SELECTED PAPERS IN STATISTICS AND PROBABILITY BY ABRAHAM WALD

A. Wald

SELECTED PAPERS IN STATISTICS AND PROBABILITY BY ABRAHAM WALD

Edited for the Institute of Mathematical Statistics

T. W. ANDERSON, *Committee Chairman*

H. CRAMÉR E. L. LEHMANN

H. A. FREEMAN A. M. MOOD

J. L. HODGES, JR. C. M. STEIN

McGRAW-HILL BOOK COMPANY, INC.

New York Toronto London

1955

SELECTED PAPERS IN STATISTICS AND PROBABILITY BY ABRAHAM WALD

Library of Congress Catalog Card Number 54-8796

PREFACE

On December 13, 1950, Abraham Wald met his sudden and untimely death. During his relatively brief career as a statistician, he made prolific contributions to the field; probably the work of no other person has had such an impact on mathematical statistics during the last decade. In this volume is collected most of Wald's research in statistics and probability except for the work included in the books under his authorship (references {47},[1] {76}, and {94}). The publication of this volume thus makes practically all of Wald's contributions to statistics available in book form.

Both the wide range of areas in which Wald made original contributions and the penetrating quality of his work are impressive. The introduction to this volume attempts to give some perspective to Wald's work by relating his different papers to each other and to the work of others. Following a brief biographical sketch, there are surveyed Wald's papers in decision function theory; the development of this theory was his most important contribution to the field of statistics. Since the later developments in sequential analysis were made as part of this general theory, comment on such papers is included in the first section of the survey. In the second section work in probability is covered. A number of other papers reported research in nonparametric inference; these are considered in the third section. The fourth deals with papers on topics in the Neyman-Pearson system of statistical inference; the fifth treats papers in the theory of regression and analysis of variance. Because of the broad scope of Wald's work, a number of papers cannot be subsumed under any particular heading; several of these are commented on in the sixth section. The committee wishes to express its appreciation to Miss Elizabeth Scott for assistance in the discussion of one of these papers.

The paper "Testing the Difference between the Means of Two Normal Populations with Unknown Standard Deviations" has not been published previously and appears here for the first time.

Acknowledgment is due to the following for permission to reprint papers in this volume: American Mathematical Society, *Annals of Mathematics*, Cowles Commission for Research in Economics, Econometric Society, Hermann et Cie, John Wiley and Sons, Notre Dame University, University of California Press.

COMMITTEE ON THE WALD MEMORIAL VOLUME

T. W. ANDERSON, *Chairman*

H. CRAMÉR J. L. HODGES, JR. A. M. MOOD

H. A. FREEMAN E. L. LEHMANN C. M. STEIN

[1] Numbers in braces refer to the complete bibliography of Wald, reprinted from the *Annals of Mathematical Statistics*, beginning on page 20.

CONTENTS

THE LIFE OF ABRAHAM WALD

Abraham Wald was born October 31, 1902, in Cluj, Rumania. His entire elementary and secondary school education was obtained at home, mainly under the direction of an older brother. After graduating from the local university in Cluj, he went to Vienna, entering the University of Vienna in 1927. Here he studied mathematics, particularly geometry, and contributed to the colloquium led by Karl Menger. After receiving his Ph.D. in 1931, Wald could not find an academic post because of religious prejudice; he began work in econometrics first with Karl Schlesinger, a banker and economist, and then with Oskar Morgenstern at the Business Cycle Research Institute.

By 1938 the rise of Nazism made life in Vienna untenable for Wald. In the summer of that year he came to the United States as a fellow of the Cowles Commission for Research in Economics, and in the fall he came to Columbia University on a fellowship of the Carnegie Corporation. He began intensive study of modern statistical inference under the tutelage of Harold Hotelling. It was not long before Wald began research in statistics; in fact, his path-breaking paper in decision theory {37} was written during this academic year. In the following academic year (1939–1940) Wald lectured in statistics while Hotelling was absent on sabbatical leave; this was the beginning of Wald's pedagogical career at Columbia. In 1941 he was made an assistant professor of economics, in 1943 an associate professor, and in 1944 a professor. In 1946 when Hotelling was called to the University of North Carolina, Columbia met the possibility of Wald also leaving by setting up a Department of Mathematical Statistics with Wald as executive officer.

In 1941 Wald married Lucille Lang. A daughter, Betty, was born in 1943, and a son, Robert, was born in 1947. In the fall of 1950, Wald and his wife traveled to India for a lecture tour. It was on one leg of this trip that they met their deaths in a plane crash.

A fuller biography of Wald is given in papers by J. Wolfowitz and Karl Menger in the March issue of the 1952 (Wald Memorial) volume of the *Annals of Mathematical Statistics.*[1] These papers also survey Wald's work in statistics and geometry; another paper in that issue by G. Tintner[2] surveys Wald's contributions to econometrics.

[1] J. Wolfowitz, "Abraham Wald, 1902–1950," *Annals of Math. Stat.*, Vol. 23 (1952), pp. 1–13.

K. Menger, "The formative years of Abraham Wald and his work in geometry," *Annals of Math. Stat.*, Vol. 23 (1952), pp. 14–20.

[2] G. Tintner, "Abraham Wald's contributions to econometrics," *Annals of Math. Stat.*, Vol. 23 (1952), pp. 21–28.

DISCUSSION OF PAPERS

1. Statistical decision function theory and sequential analysis. One of Wald's first papers in mathematical statistics (published in 1939) already contains an outline of what was to become his main work. In {37} he defines a general class of decision problems and the notion of a loss function, introduces the minimax principle and points out that minimax solutions can be obtained through the use of Bayes solutions; more specifically, that under suitable restrictions it is the Bayes solution corresponding to the "least favorable" a priori distribution.

Not all of the definitions appear yet in their final form, and some elements of the general theory are still missing (the complete class concept, the notion of sequential experimentation, and the choice of design as part of the decision problem). Also in some of the ideas Wald had been anticipated by earlier authors. However, here is the first realization that these ideas can be used to construct a unified general theory of statistics and, equally important, that they provide the tools for solving explicit problems.

At the time he wrote this paper Wald had not yet familiarized himself sufficiently with mathematical statistics to carry out this program, and he did not seriously return to it until 1946. The main development of the general theory of decision functions was then carried out by him in a series of papers which are not included in the present volume since Wald gave a presentation of the whole theory in his book *Statistical Decision Functions*, published a year before his death. There is included here, however, Wald's last paper on the subject {102}, which constitutes his address to the International Congress of Mathematicians at Harvard in 1950 and in which he gives an outline of the main ideas, exemplified in the case of a finite number of decisions.

There are also included three papers (developed since the book was written) dealing with the following special aspect of the general decision theory. In a decision problem one may decide after each observation whether or not to continue taking observations, and this as well as the final decision may be carried out with the help of a random mechanism, which chooses between various possibilities at random according to probabilities specified by the statistician. The possibility of leaving some of the choices to chance is somewhat disturbing and in practice rather complicated. In the three papers {96}, {97}, and {99} Wald considers the extent to which randomization can be avoided. A joint paper with J. Wolfowitz {99} shows that the same results as with randomization at each step can be achieved by performing at the beginning of the experiment a single randomization over the class of pure (that is, nonrandomized) strategies. The result is analogous to that of reducing a game to normal form. In the papers {96} and {97}, written jointly with A. Dvoretzky and J. Wolfowitz, it is proved that when the parameter space is finite and the distributions are atomless, one can dispense with randomization altogether even in sequential problems. The principal tool in the proof is Lyapunov's theorem concerning the range of a vector measure.

The other main strand of Wald's work was the development of sequential methods of taking observations. This was begun during the war, the first problems being suggested to Wald by Milton Friedman and W. A. Wallis. As a solution to the problem of testing a simple hypothesis against a simple alternative, Wald proposed the sequential probability ratio test. Most of his investigations into the power of those tests and the expected number of observations that they require are contained in his book *Sequential Analysis*, and the papers covered by this volume are omitted here. Certain refinements of these results given there are to be found in {72} and {73}. The same methods were also employed in a joint paper with Milton Sobel {89} to give a sequential treatment of a three-decision problem.

As was mentioned above, the approach developed in the general theory, particularly the method of Bayes solutions, provides a tool for solving specific decision problems. The main difficulty that remains is the explicit determination of the Bayes solutions. Especially in the sequential case this constitutes a formidable problem. The first specific problem of this type was considered in a joint paper by Charles Stein and Wald {77}, in which they show that, for estimating the mean of a normal distribution with known variance, the usual nonsequential procedure provides confidence intervals of fixed length and confidence coefficients that minimize the maximum expected number of observations. Perhaps the greatest achievement of the method to date is that contained in a joint paper with Wolfowitz {84}, where it was shown that the Bayes solutions for testing a simple hypothesis against a simple alternative are, when the cost per observation is constant, sequential probability ratio tests. From this the authors were able to derive a strong optimum property of sequential probability ratio tests, which Wald had conjectured much earlier. Another special case was treated in {101}, also written jointly with Wolfowitz, in which the authors give an explicit characterization of the smallest complete class in the case that both the number of distributions and of possible decisions is finite. Because of the great difficulty involved in the determination of minimax solutions in the sequential case, Wald also considered the problem from a large sample point of view and in {100} used the maximum likelihood method to obtain sequential solutions for certain estimation problems, which are asymptotically minimax.

The three lines of research indicated above, the development of a general theory of decision functions, the construction of usable sequential procedures and the evaluation of their performance characteristics, and the determination of minimax (or other optimum solutions of specific decision problems), which were initiated by Wald, have been furthered in various ways since the appearance of Wald's original papers that are here reproduced. The following examples, while in no way complete, are intended to indicate some of these developments.

(i) With regard to the general theory, there have been investigations into the possibility of weakening the conditions imposed by Wald.[1] It should be mentioned here that Wald himself in {87} has given a condition for the case of discrete variables that is weaker than the corresponding condition in his book.

(ii) Complete classes of decision procedures are being worked out for various classes of problems.[2] Unfortunately, the complete classes will nearly always be infinite. In this connection, Wolfowitz[3] has introduced the notion of ϵ-complete class, a class τ of decision procedures being ϵ-complete if to any procedure δ there exists δ' in τ such that the risk function of δ' never exceeds that of δ by more than ϵ. It has been shown by Wolfowitz that finite ϵ-complete classes exist under mild conditions.

(iii) The possibility, envisaged in Wald's general theory, of the statistician deciding what observations to take and thereby selecting his experiment, suggests that one may at various levels of generality compare different experiments that would serve the same purpose and select among them the most efficient one. In fact, in {54}, Wald considered one such problem. He proved there that among a certain class of experimental designs the Latin square is, in a certain sense, most efficient. This problem has been taken up again recently by Ehrenfeld[4] who reached similar conclusions.

A general theory of the relative informativeness of competitive experiments has been developed by Blackwell[5] in analogy to related work in the theory of games by Bohnenblust, Shapley, and Sherman.

[1] M. N. Ghosh, "An extension of Wald's decision theory to unbounded weight functions," *Sankhyā*, Vol. 12 (1952), pp. 8–26.

S. Karlin, "Operator treatment of minimax principle," *Contributions to the Theory of Games*, Princeton University Press, 1950, pp. 133–154; "The theory of infinite games," *Annals of Mathematics*, 2d ser., Vol. 58 (1953), pp. 371–401.

J. Kiefer, "On Wald's complete class theorems," *Annals of Math. Stat.*, Vol. 24 (1953), pp. 70–75.

L. LeCam, "On the completeness of classes of Bayes solutions," Abstract, *Annals of Math. Stat.*, Vol. 24 (1953), pp. 492–493.

[2] A. Birnbaum, "Admissible tests for the mean of a rectangular distribution," *Annals of Math. Stat.*, Vol. 25 (1954), pp. 157–161; "Characterizations of complete classes of tests of some multiparameter hypotheses, with applications to likelihood ratio tests," Abstract, *ibid.*, Vol. 24 (1953), p. 490.

M. Sobel, "An essentially complete class of decision functions for certain standard sequential problems," *Annals of Math. Stat.*, Vol. 24 (1953), pp. 319–337.

[3] J. Wolfowitz, "On ϵ-complete classes of decision functions," *Annals of Math. Stat.*, Vol. 22 (1951), pp. 461–464.

[4] S. Ehrenfeld, "On increasing efficiency by proper choice of design," submitted to the *Annals of Math. Stat.*

[5] D. Blackwell, "Comparison of experiments," *Proceedings of the Second Berkeley Symposium on Mathematical Statistics and Probability*, University of California Press, 1951, pp. 93–102; "Equivalent comparisons of experiments," *Annals of Math. Stat.*, Vol. 24 (1953), pp. 265–272.

(iv) The central role that the minimax principle plays in the general theory has led various authors to study its consequences. In many cases the minimax solutions turn out to be just those procedures that one might choose on intuitive grounds,[1] but there are also many others in which the resulting procedures are quite unsatisfactory.[2] These latter have led to various modifications of the principle. We mention here the notion of minimizing the maximum regret which has been discussed by Savage,[3] and that of restricted Bayes solutions which is an attempt to utilize previous experiment in decision making.[4] Another result of these and similar difficulties is an axiomatic investigation of the "rational selection of decision functions."[5]

(v) While the decision theoretic point of view has not yet penetrated all of statistics, various fields of statistical activity are in the process of being investigated in this light. Some progress has been made in particular in the theory of sampling human populations.[6]

2. Probability. *On Kollektivs.* The papers {29} and {31} form a contribution to the axiomatization of probability theory according to the system introduced by R. von Mises. The main question at issue is the existence of kollektivs satisfying the two fundamental conditions of von Mises, which is proved by Wald under various conditions. From the point of view of the measure-theoretic approach to the subject, however, the main results of Wald are consequences of the strong law of large numbers, so that the interest of these papers is now mainly historical. Accordingly, this collection includes

[1] J. Wolfowitz, "Minimax estimates of the mean of a normal distribution with known variance," *Annals of Math. Stat.*, Vol. 21 (1950), pp. 218–230.

M. A. Girshick and L. J. Savage, "Bayes and minimax estimates for quadratic loss functions," *Proceedings of the Second Berkeley Symposium on Mathematical Statistics and Probability*, University of California Press, 1951, pp. 53–74.

E. Paulson, "An optimum solution to the k-sample slippage problem for the normal distribution," *Annals of Math. Stat.*, Vol. 23 (1952), pp. 610–616.

[2] J. L. Hodges, Jr., and E. L. Lehmann, "Some problems in minimax point estimation," *Annals of Math. Stat.*, Vol. 21 (1950), pp. 182–197.

H. Robbins, "Asymptotically subminimax solutions of compound statistical decision problems," *Proceedings of the Second Berkeley Symposium on Mathematical Statistics and Probability*, University of California Press, 1951, pp. 131–148.

P. Frank and J. Kiefer, "Almost subminimax and biased minimax procedures," *Annals of Math Stat.*, Vol. 22 (1951), pp. 465–468.

J. Laderman, "Asymptotically subminimax and asymptotically admissible statistical decision procedures," Abstract, *Annals of Math. Stat.*, Vol. 23 (1952), p. 476.

[3] L. J. Savage, "The theory of statistical decision," *Jour. Amer. Stat. Assn.*, Vol. 46 (1951), pp. 55–67.

[4] J. L. Hodges, Jr., and E. L. Lehmann, "The use of previous experience in reaching statistical decisions," *Annals of Math. Stat.*, Vol. 23 (1952), pp. 396–407.

[5] H. Chernoff, "Remarks on the rational selection of a decision function," Abstract, *Econometrica*, Vol. 18 (1950), p. 183.

[6] Om. P. Aggarwal, "Minimax sampling and estimation in finite populations," Abstract, *Annals of Math. Stat.*, Vol. 23 (1952), p. 640.

J. Putter, "Sur une méthode de double echantillonage pour estimer la moyenne d'une population Laplacienne stratifié," *Revue de l'Institut International de Statistique*, 1951.

only {31}, which gives a survey without detailed proofs of the results reached in the more elaborate paper {29}.

We shall give a brief review of the question, considering only the particular case of a variable x having the only possible values 0 and 1. Let $X = (x_1, x_2, \cdots)$ be an infinite sequence of values of x, so that each x_i is equal to 0 or 1. The following conditions A and B are then fundamental in the von Mises theory.

Condition A
$$\lim_{n \to \infty} \frac{1}{n} \sum_{1}^{n} x_i = p \text{ exists}$$

The limit p is, by definition, the probability of the event $x = 1$. Further, let $F = (f_1, f_2, \cdots)$ be a sequence of functions, each of which may only take the values 0 and 1, and such that $f_i = f_i(x_1, x_2, \cdots, x_{i-1})$ for $i \geqq 2$, while f_1 is a constant. The Condition B then requires that

Condition B
$$\lim_{n \to \infty} \frac{\sum_{1}^{n} f_i x_i}{\sum_{1}^{n} f_i} = p \text{ exists}$$

Clearly B reduces to A in the particular case when all f_i are identically equal to 1, so that it is only a matter of convenience that the conditions are stated separately. Condition A postulates the *existence of a limit* of the frequency ratio $\frac{1}{n} \sum_{1}^{n} x_i$ of 1's while B postulates the *invariance of this limit* against the *place selection* defined by F. This invariance can be interpreted as showing the *impossibility of a successful gambling system* founded on the place selection F.

One of the main theorems of Wald states, for the particular case considered here, that, to any given probability p and to any sequence of place selections $F_1, F_2, \cdots$, there exist continuously many sequences $X = (x_1, x_2, \cdots)$ satisfying A and B for all place selections F_i.

In the measure-theoretic treatment of the subject, a measure is introduced in the infinite-dimensional space of the sequences $X = (x_1, x_2, \cdots)$ in the obvious way, regarding the x_i as independent random variables, each with the probability p of taking the value 1 and the probability $1 - p$ of taking the value 0. It is then a consequence of the strong law of large numbers[1] that the above-mentioned result of Wald holds *almost everywhere* in the space of sequences X.

[1] See W. Feller, "Ueber die Existenz von sogenannten Kollektiven," *Fund. Math.*, Vol. 32 (1939), p. 87, which covers also the case of nonmeasurable place selections. See also J. L. Doob, "Probability as measure," *Annals of Math. Stat.*, Vol. 12 (1941), p. 206, and the subsequent discussion between Doob and von Mises, *ibid.*, p. 215.

Use of moments. In the paper {35}, which forms an extension of the preliminary investigations published in {32}, Wald is concerned with the relations between a distribution function and its absolute moments. If a finite number of absolute moments of the distribution function $F(x)$ are given, say the absolute moments of orders $i_1, i_2, \cdots, i_r$, what can be said about $F(x)$? In particular it is required to find sharp upper and lower limits of the difference $F(x) - F(-x)$ for any positive x. Further, if r real numbers are given, under what conditions will it be possible to find a distribution function $F(x)$ having the given numbers for absolute moments of orders $i_1, \cdots, i_r$? These problems are first reduced to the equivalent problems concerning ordinary moments of a distribution function $F(x)$ such that $F(-0) = 0$. Wald then proceeds to give the general solutions of the problems, thus generalizing previous work of Tchebychef, Markoff, Stieltjes, Cantelli, Guldberg, and others. This paper represents perhaps the main contribution of Wald to purely mathematical analysis. It shows his remarkable mathematical skill, and its results contain the complete solution of long outstanding and intricate problems. For the literature of the subject, we may refer to the monograph of Shohat and Tamarkin.[1]

Limit theorems. In many statistical applications of probability theory, questions of convergence of sequences of distributions or of random variables occur. A large number of particular results in this direction have been given by various authors, in connection with special applications. In the paper {56}, written in collaboration with H. B. Mann, the subject has been treated in a systematic way, with restriction to convergence in probability for sequences of random variables, and convergence in every continuity point of the limiting distribution for sequences of distributions. A great number of general and practically useful results are given. One of the most important results is perhaps the following: Let $x_1, x_2, \cdots$ and x be r-dimensional random vectors, with the distribution functions $F_1, F_2, \cdots$, and suppose that $F_n \to F$ in every continuity point of F. Let $g(y)$ be a Borel measurable function of the r-dimensional vector y, whose set of discontinuity points is closed and has F-measure zero. Then the distribution function of $g(x_n)$ tends to the distribution function of $g(x)$ in every continuity point of the latter.

Cumulative sums. The papers {75} and {82} are closely connected with Wald's work on sequential analysis. A double sequence of independent random variables:

$$\begin{array}{l} x_{11}, \\ x_{21}, x_{22}, \\ \cdots\cdots\cdots\cdots \\ x_{n1}, x_{n2}, \cdots, x_{nn}, \\ \cdots\cdots\cdots\cdots \end{array}$$

[1] J. A. Shohat and J. D. Tamarkin, *The Problem of Moments*, American Mathematical Society, New York, 1943.

is considered, and the limiting distribution (as $n \to \infty$) of the variable

$$M_n = \max_{r=1, \cdots, n} \sum_{i=1}^{r} x_{ni}$$

is studied under various assumptions. The results obtained include as particular cases some previous results by Erdos and Kac.[1] Some further results in the same direction have been given by Chung.[2]

Consistency of maximum likelihood estimates. The note {88} gives a simple proof of the consistency of the maximum likelihood estimate, based on the strong law of large numbers. In a subsequent note by Wolfowitz,[3] the proof is modified so as to use only the weak law of large numbers and thus apply to a wider class of random variables.

3. Nonparametric inference. Since about 1940 there has been considerable development of methods which do not require assumptions of normality. A number of nonparametric techniques had already been proposed for special problems, but the first to envisage (in 1938–1939) the possibility of a general body of nonparametric statistical theory with techniques suitable for any situation in which the functional form of the distribution is unknown was S. S. Wilks. It was also he who suggested the two-sample problem discussed in {40}. Wald took an early interest in the development of nonparametric methods and made extensive contributions to it, which included both the proposal of important new techniques and evaluation of the performance of such procedures.

Confidence belts for distribution functions. The paper {34} and note {42} deal with the problem of using an empirical distribution to determine a confidence belt for the graph of a population distribution. The main theorem of the paper states that confidence belts can be constructed in a manner such that the confidence coefficient is independent of the form of the population distribution (provided only that it is continuous). A method for constructing such belts is presented and illustrated.

However, this work had already been done essentially by Kolmogorov[4] who also derived the important large sample distribution of the largest absolute deviation between the sample and population distributions. His work was brought to the attention of Wald and Wolfowitz after their paper was published and they then published the note {42}. Later Kolomogorov[5] published a short

[1] P. Erdos and M. Kac, "On certain limit theorems of the theory of probability," *Bull. Amer. Math. Soc.*, Vol. 52 (1946), pp. 292–302.

[2] K. L. Chung, "Asymptotic distribution of the maximum cumulative sum of independent random variables," *Bull. Amer. Math. Soc.*, Vol. 54 (1948), pp. 1162–1170.

[3] J. Wolfowitz, "On Wald's proof of the consistency of the maximum likelihood estimate," *Annals of Math. Stat.*, Vol. 20 (1949), pp. 601–602.

[4] A. N. Kolmogorov, "Sulla determinatione empirica di una legge di distributione," *Giorn. Inst. Ital. Attuari*, Vol. 4 (1933), pp. 1–11.

[5] A. N. Kolmogorov, "Confidence limits for an unknown distribution function," *Annals of Math. Stat.*, Vol. 12 (1941), pp. 461–464.

summary of his work in English and showed how his basic distribution made the construction of confidence belts a simple matter.

A two-sample test and some asymptotic theory. One of the classic papers in the development of nonparametric inference was {40}, which proposed a method for testing whether two samples are from populations with the same distribution. The test criterion is the number of runs in the combined ordered samples; it is shown to be distribution-free, asymptotically normal under the null hypothesis, "consistent" in a certain sense, and thereby superior in this sense to an earlier two-sample nonparametric test presented by Thompson.[1] This appears to be the first theoretical result concerning the power of a nonparametric test.

The value of the paper is in showing how one could come to grips with nonparametric tests not only in defining a criterion and solving the distribution problem but in evaluating such a test and comparing it with other tests. Later studies have raised questions as to the efficacy of this particular test. (See, for example, Wolfowitz[2] and Lehmann.[3]) Also, the evaluation criterion (consistency) itself is not very discriminating. Nevertheless, the paper is one of the significant landmarks of research in nonparametric methods.

The papers {59} and {64} demonstrate the asymptotic normality of certain test criteria based on permutations of the observations. The basic idea for the criteria is due to Fisher and is a source of some excellent nonparametric tests. Pitman[4] was an early protagonist of these tests and many other authors used the idea, but the large sample distributions of the test criteria were generally unavailable or were unproved until supplied via a theorem presented in {64}. The other paper investigates the serial correlation on a randomization basis rather than as a function of normally distributed variates. These two papers initiated several important pieces of research.[5]

[1] W. R. Thompson, "Biological applications of normal range and associated significance tests in ignorance of original distribution forms," *Annals of Math. Stat.*, Vol. 9 (1938), pp. 281–287.

[2] J. Wolfowitz, "Nonparametric statistical inference," *Proceedings of the Berkeley Symposium on Mathematical Statistics and Probability*, University of California Press, 1949, pp. 93–113.

[3] E. L. Lehmann, "The power of rank tests," *Annals of Math. Stat.*, Vol. 24 (1953), pp. 23–43.

[4] E. J. G. Pitman, "Significance tests which may be applied to samples from any populations," *Suppl. Jour. Roy. Stat. Soc.*, Vol. 4 (1937), pp. 117–130; "Significance tests which may be applied to samples from any populations II," *ibid.*, pp. 225–232; "Significance tests which may be applied to samples from any populations III," *Biometrika*, Vol. 29 (1938), pp. 322–335.

[5] G. Noether, "On a theorem by Wald and Wolfowitz," *Annals of Math. Stat.*, Vol. 20 (1949), pp. 455–458; "Asymptotic properties of the Wald-Wolfowitz test of randomness," *ibid.*, Vol. 21 (1950), pp. 231–246.

W. Hoeffding, "A combinatorial central limit theorem," *Annals of Math. Stat.*, Vol. 22 (1951), pp. 558–566; "The large sample power of tests based on permutations of the observations," *ibid.*, Vol. 23 (1952), pp. 169–192.

As is now well-known, these tests based on permutations and the asymptotic normality of the criteria are at the heart of Fisher's requirement for randomization of experimental designs. The randomization permits probability theory to be used in the interpretation of the data and often allows normal theory to be applied without assuming individual observations are from normal populations. (See references in {64}.)

The χ^2 *test of goodness of fit.* In setting up a goodness of fit test using chi square, how many and what size class intervals should be used? Using sensible assumptions and large sample approximations, the authors show in {49} that the number of intervals k for samples of size N should be roughly $k = 3.5N^{2/5}$, and the intervals should be so defined that each one has probability $1/k$ associated with it under the null hypothesis. The authors feel that this result is probably valid down to $N = 200$, which corresponds to an expected number of observations of only seven per interval.

There does not yet exist in the literature a comparable study for moderate sample sizes despite the manifest importance of the problem. However the results of Mann and Wald have been extended to show that k can be taken smaller than they recommended without much loss of efficiency.[1]

Tolerance limits. Of three papers on tolerance limits, {52}, {50}, and {71}, the first is surely the most significant. Wald's ingenious extension of one-dimensional nonparametric tolerance limits was second only to Wilks' original idea[2] in stimulating an important sequence of theoretical and applied developments in nonparametric inference.

Suppose four observations (x_i,y_i) are drawn from a bivariate population with a continuous distribution function and the corresponding points plotted in the euclidean plane. (The same four points are plotted three times in the figure.) Using Tukey's characterization,[3] the four lines $y = y_i$ shown in (a)

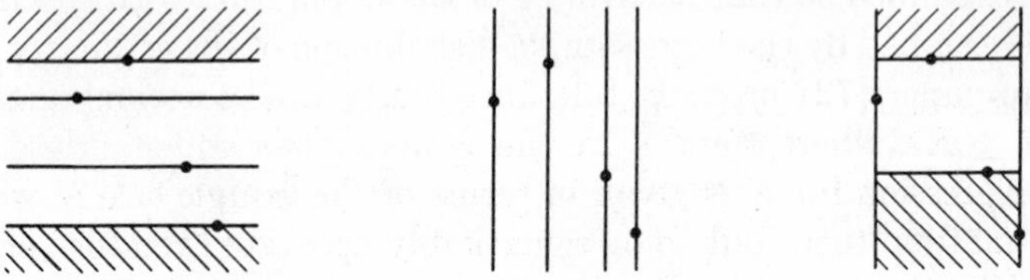

(a) (b) (c)

divide the plane into five statistically equivalent regions as is obvious by applying Wilks' result to the marginal distribution of y. The proportion of the population covered by any one of the regions has the same distribution; each region covers one-fifth of the population on the average. The joint distribu-

[1] C. Arthur Williams, Jr., "On the choice of the number and width of classes for the chi-square test of goodness of fit," *Jour. Amer. Stat. Assn.*, Vol. 45 (1950), pp. 77–86.

[2] S. S. Wilks, "Determination of sample sizes for setting tolerance limits," *Annals of Math. Stat.*, Vol. 12 (1941), pp. 91–96.

[3] J. W. Tukey, "Statistically equivalent blocks and tolerance regions—the continuous case," *Annals of Math. Stat.*, Vol. 18 (1947), pp. 529–539.

tion of the proportions of the population covered by the regions does not depend on the population distribution. Similarly, considering the marginal distribution of x, the four lines $x = x_i$ of (*b*) divide the plane into five statistically equivalent regions. Wald's paper shows, for example, that the five regions of (*c*) are also statistically equivalent in exactly the same sense.

As mentioned above, Tukey formulated the idea of using a sample of n to divide the plane into $n + 1$ statistically equivalent regions; in the same paper he showed that arbitrary curves could be employed instead of the lines parallel to the coordinate axes, and he provided methods for obtaining (roughly) domains of a desired shape in place of the rectangle (or collection of rectangles) given by Wald. Later Tukey[1] removed the restriction that the distribution functions be continuous. Fraser[2] has further generalized these techniques by studying the extent to which curves used to cut off regions may be chosen sequentially instead of prescribed in advance. Also Fraser[3] studied methods of cutting the set of $n + 1$ elementary regions into groups to be truncated later by other curves.

The second paper of this group {50} is a straightforward application of large sample theory to the parametric tolerance limit problem. Let $f(x,\theta)$ be a density of x with parameter θ and let $\phi(\theta)$ and $\psi(\theta)$ be such that

$$\int_{\phi(\theta)}^{\psi(\theta)} f(x,\theta)\, dx = \xi$$

then

$$\phi(\theta) < x < \psi(\theta)$$

is an exact ξ tolerance interval (x, ϕ, ψ may be vectors). In this paper θ (which may also be a vector) is estimated from a sample and the estimator substituted for θ in ϕ and ψ to calculate a random tolerance interval; this interval is suitably expanded so that it covers ξ of the population with specified probability as determined by the large sample distribution of the estimator.

The third paper {72} provides tolerance limits for the normal distribution of the form $\bar{x} \pm Ks$, where $\bar{x}$ and s are the sample mean and standard deviation. A simple expression for K is given in terms of the sample size N which, while an approximation, turns out to be remarkably accurate even for very small N. Tables of K are provided by Bowker.[4]

4. Theory of testing hypotheses and estimation. *Asymptotic theory.* In the early 1940's Wald published four papers, {43}, {46}, {48}, and {58}, which were concerned with devising definitions of optimum performance for a

[1] J. W. Tukey, "Statistically equivalent blocks and multivariate tolerance regions—the discontinuous case," *Annals of Math. Stat.*, Vol. 19 (1948), pp. 30–39.

[2] D. A. S. Fraser, "Sequentially determined statistically equivalent blocks," *Annals of Math. Stat.*, Vol. 22 (1951), pp. 372–381.

[3] D. A. S. Fraser, "Nonparametric tolerance regions," *Annals of Math. Stat.*, Vol. 24 (1953), pp. 44–45.

[4] Statistical Research Group of Columbia University, *Selected Techniques of Statistical Analysis*, McGraw-Hill Book Co., 1947, pp. 97–110.

sequence of tests based on increasing numbers of observations, and with finding conditions on the distributions under which the classical methods of maximum likelihood and likelihood ratio would yield optimum sequences. In all of these papers, the method of attack is to show that in the limit the distributions involved behave like normal distributions, for which the optimum problem is relatively easy.

Several different notions of asymptotically optimum sequences of tests were presented in these papers. The first ({43} and {46}) is that of an asymptotically most powerful (AMP) sequence of tests. Roughly speaking, a sequence of tests is AMP if the amount by which its power can be anywhere increased tends to zero as the number n of observations is increased. This notion has been examined critically by Lehmann.[1] We note that for a sequence of tests which is uniformly consistent outside of every neighborhood of the hypothesis, the AMP property becomes a local one which is therefore closely related to the notions of locally most powerful sequences.

In {43} and {46} Wald showed, under certain restrictions on the distributions, and for one-sided and unbiased tests of a simple hypothesis about a single real parameter, that the tests based on the maximum likelihood estimates are AMP. The main difference between these papers is that in the former the maximum likelihood estimate is considered directly, whereas in the latter Wald studies the estimate based on the derivative of the logarithm of the likelihood function. Because of the connection between tests of hypotheses and the theory of confidence intervals, which is particularly close when the classical Neyman definition of "shortest" intervals is used, Wald was able to extend the results to the confidence interval problem rather easily. This is done in {48}, where the main novel point lies in verifying that the confidence set obtained is an interval.

Wald's work on asymptotic theory of tests culminated in {58}, which generalizes the preceding work by considering random vectors instead of random variables, by allowing the parameter space to be multidimensional, and by dealing with composite as well as simple hypotheses. The paper falls naturally into two parts: first, the general problem is reduced to the normal case, and then several optimum properties are obtained for tests on normal distributions. The latter work is in some ways an extension of {51}.

The reduction of the problem to normal distributions is effected by the remarkable Lemma 1 ({58}, page 433). The proof of this lemma, which takes more than ten pages, is a good example of Wald's power with complicated analytic problems. In the terminology of LeCam,[2] this result establishes the

[1] E. L. Lehmann, "Some comments on large sample tests," *Proceedings of the Berkeley Symposium on Mathematical Statistics and Probability,* University of California Press, 1949, pp. 451–457.

[2] Lucien LeCam, "On some asymptotic properties of maximum likelihood estimates and related Bayes estimates," *Univ of Calif. Publications in Statistics,* Vol. 1, No. 11 (1953), pp. 277–330.

"uniform asymptotic sufficiency" of the maximum likelihood estimate. It is shown that we may replace each Borel set of the original sample space by another Borel set depending only on the maximum likelihood estimate, in such a way that the probability of the original set will differ from the probability of the new set by an amount tending to zero uniformly in both the parameter and the original set.

Wald considers three optimum properties of tests in {58}, together with the asymptotic versions of each: best constant power, best average power, and most stringency. The first two notions have clear relations to the properties of the F-test established by Hsu and Wald. The stringency idea seems to have been introduced into the literature by Wald ({47}, page 30), and has had an important influence.[1] Briefly, we may define the envelope power function of a family of tests as the supremum at each parameter point θ of the powers of the tests of the family, and define the "shortcoming" of a test at a point θ as the amount by which the power of the test falls short of the envelope power there. We may then define the maximum shortcoming of the test as the supremum over θ of its shortcoming. A test is most stringent if it has the smallest possible maximum shortcoming. A sequence of tests is asymptotically most stringent if the maximum amount by which its maximum stringency can be reduced tends to zero as n increases. Note that an asymptotically most powerful test {43} is also asymptotically most stringent.

The stringency idea is clearly a minimax idea, or more specifically an illustration of what is sometimes called the *minimax regret* approach. The shortcoming measures the failure of the test compared to what would have been possible under the circumstances, and this may be called the *regret*. The most stringent test is the test which minimizes the maximum regret. A theory of most stringent tests has been developed by Gilbert Hunt and Charles Stein. Their work has not been published but is summarized by Lehmann.[2]

Existence of most powerful tests. In {98}, the results of Lyapunov on the range of a vector measure are applied to complete the Neyman-Pearson lemma on most powerful tests satisfying certain integral side conditions. The Neyman-Pearson lemma, while of fundamental usefulness in verifying that given tests were optimum, did not assert the existence of optimum tests, nor was it known that the conditions of the lemma were necessary. These problems had been considered separately by Wald and by George B. Dantzig, who had presented in his doctoral dissertation (University of California, 1946) a solution based on variational arguments. Using Lyapunov's theorem, it is

[1] The term "most stringent test" had been used earlier by Robert W. B. Jackson, "Tests of statistical hypotheses in the case when the set of alternatives is discontinuous, illustrated on some genetical problems," *Statistical Research Memoirs*, Vol. 1 (1936), pp. 138–161. However, Jackson defines a test to be most stringent if it minimizes the maximum probability of error, so that in Wald's terminology the Jackson notion (which that author credits to Neyman) is a minimax one in the ordinary sense.

[2] E. L. Lehmann, "Some principles of the theory of testing hypotheses," *Annals of Math. Stat.*, Vol. 21 (1950), pp. 1–26.

immediate that there will always exist most powerful tests satisfying the side conditions whenever there exist any tests which do. Further, if the side conditions are not of an extreme nature, the Neyman-Pearson sufficient conditions are also necessary (Theorem 3.1). In the extreme case, a further condition is required (Theorem 3.2). These considerations provided the basis for an extended result concerning the maximum of a function of several integrals, given by Chernoff and Scheffé.[1]

It may be noted that the Lyapunov theorem has had other interesting statistical consequences, as in {91} and {96}, and in a paper by Blackwell.[2]

Distinct hypotheses. Paper {86}, written with Agnes Berger, is also concerned with clearing up a point in the classical theory of tests, which had been raised by J. Neyman. The introduction of the concept of distinct hypotheses ({86}, page 104) is motivated by the fact that there can exist different sets H_0 and H_1 of distributions such that for every test ϕ, the power function $\beta_\phi(\theta)$ of the test ϕ has identical range in H_0 and H_1. Berger and Wald found necessary and sufficient conditions for ruling out such unpleasant possibilities.

5. Analysis of variance. *Variance component analysis.* In the three papers {38}, {45}, and {80} Wald treated, with increasing generality, a problem in what is now usually known as variance component analysis or Model II analysis of variance.[3] If the observations in groups are normally distributed about the group expectation with variance σ^2, while the group expectations were themselves sampled from a normal population with variance σ'^2, we may be interested in the parameter $\lambda = \sigma'^2/\sigma^2$. The problem of testing the hypothesis that $\lambda = 0$, which is equivalent to the hypothesis that the intraclass correlation is 0, had been dealt with by R. A. Fisher[4] for the special case in which the number of observations in each group is the same. Wald showed how, by using appropriately weighted sums of squares and weighted means, the F distribution could also be used to obtain confidence intervals (and hence of course tests) for λ even with unequal frequencies. Unfortunately, the actual calculation of the confidence intervals would involve somewhat troublesome numerical work.

Papers {45} and {80} are generalizations, first to the r-way classification and then to the most general linear regression problem. The method does not seem yet to have come into wide use, but it may be noted that in a recent survey paper on variance component analysis[5] it was stated: "Wald's approach to the particular problem of confidence limits on the ratios of variance com-

[1] Herman Chernoff and Henry Scheffé, "A generalization of the Neyman-Pearson fundamental lemma," *Annals of Math. Stat.*, Vol. 23 (1952), pp. 213–225.

[2] David Blackwell, "On a theorem of Lyapunov," *Annals of Math. Stat.*, Vol. 22 (1951), pp. 112–114.

[3] Churchill Eisenhart, "The assumptions underlying the analysis of variance," *Biometrics*, Vol. 3 (1947), pp. 1–21.

[4] R. A. Fisher, *Statistical Methods for Research Workers*, Chap. 7, Oliver and Boyd, Edinburgh.

[5] S. Lee Crump, "The present status of variance component analysis," *Biometrics*, Vol. 7 (1951), pp. 1–16.

ponents is very general within the framework of Model II. It might well serve as a model for other work directed toward other problems within Model II."

Optimum properties. In {51} Wald made an important contribution to the theory of optimum tests by introducing the notion of best average power. The power function of the F-test depends on a single real parameter λ, and it was shown in 1940 by P. L. Hsu[1] that the F-test is uniformly most powerful among all those tests whose power depends only on λ. Wald felt that this condition was too restrictive and wanted instead to compare the performance of all tests of the given size in some global sense. He considered the "spheres" in the parameter space where λ has a specified value, and proved that the F-test is uniformly best in the sense of having on each such sphere the greatest possible average power. In the language of decision theory, the F-test is the Bayes solution corresponding to a uniform a priori distribution over the sphere.

Wald's results on the power of the F-test were further extended and simplified by Hsu[2] and Wolfowitz.[3] The notions of best constant power and best average power were generalized by Wald in {58}. In the proof of the main lemma of {51} there was an error, which is corrected in {61}.

6. Other topics. *Stochastic difference equations.* In {57}, {92}, and {93} Wald deals with problems of statistical inference in models where the observed variables satisfy stochastic difference equations. Haavelmo[4] studied the usefulness of such models in econometrics, pointed out new statistical problems, and solved some of them. Reading early drafts of Haavelmo's papers and discussion with Haavelmo stimulated Wald to work with H. B. Mann on these problems.

In {57} Mann and Wald first consider a single stochastic difference equation

$$x_t + \alpha_1 x_{t-1} + \cdots + \alpha_p x_{t-p} + \alpha_0 = \epsilon_t \qquad t = 1, 2, \cdots N$$

where $\{x_t\}$ are the observables and $\{\epsilon_t\}$ are independently and identically distributed with $E(\epsilon_t) = 0$ and $E(\epsilon_t^2) = \sigma^2$. The authors find the maximum likelihood estimates of $\alpha_0, \alpha_1, \cdots, \alpha_p$ and σ^2 under certain assumptions and show that they are consistent and are asymptotically normally distributed. The second part of the paper deals with a system of stochastic difference equations

$$\sum_{j=1}^{r} \sum_{k=0}^{p_{ij}} \alpha_{ijk} x_{j,t-k} + \alpha_i = \epsilon_{it} \qquad i = 1, \cdots, r$$
$$t = 1, \cdots, N$$

[1] P. L. Hsu, "Analysis of variance from the power function standpoint," *Biometrika,* Vol. 32 (1940), pp. 62–69.

[2] P. L. Hsu, "On the power functions of the E^2-test and the T^2-test," *Annals of Math. Stat.,* Vol. 16 (1945), pp. 278–286.

[3] J. Wolfowitz, "The power of the classical tests associated with the normal distribution," *Annals of Math. Stat.,* Vol. 20 (1949), pp. 540–551.

[4] T. Haavelmo, "The statistical implications of a system of simultaneous equations," *Econometrica,* Vol. 11 (1943), pp. 1–12; "The probability approach in econometrics," *ibid.,* Vol. 12 (1944), Supplement.

The maximum likelihood estimates are derived and are shown to be consistent and asymptotically normally distributed under certain conditions.

The multivariate stochastic difference equation as stated above raises a question of "identification." Suppose $p_{11} = p_{12} = \cdots = p_{rr}$. If the parameters are unrestricted there is an indeterminancy; we can multiply the above equation by $\beta_{hi}(h, i = 1, \cdots, r)$ and sum over i to obtain a new set of equations. If we let $\alpha^*_{hjk} = \Sigma_i\beta_{hi}\alpha_{ijk}$, $\epsilon^*_{ht} = \Sigma_i\beta_{hi}\epsilon_{it}$, and $\sigma^*_{hg} = \Sigma\beta_{hi}\sigma_{ij}\beta_{gj}$, then the new model has exactly the same form as before (that is, the distribution of the observed x_{it} is the same), but the values of the parameters are different. In order that the parameters be identified (that is, that the representation be unique) it is necessary that a priori restrictions be put on the parameters, such as $\alpha_{ij0} = 0, i \neq j, \alpha_{ii} = 1$. The authors consider the statistical problems under these restrictions and under other sets of restrictions.

This work on stochastic difference equations has been extended by Koopmans, Leipnik, Rubin, and others.[1] Identification problems have been studied extensively; explicit estimation equations (and computational procedures) have been given under rather general identification restrictions; provision for addition of fixed variables into the equations have been made, and consistency and asymptotic normality of estimates have been proved under much more general assumptions.

In {92} Wald gives some conditions for identification of parameters. However, the conditions are stated very generally and they cannot be easily verified in any given case. The work of Koopmans, Leipnik, and Rubin (written later than Wald's contribution) gives much more specific conditions which are easily verified.

In {93} Wald considers the question of estimating some of the parameters in (2) when not all of the component equations are given. He suggests a general method of deriving confidence regions but points out that unless one knows something about the missing component equations one cannot be sure that the confidence regions will be very effective (for example, that the regions will shrink to a point when the sample size increases). Anderson and Rubin[2] have gone further with this problem and have obtained useful estimates and confidence regions under fairly general conditions (using a minimum of knowledge about the missing equations).

The fitting of straight lines if both variables are subject to error. In {41} Wald considers the problem of fitting a straight line when both variables x and y are subject to "error." This problem arises in empirical sciences when we deduce laws from observations of a limited number (here, two) of variables which either are observed with error or are influenced by the disturbances of other variables.

Consider N pairs of observations (x_i, y_i) such that the random errors

$$\epsilon_i = x_i - E(x_i)$$

[1] T. C. Koopmans (ed.), *Statistical Inference in Dynamic Economic Models*, Cowles Commission Monograph 10, John Wiley and Sons, 1950.

[2] T. W. Anderson and Herman Rubin, "Estimation of the parameters of a single stochastic difference equation in a complete system," *Annals of Math. Stat.*, Vol. 20 (1949), pp. 46–63.

are identically distributed, have finite variance σ_ϵ^2, and are uncorrelated, with similar assumptions about $\eta_i = y_i - E(y_i)$, and assume also $E(\epsilon_i\eta_i) = 0$ for $i = 1, 2, \cdots, N; j = 1, 2, \cdots, N$. Finally, assume that there exists a single linear relation $E(y_i) = \alpha E(x_i) + \beta$, called the *structural relation* between y and x.

Wald provides consistent estimates of the parameters α, β, σ_ϵ and σ_η depending only on the observable x_i, y_i without making any assumption about σ_ϵ or σ_η or their ratio. He also finds a confidence interval for α, for β given α, and then a confidence region for α and β jointly.

The method consists in dividing the observations into two equal groups, computing the center of gravity of each group, and then connecting the two centers by a straight line which is the estimated structural relation. The method of grouping is arbitrary except that it must be entirely independent of errors. Wald notes that the most advantageous grouping is to put a point (x_k, y_k) into the first group only if the number of elements x_j for which $x_j \leqq x_k$ is less than $N/2$. However, this grouping will generally not be independent of the errors. The consequence of ignoring this fact is discussed by Neyman and Scott.[1] Wald discusses other methods of grouping which will be independent of the errors. He points out and illustrates the fact that the structural relation between y and x is not generally appropriate for estimating one variable from the other.

Wald's idea for estimating the structural relation was extended to the forming of three groups by Nair and Banerjee[2] and by Bartlett.[3]

Sampling inspection. Paper {65} describes an inspection plan for continuous production which consists of grouping the items produced into blocks of size $n(= 1/f)$ and inspecting either one item or all items of a given block according to certain criteria. It is shown that such a procedure has two interesting properties:

a. The AOQL (average outgoing quality limit) is guaranteed no matter what the vagaries of the production process.

b. When the process is in statistical control the procedure minimizes the amount of inspection required to guarantee the AOQL.

An example is constructed to show that an inspection plan presented by Dodge[4] does not guarantee the claimed AOQL when the production process is not in control.

The procedure described in the present paper might be made the basis of a useful inspection procedure with further investigation and development. The

[1] J. Neyman and E. L. Scott, "On certain methods of estimating the linear structural relation," *Annals of Math. Stat.*, Vol. 22 (1951), pp. 352–361.

[2] K. R. Nair and K. S. Banerjee, "A note on fitting of straight lines if both variables are subject to error," *Sankhyā*, Vol. 6 (1942), p. 33.

[3] M. S. Bartlett, "Fitting a straight line when both variables are subject to error," *Biometrics*, Vol. 5 (1949), pp. 207–212.

[4] H. F. Dodge, "A sampling plan for continuous production," *Annals of Math. Stat.*, Vol. 14 (1943), pp. 264–279.

18 ABRAHAM WALD

property (*a*) just mentioned is only asymptotically true; in the short run unpleasant things may happen:

1. After a long good run a bad run will be allowed to continue for some time without interruption. The authors suggest one way of reducing this effect.

2. The procedure essentially estimates the proportion of defects in partially inspected blocks and requires enough blocks to be completely inspected to keep the estimated proportion of defects below the AOQL. The sampling variation of the estimate is ignored, however, so that with bad luck the AOQ of a finite set of items could seriously violate the AOQL.

The property (*b*) is also an asymptotic property and is proved under the restriction that n be fixed. An economical inspection procedure would probably have to alter n according to the current behavior of the production process.

Two problems of multivariate statistics. In {44} Wald and R. J. Brookner deal with the distribution of the likelihood ratio criterion for independence of sets of variates (where the variates are normally distributed) derived by Wilks.[1] The authors derive the exact distribution of this criterion (under the null hypothesis) when the numbers in each set are even and when all but one of these numbers are even. This treatment extends the cases given by Wilks. The second result in this paper is a series expansion of the cumulative distribution function of -2 times the logarithm of the likelihood ratio criterion. The terms of the series are incomplete Γ functions (equivalent to the χ^2 cumulative distribution functions). Since the terms of the series are tabulated and since usually not many terms are needed, this result is useful for carrying out the tests of independence. In the case of two sets of variates this criterion is equivalent to the criterion for the multivariate general linear hypothesis. Rao[2] used the result of Wald and Brookner to obtain an expansion of the distribution of a multiple of the logarithm; the multiple was chosen so that the second term in the series vanishes. This leads to a considerable simplification of the actual use of the series. Box[3] has given a more general treatment of the expansion of distributions of multivariate test criteria.

In {60} Wald deals with what is called a classification problem. One wants to classify an observed individual into one of two populations. If the two populations are known, the problem is essentially that of testing one simple hypothesis against another except that here the two hypotheses are treated symmetrically. In the case of two normal populations with a common covariance matrix, Wald uses the Neyman-Pearson fundamental lemma to show that the best criterion is a constant plus the population analogue of the dis-

[1] S. S. Wilks, "On the independence of k sets of normally distributed variables," *Econometrica*, Vol. 3 (1935), pp. 309–326.

[2] C. R. Rao, "Tests of significance in multivariate analysis," *Biometrika*, Vol. 35 (1948), pp. 58–79.

[3] G. E. P. Box, "A general distribution theory for a class of likelihood criteria," *Biometrika*, Vol. 36 (1949), pp. 317–346.

criminant function. Von Mises[1] derived similar criteria when there are more than two populations. Now suppose that the parameters of the two populations are unknown but that a sample from each population is given. Then Wald suggests that in the criterion above one replace the parameters by estimates from the sample and use the first of the two terms; this gives the discriminant function. If the samples are sufficiently large this criterion can be treated as the one with known parameters; that is, normally distributed. Wald uses the ideas of decision function theory to suggest the best procedure. For small sample sizes Wald attempted to derive the distribution of his criterion but did not carry the derivation to the point of giving an explicit result. If one modifies the discriminant function by adding a function of the two samples, one obtains a criterion that is more useful and whose distribution has only one parameter instead of several. Then it is possible to continue Wald's development to the point of giving the distribution except for the normalizing constant.[2] Wald's development is extremely involved because he used geometric methods; Sitgreaves[3] has derived the distribution by analytical means.

[1] R. von Mises, "On the classification of observation data into distinct groups," *Annals of Math. Stat.*, Vol. 16 (1945), pp. 68–73.

[2] T. W. Anderson, "Classification by multivariate analysis," *Psychometrika*, Vol. 16 (1951), pp. 31–50.

[3] Rosedith Sitgreaves, "On the distribution of two random matrices used in classification procedures," *Annals of Math. Stat.*, Vol. 23 (1952), pp. 263–270.

Reprinted from The Annals of Mathematical Statistics
Vol. 23, No. 1, March, 1952
Printed in U.S.A.

THE PUBLICATIONS OF ABRAHAM WALD

1931

{1} "Axiomatik des Zwischenbegriffes in metrischen Räumen," *Wiener Akademischer Anzeiger*, No. 16, pp. 1–3.

{2} "Axiomatik des Zwischenbegriffes in metrischen Räumen," *Math. Ann.*, Vol. 104, pp. 476–484.

{3} "Axiomatik des metrischen Zwischenbegriffes," *Ergebnisse eines Math. Kolloquiums*, Vol. 2, p. 17.

{4} "Über den allgemeinen Raumbegriff," *Ergebnisse eines Math. Kolloquiums*, Vol. 3, pp. 6–11.

{5} "Über das Hilbertsche Axiomensystem der Geometrie," *Ergebnisse eines Math. Kolloquiums*, Vol. 3, pp. 23–24.

1933

{6} "Zur Axiomatik der Verknüpfungsbeziehungen," *Ergebnisse eines Math. Kolloquiums*, Vol. 4, p. 6.

{7} "Zur Axiomatik des Zwischenbegriffes," *Ergebnisse eines Math. Kolloquiums*, Vol. 4, pp. 23–24.

{8} "Über die Volumsdeterminante," *Ergebnisse eines Math. Kolloquiums*, Vol. 4, pp. 25–28.

{9} "Über das Volumen der Euklidischen Simplexe," *Ergebnisse eines Math. Kolloquiums*, Vol. 4, pp. 32–33.

{10} "Halbmetrische Räume und Konvexifizierbarkeit," *Ergebnisse eines Math. Kolloquiums*, Vol. 5, p. 7.

{11} "Vereinfachter Beweis des Steinitzschen Satzes über Vektorenreihen im R_n," *Ergebnisse eines Math. Kolloquiums*, Vol. 5, pp. 10–13.

{12} "Bedingt konvergente Reihen von Vektoren im R_ω," *Ergebnisse eines Math. Kolloquiums*, Vol. 5, pp. 13–14.

{13} "Reihen in topologischen Gruppen," *Ergebnisse eines Math. Kolloquiums*, Vol. 5, pp. 14–16.

{14} "Komplexe und indefinite Räume," *Ergebnisse eines Math. Kolloquiums*, Vol. 5, pp. 32–42.

1935

{15} "Über die eindeutige positive Lösbarkeit der neuen Produktionsgleichungen (Mitteilung I)," *Ergebnisse eines Math. Kolloquiums*, Vol. 6, pp. 12–18.

{16} "Über die Anzahl der abgerundeten Mengen," *Ergebnisse eines Math. Kolloquiums*, Vol. 6, p. 26.

{17} "Eine Charakterisierung des Lebesgueschen Masses," *Ergebnisse eines Math. Kolloquiums*, Vol. 6, pp. 27–29.

{18} "Zur Differentialgeometrie der Flächen: I. Eine neue Definition der Flächenkrümmung," *Ergebnisse eines Math. Kolloquiums*, Vol. 6, pp. 29–39.

{19} "Sur la courbure des surfaces," *C. R. Acad. Sci. Paris*, Vol. 201, pp. 918–920.

1936

{20} "Sur la notion de collectif dans le calcul des probabilités," *C. R. Acad. Sci. Paris*, Vol. 202, pp. 180–183.

{21} "Über die Produktionsgleichungen der ökonomischen Wertlehre (Mitteilung II)," *Ergebnisse eines Math. Kolloquiums*, Vol. 7, pp. 1–6.

{22} "Begründung einer koordinatenlosen Differentialgeometrie der Flächen," *Ergebnisse eines Math. Kolloquiums*, Vol. 7, pp. 24–46.

{23} "Über einige Gleichungssysteme der mathematischen Ökonomie," *Zeitschrift für Nationalökonomie*, Vol. 7, pp. 637–670. English translation by Otto Eckstein, "On some systems of equations of mathematical economics," *Econometrica*, Vol. 19 (1951), pp. 368–403.

{24} *Berechnung und Ausschaltung von Saisonschwankungen, Beiträge zur Konjunkturforschung, Vol. 9*, Verlag von Julius Springer, Vienna.

1937

{25} "Ein Streckenbild, für welches λ_φ nicht existiert, obwohl λ_φ für jeden Anfangabschnitt existiert," *Ergebnisse eines Math. Kolloquiums*, Vol. 8, pp. 34–35.

{26} "Die ϕ-Länge im Hilbertschen Raum," *Ergebnisse eines Math. Kolloquiums*, Vol. 8, pp. 36–37.

{27} "Grundsätzliches zur Berechnung des Produktionsindex," *Monatsberichte des Österreichischen Institutes für Konjunkturforschung*, Vol. 11, Appendix 6 of February issue, pp. i–vii.

{28} "Zur Theorie der Preisindexziffern," *Zeitschrift für Nationalökonomie*, Vol. 8, pp. 179–219.

{29} "Die Widerspruchsfreiheit des Kollektivbegriffes der Wahrscheinlichkeitsrechnung," *Ergebnisse eines Math. Kolloquiums*, Vol. 8, pp. 38–72.

{30} "Extrapolation des gleitenden 12-Monatsdurchschnittes," *Monatsberichte des Österreichischen Institutes für Konjunkturforschung*, Vol. 11, Appendix 8 of November issue, pp. i–vii.

1938

{31} "Die Widerspruchsfreiheit des Kollektivbegriffes," *Actualités Scientifiques et Industrielles, No. 735, Colloque Consacré à la Théorie des Probabilités*, Hermann et Cie., Paris, pp. 79–99.

{32} "Generalization of the inequality of Markoff," *Annals of Math. Stat.*, Vol. 9, pp. 244–255.

1939

{33} "Long cycles as a result of repeated integration," *Am. Math. Monthly*, Vol. 46, pp. 136–141.

{34} "Confidence limits for continuous distribution functions" (with J. Wolfowitz), *Annals of Math. Stat.*, Vol. 10, pp. 105–118.

{35} "Limits of a distribution function determined by absolute moments and inequalities satisfied by absolute moments," *Trans. Am. Math. Soc.*, Vol. 46, pp. 280–306.

{36} "A new formula for the index of cost of living," *Econometrica*, Vol. 7, pp. 319–331.

{37} "Contributions to the theory of statistical estimation and testing hypotheses," *Annals of Math. Stat.*, Vol. 10, pp. 299–326.

1940

{38} "A note on the analysis of variance with unequal class frequencies," *Annals of Math. Stat.*, Vol. 11, pp. 96–100.

{39} "The approximate determination of indifference surfaces by means of Engel curves," *Econometrica*, Vol. 8, pp. 144–175.

{40} "On a test whether two samples are from the same population" (with J. Wolfowitz), *Annals of Math. Stat.*, Vol. 11, pp. 147–162.

{41} "The fitting of straight lines if both variables are subject to error," *Annals of Math. Stat.*, Vol. 11, pp. 284–300.

1941

{42} "Note on confidence limits for continuous distribution functions" (with J. Wolfowitz), *Annals of Math. Stat.*, Vol. 12, pp. 118–119.

{43} "Asymptotically most powerful tests of statistical hypotheses," *Annals of Math. Stat.*, Vol. 12, pp. 1–19.

{44} "On the distribution of Wilks' statistic for testing the independence of several groups of variates" (with R. J. Brookner), *Annals of Math. Stat.*, Vol. 12, pp. 137–152.

{45} "On the analysis of variance in case of multiple classifications with unequal class frequencies," *Annals of Math. Stat.*, Vol. 12, pp. 346–350.

{46} "Some examples of asymptotically most powerful tests," *Annals of Math. Stat.*, Vol. 12, pp. 396–408.

1942

{47} *On the Principles of Statistical Inference, Notre Dame Mathematical Lectures, No. 1*, University of Notre Dame.

{48} "Asymptotically shortest confidence intervals," *Annals of Math. Stat.*, Vol. 13, pp. 127–137.

{49} "On the choice of the number of class intervals in the application of the chi-square test" (with H. B. Mann), *Annals of Math. Stat.*, Vol. 13, pp. 306–317.

{50} "Setting of tolerance limits when the sample is large," *Annals of Math. Stat.*, Vol. 13, pp. 389–399.

{51} "On the power function of the analysis of variance test," *Annals of Math. Stat.*, Vol. 13, pp. 434–439.

1943

{52} "An extension of Wilks' method for setting tolerance limits," *Annals of Math. Stat.*, Vol. 14, pp. 45–55.

{53} "On a statistical generalization of metric spaces," *Proc. Nat. Acad. Sci.*, Vol 29, pp. 196–197.

{54} "On the efficient design of statistical investigations," *Annals of Math. Stat.*, Vol 14, pp. 134–140.

{55} *Sequential Analysis of Statistical Data: Theory*, Statistical Research Group, Columbia University.

{56} "On stochastic limit and order relationships" (with H. B. Mann), *Annals of Math. Stat.*, Vol. 14, pp. 217–226.

{57} "On the statistical treatment of linear stochastic difference equations" (with H. B. Mann), *Econometrica*, Vol. 11, pp. 173–220.

{58} "Tests of statistical hypotheses concerning several parameters when the number of observations is large," *Trans. Am. Math. Soc.*, Vol. 54, pp. 426–482.

{59} "An exact test for randomness in the nonparametric case based on serial correlation" (with J. Wolfowitz), *Annals of Math. Stat.*, Vol. 14, pp. 378–388.

1944

{60} "On a statistical problem arising in the classification of an individual into one of two groups," *Annals of Math. Stat.*, Vol. 15, pp. 145–162.

{61} "Note on a lemma," *Annals of Math. Stat.*, Vol. 15, pp. 330–333.

{62} "On cumulative sums of random variables," *Annals of Math. Stat.*, Vol. 15, pp. 283–296.

{63} "On a statistical generalization of metric spaces," *Rep. Math. Colloq., Notre Dame*, Ser. 2, Vol. 5-6, pp. 76–79.

{64} "Statistical tests based on permutations of the observations" (with J. Wolfowitz), *Annals of Math. Stat.*, Vol. 15, pp. 358–372.

1945

{65} "Sampling inspection plans for continuous production which insure a prescribed limit on the outgoing quality" (with J. Wolfowitz), *Annals of Math. Stat.*, Vol. 16, pp. 30–49.

{66} "Statistical decision functions which minimize the maximum risk," *Annals of Mathematics*, Vol. 46, pp. 265–280.

{67} "Generalization of a theorem by v. Neumann concerning zero-sum two-person games," *Annals of Mathematics*, Vol. 46, pp. 281–286.

{68} "Sequential tests of statistical hypotheses," *Annals of Math. Stat.*, Vol. 16, pp. 117–186.

{69} "Sequential method of sampling for deciding between two courses of action," *Jour. Am. Stat. Assn.*, Vol. 40, pp. 277–306.

{70} "Some generalizations of the theory of cumulative sums of random variables," *Annals of Math. Stat.*, Vol. 16, pp. 287–293.

1946

{71} "Tolerance limits for a normal distribution" (with J. Wolfowitz), *Annals of Math. Stat.*, Vol. 17, pp. 208–215.

{72} "Some improvements in setting limits for the expected number of observations required by a sequential probability ratio test," *Annals of Math. Stat.*, Vol. 17, pp. 466–474.

{73} "Differentiation under the expectation sign in the fundamental identity of sequential analysis," *Annals of Math. Stat.*, Vol. 17, pp. 493–497.

1947

{74} "A review of J. von Neumann and O. Morgenstern, *Theory of Games and Economic Behavior*," *Rev. Econ. Stat.*, Vol. 29, pp. 47–51.

{75} "Limit distribution of the maximum and minimum of successive cumulative sums of random variables," *Bull. Am. Math. Soc.*, Vol. 53, pp. 142–153.

{76} *Sequential Analysis*, John Wiley and Sons.

{77} "Sequential confidence intervals for the mean of a normal distribution with known variance" (with C. Stein), *Annals of Math. Stat.*, Vol. 18, pp. 427–433.

{78} "Foundations of a general theory of sequential decision functions," *Econometrica*, Vol. 15, pp. 279–313.

{79} "An essentially complete class of admissible decision functions," *Annals of Math. Stat.*, Vol. 18, pp. 549–555.

{80} "A note on regression analysis," *Annals of Math. Stat.*, Vol. 18, pp. 586–589.

1948

{81} "Asymptotic properties of the maximum likelihood estimate of an unknown parameter of a discrete stochastic process," *Annals of Math. Stat.*, Vol. 19, pp. 40–46.

{82} "On the distribution of the maximum of successive cumulative sums of independently but not identically distributed chance variables," *Bull. Am. Math. Soc.*, Vol. 54, pp. 422–430.

{83} "Estimation of a parameter when the number of unknown parameters increases indefinitely with the number of observations," *Annals of Math. Stat.*, Vol. 19, pp. 220–227.

{84} "Optimum character of the sequential probability ratio test" (with J. Wolfowitz), *Annals of Math. Stat.*, Vol. 19, pp. 326–339.

1949

{85} "Bayes solutions of sequential decision problems" (with J. Wolfowitz), *Proc. Nat. Acad. Sci.*, Vol. 35, pp. 99–102.

{86} "On distinct hypotheses" (with Agnes Berger), *Annals of Math. Stat.*, Vol. 20, pp. 104–109.

{87} "Statistical decision functions," *Annals of Math. Stat.*, Vol. 20, pp. 165–205.

{88} "Note on the consistency of the maximum likelihood estimate," *Annals of Math. Stat.*, Vol. 20, pp. 595–601.

{89} "A sequential decision procedure for choosing one of three hypotheses concerning the unknown mean of a normal distribution" (with Milton Sobel), *Annals of Math. Stat.*, Vol. 20, pp. 502–522.

1950

{90} "Bayes solutions of sequential decision problems" (with J. Wolfowitz), *Annals of Math. Stat.*, Vol. 21, pp. 82–89.

{91} "Elimination of randomization in certain problems of statistics and of the theory of games" (with A. Dvoretzky and J. Wolfowitz), *Proc. Nat. Acad. Sci.*, Vol. 36, pp. 256–260.

{92} "Note on the identification of economic relations," *Statistical Inference in Dynamic Economic Models, Cowles Commission Monograph No. 10*, John Wiley and Sons, pp. 238–244.

{93} "Remarks on the estimation of unknown parameters in incomplete systems of equations," *Statistical Inference in Dynamic Economic Models, Cowles Commission Monograph No. 10*, John Wiley and Sons, pp. 305–310.

{94} *Statistical Decision Functions*, John Wiley and Sons.

{95} "Note on zero-sum two-person games," *Annals of Mathematics*, Vol. 52, pp. 739–742.

1951

{96} "Relations among certain ranges of vector measures" (with A. Dvoretzky and J. Wolfowitz), *Pacific Jour. Math.*, Vol. 1, pp. 59–74.

{97} "Elimination of randomization in certain statistical decision procedures and zero-sum two-person games" (with A. Dvoretzky and J. Wolfowitz), *Annals of Math. Stat.*, Vol. 22, pp. 1–21.

{98} "On the fundamental lemma of Neyman and Pearson" (with G. B. Dantzig), *Annals of Math. Stat.*, Vol. 22, pp. 87–93.

{99} "Two methods of randomization in statistics and the theory of games" (with J. Wolfowitz), *Annals of Mathematics*, Vol. 53, pp. 581–586.

{100} "Asymptotic minimax solutions of sequential point estimation problems," *Proceedings of the Second Berkeley Symposium on Mathematical Statistics and Probability*, University of California Press.

{101} "Characterization of the minimal complete class of decision functions when the number of distributions and decisions is finite" (with J. Wolfowitz), *Proceedings of the Second Berkeley Symposium on Mathematical Statistics and Probability*, University of California Press.

1952

{102} "Basic ideas of a general theory of statistical decision rules," *Proceedings of the International Congress of Mathematicians*, Harvard University Press.

{103} "On a relation between changes in demand and price changes," *Econometrica*, Vol. 20 (1952), pp. 304–305.

1954

{104} "Testing the difference between the means of two normal populations with unknown standard deviations," *Selected Papers in Statistics and Probability by Abraham Wald*, McGraw-Hill Book Company.

VI

DIE WIDERSPRUCHSFREIHEIT DES KOLLEKTIVBEGRIFFES *

Von A. WALD
(Wien)

R. v. Misès (1) führt zur Grundlegung der Wahrscheinlichkeitsrechnung den Begriff des Kollektivs ein. In der Literatur wurde die Mises'sche Theorie vielfach diskutiert und auch verschiedene Einwände wurden dagegen erhoben. Insbesondere war die Frage der Widerspruchsfreiheit des Kollektivbegriffes umstritten. Ich möchte hier über die Ergebnisse meiner Arbeit (2) kurz referieren, in der der Kollektivbegriff präzisiert und seine Widerspruchsfreiheit bewiesen wird.

R. v. Misès geht von der Betrachtung einer unendlichen Folge $\{a_i\}$ ($i = 1, 2, \ldots$ ad inf.) gleichartiger Einzelerscheinungen aus. Diese Erscheinungen werden im einzelnen durch gewisse beobachtbare Merkmale unterschieden. Eine solche Erscheinungsreihe ist z. B. das unbegrenzt fortgesetzte Spiel mit einer Münze. Die i-te Erscheinung ist in diesem Falle der i-te Wurf mit der Münze, und diese Erscheinungen werden durch zwei Merkmale unterschieden, nämlich dadurch, ob Kopf oder Adler erscheint. Die Menge M aller Merkmale nennt man auch Merkmalraum. Der Merkmalraum M kann aus beliebig (endlich oder unendlich) vielen Elementen bestehen. Wesentlich ist, dass jeder Erscheinung a_i ($i = 1, 2, 3, \ldots$ ad inf.) ein bestimmtes Merkmal m_i aus dem Merkmalraum M

(1) R. v. Mises : Wahrscheinlichkeitsrechnung und ihre Anwendungen in der Statistik und theor. *Physik. Deuticke*, Leipzig Wien 1931. Die wesentlichen Grundzüge seiner Theorie hat Mises bereits 1919 in der *Mathem. Zeitschrift* Bd. 5 publiziert.

(2) Ergebnisse eines mathematischen Kolloquiums, Heft 8, Wien 1937.

* Reprinted from *Actualités scientifiques et industrielles:* No. 735, *Colloque consacré à la théorie des probabilités*, 1938.

zugeordnet ist, nämlich das Merkmal, das a_i aufweist. Da die Erscheinung a_i selbst keine Rolle spielt und nur das ihr zugeordnete Merkmal m_i in die Theorie eingeht, so genügt es für die Begründung des Kollektivbegriffes, bloss von einer abstrakten Menge M auszugehen, deren Elemente Merkmale oder Merkmalpunkte genannt werden, und Merkmalfolgen, d. h. Folgen von Elementen von M zu betrachten. Besteht der Merkmalraum z. B. bloss aus den zwei Elementen 0 und 1, so ist eine Merkmalfolge $\{ m_i \}$ ($i = 1,2, \ldots$ ad inf.) eine Folge von 0-en und 1-en, d. h. für jedes i gilt $m_i = 1$ oder O.

Unter « Auswahlvorschrift » verstehen wir eine Vorschrift, die jeder Merkmalfolge $\{ m_i \}$ eine Teilfolge (Auswahlfolge) zuordnet, wobei aber die Zugehörigkeit bezw. Nichtzugehörigkeit eines Elementes m_i zur Auswahlfolge bloss auf Grund des Wertes i und der vorangehenden Merkmalwerte m_1, $m_2 \ldots$, m_{i-1}, bestimmt wird. Es kann also die Auswahl eines Elementes nicht von dem Merkmalwert desselben Elementes oder von Merkmalwerten nachfolgender Elemente abhängig gemacht werden. Besteht der Merkmalraum bloss aus den zwei Elementen 0 und 1, so ist eine Auswahlvorschrift etwa die folgende Vorschrift : das Element m_i wird dann und nur dann ausgewählt, falls i eine Primzahl und $m_{i-1} = 0$ ist. Dagegen ist die Vorschrift, dass m_i dann und nur dann ausgewählt wird, falls $m_{i+1} = 1$ ist, keine Auswahlvorschrift. Der so definierte Begriff der Auswahlvorschrift stimmt mit dem Mises'schen Bergiff der « Stellenauswahl » überein. Eine Auswahlfolge kann unendlich, endlich oder leer sein.

Ist L eine Teilmenge des Merkmalraumes M und $A = \{ m_i \}$ eine Folge von Merkmalen, so verstehen wir unter der relativen Häufigkeit H_n (L, A) von L in dem Anfangsabschnitt $m_1, \ldots m_n$ die durch n dividierte Anzahl der zu L gehörigen Elemente dieses Anfangsabschnittes.

Die Mises'sche Fassung des Kollektivbegriffes lässt sich dann so formulieren : Ist M ein Merkmalraum und $K = \{ m_i \}$ ($i = 1,2, \ldots$ ad inf.) eine Folge von Merkmalen, so nennen wir K ein Kollektiv, falls die folgenden zwei Bedingungen erfüllt sind :

1) Das Grenzwertaxiom. — Für jede Teilmenge L von M konvergiert die Folge der relativen Häufigkeiten H_n (L, K) mit wachsendem n gegen einen Grenzwert. Diesen Grenzwert nennt

man auch die Wahrscheinlichkeit von L in bezug auf das Kollektiv K und wir bezeichnen sie mit μ (L).

2) Das Regellosigkeitsaxiom. — Ist A eine unendliche Teilfolge von K, die nach irgendeiner Auswahlvorschrift aus K ausgewählt wurde, ist ferner L eine beliebige Teilmenge von M, so konvergiert die Folge der relativen Häufigkeiten H_n (L, A) mit wachsendem n ebenfalls gegen den Grenzwert μ(L).

Die Wahrscheinlichkeit ist also eine nicht negative additive Mengenfunktion μ(L), definiert für alle Teilmengen von M. Diese Mengenfunktion μ(L) wird auch Verteilungsfunktion des Kollektivs K genannt.

Die Einwände gegen die Mises'sche Theorie richteten sich hauptsächlich gegen die Formulierung des Regellosigkeitsaxioms. In der Tat führt die obige Formulierung zu Widersprüchen. Wird nämlich zur Bildung einer Auswahlfolge A ohne Einschränkung jede Auswahlvorschrift zugelassen, dann kann die Regellosigkeitsforderung für keine Folge K erfüllt sein. Dies sieht man wie folgt ein : eine Folge $\{ i_j \}$ ($j = 1, 2,...$ ad inf.) von natürlichen Zahlen definiert offenbar eine Auswahlvorschrift, indem man festsetzt, das n-te Element der Merkmalfolge $\{ m_i \}$ ($i = 1, 2...$ ad inf.) solle dann und nur dann ausgewählt werden, falls n ein Element der Folge $\{ i_j \}$ ist. Wir bezeichnen mit γ das System der Auswahlvorschriften, das dem System aller natürlichen Zahlenfolgen entspricht. Ist nun K $= \{ m_i \}$ ein Kollektiv, so betrachten wir eine Teilmenge L des Merkmalraumes M, deren Wahrscheinlichkeit μ (L) $\neq 0$ und $\neq 1$ ist. (Den trivialen Fall, dass μ(L) für jede Teilmenge L von M gleich 0 oder gleich 1 ist, wollen wir ausschliessen). Aus μ(L) $\neq 0$ folgt, dass unendlich viele Elemente aus $\{ m_i \}$ in L enthalten sind. Daraus ergibt sich aber, dass es eine Auswahlvorschrift in γ gibt, die aus $\{ m_i \}$ eine unendliche Auswahlfolge liefert, deren sämtliche Elemente in L enthalten sind. Damit sind wir zu einem Widerspruch gelangt un haben gezeigt, dass kein Kollektiv K existiert.

Offenbar meinte jedoch Mises, dass nicht alle beliebigen Auswahlvorschriften zugelassen seien, sondern nur solche, die durch mathematische Gesetze definierbar sind. Dann müsste allerdings der Begriff des mathematischen Gesetzes präzisiert werden. Dies tut aber Mises nicht und solange dies nicht geschieht, kann man die Frage der Widerspruchsfreiheit des Mises'schen Kollektivbegriffes gar nicht sinnvoll stellen. Es sei hier noch bemerkt, dass Mises die mit dem Wort « alle » verbundenen Schwierigkeiten auch dadurch zu beseitigen versucht, dass er seiner Regellosigkeitsforderung wörtlich folgende Deutung gibt [1] : « Die Festsetzung, dass in einem Kollektiv jede Stellenauswahl (in unserer Ausdrucksweise, Auswahl nach einer « Auswahlvorschrift ») die Grenzhäufigkeit unverändert lässt, besagt nichts anderes als dies : Wir verabreden, dass, wenn in einer konkreten Aufgabe das Kollektiv einer bestimmten Stellenauswahl unterworfen wird, wir annehmen wollen, diese Stellenauswahl ändere nichts an den Grenzwerten der relativen Häufigkeiten ». Gibt man der Regellosigkeitsforderung dieseDeutung, so wird der Kollektivbegriff sozusagen bezüglich der vorliegenden Aufgabe relativiert, indem nur das System der in der Aufgabe tatsächlich vorkommenden Auswahlvorschriften für die Bildung von Auswahlfolgen zugelassen wird. Liegt dann eine konkrete Aufgabe vor, so kann man zwar die Frage, ob Kollektivs dieser Art existieren, sinnvoll stellen, sie muss jedoch noch gelöst werden, d. h. es muss für jede konkret vorliegende Aufgabe gezeigt werden, dass entsprechende Kollektivs existieren.

Falls der Merkmalraum M unendlich ist, so führt auch die For derung, dass alle Teilmengen von M eine Wahrscheinlichkeit haben, zu Widersprüchen. Es gilt nämlich : falls die Verteilungsfunktion μ für jeden Merkmalpunkt p den Wert $\mu\ (p) = 0$ hat, so gibt es kein Kollektiv K, in dem jede Teilmenge des Merkmalraumes eine Wahrscheinlichkeit hat. Es sei bemerkt, dass diese Tatsache in keinem Zusammenhang mit dem bekannten Satz steht, dass in einer unendlichen Menge N keine totaladditive nichtnegative Mengenfunktion ν für alle Teilmengen von N existieren kann, falls $\nu\,(\mathrm{N}) > 0$ und für jeden Punkt p von N $\nu(p) = 0$ ist. Die Verteilungsfunktion μ muss nämlich nicht totaladditiv sein, sie ist nur additiv, und es gibt tatsächlich nichtnegative additive Mengenfunktionen ν, die für alle Teilmengen einer unendlichen Menge N definiert sind und für die für jeden Punkt p von N $\nu\ (p) = 0$ und $\nu\,(\mathrm{N}) > 0$ gilt. Dass eine solche nichtnegative additive Mengenfunktion nicht als Verteilungsfunktion eines Kollektivs K realisierbar ist, liegt daran, dass es keine Merkmalfolge gibt, in der die Grenz-

---

[1] Mises : Wahrscheinlichkeit, Statistik und Wahrheit, vierte Auflage Wien, Julius Springer 1936, S. 119.

werte der relativen Häufigkeiten, gebildet für jede Teilmenge des Merkmalraumes, eine solche Mengenfunktion ergeben würden.

Die obigen Ueberlegungen zeigen also, dass man, wenn man Widersprüche vermeiden will, sowohl bezüglich des Systems der zulässigen Auswahlvorschriften als auch bezüglich des Systems der Teilmengen von M, die eine Wahrscheinlichkeit haben sollen, gewisse Einschränkungen machen muss. Wir werden daher den Kollektivbegriff relativieren.

1. Definition des Kollektivs. — Es sei M ein Merkmalraum, $\mathfrak{M}$ ein System von Teilmengen von M, γ ein System von Auswahlvorschriften und $K = \{m_i\}$ $(i = 1,2,...$ ad inf.) eine Folge von Merkmalen. Wir sagen, *dass* K *ein Kollektiv bezüglich* γ *und* $\mathfrak{M}$ *ist, in Zeichen* K (γ, μ), *falls folgende zwei Bedingungen erfüllt sind* :

1) Grenzwertaxiom. — *Für jedes Element* L *aus* $\mathfrak{M}$ *konvergieren die relativen Häufigkeiten* H_n (L, K) *mit wachsendem n gegen einen Grenzwert* μ (L), *genannt die Wahrscheinlichkeit von* L.

2) Regellosigkeitsaxiom. — *Ist* A *eine unendliche Teilfolge von* K, *die durch eine Vorschrift des Systems* γ *aus K ausgewählt wurde, und ist* L *ein Element von* $\mathfrak{M}$, *so konvergieren die relativen Häufigkeiten* H_n (L, A) *mit wachsendem n gegen die Wahrscheinlichkeit* μ (L) *von* L.

Die Wahrscheinlichkeit ist also eine nichtnegative additive Mengenfunktion, definiert für alle Elemente aus $\mathfrak{M}$. Diese Mengenfunktion nennt man auch Verteilungsfunktion des Kollektivs K $(\gamma, \mathfrak{M})$.

2. Formulierung der Hauptergebnisse. — Anknüpfend an die obige Definition des Kollektivs formulieren wir zunächst das.

Problem I. — *Sei* M *ein gegebener Merkmalraum. Welchen Bedingungen muss das System* γ *von Auswahlvorschriften, das System* $\mathfrak{M}$ *von Teilmengen von* M, *und die für alle Elemente von* $\mathfrak{M}$ *erklärte Mengenfunktion* μ *genügen, damit ein Kollektiv* K $(\gamma, \mathfrak{M})$ *existiert, dessen Verteilungsfunktion mit* μ *identisch ist* ?

Die Lösung dieses Problems wird durch die folgenden vier Theoreme gegeben :

Theorem I. — *Ist der Merkmalraum* M *eine endliche Menge* (*von einer Mächtigkeit* > 1), γ *ein System von abzählbar vielen Auswahl-*

vorschriften, $\mathfrak{M}$ *das System aller Teilmengen von* M, *und* μ *eine nichtnegative additive Mengenfuntion, definiert für alle Elemente von* $\mathfrak{M}$, *wobei* μ (M) = 1 *gilt, so gibt es kontinuierlich viele Kollektivs* K (γ, $\mathfrak{M}$), *deren Verteilungsfunktion gleich* μ *ist.*

Theorem II. — *Ist der Merkmalraum* M *eine unendliche (abzählbare oder unabzählbare) Menge,* γ *ein System von abzählbar vielen Auswahlvorschriften,* $\mathfrak{M}$ *das System aller Teilmengen von* M, *und* μ *eine nichtnegative additive Mengenfunktion, definiert für alle Elemente von* $\mathfrak{M}$, *wobei* μ (M) = 1 *gilt, dann ist die notwendige und hinreichende Bedingung dafür, dass ein Kollektiv* K (γ, $\mathfrak{M}$) *mit der Verteilungsfunktion* μ *existiere, dass es eine Folge* $\{m_i\}$ ($i = 1, 2...$ *ad inf.*) *von paarweise verschiedenen Merkmalen gibt, für welche*

$$\sum_{i=1}^{\infty} \mu(m_i) = 1 \text{ ist.}$$

In der klassischen Theorie wird im Falle der sogen, geometrischen Wahrscheinlichkeiten (wo also der Merkmalraum aus kontinuierlich vielen Punkten besteht) stets die Annahme gemacht, dass die einzelnen Merkmalpunkte die Wahrscheinlichkeit 0 haben. Aus Theorem II ergibt sich daher, dass in allen diesen Fällen *keine* Kollektivs bezüglich aller Teilmengen von M existieren. Wir können also höchstens von Kollektivs bezüglich gewisser Teilmengen von M sprechen. Das folgende Theorem zeigt, dass es jedoch möglich ist, sehr umfassende für die Anwendungen völlig ausreichende Systeme von Teilmengen von M anzugeben, bezüglich welcher Kollektivs existieren.

Theorem III. — *Voraussetzungen* :

1. $\mathfrak{M}$ *ist ein unendlicher Merkmalraum von beliebiger Mächtigkeit.*

2. $\mathfrak{K}$ *ist ein abzählbarer Mengenkörper, dessen Elemente Teilmengen von* M *sind und der* M *selbst als Element enthält.*

3. μ *ist eine nichtnegative additive Mengenfunktion, erklärt für alle Elemente aus* $\mathfrak{K}$, *wobei* μ (M) = 1 *gilt.*

4. $\mathfrak{M}$ *sei das System aller in bezug auf* $\mathfrak{K}$ μ*-messbaren Teilmengen von* M. Dabei heisst eine Teilmenge A von M in bezug auf $\mathfrak{K}$ μ-messbar, falls das äussere μ-Mass von A, d. i. die untere Schranke der Werte von μ für alle Elemente aus $\mathfrak{K}$, die A enthalten, und das

innere μ-Mass von A, d. i. die obere Schranke der Werte von μ für alle Elemente aus $\mathcal{K}$, die in A enthalten sind, einander gleich sind ; dieser gemeinsame Wert heisst $\mu(A)$, das μ-Mass von A.

5. *γ sei ein System von abzählbar vielen Auswahlvorschriften.* *Behauptung : Es gibt kontinuierlich viele Kollektivs* K $(\gamma, \mathfrak{M})$, *deren Verteilungsfunktion gleich* $\mathfrak{M}$ *ist.*

In den Anwendungen werden nur Merkmalmengen M betrachtet, die entweder endlich, oder im Peanoschen Sinne messbare Teilmengen des n-dimensionalen euklidischen Raumes R_n sind. Im letzteren Fall sind nur die Wahrscheinlichkeiten der im Peanoschen Sinne messbaren Teilmengen von M für die Anwendungen von Bedeutung. Ferner handelt es sich hier ausschliesslicht um Verteilungsfunktionen μ, für welche folgendes gilt : Ist $\{M_i\}$ $(i = 1, 2, \ldots$ ad inf.) eine Folge von Teilmengen von M, wobei das Peanosche Mass μ_P (M_i) von M_i mit wachsendem i gegen 0 konvergiert, so gilt auch $\lim \mu(M_i) = 0$. Für die Zwecke der Anwendungen reicht daher vollkommen aus das

Theorem IV. — *Ist der Merkmalraum* M *eine beschränkte im Peanoschen Sinne messbare Teilmenge des n-dimensionalen euklidischen Raumes* R_n, *ist ferner* μ *eine nichtnegative additive Mengenfunktion, erklärt für alle im Peanoschen Sinne messbaren Teilmengen von* M, *wobei* $\mu(M) = 1$ *gilt und* $\lim_{i=\infty} \mu(M_i) = 0$ *ist, falls* $\{M_i\}$ *eine Folge von Peanosch-messbaren Teilmengen bedeutet, für welche das Peanosche Mass* $\mu_P(M_i)$ *mit wachsendem i gegen* 0 *konvergiert, und ist schliesslich* γ *ein System von abzählbar vielen Auswahlvorschriften, so gibt es kontinuierlich viele Kollektivs bezüglich* γ *und aller im Peanoschen Sinne messbaren Teilmengen von* M, *deren Verteilungsfunktion identisch gleich* μ *ist.*

Theorem IV ist als Spezialfall in Theorem III enthalten, denn wählt man den abzählbaren Mengenkörper $\mathcal{K}$ so, dass er jede Teilmenge von M, die Durchschnitt von M mit einem rationalen Intervall des R_n ist, enthält, so ist jede im Peanoschen Sinne messbare Teilmenge von M zugleich μ-messbar in bezug auf den Körper $\mathcal{K}$.

Die Theoreme I-IV geben Bedingungen über γ, $\mathfrak{M}$ und μ an, die hinreichend, aber nicht notwendig sind, damit Kollektivs K $(\gamma, \mathfrak{M})$ mit der Verteilungsfunktion μ existieren. Doch sind diese

Bedingungen schon so wenig einschränkend, dass eine weitere Abschwächung derselben kaum von Interesse wäre. Hinsichtlich der Bedingungen für $\mathfrak{M}$ und μ haben wir dies bereits gesehen. Ist nämlich M endlich, so haben wir üver μ und M gar keine einschränkende Annahmen gemacht (Theorem I). Ist M unendlich, so haben wir bloss die Bedingung gestellt, dass die Elemente von $\mathfrak{M}$ in gewissem Sinne messbare Teilmengen von M seien (Theorem III und IV). Nun wollen wir klar machen, dass auch die an γ gestellte Bedingung, nämlich dass γ nur abzählbar viele Auswahlvorschriften enthält, so schwach ist, dass eine weitere Abschwächung für die Anwendungen nicht von Interesse wäre. Sogar der Mises'sche Kollektivbegriff genügt nämlich, wenn man den Begriff des mathematischen Gesetzes im Sinne der modernen Logik präzisiert, dieser Bedingung. Der Begriff des mathematischen Gesetzes kann ja nur innerhalb eine formalisierten Logik sinnvoll definiert werden und mithin gibt es offenbar nur abzählbar viele mathematische Gesetze. Man kann z. B. die von Whitehead und Russell in den Principia Mathematica entwickelte formalisierte Logik (innerhalb deren das gesamte Gebiet der Mathematik dargestellt werden kann) zugunde legen, und γ als das System aller Auswahlvorschriften annehmen, die durch mathematische Gesetze dieser formalisierten Logik definiert werden können. Man sieht also, dass die Bedingung der Abzählbarkeit von γ so wenig einschränkend ist, dass eine weitere Abschwächung für die Anwendungen von Bedeutung wäre. Besonders klar wird dies, wenn man der Regellosigkeitsforderung die erwähnte Mises'sche Deutung gibt, welche die Regellosigkeitsforderung bezüglich der vorliegenden Aufgabe relativiert, indem bloss gefordert wird, dass die Auswahlen, die nach in der Aufgabe tatsächlich vorkommenden Auswahlvorschriften gebildet werden, an den Grenzwerten der relativen Häufigkeiten nichts ändern. Denn es ist klar, dass in jeder konkreten Aufgabe der Wahrscheinlichkeitsrechnung höchstens abzählbar viele Auswahlvorschriften tatsächlich verwendet werden. Die Theoreme I-IV liefern also auch dann einen Widerspruchsfreiheitsbeweis für den Mises'schen Kollektivbegriff, wenn man der Regellosigkeitsforderung die obige Deutung gibt. Bezeichnet man ein Kollektiv K(γ, $\mathfrak{M}$) als Misessches Kollektiv bezüglich einer formalisierten Logik L, falls γ das System aller Auswahlvorschriften bedeutet, die durch mathematische Gesetze der formalisierten

Logik L definiert sind, und als Mises'sches Kollektiv bezüglich einer Aufgabe A, falls γ das System der in der Aufgabe A tatsächlich vorkommenden Auswahlvorschriften bedeutet, so gilt folgendes.

Korollar. — *Ist* M *ein Merkmalraum,* μ *eine nichtnegative additive Mengenfunktion definiert in einem abzählbaren Körper* $\mathfrak{K}$ *von Teilmengen von* M (*falls* M *endlich ist, so sei* μ *für alle Teilmengen von* M *definiert*), *wobei* $\mathfrak{K}$ *die Menge* M *als Element enthält und* μ (M) = 1 *gilt, so gibt es Mises'sche Kollektivs sowohl bezüglich jeder formalisierten Logik* L, *als auch bezüglich jeder Aufgabe* A, *die nur abzählbar viele Auswahlvorschriften verwendet* (*und nur solche kommen in den Anwendungen vor*), *derart, dass jede in bezug auf den Körper* $\mathfrak{K}$ μ*-messbare Teilmenge* N (*falls* M *endlich ist jede Teilmenge* N) *eine Wahrscheinlichkeit* W (N) *besitzt, die gleich* μ (N) *ist.*

Die Theoreme I-IV stellen also eine für die Anwendungen völlig ausreichende Lösung des Problems 1 dar. Es sei hier noch bemerkt, dass, falls die Wahrscheinlichkeitsrechnung als ein Teil der Mathematik in einer formalisierten Logik L entwickelt und dargestellt wird, es am zweckmässigsten ist, das Misessche Kollektiv bezüglich der formalisierten Logik L an die Spitze zu stellen. Das Misessche Kollektiv bezüglich der formalisierten Logik L wird in dieser Theorie gleichzeitig auch ein Misessches Kollektiv bezüglich jeder Aufgabe A sein, da in jeder Aufgabe nur Auswahluorschriften auftreten, die durch mathematische Gesetze der formalisierten Logik L definiert sind.

Ein weiteres Problem knüpft an folgende Definition an: Die Merkmalfolge $\{m_i\}$ (i = 1,2..., ad inf.) heisse konstruktiv definiert, falls ein Verfahren vorliegt, das für jede natürliche Zahl i den Merkmalwert m_i in endlich vielen Schritten tatsächlich zu berechnen gestattet. Z. B. ist die Merkmalfolge $\{m_i\}$ (i = 1,2..., ad inf.) wo m_i = 1 oder = 0 ist, je nachdem ob n eine Primzahl ist oder nicht, konstruktiv definiert, da man für jede natürliche Zahl n in endlich vielen Schritten berechnen kann, ob n eine Primzahl ist oder nicht. Dageden ist die Folge $\{m_i\}$ (i = 1, 2,... ad inf.), wo m_i gleich 1 oder 0 ist, je nachdem der grosse Fermatsche Satz für i gilt oder nicht, nicht konstruktiv definiert, da wir nicht für jedes i in endlich vielen Schritten feststellen können, ob der grosse Fermatsche Satz gilt oder nicht.

PROBLEM 2. — *Welchen Bedingungen muss das abzählbare System* γ *von Auswahlvorschriften, das System* $\mathfrak{M}$ *von Teilmengen des Merkmalraumes* M, *und die für die Elemente von* $\mathfrak{M}$ *erklärte Mengenfunktion* μ *genügen, damit Kollektivs* K (γ, $\mathfrak{M}$) *mit der Verteilungsfunktion* μ *konstruktiv definiert werden können ? Ein entsprechendes Konstruktionsverfahren für solche Kollektivs* K (γ, $\mathfrak{M}$) *soll effektiv angegeben werden.*

Eine Folge $\gamma = \{V_i\}$ ($i = 1,2...$ ad inf.) von Auswahlvorschriften heisse konstruktiv definiert, falls für jedes natürliche Zahlanpaar n und i und für ein beliebiges geordnetes n-1-Tupel $= m_1,... m_{n-1}$ von Merkmalen in endlich vielen Schritten berechnet werden kann, ob das n-te Element einer Merkmalfolge, in der die $n - 1$ vorangehenden Merkmale mit dem gegebenen $n - 1$ Tupel von Merkmalen übereinstimmen, nach der Vorschrift V_i ausgewhält wird oder nicht.

Ist $\mathfrak{a} = \{A_i\}$ ($i = 1,2...$) eine endliche oder unendliche Folge von Teilmengen von M, so sagen wir, dass die Mengenfunktion μ für die Elemente von $\mathfrak{a}$ konstruktiv definiert ist, falls für jede natürliche Zahl i der Funktionswert μ (A_i) in endlich vielen Schritten berechnet werden kann. Ist der Merkmalraum endlich, so ist μ in den Anwendungen stets konstruktiv definiert, da man für jeden Merkmalpunkt von M den Wert von μ explizit angibt. Die Lösung des Problems 2 wird dann durch die folgenden zwei Theoreme gegeben :

THEOREM V. — *Ist* M *ein endlicher Merkmalraum, ist* μ *eine nichtnegative additive Mengenfunktion, definiert für alle Teilmengen von* M, *wobei* $\mu(M) = 1$ *ist, und ist* γ *ein konstruktiv definiertes abzählbares System von Auswahlvorschriften, so kann man Kollektivs bezüglich* μ *und aller Teilmengen von* M *mit der Verteilungsfunktion* μ *konstruktiv definieren.*

THEOREM VI. — *Ist* M *ein unendlicher Merkmalraum, ist* μ *eine nichtnegative additive Mengenfunktion, die konstruktiv definiert ist für die Elemente eines abzählbaren Körpers* $\mathfrak{K}$ *von Teilmengen von* M, *wobei* M *ein Element von* $\mathfrak{K}$ *ist und* m (M) $= 1$ *gilt, und ist* γ

ein konstruktiv definiertes abzählbares System von Auswahlvorschriften, so kann man Kollektivs K (γ, $\mathfrak{M}$) *mit der Verteilungsfunktion* μ *konstruktiv definieren, wobei* $\mathfrak{M}$ *das System aller in bezug auf* $\mathcal{K}$ μ*-messbaren Teilmengen von* M *bedeutet.*

Für Kollektivs, die den Voraussetzungen des Theorems V bezw. VI genügen, gebe ich in meinen Arbeit l. c. ein Konstruktionsverfahren effektiv an. Damit ist das Problem 2 vollständig gelöst. In den konkreten Aufgaben werden wohl nur Kollektivs betrachtet, bei denen die Verteilungsfunktion μ und das System der Auswahlvorschriften, die in der gegebenen Aufgabe zur Bildung von Auswahlfolgen tatsächlich verwendet werden, konstruktiv definiert sind. Unsere Theorie liefert also ein Konstruktionsverfahren für Misessche Kollektivs in bezug auf jede praktisch vorkommende Aufgabe. Damit dürften auch die Einwände der Intuitionisten wegfallen, da nach dem intuitionistischen Standpunkt erst recht nur Aufgaben gestellt werden, bei denen die obgenannten Bedingungen erfüllt sind.

Es sei noch einiges über die Beziehungen dieser Ergebnisse zu den Untersuchungen von H. Reichenbach, K. Popper und A. H. Copeland bemerkt. H. Reichenbach [1] und neuerdings K. Popper [2] haben Merkmalfolgen untersucht, die einer stark eingeschränkten Regellosigkeitsforderung genügen, und die « normale » bezw, « nachwirkungsfreie » Folgen genannt werden. Ist M endlich, so ist eine nachwirkungsfreie Folge $\{ m_i \}$ in unserer Ausdrucksweise ein Kollektiv K (γ, $\mathfrak{M}$), wobei $\mathfrak{M}$ das System aller Teilmengen von M bedeutet und γ alle und nur die Auswahlvorschriften von folgender Art enthält : ein Element m_n aus $\{ m_i \}$ wird dann und nur dann ausgewählt, falls das diesem Element unmittelbar vorangehende k-Tupel (m_{n-k},..., m_{n-1}) von Merkmalen einem festgegebenen k-Tupel von Merkmalen (a_1,..., a_k) gleich ist. Besteht der Merkmalraum M bloss aus den zwei Elementen 0 und 1, so sind Auswahlvorschriften solcher Art z. B. die folgenden : ein Element wird dann und nur dann ausgewählt, falls ihm das Element 0 unmittelbar vorangeht ; oder : ein Element wird dann und nur dann ausgewählt, falls ihm die Kombination 0 0 unmittelbar vorangeht, usw. Reichenbach formuliert l. c. als ungelöstes

---

(1) Reichenbach, *Mathem. Zeitschrift*, Bd. 34.

(2) K. Popper, *Logik der Forschung*, Wien 1935.

Problem die Frage, ob nachwirkungsfreie Folgen durch eine Rechenvorschrift konstruktiv definiert werden können. Diese Frage hat K. Popper l. c. für den Fall, dass der Merkmalraum M bloss aus zwei Elementen besteht, positiv entschieden, indem er ein Konstruktionsverfahren für nachwirkungsfreie Folgen angibt. Diese Ergebnisse sind als Spezialfall in unseren Theoremen enthalten. Man kann nämlich leicht zeigen, dass die normalen bezw. nachwirkungsfreien Folgen Kollektivs in bezug auf ein bestimmtes System γ_0 von Auswahlvorschriften sind. Dieses System γ_0 ist sogar konstruktiv definiert und mithin sind die normalen Folgen Kollektivs, die auch den Voraussetzungen des Theorems V genügen. Also liefert unsere Theorie auch Konstruktionsverfahren für normale bezw. nachwirkungsfreie Folgen.

Die interessanten Copeland' (1) schen Untersuchungen gehen weit über die oben erwähnten Ergebnisse von Popper und Reichenbach hinaus. Diese Arbeit weist ebenfalls gewisse Berührungspunkte mit der unseren auf. Insbesondere ist Copelands Theorem I (l. c. S. 185) als Spezialfall in unserem Theorem I enthalten. Copeland betrachtet speziell einen Merkmalraum, der bloss aus zwei Merkmalen besteht, und Auswahlvorschriften von folgender speziellen Art : Es wird das i_k-te Element ($k = 1,2...$ ad inf.) ausgewählt, wobei $\{ i_k \}$ eine monoton wachsende Folge von natürlichen Zahlen ist. (Insbesondere hängt also in diesem Falle die Auswahl des Merkmales m_i aus einer Merkmalfolge bloss von i ab, nicht von den m_i vorangehenden Merkmalwerten). Copelands Theorem I besagt dann, ohne übrigens Konstruktionsvorschriften zu enthalten : Ist D ein abzählbares System von Auswahlvorschriften obiger Art, so existieren kontinuierlich viele « admissible numbers », von denen jede die Eigenschaft hat, dass jede aus ihr nach einer Vorschrift des Systems D gebildete Auswahlfolge ebenfalls ein « admissible number » ist.

Nun ist ein « admissible number » ein Kollektiv bezüglich eines abzählbaren (und sogar konstruktiv definierten) Systems γ_0 von Auswahlvorschriften. Um zu zeigen, dass Copelands Theorem I in unserem Theorem I enthalten ist, definieren wir zunächst als Produkt von endlich vielen Auswahlvorschriften $V_1, ... V_k$ folgende

(1) A. H. Copeland : Point set theory applied to the random selection of the digits of an admissible number. *Amer. Journ. of math.* 58, 1936.

Vorschrift V : Es wird aus der Merkmalfolge $\{m_i\}$ ($i = 1,2,...$ ad inf.) zuerst eine Auswahlfolge $\{m_i^{(1)}\}$ nach Vorschrift V_1 gebildet, dann aus der Folge $\{m_i^{(1)}\}$ nach Vorschrift V_2 eine Auswahlfolge $\{m_i^{(2)}\}$ gebildet, ...und schliesslich wird aus $\{m_i^{(k-1)}\}$ ($i = 1,2...$ ad inf.) eine Auswahlfolge $\{m_i^{(k)}\}$ ($i = 1, 2,...$ ad inf.) nach Vorschrift V_k gebildet. Die Vorschrift V ordnet also einer Merkmalfolge $\{m_i\}$ ($i = 1, 2...$ ad inf.) die Auswahlfolge $\{m_i^{(k)}\}$ ($i = 1, 2,...$ ad inf.) zu. Man bestätigt leicht, dass das Produkt von k Auswahlvorschriften ebenfalls eine Auswahlvorschrift ist. Wir nennen ein System von Auswahlvorschriften, das zu je zwei Auswahlvorschriften auch ihr Produkt enthält, eine Gruppe. Es sei nun γ_1 die kleinste Gruppe, die die Auswahlvorschriften von γ_0 und D enthält. Offenbar ist γ_1 abzählbar und jedes Kollektiv bezüglich γ_1, genügt der Behauptung des Copelandschen Theorems I.

3. Ueber einige Einwande gegen den Kollektivbegriff. — Nachdem die Widerspruchsfreiheit des Kollektivbegriffes bewiesen ist, fallen alle Einwände *logischer* Natur weg; der Aufbau der Wahrscheinlichkeitsrechnung auf der Grundlage des Kollektivbegriffes kann widerspruchsfrei durchgeführt werden. Es können gegen diese Begriffsbildung höchstens Einwände erhoben werden, die ihre Zweckmässigkeit oder Fruchtbarkeit betreffen. Man kann etwa fragen, ob die so aufgebaute Theorie das klassische Lehrgebäude der Wahrscheinlichkeitsrechnung umfasst, d. h., ob alle Sätze der klassischen Theorie auch in der neuen Theorie beweisbar sind, und insbesondere ob die neue Theorie nicht Sätze enthält, die den Sätzen der klassischen Theorie widersprechen ? Man kann auch fragen, ob nicht einfachere Begründungen der Wahrscheinlichkeitsrechnung möglich sind, die aber dasselbe leisten. Auf diese und ähnliche Fragen möchte ich hier nicht ausführlich eingehen. Ich werde mich vielmehr mit den zwei Einwänden beschäftigen, die Herr Fréchet in seinem Vortrage gegen den Kollektivbegriff geltend gemacht hat, und werde dabei auch die erwähnten Fragen kurz behandeln.

Der erste Einwand des Herrn Fréchet richtete sich gegen das Grenzwertaxiom. Nach dem Grenzwertaxiom wird die Existenz des Grenzwertes der relativen Häufigkeiten gefordert und die

Wahrscheinlichkeit wird diesem Grenzwert gleichgesetzt. Nun ist es aber offenbar *möglich*, dass man auch mit einer richtigen Münze (wobei « Kopf » und « Adler » die gleiche Wahrscheinlichkeit haben) stets nur « Kopf » wirft, d. h. dass man die Reihe.

(*) 1111...

erhält, wobei das Merkmal « Kopf » durch 1 bezeichnet ist. Diese Möglichkeit wird aber ausgeschlossen, wenn man die Wahrscheinlichkeit im Sinne des Kollektivbegriffes definiert, und dies sei ein Mangel der Theorie.

Es scheint mir jedoch, dass der obige Umstand keinen Nachteil bedeutet, u. zw. aus folgenden Gründen :

Die Wahrscheinlichkeitsrechnung dient dazu, um auf die realen Erscheinungen der Aussenwelt angewendet zu werden. Macht man eine genügend lange Reihe von Versuchen und findet man, dass *immer* das Ereignis 1 eingetreten ist, so wird kein Statistiker den Ansatz machen, dass die Münze eine richtige ist. Es würde zwar keinen Widerspruch bedeuten diese Annahme zu machen, denn auch in diesem Falle ist es *möglich*, dass die gegebene Versuchsreihe entsteht, es wäre nur ein sehr unwahrscheinlicher Fall; jedoch wird dies kein Statistiker tun. Ich glaube, dass es keinen Nachteil bedeutet, wenn man Fälle, die von den Statistikern ausgeschlossen werden, auch theoretisch ausschliesst. Allerdings könnte man dann sagen, wie Herr DE FINETTI bemerkt hat, dass man z. B. auch den Fall, dass die ersten 1000 Würfe mit einer rich-, tigen Münze stets das Resultat 1 ergeben, ausschliessen könnte denn kein Statistiker wird in diesem Falle die Annahme machen, dass die Münze eine richtige ist. Dies könnte man in der Tat tun, wenn es keine Schwierigkeiten gäbe, wo man diese Grenze festlegen soll, bei 100, oder bei 1000, oder erst bei 10000, denn eine scharfe Grenze kann man ja nicht angeben. Gegen den Ausschluss des Grenzfalles, dass alle Versuchsergebnisse ad inf. stets gleich 1 sind, bestehen jedoch solche Bedenken nicht.

Uebrigens handelt es sich bei diesem Grenzfall um einen idealisierten Fall, der in der Wirklichkeit nie auftreten kann, da man nicht unendlich viele Versuche machen kann. Es besteht dafür bloss eine logische Möglichkeit. Ob der Ausschluss dieser logischen Möglichkeit einen Nachteil bedeutet, kann einzig und allein von Zweckmässigkeitsgründen abhängen, nämlich davon, ob dadurch in dem Aufbau der Theorie gewisse Schwierigkeiten sich ergeben.

Es scheint mir aber, dass dies nicht der Fall ist. Man kann nämlich zeigen, dass die auf den Kollektivbegriff aufgebaute Theorie der klassischen nicht widerspricht, und sie ist auch nicht enger als die klassische, d. h. man kann jeden klassischen Satz auch in der Theorie des Kollektivs beweisen. Wir wollen uns dies etwas klarer machen. In der Wahrscheinlichkeitstheorie werden zwei Arten von Sätzen ausgesprochen. Es gibt Sätze, die Wahrscheinlichkeitsaussagen über endliche Versuchsreihen, und solche, die Wahrscheinlichkeitsaussagen über unendliche Versuchsreihen machen. Die letzteren werden aus den ersteren durch Grenzprozesse hergeleitet. Um die Wahrscheinlichkeitsaussagen über unendliche Versuchsreihen aus denen über endliche Versuchsreihen herleiten zu können, muss man nur die Totaladditivität der Wahrscheinlichkeit postulieren. Gemäss dem Kollektivbegriffe haben jedoch nur Wahrscheinlichkeitsaussagen über endliche Versuchsreihen einen wirklichen Sinn, da eine unendliche Versuchsreihe keinen Wiederholungsvorgang bildet. Es ist bekannt, dass man alle Wahrscheinlichkeitsaussagen über endliche Versuchsreihen, die in der klassischen Theorie beweisbar sind, auch in der Theorie des Kollektivs beweisen kann. Was die Wahrscheinlichkeitsaussagen über unendliche Versuchsreihen anbelangt, so haben diese in der Mises'schen Theorie zwar keinen unmittelbaren Sinn, jedoch können wir sie auch in der Mises'schen Theorie rein konventionell einführen. Um dies klarer auszudrücken, betrachten wir den einfachen Fall, dass der Merkmalraum bloss aus den Elementen 0 und 1 besteht und dass die beiden Merkmale die Wahrscheinlichkeit 1/2 haben. Einer unendlichen Versuchsreihe, d. h. einer unendlichen Folge $\{ m_i \}$ ($i = 1, 2, \ldots$ ad inf.) von Nullen und Einsen kann die zwischen 0 und 1 liegende dyadische Zahl $0, m_1, m_2 \ldots$ eindeutig zugeordnet werden. Eine Folge $\{ m_i \}$ von Merkmalen ist dann durch einen Punkt des Zahlenintervalles $J = [0, 1]$ dargestellt. Eine Eigenschaft E der Folge $\{ m_i \}$ ist durch die Menge aller Punkte von J, die die Eigenschaft E haben, charakterisiert. Wenn das Erfülltsein bezw. Nichterfülltsein der Eigenschaft E stets schon durch die n ersten Merkmalwerte $m_1 \ldots, m_n$ eindeutig bestimmt ist, d. h. E eine Eigenschaft einer endlichen Versuchsreihe ist, so wird E durch eine Summe von endlich vielen Teilintervallen von J dargestellt. In der Theorie des

Kollektivs haben nur die Wahrscheinlichkeiten von Summen endlich vieler Teilintervalle von J einen eigentlichen Sinn, da nur diese als Grenzwerte von relativen Häufigkeiten interpretiert werden können (es ist ja nur eine endliche und keine unendliche Versuchsreihe ein Wiederholungsvorgang). Es besteht aber gar keine Schwierigkeit, rein konventionell auch solchen Teilmengen von J eine Wahrscheinlichkeit zuzuordnen, die nicht Summe von endlich vielen Intervallen sind. Wir haben für die Bestimmung der Wahrscheinlichkeiten von Teilmengen von J, die nicht Summe von endlich vielen Intervallen sind, bloss folgende zwei Konventionen zu treffen :

(1) Ist $\{ A_i \}$ $(i = 1, 2, \ldots$ ad inf.) eine Folge von paarweisen fremden Teilintervallen von J, so gilt

$$W(A) = \sum_{i=1}^{\infty} W(A_i),$$

wobei $A = \sum_{i=1}^{\infty} A_i$, W(A) die Wahrscheinlichkeit von A, und $W(A_i)$ die Wahrscheinlichkeit von A_i bedeutet.

(2) Sind A und B zwei Teilmengen von J und ist A in B enthalten, so gilt $W(A) \leqq W(B)$.

Die Konvention (2) ist notwendig, falls A oder B nicht Summe von abzählbar vielen Intervallen ist. Ohne die Konvention (2) könnte man die Wahrscheinlichkeit nur für Teilmengen bestimmen, die Summe von abzählbar vielen Intervallen sind.

Auf Grund der Konventionen (1) und (2) wird dann jeder im Lebesgue'schen Sinne messbaren Teilmenge von J eindeutig eine Wahrscheinlichkeit zugeordnet. Auf diese Weise kann man dann jeden Satz der klassischen Theorie auch in der Theorie des Kollektivs beweisen. Man muss sich nur vor Augen halten, dass bloss die Wahrscheinlichkeiten von Summen endlich vieler Intervalle einen eigentlichen Sinn und die Wahrscheinlichkeiten der übrigen Teilmengen rein konventionellen Charakter haben, indem sie aus den Wahrscheinlichkeiten erster Art auf Grund der Konventionen (1) und (2) hergeleitet werden. Man sieht unmittelbar, dass auf Grund der Konventionen (1) und (2) der Satz gilt : Die Wahrscheinlichkeit dafür, dass man mit einer richtigen Münze bei unbegrenzt fortgesetztem Spiel stets nur das Resultat 1 erhält, ist gleich Null.

Der wirkliche Inhalt der obigen Aussage besteht darin, dass die Wahrscheinlichkeit, mit einer richtigen Münze n mal nacheinander 1 zu werfen, *mit wachendem n gegen* 0 *konvergiert*, und ohne die Konvention (2) könnte man nicht mehr als dies aussagen. Die Aussage steht weder in Widerspruch zum Grenzwertaxiom, noch ist sie eine Folge desselben.

Der zweite Einwand des Herrn FRÉCHET knüpft an einen Satz an, den A. VILLE bewiesen hat. Dieser Satz klautet folgendermassen : *Besteht der Merkmalraum aus den Elementen* 0 *und* 1 *und haben die beiden Merkmale die Wahrscheinlichkeit* 1 /2, *ist* γ *ferner ein abzählbares System von Auswahlvorschriften, so gibt es stets ein Kollektiv* $K = \{ m_i \}$ $(i = 1, 2, \ldots$ *ad inf.*) *bezüglich des Systems* γ, *in dem die beiden Merkmale die Wahrscheinlichkeit* 1 /2 *haben und die folgende merkwürdige Eigenschaft besteht : für jede natürliche Zahl n gilt, dass die relative Hufigkeit der Eins in demAnfangsabschnitt* $m_1, \ldots, m_n$ *grösser als der Grenzwert der relativen Häufigkeiten, also grösser als* 1 /2 *ist.* Wir wollen diese Eigenschaft mit E bezeichnen. Man kann leicht zeigen, dass die Wahrscheinlichkeit W (E) von E gleich 0 ist. Bezeichnet man nämlich mit E_n die Eigenschaft, dass für jede natürliche Zahl $r \leqq n$ die relative Häufigkeit der Eins in dem Anfangsabschnitt $m_1, \ldots, m_r$ grösser als 1 /2 ist, so lässt sich leicht zeigen, dass die Wahrscheinlichkeit W (E_n) von E_n mit wachsendem n gegen 0 konvergiert. Der Einwand bestenht nun darin, dass, wie man auch das System γ von Auswahlvorschriften wählt, stets Kollektivs existieren, die die Eigenschaft E aufweisen, obwohl W (E) = 0 ist.

Ich möchte zunächst bemerken, dass diese Tatsache weder einen Widerspruch noch die Unbeweisbarkeit des Satzes W(E) = 0 in der Theorie des Kollektivs bedeutet. Dies geht schon aus unseren früheren Ueberlegungen hervor. Die Eigenschaft E ist nämlich eine Eigenschaft einer unendlichen Versuchsreihe und über die Wahrscheinlichkeit von E zu sprechen, ist nur eine Angelegenheit konventioneller Natur. Aus den Konventionen (1) und (2) folgt unmittelbar, dass W (E) = 0 ist. Der wirkliche Inhalt dieser Aussage ist bloss die Behauptung, dass $\lim_{n=\infty} W(E_n) = 0$ ist, und diese Behauptung lässt sich auch ohne die Konventionen (1) und (2) in der Theorie des Kollektivs beweisen. Aus $\lim W(E_n) = 0$ folgt keinesfalls, dass kein Kollektiv die Eigenschaft E hat.

Der eigentliche Sinn des Einwandes besteht vielleicht darin, dass vom Kollektivbegriff gefordert wird, es solle jede asymptotische Eigenschaft, die die klassische Wahrscheinlichkeit 1 besitzt, im Kollektiv immer erfüllt sein, oder was dasselbe bedeutet, es solle keine asymptotische Eigenschaft mit der Wahrscheinlichkeit 0 im Kollektiv jemals bestehen. Die in dem Grenzwertaxiom ausgedrückte Eigenschaft, dass die relativen Häufigkeiten gegen einen Grenzwert streben, ist ja im Sinne der klassischen Theorie ebenfalls nichts anderes als eine asymptotische Eigenschaft mit der Wahrscheinlichkeit 1. Es wird nun der Einwand gemacht : Warum will man eben *diese* asymptotische Eigenschaft auszeichnen und nicht lieber fordern, dass *alle* asymptotischen Eigenschaften mit der Wahrscheinlichkeit 1 im Kollektivs stets erfüllt seien ?

Es scheint mir zwar, dass man nicht jeder asymptotischen Eigenschaft dieselbe Bedeutung beilegen kann, wie der in dem Grenzwertaxiom ausgedrückten Eigenschaft, jedoch glaube ich, dass eine Verschärfung des Kollektivbegriffes in diesem Sinne tatsächlich von Interesse wäre. Eine solche lässt sich in der Tat leicht durchführen. Allerdings muss man auch hier das Wort « alle » einer Einschränkung unterwerfen. Genau so wie die Regellosigkeitsforderung nur für ein abzählbares System von Auswahlvorschriften postuliert wurde, ebenso müssen wir uns auf ein abzählbares System von asymptotischen Eigenschaften beschränken. Diese Beschränkung bedeutet keinen Nachteil, denn in jeder formalisierten Theorie können nur abzählbar viele asymptotische Eigenschaften formuliert werden und wir können uns auf das System der in der Theorie ausdrückbaren asymptotischen Eigenschaften beschränken. Ein umfassenderes System zu betrachten, hat nicht viel Interesse.

Wir wollen hier den verschärften Kollektivbegriff für den Fall angeben, dass der Merkmabraun bloss aus den Elementen 0 und 1 besteht und die beiden Merkmalpunkte die gleiche Wahrscheinlichkeit haben. Es sei auch hier jede Merkmalfolge $\{ m_i \}$ als eine dyadische Zahl zwischen 0 und 1 aufgefasst. Jeder Merkmalfolge entspricht dann ein Punkt des Intervalles $J = [0,1]$. Eine Auswahlvorschrift V heisse *singulär*, falls die Menge der Punkte von J, aus denen nach Vorschrift V eine unendliche Teilfolge von Merkmalen ausgewählt wird, eine Nullmenge im Lebesgue'schen Sinne ist.

Ist K = { m_i } eine Merkmalfolge und γ ein System von Auswahlvorschriften, so sagen wir, dass K ein Kollektiv im schärferen Sinne in bezug auf γ ist, falls K ein Kollektiv im früheren Sinne in bezug auf γ ist, und nach jeder singulären Vorschrift des Systems γ nur endlich viele Elemente aus K ausgewählt werden [1].

Für diesen verschärften Begriff des Kollektivs gelten die folgenden zwei Sätze :

I) *Ist γ ein System von abzählbar vielen Auswahlvorschriften, so gibt es Kollektivs im schärferen Sinne in bezug auf γ.*

II) *Zu jedem System $\mathfrak{v}$ von abzhälbar vielen asymptotischen Eigenschaften gibt es ein System γ von abzahlbar vielen Auswahlvorschriften derart, dass für jedes Kollektiv im schärferen Sinne in bezug auf γ, folgendes gilt : jede asymptotische Eigenschaft* E, *die Element von $\mathfrak{v}$ ist, ist für das Kollektiv erfüllt bezw, nicht erfüllt, je nachdem ob die Wahrscheinlichkeit von* E *gleich* 1 *oder* 0 *ist.*

Wird die Wahrscheinlichkeitserechnung innerhalb einer formalisierten Logik entwickelt und bedeutet $\mathfrak{v}$ das System der asymptotischen Eigenschaften, die innerhalb dieser Theorie formuliert werden können, bedeutet ferner γ ein System von abzählbar vielen Auswahlvorschriften, für welches Satz II gilt, dann ist jede Merkmalfolge, die ein Kollektiv im schärferen Sinne in bezug auf γ ist, zugleich ein vollständiges häufigkeitstheoretisches Modell der Wahrscheinlichkeitsrechnung.

Zum Schluss möchte ich noch eine Bemerkung über die sogenannten « normalen » Merkmalfolgen machen. Wie bereits erwähnt wurde, sind die normalen Folgen Kollektivs in bezug auf ein speziell gewähltes System γ_0 von Auswahlvorschriften. Es sei nun K = { m_i } eine normale Folge, und E eine Eigenschaft eines n-Tupels von Versuchsergebnissen, d. h. eine Eigenschaft eines geordneten n-Tupels von Merkmalen. Zerlegt man die normale

[1] Es ist hier nicht bezwecht diese Verschärfung des Kollektivbegriffes zu empfehlen ; wir wollen damit bloss veranschanlicher, dass überhaupt eine Verschärfung des Kollektivbegriffes, wodurch die erwähnten Einwände hinfällig werden, moglich und leicht durchuführbar ist.

Folge K in aufeinanderfolgende Abschnitte der Länge n, so hat die relative Häufigkeit jener Abschnitte, welche die Eigenschaft E besitzen, einen Grenzwert, der gleich der Wahrscheinlichkeit W(E) von E ist. Man kann also schon in der Theorie der normalen Folgen die Wahrscheinlichkeit von E richtig bestimmen, falls man festsetzt, dass die Elemente des Kollektivs, in bezug auf welches die Wahrscheinlichkeit von E bestimmt wird, die in der normalen Folge K unmittelbar aufeinanderfolgende Abschnitte der Länge n sind. Macht man noch die Konvention (1) und (2), so kann man schon alle Wahrscheinlichkeitsaussagen der klassischen Theorie herleiten. Man könnte dann fragen, ob nicht die Theorie der normalen Folgen für den Aufbau der Wahrscheinlichkeitsrechnung bereits ausreicht ? Eine solche Theorie eignet sich jedoch nicht, wie auch von Herrn v. Mises betont wird ([1]), zur Beschreibung der Glücksspiele. Um dies zu veranschaulichen, betrachten wir wieder den Fall, dass der Merkmalraum bloss aus den Elementen 0 und 1 besteht. Man kann dann eine normale Merkmalfolge angeben (sogar durch eine konstruktive Rechenvorschrift), in der der Grenzwert der relativen Häufigkeiten des Merkmals 1 gleich 1/2 ist und an jeder Primzahlstelle das Merkmal 1 auftritt. Eine solche Merkmalfolge ist sicherlich nicht geeignet für die Beschreibung des Glücksspieles mit einer richtigen Münze. Es gibt nämlich in Wirklichkeit kein Spielsystem, bei dem man immer gewinnt. Bei der erwähnten Merkmalfolge wäre aber ein solches Spielsystem möglich, indem man sich nur an den primzahlstelligen Spielen beteiligt und immer auf das Eintreffen des Ereignisses 1 setzt.

Trifft man in der Theorie der normalen Folgen die Festsetzung, dass für die Bestimmung der Wahrscheinlichkeit einer Eigenschaft eines n-Tupels von Versuchen das Kollektiv betrachtet wird, dessen Elemente die in der normalen Folge unmittelbar aufeinanderfolgenden Abschnitte der Länge n sind, macht man ferner die Konventionen (1) und (2), so kann man auch in der Theorie der normalen Folgen den folgenden Satz beweisen : Wählt man aus der Merkmalfolge $\{ m_i \}$ jedes primzahlstellige Element aus, so ist die Wahrscheinlichkeit dafür, dass auch innerhalb der Auswahlfolge der Grenzwert der relativen Häufigkeiten exis-

---

([1]) R. v. Mises : Wahrscheinlichkeit, *Statistik und Wahrheit*, Wien Verlag von Julius Springer 1936, Seite 116.

tiert und mit dem entsprechenden Grenzwert in der ursprünglichen Folge übereinstimmt, gleich 1. Wie man sieht, ist diese Aussage vereinbar damit, dass in der ursprünglichen Merkmalfolge an jeder Primzahlstelle das Merkmal 1 steht. Für die richtige Beschreibung des Glücksspieles müssen wir aber von einem Kollektiv $K = \{ m_i \}$ fordern, dass auch die Auswahlfolge tatsächlich ein Kollektiv sei, was hier nicht der Fall ist.

Imprimerie R. Bussière. — Saint-Amand (Cher). France. — 13-10-1938

Reprinted from THE ANNALS OF MATHEMATICAL STATISTICS
Vol. X, No. 2, June, 1939

CONFIDENCE LIMITS FOR CONTINUOUS DISTRIBUTION FUNCTIONS[1]

BY A. WALD[2] AND J. WOLFOWITZ

1. **Introduction.** The theory of confidence limits for unknown parameters of distribution functions has been considerably developed in recent years. This theory assumes that there is given a family F of systems of n stochastic variables $X_1(\theta_1, \cdots, \theta_k), \cdots, X_n(\theta_1, \cdots, \theta_k)$ depending upon k parameters $\theta_1, \cdots, \theta_k$ and such that the distribution function of every element of F is known.

For the case $k = 1$, for example, this theory proceeds as follows:

Denote by E an n-tuple $x_1, \cdots, x_n$ of observed values of the stochastic variables $X_1(\theta), \cdots, X_n(\theta)$ of which we know only that they constitute a system which is an element of F. E can be represented as the point $x_1, \cdots, x_n$ in an n-dimensional Euclidean space. Let there be given a positive number α, $0 < \alpha < 1$. Then to each pair E, α there is constructed a θ-interval, $[\underline{\theta}(E, \alpha), \bar{\theta}(E, \alpha)]$ with the following property: If we were to draw a sample from the system $X_1(\theta), \cdots, X_n(\theta)$, the probability is exactly α that we shall get a system of observations $E = x_1, \cdots, x_n$ such that the interval corresponding to E, α will include θ (i.e., that $\underline{\theta}(E, \alpha) \leq \theta \leq \bar{\theta}(E, \alpha)$).

In this paper we do not limit ourselves to a family of systems of n stochastic variables depending upon a finite number of parameters, but consider the family G of all systems of n stochastic variables $X_1, \cdots, X_n$ subject only to the condition that $X_1, \cdots, X_n$ are independently distributed with the same continuous distribution function.

Let E be the point in an n-dimensional Euclidean space which corresponds to the observed values $x_1, \cdots, x_n$ of the n stochastic variables $X_1, \cdots, X_n$ of which we know only that they constitute an element of the family G, i.e., that they are independently distributed with the same continuous distribution function. Let us denote their distribution function by $f(x)$; the probability that $X_i < x$ is $f(x)$, $i = 1, \cdots, n$. Let α be a number such that $0 < \alpha < 1$. To each pair E, α we shall construct two functions, $\bar{l}_{E,\alpha}(x)$ and $\underline{l}_{E,\alpha}(x)$, with the following property: The probability is α that, if we were to draw a sample from the system $X_1, \cdots, X_n$, we would get a system of observations $E = x_1, \cdots, x_n$ such that $f(x)$ lies entirely between $\bar{l}_{E,\alpha}(x)$ and $\underline{l}_{E,\alpha}(x)$ (i.e., that $\underline{l}_{E,\alpha}(x) \leq f(x) \leq \bar{l}_{F,\alpha}(x)$ for all x). We shall call $\bar{l}_{E,\alpha}(x)$ and $\underline{l}_{E,\alpha}(x)$ the upper and lower confidence limits, respectively, corresponding to the confidence coefficient α.

---

[1] Presented to the American Mathematical Society at New York, February 25, 1939.

[2] Research under a grant-in-aid from the Carnegie Corporation of New York.

All the stochastic variables considered hereafter in this paper are to have continuous distribution functions.

2. **A theorem on continuous distribution functions.** Let $f(x)$ be the continuous distribution function of a stochastic variable X whose range is from $-\infty$ to $+\infty$. Let $\delta_1(x)$ and $\delta_2(x)$ be two functions defined for $0 \leq x \leq 1$ and satisfying the following requirements:

(a) $\delta_1(x)$ and $\delta_2(x)$ are non-negative and continuous for $0 \leq x \leq 1$.

(b) $l_1(x)$ and $l_2(x)$ are monotonically non-decreasing for all x, where

$$l_1(x) \equiv f(x) + \delta_1(f(x))$$

$$l_2(x) \equiv f(x) - \delta_2(f(x)).$$

(c) There exists a number h, such that $f(h) < 1$ and $l_1(h) = 1$.

(d) There exists a number h', such that $f(h') > 0$ and $l_2(h') = 0$.

(e) $l_1(x) \leq 1$ for all x
$l_2(x) \geq 0$ for all x

(f) $\delta_1(x) + \delta_2(x) \geq \frac{1}{n}$ for all x, where n is the number of random, independent observations of the stochastic variable X.

Let $\varphi(x)$ be the distribution function of such a system of observations, i.e., the ratio, to n, of the number of observations $<x$ is $\phi(x)$. $\varphi(x)$ is, of course, a multiple of $\frac{1}{n}$ for all x.

We shall consider the following problem:

What is the probability P that

$$l_2(x) \leq \varphi(x) \leq l_1(x) \tag{1}$$

for all x?

The reasons for restrictions (b), (c), (d), (e), and (f) on $\delta_1(x)$ and $\delta_2(x)$ are now apparent. If there exist two numbers $q_1 < q_2$, such that, for $q_1 < x < q_2$, $l_1(x) > l_1(q_2)$ and $l_1(q_1) = l_1(q_2)$, then, if we change $l_1(x)$ so that $l_1(x) = l_1(q_2)$ for $q_1 \leq x \leq q_2$, P will remain unchanged. An analogous process leads to a similar conclusion for $l_2(x)$. Hence $l_1(x)$ and $l_2(x)$ are to be monotonically non-decreasing. If there did not exist a number h or h', P would be 0. Hence requirements (c) and (d). Since $0 \leq \varphi(x) \leq 1$, there is no point to considering functions which do not satisfy (e). $\varphi(x)$ is a step-function whose saltuses are $\geq \frac{1}{n}$. If, for all x.

$$\delta_1(x) + \delta_2(x) < \frac{1}{n}$$

then $P = 0$. If there is an interval $[\beta, \gamma]$ within which $\delta_1(x) + \delta_2(x) < \frac{1}{n}$, then all samples in which one of the observed values lies in this interval are such that (1) does not hold for all x. For the sake of simplicity and because the situation described in (f) is the one of importance, we make the latter requirement.

It would appear that P depends upon $f(x)$, $\delta_1(x)$, $\delta_2(x)$, and n.

THEOREM: *P is independent of $f(x)$ and depends only upon $\delta_1(x)$, $\delta_2(x)$, and n.*

PROOF: Let $Y = f(X)$. Then Y is a stochastic variable distributed in the range 0 to 1 with a distribution function $\equiv x$. By this transformation $l_1(x)$ and $l_2(x)$ become respectively

$$\left.\begin{aligned} l_1'(x) &= x + \delta_1(x) \\ l_2'(x) &= x - \delta_2(x) \end{aligned}\right\} 0 \leq x \leq 1. \tag{2}$$

Then P is the probability that the distribution function $\varphi(x)$ of a random sample of n of the stochastic variable Y shall be such that $l_2'(x) \leq \varphi(x) \leq l_1'(x)$ and is therefore independent of $f(x)$.

3. **Computation of** P. From the previous section it follows that, in computing P, we may confine ourselves to a stochastic variable X whose range is from 0 to 1 and whose distribution function $\equiv x$. Let $l_1(x)$ and $l_2(x)$ be the upper and lower limits, respectively, which are set for $\varphi(x)$. $l_1(x)$ and $l_2(x)$ are defined in (2), if the accents are omitted.

Consider the equations:

$$l_1(x) = \frac{i}{n} \qquad (i = 1, 2, \cdots, n; 0 \leq x \leq 1). \tag{3}$$

If, for a certain i, the corresponding equation possesses one or more solutions in x, let a_i be the minimum of these solutions. If the first r of these equations (3) have no solutions, let

$$a_i = 0 \qquad (i = 1, \cdots, r).$$

If the ith, say, of the equations

$$l_2(x) = \frac{i-1}{n} \qquad (i = 1, \cdots, n; 0 \leq x \leq 1) \tag{4}$$

possesses one or more solutions in x, let b_i be the maximum of these. If the last $n - s$ of the equations (4) have no solutions, let

$$b_i = 1 \qquad (i = s + 1, \cdots, n).$$

Obviously

$$a_i \leq a_{i+1}, \qquad b_i \leq b_{i+1}, \qquad a_i \leq b_i .$$

From restrictions, (e) and (f) on $l_1(x)$ and $l_2(x)$, it follows that $a_1 = 0$, $b_n = 1$.

Suppose the sample $E = x_1, \cdots, x_n$ has been obtained. Arrange the x's in ascending order, thus: $x_{p_1}, x_{p_2}, \cdots, x_{p_n}$ where $x_{p_1} \leq x_{p_2} \leq \cdots \leq x_{p_n}$. Then necessary and sufficient conditions that (1) hold are:

$$a_i \leq x_{p_i} \leq b_i \qquad (i = 1, \cdots, n). \tag{5}$$

Let $P_k(t, \Delta t)$, $(k = 0, 1, \cdots, (n-1); a_{k+1} \leq t \leq b_{k+1})$ be the probability that a sample $E = x_1, \cdots, x_n$ shall fulfill the following conditions:

(a) $x_1 \leq x_2 \leq \cdots \leq x_{k+1}$,

(b) $x_1, \cdots, x_k$ satisfy the first k inequalities (5),

(c) $t \leq x_{k+1} \leq t + \Delta t$.

Let

$$P_k(t) = \lim_{\Delta t \to 0} \frac{P_k(t, \Delta t)}{\Delta t}.$$

Since $f(x) \equiv x$, we get easily

$$P_0(t) \equiv 1. \tag{6}$$

We shall now develop a recursion formula for $P_{k+1}(t)$. For this purpose let us consider the following composite event: The observations $x_1, \cdots, x_n$ satisfy the conditions (a), (b), and

$$t' \leq x_{k+1} \leq t' + \Delta t'$$

and

$$t \leq x_{k+2} \leq t + \Delta t.$$

If $a_{k+1} \leq t' \leq b_{k+1}$, the probability of this event is $P_k(t', \Delta t')\Delta t$. Now

$$\lim_{\substack{\Delta t' \to 0 \\ \Delta t \to 0}} \frac{P_k(t', \Delta t')\Delta t}{\Delta t' \cdot \Delta t} = P_k(t').$$

$P_k(t')$ is obviously the probability density of the bivariate distribution of t' and t. In order to obtain $P_{k+1}(t)$ we have to integrate $P_k(t')\, dt'$ over the region defined by the two inequalities

$$t' \leq t$$

$$a_{k+1} \leq t' \leq b_{k+1}.$$

Hence, omitting the now unnecessary accent, if

$$t \leq b_{k+1} \tag{7}$$

then

$$P_{k+1}(t) = \int_{a_{k+1}}^{t} P_k(t)\, dt \qquad (k = 0, 1, \cdots, (n-2)), \tag{8}$$

and if

$$t > b_{k+1} \tag{9}$$

then

$$P_{k+1}(t) = \int_{a_{k+1}}^{b_{k+1}} P_k(t)\,dt \quad (k = 0, 1, 2, \cdots (n-2)). \tag{10}$$

Now, to obtain P, we cannot confine ourselves only to cases where $x_1 \leq x_2 \leq \cdots \leq x_n$, but have to consider all the $n!$ permutations of the n x's. Hence

$$P = n! \int_{a_n}^{b_n} P_{n-1}(t)\,dt. \tag{11}$$

The fact that there are two forms of the recursion formula corresponding to the two possible cases (7) and (9) makes actual calculation very cumbersome for n of any considerable size. We shall therefore give an approximation formula which is considerably easier to apply to practical calculations.

4. **Computation of $\overline{P}$ and $\underline{P}$.** Let $\overline{P}$ be the probability that, for a sample of n, $l_1(x) \geq \varphi(x)$ for all x. Let $\underline{P}$ be the probability that, for a sample of n, $\varphi(x) \geq l_2(x)$ for all x.

Consider the inequalities

$$\left.\begin{aligned} x_i &\geq a_i \quad (12)\\ x_i &\leq b_i \quad (13) \end{aligned}\right\} \qquad (i = 1, 2, \cdots, n)$$

Let

$$\overline{P}_k(t, \Delta t), \qquad (k = 0, 1, \cdots, (n-1);\ t \geq a_{k+1})$$

be the probability that a sample $E = x_1, \cdots, x_n$ of the stochastic variable X should fulfill the following conditions:

(a) $x_1 \leq x_2 \leq \cdots \leq x_{k+1}$

(b) $x_1, \cdots, x_k$ satisfy the first k inequalities (12)

(c) $t \leq x_{k+1} \leq t + \Delta t$.

Let

$$\overline{P}_k(t) = \lim_{\Delta t \to 0} \frac{\overline{P}_k(t, \Delta t)}{\Delta t}.$$

Then, by an argument like that employed in the preceding section, we obtain

$$\overline{P}_0(t) \equiv 1, \tag{14}$$

and the recursion formula

$$\overline{P}_{k+1}(t) = \int_{a_{k+1}}^{t} \overline{P}_k(t)\,dt. \tag{15}$$

Let $\overline{P}_n(t)$ be defined formally by (15). Then, in the same way in which we obtained (11), we get

$$\overline{P} = n!\,\overline{P}_n(1). \tag{16}$$

In the same manner we shall obtain an expression for $\underline{P}$.

Let $\underline{P}_k(t, \Delta t)$, $(k = 0, 1, \cdots (n - 1); t \leq b_{n-k})$ be the probability that a sample $E = x_1, \cdots, x_n$ of the stochastic variable X should fulfill the following conditions:

(a) $x_{n-k} \leq x_{n-k+1} \leq \cdots \leq x_n$,
(b) $x_{n-k+1}, \cdots, x_n$ satisfy the last k inequalities (13),
(c) $t \leq x_{n-k} \leq t + \Delta t$.

Let

$$\underline{P}_k(t) = \lim_{\Delta t \to 0} \frac{\underline{P}_k(t, \Delta t)}{\Delta t}.$$

Then

$$\underline{P}_0(t) \equiv 1 \tag{17}$$

and by an argument very similar to that employed above,

$$\underline{P}_{k+1}(t) = \int_t^{b_{n-k}} \underline{P}_k(t)\, dt. \tag{18}$$

Let $\underline{P}_n(t)$ be defined formally by (18). Then

$$\underline{P} = n!\, \underline{P}_n(0). \tag{19}$$

The $\overline{P}_i(t)$ and $\underline{P}_i(t)$ are polynomials in t. Denote by c_j the constant term of $\overline{P}_j(t)$ and by d_j the constant term of $(-1)^j \underline{P}_j(t)$. Obviously

$$c_0 = 1 \tag{20}$$

$$d_0 = 1 \tag{21}$$

and

$$\overline{P}_i(t) = \frac{c_0}{i!} t^i + \frac{c_1}{(i-1)!} t^{i-1} + \cdots + c_{i-1} t + c_i \tag{22}$$

$$\underline{P}_i(t) = (-1)^i \left(\frac{d_0}{i!} t^i + \frac{d_1}{(i-1)!} t^{i-1} + \cdots + d_{i-1} t + d_i \right). \tag{23}$$

Since

$$\overline{P}_i(a_i) = 0, \qquad \underline{P}_i(b_{n-i+1}) = 0$$

we obtain

$$c_0 \frac{a_i^i}{i!} + c_1 \frac{a_i^{i-1}}{(i-1)!} + \cdots + c_{i-1} a_i + c_i = 0 \qquad (i = 1, 2, \cdots, n) \tag{24}$$

and

$$\frac{d_0}{i!} b_{n-i+1}^i + \frac{d_1}{(i-1)!} b_{n-i+1}^{i-1} + \cdots + d_{i-1} b_{n-i+1} + d_i = 0 \tag{25}$$

$$(i = 1. 2. \cdots, n)$$

The determinant of (20) and the first j equations (24) ($j = 1, \cdots, n$) considered as equations in $c_0, c_1, \cdots, c_j$ equals 1, since all the elements of the principal diagonal are 1 and all the elements above the principal diagonal are 0. Then

$$c_i = \begin{vmatrix} 1 & 0 & 0 & \cdots & 0 & 1 \\ a_1 & 1 & 0 & \cdots & 0 & 0 \\ \dfrac{a_2^2}{2!} & a_2 & 1 & \cdots & 0 & 0 \\ \cdots & \cdots & \cdots & \cdots & \cdots & \cdots \\ \dfrac{a_i^i}{i!} & \dfrac{a_i^{i-1}}{(i-1)!} & \dfrac{a_i^{i-2}}{(i-2)!} & \cdots & a_i & 0 \end{vmatrix} \tag{26}$$

$$= (-1)^i \begin{vmatrix} a_1 & 1 & 0 & \cdots & 0 \\ \dfrac{a_2^2}{2!} & a_2 & 1 & \cdots & 0 \\ \cdots & \cdots & \cdots & \cdots & \cdots \\ \dfrac{a_i^i}{i!} & \dfrac{a_i^{i-1}}{(i-1)!} & \dfrac{a_i^{i-2}}{(i-2)!} & \cdots & a_i \end{vmatrix}.$$

From (16) and (22) for $i = n$, we get

$$\bar{P} = c_0 + nc_1 + n(n-1)c_2 + \cdots + n(n-1)\cdots(3)(2)c_{n-1} + n!\, c_n$$

$$= \begin{vmatrix} \dfrac{n!}{n!} & \dfrac{n!}{(n-1)!} & \dfrac{n!}{(n-2)!} & \cdots & \dfrac{n!}{1!} & \dfrac{n!}{0!} \\ a_1 & 1 & 0 & \cdots & 0 & 0 \\ \dfrac{a_2^2}{2!} & a_2 & 1 & \cdots & 0 & 0 \\ \cdots & \cdots & \cdots & \cdots & \cdots & \cdots \\ \dfrac{a_n^n}{n!} & \dfrac{a_n^{n-1}}{(n-1)!} & \dfrac{a_n^{n-2}}{(n-2)!} & \cdots & a_n & 1 \end{vmatrix}. \tag{27}$$

In the same way, we obtain

$$d_i = \begin{vmatrix} 1 & 0 & 0 & \cdots & 0 & 1 \\ b_n & 1 & 0 & \cdots & 0 & 0 \\ \dfrac{b_{n-1}^2}{2!} & b_{n-1} & 1 & \cdots & 0 & 0 \\ \cdots & \cdots & \cdots & \cdots & \cdots & \cdots \\ \dfrac{b_{n-i+1}^i}{i!} & \dfrac{b_{n-i+1}^{i-1}}{(i-1)!} & \dfrac{b_{n-i+1}^{i-2}}{(i-2)!} & \cdots & b_{n-i+1} & 0 \end{vmatrix} \tag{28}$$

$$= (-1)^i \begin{vmatrix} b_n & 1 & 0 & \cdots & 0 \\ \frac{b_{n-1}^2}{2!} & b_{n-1} & 1 & \cdots & 0 \\ \cdots & \cdots & \cdots & \cdots & \cdots \\ \frac{b_{n-i+1}^i}{i!} & \frac{b_{n-i+1}^{i-1}}{(i-1)!} & \frac{b_{n-i+1}^{i-2}}{(i-2)!} & \cdots & b_{n-i+1} \end{vmatrix}$$

and from (19) and (23) for $i = n$,

(29) $$\underline{P} = (-1)^n n!\, d_n .$$

Perhaps if the determinants in (27) and (28) were to be simplified it might be easier to calculate $\overline{P}$ and $\underline{P}$ that way than by the recursion formulas.

5. **The approximation of P.** Let J be the probability that, for a sample of n, there exists at least one pair of numbers ω_1, ω_2, such that

$$0 \leq \omega_i \leq 1 \qquad (i = 1, 2)$$

$$\varphi(\omega_1) > l_1(\omega_1)$$
$$\varphi(\omega_2) < l_2(\omega_2).$$

Recalling the definitions of P, $\overline{P}$, and $\underline{P}$, it is obvious that

(30) $$1 - P = (1 - \overline{P}) + (1 - \underline{P}) - J.$$

Now if

(31) $$J \leq (1 - \overline{P})(1 - \underline{P})$$

and $(1 - P)$ is small, the right member of (30) with J omitted furnishes an excellent approximation to $(1 - P)$. Suppose, for example, that it were desired to give upper and lower limits $l_1(x)$ and $l_2(x)$ such that $P = .95$. Choose $l_1(x)$ and $l_2(x)$ so that, for example, $\overline{P} = \underline{P} = .975$. Then P cannot differ from .95 by more than .000625. Even if

(32) $$J \leq K(1 - \overline{P})(1 - \underline{P})$$

where K is a small factor, say 10, the approximation would still be excellent. It seems very plausible that even (31) holds. However, we have not yet succeeded in obtaining a rigorous proof.

6. **The construction of confidence limits.** We now proceed to the construction of $\overline{l}_{E,\alpha}(x)$ and $\underline{l}_{E,\alpha}(x)$ which were defined in Section I of this paper.

A confidence coefficient $\alpha (0 < \alpha < 1)$ is selected to which it is desired that the confidence limits correspond. Functions $\delta_1(x)$ and $\delta_2(x)$ are chosen to be as defined in Section 2 and also to be such as to make $P = \alpha$. This can be done by application of the formulas for the evaluation of P.

The functions $\overline{l}_{E,\alpha}(x)$ and $\underline{l}_{E,\alpha}(x)$ are to be known when E and α are known.

Since α is given, $\bar{l}_{E,\alpha}(x)$ and $\underline{l}_{E,\alpha}(x)$ depend upon the outcome of the experiment which yields observed values of the stochastic variable X. Let $E = x_1, \cdots, x_n$ be this system of values and let $\varphi(x)$ be its distribution function. Consider the equations

$$\delta_2(\varphi(x) + \Delta_1(x)) = \Delta_1(x) \tag{33}$$

$$\delta_1(\varphi(x) - \Delta_2(x)) = \Delta_2(x). \tag{34}$$

For a fixed but arbitrary x, $-\infty < x < +\infty$, $\varphi(x)$ is known and (33) and (34) are equations in $\Delta_1(x)$ and $\Delta_2(x)$. If, for a certain x, (33) has one or more solutions, let $\epsilon_1(x)$ be the maximum of the set of solutions (for this x, of course). Similarly, if for a certain x, (34) has one or more solutions, let $\epsilon_2(x)$ be the maximum of the set of solutions.

We can now give $\bar{l}_{E,\alpha}(x)$ and $\underline{l}_{E,\alpha}(x)$ as follows:

For an x such that (33) has at least one solution,

$$\bar{l}_{E,\alpha}(x) = \varphi(x) + \epsilon_1(x). \tag{35}$$

For an x such that (33) has no solutions,

$$\bar{l}_{E,\alpha}(x) = 1. \tag{36}$$

For an x such that (34) has at least one solution,

$$\underline{l}_{E,\alpha}(x) = \varphi(x) - \epsilon_2(x). \tag{37}$$

For an x such that (34) has no solution,

$$\underline{l}_{E,\alpha}(x) = 0. \tag{38}$$

We recapitulate briefly the meaning of $\bar{l}_{E,\alpha}(x)$ and $\underline{l}_{E,\alpha}(x)$ which were defined in Section 1. These are two functions defined for $-\infty < x < +\infty$ which may be constructed as above after a confidence coefficient α has been assigned and after the outcome of the physical experiment which determines the stochastic point E is known. These functions have the following property: No matter what the distribution function $f(x)$ of each of n stochastic independent variables $X_1, \cdots, X_n$ may be, provided only that $f(x)$ is continuous and the same for each $X_1, \cdots, X_n$, the probability is exactly α that, if we were to perform the physical experiment which gives a set of values E of the stochastic system $X_1, \cdots, X_n$ and were then to construct $\bar{l}_{E,\alpha}(x)$ and $\underline{l}_{E,\alpha}(x)$, the inequality

$$\underline{l}_{E,\alpha}(x) \leq f(x) \leq \bar{l}_{E,\alpha}(x) \tag{39}$$

would hold for all x.

A less precise but more intuitive statement of the above result is as follows: If, in many experiments we were to proceed as above to construct $\bar{l}_{E,\alpha}(x)$ and $\underline{l}_{E,\alpha}(x)$ and then, in each instance, we were to predict that the unknown $f(x)$ (which need not be the same in all experiments) satisfies (39), the relative frequency of correct predictions would be α.

The formal proof of this result is exceedingly simple. For any continuous $f(x)$, the probability is α that

(40) $$l_2(x) \leq \varphi(x) \leq l_1(x)$$

will hold for all x. This is so because of the way in which $\delta_1(x)$ and $\delta_2(x)$ were chosen. To prove the required result it would therefore be sufficient to show that, if (39) holds for all x, (40) holds for all x and conversely.

Let x be fixed but arbitrary. We shall show that

(41) $$f(x) \leq l_{E,\alpha}(x)$$

implies

(42) $$l_2(x) \leq \varphi(x)$$

and conversely.

If (33) has no solution, $\varphi(x) > l_2(1) \geq l_2(x)$, $l_{E,\alpha}(x) = 1$, and (41) and (42) are trivial. Assume therefore that (33) has at least one solution. For this situation, then, we have to show that

(43) $$f(x) \leq \varphi(x) + \epsilon_1(x)$$

implies

(44) $$l_2(x) \leq \varphi(x)$$

and conversely.

With x and hence $\varphi(x)$ and $\epsilon_1(x)$ fixed, consider the equation in x':

(45) $$l_2(x') = \varphi(x).$$

Since $\varphi(x) \leq l_2(1)$, (45) has at least one solution. Let x'_m be the maximum of these solutions for a fixed x. Then from the definition of $\epsilon_1(x)$ it follows that

(46) $$f(x'_m) - l_2(x'_m) = \epsilon_1(x),$$

or, on account of the definition of x'_m,

(47) $$f(x'_m) = \varphi(x) + \epsilon_1(x).$$

Now, if (43) holds, $x \leq x'_m$ because of (47). Then, from the definition of x'_m and the fact that $l_2(x')$ is monotonically non-decreasing (44) follows.

If (44) holds, then $x \leq x'_m$ (by the definition of x'_m and the monotonic character of $l_2(x')$). Hence, because of (47), (43) is true. This shows the equivalence of (43) and (44).

In a similar manner, it may be shown that

(48) $$l_{E,\alpha}(x) \leq f(x)$$

implies

(49) $$\varphi(x) \leq l_1(x)$$

and conversely. This completes the proof.

7. **Miscellaneous remarks.** An expedient way of choosing $\delta_1(x)$ and $\delta_2(x)$ is such that, with c a constant,

$$
\begin{aligned}
x + \delta_1(x) &\equiv \min\,[x + c, 1] \\
x - \delta_2(x) &\equiv \max\,[x - c, 0].
\end{aligned}
\qquad 0 \leq x \leq 1 \tag{50}
$$

Tables of double entry could be constructed giving the c corresponding to specified α and n. With such tables available the construction of confidence limits would be quick and simple in practice. In this case, $\epsilon_1(x) = \epsilon_2(x) = c$.

Another expedient and plausible way of choosing $\delta_1(x)$ and $\delta_2(x)$ might be to choose them so that

$$
\begin{aligned}
x + \delta_1(x) &\equiv \min\,[px + q, 1] \\
x - \delta_2(x) &\equiv \max\,[p'x + q', 0]
\end{aligned}
\qquad 0 \leq x \leq 1 \tag{51}
$$

where p, p', q, and q' are constants. The actual construction of confidence limits could then be handled with dispatch if similar tables were constructed.

$$\bar{l}_{E,\alpha}(x) \quad \text{and} \quad \underline{l}_{E,\alpha}(x)$$

are, like $\varphi(x)$, step-functions. The situation may occur where, for $x = e$,

$$\lim_{(x<e),x\to e} \bar{l}_{E,\alpha}(x) < \lim_{(x>e),x\to e} \underline{l}_{E,\alpha}(x).$$

This would give a prediction, corresponding to the confidence coefficient α, that $f(x)$ is not continuous. If $f(x)$ is continuous the probability of such a situation is 0.

8. **Further problems.** Even with α fixed, the functions $\delta_1(x)$ and $\delta_2(x)$ may be chosen in many ways. Each different choice gives, in general, different confidence limits. Which is to be preferred? This very problem also arose in the theory of parameter estimation and the testing of hypotheses and gave rise to the Neyman-Pearson theory. It would be desirable to develop such a theory for the confidence limits discussed in this paper.

We have treated here only the case where $f(x)$ is continuous. A similar theory is needed for the case where $f(x)$ is not continuous.

It would be of practical value to construct tables such as those described in Section 7. The construction of tables could be greatly facilitated if the formulas for P or $\bar{P}$ and $\underline{P}$ could be simplified so as to render them more practical for calculation or else if they were to be replaced by asymptotic expansions.

9. **An example.** To illustrate the method we shall consider an example for the case of samples of size 6, i.e. $n = 6$.

Let $\delta_1(x)$ and $\delta_2(x)$ be given as follows:

$$\delta_1(x) = d \quad \text{for} \quad 0 \leq x \leq 1 - .d,$$

$$\delta_1(x) = 1 - x \quad \text{for} \quad 1 - d < x \leq 1,$$

$$\delta_2(x) = x \quad \text{for} \quad 0 \leq x \leq d,$$

and

$$\delta_2(x) = d \quad \text{for} \quad d < x \leq 1.$$

Denote by $\bar{P}$ the probability that

$$\varphi(x) \leq f(x) + \delta_1[f(x)],$$

by $\underline{P}$ the probability that

$$\varphi(x) \geq f(x) - \delta_2[f(x)]$$

and by P the probability that

$$f(x) - \delta_2[f(x)] \leq \varphi(x) \leq f(x) + \delta_1[f(x)].$$

$\varphi(x)$ denotes the sample distribution and $f(x)$ denotes the population distribution.

Since $\delta_2(x) = \delta_1(1 - x)$, we obviously have

$$\bar{P} = \underline{P}.$$

Let us calculate $\bar{P} = \underline{P}$ in case $d = \frac{1}{2}$. According to (3) we have

$$a_1 = a_2 = a_3 = 0, \qquad a_4 = \tfrac{1}{6}, \qquad a_5 = \tfrac{1}{3}, \qquad a_6 = \tfrac{1}{2}.$$

According to (16)

$$\bar{P} = 6!\bar{P}_6(1)$$

where

$$\bar{P}_0(t) \equiv 1,$$

$$\bar{P}_k(t) \equiv \int_{a_k}^{t} \bar{P}_{k-1}(t)\,dt \qquad (k = 1, \cdots, 6).$$

Applying this recursion formula we get

$$\bar{P}_1(t) = t; \qquad \bar{P}_2(t) = \frac{t^2}{2}, \qquad \bar{P}_3(t) = \frac{t^3}{6},$$

$$\bar{P}_4(t) = \frac{t^4}{24} - \frac{1}{2^7 \cdot 3^5}$$

$$\bar{P}_5(t) = \frac{t^5}{120} - \frac{t}{2^7 \cdot 3^5} - \frac{11}{3^6 \cdot 2^7 \cdot 5}$$

$$\bar{P}_6(t) = \frac{t^6}{720} - \frac{t^2}{2^8 \cdot 3^5} - \frac{11t}{3^6 \cdot 2^7 \cdot 5} - \frac{11}{2^9 \cdot 3^6 \cdot 5}.$$

Hence

$$\overline{P} = \underline{P} = 6!\overline{P}_6(1) = 1 - \frac{85}{2592} = 0.967.$$

Let us now calculate $\overline{P} = \underline{P}$ in case $d = \frac{1}{3}$. We have

$$a_1 = a_2 = 0, \qquad a_3 = \tfrac{1}{6}, \qquad a_4 = \tfrac{1}{3}, \qquad a_5 = \tfrac{1}{2} \text{ and } a_6 = \tfrac{2}{3}.$$

Applying the recursion formula we get

$$\overline{P}_0(t) = 1, \qquad \overline{P}_1(t) = t, \qquad \overline{P}_2(t) = \frac{t^2}{2}, \qquad \overline{P}_3(t) = \frac{t^3}{6} - \frac{1}{2^4 \cdot 3^4},$$

$$\overline{P}_4(t) = \frac{t^4}{24} - \frac{t}{2^4 \cdot 3^4} - \frac{1}{2^4 \cdot 3^5},$$

$$\overline{P}_5(t) = \frac{t^5}{120} - \frac{t^2}{2^5 \cdot 3^4} - \frac{t}{2^4 \cdot 3^5} - \frac{11}{2^8 \cdot 3^5 \cdot 5},$$

$$\overline{P}_6(t) = \frac{t^6}{720} - \frac{t^3}{2^5 \cdot 3^5} - \frac{t^2}{2^5 \cdot 3^5} - \frac{11t}{2^8 \cdot 3^5 \cdot 5} - \frac{13}{2^7 \cdot 3^8 \cdot 5}.$$

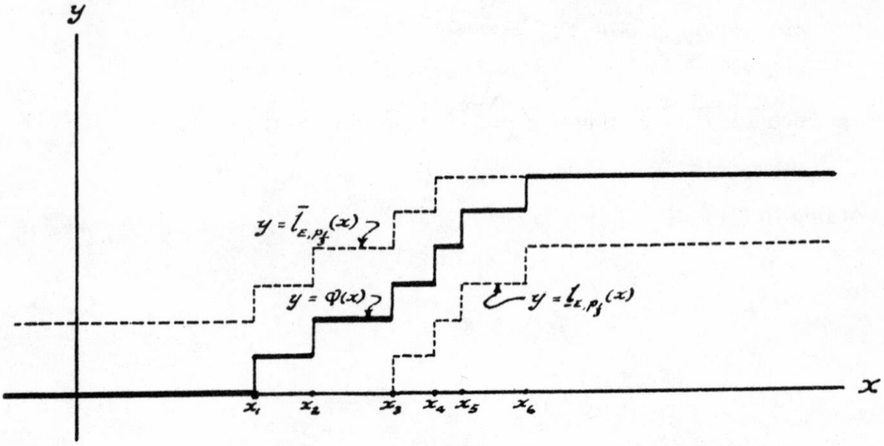

FIG. 1

Hence

$$\overline{P} = \underline{P} = 6!\overline{P}_6(1) = 1 - \frac{2483}{11664} = 0.787.$$

It is obvious that

$$1 - P = (1 - \overline{P}) + (1 - \underline{P}) - J,$$

where J denotes the probability that $\varphi(x)$ violates both limits. In case $d = \frac{1}{2}$ no $\varphi(x)$ exists which violates both limits, and therefore $J = 0$. If $d = \frac{1}{3}$,

J is not zero but so small that it can be neglected. Hence

$$P = 0.934 \quad \text{if} \quad d = \tfrac{1}{2}$$

and

$$P = 0.574 \quad \text{if} \quad d = \tfrac{1}{3}$$

P increases monotonically from 0.574 to 0.934 if d increases from $\frac{1}{3}$ to $\frac{1}{2}$. Denote by P_d the probability corresponding to d. According to (33)–(38), the confidence limits corresponding to the probability level P_d are given as follows:

$$\begin{aligned} l_{E,P_d}(x) &= \varphi(x) + d \text{ if } \varphi(x) + d \leq 1, \\ l_{E,P_d}(x) &= 1 \text{ if } \varphi(x) + d > 1, \\ \underline{l}_{E,P_d}(x) &= \varphi(x) - d \text{ if } \varphi(x) - d \geq 0 \end{aligned}$$

and

$$\underline{l}_{E,P_d}(x) = 0 \text{ if } \varphi(x) - d < 0.$$

Substituting for d the numbers $\frac{1}{2}$ and $\frac{1}{3}$, we get the confidence limits corresponding to the probability levels 0.934 and 0.574 respectively. The upper and lower confidence limits for the *population* distribution corresponding to the probability level 0.574 are represented geometrically in Figure 1 by the upper and lower dotted broken lines for a sample of 6 having the values $x_1, x_2, \cdots x_6$. The sample distribution is represented by the solid broken line.

Columbia University
and
Brooklyn, N. Y.

Reprinted from the
TRANSACTIONS OF THE AMERICAN MATHEMATICAL SOCIETY
Vol. 46, No. 2, pp. 280-306
September, 1939

LIMITS OF A DISTRIBUTION FUNCTION DETERMINED BY ABSOLUTE MOMENTS AND INEQUALITIES SATISFIED BY ABSOLUTE MOMENTS*

BY
ABRAHAM WALD†

1. **Introduction.** Denote by X a chance variable and by $P(\alpha<X<\beta)$ the probability that $\alpha<X<\beta$. Similarly, denote by $P(X<\beta)$ the probability that $X<\beta$ and by $P(X=\beta)$ the probability that $X=\beta$. For any positive integer r the expected value $E|X-x_0|^r$ of $|X-x_0|^r$ is called the absolute moment of order r about x_0, where x_0 denotes a certain real value. If the absolute moments $M_{i_1}=E|X-x_0|^{i_1}, \cdots, M_{i_j}=E|X-x_0|^{i_j}$ of a chance variable X are given (and no further data about X are known), then we shall say for any positive number d that a_d is the sharp lower limit of $P(-d<X-x_0<d)$ if the following two conditions are fulfilled:

(1) For each chance variable Y for which $E|Y-x_0|^{i_\nu}=E|X-x_0|^{i_\nu}$ $(\nu=1, \cdots, j)$ the inequality $P(-d<Y-x_0<d)\geqq a_d$ holds.

(2) To each $\epsilon>0$ a chance variable Y can be given such that $E|Y-x_0|^{i_\nu}=E|X-x_0|^{i_\nu}$ $(\nu=1, \cdots, j)$ and $P(-d<Y-x_0<d)<a_d+\epsilon$.

In other words, a_d is the greatest lower bound of the probabilities $P(-d<Y-x_0<d)$ formed for all chance variables Y for which the i_νth absolute moment about x_0 is equal to the i_νth absolute moment of X about x_0 $(\nu=1, \cdots, j)$.

Similarly we shall say that b_d is the sharp upper limit of $P(-d<X-x_0<d)$ if b_d is the least upper bound of the probabilities $P(-d<Y-x_0<d)$ formed for all chance variables Y for which the i_νth absolute moment about x_0 is equal to the i_νth absolute moment of X about x_0 $(\nu=1, \cdots, j)$.

In this paper we shall give the solution of the following two problems:

PROBLEM 1. *The absolute moments of the order $i_1, \cdots, i_j$ of a chance variable X are given about the point x_0, where $i_1, \cdots, i_j$ denote any positive integers. It is required to determine the sharp lower and sharp upper limit of $P(-d<X-x_0<d)$ for any positive value d.*

PROBLEM 2. *A real value x_0 and a system of j positive integers $i_1, \cdots, i_j$ are given. What are the necessary and sufficient conditions which must be satisfied*

* Presented to the Society, February 25, 1939; received by the editors February 15, 1939.
† This research was done under a grant-in-aid from the Carnegie Corporation of New York.

by j positive numbers $a_1, \cdots, a_j$ *that a chance variable X exists for which the* i_ν*th moment about* x_0 *is equal to* a_ν $(\nu=1, \cdots, j)$?

The solution of Problem 1 is a generalization of the inequality of Markoff. In fact, the inequality of Markoff can be written as follows:

$$(1) \qquad P(-d < X - x_0 < d) \geqq 1 - M_r/d^r,$$

where d denotes an arbitrary positive value and M_r denotes the rth absolute moment of X about x_0. As is well known, the inequality (1) cannot be improved for $d \geqq M_r^{1/r}$, that is to say that $1-M_r/d^r$ is the sharp lower limit of $P(-d<X-x_0<d)$ for $d \geqq M_r^{1/r}$. The generalization in our Problem 1 consists in the circumstance that instead of a single moment M_r we consider a finite number of moments $M_{i_1}, \cdots, M_{i_j}$, and besides the sharp lower limit of $P(-d<X-x_0<d)$ also its sharp upper limit is to be determined. The inequality (1) is called for $r=2$ also the inequality of Tshebysheff.

Some results concerning the case when two moments M_r and M_s are given, have been obtained by different authors. A. Guldberg* gave the following formula:

$$(2) \qquad P(|X - x_0| < \lambda M_r^{1/r}) \geqq 1 - \frac{1}{\lambda^s}\left(\frac{M_s^{1/s}}{M_r^{1/r}}\right)^s.$$

If we substitute $2k$ for s, and 2 for r, we get the inequality of K. Pearson.† By other substitutions we get the formula of E. Lurquin.‡ It is easy to show that the limit given in (2) is not sharp.

P. Cantelli§ gave a formula in case that $s=2r$. His formula can be written as follows:

(3a) If $M_r/d^r \leqq M_{2r}/d^{2r}$, then $P(|X-x_0|<d) \geqq 1-M_r/d^r$.

(3b) If $M_r/d^r > M_{2r}/d^{2r}$, then

$$P(|X - x_0| < d) \geqq 1 - \frac{M_{2r} - M_r^2}{(d^r - M_r)^2 + M_{2r} - M_r^2}.$$

The writer of this article gave in a previous paper‖ some results concerning the general case and the sharp lower limit of $P(-d<X-x_0<d)$ if two

---

* A. Guldberg, Comptes Rendus de l'Académie des Sciences, Paris, vol. 175, p. 679.

† K. Pearson, Biometrika, vol. 12 (1918–1919).

‡ E. Lurquin, Comptes Rendus de l'Académie des Sciences, Paris, vol. 175, p. 681.

§ Cantelli's formula and its demonstration are given in the book of M. Fréchet, *Recherches Théoriques Modernes sur la Théorie des Probabilités*, Paris, 1937, pp. 123–126.

‖ A. Wald, *A generalization of Markoff's inequality*, Annals of Mathematical Statistics, December, 1938.

moments M_r and M_s are given, where r and s denote arbitrary positive integers. If $s=2r$ the formula reduces to Cantelli's formula.

In case of consecutive algebraic moments, that is to say, if $M_1, \cdots, M_j$ are given and $M_i=E(X-x_0)^i$ $(i=1, \cdots, j)$, Tshebysheff determined the sharp lower and sharp upper limit of the distribution function $P(X<d)$. These inequalities are called Tshebysheff's inequalities. The first proof of these inequalities was given by Markoff in 1884 and the same proof was discovered almost at the same time by Stieltjes.*

The solution of Problem 2 is well known† if $i_1, \cdots, i_j$ are consecutive integers, that is to say, if $i_\nu=\nu$ $(\nu=1, \cdots, j)$ and if a_ν $(\nu=1, \cdots, j)$ is the νth algebraic moment, that is to say, $a_\nu=E(X-x_0)^\nu$. In this paper we shall give the solution for absolute moments and for arbitrary positive integers $i_1, \cdots, i_j$.

2. **Reduction of the problem to the case of nonnegative chance variables.** We shall call a chance variable X nonnegative if $P(X<0)=0$. Since the moments of the nonnegative chance variable $Y=|X-x_0|$ about the origin are equal to the absolute moments of X about x_0 and since

$$P(Y < d) = P(- d < X - x_0 < d),$$

the following proposition holds true:

PROPOSITION 1. *Denote by $M_{i_1}, \cdots, M_{i_j}$ the absolute moments of order $i_1, \cdots, i_j$ of a certain chance variable X about the point x_0. There exists a nonnegative chance variable Y such that the i_νth moment of Y about the origin is equal to M_{i_ν} $(\nu=1, \cdots, j)$. The greatest lower (least upper) bound of the probabilities $P(-d<Z-x_0<d)$ is equal to the greatest lower (least upper) bound of the probabilities $P(Z'<d)$, where $P(-d<Z-x_0<d)$ is formed for all chance variables Z for which the i_νth absolute moment about x_0 is equal to M_{i_ν} and $P(Z'<d)$ is formed for all nonnegative chance variables Z' for which the i_νth moment about the origin is equal to M_{i_ν} $(\nu=1, \cdots, j)$.*

On account of Proposition 1 we can restrict ourselves to the consideration of nonnegative chance variables and of the moments about the origin. Throughout the following developments we shall understand by a chance variable a nonnegative chance variable and by moments the moments about the origin.

3. **Some definitions and propositions.** Let us begin with some definitions.

* See, for instance, J. Uspensky, *Introduction to Mathematical Probability*, New York, McGraw-Hill, 1937, pp. 373–380.

† See, for instance, R. von Mises, *Wahrscheinlichkeitsrechnung und ihre Anwendung in der Statistik und theoretischen Physik*, Deuticke, Leipzig, 1931, pp. 247–248.

DEFINITION 1. *A chance variable X is said to be an arithmetic chance variable, if there exist a finite system of different numbers* $x_1, \cdots, x_k$ *such that* $\sum_{i=1}^{k} P(X=x_i)=1$.

DEFINITION 2. *A chance variable X for which k different positive values* $x_1, \cdots, x_k$ *exist such that* $P(X=x_i)>0$ $(i=1, \cdots, k)$ *and* $\sum_{i=1}^{k} P(X=x_i)=1$, *is called an arithmetic chance variable of the degree k.*

DEFINITION 3. *A chance variable X is said to be an arithmetic chance variable of the degree* $k+1/2$ *if* $P(X=0)>0$ *and if there exist k different positive values* $x_1, \cdots, x_k$ *such that* $P(X=x_i)>0$ $(i=1, \cdots, k)$ *and* $\sum_{i=1}^{k} P(X=x_i)+P(X=0)=1$.

DEFINITION 4. *Denote by* $M_{i_1}, \cdots, M_{i_j}$ *the moments of the order* $i_1, \cdots, i_j$ *of a certain chance variable X. A chance variable Y is said to be characteristic relative to* $M_{i_1}, \cdots, M_{i_j}$ *if the* i_ν*th moment of Y is equal to* M_{i_ν} $(\nu=1, \cdots, j)$ *and Y is an arithmetic chance variable of the degree less than or equal to* $(j+1)/2$. *A characteristic chance variable is said to be degenerate if its degree is less than* $(j+1)/2$.

DEFINITION 5. *We shall say that the numbers* $M_{i_1}, \cdots, M_{i_j}$ *can be realized as moments of the order* $i_1, \cdots, i_j$ *if there exists a chance variable X such that the* i_ν*th moment of X is equal to* M_{i_ν} $(\nu=1, \cdots, j)$.

DEFINITION 6. *A function* $f(x)$ *defined for all real values x is said to change its sign at the point* $x=\alpha$ *if the following conditions are fulfilled:*

(1) *If* $f(x)=0$ *for all values* $x<\alpha$, *then any open interval containing* α *must contain at least one value* α' *such that* $f(\alpha)f(\alpha')<0$.

(2) *If* $f(x)$ *is not identically zero for* $x<\alpha$, *then any open interval which contains* α *and a point* $\beta<\alpha$ *for which* $f(\beta)\neq 0$, *must also contain two points* α_1 *and* α_2 *such that* $\alpha_1\leqq\alpha$, $\alpha_2\geqq\alpha$ *and* $f(\alpha_1)f(\alpha_2)<0$.

By the number of changes in sign of $f(x)$ we shall understand the number of points at which $f(x)$ changes its sign. Similarly we shall understand by the number of changes in sign in an (open or closed) interval A, the number of points of A at which $f(x)$ changes its sign.

It is easy to prove that if $f(\alpha_1)f(\alpha_2)<0$ then there exists at least one point of the closed interval $[\alpha_1, \alpha_2]$ at which $f(x)$ changes its sign. In order to prove this, let us assume that $\alpha_1<\alpha_2$ and denote by α the greatest lower bound of all values γ of the interval $[\alpha_1, \alpha_2]$ for which $f(\alpha_1)f(\gamma)<0$. It is obvious that $\alpha_1\leqq\alpha\leqq\alpha_2$. We shall show that $f(x)$ changes its sign at α. If $\alpha=\alpha_1$ then from the definition of α it follows that any open interval containing α contains also a point α' such that

$$f(\alpha_1)f(\alpha') = f(\alpha)f(\alpha') < 0.$$

Hence $f(x)$ changes its sign at α. If $\alpha > \alpha_1$ then for any value $\delta \geqq \alpha_1$ and less than α, $f(\delta)$ has the same sign as $f(\alpha_1)$ or is equal to zero. From this fact it follows easily that any open interval which contains α and a value $\beta < \alpha$ for which $f(\beta) \neq 0$, contains also two points β_1 and β_2 such that $\beta_1 \leqq \alpha$, $\beta_2 \geqq \alpha$ and $f(\beta_1)f(\beta_2) < 0$. Hence $f(x)$ changes its sign at α in any case.

If $f(x)$ does not change its sign at any point of the (open or closed) interval I, then $f(\alpha)f(\beta) \geqq 0$ for any two points α, β of I. In fact if I should contain two points α, β such that $f(\alpha)f(\beta) < 0$, then $[\alpha, \beta]$ and therefore also I must contain a point γ at which $f(x)$ changes its sign, in contradiction to our assumption.

We shall prove now the

PROPOSITION 2. *If X denotes an arithmetic chance variable of degree k and Y denotes an arbitrary chance variable, then the number of changes in sign of $D(x) = P(X < x) - P(Y < x)$ is less than or equal to $2k-1$.*

Let us first consider the case that k is integral. In this case there are k different positive values $\alpha_1, \cdots, \alpha_k$ such that $P(X = \alpha_i) > 0$ $(i = 1, \cdots, k)$ and $\sum_{i=1}^{k} P(X = \alpha_i) = 1$. It is obvious that at most one change in sign of $D(x)$ can take place in the interior of the interval $I_i = [\alpha_i, \alpha_{i+1}]$ $(i = 1, \cdots, k-1)$. Besides the changes in sign in the interior of the intervals $I_1, \cdots, I_{k-1}$ a change in sign can only occur at the points $\alpha_1, \cdots, \alpha_k$. Hence the total number of changes in sign cannot exceed $(k-1) + k = 2k-1$.

If $k = k' + 1/2$, where k' denotes a nonnegative integer, then $P(X = 0) > 0$ and there exist k' different positive values $\alpha_1, \cdots, \alpha_{k'}$ such that $P(X = \alpha_i) > 0$ $(i = 1, \cdots, k')$ and $\sum_{i=1}^{k'} P(X = \alpha_i) + P(X = 0) = 1$. Let us denote the point 0 by α_0. It is obvious that in the interior of the interval $I_i = [\alpha_i, \alpha_{i+1}]$ $(i = 0, 1, \cdots, k'-1)$ at most one change in sign of $D(x)$ can take place. Further changes in sign can occur only at the points $\alpha_1, \cdots, \alpha_{k'}$. Hence the total number of changes in sign cannot exceed $2k' = 2k-1$.

PROPOSITION 3. *If X and Y denote two arithmetic chance variables of degree less than or equal to k, then the number of changes in sign of $D(x) = P(X < x) - P(Y < x)$ is less than or equal to $2k-2$.*

First let us consider the case that both chance variables X and Y are of the degree k. If k is a positive integer, then there exist two systems of k positive values $\alpha_1, \cdots, \alpha_k$ and $\beta_1, \cdots, \beta_k$ such that

$$\sum_{i=1}^{k} P(X = \alpha_i) = \sum_{i=1}^{k} P(Y = \beta_i) = 1.$$

We may assume that $\alpha_1 \leqq \beta_1$. (If it happens that $\beta_1 < \alpha_1$, we can change the notation.) Hence $D(x)$ has no change in sign at the point α_1. Since in the interior of the interval $[\alpha_i, \alpha_{i+1}]$ $(i=1, \cdots, k-1)$ at most one change in sign can take place and further changes in sign can occur only at the points $\alpha_2, \alpha_3, \cdots, \alpha_k$, the total number of changes in sign cannot exceed $(k-1)+(k-1)=2k-2$. If $k=k'+1/2$, where k' denotes a nonnegative integer, then $P(X=0)>0$, $P(Y=0)>0$ and there exist two systems of k' positive numbers $\alpha_1, \cdots, \alpha_{k'}; \beta_1, \cdots, \beta_{k'}$ such that $P(X=\alpha_i)>0$, $P(Y=\beta_i)>0$ $(i=1, \cdots, k')$ and

$$\sum_{i=1}^{k'} P(X = \alpha_i) + P(X = 0) = \sum_{i=1}^{k'} P(Y = \beta_i) + P(Y = 0) = 1.$$

We may assume that $\alpha_1 \leqq \beta_1$. It is obvious that $D(x)$ has no change in sign in the interior of the interval $[0, \alpha_1]$. Since $D(x)$ has at most one change in sign in the interior of the interval $[\alpha_i, \alpha_{i+1}]$ $(i=1, \cdots, k'-1)$, and since further changes in sign can occur only at the points $\alpha_1, \cdots, \alpha_{k'}$, the total number of changes in sign cannot exceed $2k'-1=2k-2$. Hence Proposition 3 is proved if X and Y are of the degree k.

Let us now consider the case that X or Y or both are of degree less than k. Let for instance the degree of X be less than k. Hence the degree of X is less than or equal to $k-1/2$ and therefore on account of Proposition 2 the number of changes in sign of $D(x)$ cannot exceed $2(k-1/2)-1=2k-2$.

PROPOSITION 4. *If X and Y denote two arithmetic chance variables of degree less than or equal to $k>1$ and if there exists a positive number α such that $P(X=\alpha)>0$ and $P(Y=\alpha)>0$, then the number of changes in sign of $D(x)=P(X<x)-P(Y<x)$ is less than or equal to $2k-3$.*

We may assume that $P(X<\alpha) \leqq P(Y<\alpha)$. Consider first the case that $P(Y<\alpha)>0$ and denote by α' the greatest value less than α for which $P(Y=\alpha')>0$. It is obvious that $D(x)$ has no change in sign in the interior of the interval $[\alpha', \alpha]$. If $D(x)$ is identically zero in the interior of $[\alpha', \alpha]$, then $D(x)$ has no change in sign at α'. If $D(x)$ is not identically zero in the interior of $[\alpha', \alpha]$ and if $P(X \leqq \alpha) \leqq P(Y \leqq \alpha)$, then $D(x)$ has no change in sign at α. Finally if $P(X \leqq \alpha) > P(Y \leqq \alpha)$ and α'' denotes the smallest value greater than α for which $P(Y=\alpha'')>0$, then $D(x)$ has no change in sign in the interior of the interval $[\alpha, \alpha'']$. Hence in any case the number of changes in sign of $D(x)$ cannot exceed $(2k-1)-2=2k-3$. Now we have to prove Proposition 4 if $P(Y<\alpha)=0$. Since $P(X<\alpha) \leqq P(Y<\alpha)=0$, $D(x)$ has no change in sign at α. If $P(X \leqq \alpha)=P(Y \leqq \alpha)=1$, then $D(x)$ has no change in sign at all and Proposition 4 is proved. We have to consider only the case that at least one of the values $P(X \leqq \alpha)$, $P(Y \leqq \alpha)$ is less than 1. Let us assume that $P(X \leqq \alpha) \geqq P(Y \leqq \alpha)$. The probability $P(Y \leqq \alpha)$ must be less than 1, since otherwise also $P(X \leqq \alpha)$ would be equal to 1, in contradiction to our assumption. Denote by β the smallest value greater than α for which $P(Y=\beta)>0$. Then $D(x)$ has obviously no change in sign in the interior of $[\alpha, \beta]$ and therefore the total number of changes in sign cannot exceed $2k-3$. If $P(X \leqq \alpha)<P(Y \leqq \alpha)$, then denote by β the smallest value greater than α for which $P(X=\beta)>0$. The function $D(x)$ has no change in sign in the interior of $[\alpha, \beta]$ and therefore also in this case the total number of changes in sign of $D(x)$ cannot exceed $2k-3$.

PROPOSITION 5. *Denote by X and Y two chance variables. If the i_νth moment ($\nu=1, \cdots, j$) of X is finite and equal to the i_νth moment of Y, then $D(x)=P(X<x)-P(Y<x)$ must have at least j changes in sign, unless $D(x)$ is identically zero.*

Denote $P(X<x)$ by $V_1(x)$ and $P(Y<x)$ by $V_2(x)$. Since the i_νth moment of X is equal to the i_νth moment of Y ($\nu=1, \cdots, j$), the Stieltjes integral

$$I=\int_0^\infty (a_1x^{i_1}+\cdots+a_jx^{i_j})d[V_1(x)-V_2(x)]=0 \tag{4}$$

for arbitrary real values $a_1, \cdots, a_j$. Denote the integral

$$\int_0^\lambda (a_1x^{i_1}+\cdots+a_jx^{i_j})d[V_1(x)-V_2(x)]$$

by I_λ. It is obvious that

$$\lim_{\lambda=\infty} I_\lambda = I = 0. \tag{5}$$

We get by integration by parts

$$\begin{aligned} I_\lambda &= (a_1\lambda^{i_1}+\cdots+a_j\lambda^{i_j})[V_1(\lambda)-V_2(\lambda)] \\ &\quad -\int_0^\lambda (i_1a_1x^{i_1-1}+\cdots+i_ja_jx^{i_j-1})[V_1(x)-V_2(x)]dx. \end{aligned} \tag{6}$$

Now we shall show that

$$\lim_{\lambda=\infty} \lambda^{i_\nu}[V_1(\lambda)-V_2(\lambda)]=0, \qquad \nu=1, \cdots, j. \tag{7}$$

Since

$$\lambda^{i_\nu}[V_1(\lambda)-V_2(\lambda)]=\lambda^{i_\nu}[1-V_2(\lambda)]-\lambda^{i_\nu}[1-V_1(\lambda)]$$

we have only to show that

$$\lim \lambda^{i_\nu}[1-V_r(\lambda)]=0, \qquad r=1, 2. \tag{8}$$

It is obvious that for any $\lambda>0$

$$\lambda^{i_\nu}[1 - V_r(\lambda)] = \int_\lambda^\infty \lambda^{i_\nu} dV_r(x) \leqq \int_\lambda^\infty x^{i_\nu} dV_r(x).$$

Since the i_νth moments of X and Y are finite, we have

$$\lim_{\lambda=\infty} \int_\lambda^\infty x^{i_\nu} dV_r(x) = 0,$$

and therefore (8) and (7) must hold. Then we get from the equations (4), (5) and (6)

$$(9) \qquad \int_0^\infty (i_1 a_1 x^{i_1 - 1} + \cdots + i_j a_j x^{i_j - 1})[V_1(x) - V_2(x)]dx = 0.$$

Let us suppose that the number of changes in sign of $D(x) = V_1(x) - V_2(x)$ is less than j and denote by $\alpha_1 < \alpha_2 < \cdots < \alpha_k$ $(k < j)$ the points at which $D(x)$ changes its sign. It is obvious that $\alpha_1 > 0$. Consider the intervals

$$I_1 = [0, \alpha_1], I_2 = [\alpha_1, \alpha_2], \cdots, I_k = [\alpha_{k-1}, \alpha_k], I_{k+1} = [\alpha_k, \infty].$$

$D(x)$ is in the interior of the interval I_ν either everywhere nonnegative or everywhere nonpositive, and if $D(x) \geqq 0$ $(\leqq 0)$ in the interior of I_ν then $D(x) \leqq 0$ $(\geqq 0)$ in the interior of $I_{\nu-1}$ $(\nu = 2, 3, \cdots, k+1)$. We put $a_{k+2} = a_{k+3} = \cdots = a_i = 0$ and consider the k equations

$$i_1 a_1 \alpha_1^{i_1 - 1} + \cdots + i_{k+1} a_{k+1} \alpha_1^{i_{k+1} - 1} = 0,$$

$$\cdots\cdots\cdots\cdots\cdots\cdots\cdots\cdots\cdots,$$

$$i_1 a_1 \alpha_k^{i_1 - 1} + \cdots + i_{k+1} a_{k+1} \alpha_k^{i_{k+1} - 1} = 0.$$

There exists a system of roots $a_1 = a_1', \cdots, a_{k+1} = a_{k+1}'$ such that at least one among them is not equal to zero. Denote the polynomial

$$i_1 a_1' x^{i_1 - 1} + \cdots + i_{k+1} a_{k+1}' x^{i_{k+1} - 1}$$

by $Q(x)$. It is obvious that $\alpha_1, \cdots, \alpha_k$ are roots of $Q(x)$. Since the number of changes in sign in the sequence of the coefficients of $Q(x)$ is less than or equal to k, $Q(x)$ has at most k positive roots. Hence $\alpha_1, \cdots, \alpha_k$ must be simple roots of $Q(x)$ and therefore the sign of $Q(x)$ in the interval $I_\nu = [\alpha_{\nu-1}, \alpha_\nu]$ is opposite to the sign of $Q(x)$ in the interval $I_{\nu-1}$ $(\nu = 2, \cdots, k+1)$. From this fact it follows that the product $Q(x)[V_1(x) - V_2(x)]$ has no change of sign at all. Hence the integral

$$\int_0^\infty Q(x)[V_1(x) - V_2(x)]dx$$

can vanish only if $V_1(x) - V_2(x)$ is identically zero. This proves Proposition 5.

From Propositions 2 and 5 follows easily

PROPOSITION 6. *If X denotes an arithmetic chance variable of degree k and Y denotes a chance variable such that $2k$ different moments of Y are equal to the corresponding moments of X, then $P(Y<x)$ is identically equal to $P(X<x)$.*

From Propositions 3 and 5 we get the

PROPOSITION 7. *For each system of moments $M_{i_1}, \cdots, M_{i_j}$ there exists at most one chance variable which is characteristic relative to $M_{i_1}, \cdots, M_{i_j}$.*

We shall now prove the

PROPOSITION 8. *If the chance variable X is characteristic relative to $M_{i_1}, \cdots, M_{i_j}$, and M_r' is the rth moment of X, where $r>i_1, \cdots, i_j$, then $M_{i_1}, \cdots, M_{i_j}, M_r$ for $M_r<M_r'$ cannot be realized as moments of the orders $i_1, \cdots, i_j, r$.*

Let us suppose that there exists a chance variable Y with the moments $M_{i_1}, \cdots, M_{i_j}, M_r$ where $M_r<M_r'$. We shall deduce a contradiction from this assumption. We can assume that Y is an arithmetic chance variable, because according to a well known theorem a finite system of moments can always be realized by an arithmetic chance variable. On account of Proposition 5, $D(x)=P(Y<x)-P(X<x)$ must have at least j changes in sign. Since X is a characteristic chance variable, the number of changes in sign of $D(x)$ cannot exceed j; hence the number of changes in sign must be equal to j. It is easy to see that the number of changes in sign can be equal to j only if the greatest value x' for which $P(Y=x')>0$ is greater than the greatest value x'' for which $P(X=x'')>0$. We denote by Y_d the arithmetic chance variable defined as follows:

$$P(Y_d = d) = \frac{M_r' - M_r}{d^r}, \qquad P(Y_d = x') = P(Y = x') - \frac{M_r' - M_r}{d^r},$$

$$P(Y_d = x) = P(Y = x), \qquad \text{for } x \neq d, x',$$

where $d>x'$ and $P(Y=x')>(M_r'-M_r)/d^r$. The differences between the moments (of the orders $i_1, \cdots, i_j, r$) of X and the corresponding moments of Y_d become arbitrarily small if we choose d sufficiently large. It is obvious that $P(X<x)-P(Y_d<x)$ has always the same sign as $P(X<x)-P(Y<x)$. Since the number of changes in sign of $D(x)$ is equal to j, a polynomial $P(x)=a_1x^{i_1}+ \cdots +a_jx^{i_j}+a_rx^r$ can be given such that $P'(x)=i_1a_1x^{i_1-1}+ \cdots +i_ja_jx^{i_j-1}+ra_rx^{r-1}$ has always the same sign as that of $P(X<x)-P(Y<x)$ and therefore has also the same sign as that of $P(X<x)-P(Y_d<x)$ for any d. Since $P(Y_d<x)=P(Y<x)$ for any $x<x'$ and since $P(Y<x)-P(X<x)$ is not equal to zero for any $x<x'$, the integral

$$\int_0^\infty P'(x)[P(Y_d < x) - P(X < x)]dx$$

cannot converge towards zero if $d \to \infty$. But on the other hand the moments of the order $i_1, \cdots, i_j, r$ of Y_d converge towards the corresponding moments of X if $d \to \infty$ and therefore, as can easily be shown, the above integral must converge towards zero. Hence we get a contradiction and our proposition is proved.

DEFINITION 7. *A sequence* $\{X_i\}$ *of chance variables is said to be convergent towards the chance variable* X, *in symbols* $\lim_{i=\infty} X_i = X$, *if* $\{P(X_i < x)\}$ $(i = 1, 2, \cdots, ad\ inf.)$ *converges uniformly towards* $P(X < x)$ *in any closed set of values of* x *which does not contain any point of discontinuity of* $P(X < x)$.

In the following development we shall understand by "X is equal to Y," in symbols $X = Y$, that $P(X < x)$ is identically equal to $P(Y < x)$.

For any integer r we shall denote the rth moment of a chance variable X also by $M_r(X)$.

DEFINITION 8. *A chance variable* X_α *defined for any point* α *of a domain* D *is said to be a continuous function of* α *in* D, *if for any* α *in* D *and for any sequence of points* $\{\alpha_i\}$ *in* D *which converges towards* α, $\lim_{i=\infty} X_{\alpha_i} = X_\alpha$.

Now we shall prove

PROPOSITION 9. *If* $\{X_i\}$ $(i = 1, 2, \cdots, ad\ inf.)$ *denotes a sequence of arithmetic chance variables of degree less than a certain integer* n *which converges towards the chance variable* X, *and if for a certain positive integer* r, $\{M_r(X_i)\}$ $(i = 1, 2, \cdots, ad\ inf.)$ *is bounded, then* $\lim_{i=\infty} M_s(X_i) = M_s(X)$ *for any positive integer* $s < r$.

It is obvious that X is an arithmetic chance variable of degree less than n. Denote by $\epsilon_s(X_i, t)$ the Stieltjes integral $\int_t^\infty x^s dP(X_i < x)$ where $t > 0$. It is obvious that for any positive value t for which $P(X \geq t) = 0$

$$\lim_{i=\infty} [M_k(X_i) - \epsilon_k(X_i, t)] = M_k(X), \qquad k = 1, 2, \cdots, \text{ad inf.}$$

Suppose that $\{M_r(X_i)\}$ is bounded for a certain r. Since $\epsilon_r(X_i, t) \leq M_r(X_i)$ $(i = 1, \cdots, \text{ad inf.})$, $\{\epsilon_r(X_i, t)\}$ must also be bounded. That is to say, there exists a positive value N such that $\epsilon_r(X_i, t) < N$ for any integer i and for any positive value t. Hence $\epsilon_s(X_i, t) < N/t$ for $s = 1, 2, \cdots, r-1$. Let us now suppose that for a certain $s < r$, $M_s(X_i)$ does not converge towards $M_s(X)$. Then a subsequence $\{X_{i_j}\}$ $(j = 1, \cdots, \text{ad inf.})$ can be given such that $M_s(X_{i_j})$ converges with increasing j towards a value $M_s' \neq M_s(X)$. We choose a value t for which $P(X < t) = 1$ and $N/t < |M_s' - M_s(X)|/2$. It is obvious that for this t, $M_s(X_i)-\epsilon_s(X_i,\, t)$ cannot converge towards $M_s(X)$. Hence we have a contradiction and the assumption that $M_s(X_i)$ does not converge towards $M_s(X)$ is proved to be an absurdity.

PROPOSITION 10. *If* $\{X_i\}$ $(i=1, 2, \cdots, ad\ inf.)$ *denotes a sequence of arithmetic chance variables of degree less than or equal to* k, *and if there exists an integer* r *for which* $\{M_r(X_i)\}$ $(i=1, \cdots, ad\ inf.)$ *is bounded, then there exists a subsequence* $\{X_{i_j}\}$ $(j=1, \cdots, ad\ inf.)$ *which is convergent.*

Since X_i is of degree less than or equal to k, there exists a subsequence $\{X_i'\}$ of $\{X_i\}$ such that the number of different values with positive probability is the same for each element of the sequence $\{X_i'\}$ $(i=1, \cdots, \text{ad inf.})$. Denote this number by s. Denote by $\alpha_{i,1}< \cdots <\alpha_{i,s}$ the values for which $P(X_i'=\alpha_{i,m})>0$ $(m=1, \cdots, s)$. It is obvious that there exists a subsequence $\{X_{i_j}'\}$ $(j=1, \cdots, \text{ad inf.})$ of the sequence $\{X_i'\}$ such that $\lim P(X_{i_j}'=\alpha_{i_j,m})$ exists for $m=1, \cdots, s$ and the sequence $\{\alpha_{i_j,m}\}$ $(j=1, \cdots, \text{ad inf.})$ converges for each $m\leqq s$ towards a finite value or towards infinity. Since $\{M_r(X_i)\}$ $(i=1, \cdots, \text{ad inf.})$ is bounded, $\lim_{j=\infty} P(X_{i_j}'=\alpha_{i_j,m})=0$ for all m for which $\lim_{j=\infty}\alpha_{i_j,m}=\infty$. Since $\alpha_{i,1}<\alpha_{i,2}< \cdots <\alpha_{i,s}$, we have $\lim \alpha_{i_j,m}=\infty$ if $\lim \alpha_{i_j,m-1}=\infty$. Hence there exists an integer $m'\leqq s$ such that for any integer $m>m'$ and less than or equal to s, $\lim \alpha_{i_j,m}=\infty$ and for any integer $m\leqq m'$, $\lim \alpha_{i_j,m}$ is finite. Denote $\lim \alpha_{i_j,m}$ by α_m and $\lim P(X_{i_j}'=\alpha_{i_j,m})$ by p_m for any $m\leqq m'$. It is obvious that $\sum_{m=1}^{m'} p_m=1$ and $\{X_{i_j}'\}$ converges towards the arithmetic chance variable X defined as follows: $P(X=\alpha_m)=p_m$ for $m\leqq m'$ and $P(X=\alpha)=0$ for any $\alpha\neq\alpha_1, \cdots, \alpha_{m'}$. Hence Proposition 10 is proved.

PROPOSITION 11. *Denote by* $\{X_i\}$ $(i=1, \cdots, ad\ inf.)$ *a sequence of arithmetic chance variables of degree less than or equal to* k *for which* $\{M_r(X_i)\}$ *is bounded for a certain integer* r. *If* $\{X_i\}$ *does not converge towards the chance variable* X, *then there exists a convergent subsequence* $\{X_{i_j}\}$ *such that* $\lim_{j=\infty} X_{i_j}=Y\neq X$.

Since $\{X_i\}$ does not converge towards X, there exists a positive ϵ, a sequence of numbers $\{\alpha_i\}$ contained in a closed set which does not contain any discontinuity point of $P(X<x)$, and a subsequence $\{X_i'\}$ of $\{X_i\}$ such that $|P(X_i'<\alpha_i)-P(X<\alpha_i)|>\epsilon$ for $i=1, \cdots$, ad inf. Hence no subsequence of the sequence $\{X_i'\}$ can converge towards X. On account of Proposition 10 there exists a convergent subsequence $\{X_i''\}$ of the sequence $\{X_i'\}$. Hence $\lim X_i''$ must be different from X and our proposition is proved.

PROPOSITION 12. *Denote by* $\{X_n\}$ $(n=1, \cdots, ad\ inf.)$ *a sequence of arithmetic chance variables of the degree less than or equal to* $(j+1)/2$, *where* j *denotes* *a nonnegative integer. Denote further by X an arithmetic chance variable of degree less than or equal to $(j+1)/2$ for which $M_{i_1}(X), \cdots, M_{i_j}(X)$ are finite and $i_1<i_2<\cdots<i_j$ denote certain integers. If* $\lim_{n=\infty} M_{i_\nu}(X_n)=M_{i_\nu}(X)$ $(\nu=1, \cdots, j)$, *then* $\lim X_n=X$.

Let us suppose that $\{X_n\}$ does not converge towards X but $\lim M_{i_\nu}(X_n) = M_{i_\nu}(X)$ $(\nu=1, \cdots, j)$. According to Proposition 11 there exists a subsequence $\{X_{n_m}\}$ $(m=1, \cdots$, ad inf.) such that $\lim_{m=\infty} X_{n_m}=X'\neq X$. It is obvious that X' is of degree less than or equal to $(j+1)/2$. Consider now the case that there exists an integer $r>i_j$ such that $\{M_r(X_{n_m})\}$ $(m=1, \cdots$, ad inf.) is bounded. Then we have on account of Proposition 9, $M_{i_\nu}(X)=\lim M_{i_\nu}(X_{n_m}) = M_{i_\nu}(X')$. From Proposition 5 it follows that $D(x)=P(X<x)-P(X'<x)$ must have at least j changes in sign. But this is not possible since on account of Proposition 3, $D(x)$ cannot have more than $2(j+1)/2-2=j-1$ changes in sign. Hence for any integer $r>i_j$, $\{M_r(X_{n_m})\}$ is not bounded. Hence there exists a subsequence $\{X'_{n_m}\}$ of $\{X_{n_m}\}$ such that $\lim_{m=\infty} M_r(X'_{n_m})=\infty$. Denote by α_m the greatest value for which $P(X'_{n_m}=\alpha_m)>0$. Obviously $\lim \alpha_m=\infty$. Since $\lim X'_{n_m}=X'$, $\lim P(X'_{n_m}=\alpha_m)$ must be equal to zero. From this fact it follows easily that the degree of X' must be less than or equal to $(j+1)/2-1=(j-1)/2$. From Proposition 9 we get that

$$M_{i_\nu}(X') = \lim M_{i_\nu}(X'_{n_m}) = M_{i_\nu}(X), \quad \nu = 1, \cdots, j-1.$$

Hence according to Proposition 5, $D(x)=P(X<x)-P(X'<x)$ must have at least $j-1$ changes in sign. But this is not possible, because the degree of X' is less than or equal to $(j-1)/2$ and therefore on account of Proposition 2 the number of changes in sign of $D(x)$ is less than or equal to $2(j-1)/2-1=j-2$. Hence we obtain a contradiction and our assumption that $\{X_n\}$ does not converge towards X is proved to be an absurdity.

PROPOSITION 13. *Denote by $M_{i_1}, \cdots, M_{i_j}$ the moments of the orders $i_1<i_2<\cdots<i_j$ of a certain chance variable X. There exists always a chance variable X' which is characteristic relative to $M_{i_1}, \cdots, M_{i_j}$.*

We shall prove this proposition by mathematical induction. Proposition 13 is obviously true for $j=1$. We shall suppose that 13 is true for any integer $r\leqq k$. That is to say, we shall make the

ASSUMPTION A_k. *Denote by $M_{i_1}, \cdots, M_{i_r}$ the moments of the orders $i_1<\cdots<i_r$ of a certain chance variable X, where $r\leqq k$. There exists a chance variable X' which is characteristic relative to $M_{i_1}, \cdots, M_{i_r}$.*

In order to prove A_{k+1}, we shall first prove by means of A_k the

LEMMA B_k. *If the chance variable which is characteristic relative to the moments* $M_{i_1}, \cdots, M_{i_r}$ $(r \leq k)$ *is not degenerate, then there exists a positive* δ *such that any r-tuple* $M'_{i_1}, \cdots, M'_{i_r}$ *can be realized as moments for which*

$$|M_{i_1} - M'_{i_1}| < \delta, \cdots, |M_{i_{r-1}} - M'_{i_{r-1}}| < \delta$$

and $M'_{i_r} > M_{i_r} - \delta$.

We shall say that an n-tuple $y_1, \cdots, y_n$ lies in the ϵ-neighborhood of the n-tuple $x_1, \cdots, x_n$ if $|x_1 - y_1| < \epsilon, \cdots, |x_n - y_n| < \epsilon$.

B_k is obviously true for $r=1$. We shall prove B_k for r by assuming that it is true for $r-1$. Denote by X the chance variable which is characteristic relative to $M_{i_1}, \cdots, M_{i_r}$ and suppose that X is not degenerate. That is to say, the degree of X is equal to $(r+1)/2$. According to A_k there exists a chance variable Y which is characteristic relative to $M_{i_1}, \cdots, M_{i_{r-1}}$. The chance variable Y is also not degenerate. In fact, if Y were degenerate, that is to say, if the degree of Y were less than or equal to $(r-1)/2$, then according to Proposition 6, $P(X<x)$ would be identically equal to $P(Y<x)$ and therefore also X would be degenerate, in contradiction to our assumption. Hence the degree of Y is equal to $r/2$. From Propositions 2 and 5 it follows that $M_{i_r}(Y) \neq M_{i_r}(X)$. Hence on account of Proposition 8, $M_{i_r}(Y) < M_{i_r}(X)$. Since B_k is assumed to be true for $r-1$, there exists a positive ϵ such that any $(r-1)$-tuple $M'_{i_1}, \cdots, M'_{i_{r-1}}$ in the ϵ-neighborhood of $M_{i_1}, \cdots, M_{i_{r-1}}$ can be realized as moments. Hence according to A_k, for each point $M' = M'_{i_1}, \cdots, M'_{i_{r-1}}$ of the ϵ-neighborhood of $M = M_{i_1}, \cdots, M_{i_{r-1}}$, a chance variable (and only one) exists which is characteristic relative to M'. Denote by $X(M')$ the chance variable which is characteristic relative to M'. From Proposition 12 it follows that $X(M')$ is a continuous function of M' in the ϵ-neighborhood of M. For each point $M' = M'_{i_1}, \cdots, M'_{i_{r-1}}$ of the ϵ-neighborhood of M the degree of $X(M')$ must be equal to $r/2$. In fact, if $X(M')$ were of degree less than $r/2$, then $X(M')$ would be characteristic also relative to $M'_{i_1}, \cdots, M'_{i_{r-2}}$ and therefore on account of Proposition 8 not every point of a neighborhood of M' could be realized, in contradiction to the statement that every point of the ϵ-neighborhood of M can be realized. Hence the degree of $X(M')$ is equal to $r/2$ for any point M' of the ϵ-neighborhood of M. From this fact it follows easily that for any integer n the nth moment of $X(M')$ is a continuous function of M' in the ϵ-neighborhood of M. Since $X(M) = Y$ and $M_{i_r}(Y) < M_{i_r}(X)$, there exists a positive value $\delta < \epsilon$ such that for any point M' of the δ-neighborhood of M, the i_rth moment of $X(M')$ is less than $M_{i_r}(X) - \delta$. Consider a certain point $M' = M'_{i_1}, \cdots, M'_{i_{r-1}}$ of the δ-neighborhood of M and the $(r-1)$-tuple $M_{i_1}(d, \eta), \cdots, M_{i_{r-1}}(d, \eta)$ defined as follows:

$$M_{i_1}(d, \eta) = \frac{M'_{i_1} - d^{i_1}\eta}{1 - \eta}, \cdots, M_{i_{r-1}}(d, \eta) = \frac{M'_{i_{r-1}} - d^{i_{r-1}} \cdot \eta}{1 - \eta},$$

where d and η are positive numbers such that the $(r-1)$-tuple $M_{i_1}(d, \eta), \cdots, M_{i_{r-1}}(d, \eta)$ is contained in the δ-neighborhood of M. Denote by $X(d, \eta)$ the chance variable which is characteristic relative to $M_{i_1}(d, \eta), \cdots, M_{i_{r-1}}(d, \eta)$. Denote further by $Y(d, \eta)$ the arithmetic chance variable defined as follows:

$$P[Y(d, \eta) = x] = P[X(d, \eta) = x] \cdot (1 - \eta), \qquad \text{for } x \neq d,$$

$$P[Y(d, \eta) = d] = P[X(d, \eta) = d] \cdot (1 - \eta) + \eta.$$

It is obvious that the i_νth moment of $Y(d, \eta)$ is equal to M'_{i_ν} $(\nu = 1, \cdots, r-1)$. The i_rth moment of $Y(d, \eta)$ is a continuous function of d and η. For $\eta = 0$, $Y(d, \eta)$ is equal to $X(M')$ and therefore the i_rth moment of $Y(d, 0)$ is less than $M_{i_r}(X) - \delta$. Let us now consider two sequences of positive numbers $\{d_\nu\}$ and $\{\eta_\nu\}$ $(\nu = 1, \cdots, \text{ad inf.})$ such that $\lim d_\nu = \infty$, $\lim \eta_\nu = 0$, $\lim d_\nu^{i_{r-1}} \eta_\nu = 0$, and $\lim d_\nu^{i_r} \eta_\nu = \infty$. It is obvious that $\lim_{\nu=\infty} M_{i_n}(d_\nu, \eta_\nu) = M'_{i_n}$ for $n = 1, \cdots, r-1$. Hence for sufficiently large ν the $(r-1)$-tuple $M_{i_1}(d_\nu, \bar{\eta}_\nu), \cdots, M_{i_{r-1}}(d_\nu, \bar{\eta}_\nu)$ lies in the δ-neighborhood of M for any positive $\bar{\eta}_\nu \leq \eta_\nu$. On the other hand the i_rth moment of $Y(d_\nu, \eta_\nu)$ converges towards infinity. If α denotes an arbitrary number greater than $M_{i_r}(X) - \delta$, then for sufficiently large ν the i_rth moment of $Y(d_\nu, \eta_\nu)$ will be greater than α. Since the i_rth moment of $Y(d_\nu, 0)$ is less than α, there exists a number $\bar{\eta}_\nu < \eta_\nu$ such that the i_rth moment of $Y(d_\nu, \bar{\eta}_\nu)$ is equal to α. This proves the Lemma B_k.

Now we shall prove A_{k+1} by means of A_k and B_k. Denote by $M_{i_1}, \cdots, M_{i_{k+1}}$ the moments of the orders $i_1 < \cdots < i_{k+1}$ of a certain chance variable X. According to A_k there exists a chance variable Y which is characteristic relative to $M_{i_1}, \cdots, M_{i_k}$. If Y is degenerate, then according to Proposition 6, X must be equal to Y and Y is therefore characteristic also relative to $M_{i_1}, \cdots, M_{i_{k+1}}$. Hence in this case A_{k+1} is proved. We have to consider only the case that Y is not degenerate. Hence the degree of Y is equal to $(k+1)/2$. On account of Proposition 8, $M_{i_{k+1}}(Y) \leq M_{i_{k+1}}$. If $M_{i_{k+1}}(Y) = M_{i_{k+1}}$, then Y is characteristic relative to $M_{i_1}, \cdots, M_{i_{k+1}}$ and A_{k+1} is proved. We have to deal only with the case that $M_{i_{k+1}}(Y) < M_{i_{k+1}}$. Denote by d_0 the greatest positive value for which $P(Y = d_0) > 0$. Consider the chance variable $Y_{d,\epsilon}$ which is characteristic relative to

$$M_{i_1}(d, \epsilon) = \frac{M_{i_1} - d^{i_1}\epsilon}{1 - \epsilon}, \cdots, M_{i_k}(d, \epsilon) = \frac{M_{i_k} - d^{i_k} \cdot \epsilon}{1 - \epsilon},$$

where $d > d_0$. On account of B_k, $Y_{d,\epsilon}$ exists for sufficiently small ϵ. According to Proposition 12, $\lim_{\epsilon=0} Y_{d,\epsilon} = Y$. Hence for sufficiently small values of ϵ,

$Y_{d,\epsilon}$ is not degenerate. From Proposition 12 and B_k it follows easily that for any given d the set of values ϵ for which $Y_{d,\epsilon}$ exists and is not degenerate is an open set. Hence there exists a smallest value $\epsilon(d)$ for which $Y_{d,\epsilon(d)}$ is degenerate or does not exist. First we shall prove

LEMMA 1. *$P(Y_{d,\epsilon}=d)=0$ for $d>d_0$ and for any ϵ for which $Y_{d,\epsilon}$ exists.*

Let us suppose that there exists a value $d>d_0$ and a positive ϵ such that $Y_{d,\epsilon}$ exists and $P(Y_{d,\epsilon}=d)>0$. Consider the chance variable $\overline{Y}_{d,\epsilon}$ defined as follows:

$$P(\overline{Y}_{d,\epsilon} = d) = P(Y_{d,\epsilon} = d)\cdot(1 - \epsilon) + \epsilon;$$

$$P(\overline{Y}_{d,\epsilon} = x) = P(Y_{d,\epsilon} = x)\cdot(1 - \epsilon), \qquad \text{for } x \neq d.$$

It is obvious that $M_{i_\nu}(\overline{Y}_{d,\epsilon})=M_{i_\nu}$ $(\nu=1, \cdots, k)$ and the degree of $\overline{Y}_{d,\epsilon}$ is not greater than the degree of $Y_{d,\epsilon}$. Hence $\overline{Y}_{d,\epsilon}$ is characteristic relative to $M_{i_1}, \cdots, M_{i_k}$. According to Proposition 7, $\overline{Y}_{d,\epsilon}$ must be equal to Y, which is not the case, since $P(\overline{Y}_{d,\epsilon}=d)>0$ and $P(Y=d)=0$. Hence we have a contradiction and the assumption $P(Y_{d,\epsilon}=d)>0$ is proved to be an absurdity.

We shall now prove the

LEMMA 2. *If $d>d_0$ then for each $\epsilon<\epsilon(d)$, $P(Y_{d,\epsilon}\geqq d)=0$.*

In fact $Y_{d,0}=Y$ and therefore $P(Y_{d,0}\geqq d)=0$. On account of Proposition 12, $Y_{d,\epsilon}$ is a continuous function of ϵ in the half open interval $[0, \epsilon(d))$. Hence if there exists a value $\epsilon'<\epsilon(d)$ for which $P(Y_{d,\epsilon'}\geqq d)>0$, then $P(Y_{d,\epsilon}=d)>0$ must hold for a certain value $\epsilon=\epsilon''\leqq\epsilon'$, in contradiction to Lemma 1.

LEMMA 3. *For $d>d_0$, $Y_{d,\epsilon(d)}$ exists and $P(Y_{d,\epsilon(d)}\geqq d)=0$.*

Denote by $\{\epsilon_n\}$ a sequence of positive numbers for which $\epsilon_n<\epsilon(d)$ and $\lim \epsilon_n=\epsilon(d)$. Consider the corresponding sequence $\{Y_{d,\epsilon_n}\}$ of chance variables. On account of Proposition 10 there exists a convergent subsequence $\{Y_{d,\epsilon_n'}\}$ of the sequence $\{Y_{d,\epsilon_n}\}$. Denote $\lim_{n=\infty} Y_{d,\epsilon_n'}$ by Y_d. Since according to Lemma 1, $P(Y_{d,\epsilon}\geqq d)=0$, $\{M_r(Y_{d,\epsilon_n'})\}$ is bounded for any integer r. Hence we have on account of Proposition 9

$$\lim_{n=\infty} M_r(Y_{d,\epsilon_n'}) = M_r(Y_d)$$

for any integer r. Then from

$$M_{i_\nu} = \frac{M_{i_\nu} - d^{i_\nu}\cdot\epsilon_n}{1 - \epsilon_n}, \qquad \nu = 1, \cdots, k;\ n = 1, \cdots, \text{ad inf.},$$

it follows that

$$M_{i_\nu}(Y_d) = \lim \frac{M_{i_\nu} - d^{i_\nu}\cdot\epsilon_n'}{1 - \epsilon_n'} = \frac{M_{i_\nu} - d^{i_\nu}\epsilon(d)}{1 - \epsilon(d)}, \qquad \nu = 1, \cdots, k.$$

Since Y_d is characteristic relative to the above moments, $Y_{d,\epsilon(d)}$ exists and is equal to Y_d. From $P(Y_{d,\epsilon_n}\geqq d)=0$ and $\lim Y_{d,\epsilon_n'}=Y_{d,\epsilon(d)}$ it follows that $P(Y_{d,\epsilon(d)}>d)=0$. Since on account of Lemma 1, $P(Y_{d,\epsilon(d)}=d)=0$, we have $P(Y_{d,\epsilon(d)}\geqq d)=0$.

Now we are able to prove

Lemma 4. *Besides $\epsilon(d)$ no other value ϵ' can be given for which $Y_{d,\epsilon'}$ exists and is degenerate provided $d>d_0$.*

Let us suppose that there exists a positive $\epsilon'\neq\epsilon(d)$ for which $Y_{d,\epsilon'}$ exists and is degenerate. Consider the chance variables $\overline{Y}_{d,\epsilon'}$ and $\overline{Y}_{d,\epsilon(d)}$ defined as follows:

$$P(\overline{Y}_{d,\epsilon} = d) = \epsilon;\ P(\overline{Y}_{d,\epsilon} = x) = P(Y_{d,\epsilon} = x)\cdot(1-\epsilon),\ \text{for } x\neq d;\ \epsilon = \epsilon',\ \epsilon(d).$$

Since $Y_{d,\epsilon'}$ and $Y_{d,\epsilon(d)}$ are degenerate, their degrees are less than or equal to $(k+1)/2-1/2=k/2$. The degree of $\overline{Y}_{d,\epsilon'}$ and that of $\overline{Y}_{d,\epsilon(d)}$ are obviously not greater than $k/2+1$. Hence on account of Proposition 4,

$$D(x) = P(\overline{Y}_{d,\epsilon'} < x) - P(\overline{Y}_{d,\epsilon(d)} < x)$$

has at most $2(k/2+1)-3=k-1$ changes in sign. Since $M_{i_\nu}(\overline{Y}_{d,\epsilon'}) = M_{i_\nu}(\overline{Y}_{d,\epsilon(d)})=M_{i_\nu}$ $(\nu=1,\cdots,k)$, $D(x)$ must be identically equal to zero on account of Proposition 5. But $D(x)$ cannot be identically equal to zero since $P(\overline{Y}_{d,\epsilon'}=d)=\epsilon'$, $P(\overline{Y}_{d,\epsilon(d)}=d)=\epsilon(d)$ and $\epsilon'\neq\epsilon(d)$. Hence the assumption that there exists an $\epsilon'\neq\epsilon(d)$ for which $Y_{d,\epsilon'}$ exists and is degenerate is proved to be an absurdity.

Let us consider a sequence of numbers $\{d_n\}$ $(n=1,\cdots,\text{ad inf.})$ for which $d_n>d_0$ and $\lim d_n=d>d_0$. We shall show that $\lim \epsilon(d_n)=\epsilon(d)$ and $\lim Y_{d_n,\epsilon(d_n)} = Y_{d,\epsilon(d)}$. In order to prove $\lim Y_{d_n,\epsilon(d_n)}=Y_{d,\epsilon(d)}$, we have only to show on account of Proposition 11 that for each convergent subsequence $\{Y_{d_n',\epsilon(d_n')}\}$ of the sequence $\{Y_{d_n,\epsilon(d_n)}\}$

$$\lim Y_{d_n',\epsilon(d_n')} = Y_{d,\epsilon(d)}.$$

Denote $\lim Y_{d_n',\epsilon(d_n')}$ by Y^*. Since $P[Y_{d_n',\epsilon(d_n')}>d_n']=0$, $\{M_r(Y_{d_n',\epsilon(d_n')})\}$ is bounded for any r. Hence we have on account of Proposition 9

$$\lim_{n=\infty} M_r(Y_{d_n',\epsilon(d_n')}) = M_r(Y^*)$$

for any positive integer r. Since

$$M_{i_\nu}(Y_{d_n',\epsilon(d_n')}) = \frac{M_{i_\nu} - (d_n')^{i_\nu}\cdot\epsilon(d_n')}{1-\epsilon(d_n')}$$

converges with increasing n and since $\lim (d_n')^{i_\nu}=d^{i_\nu}>d_0^{i_\nu}>M_{i_\nu}$, the sequence

$\{\epsilon(d_n')\}$ must also converge. Denote lim $\epsilon(d_n')$ by ϵ^*. Then Y^* is characteristic relative to

$$M_{i_1}(d, \epsilon^*), \cdots, M_{i_k}(d, \epsilon^*);$$

that is to say, Y^* is equal to Y_{d,ϵ^*}. Since $Y_{d,\epsilon^*} = \lim Y_{d_n',\epsilon(d_n')}$ and $Y_{d_n',\epsilon(d_n')}$ is degenerate, Y_{d,ϵ^*} must also be degenerate. Then according to Lemma 4, $\epsilon^* = \epsilon(d)$ and therefore Y_{d,ϵ^*} is equal to $Y_{d,\epsilon(d)}$. Hence our statement that $\lim Y_{d_n,\epsilon(d_n)} = Y_{d,\epsilon(d)}$ is proved. Since according to Lemma 3, $P(Y_{d_n,\epsilon(d_n)} \geq d_n) = 0$ and therefore $M_r(Y_{d_n,\epsilon(d_n)})$ is bounded for any integer r, we have on account of Proposition 9

$$\lim_{n=\infty} M_r(Y_{d_n,\epsilon(d_n)}) = M_r(Y_{d,\epsilon(d)}).$$

From this it follows that $\lim \epsilon(d_n) = \epsilon(d)$ and that the moments of $Y_{d,\epsilon(d)}$ are continuous functions of d for $d > d_0$.

Denote by $\overline{Y}_{d,\epsilon}$ the chance variable defined as follows:

$$P(\overline{Y}_{d,\epsilon} = d) = \epsilon; \quad P(\overline{Y}_{d,\epsilon} = x) = P(Y_{d,\epsilon} = x)\cdot(1 - \epsilon), \qquad \text{for } x \neq d.$$

It is obvious that

$$M_{i_\nu}(\overline{Y}_{d,\epsilon}) = M_{i_\nu}, \qquad \nu = 1, \cdots, k.$$

In order to show that $\lim_{d=\infty} M_{i_{k+1}}(\overline{Y}_{d,\epsilon(d)}) = \infty$ we have only to show that for any sequence $\{d_n\}$ for which $\lim d_n = \infty$, $d_n^{i_k}\cdot\epsilon(d_n)$ does not converge towards zero. In order to prove the latter statement, let us assume that $\lim d_n^{i_k}\epsilon(d_n) = 0$ and $\lim d_n = \infty$. It is obvious that $\lim d_n^{i_\nu}\epsilon(d_n) = 0$ for $\nu = 1, 2, \cdots, k$. Hence

$$\lim_{d=\infty} M_{i_\nu}(Y_{d,\epsilon(d)}) = M_{i_\nu}, \qquad \nu = 1, \cdots, k.$$

Since Y is characteristic relative to $M_{i_1}, \cdots, M_{i_k}$, we have, on account of Proposition 12, $\lim Y_{d_n,\epsilon(d_n)} = Y$. But this is not possible since $Y_{d_n,\epsilon(d_n)}$ is degenerate and therefore $\lim Y_{d_n,\epsilon(d_n)}$ must also be degenerate and consequently cannot be equal to Y which is not degenerate. Hence we have $\lim M_{i_{k+1}}(\overline{Y}_{d_n,\epsilon(d_n)}) = \infty$.

On account of Proposition 10 there exists a sequence $\{d_n\}$ such that $d_n > d_0$, $\lim d_n = d_0$, and the sequence $\{\overline{Y}_{d_n,\epsilon(d_n)}\}$ is convergent. Denote $\lim \overline{Y}_{d_n,\epsilon(d_n)}$ by Y^*. Since $M_{i_\nu}(\overline{Y}_{d_n,\epsilon(d_n)}) = M_{i_\nu}$ $(\nu = 1, \cdots, k)$ and $P(\overline{Y}_{d_n,\epsilon(d_n)} > d_n) = 0$, we have, on account of Proposition 9, $M_{i_\nu}(Y^*) = M_{i_\nu}$ $(\nu = 1, \cdots, k)$. The degree of $Y_{d,\epsilon(d)}$ is less than or equal to $k/2$ and therefore the degree of $\overline{Y}_{d,\epsilon(d)}$ is less than or equal to $k/2+1$. Hence also the degree of Y^* is less than or equal to $k/2+1$. Now we shall show that $P(Y^* = d_0) > 0$. Let us assume that $P(Y^* = d_0) = 0$. Then $\lim \epsilon(d_n)$ must be equal to zero. Hence $\lim M_{i_\nu}(Y_{d_n,\epsilon(d_n)})$

$=M_{i_\nu}$ $(\nu=1,\cdots,k)$, and then on account of Proposition 12, $Y_{d_n,\epsilon(d_n)}$ must converge towards Y which cannot be the case since $Y_{d_n,\epsilon(d_n)}$ is degenerate and Y is not degenerate. Hence $P(Y^*=d_0)>0$ is proved. Since $P(Y=d_0)>0$, $P(Y^*=d_0)>0$, we get from Proposition 4 that the number of changes in sign of $D(x)=P(Y^*<x)-P(Y<x)$ is less than or equal to $2(k/2+1)-3=k-1$. Since $M_{i_\nu}(Y^*)=M_{i_\nu}(Y)$ $(\nu=1,\cdots,k)$, on account of Proposition 5, $D(x)$ must be identically equal to zero, that is to say, $Y^*=Y$. Hence

$$\lim M_{i_{k+1}}(\overline{Y}_{d_n,\epsilon(d_n)}) = M_{i_{k+1}}(Y^*) = M_{i_{k+1}}(Y) < M_{i_{k+1}}.$$

Since $\lim_{d=\infty} M_{i_{k+1}}(\overline{Y}_{d,\epsilon(d)})=\infty$ and $M_{i_{k+1}}(\overline{Y}_{d,\epsilon(d)})$ is a continuous function of d, there exists a value d' such that $M_{i_{k+1}}(\overline{Y}_{d',\epsilon(d')})=M_{i_{k+1}}$. The degree of $\overline{Y}_{d',\epsilon(d')}$ is less than or equal to $k/2+1$, and therefore $\overline{Y}_{d',\epsilon(d')}$ is characteristic relative to $M_{i_1},\cdots,M_{i_{k+1}}$. This proves A_{k+1}, and therefore Proposition 13 is also proved.

Since Proposition 13 is proved, B_k is also proved for any positive integer k. Hence we can formulate

PROPOSITION 14. *If the chance variable which is characteristic relative to the moments $M_{i_1},\cdots,M_{i_k}$ is not degenerate, then there exists a positive δ such that any k-tuple $M'_{i_1},\cdots,M'_{i_k}$ in the δ-neighborhood of the k-tuple $M_{i_1},\cdots,M_{i_k}$ can be realized as moments of the orders $i_1,\cdots,i_k$.*

4. **Solution of Problem 1.** Denote by $M_{i_1},\cdots,M_{i_k}$ the moments of the order $i_1<i_2<\cdots<i_k$ of a certain chance variable X. Denote by X' the characteristic chance variable relative to $M_{i_1},\cdots,M_{i_k}$. If X' is degenerate, then according to Proposition 6 no chance variable $Y\neq X'$ exists for which $M_{i_\nu}(Y)=M_{i_\nu}(X')$ $(\nu=1,\cdots,k)$. Hence the sharp lower and the sharp upper limits of $P(X<d)$ are equal to $P(X'<d)$ and our problem is solved. Throughout the following development we shall suppose that X' is not degenerate.

Consider the k-tuple of values

$$M_{i_\nu}(d,\lambda) = \frac{M_{i_\nu} - d^{i_\nu}\cdot\lambda}{1-\lambda}, \qquad \nu = 1,\cdots,k,$$

where $d>0$, $0\leqq\lambda<1$. According to Proposition 14 the k-tuple $M_{i_1}(d,\lambda),\cdots,$ $M_{i_k}(d,\lambda)$ can be realized as moments for sufficiently small values of λ. Denote by $Y(d,\lambda)$ the characteristic chance variable relative to the moments $M_{i_1}(d,\lambda),\cdots,M_{i_k}(d,\lambda)$. Denote further by $\overline{Y}(d,\lambda)$ the arithmetic chance variable defined as follows:

$$P[\overline{Y}(d,\lambda) = d] = P[Y(d,\lambda) = d](1-\lambda) + \lambda,$$
$$P[\overline{Y}(d,\lambda) = x] = P[Y(d,\lambda) = x](1-\lambda), \qquad \text{for } x \neq d.$$

It is obvious that

$$M_{i_\nu}[\overline{Y}(d, \lambda)] = M_{i_\nu}, \qquad \nu = 1, \cdots, k.$$

From Proposition 14 it follows that for any given $d>0$ the set Ω of values of λ for which the characteristic chance variable relative to the moments $M_{i_1}(d, \lambda), \cdots, M_{i_k}(d, \lambda)$ exists and is not degenerate is an open set. Denote by λ_d the smallest positive value not belonging to Ω.

As is well known, $M_r^{s/r} \leqq M_s$ for any integer $r<s$, and the equality sign holds only if the chance variable is of the degree less than or equal to 1. Since for $\lambda < \lambda_d$ the characteristic chance variable $Y(d, \lambda)$ is not degenerate, we have

$$[M_{i_\mu}(d, \lambda)]^{i_\nu/i_\mu} < M_{i_\nu}(d, \lambda), \qquad \text{for } \mu < \nu, \mu < k, \nu \leqq k.$$

From these inequalities and from the fact that negative moments are not possible, it follows easily that if $k \geqq 3$, $\lambda_d < 1$ for any positive value d. If $k=2$, λ_d can be equal to 1 only if $d = M_{i_1}^{1/i_1}$.

Now we shall prove

PROPOSITION 15. *Denote by* $\{\lambda_n\}$ $(n=1, 2, \cdots,$ *ad inf.) a sequence of positive values such that* $\lambda_n < \lambda_d$ *and* $\lim \lambda_n = \lambda_d$. *Then* $\lim Y(d, \lambda_n)$ *exists and is equal to the chance variable* Y_d *which is characteristic relative to the* $k-1$ *moments* $M_{i_1}(d, \lambda_d), \cdots, M_{i_{k-1}}(d, \lambda_d)$. *If* Y_d *is not degenerate, then* Y_d *is characteristic also relative to the k moments* $M_{i_1}(d, \lambda_d), \cdots, M_{i_k}(d, \lambda_d)$.

According to Proposition 10 there exists a convergent subsequence of the sequence $\{Y(d, \lambda_n)\}$ $(n=1, \cdots,$ ad inf.). Denote by $\{Y(d, \lambda_n')\}$ a convergent subsequence of $\{Y(d, \lambda_n)\}$ and denote $\lim Y(d, \lambda_n')$ by Y^*. If $Y(d, \lambda_d)$ exists, then according to Proposition 12, Y^* must be equal to $Y(d, \lambda_d)$. Since $Y(d, \lambda_d)$ is degenerate, $Y(d, \lambda_d)$ is characteristic also relative to the $k-1$ moments $M_{i_1}(d, \lambda_d), \cdots, M_{i_{k-1}}(d, \lambda_d)$. Hence $Y^* = Y(d, \lambda_d) = Y_d$. We have now to consider the case that $Y(d, \lambda_d)$ does not exist, that is to say, the k-tuple $M_{i_1}(d, \lambda_d), \cdots, M_{i_k}(d, \lambda_d)$ cannot be realized as moments. On account of Proposition 9,

$$M_{i_\nu}(Y^*) = M_{i_\nu}(d, \lambda_d), \qquad \nu = 1, \cdots, k-1.$$

Since the k-tuple $M_{i_1}(d, \lambda_d), \cdots, M_{i_k}(d, \lambda_d)$ cannot be realized,

$$M_{i_k}(Y^*) \neq M_{i_k}(d, \lambda_d).$$

From this it follows on account of Proposition 9 that $\{M_r[Y(d, \lambda_n')]\}$ $(n=1, 2, \cdots,$ ad inf.) is not bounded for any integer $r > i_k$. Hence there exists a subsequence $\{Y(d, \lambda_n'')\}$ such that $\lim_{n=\infty} M_r[Y(d, \lambda_n'')] = \infty$. Denote by α_n the greatest positive value for which $P[Y(d, \lambda_n'') = \alpha_n] > 0$. It is obvious that $\lim \alpha_n = \infty$ and

$$\lim P[Y(d, \lambda_n'') = \alpha_n] = 0.$$

Hence the degree of Y^* must be less than or equal to $(k+1)/2-1=(k-1)/2$. Since $M_{i_\nu}(Y^*)=M_{i_\nu}(d, \lambda_d)$ $(\nu=1, \cdots, k-1)$, Y^* is characteristic and degenerate relative to $M_{i_1}(d, \lambda_d), \cdots, M_{i_{k-1}}(d, \lambda_d)$. That is to say, $Y^*=Y_d$ and Y_d is degenerate.

Hence we have proved that in any case the limit of a convergent subsequence of $\{Y(d, \lambda_n)\}$ is equal to Y_d. From this fact it follows on account of Proposition 11 that $\lim Y(d, \lambda_n)=Y_d$. As we have shown, $Y_d=Y(d, \lambda_d)$ if $Y(d, \lambda_d)$ exists, and Y_d is degenerate if $Y(d, \lambda_d)$ does not exist. Hence Proposition 15 is proved.

PROPOSITION 16. *$P(Y_d=d)=0$, where Y_d denotes the characteristic chance variable relative to the $k-1$ moments $M_{i_1}(d, \lambda_d), \cdots, M_{i_{k-1}}(d, \lambda_d)$.*

Let us suppose $P(Y_d=d)>0$. Denote by $\overline{Y}_d$ the chance variable defined as follows:

$$P(\overline{Y}_d = d) = P(Y_d = d)\cdot(1 - \lambda_d) + \lambda_d;$$

$$P(\overline{Y}_d = x) = P(Y_d = x)\cdot(1 - \lambda_d), \qquad \text{for } x \neq d.$$

If $Y(d, \lambda_d)$ exists, then $M_{i_\nu}(Y_d)=M_{i_\nu}(d, \lambda_d)$ $(\nu=1, \cdots, k)$ and therefore $M_{i_\nu}(\overline{Y}_d)=M_{i_\nu}$ $(\nu=1, \cdots, k)$. The degree of $\overline{Y}_d$ is equal to the degree of Y_d. Hence $\overline{Y}_d$ is characteristic and degenerate relative to $M_{i_1}, \cdots, M_{i_k}$, in contradiction to our assumption that the characteristic chance variable relative to $M_{i_1}, \cdots, M_{i_k}$ is not degenerate. If $Y(d, \lambda_d)$ does not exist, then Y_d is degenerate. That is to say, the degree of Y_d is less than or equal to $(k-1)/2$. Since $M_{i_\nu}(\overline{Y}_d)=M_{i_\nu}$ $(\nu=1, \cdots, k-1)$ and the degree of $\overline{Y}_d$ is equal to the degree of Y_d, $\overline{Y}_d$ is characteristic and degenerate relative to the $k-1$ moments $M_{i_1}, \cdots, M_{i_{k-1}}$. But on account of our assumption that the characteristic chance variable relative to $M_{i_1}, \cdots, M_{i_k}$ is not degenerate, from Proposition 6 it follows that the characteristic chance variable relative to $M_{i_1}, \cdots, M_{i_{k-1}}$ also cannot be degenerate. Hence we have a contradiction and the assumption that $P(Y_d=d)>0$ is proved to be an absurdity.

PROPOSITION 17. *Denote by $\{d_n\}$ and $\{\lambda_n\}$ $(n=1, 2, \cdots$, ad inf.) two sequences of positive values such that* $\lim d_n=d>0$, $\lim \lambda_n=\lambda<\lambda_d$. *Then* $\lim Y(d_n, \lambda_n)=Y(d, \lambda)$.

On account of Proposition 14, $Y(d_n, \lambda_n)$ exists for almost every n. Since $\lim M_{i_\nu}[Y(d_n, \lambda_n)]=M_{i_\nu}[Y(d, \lambda)]$, we have on account of Proposition 12 that $\lim Y(d_n, \lambda_n)=Y(d, \lambda)$.

PROPOSITION 18. *The sharp lower limit a_d of $P(X<d)$ is equal to $P(\overline{Y}_d<d)$, and the sharp upper limit b_d of $P(X<d)$ is equal to $P(\overline{Y}_d\leqq d)$ where $\overline{Y}_d$ denotes*

the arithmetic chance variable defined as follows:

$$P(\overline{Y}_d = d) = \lambda_d; \qquad P(\overline{Y}_d = x) = P(Y_d = x)\cdot(1 - \lambda_d), \qquad \text{for } x \neq d.$$

We shall consider two cases.

(1) *Y_d is not degenerate.* Hence the degree of Y_d is equal to $k/2$. According to Proposition 15, Y_d is characteristic also relative to $M_{i_1}(d, \lambda_d), \cdots, M_{i_k}(d, \lambda_d)$. Hence

$$M_{i_\nu}(\overline{Y}_d) = M_{i_\nu}, \qquad \nu = 1, \cdots, k.$$

Since, according to Proposition 16, $P(Y_d = d) = 0$, the degree of $\overline{Y}_d$ is obviously equal to $k/2+1$. Let us suppose that there exists a chance variable X such that $M_{i_\nu}(X) = M_{i_\nu}$ $(\nu = 1, \cdots, k)$ and $P(X < d) < P(\overline{Y}_d < d)$. Denote by α the greatest number less than d for which $P(\overline{Y}_d = \alpha) > 0$. It is obvious that $D(x) = P(X < x) - P(\overline{Y}_d < d)$ has no change in sign in the interior of the interval $[\alpha, d]$. If $D(x)$ is identically zero in the interior of $[\alpha, d]$, then $D(x)$ has no change in sign at α. If $D(x)$ is not identically zero in the interior of $[\alpha, d]$ and if $P(X \leqq d) \leqq P(\overline{Y}_d \leqq d)$, then $D(x)$ has no change in sign at d. Finally if $P(X \leqq d) > P(\overline{Y}_d \leqq d)$ and if β denotes the smallest value greater than d for which $P(\overline{Y}_d = \beta) > 0$, then $D(x)$ has no change in sign in the interior of the interval $[d, \beta]$. From this fact it follows easily that the number of changes in sign of $D(x)$ cannot exceed $2(k/2+1) - 3 = k - 1$. Since $M_{i_\nu}(X) = M_{i_\nu}(\overline{Y})$ $(\nu = 1, \cdots, k)$, this is in contradiction to Proposition 5. Hence the assumption $P(X < d) < P(\overline{Y}_d < d)$ is proved to be an absurdity. Now let us assume that there exists a chance variable X such that $M_{i_\nu}(X) = M_{i_\nu}$ $(\nu = 1, \cdots, k)$ and $P(X < d) > P(\overline{Y}_d \leqq d)$. Denote by β the smallest number greater than d for which $P(\overline{Y}_d = \beta) > 0$. It is obvious that $D(x) = P(X < x) - P(\overline{Y}_d < x)$ has no change in sign at the point d and also no change in sign in the interior of the interval $[d, \beta]$. Hence the number of changes in sign of $D(x)$ cannot exceed $2(k/2+1) - 3 = k - 1$. But this is in contradiction to Proposition 5, and the assumption $P(X < d) > P(\overline{Y}_d \leqq d)$ therefore is proved to be an absurdity.

We now have to show that the limits $P(\overline{Y}_d < d)$ and $P(\overline{Y}_d \leqq d)$ are sharp. Since $M_{i_\nu}(\overline{Y}_d) = M_{i_\nu}$ $(\nu = 1, \cdots, k)$, the lower limit $P(\overline{Y}_d < d)$ is evidently sharp. Denote by $\{d_n\}$ $(n = 1, 2, \cdots$, ad inf.) a sequence of positive numbers for which $d_n < d$ and $\lim d_n = d$. Denote by λ some value less than λ_d. It is obvious that $Y(d_n, \lambda)$ exists for almost every n and that on account of Proposition 12, $\lim_{n=\infty} Y(d_n, \lambda) = Y(d, \lambda)$. Since $P(Y_d = d) = 0$ the function $P(Y_d < x)$ is constant in the neighborhood of $x = d$. Then from $\lim_{\lambda=\lambda_d} Y(d, \lambda) = Y_d$ it follows that there exists a positive η such that

$$\lim_{\lambda=\lambda_d} P[Y(d, \lambda) < d - \eta] = P(Y_d < d).$$

Hence to an arbitrarily small positive ϵ a value $\lambda_\epsilon < \lambda_d$ can be given such that

$$P[Y(d, \lambda) < d - \eta] > P(Y_d < d) - \epsilon$$

for any λ greater than λ_ϵ and smaller than λ_d. Since $\lim_{n=\infty} Y(d_n, \lambda) = Y(d, \lambda)$,

$$\text{(a)} \qquad P[Y(d_n, \lambda) < d] > P(Y_d < d) - 2\epsilon$$

for almost every n. On account of $d_n < d$,

$$\text{(b)} \qquad P[\overline{Y}(d_n, \lambda) < d] = (1 - \lambda)P[Y(d_n, \lambda) < d] + \lambda,$$

and on account of $P(Y_d = d) = 0$,

$$\text{(c)} \qquad P(\overline{Y}_d \leqq d) = (1 - \lambda_d)P(Y_d < d) + \lambda_d.$$

From (a), (b), and (c), it follows that if we choose λ sufficiently near to λ_d, we have

$$P[\overline{Y}(d_n, \lambda) < d] > P(\overline{Y}_d \leqq d) - 3\epsilon.$$

Since $M_{i_\nu}[\overline{Y}(d_n, \lambda)] = M_{i_\nu}$ $(\nu = 1, \cdots, k)$ and since ϵ can be chosen arbitrarily small, the upper limit $P(\overline{Y}_d \leqq d)$ is proved to be sharp. Hence Proposition 18 is proved if Y_d is not degenerate.

(2) *Y_d is degenerate.* Denote the degree of Y_d by $k'/2$ where k' denotes a positive integer. It is obvious that $k' \leqq k - 1$. Since the characteristic chance variable relative to $M_{i_1}, \cdots, M_{i_k}$ is not degenerate, from Proposition 6 it follows that also the characteristic chance variable relative to $M_{i_1}, \cdots, M_{i_{k'}}$ is not degenerate. Considering only the moments of the orders $i_1, \cdots, i_{k'}$ we have case (1) since Y_d is obviously characteristic and degenerate relative to the moments $M_1(d, \lambda_d), \cdots, M_{i_{k'}}(d, \lambda_d)$. Hence $P(\overline{Y}_d < d)$ is the greatest lower and $P(\overline{Y}_d \leqq d)$ is the least upper bound of $P(Z < d)$ where $P(Z < d)$ is formed for all chance variables Z for which $M_{i_\nu}(Z) = M_{i_\nu}$ $(\nu = 1, \cdots, k')$.

In order to show that the lower limit $P(\overline{Y}_d < d)$ is sharp consider the sequence $\{Y(d, \lambda_n)\}$ of chance variables where $\lambda_n < \lambda_d$ and $\lim \lambda_n = \lambda_d$. Since $\lim Y(d, \lambda_n) = Y_d$ and $P(Y_d = d) = 0$, we have

$$\lim_{n=\infty} P[Y(d, \lambda_n) < d] = P(Y_d < d).$$

On account of the fact that

$$P[\overline{Y}(d, \lambda_n) < d] = (1 - \lambda_n)P[Y(d, \lambda_n) < d],$$

and that

$$P(\overline{Y}_d < d) = P(Y_d < d) \cdot (1 - \lambda_d),$$

we have

$$\lim_{n=\infty} P[\overline{Y}(d, \lambda_n) < d] = P(\overline{Y}_d < d).$$

Since $M_{i_\nu}[\overline{Y}(d, \lambda_n)] = M_{i_\nu}$ $(\nu = 1, \cdots, k)$ the lower limit $P(\overline{Y}_d < d)$ is proved to be sharp. The proof of the fact that also the upper limit $P(\overline{Y}_d \leqq d)$ is sharp is quite analogous to that given in case (1). Hence Proposition 18 is proved.

We can summarize our results in the following

THEOREM 1. *The moments $M_{i_1}, \cdots, M_{i_j}$ of the orders $i_1, \cdots, i_j$ of a certain chance variable X are given. If the chance variable X' which is characteristic relative to $M_{i_1}, \cdots, M_{i_j}$ is degenerate, then the sharp lower limit a_d and the sharp upper limit b_d are equal to $P(X' < d)$. If X' is not degenerate, we have to consider the chance variable Y_d which is characteristic relative to $M_{i_1}(d, \lambda_d), \cdots, M_{i_{j-1}}(d, \lambda_d)$ where*

$$M_{i_\nu}(d, \lambda) = \frac{M_{i_\nu} - d^{i_\nu}\lambda}{1 - \lambda}, \qquad \nu = 1, \cdots, j,$$

and λ_d denotes the smallest value λ for which $M_{i_1}(d, \lambda), \cdots, M_{i_j}(d, \lambda)$ cannot be realized as moments, or the characteristic chance variable relative to them is degenerate. The sharp lower limit a_d is equal to $P(\overline{Y}_d < d)$ and the sharp upper limit b_d is equal to $P(\overline{Y}_d \leqq d)$, where $\overline{Y}_d$ denotes the arithmetic chance variable defined as follows:

$$P(\overline{Y}_d = d) = P(Y_d = d) \cdot (1 - \lambda_d) + \lambda_d,$$

$$P(\overline{Y}_d = x) = P(Y_d = x) \cdot (1 - \lambda_d) \qquad \text{for} \qquad x \neq d.$$

5. **Solution of Problem 2.** Denote by $M_{i_1}, \cdots, M_{i_k}$ the moments of the orders $i_1 < i_2 < \cdots < i_k$ of a certain chance variable X. Consider an integer $i_{k+1} > i_k$ and a number $M_{i_{k+1}}$. First we shall deal with the question: what conditions must be satisfied by $M_{i_{k+1}}$ in order that $M_{i_1}, \cdots, M_{i_{k+1}}$ can be realized as moments of the orders $i_1, \cdots, i_{k+1}$.

If the chance variable Y which is characteristic relative to $M_{i_1}, \cdots, M_{i_k}$ is degenerate, then on account of Proposition 6 no chance variable $Z \neq Y$ exists such that $M_{i_\nu}(Z) = M_{i_\nu}(Y)$ $(\nu = 1, \cdots, k)$. Hence $M_{i_1}, \cdots, M_{i_{k+1}}$ can be realized if and only if $M_{i_{k+1}} = M_{i_{k+1}}(Y)$.

Let us consider the case that Y is not degenerate. Denote by $\{d_n\}$ and $\{\epsilon_n\}$ $(n = 1, \cdots, \text{ad inf.})$ two sequences of positive numbers such that $\lim d_n = \infty$, $\lim d_n^{i_\nu} \cdot \epsilon_n = 0$ for $\nu \leqq k$, and $\lim d_n^{i_{k+1}} \cdot \epsilon_n = \infty$. Consider the k-tuple of values

$$M_{i_\nu}(d, \epsilon) = \frac{M_{i_\nu} - d^{i_\nu} \cdot \epsilon}{1 - \epsilon}, \qquad \nu = 1, \cdots, k.$$

If $M_{i_1}(d, \epsilon), \cdots, M_{i_k}(d, \epsilon)$ can be realized as moments of the orders $i_1, \cdots, i_k$, then we shall denote by $Y(d, \epsilon)$ the characteristic chance variable relative to these moments, and by $\overline{Y}(d, \epsilon)$ the arithmetic chance variable defined as follows:

$$P[\overline{Y}(d, \epsilon) = d] = P[Y(d, \epsilon) = d](1 - \epsilon) + \epsilon,$$

$$P[\overline{Y}(d, \epsilon) = x] = P[Y(d, \epsilon) = x](1 - \epsilon), \qquad \text{for } x \neq d.$$

It is obvious that

$$M_{i_\nu}[\overline{Y}(d, \epsilon)] = M_{i_\nu}, \qquad \nu = 1, \cdots, k.$$

Since $\lim \epsilon_n = \lim d_n^{i_\nu}\epsilon_n = 0$ ($\nu = 1, \cdots, k$), from Propositions 14 and 8 it follows easily that for almost every n, $Y(d_n, \epsilon)$ exists and is not degenerate for any nonnegative value $\epsilon \leqq \epsilon_n$. On account of Proposition 12, $Y(d_n, \epsilon)$ is a continuous function of ϵ in the interval $[0, \epsilon_n]$. Since $Y(d_n, \epsilon)$ is not degenerate for $0 \leqq \epsilon \leqq \epsilon_n$, also $M_r[Y(d_n, \epsilon)]$ is a continuous function of ϵ for any positive integer r. From this it follows that also $M_r[\overline{Y}(d_n, \epsilon)]$ is a continuous function of ϵ in the interval $[0, \epsilon_n]$. Since

$$M_{i_\nu}[\overline{Y}(d_n, \epsilon)] = M_{i_\nu} \quad (\nu = 1, \cdots, k), \qquad M_{i_{k+1}}[\overline{Y}(d_n, 0)] = M_{i_{k+1}}(Y),$$

we get that $M_{i_1}, \cdots, M_{i_{k+1}}$ can be realized as moments if

$$M_{i_{k+1}}(Y) \leqq M_{i_{k+1}} \leqq M_{i_{k+1}}[\overline{Y}(d_n, \epsilon_n)].$$

Because $\lim d_n^{i_{k+1}}\epsilon_n = \infty$ we obtain easily that $\lim M_{i_{k+1}}[\overline{Y}(d_n, \epsilon_n)] = \infty$ and therefore $M_{i_1}, \cdots, M_{i_{k+1}}$ can be realized as moments if $M_{i_{k+1}} \geqq M_{i_{k+1}}(Y)$. From Proposition 8 it follows that this condition is also necessary. Hence we have proved

PROPOSITION 19. *Denote by $M_{i_1}, \cdots, M_{i_k}$ k numbers which can be realized as moments of the orders $i_1 < i_2 < \cdots < i_k$. Denote by i_{k+1} an integer greater than i_k and by $M_{i_{k+1}}$ a certain number. If the chance variable Y which is characteristic relative to $M_{i_1}, \cdots, M_{i_k}$ is degenerate, then $M_{i_1}, \cdots, M_{i_{k+1}}$ can be realized as moments of the orders $i_1, \cdots, i_{k+1}$ if and only if $M_{i_{k+1}} = M_{i_{k+1}}(Y)$. If Y is not degenerate, then $M_{i_1}, \cdots, M_{i_{k+1}}$ can be realized as moments if and only if $M_{i_{k+1}} \geqq M_{i_{k+1}}(Y)$.*

If $M_{i_{k+1}} = M_{i_{k+1}}(Y)$, the characteristic chance variable relative to $M_{i_1}, \cdots, M_{i_{k+1}}$ is obviously equal to Y and therefore is degenerate. Since M_{i_1} can be realized as a moment of the order i_1 if and only if $M_{i_1} \geqq 0$, we get from Proposition 19

THEOREM 2. *Denote by $i_1 < i_2 < \cdots < i_k$ positive integers and by $M_{i_1}, \cdots, M_{i_k}$ some numbers. The values $M_{i_1}, \cdots, M_{i_k}$ can be realized as moments of the orders $i_1, \cdots, i_k$ if and only if*

$$M_{i_1} \geqq 0, M_{i_2} \geqq M_{i_2}(X_1), \cdots, M_{i_k} \geqq M_{i_k}(X_{k-1}),$$

where X_r denotes the characteristic chance variable relative to $M_{i_1}, \cdots, M_{i_r}$; if in one of the above relations the equality sign holds, then in all subsequent relations the equality sign must hold.

This theorem gives the solution of Problem 2, since $M_{i_r}(X_{r-1})$ is a function of $M_{i_1}, \cdots, M_{i_{r-1}}$ which can be calculated.

6. **Some applications of Theorems 1 and 2.** Let us calculate by means of Theorem 2 the inequalities which must be satisfied by the numbers M_r, M_s, M_t if they can be realized as moments of the orders r, s, t, where $r<s<t$.

According to Theorem 2 the necessary and sufficient conditions are given by

$$(10) \qquad M_r \geqq 0, \qquad M_s \geqq M_s(X_1), \qquad M_t \geqq M_t(X_2),$$

where X_1 denotes the characteristic chance variable relative to M_r, and X_2 denotes the characteristic chance variable relative to M_r and M_s. The degree of X_1 is less than or equal to 1. Hence there exists only a single point a with positive probability and therefore $M_r = M_r(X_1) = a^r$. Hence $a = M_r^{1/r}$. It is obvious that

$$(11) \qquad M_s(X_1) = a^s = M_r^{s/r}.$$

Let us now calculate the chance variable X_2. The degree of X_2 is less than or equal to 3/2. Hence only the origin and a single positive value b can have positive probability. The value of b and the probability $P(X_2=b)$ are determined by the equations

$$M_r(X_2) = b^r P(X_2 = b) = M_r; \qquad M_s(X_2) = b^s P(X_2 = s) = M_s.$$

From these equations we obtain

$$P(X_2 = b) = \frac{M_r}{(M_s/M_r)^{r/(s-r)}}, \qquad b = \left(\frac{M_s}{M_r}\right)^{1/(s-r)}.$$

Hence

$$(12) \qquad M_t(X_2) = b^t P(X_2 = b) = M_r \left(\frac{M_s}{M_r}\right)^{(t-r)/(s-r)}.$$

From (10), (11), and (12) we get

$$(13) \qquad M_r \geqq 0, \qquad M_s \geqq M_r^{s/r}, \qquad M_t \geqq M_r \left(\frac{M_s}{M_r}\right)^{(t-r)/(s-r)}.$$

If in one of the relations (13) the equality sign holds, then in all subsequent relations the equality sign must hold. These relations are necessary and sufficient in order that M_r, M_s, M_t can be realized as moments of the orders r, s, t.

As an application of Theorem 1 let us calculate the sharp lower limit a_d and the sharp upper limit b_d if two moments M_r and M_s are given, where $r<s$. According to the relations (13) we have

$$M_r \geqq 0, \qquad M_s \geqq M_r^{s/r}.$$

If $M_r=0$ (and therefore also $M_s=0$), or if $M_r>0$ and $M_s=M_r^{s/r}$, the chance variable X which is characteristic relative to M_r and M_s is degenerate and we have $a_d=b_d=P(X<d)$. Since $P(X<x)=0$ for $x\neq M_r^{1/r}$ and $P(X=M_r^{1/r})=1$, we have

$$a_d = b_d = 1, \qquad \text{for } d > M_r^{1/r},$$

$$a_d = b_d = 0, \qquad \text{for } d \leqq M_r^{1/r}.$$

Now we have to consider the case that

$$M_r > 0, \qquad M_s > M_r^{s/r}. \tag{14}$$

In order to calculate λ_d we have to consider the expressions:

$$M_r(d, \lambda) = \frac{M_r - d^r\lambda}{1-\lambda}, \qquad M_s(d, \lambda) = \frac{M_s - d^s\lambda}{1-\lambda}.$$

From Theorem 2 it follows that for any λ for which $M_r(d, \lambda)>0$ and $M_s(d, \lambda)>[M_r(d, \lambda)]^{s/r}$, $M_r(d, \lambda)$ and $M_s(d, \lambda)$ can be realized as moments of the orders r, s, and the corresponding characteristic chance variable is not degenerate. Hence either $M_r(d, \lambda_d)=0$ or $M_s(d, \lambda_d)=[M_r(d, \lambda_d)]^{s/r}$ must hold. That is to say, λ_d is either equal to M_r/d^r or is the root of the equation

$$\frac{M_s - d^s\lambda}{1-\lambda} = \left[\frac{M_r - d^r\lambda}{1-\lambda}\right]^{s/r}. \tag{15}$$

We have $\lambda_d=M_r/d^r$ if and only if the smallest positive root of (15) is greater than or equal to M_r/d^r. It is easy to show that this is the case if $M_r/d^r \leqq M_s/d^s$. Hence we have:

If $M_r/d^r\leqq M_s/d^s$ then $\lambda_d=M_r/d^r$, and if $M_r/d^r>M_s/d^s$ then λ_d is equal to the smallest positive root of (15).

If $\lambda_d=M_r/d^r$ then the chance variable Y_d which is characteristic relative to $M_r(d, \lambda_d)$ is given as follows: $P(Y_d=0)=1$ and $P(Y_d=x)=0$ for $x\neq 0$. Hence the chance variable $\overline{Y}_d$ is given as follows:

$$P(\overline{Y}_d = 0) = 1 - \lambda_d = 1 - M_r/d^r,$$

$$P(\overline{Y}_d = d) = \lambda_d = M_r/d^r.$$

Hence

(16) $a_d = P(\overline{Y}_d < d) = 1 - M_r/d^r;\ b_d = P(\overline{Y}_d \leqq d) = 1,\quad M_r/d^r \leqq M_s/d^s.$

Let us now consider the case that $M_r/d^r > M_s/d^s$. Then λ_d is the smallest positive root of (15). The chance variable Y_d which is characteristic relative to $M_r(d, y_d)$ is given as follows: $P(Y_d = \delta) = 1$ where

$$(17)\qquad \delta^r = \frac{M_r - d^r \cdot \lambda_d}{1 - \lambda_d}, \quad \text{or} \quad \frac{\delta^r}{d^r} = \frac{M_r/d^r - \lambda_d}{1 - \lambda_d}.$$

The chance variable $\overline{Y}_d$ is given as follows:

$$P(\overline{Y}_d = \delta) = P(Y_d = \delta)\cdot(1 - \lambda_d) = (1 - \lambda_d), \qquad P(\overline{Y}_d = d) = \lambda_d.$$

We shall show that $\delta < d$. One can easily see that $M_r/d^r < 1$. In fact, if $M_r/d^r > 1$, then

$$\left(\frac{M_r}{d^r}\right)^{s/r} > \frac{M_r}{d^r} > \frac{M_s}{d^s}$$

and therefore $M_r^{s/r} > M_s$ which is not possible. The inequality $\delta/d < 1$ follows from (17) on account of $M_r/d^r < 1$. Hence we have

$$(18)\qquad \begin{aligned} a_d = P(\overline{Y}_d < d) = P(\overline{Y}_d = \delta) = 1 - \lambda_d; \qquad b_d = P(\overline{Y}_d \leqq d) = 1,& \\ M_r/d^r > M_s/d^s.& \end{aligned}$$

The equations (16) and (18) give the complete formulas for a_d and b_d if two moments M_r and M_s are given.

If $s = 2r$ the root λ_d of (15) is given by the expression:

$$\lambda_d = \frac{M_{2r} - M_r^2}{(d^r - M_r)^2 + (M_{2r} - M_r^2)}.$$

Hence we get

$$(19)\qquad \begin{aligned} a_d &= 1 - \lambda_d = 1 - \frac{M_{2r} - M_r^2}{(d^r - M_r)^2 + (M_{2r} - M_r^2)}, \\ b_d &= 1, \qquad\qquad M_r/d^r > M_s/d^s, \end{aligned}$$

The sharp lower limits given in the formulas (16) and (19) are identical with the lower limits in the formulas (3) given by Cantelli.

COLUMBIA UNIVERSITY,
NEW YORK, N. Y.

Reprinted from THE ANNALS OF MATHEMATICAL STATISTICS
Vol. X, No. 4, December, 1939

CONTRIBUTIONS TO THE THEORY OF STATISTICAL ESTIMATION AND TESTING HYPOTHESES[1]

BY ABRAHAM WALD

1. **Introduction.** Let us consider a family of systems of n variates $X_1(\theta^{(1)}, \cdots, \theta^{(k)}), \cdots, X_n(\theta^{(1)}, \cdots, \theta^{(k)})$ depending on k parameters $\theta^{(1)}, \cdots, \theta^{(k)}$. A system of k values $\theta^{(1)}, \cdots, \theta^{(k)}$ can be represented in the k-dimensional parameter space by the point θ with the co-ordinates $\theta^{(1)}, \cdots, \theta^{(k)}$. Denote by Ω the set of all possible points θ. For any point θ of Ω we shall denote by $P(E \epsilon w|\theta)$ the probability that the sample point $E = (x_1, \cdots, x_n)$ falls into the region w of the n-dimensional sample space, where x_j denotes the observed value of the variate $X_j(\theta)(j = 1, \cdots, n)$. The distribution $P(E \epsilon w|\theta)$ is supposed to be known for any point θ of Ω. In the theory of testing hypotheses and of statistical estimation we have to deal with problems of the following type: A sample point $E = (x_1, \cdots, x_n)$ of the n-dimensional sample space is given. We know that x_j is the observed value of $X_j(\theta)$ but we do not know the parameter point θ, and we have to draw inferences about θ by means of the sample point observed. The assumption that θ belongs to a certain subset ω of Ω is called a hypothesis. We shall deal in this paper with the following general problem: Let us consider a system S of subsets of Ω. Denote by H_ω the hypothesis corresponding to the element ω of S, and by H_S the system of all hypotheses corresponding to all elements of S. We have to decide by means of the observed sample point E which hypothesis of the system H_S should be accepted. That is to say for each H_ω we have to determine a region of acceptance M_ω in the n-dimensional sample space. The hypothesis H_ω will be accepted if and only if the sample point E falls in the region M_ω. M_ω and $M_{\omega'}$ are disjoint if $\omega \neq \omega'$. The statistical problem is the question as to how the system M_S of all regions M_ω should be chosen.

The problem in this formulation is very general. It contains the problems of testing hypotheses and of statistical estimation treated in the literature.[2] For instance if we want to test the hypothesis H_ω corresponding to a certain subset ω of Ω, the system of hypotheses H_S consists only of the two hypotheses H_ω and $H_{\bar{\omega}}$ where $\bar{\omega}$ denotes the subset of Ω complementary to ω. If we want to estimate θ by a unique point, then S is the system of all points of Ω. In the theory of confidence intervals we estimate one of the parameter co-ordinates $\theta^{(1)}, \cdots, \theta^{(k)}$,

[1] Research under a grant-in-aid from the Carnegie Corporation of New York.

[2] See, for instance, J. Neyman, "Outline of a Theory of Statistical Estimation Based on the Classical Theory of Probability," *Phil. Transactions of the Royal Society*, London, Vol. 231 (1937), pp. 333–380.

say $\theta^{(1)}$, by an interval. In this case S is a certain system of subsets ω of the following type: ω is the set of all points $\theta = (\theta^{(1)}, \cdots, \theta^{(k)})$ for which $\theta^{(1)}$ lies in a certain interval $[a, b]$. The problem in our formulation covers also cases which, as far as I know, have not yet been treated. Consider for instance 3 subsets ω_1, ω_2 and ω_3 of Ω such that the sum of them is equal to Ω. It may be that we are interested only to know in which of the subsets ω_1, ω_2, ω_3 the unknown parameter point lies. In this case the system of hypotheses H_S consists only of the 3 hypotheses H_{ω_1}, H_{ω_2} and H_{ω_3}. Cases like this might be of practical interest.

For the determination of the "best" system (in a certain sense) of regions of acceptance we shall use methods and principles which are closely related to those of the Neyman-Pearson theory of testing hypotheses. In the Neyman-Pearson theory two types of error are considered. Let $\theta = \theta_1$ be the hypothesis to be tested, where θ_1 denotes a certain point of the parameter space. Denote this hypothesis by H_1 and the hypothesis $\theta \neq \theta_1$ by $\bar{H}$. The type I error is that which is made by rejecting H_1 when it is true. The type II error is made by accepting H_1 when it is false. The fundamental principle in the Neyman-Pearson theory can be formulated as follows: among all critical regions (regions of rejection of H_1, i.e. regions of acceptance of $\bar{H}$) for which the probability of type I error is equal to a given constant α, we have to choose that region for which the probability of type II error is a minimum. The difficulty which arises here lies in the circumstance that the probability of type II error depends on the true parameter point θ. That is to say, if the critical region is given the probability of type II error will be a function of the true parameter point θ. Since we do not know the true parameter point θ, we want to have a critical region which minimizes the probability of type II error with respect to any possible alternative hypothesis $\theta = \theta_2 \neq \theta_1$. If such a common best critical region exists, then the problem is solved. But such cases are rather exceptional. If a common best critical region does not exist, Neyman and Pearson consider unbiased critical regions of different types,[3] which minimize the type II error locally, that is to say with respect to alternative hypotheses in the neighborhood of the hypothesis considered. In this paper we develop methods for the determination of a system of regions of acceptance taking in account type II errors also relative to alternative hypotheses not lying in the neighborhood of the hypothesis to be tested.

2. Some Definitions. Let us denote by Ω the set of all possible parameter points θ and by S a system of subsets of Ω. If ρ denotes the sum of the elements of a subset σ of S, then we shall denote ΣM_ω by M_ρ, where M_ω denotes the

[3] J. Neyman and E. S. Pearson: *Statistical Research Memoirs*, Volumes I and II. The authors consider also unbiased regions of type A_1 for which the probability of type II error with respect to every alternative hypothesis is not greater than for any other unbiased region of the same size. However regions of type A_1 do not always exist (the existence of such regions has been proved for a special but important class of cases).

region of acceptance of H_ω and the summation is to be taken over all elements ω of σ.

Definition 1. Denote by M_S and M'_S two different systems of regions of acceptance corresponding to the same system H_S of hypotheses. The systems M_S and M'_S are said to be equivalent if for each point θ of Ω and for every ρ which is a sum of elements of S which does not contain θ, the equation

$$P(E \epsilon M'_\rho \mid \theta) = P(E \epsilon M_\rho \mid \theta)$$

holds, where M'_ρ denotes the region according to the system M'_S and M_ρ denotes the region according to the system M_S.

Definition 2. Denote by M_S and M'_S two different systems of regions of acceptance corresponding to the same system of hypotheses. The system M'_S is said to be absolutely better than the system M_S if they are not equivalent and if for each θ and for every ρ which is a sum of elements of S which does not contain θ the inequality

$$P(E \epsilon M'_\rho \mid \theta) \leq P(E \epsilon M_\rho \mid \theta)$$

holds.

Definition 3. A system M_S of regions of acceptance is said to be admissible if no absolutely better system of regions exists.

3. **The problem of the choice of M_S.** The choice of M_S will in general be affected by the following two circumstances:

(1) We do not attribute the same importance to each error. For instance the acceptance of the hypothesis that θ lies in a certain interval I has in general more serious consequences if θ is far from I than if θ is near to I. The choice of M_S will in general depend on the relative importance of the different possible errors.

(2) In some cases we have a priori more confidence that the true parameter point lies in a certain interval I than in some other cases. The choice of M_S will in general be affected also by this fact. Let us illustrate this by an example. We have two coins, a new and an old one and we want to test for both coins whether the probability p of tossing head is equal to $\frac{1}{2}$. Let us assume that we make 100 tosses with each of the coins and we get head 40 times in each case. Since we have a priori no very great confidence that the old coin is unbiased, the fact that head occured only 40 times will suffice to reject the hypothesis that for the old coin $p = \frac{1}{2}$. But in the case of the new coin, having much greater a priori confidence that it is unbiased, we shall perhaps not reject the hypothesis $p = \frac{1}{2}$ and we shall rather assume that a somewhat improbable event occurred. That is to say, we do not choose the same critical region in both cases due to the fact that our a priori confidence for $p = \frac{1}{2}$ is in the case of the new coin greater than in the case of the old one.

In order to study the dependence of the choice of M_S on the two circumstances mentioned, let us introduce a weight function for the possible errors and an a priori probability distribution for the unknown parameter θ. The weight function $W(\theta, \omega)$ is a real valued non-negative function defined for all points θ of Ω and for all elements ω of S, which expresses the relative importance of the error committed by accepting H_ω when θ is true. If θ is contained in ω, $W(\theta, \omega)$ is, of course, equal to zero. The question as to how the form of the weight function $W(\theta, \omega)$ should be determined, is not a mathematical or statistical one. The statistician who wants to test certain hypotheses must first determine the relative importance of all possible errors which will entirely depend on the special purposes of his investigation. If that is done, we shall in general be able to give a more satisfactory answer to the question as to how the system of regions of acceptance should be chosen. In many cases, especially in statistical questions concerning industrial production, we are able to express the importance of an error in terms of money, that is to say, we can express the loss caused by the error considered in terms of money. We shall also say that $W(\theta, \omega)$ is the loss caused by accepting H_ω when θ is true.

The situation regarding the introduction of an a priori probability distribution of θ is entirely different. First, the objection can be made against it, as Neyman has pointed out, that θ is merely an unknown constant and not a variate, hence it makes no sense to speak of the probability distribution of θ. Second, even if we may assume that θ is a variate, we have in general no possibility of determining the distribution of θ and any assumptions regarding this distribution are of hypothetical character. On account of these facts the determination of the system of regions of acceptance should be independent of any a priori probability considerations. The "best" system of regions of acceptance, which we shall define later, will depend only on the weight function of the errors. The reason why we introduce here a hypothetical probability distribution of θ is simply that it proves to be useful in deducing certain theorems and in the calculation of the best system of regions of acceptance.

Let us denote by $f(\theta)$ a distribution function of θ. For the sake of simplicity let us assume that the probability density of the distribution $P(E \epsilon w \mid \theta)$ exists in any point E of the sample space for any θ and denote it by $p(E \mid \theta)$. The expected value of the loss is given by

$$I = \int_M \int_\Omega W(\theta, \omega_E) p(E \mid \theta)\, df(\theta)\, dE \tag{1}$$

where ω_E denotes the element of S corresponding to E (that is to say, ω_E is that element of S for which E is a point of the region of acceptance M_{ω_E}), and the integral is to be taken over the product of the sample space M with the parameter space Ω. The expected value I of the loss depends on the system M_S of regions of acceptance. The system M_S for which I becomes a minimum, can be regarded as the best system of regions relative to the given weight function and to the given a priori distribution of θ.

One can easily show the following: If M'_S is an absolutely better system of regions (in sense of the definition 2) than the system M_S, then for any weight function $w(\theta, \omega)$ and for any a priori distribution $f(\theta)$ the expected value I' of the loss corresponding to M'_S is less than the expected value I of the loss corresponding to M_S. (For some exceptional weight and a priori distribution functions I' may be equal to I.)

Hence we can give the following rule: *We have to choose an admissible system of regions of acceptance.*

Now let us consider the question whether besides admissibility further restrictions upon the choice of M_S can be made. In order to see this, let us consider two admissible systems of regions M_S and M'_S which are not equivalent. One can easily show that there exist two weight functions $W_1(\theta, \omega)$, $W_2(\theta, \omega)$ and two a priori distributions $f_1(\theta)$ and $f_2(\theta)$ such that for $W_1(\theta, \omega)$ and $f_1(\theta)$ the expected value of the loss corresponding to M_S is less than that corresponding to M'_S, and for $W_2(\theta, \omega)$ and $f_2(\theta)$ the expected value of the loss corresponding to M_S is greater than that corresponding to M'_S. Hence no absolute criteria can be given as to which of the systems M_S and M'_S should be chosen. In order to be able to make further restrictions upon the choice of M_S, we have to make assumptions regarding the form of the weight function. We shall deal with this question in section 6.

4. **Calculation of admissible systems of regions.** As we have seen, we have to choose an admissible system of regions. The question arises as to how we can find admissible systems of regions.

Provided that $p(E \mid \theta)$ is continuous in E and θ jointly, one can easily show that M'_S is an admissible system of regions if there exists a bounded, uniformly continuous and everywhere positive (except if θ is contained in ω) weight function $W(\theta, \omega)$ and an a priori distribution $f(\theta)$ such that every open subset of Ω has a positive probability and the expected value of the loss

$$I(M_S) = \int_M \int_\Omega W(\theta, \omega_E) p(E \mid \theta)\, df(\theta)\, dE. \tag{2}$$

becomes a minimum for $M_S = M'_S$. (ω_E denotes that element of S for which M_{ω_E} contains E). In fact if there existed an absolutely better system M''_S of regions, then $I(M''_S)$ would be less than $I(M'_S)$ in contradiction to our assumption that $I(M_S)$ becomes a minimum for $M_S = M'_S$.

In order to obtain an admissible system M_S we may choose any bounded, uniformly continuous and everywhere positive (except if θ is contained in ω) weight function $W(\theta, \omega)$ and any arbitrary a priori distribution $f(\theta)$ (subject only to the condition that every open subset of Ω should have a positive probability) and then the system M_S which makes

$$I(M_S) = \int_M \int_\Omega W(\theta, \omega_E) p(E \mid \theta)\, df(\theta)\, dE$$

a minimum is an admissible one. In order to determine M_S we have only to determine for each E the corresponding element ω_E of S. Let us consider the integral

$$I_E = \int_\Omega W(\theta, \omega) p(E \mid \theta)\, df(\theta).$$

The integral I_E is for a fixed E only a function of ω. It is obvious that ω_E must be that element of S for which I_E becomes a minimum.

5. **Admissible systems M_S and the Neyman-Pearson best critical regions.** Let us consider the case that the system H_S of hypotheses consists only of the following two hypotheses: 1) $\theta = \theta_0$ where θ_0 is a certain point of Ω. 2) θ belongs to the set complementary to θ_0. Let us denote by ω_1 the set consisting only of the point θ_0, and by ω_2 the set complementary to ω_1. S consists in this case only of two elements ω_1 and ω_2. The system M_S of regions consists of two regions of acceptance M_{ω_1} and M_{ω_2} corresponding to the hypotheses H_{ω_1} and H_{ω_2}. If a common best critical region in the sense of Neyman-Pearson exists and if M_S is admissible, then M_{ω_2} is obviously a common best critical region. This leads to the following remarkable conclusion: If a common best critical region exists and if the system M_S of regions consisting of the two regions M_{ω_1} and M_{ω_2} minimizes the expectation of the loss (formula 2) for a weight function and for an a priori distribution subject to some weak conditions mentioned in paragraph 4, then M_{ω_2} is a common best critical region. That is to say, the form of the weight function and of the a priori distribution affects only the size of the region M_{ω_2} but it will always be a common best critical region.

6. **The choice of M_S if a weight function is given.** We shall now consider the case in which a weight function $W(\theta, \omega)$ is given and we shall deal with the question as to how M_S in this case is to be chosen.

If the parameter point is an unknown constant and if θ denotes the true parameter point, then the expected value of the loss is given by

$$r(\theta) = \int_M W(\theta, \omega_E)p(E \mid \theta)\, dE \tag{3}$$

where the integration is to be taken over the whole sample space M and H_{ω_E} denotes the hypothesis accepted if E is the observed sample point. That is to say ω_E is that element of S for which E is contained in the region of acceptance M_{ω_E}. We shall call the expression (3) the risk of accepting a false hypothesis if θ is the true parameter point. Since we do not know the true parameter point θ, we shall have to study the risk $r(\theta)$ as a function of θ. We shall call this function the risk function. The form of the risk function depends on the system M_S of regions and on the form of the weight function. In order to express this fact, we shall denote the risk function corresponding to the system M_S and to the weight function $W(\theta, \omega)$ also by

$$r[\theta \mid M_S, W(\theta, \omega)].$$

Definition 4. Denote by M_S and M'_S two systems of regions of acceptance corresponding to the same system H_S of hypotheses. We shall say that M_S and M'_S are equivalent relative to the weight function $W(\theta, \omega)$ if the risk function

$r[\theta \mid M_S, W(\theta, \omega)]$ is identically equal to the risk function $r[\theta \mid M'_S, W(\theta, \omega)]$, that is to say if for each point θ,

$$r[\theta \mid M'_S, W(\theta, \omega)] = r[\theta \mid M_S, W(\theta, \omega)].$$

Definition 5. Denote by M_S and M'_S two systems of regions corresponding to the same system H_S of hypotheses. We shall say that M_S is uniformly better than M'_S relative to the weight function $W(\theta, \omega)$ if M_S and M'_S are not equivalent and for each θ

$$r(\theta \mid M_S, W(\theta, \omega)] \leq r[\theta \mid M'_S, W(\theta, \omega)].$$

Definition 6. A system M_S of regions of acceptance is said to be admissible relative to the weight function $W(\theta, \omega)$ if no uniformly better system of regions exists relative to the weight function considered.

It is obvious that *we have to choose a system M_S of regions which is admissible relative to the weight function considered.*

There exist in general many systems M_S which are admissible relative to the weight function given. The question arises as to how can we distinguish among them. Denote by r_{M_S} the maximum of the risk function corresponding to the system M_S of regions and to the given weight function. If we do not take into consideration a priori probabilities of θ, then it seems reasonable to choose that system M_S for which r_{M_S} becomes a minimum. We shall see in section 8 that the system M_S for which r_{M_S} becomes a minimum has some important properties which justify the distinction of this particular system of regions among all admissible systems.

Definition 7. We shall call an admissible system M'_S of regions for which r_{M_S} becomes a minimum a best system of regions of acceptance relative to the weight function given.[4]

Now we shall have to deal with the question of determining a best system M_S of regions and what special properties this system M_S has.

7. **Reduction of the problem to the case when the system H_S of hypotheses is the system of all simple hypotheses.** A hypothesis H_ω is said to be a simple hypothesis if ω contains exactly one point of the parameter space Ω. We assume that each element ω of S is a closed subset of Ω. Hence the power of S is not greater than the power of the continuum and therefore we can always set up a correspondence between the elements ω of S and the points θ of Ω such that to each point θ corresponds a certain element ω_θ of S and to each element ω of S at least one point θ exists for which $\omega_\theta = \omega$. For instance if S consists of the two elements ω_1 and ω_2 then we can set up a correspondence as follows: the element ω_θ of S corresponding to θ is ω_1 if θ is contained in ω_1 and ω_2 otherwise.

[4] As we shall see later (Theorem 3), the best system of regions is uniquely determined if some regularity conditions are fulfilled.

If Ω is one dimensional and S is the system of all intervals of a certain length ϵ then we can define the interval ω_θ corresponding to θ as the interval of which the initial point is θ and the terminal point $\theta + \epsilon$.

Let us denote the weight function by $W(\theta, \omega)$ defined for all values of θ and for all elements ω of S. Consider the system $H_{\bar{S}}$ of all simple hypotheses and the following weight function

$$W(\theta, \bar{\theta}) = W(\theta, \omega_{\bar{\theta}}) \tag{4}$$

where θ denotes the true parameter point and $\bar{\theta}$ denotes the estimated point. A system $M_{\bar{S}}$ of regions of acceptance for $H_{\bar{S}}$ is given by a vector function $\bar{\theta}(E)$ of the observations such that to each point $E = (x_1, \cdots, x_n)$ of the sample space M corresponds a certain point $\bar{\theta}(E)$ of the parameter space. For each point θ_0 the region M_{θ_0} of the acceptance of the hypothesis $\theta = \theta_0$ is given by the equation $\bar{\theta}(E) = \theta_0$. We shall call the function $\bar{\theta}(E)$ an estimate of θ, the system of regions $M_{\bar{S}}$ is uniquely determined by the estimate. We shall call $\bar{\theta}(E)$ a best estimate relative to a given weight function if the system of regions determined by $\bar{\theta}(E)$ is a best system of regions relative to the weight function considered.

Let us denote by $\bar{\theta}(E)$ a best estimate of θ relative to the weight function $W(\theta, \bar{\theta})$ defined in (4). A best system M_S of regions of acceptance in the original problem can obviously be obtained in the following way: Denote by ω an element of S. The region M_ω of acceptance of the hypothesis H_ω consists of the points E for which

$$\omega_{\bar{\theta}(E)} = \omega.$$

Hence we can restrict our considerations to the case when the system of hypotheses is the system of all simple hypotheses. We shall deal with the problem of how a best estimate of θ can be found and what properties this estimate has.

8. **Some theorems concerning the best estimate.** In order to study the properties of a best estimate $\bar{\theta}(E)$ it is useful to consider hypothetical a priori distributions of θ. We shall especially consider point distributions of θ, that is to say, distributions where a finite number of points $\theta_1, \cdots, \theta_s$ of the parameter space Ω exist such that the probability of any subset of Ω not containing any of the points $\theta_1, \cdots, \theta_s$ is zero. If $\theta_1, \cdots, \theta_s$ are given, a point distribution is characterized by a vector $\rho = (\rho_1, \cdots, \rho_s)$ where ρ_i denotes the probability of θ_i and $\Sigma\rho_i = 1$.

If $\theta(E)$ denotes an estimate of θ and if $f(\theta)$ denotes a distribution function of θ then the expected value of the loss, that is to say the expected value of the weight function $W[\theta, \theta(E)]$ is obviously given by

$$\int_M \int_\Omega W[\theta, \theta(E)] p(E \mid \theta)\, df(\theta)\, dE \tag{5}$$

where $p(E \mid \theta)$ denotes the probability density in E if θ is the true parameter point and the integration is to be taken over the product of the sample space M and parameter space Ω.

Let us assume that for every sample point E there exists a parameter point $\theta_f(E)$ such that the expression

$$\int_\Omega W(\theta, \bar{\theta})p(E \mid \theta)\, df(\theta) \tag{6}$$

becomes a minimum with respect to $\bar{\theta}$ for $\bar{\theta} = \theta_f(E)$. We shall call the estimate $\theta_f(E)$ a minimum risk estimate with respect to the distribution $f(\theta)$, since also the expression (5) becomes a minimum for the estimate $\theta_f(E)$.

We shall make the following assumptions:

Assumption 1. The parameter space is a bounded and closed subset of the k-dimensional Euclidean space.

Assumption 2. The weight function $W(\theta, \bar{\theta})$ is continuous in θ and $\bar{\theta}$ jointly.

Assumption 3. The probability density $p(E \mid \theta)$ is continuous in E and θ jointly. That is to say if $\lim E_i = E$ and $\lim \theta_i = \theta$ then $\lim p(E_i \mid \theta_i) = p(E \mid \theta)$.

Assumption 4. For any distribution $f(\theta)$ of θ there exists at most one minimum risk estimate $\theta_f(E)$.[5]

Assumption 5. If $f(\theta)$ and $f'(\theta)$ denote two different point distributions of θ and if $\theta_f(E)$ and $\theta_{f'}(E)$ are minimum risk estimates corresponding to $f(\theta)$ and $f'(\theta)$ respectively, then $\theta_f(E)$ is not identically equal to $\theta_{f'}(E)$.

The assumptions 1–5, with addition of an assumption 6 which we shall formulate later, enables us to deduce important properties of the best estimate $\bar{\theta}(E)$. First we shall prove some Lemmas by means of the assumptions 1–5.

LEMMA 1. *For any a priori distribution $f(\theta)$ of θ there exists exactly one minimum risk estimate $\theta_f(E)$.*

According to Assumption 2 $W(\theta, \bar{\theta})$ is continuous. Since the parameter space Ω is compact on account of Assumption 1, $W(\theta, \bar{\theta})$ is uniformly continuous. According to Assumption 3 $p(E \mid \theta)$ is continuous; hence for any fixed sample point E, $p(E \mid \theta)$ is bounded. From these facts it follows easily that the expression (6) is a continuous function of $\bar{\theta}$ for any fixed sample point E. Hence there exists at least one parameter point $\theta_f(E)$ such that (6) becomes a minimum for $\bar{\theta} = \theta_f(E)$. Since, according to Assumption 4, at most one parameter point exists for which (6) becomes a minimum, Lemma 1 is proved.

If a distribution $f(\theta)$ of θ is given then the distribution of each of the components $\theta^{(1)}, \cdots, \theta^{(k)}$ of θ can be found. Denote by Q_j the set of real numbers which are discontinuities of the distribution of the component $\theta^{(j)}$ $(j = 1, \cdots, k)$ and form the set $Q = Q_1 + \cdots + Q_k$. As is well known, Q is at most denumerable. A k-dimensional interval J of the parameter space given by

$$a_j \leq \theta^{(j)} \leq b_j \qquad (j = 1, \cdots, k)$$

is called a continuity interval of the distribution $f(\theta)$ if no a_j and no b_j belongs to Q. A sequence $\{f_n(\theta)\}$ of distributions is said to be convergent towards the

[5] As will be shown in Section 10, Assumption 4 is not as restrictive as it would appear. It will be satisfied in the great majority of practical cases.

distribution $f(\theta)$, i.e. in symbols $\lim f_n(\theta) = f(\theta)$, if for any continuity interval J of $f(\theta)$ the probability of J corresponding to the distribution $f_n(\theta)$ converges with increasing n towards the probability of J corresponding to the distribution $f(\theta)$.

LEMMA 2. *If* $\{f_n(\theta)\}$ ($n = 1, \cdots$, ad inf.) *denotes a sequence of distributions, then there exists a subsequence* $\{f_{n_m}(\theta)\}$ ($m = 1, \cdots$, ad inf.) *which converges towards a distribution.*

As is well known, there exists a completely additive set function $P(\omega)$ defined for all Borel measurable subsets ω of Ω and a subsequence $\{n_m\}$ of $\{n\}$, such that for any continuity interval J of $P(\omega)$ the probability of J corresponding to the distribution $f_{n_m}(\theta)$ converges with increasing m towards $P(J)$. Since Ω is bounded, there exists a continuity interval J such that for all n the probability of J according to $f_n(\theta)$ is equal to 1. Hence $P(\Omega) = 1$, that is to say, $P(\omega)$ is a probability set function which proves Lemma 2.

LEMMA 3. *If* $\{f_n(\theta)\}$ ($n = 1, \cdots$, ad inf.) *denotes a sequence of distributions which converges towards the distribution* $f(\theta)$ *and if* $\lim E_n = E$ *then*

$$\lim_{n=\infty} \theta_{f_n}(E_n) = \theta_f(E),$$

where $\theta_{f_n}(E)$ *denotes the minimum risk estimate corresponding to* $f_n(\theta)$ *and* $\theta_f(E)$ *denotes the minimum risk estimate corresponding to* $f(\theta)$.

If $\{\varphi_n(\theta)\}$ denotes a sequence of real valued functions which converges uniformly towards a continuous function $\varphi(\theta)$ then

$$\lim \int_\Omega \varphi_n(\theta)\, df_n(\theta) = \int_\Omega \varphi(\theta)\, df(\theta). \tag{7}$$

Since $\{\varphi_n(\theta)\}$ converges uniformly towards $\varphi(\theta)$, (7) is obviously true if

$$\lim \int_\Omega \varphi(\theta)\, df_n(\theta) = \int_\Omega \varphi(\theta)\, df(\theta)$$

holds. The latter equality follows easily from the fact that Ω is compact.

Consider a subsequence $\{n_m\}$ of $\{n\}$ such that $\lim_{m=\infty} \theta_{f_{n(m)}}(E_{n_m})$ exists. Denote this limit by θ^*. In order to prove Lemma 3, we have only to show that $\theta^* = \theta_f(E)$. If $\theta_f(E) \neq \theta^*$ then on account of Assumption 4

$$\int_\Omega W[\theta, \theta_f(E)]p(E \mid \theta)\, df(\theta) < \int_\Omega W(\theta, \theta^*)p(E \mid \theta)\, df. \tag{8}$$

$W(\theta, \bar{\theta})$ is uniformly continuous since Ω is compact. On account of Assumption 3 also $p(E \mid \theta)$ is uniformly continuous in the product of Ω with a bounded subset of the sample space. Hence

$$W[\theta, \theta_{f_{n(m)}}(E_{n_m})]p(E_{n_m} \mid \theta)$$

converges uniformly in θ towards

$$W(\theta, \theta^*)p(E \mid \theta)$$

and we have on account of (7) and (8)

$$(9)\qquad \begin{aligned}\lim_{m=\infty}\int_\Omega W[\theta,\, \theta_{f_{n(m)}}(E_{n_m})]p(E_{n_m}\mid\theta)\,df_{n_m} &= \int_\Omega W(\theta,\,\theta^*)p(E\mid\theta)\,df \\ &> \int_\Omega W[\theta,\,\theta_f(E)]p(E\mid\theta)\,df,\end{aligned}$$

and

$$(10)\qquad \lim_{m=\infty}\int_\Omega W[\theta,\,\theta_f(E)]p(E\mid\theta)\,df_{n_m} = \int_\Omega W[\theta,\,\theta_f(E)]p(E\mid\theta)\,df.$$

From (9) and (10) it follows that there exists a positive δ such that for sufficiently large m

$$\int_\Omega W[\theta,\,\theta_{f_{n(m)}}(E_{n_m})]p(E_{n_m}\mid\theta)\,df_{n_m} > \int_\Omega W[\theta,\,\theta_f(E)]p(E\mid\theta)\,df_{n_m} + \delta.$$

Since the sequence of functions $\{p(E_n \mid \theta)\}$ converges uniformly in θ towards $p(E \mid \theta)$, we have for sufficiently large m

$$\int_\Omega W[\theta,\,\theta_{f_{n(m)}}(E_{n_m})]p(E_{n_m}\mid\theta)\,df_{n_m} > \int_\Omega W[\theta,\,\theta_f(E)]p(E_{n_m}\mid\theta)\,df_{n_m}.$$

But this is a contradiction, since $\theta_{f_n}(E)$ is a minimum risk estimate. Hence the assumption $\theta^* \neq \theta_f(E)$ is proved to be an absurdity. This proves Lemma 3.

LEMMA 4. *To each positive ϵ a bounded and closed subset M_ϵ of the sample space M can be given such that*

$$\int_{M_\epsilon} p(E\mid\theta)\,dE \geq 1 - \epsilon$$

for every point θ of the parameter space Ω.

Let us assume that Lemma 4 is not true and we shall deduce a contradiction. Denote by $M_\nu(\nu = 1, 2, \cdots, \text{ad inf.})$ the sphere in the sample space M whose center is the origin and whose radius is equal to ν. Since Lemma 4 is supposed to be not true, to each ν there exists a parameter point θ_ν such that

$$(11)\qquad \int_{M_\nu} p(E\mid\theta_\nu)\,dE < 1 - \epsilon \qquad (\nu = 1, \cdots, \text{ad inf.}).$$

Since Ω is compact, there exists a subsequence $\{\theta_{\nu_\mu}\}$ of the sequence $\{\theta_\nu\}$ such that $\lim_{\mu=\infty}\theta_{\nu_\mu}$ exists. Denote $\lim\theta_{\nu_\mu}$ by θ. Since

$$\int_M p(E\mid\theta)\,dE = 1$$

there exists a positive integer ν' such that

$$\int_{M_{\nu'}} p(E\mid\theta)\,dE > 1 - \frac{\epsilon}{2}.$$

On account of Assumption 3 we get easily

$$\lim_{\mu=\infty} \int_{M_{\nu'}} p(E \mid \theta_{\nu_\mu})\, dE = \int_{M_{\nu'}} p(E \mid \theta)\, dE.$$

Hence for sufficiently large μ we get

$$\int_{M_{\nu_\mu}} p(E \mid \theta_{\nu_\mu})\, dE \geq \int_{M_{\nu'}} p(E \mid \theta_{\nu_\mu})\, dE > 1 - \epsilon,$$

in contradiction to (11). This proves Lemma 4.

For any estimate $\theta(E)$ we shall call the integral

$$r(\theta) = \int_M W[\theta, \theta(E)] p(E \mid \theta)\, dE$$

the risk function of the estimate $\theta(E)$. The value of the risk function $r(\theta)$ is for any θ equal to the expected value of the loss (of the weight function) if θ is the true parameter point.

LEMMA 5. *To any positive η a positive δ can be given such that for any estimate $\theta(E)$ and for any pair θ, θ' of parameter points whose Euclidean distance is less than δ the inequality*

$$|r(\theta) - r(\theta')| = \left| \int_M W[\theta, \theta(E)] p(E \mid \theta)\, dE - \int_M W[\theta', \theta(E)] p(E \mid \theta')\, dE \right| < \eta$$

holds.

Since $W(\theta, \bar{\theta})$ is uniformly continuous, to any $\epsilon > 0$ a positive δ can be given such that for any pair of points θ, θ' whose Euclidean distance is less than δ the relation

$$(12) \qquad |W(\theta, \bar{\theta}) - W(\theta', \bar{\theta})| < \epsilon$$

holds for every $\bar{\theta}$. On account of Assumption 3 δ can be chosen in such a way that also the inequality

$$(13) \qquad |p(E \mid \theta) - p(E \mid \theta')| < \epsilon$$

is satisfied for any sample point E of a bounded subset M' of M and for any pair θ, θ' whose Euclidean distance is less than δ.

Since $W(\theta, \bar{\theta})$ is continuous and Ω is compact, $W(\theta, \bar{\theta})$ must be bounded. Denote by A an upper bound of $W(\theta, \bar{\theta})$. According to Lemma 4 there exists a bounded and closed subset M' of the sample space M such that

$$\int_{M'} p(E \mid \theta)\, dE \geq 1 - \frac{\eta}{2A} \text{ for any } \theta.$$

It is obvious that

$$\left| \int_{M-M'} W[\theta, \theta(E)] p(E \mid \theta)\, dE - \int_{M-M'} W[\theta', \theta(E)] p(E \mid \theta')\, dE \right| \leq \frac{\eta}{2}.$$

In order to prove Lemma 5 we have only to show that

$$(14) \qquad \left| \int_{M'} W[\theta, \theta(E)]p(E \mid \theta)\, dE - \int_{M'} W[\theta', \theta(E)]p(E \mid \theta')\, dE \right| < \frac{\eta}{2}.$$

On account of (12) and (13), (14) is certainly true for sufficiently small ϵ. Hence Lemma 5 is proved.

LEMMA 6. *If the sequence $\{f_n(\theta)\}$ of distributions converges towards the distribution $f(\theta)$ and if $r_{f_n}(\theta)$ denotes the risk function of the minimum risk estimate $\theta_{f_n}(E)$ then $\{r_{f_n}(\theta)\}$ converges uniformly towards the risk function $r_f(\theta)$ of the minimum risk estimate $\theta_f(E)$.*

According to Lemma 4 to any positive ϵ a bounded and closed subset M_ϵ of M can be given such that

$$(15) \qquad \int_{M_\epsilon} p(E \mid \theta)\, dE \geq 1 - \epsilon$$

for every θ. From Lemma 3 it follows easily that $\{\theta_{f_n}(E)\}$ converges uniformly towards $\theta_f(E)$ in M_ϵ. Hence

$$\lim_{n=\infty} \int_{M_\epsilon} W[\theta, \theta_{f_n}(E)]p(E \mid \theta)\, dE = \int_{M_\epsilon} W[\theta, \theta_f(E)]p(E \mid \theta)\, dE$$

holds for every θ and for every positive ϵ. Since $W(\theta, \bar{\theta})$ is bounded and ϵ can be chosen arbitrarily small, we get on account of (15) that

$$\lim_{n=\infty} \int_{M} W[\theta, \theta_{f_n}(E)]p(E \mid \theta)\, dE = \int_{M} W[\theta, \theta_f(E)]p(E \mid \theta)\, dE,$$

that is to say

$$\lim r_{f_n}(\theta) = r_f(\theta).$$

The uniformity of the convergence follows easily from Lemma 5.

In the following argument we shall consider an arbitrary but fixed system of s parameter points $\theta_1, \cdots, \theta_s$, and point distributions such that no point $\theta \neq \theta_1, \cdots, \theta_s$ has positive probability. Such a point distribution is characterized by a vector $\rho = (\rho_1, \cdots, \rho_s)$ where ρ_i denotes the probability of θ_i $(i = 1, \cdots, s)$ and $\Sigma\rho_i = 1$. The points $\theta_1, \cdots, \theta_s$ are kept constant and only ρ will vary. Hence if we speak about different distributions $\rho = (\rho_1, \cdots, \rho_s)$, $\rho' = (\rho_1', \cdots, \rho_s')$ they are always related to the same points $\theta_1, \cdots, \theta_s$ unless we state explicitly the contrary.

LEMMA 7. *If $\rho = (\rho_1, \cdots, \rho_s)$ and $\rho' = (\rho_1 + \Delta\rho_1, \cdots, \rho_s + \Delta\rho_s)$ denote two different distributions then*

$$\sum_{i=1}^{s} [(\lambda - 1)\rho_i + \lambda\Delta\rho_i][r_i(\rho') - r_i(\rho)] < 0$$

holds for any positive λ, *where*

$$r_i(\rho) = \int_M W[\theta_i, \theta_\rho(E)]p(E \mid \theta_i)\, dE \qquad (i = 1, \cdots, s),$$

$$r_i(\rho') = \int_M W(\theta_i, \theta_{\rho'}(E)]p(E \mid \theta_i)\, dE,$$

and $\theta_\rho(E)$ *and* $\theta_{\rho'}(E)$ *denote the minimum risk estimates corresponding to* ρ *and* ρ' *respectively.*

We have

$$\sum_i (\rho_i + \Delta\rho_i)r_i(\rho) = \int_M \sum_i W[\theta_i, \theta_\rho(E)]\rho_i' p(E \mid \theta_i)\, dE = I_1$$

and

$$\sum_i (\rho_i + \Delta\rho_i)r_i(\rho') = \int_M \sum W[\theta_i, \theta_{\rho'}(E)]\rho_i' p(E \mid \theta_i)\, dE = I_2.$$

Since $\theta_{\rho'}(E)$ is the minimum risk estimate corresponding to ρ', we have $I_1 \geq I_2$. We shall show that $I_1 > I_2$. According to Assumption 5 $\theta_\rho(E)$ is not identically equal to $\theta_{\rho'}(E)$. Hence there exists a point E' such that $\theta_\rho(E') \neq \theta_{\rho'}(E')$. On account of Assumption 4

$$\Sigma W[\theta_i, \theta_\rho(E')]\rho_i' p(E' \mid \theta_i) > \Sigma W[\theta_i, \theta_{\rho'}(E')]\rho_i' p(E' \mid \theta_i).$$

From Lemma 3 it follows that $\theta_\rho(E)$ and $\theta_{\rho'}(E)$ are continuous functions of E. Hence there exists a positive δ and a sphere s with center in E' such that

$$\Sigma W[\theta_i, \theta_\rho(E)]\rho_i' p(E \mid \theta_i) > \Sigma W[\theta_i, \theta_{\rho'}(E)]\rho_i' p(E \mid \theta_i) + \delta$$

for every point E of S. Since $\theta_{\rho'}(E)$ is the minimum risk estimate corresponding to ρ' we have

$$\Sigma W[\theta_i, \theta_\rho(E)]\rho_i' p(E \mid \theta_i) \geq \Sigma W[\theta_i, \theta_{\rho'}(E)]\rho_i' p(E \mid \theta_i)$$

for every point E outside S. Hence $I_1 > I_2$ that is to say

(16) $$\Sigma(\rho_i + \Delta\rho_i)r_i(\rho) > \Sigma(\rho_i + \Delta\rho_i)r_i(\rho').$$

Analogously we get

(17) $$\Sigma\rho_i r_i(\rho) < \Sigma\rho_i r_i(\rho').$$

Multiplying (16) by an arbitrary positive value λ and subtracting (17) we get

$$\Sigma[\lambda(\rho_i + \Delta\rho_i) - \rho_i]r_i(\rho) > \Sigma[\lambda(\rho_i + \Delta\rho_i) - \rho_i]r_i(\rho').$$

Hence

$$\Sigma[(\lambda - 1)\rho_i + \lambda\Delta\rho_i][r_i(\rho') - r_i(\rho)] < 0.$$

Let us denote for any ρ the maximum of the numbers

$$r_1(\rho), \cdots, r_s(\rho)$$

by $r(\rho)$. We shall call a distribution ρ for which $r(\rho)$ becomes a minimum, a risk-minimizing distribution. We shall say that the risk-minimizing distribution $\rho = (\rho_1, \cdots, \rho_s)$ is not degenerate if $\rho_1 > 0, \cdots, \rho_s > 0$. Otherwise we shall say that ρ is degenerate.

LEMMA 8. *There exists at least one risk-minimizing distribution* ρ.

From Lemma 6 it follows that $r_1(\rho), \cdots, r_s(\rho)$ are continuous functions of ρ. Hence also $r(\rho)$ is continuous. Since the set of all possible distributions ρ is bounded and closed, there must be at least one distribution ρ for which $r(\rho)$ becomes a minimum.

LEMMA 9. *If* $\rho = (\rho_1, \cdots, \rho_s)$ *denotes a risk-minimizing distribution which is not degenerate then*

$$r_1(\rho) = r_2(\rho) = \cdots = r_s(\rho).$$

Let us assume that there are two integers i and j, for instance 1 and 2, such that $r_1(\rho) < r_2(\rho)$. We shall deduce a contradiction from this assumption. Let us consider two different distributions $\rho' = (\rho_1', \cdots, \rho_s')$ and $\rho'' = (\rho_1'', \cdots, \rho_s'')$ where $\rho_1'' > 0$. Hence at least one of the quantities

$$(\rho_1' - \rho_1''), \cdots, (\rho_s' - \rho_s'')$$

is unequal to zero. Since $\Sigma\rho_i' = \Sigma\rho_i'' = 1$, also at least one of the quantities

$$(\rho_2' - \rho_2''), \cdots, (\rho_s' - \rho_s'')$$

must be unequal to zero. On account of Lemma 7 we have

$$\sum_{i=1}^{s} [(\lambda - 1)\rho_i' + \lambda(\rho_i'' - \rho_i')][r_i(\rho'') - r_i(\rho')] < 0.$$

If we put $\lambda = \frac{\rho_1'}{\rho_1''}$, we get

$$\sum_{i=2}^{s} \left[\left(\frac{\rho_1'}{\rho_1''} - 1\right)\rho_i' + \frac{\rho_1'}{\rho_1''}(\rho_i'' - \rho_i')\right][r_i(\rho'') - r_i(\rho')] < 0.$$

Hence at least one of the quantities

$$r_2(\rho'') - r_2(\rho'), \cdots, r_s(\rho'') - r_s(\rho')$$

must be unequal to zero.

Since $\rho_1 > 0$, there exists a closed sphere S_ρ with center at ρ such that for any point ρ' of S_ρ $\rho_1' > 0$. Hence for any two different points ρ' and ρ'' of S_ρ at least one of the quantities

$$r_2(\rho'') - r_2(\rho'), \cdots, r_s(\rho'') - r_s(\rho')$$

is unequal to zero. Denote by $\bar{S}_\rho$ the projection of S_ρ on the $s - 1$ dimensional space given by $\rho_1 = 0$. Consider the transformation according to which the image of the point $\bar{\rho}' = (\rho_2', \cdots, \rho_s')$ of $\bar{S}_\rho$ is the point $\bar{q}(\bar{\rho}') = [r_2(\rho'), \cdots, r_s(\rho')]$. It is obvious that the images of two different points of $\bar{S}_\rho$ are different.

Since $r_i(\rho)$ $(i = 1, \cdots, s)$ is continuous, the transformation is continuous and therefore topological. Denote the image of $\bar{S}_\rho$ by $\bar{R}_\rho$. Since $\bar{\rho} = (\rho_2, \cdots, \rho_s)$ is an interior point of $\bar{S}_\rho$, according to the Brouwer-Jordan theorem[6] on domain invariance the image $\bar{q}(\bar{\rho}) = [r_2(\rho), \cdots, r_s(\rho)]$ of $\bar{\rho}$ must also be an interior point of $\bar{R}_\rho$. Hence for sufficiently small $\epsilon > 0$ the point

$$t(\epsilon) = [r_2(\rho) - \epsilon, \cdots, r_s(\rho) - \epsilon]$$

is contained in $\bar{R}_\rho$. Denote by $\bar{\rho}(\epsilon) = [\rho_2(\epsilon), \cdots, \rho_s(\epsilon)]$ the point of $\bar{S}_\rho$ whose image is $t(\epsilon)$. It is obvious that

$$\lim_{\epsilon=0} \bar{\rho}(\epsilon) = \bar{\rho} = (\rho_2, \cdots, \rho_s). \tag{18}$$

Consider the point $\rho(\epsilon)$ of S_ρ whose projection is $\bar{\rho}(\epsilon)$ that is to say $\rho(\epsilon)$ has the co-ordinates $1 - \Sigma\bar{\rho}_i(\epsilon), \bar{\rho}_2(\epsilon), \cdots, \bar{\rho}_s(\epsilon)$. From (18) it follows that also

$$\lim_{\epsilon=0} \rho(\epsilon) = \rho = (\rho_1, \rho_2, \cdots, \rho_s). \tag{19}$$

Since $r_1[\rho(\epsilon)], \cdots, r_s[\rho(\epsilon)]$ are continuous functions of ϵ and since $r_1(\rho) < r_2(\rho)$, for sufficiently small ϵ the maximum of the numbers

$$r_1[\rho(\epsilon)], r_2[\rho(\epsilon)] = r_2(\rho) - \epsilon, \cdots, r_s[\rho(\epsilon)] = r_s(\rho) - \epsilon$$

is certainly smaller than the maximum $r(\rho)$ of the numbers

$$r_1(\rho), \cdots, r_s(\rho),$$

in contradiction to our assumption that ρ is a risk minimizing distribution. Hence the assumption $r_1(\rho) < r_2(\rho)$ is proved to be an absurdity and Lemma 9 is proved.

In the previous arguments we have considered an arbitrary but fixed system of s parameter points $\theta_1, \cdots, \theta_s$ and all distributions ρ were related to these points. In the following arguments we shall vary the points $\theta_1, \cdots, \theta_s$ and therefore we shall have to state the parameter points to which the distribution ρ is related.

Let us consider a sequence $\{\theta_\nu\}$ $(\nu = 1, \cdots, \text{ad inf.})$ of parameter points which is dense in Ω. We say that a subset ω of Ω is dense in Ω if for each point θ of Ω any arbitrarily small open neighborhood of θ contains at least one point of ω. Since Ω is compact, a sequence $\{\theta_\nu\}$ which is dense in Ω certainly exists. Let us consider the first s points $\theta_1, \cdots, \theta_s$ of the sequence $\{\theta_\nu\}$. According to Lemma 8 there exists for any s a risk-minimizing distribution $\rho(s) = [\rho_1(s), \cdots, \rho_s(s)]$ related to $\theta_1, \cdots, \theta_s$.

Assumption 6. There exists a sequence $\{\theta_s\}$ $(s = 1, \cdots, \text{ad inf.})$ of parameter points which is dense in Ω and such that for almost any s[7] the risk-minimizing

[6] See for instance Alexandroff and Hopf, *Topologie*, Berlin 1935, p. 396.

[7] By "almost any s" we understand "for all s greater than a sufficiently large integer."

distribution $\rho(s) = [\rho_1(s), \cdots, \rho_s(s)]$ related to the first s points $\theta_1, \cdots, \theta_s$, is not degenerate.

LEMMA 10. *Denote by* $\{\theta_s\}$ ($s = 1, 2, \cdots$, ad inf.) *a sequence of parameter points for which the conditions of Assumption* 6 *are fulfilled. Denote by* $\rho(s) = [\rho_1(s), \cdots, \rho_s(s)]$ *the risk-minimizing distribution related to the first s points* $\theta_1, \cdots, \theta_s$. *Then there exists a non-negative constant c such that for any arbitrarily small positive* ϵ *the inequality*

$$c - \epsilon \leq \int_M W[\theta, \theta_{\rho(s)}(E)]p(E \mid \theta)\, dE \leq c + \epsilon$$

holds identically in θ *for almost every s. That is to say the risk function of the minimum risk estimate* $\theta_{\rho(s)}(E)$ *lies entirely between* $c - \epsilon$ *and* $c + \epsilon$ *for almost every s.*

Denote the risk function

$$\int_M W[\theta, \theta_{\rho(s)}(E)]p(E \mid \theta)\, dE$$

of the estimate $\theta_{\rho(s)}(E)$ by $r(\theta, s)$. First we shall prove that there exists a sequence $\{c_s\}$ ($s = 1, \cdots$, ad inf.) of non-negative numbers such that for every $\epsilon > 0$ the inequality

$$c_s - \epsilon \leq r(\theta, s) \leq c_s + \epsilon \tag{20}$$

holds for almost every s. In fact to any positive η a positive integer s_η can be given such that for any $s > s_\eta$ the points $\theta_1, \cdots, \theta_s$ are η-dense in Ω. That is to say every point θ of Ω lies in a sphere with radius η and center in one of the points $\theta_1, \cdots, \theta_s$. Since for sufficiently large s $\rho(s)$ is not degenerate, we have on account of Lemma 9 for sufficiently large s

$$r(\theta_1, s) = \cdots = r(\theta_s, s) = c_s. \tag{21}$$

Since for sufficiently large s $\theta_1, \cdots, \theta_s$ is η-dense in Ω, we get easily from Lemma 5 that (20) holds for any positive ϵ for almost every s.

In order to prove Lemma 10 we have only to show that $\lim_{s=\infty} c_s$ exists and is finite. First we see that for no estimate $\theta(E)$ can the corresponding risk function

$$r(\theta) = \int_M W[\theta, \theta(E)]p(E \mid \theta)\, dE$$

lie entirely below $r(\theta, s)$ that is to say

$$r(\theta) < r(\theta, s) \tag{22}$$

cannot hold for any θ. In fact if (22) were true for a certain estimate $\theta(E)$ then

$$\Sigma\rho_i(s)r(\theta_i) = \int_M \Sigma W[\theta_i, \theta(E)]\rho_i(s)p(E \mid \theta_i)\, dE < \Sigma\rho_i(s)r(\theta_i, s)$$

$$= \int_M \Sigma W[\theta_i, \theta_{\rho(s)}(E)]\rho_i(s)p(E \mid \theta_i)\, dE,$$

which is not possible since $\theta_{\rho(s)}(E)$ is a minimum risk estimate. Hence (22) cannot hold for any θ. From this fact follows easily that lim c_s exists and is finite. This proves Lemma 10.

LEMMA 11. *Denote $f(\theta)$ a distribution of θ and let $\theta_f(E)$ be the corresponding minimum risk estimate. If $\theta(E)$ denotes an arbitrary estimate then*

$$r(\theta) \equiv r_f(\theta)$$

if $\theta_f(E) \neq \theta(E)$ only in a set of measure 0, *and*

$$\int_\Omega r(\theta)\, df(\theta) > \int_\Omega r_f(\theta)\, df(\theta)$$

if $\theta_f(E) \neq \theta(E)$ in a set of positive measure. $r(\theta)$ denotes the risk function of $\theta(E)$ and $r_f(\theta)$ denotes the risk function of $\theta_f(E)$.

If $\theta_f(E) \neq \theta(E)$ only in a set of measure zero, then we have obviously $r(\theta) \equiv r_f(\theta)$. Consider the case that $\theta_f(E) \neq \theta(E)$ in a set M' of positive measure. According to Assumption 4 we have

$$\int_\Omega W[\theta, \theta(E)]\, p(E \mid \theta)\, df(\theta) > \int_\Omega W[\theta, \theta_f(E)]\, p(E \mid \theta)\, df(\theta)$$

for any point E of M'. Since

$$\int_\Omega W[\theta, \theta(E)]\, p(E \mid \theta)\, df(\theta) = \int_\Omega W[\theta, \theta_f(E)]\, p(E \mid \theta)\, df(\theta)$$

for any other point E of the sample space M, we get

$$\int_\Omega r(\theta)\, df = \int_M \int_\Omega W[\theta, \theta(E)]\, p(E \mid \theta)\, df\, dE$$

$$> \int_M \int_\Omega W[\theta, \theta_f(E)]\, p(E \mid \theta)\, df\, dE = \int_\Omega r_f(\theta)\, df.$$

Hence Lemma 11 is proved.

We are now able to prove some theorems about the best estimate $\bar{\theta}(E)$ relative to a given weight function. An estimate $\bar{\theta}(E)$ is a best estimate according to our definition 7, if the maximum of the risk function of $\bar{\theta}(E)$ is less than or equal to the maximum of the risk function of any other estimate $\theta(E)$ and if $\bar{\theta}(E)$ is an admissible estimate (that is to say there exists no estimate $\theta(E)$ such that the risk function $r(\theta)$ of $\theta(E)$ is not identical to the risk function $\bar{r}(\theta)$ of $\bar{\theta}(E)$ and in every point θ $\bar{r}(\theta) \geq r(\theta)$.

THEOREM 1. *If $\bar{\theta}(E)$ is a best estimate and if the Assumptions* 1–6 *are fulfilled then the risk function $\bar{r}(\theta)$ of $\bar{\theta}(E)$ is constant, that is to say*

$$\bar{r}(\theta) \equiv c.$$

According to Assumption 6 there exists a sequence $\{\theta_s\}$ ($s = 1, \cdots$, ad inf.) of parameter points such that $\{\theta_s\}$ is dense in Ω and for almost every s the risk-minimizing distribution $\rho(s)$ related to $\theta_1, \cdots, \theta_s$ is not degenerate. On account of Lemma 10 there exists a non-negative constant c such that for any $\epsilon > 0$ the inequality

$$c - \epsilon \leq r(\theta, s) \leq c + \epsilon \tag{23}$$

holds for almost every s. $r(\theta, s)$ denotes the risk function of the estimate $\theta_{\rho(s)}(E)$. According to Lemma 2 there exists a subsequence $\{s_n\}$ $(n = 1, \cdots,$ ad inf.) of integers such that the sequence $\{\rho(s_n)\}$ of distributions converges towards a distribution $f(\theta)$. From Lemma 6 it follows that

$$\lim_{n=\infty} r(\theta, s_n) = r_f(\theta)$$

where $r_f(\theta)$ denotes the risk function of the minimum risk estimate $\theta_f(E)$. On account of (23) we have

$$r_f(\theta) \equiv c.$$

From Lemma 11 it follows that for any other estimate $\theta(E)$ either

$$r(\theta) \equiv r_f(\theta) \equiv c$$

or

$$\int_\Omega r(\theta)\, df > \int_\Omega r_f(\theta)\, df,$$

where $r(\theta)$ denotes the risk function of $\theta(E)$. In the latter case there exists at least one point θ for which $r(\theta) > r_f(\theta)$. Hence $\theta_f(E)$ is a best estimate. If $\bar{\theta}(E)$ is also a best estimate, we get on account of Lemma 11 that $\bar{\theta}(E)$ can differ from $\theta_f(E)$ only in a set of measure 0 and the risk function of $\bar{\theta}(E)$ is identically equal to c. Hence we have proved Theorem 1 and also the following Theorems 2–3:

THEOREM 2. *If the Assumptions* 1–6 *are fulfilled there exists a distribution $f(\theta)$ of θ such that the corresponding minimum risk estimate $\theta_f(E)$ is a best estimate.*

THEOREM 3. *If Assumptions* 1–6 *are fulfilled and $\bar{\theta}(E)$, $\theta^*(E)$ are best estimates, then $\bar{\theta}(E) = \theta^*(E)$ almost everywhere and the corresponding risk functions are identically equal.*

Now we shall prove (without making the Assumptions 1–6)

THEOREM 4. *If $W(\theta, \bar{\theta})$ and $p(E \mid \theta)$ are continuous and Ω is compact, and if $f(\theta)$ denotes a distribution of θ such that any open set has a positive probability, then the minimum risk estimate $\theta_f(E)$ is a best estimate if its risk function $r_f(\theta)$ is identically equal to a constant.*

Let $r_f(\theta)$ be identically equal to c and consider an arbitrary estimate $\theta(E)$. Since $W(\theta, \bar{\theta})$ and $p(E \mid \theta)$ are continuous and Ω is compact, the risk function $r(\theta)$ of $\theta(E)$ is a continuous function of θ. Since $\theta_f(E)$ is a minimum risk estimate we have

$$\int_\Omega r(\theta)\, df \geq \int_\Omega r_f(\theta)\, df = c. \tag{24}$$

In order to prove Theorem 4, we have to show that either

$$(25) \qquad r(\theta) \equiv c$$

or there exists a point θ' such that

$$(26) \qquad r(\theta') > c.$$

If (25) does not hold there exists a point θ^* such that $r(\theta^*) \neq c$. If $r(\theta^*) > c$ our statement is proved. Consider the case $r(\theta^*) < c$. On account of the continuity of $r(\theta)$ there exists a positive δ and an open neighborhood U of θ^* such that

$$r(\theta) < c - \delta$$

for every θ in U. Since $\int_U df$ is assumed to be positive, the inequality (24) can hold only if there exists at least a point θ' for which $r(\theta') > c$. This proves Theorem 4.

9. **Determination of the best estimate $\bar{\theta}(E)$ for a certain class of distributions** $p(E \mid \theta)$.* In this paragraph we shall prove two theorems which enable us to calculate very easily the best estimate $\bar{\theta}(E)$ for a certain special but important class of distributions.

The risk function of an estimate $\bar{\theta}(E)$ is given by

$$r(\theta) = \int_M W[\theta, \bar{\theta}(E)]\, p(E \mid \theta)\, dE,$$

where the integration is to be taken over the whole sample space M. We consider the integral equation

$$(27) \qquad \int_M W[\theta, \bar{\theta}(E)]\, p(E \mid \theta)\, dE \equiv c,$$

where c denotes an arbitrary constant. If we can find an estimate $\bar{\theta}(E)$ which satisfies (27) for a certain c and which is an admissible estimate relative to the weight function considered, then $\bar{\theta}(E)$ is certainly a best estimate. If Assumptions 1–6 are fulfilled, an admissible estimate satisfying (27) certainly exists. As we shall see, a best estimate can very easily be determined by the above procedure if the conditions in the following theorem 5 are fulfilled.

THEOREM 5. *Let us assume that the following conditions are fulfilled:*

I. *The parameter space Ω is one dimensional and θ can take any real value from $-\infty$ to $+\infty$.*

II. *The probability density $p(E \mid \theta)$ depends only on the differences $x_1 - \theta$, $\cdots, x_n - \theta$, that is to say $p(E \mid \theta) = p(x_1 - \theta, \cdots, x_n - \theta)$, where $x_1, \cdots, x_n$ denote the co-ordinates of E.*

III. *The value of the weight function depends only on the difference $u = \theta - \bar{\theta}$ and is uniformly continuous in u.*

* The results of Secs. 9 and 10 are less general than stated, since the integral (30) may not be absolutely convergent. Suitable reformulation of these results is difficult; in fact, only special cases have been proved in the literature.

IV. *For any value $\bar{\theta}$ and for any sample point E the integral*

$$\psi(\bar{\theta}, E) = \int_{-\infty}^{+\infty} W(\theta - \bar{\theta}) p(E \mid \theta) \, d\theta \tag{28}$$

has a finite value.

V. *For every E there exists a finite value $\theta'(E)$ such that $\psi(\bar{\theta}, E)$ becomes a minimum for $\bar{\theta} = \theta'(E)$.*

Then there exists an estimate $\bar{\theta}(E)$ such that for any E, $\psi(\bar{\theta}, E)$ becomes a minimum for $\bar{\theta} = \bar{\theta}(E)$ and $\bar{\theta}(E'') - \bar{\theta}(E') = \lambda$ for any $E' = (x_1', \cdots, x_n')$ and $E'' = (x_1'', \cdots, x_n'')$ for which $x_1'' - x_1' = \cdots = x_n'' - x_n' = \lambda$. An estimate with these properties is a best estimate.

Let us consider two sample points $E' = (x_1', \cdots, x_n')$ and $E'' = (x_1'', \cdots, x_n'')$ such that $x_1'' - x_1' = \cdots = x_n'' - x_n' = \lambda$. From the conditions II and III follows that if $\psi(\bar{\theta}, E')$ becomes a minimum for $\bar{\theta} = \theta_1$, then $\psi(\bar{\theta}, E'')$ becomes a minimum for $\bar{\theta} = \theta_2 = \theta_1 + \lambda$. Hence there exists an estimate $\bar{\theta}(E) = \bar{\theta}(x_1, \cdots, x_n)$ such that for any E, $\psi(\bar{\theta}, E)$ becomes a minimum for $\bar{\theta} = \bar{\theta}(E)$ and $\bar{\theta}(E'') - \bar{\theta}(E') = \lambda$ if $x_1'' - x_1' = \cdots = x_n'' - x_n' = \lambda$. We shall show that such an estimate $\bar{\theta}(E)$ is a best estimate. First we shall show that the risk function

$$r(\theta) = \int_{-\infty}^{+\infty} \cdots \int_{-\infty}^{+\infty} W[\theta - \bar{\theta}(E)] \, p(x_1 - \theta, \cdots, x_n - \theta') \, dx_1 \cdots dx_n$$

is constant. Let us consider two arbitrary parameter values θ' and θ''. Then we have

$$r(\theta') = \int_{-\infty}^{+\infty} \cdots \int_{-\infty}^{+\infty} W[\theta' - \bar{\theta}(E)] \, p(x_1 - \theta', \cdots, x_n - \theta') \, dx_1 \cdots dx_n,$$

$$r(\theta'') = \int_{-\infty}^{+\infty} \cdots \int_{-\infty}^{+\infty} W[\theta'' - \bar{\theta}(E)] \, p(x_1 - \theta'', \cdots, x_n - \theta'') \, dx_1 \cdots dx_n.$$

Making in the second integral the transformation

$$y_1 = x_1 - (\theta'' - \theta'), \cdots, y_n = x_n - (\theta'' - \theta'),$$

we get

$$r(\theta'') = \int_{-\infty}^{+\infty} \cdots \int_{-\infty}^{+\infty} W\{\theta'' - \bar{\theta}[y_1 + (\theta'' - \theta'), \cdots, y_n + (\theta'' - \theta')]\} \, p(y_1 - \theta', \cdots, y_n - \theta') \, dy_1 \cdots dy_n$$

$$= \int_{-\infty}^{+\infty} \cdots \int_{-\infty}^{+\infty} W[\theta' - \bar{\theta}(y_1, \cdots, y_n)] \, p(y_1 - \theta', \cdots, y_n - \theta') \, dy_1 \cdots dy_n.$$

Hence $r(\theta') = r(\theta'')$ and our statement that $r(\theta)$ is constant is proved. In order to prove Theorem 5, we have only to show that $\bar{\theta}(E)$ is an admissible estimate. For this purpose let us consider an arbitrary estimate $\theta^*(E)$ and denote the corresponding risk function by $r^*(\theta)$. Since $\bar{\theta}(E)$ minimizes the integral (28), we have

$$(29) \qquad \psi[\theta^*(E), E] \geq \psi[\bar{\theta}(E), E)]$$

for all sample points E. Let us consider the integral

$$(30) \qquad I = \int_{-\infty}^{+\infty} \cdots \int_{-\infty}^{+\infty} \{W[\theta - \bar{\theta}(E)] - W[\theta - \theta^*(E)] \, p(E \mid \theta) \, d\theta \, dx_1 \cdots dx_n .$$

Integrating (30) with respect to θ we get

$$(31) \qquad I = \int_{-\infty}^{+\infty} \cdots \int_{-\infty}^{+\infty} \{\psi[\bar{\theta}(E), E] - \psi[\theta^*(E) - E]\} \, dx_1 \cdots dx_n .$$

Integrating (30) with respect to E, we get

$$(32) \qquad I = \int_{-\infty}^{+\infty} [r(\theta) - r^*(\theta)] \, d\theta .$$

On account of (29) and (31) we have $I \leq 0$, hence

$$(33) \qquad \int_{-\infty}^{+\infty} [r(\theta) - r^*(\theta)] \, d\theta \leq 0 .$$

From (33) it follows that if $r^*(\theta) \leq r(\theta)$ for every θ then $r^*(\theta) < r(\theta)$ can hold only for the points of a set of measure zero. In case $r^*(\theta)$ is continuous, this means that $r^*(\theta) \equiv r(\theta)$. Hence if $r^*(\theta)$ is continuous, then either $r^*(\theta) \equiv r(\theta)$ or there exists at least one point θ' such that $r^*(\theta') > r(\theta)$. The risk function $r^*(\theta)$ is continuous if the estimate $\theta^*(E)$ is uniformly continuous in the whole sample space. In fact, we have

$$r^*(\theta + t) = \int_{-\infty}^{+\infty} \cdots \int_{-\infty}^{+\infty} W[\theta + t - \theta^*(E)] \, p(x_1 - \theta - t, \cdots, x_n - \theta - t) \, dx_1 \cdots dx_n .$$

Making the transformation

$$y_i = x_i - t \qquad (i = 1, \cdots, n)$$

we get

$$r^*(\theta + t) =$$

$$\int_{-\infty}^{+\infty} \cdots \int_{-\infty}^{+\infty} W[\theta + t - \theta^*(y_1 + t, \cdots, y_n + t) \, p(y_1 - \theta, \cdots, y_n - \theta) \, dy_1 \cdots dy_n .$$

Since $W(u)$ and $\theta^*(E)$ are uniformly continuous, from the latter equation we get easily

$$\lim_{t=0} r^*(\theta + t) = r^*(\theta)$$

that is to say $r^*(\theta)$ is continuous. Considering only continuous estimates the admissibility of $\bar{\theta}(E)$, and therefore also Theorem 5, is proved. If $\theta^*(E)$ is not uniformly continuous we have only proved that if $r^*(\theta) \leq r(\theta)$ for every θ, then $r^*(\theta) < r(\theta)$ can hold only in a set of measure zero. I should like to mention without proof that even if $\theta^*(E)$ is not continuous, $r^*(\theta) \leq r(\theta)$ implies $r^*(\theta) \equiv r(\theta)$.

An estimate $\hat{\theta}(E)$ is called a maximum likelihood estimate if for any fixed E $p(E \mid \theta)$ becomes a maximum with respect to θ for $\theta = \hat{\theta}(E)$.

THEOREM 6. *Consider the following conditions:*

VI. *There exists exactly one maximum likelihood estimate $\hat{\theta}(E)$ with the following properties:*

a) *For any E $p(E \mid \theta)$ is non-decreasing with increasing θ for $\theta < \hat{\theta}(E)$ and non-increasing with increasing θ for $\theta > \hat{\theta}(E)$.*

b) *For any E $p(E \mid \theta)$ is a symmetric function of θ about $\hat{\theta}(E)$ that is to say, for for any real value λ $p[E \mid \hat{\theta}(E) - \lambda] = p[E \mid \hat{\theta}(E) + \lambda]$.*

VII. *The value of the weight function depends only on the absolute value of the difference $u = \theta - \bar{\theta}$ and $\dfrac{dw(u)}{du}$ exists, is uniformly continuous and > 0 for $u > 0$.*

If the conditions I–V of Theorem 5 and the above condition VII are fulfilled, and if $\hat{\theta}(E)$ is a maximum likelihood estimate satisfying VI, then $\hat{\theta}(E)$ is a best estimate.

Assume that the conditions I–V and VII are satisfied and that $\hat{\theta}(E)$ is a maximum likelihood estimate satisfying VI. It is obvious that $\hat{\theta}(E'') - \hat{\theta}(E') = \lambda$ for $E' = (x_1, \cdots, x_n)$ and $E'' = (x_1 + \lambda, \cdots, x_n + \lambda)$. In order to prove Theorem 6, we have, according to Theorem 5, only to show that the integral in (28)

$$\psi(\bar{\theta}, E) = \int_{-\infty}^{+\infty} W(\theta - \bar{\theta}) p(E \mid \theta)\, d\theta$$

becomes a minimum for $\bar{\theta} = \hat{\theta}(E)$. Denote $\theta - \bar{\theta}$ by u. Since $\dfrac{dW(u)}{du}$ is uniformly continuous, we have

$$\frac{\partial \psi(\bar{\theta}, E)}{\partial \bar{\theta}} = \int_{-\infty}^{+\infty} -\left[\frac{dW(u)}{du}\right] p(E \mid \theta)\, d\theta.$$

Since $\dfrac{dW(u)}{du} = -\dfrac{dW(-u)}{du}$ we have

$$\frac{\partial \psi(\bar{\theta}, E)}{\partial \bar{\theta}} = \int_{0}^{\infty} \left[\frac{dW(u)}{du}\right] [p(E \mid \bar{\theta} - u) - p(E \mid \bar{\theta} + u)]\, du. \tag{34}$$

From condition VI it follows easily that for any fixed E and $\bar{\theta}$ the function of u $(0 \leq u \leq \infty)$

$$p(E \mid \bar{\theta} - u) - p(E \mid \bar{\theta} + u)$$

does not change its sign and if $\bar{\theta} \neq \hat{\theta}(E)$ there exists an interval J such that the above expression is unequal to zero for every point u of J. Hence on account of $\frac{dW(u)}{du} > 0$ for $u > 0$, the integral in (34) vanishes only for $\bar{\theta} = \hat{\theta}(E)$. Since according to the condition V there exists a finite value θ' such that $\psi(\bar{\theta}, E)$ becomes a minimum for $\bar{\theta} = \theta'$, θ' must be equal to $\hat{\theta}(E)$. This proves Theorem 6.

The condition VI is seldom exactly fulfilled. But for large n, in the great majority of practical cases, VI will be fulfilled with good approximation and the best estimate approaches the maximum likelihood estimate with increasing n.

10. **Two examples.** As a first example we consider a normal distribution with the variance 1. The mean value θ is unknown and we have to estimate it by means of a sample $E = (x_1, \cdots, x_n)$. In this case

$$p(E \mid \theta) = \frac{1}{(2\pi)^{\frac{n}{2}}} e^{-\frac{1}{2}\Sigma(x_i-\theta)^2}.$$

It is obvious that for a very broad class of weight functions the conditions I–V of Theorem 5 are fulfilled. The maximum likelihood estimate $\hat{\theta}(x_1, \cdots, x_n) = \frac{x_1 + \cdots + x_n}{n}$ satisfies the condition VI of Theorem 6. Hence if the weight function satisfies also the condition VII, then the best estimate of θ is the maximum likelihood estimate $\hat{\theta}(x_1, \cdots, x_n) = \frac{x_1 + \cdots + x_n}{n}$.

Let us now consider a weight function defined as follows:

$$W(\theta, \bar{\theta}) = 2(\bar{\theta} - \theta) \quad \text{if} \quad \bar{\theta} \geq \theta$$

and

$$W(\theta, \bar{\theta}) = \theta - \bar{\theta} \qquad \text{if} \quad \bar{\theta} < \theta.$$

Since for this weight function, the conditions I–V satisfied, according to Theorem 5 the best estimate of θ is the value $\bar{\theta}$ for which the integral

$$\int_{-\infty}^{+\infty} W(\theta, \bar{\theta}) e^{-\frac{1}{2}\Sigma(x_i-\theta)^2}\, d\theta = \int_{-\infty}^{\bar{\theta}} 2(\bar{\theta} - \theta) e^{-\frac{1}{2}\Sigma(x_i-\theta)^2}\, d\theta + \int_{\bar{\theta}}^{\infty} (\theta - \bar{\theta}) e^{-\frac{1}{2}\Sigma(x_i-\theta)^2}\, d\theta$$

becomes a minimum. As an easy calculation shows, the estimate obtained in this way is not the arithmetic mean.

As a second example we consider the family of variates $X(\theta)$ with the probability density $f(x, \theta)$ defined as follows:

$$f(x, \theta) = 1 \quad \text{if} \quad \theta - \tfrac{1}{2} \leq x \leq \theta + \tfrac{1}{2}$$

and

$$f(x, \theta) = 0 \text{ for all other values of } x.$$

If $E = (x_1, \cdots, x_n)$ denotes a sample point where x_1 denotes the smallest and x_n denotes the greatest value in the sample, then

$$p(E \mid \theta) = \prod_{i=1}^{n} f(x_i, \theta) = 1 \quad \text{if} \quad x_n - \tfrac{1}{2} \leq \theta \leq x_1 + \tfrac{1}{2}$$

and

$$p(E \mid \theta) = 0 \quad \text{for all other values of } \theta.$$

The classical method of maximum likelihood cannot be applied here, since $p(E \mid \theta)$ is maximum for every value θ for which $x_n - \frac{1}{2} \leq \theta \leq x_1 + \frac{1}{2}$. It is obvious that for a broad class of weight functions the conditions I–V are satisfied. The estimate $\bar{\theta}(E) = \dfrac{x_1 + x_n}{2}$, where x_1 denotes the smallest and x_n the greatest value in the sample, satisfies the condition VI. Hence if the weight function satisfies also the condition VII, the best estimate of θ is given by $\bar{\theta}(E) = \dfrac{x_1 + x_n}{2}$.

Let us now calculate the best estimate of θ if the weight function is given as follows:

$$W(\theta, \bar{\theta}) = \theta - \bar{\theta} \qquad \text{if} \quad \bar{\theta} \leq \theta$$

and

$$W(\theta, \bar{\theta}) = 2(\bar{\theta} - \theta) \quad \text{if} \quad \bar{\theta} > \theta.$$

In this case the conditions I–V are satisfied but not the condition VII. We have to calculate the integral $\psi(\bar{\theta}, E)$ given in (28), which reduces in this case to

$$\psi(\bar{\theta}, E) = \int_{x_n - \frac{1}{2}}^{x_1 + \frac{1}{2}} W(\theta, \bar{\theta})\, d\theta = \int_{x_n - \frac{1}{2}}^{\bar{\theta}} 2(\bar{\theta} - \theta)\, d\theta + \int_{\bar{\theta}}^{x_1 + \frac{1}{2}} (\theta - \bar{\theta})\, d\theta$$

$$= 1.5\bar{\theta}^2 - [(x_1 + \tfrac{1}{2}) + 2(x_n - \tfrac{1}{2})]\bar{\theta} + \tfrac{1}{2}(x_1 + \tfrac{1}{2})^2 + (x_n - \tfrac{1}{2})^2.$$

This expression becomes a minimum for

$$\bar{\theta} = \frac{x_1 + 2x_n - \frac{1}{2}}{3}.$$

Hence the best estimate of θ is given by this expression.

11. **Miscellaneous remarks.** Assumptions 1–6 of paragraph 8 are sufficient but not necessary for the proof of the Theorems 1–3 (Theorems 4–6 have been deduced without Assumptions 1–6). They can be weakened in many respects. The assumption that the parameter space is bounded can be dropped if we impose certain conditions on the weight function $W(\theta, \bar{\theta})$ and the probability density $p(E \mid \theta)$. It is certainly not necessary to assume that $W(\theta, \bar{\theta})$ and $p(E \mid \theta)$ are everywhere continuous. It is however doubtful whether Theorems 1–3 remain valid in the form in which they are stated, if we admit discon-

tinuities in a set of measure zero without imposing any other restrictions. Also Assumptions 4–6 can in all probability be essentially weakened.

I should like to mention that Assumption 4 is not as restrictive as it would appear. Let us make this clear in the case that the parameter space is a one-dimensional interval $[a, b]$. If we assume that $W(\theta, \bar{\theta})$ is a polynomial of the second degree in $\bar{\theta}$ and the coefficient of $\bar{\theta}^2$ is positive for every θ, and if $p(E \mid \theta) > 0$ for every E and θ, the Assumption 4 can easily be proved. In fact,

$$\psi(\bar{\theta}, E) = \int_a^b W(\theta, \bar{\theta}) p(E \mid \theta)\, df(\theta) = A(E) + B(E)\bar{\theta} + C(E)\bar{\theta}^2.$$

Since the coefficient of $\bar{\theta}^2$ in $W(\theta, \bar{\theta})$ is positive and since $p(E \mid \theta) > 0$ for every E and θ, $C(E) > 0$ for every E and for any arbitrary distribution $f(\theta)$. From this fact follows easily that for every E there exists a value $\bar{\theta}(E)$ in the interval $[a, b]$ such that

$$\psi[\bar{\theta}(E), E] < \psi(\bar{\theta}, E)$$

for every $\bar{\theta}$ contained in $[a, b]$ and unequal to $\bar{\theta}(E)$. Hence Assumption 4 is proved.

Let us consider a system S of subsets of the parameter space Ω and the corresponding system H_S of hypotheses. The weight function $W(\theta, \omega)$ is defined for all points θ of Ω and for all elements ω of S and expresses the weight of the error committed by accepting H_ω when θ is true. If θ is an element of ω then $W(\theta, \omega)$ is of course equal to zero. Let us assume that $W(\theta, \omega)$ has the special form: $W(\theta, \omega) = 1$ if θ is not contained in ω, and $W(\theta, \omega) = 0$ if θ is an element of ω. It is obvious that in this case for any θ the value of the risk function $r(\theta)$ is equal to the probability of accepting a false hypothesis if θ is the true parameter point. Because of this fact the theory developed here has close relation to the theory of confidence intervals. Let us first make this clear for the case when the parameter space is one dimensional, that is to say θ is a real number.

In the theory of confidence intervals we estimate the unknown parameter θ by an interval $I(E)$ extending from $\theta_1(E)$ to $\theta_2(E)$ where $\theta_1(E)$ and $\theta_2(E)$ are certain functions of the sample point E. The interval $I(E)$ is defined in such a way that the following probability statement holds: If we perform an experiment, the probability that we shall obtain a sample point E such that $I(E)$ will cover the true parameter point θ, is equal to a given constant α (called confidence coefficient) and is independent of the value of θ. Let us consider a certain example of such an inference with the confidence coefficient α and denote by $I(E)$ the interval corresponding to E. We define a system S of intervals as follows: An interval I is an element of S if and only if there exists a sample point E for which $I(E) = I$. Consider the corresponding system H_S

of hypotheses and the weight function $W(\theta, I)$ defined for all values θ and all elements I of S as follows:

$$W(\theta, I) = 0 \quad \text{if} \quad \theta \text{ is a point of } I$$

$$W(\theta, I) = 1 \quad \text{if} \quad \theta \text{ is not contained in } I.$$

Denote by M_S a best system of regions of acceptance relative to the weight function defined above. Denote by $I'(E)$ the element of S which we accept according to M_S if E is the sample point. On account of the special form of the weight function, the risk is obviously equal to the probability of accepting a false interval. From the definition of the best system of regions it follows that for any θ the probability that $I'(E)$ will cover θ is greater than or equal to α. If the risk function is constant, that is to say, if the probability that $I'(E)$ will cover the true parameter point θ is independent of the value of θ, then the intervals $I'(E)$ are confidence intervals corresponding to a confidence coefficient $\alpha' \geq \alpha$.

Similar observations can be made if the parameter space is k-dimensional ($k > 1$) that is to say, θ is a system of k numbers $\theta^{(1)}, \cdots, \theta^{(k)}$. An important case is that when we have to estimate only one of the components, say $\theta^{(1)}$, by an interval. As the investigations of W. Feller[8] have shown, confidence intervals in such cases do not exist always. That is to say, it is not always possible to determine $I(E)$ such that the probability that $I(E)$ will cover $\theta^{(1)}$ is equal to a given constant α independently of the values of $\theta^{(1)}, \cdots, \theta^{(k)}$. It is of great interest to know under what conditions confidence intervals exist. I should like to mention that a further development of the theory given in paragraph 8 may contribute much to the solution of this problem. In order to make this clear, let us consider a system S_1 of one-dimensional intervals. To each element I of S_1 let there correspond the subset ω of the k-dimensional parameter space Ω consisting of all points $\theta = (\theta^{(1)}, \cdots, \theta^{(k)})$ for which $\theta^{(1)}$ lies in I. Consider the system S of subsets ω of Ω corresponding to all elements of S_1 and the system H_S of hypotheses corresponding to S. The weight function is to be chosen as follows: $W(\theta, \omega) = 1$ if θ is not an element of ω and $W(\theta, \omega) = 0$ if θ is an element of ω. Consider a best system M_S of regions of acceptance and the corresponding risk function $r(\theta)$. On account of the special definition of $W(\theta, \omega)$, $r(\theta)$ is equal to the probability of accepting a false hypothesis if θ is the true parameter point. If the risk function $r(\theta)$ is identically equal to a constant α, we have confidence intervals corresponding to the confidence coefficient α. In order to see under what conditions the risk function is constant, we have to consider an equivalent problem (see paragraph 7) where the system of hypotheses is the system of all simple hypotheses and the weight function $W(\theta, \bar{\theta})$

[8] W. Feller, "Note on Regions Similar to the Sample Space," *Statistical Research Memoirs*, Vol. II, 1938.

is given according to formula (4). If $W(\theta, \bar{\theta})$ satisfies Assumptions 1–6, the risk function of the best estimate is constant. As we have mentioned, Assumptions 1–6 can be weakened. In order to get valuable results concerning the problem of the existence of confidence intervals, we have to weaken especially Assumption 2. In fact $W(\theta, \omega)$ takes only the values 1 and 0 and therefore $W(\theta, \bar{\theta})$ cannot be continuous.

Finally I should like to mention that the most stringent test as defined by Robert W. B. Jackson[9] is contained as special case in our general definition of the best system of regions of acceptance. Jackson considers a discontinuous parameter space Ω. Consider the problem of testing the hypothesis $\theta = \theta_0$ where θ_0 denotes a point of Ω. According to Jackson's definition we have the most stringent test if the critical region w_0 satisfies the condition: the maximum of the numbers A and B

$$A = P(E \epsilon w \mid \theta_0), \quad B = \text{least upper bound of } P(E \epsilon \bar{w} \mid \theta) \text{ formed for all } \theta \neq \theta_0,$$

becomes a minimum for $w = w_0 \cdot \bar{w}$ denotes the region complementary to w. It is easy to see that Jackson's definition of the most stringent test coincides with our definition of the best system of regions of acceptance in the following special case:

1) Ω is discontinuous 2) S consists only of two elements.
3) The weight function $W(\theta, \omega)$ is equal to 1 if θ is not contained in ω.

COLUMBIA UNIVERSITY.

[9] Robert W. Jackson, "Testing Statistical Hypotheses," *Statistical Research Memoirs*, Vol. I, 1936.

Reprinted from The Annals of Mathematical Statistics
Vol. XI, No. 1, March, 1940

A NOTE ON THE ANALYSIS OF VARIANCE WITH UNEQUAL CLASS FREQUENCIES[1]

By Abraham Wald[2]

Let us consider p groups of variates and denote by m_j ($j = 1, \cdots, p$) the number of elements in the j-th group. Let x_{ij} be the i-th element of the j-th group. We assume that x_{ij} is the sum of two variates ϵ_{ij} and η_j, i.e. $x_{ij} = \epsilon_{ij} + \eta_j$, where ϵ_{ij} ($i = 1, \cdots, m_j$; $j = 1, \cdots, p$) is normally distributed with mean μ and variance σ^2, and η_j ($j = 1, \cdots, p$) is normally distributed with mean μ' and variance σ'^2. All the variates ϵ_{ij} and η_j are supposed to be distributed independently.

The intraclass correlation ρ is given by[3]

$$\rho = \frac{\sigma'^2}{\sigma^2 + \sigma'^2}.$$

Confidence limits for ρ have been derived only in case of equal class frequencies, i.e. $m_1 = m_2 = \cdots = m_p$. In this paper we shall deal with the problem of determining the confidence limits for ρ in the case of unequal class frequencies. Since ρ is a monotonic function of $\frac{\sigma'^2}{\sigma^2}$, our problem is solved if we derive confidence limits for $\frac{\sigma'^2}{\sigma^2}$.

Denote by $\bar{x}_j$ the arithmetic mean of the j-th group, i.e.

$$\bar{x}_j = \frac{\sum_{i=1}^{m_j} \epsilon_{ij}}{m_j} + \eta_j. \tag{1}$$

Hence the variance of $\bar{x}_j$ is equal to

$$\sigma_{\bar{x}_j}^2 = \frac{\sigma^2}{m_j} + \sigma'^2. \tag{2}$$

Denote $\frac{\sigma'^2}{\sigma^2}$ by λ^2. Then we have

$$\sigma_{\bar{x}_j}^2 = \sigma^2\left(\frac{1}{m_j} + \lambda^2\right) = \frac{\sigma^2}{w_j}, \tag{3}$$

---

[1] The author is indebted to Professor H. Hotelling for formulating the problem dealt with in this paper.

[2] Research under a grant-in-aid from the Carnegie Corporation at New York.

[3] See for instance R. A. Fisher, *Statistical Methods for Research Workers*, 6-th edition, p. 228.

where

$$w_j = \frac{m_j}{1 + m_j \lambda^2}. \tag{4}$$

Now we shall prove that

$$\frac{1}{\sigma^2} \sum_{j=1}^{p} \left[w_j \left(\bar{x}_j - \frac{\sum_{j=1}^{p} w_j \bar{x}_j}{\sum_{j=1}^{p} w_j} \right)^2 \right] \tag{5}$$

has the χ^2-distribution with $p - 1$ degrees of freedom. Let

$$y_j = \sqrt{w_j}\, \bar{x}_j \qquad (j = 1, \cdots, p)$$

and consider the orthogonal transformation

$$y'_1 = L_1(y_1, \cdots, y_p),$$

$$\cdots\cdots\cdots\cdots\cdots\cdots$$

$$y'_{n-1} = L_{p-1}(y_1, \cdots, y_p),$$

$$y'_p = L_p(y_1, \cdots, y_p) = \frac{\sqrt{w_1}\, y_1 + \cdots + \sqrt{w_p}\, y_p}{\sqrt{w_1 + \cdots + w_p}},$$

where $L_1(y_1, \cdots, y_p), \cdots, L_{p-1}(y_1, \cdots, y_p)$ denote arbitrary homogenous linear functions subject to the only condition that the transformation should be orthogonal.

Since the mean value of y_j is equal to $\sqrt{w_j}\,(\mu + \mu')$ and the variance of y_j is equal to σ^2, we obviously have: The mean value of y'_j $(j = 1, \cdots, p - 1)$ is equal to zero, the variance of y'_j $(j = 1, \cdots, p)$ is equal to σ^2. In order to prove our statement, we have only to show that the expression (5) is equal to $\frac{1}{\sigma^2}(y_1'^2 + \cdots + y_{p-1}'^2)$. If we substitute in (5) $\frac{y_j}{\sqrt{w_j}}$ for $\bar{x}_j$, we get

$$\frac{1}{\sigma^2} \sum_{j=1}^{p} \left\{ w_j \left[\frac{y_j^2}{w_j} - 2 \frac{y_j}{\sqrt{w_j}} \frac{\sum_{j=1}^{p} \sqrt{w_j}\, y_j}{\sum_{j=1}^{p} w_j} + \left(\frac{\sum_j \sqrt{w_j}\, y_j}{\Sigma w_j} \right)^2 \right] \right\}$$

$$= \frac{1}{\sigma^2} \left[\sum_j y_j^2 - 2 \frac{(\sum_j \sqrt{w_j}\, y_j)^2}{\sum_j w_j} + \frac{(\sum_j \sqrt{w_j}\, y_j)^2}{\sum_j w_j} \right] \tag{5'}$$

$$= \frac{1}{\sigma^2} \left[\sum_j y_j^2 - \frac{(\sum_j \sqrt{w_j}\, y_j)^2}{\sum_j w_j} \right] = \frac{1}{\sigma^2} \left[\sum_{j=1}^{p} y_j^2 - y_p'^2 \right] = \frac{1}{\sigma^2} \left[\sum_{j=1}^{p} y_j'^2 - y_p'^2 \right]$$

$$= \frac{1}{\sigma^2} (y_1'^2 + \cdots + y_{p-1}'^2).$$

Since $\frac{\Sigma\Sigma(x_{ij} - \bar{x}_j)^2}{\sigma^2}$ has the χ^2 distribution with $N - p$ degrees of freedom, the expression

$$F = \frac{N-p}{p-1} \frac{\sum_{j=1}^{p}\left\{w_j\left(\bar{x}_j - \frac{\Sigma w_j \bar{x}_j}{\Sigma w_j}\right)^2\right\}}{\Sigma\Sigma(x_{ij} - \bar{x}_j)^2} \tag{6}$$

has the analysis of variance distribution with $p - 1$ and $N - p$ degrees of freedom, where $N = m_1 + \cdots + m_p$. In case $m_1 = m_2 = \cdots = m_p = m$, we have

$$F = \frac{N-p}{p-1} \frac{\sum_{j=1}^{p}(\bar{x}_j - \bar{x})^2}{\Sigma\Sigma(x_{ij} - \bar{x}_j)^2} \cdot \frac{m}{1 + m\lambda^2} = \frac{1}{1 + m\lambda^2} F^*, \tag{6'}$$

where $\bar{x} = \frac{\Sigma\Sigma x_{ij}}{N}$ and $F^* = \frac{N-p}{p-1} \frac{m\Sigma(\bar{x}_j - \bar{x})^2}{\Sigma\Sigma(x_{ij} - \bar{x}_j)^2}$.

Hence

$$\lambda^2 = \left(\frac{F^*}{F} - 1\right)\frac{1}{m}.$$

If F_1 denotes the lower and F_2 the upper confidence limit of F, we obtain for λ^2 the confidence limits

$$\left(\frac{F^*}{F_1} - 1\right)\frac{1}{m} \quad \text{and} \quad \left(\frac{F^*}{F_2} - 1\right)\frac{1}{m}.$$

Let us now consider the general case that $m_1, \cdots, m_p$ are arbitrary positive integers. First we shall show that the set of values of λ^2, for which (6) lies between its confidence limits F_1 and F_2, is an interval. For this purpose we have only to show that

$$f(\lambda^2) \equiv \sum_{j=1}^{p}\left\{w_j\left(\bar{x}_j - \frac{\Sigma w_j \bar{x}_j}{\Sigma w_j}\right)^2\right\}$$

is monotonically decreasing with λ^2. In fact

$$\frac{df(\lambda^2)}{d\lambda^2} = \sum_{j=1}^{p} \frac{dw_j}{d\lambda^2}\left(\bar{x}_j - \frac{\Sigma w_j \bar{x}_j}{\Sigma w_j}\right)^2 - 2\frac{d}{d\lambda^2}\left(\frac{\Sigma w_j \bar{x}_j}{\Sigma w_j}\right)\left[\sum_{j=1}^{p} w_j\left(\bar{x}_j - \frac{\Sigma w_j \bar{x}_j}{\Sigma w_j}\right)\right].$$

Since

$$\sum_{j=1}^{p} w_j\left(\bar{x}_j - \frac{\Sigma w_j \bar{x}_j}{\Sigma w_j}\right) = 0,$$

we have

$$\frac{df(\lambda^2)}{d\lambda^2} = \sum_{j=1}^{p} \frac{dw_j}{d\lambda^2}\left(\bar{x}_j - \frac{\Sigma w_j \bar{x}_j}{\Sigma w_j}\right)^2 = \sum_{j=1}^{p} - w_j^2\left(\bar{x}_j - \frac{\Sigma w_j \bar{x}_j}{\Sigma w_j}\right)^2 < 0,$$

which proves our statement.

Hence the lower confidence limit λ_1^2 of λ^2 is given by the root of the equation in λ^2:

$$(7) \qquad F = \frac{N-p}{p-1} \frac{\sum_{j=1}^{p} \left\{ w_j \left(\bar{x}_j - \frac{\Sigma w_j \bar{x}_j}{\Sigma w_j} \right)^2 \right\}}{\Sigma\Sigma(x_{ij} - \bar{x}_j)^2} = F_2$$

and the upper confidence limit λ_2^2 of λ^2 is given by the root of the equation in λ^2:

$$(8) \qquad F = F_1 .$$

Since $f(\lambda^2)$ is monotonically decreasing, the equations (7) and (8) have at most one root in λ^2. If the equation (7) or (8) has no root, the corresponding confidence limit has to be put equal to zero. If neither (7) nor (8) has a root, we have to reject at least one of the hypotheses:

(1) $x_{ij} = \epsilon_{ij} + \eta_j$.

(2) The variates ϵ_{ij} and η_j $(i = 1, \cdots, m_j; j = 1, \cdots, p)$ are normally and independently distributed.

(3) Each of the variates ϵ_{ij} has the same distribution.

(4) Each of the variates η_j has the same distribution.

The equations (7) and (8) are complicated algebraic equations in λ^2. For the actual calculation of the roots of these equations, well known approximation methods can be applied making use also of the fact that the left members are monotonic functions of λ^2. In applying any approximation method it is very useful to start with two limits of the root which do not lie far apart. We shall give here a method of finding such limits.

Denote by $\bar{F}$ the function which we obtain from F (formula (6)) by substituting

$$\bar{w}_j = \frac{l_j}{1 + l_j \lambda^2} \text{ for } w_j \qquad (j = 1, \cdots, p).$$

Let f be the function obtained from f by the same process.

Denote by $\varphi(m, \lambda^2)$ the function which we obtain from $\bar{F}$ by substituting m for $l_1, \cdots, l_p$. We shall first show that $\bar{F}$ is non-decreasing with increasing l_k $(k = 1, \cdots, p)$, i.e. $\frac{\partial \bar{F}}{\partial l_k} \geq 0$. For this purpose we have only to show that $\frac{\partial \bar{f}}{\partial l_k} \geq 0$. We have:

$$\frac{\partial \bar{f}}{\partial l_k} = \sum_j \frac{\partial \bar{w}_j}{\partial l_k} \left(\bar{x}_j - \frac{\Sigma \bar{w}_j \bar{x}_j}{\Sigma \bar{w}_j} \right)^2 - 2 \frac{\partial}{\partial l_k} \left(\frac{\Sigma \bar{w}_j \bar{x}_j}{\Sigma \bar{w}_j} \right) \cdot \left[\Sigma \bar{w}_j \cdot \left(\bar{x}_j - \frac{\Sigma \bar{w}_j \bar{x}_j}{\Sigma \bar{w}_j} \right) \right]$$

$$= \sum_j \frac{\partial \bar{w}_j}{\partial l_k} \left(\bar{x}_j - \frac{\Sigma \bar{w}_j \bar{x}_j}{\Sigma \bar{w}_j} \right)^2 = \frac{1}{(1 + l_k \lambda^2)^2} \left(\bar{x}_k - \frac{\Sigma \bar{w}_j \bar{x}_j}{\Sigma \bar{w}_j} \right)^2 \geq 0.$$

Hence our statement is proved. Denote by m' the smallest and by m'' the greatest of the values $m_1, \cdots, m_p$. Then we obviously have

(9) $$\varphi(m', \lambda^2) \leq F \leq \varphi(m'', \lambda^2).$$

Denote by $\lambda_1'^2, \lambda_1''^2, \lambda_2'^2, \lambda_2''^2$ the roots in λ^2 of the following equations respectively:

$$\varphi(m', \lambda^2) = F_2 ;$$

$$\varphi(m'', \lambda^2) = F_2 ;$$

$$\varphi(m', \lambda^2) = F_1 ; \qquad \varphi(m'', \lambda^2) = F_1 .$$

Since F is monotonically decreasing with increasing λ^2, on account of (7), (8), and (9) we obviously have

$$\lambda_1'^2 \leq \lambda_1^2 \leq \lambda_1''^2$$

and

$$\lambda_2'^2 \leq \lambda_2^2 \leq \lambda_2''^2 .$$

The above inequalities give us the required limits.

COLUMBIA UNIVERSITY,
NEW YORK, N. Y.

Reprinted from THE ANNALS OF MATHEMATICAL STATISTICS
Vol. XI, No. 2, June, 1940

ON A TEST WHETHER TWO SAMPLES ARE FROM THE SAME POPULATION[1]

BY A. WALD[2] AND J. WOLFOWITZ

1. The Problem.[3] Let X and Y be two independent stochastic variables about whose cumulative distribution functions nothing is known except that they are continuous. Let $x_1, x_2, \cdots, x_m$ be a set of m independent observations on X and let $y_1, \cdots, y_n$ be a set of n independent observations on Y. It is desired to test the hypothesis (the null hypothesis) that the distribution functions of X and Y are identical.

An important step in statistical theory was made when "Student" proposed his ratio of mean to standard deviation for a similar purpose. In the problem treated by "Student" the distribution functions were assumed to be of known (normal) form and completely specified by two parameters. It is clear that in the problem to be considered here the distributions cannot be specified by any finite number of parameters.

It might nevertheless be argued that by virtue of the limit theorems of probability theory, "Student's" ratio might be used in our problem for large samples. Such a procedure is open to very serious objections. The population distributions may be of such form (e.g., Cauchy distribution) that the limit theorems do not apply. Furthermore, the distributions of X and Y may be radically different and yet have the same first two moments; clearly "Student's" ratio will not distinguish between two such distributions.

The Pearson contingency coefficient is a useful test specifically designed for the problem we are discussing here, but one which also possesses some disadvantages. The location of the class intervals is to a considerable extent arbitrary. In order to use the χ^2 distribution, the numbers in each class interval must not be small; often this can be done only by having large class intervals, thus entailing a loss of information.

2. Preliminary remarks. Denote by $P\{X < x\}$ the probability of the relation in braces. Let $f(x)$ and $g(x)$ be the distribution functions of X and Y respectively; e.g., $P\{X < x\} = f(x)$. Throughout this paper we shall assume that $f(x)$ and $g(x)$ are continuous.

Let the set of $m + n$ elements $x_1, \cdots, x_m$ and $y_1, \cdots, y_n$ be arranged in

---

[1] Presented to the Institute of Mathematical Statistics at Philadelphia, December 27, 1939.

[2] Research under a grant-in-aid from the Carnegie Corporation of New York.

[3] The authors are indebted to Prof. S. S. Wilks for proposing this problem to them.

ascending order of magnitude, and let the sequence be designated by Z, thus: $Z = z_1, z_2, \cdots, z_{m+n}$, where $z_1 < z_2 < \cdots < z_{m+n}$. ($f(x)$ and $g(x)$ were assumed to be continuous. Hence the probability is 0 that $z_i = z_{i+1}$ and therefore we may exclude this case.) Let $V = v_1, v_2, \cdots, v_{m+n}$ be a sequence defined as follows: $v_i = 0$ if z_i is a member of the set $x_1, \cdots, x_m$ and $v_i = 1$ if z_i is a member of the set $y_1, \cdots, y_n$. It is easy to show that any statistic S used to test the null hypothesis should be invariant under any continuous, reciprocally one-to-one transformation of the real axis. That is to say, if $t' = \varphi(t)$ is any such transformation, then

$$(1) \quad S(x_1, \cdots, x_m, y_1, \cdots, y_n) \equiv S(\varphi(x_1), \cdots, \varphi(x_m), \varphi(y_1), \cdots, \varphi(y_n)).$$

The reason for this requirement on S is the fact that the transformed stochastic variables $X' = \varphi(X)$ and $Y' = \varphi(Y)$ are continuous and have identical distributions if and only if X and Y have identical distributions. Hence S must be a function of V only, with the added restriction that $S(V) = S(V')$, where $V' = v_{m+n}, v_{m+n-1}, \cdots, v_1$. For if S were a function of $x_1, \cdots, x_m, y_1, \cdots, y_n$ which cannot be expressed as a function of V alone, then there exists a continuous reciprocally one-to-one transformation $t' = \varphi(t)$ such that (1) is not true. On the other hand, any continuous reciprocally one-to-one transformation of the entire line into itself is monotonic and hence either leaves V invariant or else transforms it into V'.

3. Previous results. In an interesting paper on this problem W. R. Thompson [1] proceeds as follows: Let the sets $x_1, \cdots, x_m$ and $y_1, \cdots, y_n$ be ordered in ascending order of magnitude, thus: $x_{p_1}, x_{p_2}, \cdots, x_{p_m}$ and $y_{p'_1}, y_{p'_2}, \cdots, y_{p'_n}$ where $x_{p_1} < x_{p_2} < \cdots < x_{p_m}$ and $y_{p'_1} < y_{p'_2} < \cdots < y_{p'_n}$. Let $P\{x_{p_k} < y_{p'_{k'}}\}$ denote the probability of the relation in braces under the null hypothesis ($f(x) \equiv g(x)$). This probability is shown to be independent of $f(x)$ and the relation

$$(2) \qquad P\{x_{p_k} < y_{p'_{k'}}\} = \psi(m, n, k, k')$$

holds, where the right member, which is given explicitly by Thompson, is a function only of the arguments exhibited. To make a test of the null hypothesis with, say, a 5% level of significance, this writer proposes to choose k and k' so that $\psi(m, n, k, k') = .05$. The test would then consist of noticing whether $x_{p_k} < y_{p'_{k'}}$ or not. In the former case the null hypothesis is to be considered as disproved.

It is clear that this test cannot be very efficient, ignoring as it does so many of the relations among the observations. Except under certain rather narrow restrictions on the admissible alternatives, for example, that $g(x) \equiv f(x + c)$, where c is an arbitrary constant, the test suffers the further defect of not being "consistent" in a way which will be discussed below. Hence the test suggested by Thompson can scarcely be regarded as a satisfactory solution of the problem. This criticism, of course, does not apply to those sections of Thompson's paper which deal with the question of estimating the so-called normal range.

4. The statistic U. A subsequence $v_{s+1}, v_{s+2}, \cdots, v_{s+r}$ of V (where r may also be 1) will be called a "run" if $v_{s+1} = v_{s+2} = \cdots = v_{s+r}$ and if $v_s \neq v_{s+1}$ when $s > 0$ and if $v_{s+r} \neq v_{s+r+1}$ when $s + r < m + n$. For example, $V =$ 1, 0, 0, 1, 1, 0 contains the following runs: 1; 0, 0; 1, 1; 0. The statistic[4] U defined as the number of runs in V seems a suitable statistic for testing the hypothesis that $f(x) \equiv g(x)$. In the event that the latter identity holds, the distribution of U is independent of $f(x)$. A difference between $f(x)$ and $g(x)$ tends to decrease U. U is consistent in a sense which will be discussed below.

In order to derive the distribution of U under the null hypothesis, we first note that all the $\frac{(m+n)!}{m!n!}$ $(= {}^{m+n}C_m)$ possible sequences V have the same probability $\left(= \frac{m!n!}{(m+n)!}\right)$. To see this, consider the sequence V where $v_i = 0$ $(i = 1, 2, \cdots, m)$ and $v_i = 1$ $(i = m + 1, m + 2, \cdots, m + n)$. Clearly the probability of the sequence is

$$q = \frac{m(m-1)\cdots 1 \cdot n(n-1)\cdots 1}{(m+n)(m+n-1)\cdots(n+1)n(n-1)\cdots 1}.$$

Furthermore, the probability of any other sequence is equal to the product of the factors in the numerator of q taken in a different order, divided by the product of the factors in the denominator taken in the same order. The quotient is, of course, $= q$.

Let e_0 be the number of runs in V whose elements are 0 and let e_1 be the number of runs whose elements are 1. Obviously $U = e_0 + e_1$. Let the runs of each kind be arranged in the ascending order of the indices of the v_i. Let r_{0j} be the number of elements 0 in the jth run of that kind $(j = 1, 2, \cdots, e_0)$ and let $r_{1j'}$ be the number of elements 1 in the j'^{th} run of that kind $(j' = 1, 2, \cdots, e_1)$. The following relations obviously hold:

$$\sum_{j=1}^{e_0} r_{0j} = m, \tag{3}$$

$$\sum_{j'=1}^{e_1} r_{1j'} = n, \tag{4}$$

$$1 \leq e_0 \leq m, \qquad 1 \leq e_1 \leq n, \tag{5}$$

$$| e_0 - e_1 | \leq 1. \tag{6}$$

---

[4] When this paper was already in proof, our attention was called to a paper by W. L. Stevens, entitled "Distribution of groups in a sequence of alternatives," *Annals of Eugenics*, Vol. 9 (1939). There a statistic, which is essentially the U statistic, is proposed for a problem different from that considered by us and the distribution of U is obtained in a different manner. However, the application of the U statistic for the purpose herein described, the proof of consistency and the other results of our paper are not contained in it.

Hence if $U = 2k$, then $e_0 = e_1 = k$, and if $U = 2k - 1$, then either $e_0 = k$, $e_1 = k - 1$ or $e_0 = k - 1$, $e_1 = k$. The element v_1 of V together with the numbers $r_{01}, r_{02}, \cdots, r_{0e_0}, r_{11}, r_{12}, \cdots, r_{1e_1}$, completely determines the sequence V whose probability is q.

Without loss of generality we may assume that $m \leq n$. If $U = 2k$, $1 \leq k \leq m$, $v_1 = 0$, any two sequences of k positive numbers each may constitute a sequence of $r_{01}, \cdots, r_{0e_0}, r_{11}, \cdots, r_{1e_1}$ provided only that (3) and (4) are satisfied. The number of sequences $r_{01}, r_{02}, \cdots, r_{0k}$ which satisfy (3) is the coefficient of a^m in the purely formal expansion of

$$(a + a^2 + a^3 + \cdots)^k = \left(\frac{a}{1 - a}\right)^k$$

and hence is ${}^{m-1}C_{k-1}$. Similarly the number of sequences $r_{11}, r_{12}, \cdots, r_{1k}$ which satisfy (4) is found to be ${}^{n-1}C_{k-1}$. Bearing in mind the case $U = 2k$, $v_1 = 1$, we obtain

$$(7) \qquad P\{U = 2k\} = \frac{2({}^{m-1}C_{k-1} \cdot {}^{n-1}C_{k-1})}{{}^{m+n}C_m}, \qquad (k = 1, 2, \cdots, m),$$

where the left member denotes the probability of the relation in braces under the null hypothesis. In a similar manner we obtain

$$(8) \qquad P = \{U = 2k - 1\} = \frac{({}^{m-1}C_{k-1} \cdot {}^{n-1}C_{k-2} + {}^{m-1}C_{k-2} \cdot {}^{n-1}C_{k-1})}{{}^{m+n}C_m},$$

$$(k = 2, \cdots, m + 1),$$

with the proviso that ${}^aC_b = 0$ if $a < b$.

We shall now briefly indicate a method of obtaining the mean $E(U)$ and variance $\sigma^2(U)$ of U. For example, $E(U)$ may be obtained by performing several summations of the type

$$(9) \qquad \sum_{i=0}^{m-1} i \cdot {}^{m-1}C_i \cdot {}^{n-1}C_i .$$

It is easy to verify that the expression (9) is the term free of a in the purely formal expansion in a of:

$$(10) \qquad (m - 1) \cdot (1 + a)^{m-2} \cdot a \cdot \left(1 + \frac{1}{a}\right)^{n-1},$$

and hence is

$$(11) \qquad (m - 1) \cdot {}^{m+n-3}C_{n-2} .$$

The other summations required for the mean and variance can be carried out in a similar manner. We shall omit these tedious calculations. The results are:

$$E(U) = \frac{2mn}{m+n} + 1, \tag{12}$$

$$\sigma^2(U) = \frac{2mn(2mn - m - n)}{(m+n)^2(m+n-1)}. \tag{13}$$

The critical region for testing the null hypothesis on a level of significance β is given by the inequality $U < u_0$, where u_0 is a function of m and n such that $P\{U < u_0\} = \beta$.

5. The asymptotic distribution of U. Let $m/n = \alpha$, a positive constant. Then, as $m \to \infty$,

$$E(U) \sim \frac{2m}{1+\alpha},$$

$$\sigma^2(U) \sim \frac{4\alpha m}{(1+\alpha)^3}.$$

THEOREM I. *If t is any real number, the probability of the relation* $U < \frac{2m}{1+\alpha} + 2\left[\frac{\alpha m}{(1+\alpha)^3}\right]^{\frac{1}{2}} t$ *converges uniformly in t to*

$$\frac{1}{\sqrt{2\pi}} \int_{-\infty}^{t} e^{-\frac{1}{2}w^2}\, dw$$

as $m \to \infty$.

The proof of this theorem is essentially the same as the classical proof that the binomial law converges to the normal distribution (see, for example, Fréchet [2], p. 89) and it will be unnecessary to give the details. Since the asymptotic distribution of the subpopulation of even U is the same as that of odd U, it will be sufficient to consider only the right member of (7). Let $m' = m - 1$, $n' = n - 1$, and $k' = k - 1$. We make the substitution

$$w = \frac{k' - \dfrac{m'}{1+\alpha'}}{\sqrt{m'}}, \quad \text{where} \quad \alpha' = \frac{m'}{n'}, \tag{14}$$

$$dw = \frac{1}{\sqrt{m'}}, \tag{15}$$

and evaluate the factorials by Stirling's formula. We shall give here only the results of successive simplifications. At each step we shall omit the factors free of k or w, since their product may be reconstructed from the final exponential form. Thus instead of the right member of (7) we can consider the expression:

$$^{m-1}C_{k-1} \cdot {}^{n-1}C_{k-1}. \tag{16}$$

Omitting factors free of k, we get

$$\frac{1}{(k-1)!\,(m-k)!\,(k-1)!\,(n-k)!} \tag{17}$$

and by Stirling's formula, since k and m are both large:

$$\frac{1}{k'^{(2k'+1)}(m'-k')^{(m'-k'+\frac{1}{2})}(n'-k')^{(n'-k'+\frac{1}{2})}}. \tag{18}$$

Now apply (14). We obtain

$$\begin{aligned}&\left(\sqrt{m'}\,w+\frac{m'}{1+\alpha'}\right)^{-2\sqrt{m'}w-\frac{2m'}{1+\alpha'}-1}\cdot\left(-\sqrt{m'}\,w+\frac{m'\alpha'}{1+\alpha'}\right)^{\sqrt{m'}w-\frac{m'\alpha'}{1+\alpha'}-\frac{1}{2}}\\ &\qquad\cdot\left(-\sqrt{m'}\,w+\frac{m'}{\alpha'(1+\alpha')}\right)^{\sqrt{m'}w-\frac{m'}{\alpha'(1+\alpha')}-\frac{1}{2}}.\end{aligned} \tag{19}$$

Dividing inside the parentheses by $\frac{m'}{1+\alpha'}$, $\frac{m'\alpha'}{1+\alpha'}$, $\frac{m'}{\alpha'(1+\alpha')}$, respectively, and again omitting factors free of w, we get

$$\begin{aligned}&\left(1+\frac{(1+\alpha')w}{\sqrt{m'}}\right)^{-2\sqrt{m'}w-\frac{2m'}{1+\alpha'}-1}\cdot\left(1-\frac{(1+\alpha')w}{\alpha'\sqrt{m'}}\right)^{\sqrt{m'}w-\frac{m'\alpha'}{1+\alpha'}-\frac{1}{2}}\\ &\qquad\cdot\left(1-\frac{\alpha'(1+\alpha')w}{\sqrt{m'}}\right)^{\sqrt{m'}w-\frac{m'}{\alpha'(1+\alpha')}-\frac{1}{2}}.\end{aligned} \tag{20}$$

Taking logarithms, expanding in powers of $\frac{w}{\sqrt{m'}}$ and neglecting terms in $\frac{w^3}{m'^{\frac{3}{2}}}$ and higher orders, the results are

$$\begin{aligned}&-\left(2\sqrt{m'}\,w+\frac{2m'}{1+\alpha'}+1\right)\left(\frac{(1+\alpha')w}{\sqrt{m'}}-\frac{(1+\alpha')^2w^2}{2m'}\right)\\ &-\left(\sqrt{m'}\,w-\frac{m'\alpha'}{1+\alpha'}-\frac{1}{2}\right)\left(\frac{(1+\alpha')w}{\alpha'\sqrt{m'}}+\frac{(1+\alpha')^2w^2}{2\alpha'^2m'}\right)\\ &-\left(\sqrt{m'}\,w-\frac{m'}{\alpha'(1+\alpha')}-\frac{1}{2}\right)\left(\frac{\alpha'(1+\alpha')w}{\sqrt{m'}}+\frac{\alpha'^2(1+\alpha')^2w^2}{2m'}\right)\end{aligned} \tag{21}$$

which equals

$$-\frac{w^2(1+\alpha')^3}{2\alpha'}+O(m'^{-\frac{1}{2}}). \tag{22}$$

The proof of the fact that the distribution of w converges uniformly to the normal distribution with zero mean and variance $\frac{\alpha'}{(1+\alpha')^3}$ can be carried out in the same way as the classical proof that the binomial law converges to the normal distribution.

It is obvious that

$$w^* = \frac{k - \dfrac{m}{1+\alpha}}{\sqrt{m}}$$

has the same distribution as w. From this and from the fact that $U = 2k$ or $2k - 1$ THEOREM I follows.

In using conventional tables of the Gaussian function to make tests of significance on U when m and n are large, the reader is urged not to forget that the critical region of U lies in only one tail of the curve.

6. An example. We give here a simple example illustrating the use of the statistic U and THEOREM I.

Suppose 50 observations were made on X and 50 observations on Y. Suppose further that these observations are arranged in ascending order and that the ith element of this sequence is said to have the rank i. The observations on X occupy the following ranks: 1, 5, 6, 7, 12, 13, 14, 15, 16, 17, 19, 20, 21, 25, 26, 27, 28, 31, 32, 38, 42, 43, 44, 45, 50, 51, 52, 53, 54, 56, 57, 58, 62, 63, 64, 65, 68, 69, 75, 79, 80, 81, 86, 87, 89, 90, 91, 93, 94, 95.

The observations on Y occupy the remaining ranks.

In this case, $U = 34$.

For $m = n = 50$,

$$E(U) = 51,$$
$$\sigma^2(U) = 24.747.$$

The probability of getting 34 runs or less when the distribution functions of X and Y are continuous and identical is therefore less than $5 \cdot 10^{-4}$.

7. Consistency. We shall say that a test is "consistent" if the probability of rejecting the null hypothesis when it is false (i.e., the complement of the probability of a type II error, cf. Neyman and Pearson, [3]) approaches one as the sample number approaches infinity. In the literature of statistics a function of the observations which converges stochastically to a population parameter as the sample number approaches infinity, is called a "consistent" statistic. If a test of a hypothesis about a population parameter is made by a proper use of a consistent (statistic) estimate of the parameter, the test will be consistent also according to our definition, which thus furnishes an extension of the idea of consistency to the case where the alternatives to the null hypothesis cannot be specified by a finite number of parameters.

It is obvious that consistency ought to be a minimal requirement of any good test. It is the purpose of this section to prove that, subject to some slight and from the practical statistical point of view, unimportant, restrictions on the distribution functions, the test furnished by the statistic U is consistent.

We shall say that the distribution functions $f(x)$ and $g(x)$ satisfy the condition A, if, for any arbitrarily small positive δ, there exist a finite number of closed intervals, such that the probability of the sum I of these intervals is $> 1 - \delta$ according to at least one of the distribution functions $f(x)$ and $g(x)$, and such that $f(x)$ and $g(x)$ have positive continuous derivatives $f'(x)$ and $g'(x)$ in I.

In all that follows, although m and n are considered as variables, their ratio m/n is to be a constant, denoted by α. Let $\beta > 0$ denote the level of significance on which the test is to be made, so that, if $f(x) \equiv g(x)$,

$$P\{U < u_0(m)\} = \beta \tag{23}$$

where the critical region for two samples of size m and n, respectively, is given by

$$U < u_0(m).$$

THEOREM II. *If $f(x)$ and $g(x)$ satisfy condition* A, *and if*

$$f(x) \not\equiv g(x), \tag{24}$$

then

$$\operatorname*{Lim}_{m\to\infty} P\{U < u_0(m)\} = 1. \tag{25}$$

The proof of this theorem will be given in several stages.

Let $E\left(\frac{U}{m}; f; g\right)$ and $\sigma^2\left(\frac{U}{m}; f; g\right)$ denote the mean and variance, respectively, of $\frac{U}{m}$, when X and Y have the distribution functions $f(x)$ and $g(x)$, respectively, and the sample numbers are m and n. Let the set $x_1 \cdots x_m$; $y_1 \cdots y_n$ be arranged in ascending order of magnitude, thus:

$$Z = z_1, z_2, \cdots, z_{m+n}, \tag{26}$$

where $z_1 < z_2 < \cdots < z_{m+n}$. The sequence

$$V = v_1, v_2, \cdots, v_{m+n} \tag{27}$$

is defined as follows: $v_i = 0$ if z_i is a member of the set $x_1 \cdots x_m$ and $v_i = 1$ if z_i is a member of the set $y_1 \cdots y_n$.

LEMMA 1. *If the following are fulfilled*:

a)
$$\begin{aligned} f(x) &\equiv 0 & x &< 0, \\ f(x) &\equiv x & 0 &\leq x \leq 1, \\ f(x) &\equiv 1 & x &> 1. \end{aligned}$$

b)
$$\begin{aligned} g(x) &\equiv 0 & x &\leq 0, \\ g(x) &\equiv 1 & x &\geq 1. \end{aligned}$$

c) *The derivative $g'(x)$ of $g(x)$ exists, is continuous and positive everywhere in the interval $0 \leq x \leq 1$.*

d) *k is an arbitrary but fixed positive integer. For every m, $i_{1m} < i_{2m} < \cdots < i_{km}$ are a set of k positive integers subject only to the restriction that the least upper bound γ of the sequence $\frac{i_{km}}{m+n}$ is less than* 1.

Then the expected value

$$E\left(\prod_{j=1}^{k} v_{i_{jm}}\right) \text{ of } \prod_{j=1}^{k} v_{i_{jm}}$$

satisfies the inequality

$$\left| E\left(\prod_{j=1}^{k} v_{i_{jm}}\right) - \prod_{j=1}^{k} \frac{g'(a_{\lambda_{jm}})}{\alpha+g'(a_{\lambda_{jm}})} \right| < \varphi(m) \tag{28}$$

where $\lambda_{jm} = \frac{i_{jm}}{m+n}$ and $a_{\lambda_{jm}}$ $(j = 1 \cdots k)$ *is the root of*

$$ma_{\lambda_{jm}} + ng(a_{\lambda_{jm}}) = \lambda_{jm}(m+n) \tag{29}$$

and $\varphi(m)$ depends only on m and is such that

$$\operatorname*{Lim}_{m\to\infty} \varphi(m) = 0\,. \tag{30}$$

It is easy to verify that the root $a_{\lambda_{jm}}$ of (29) exists and is unique.

PROOF: It will be sufficient to show that, for any specified set of values of

$$v_{i_{1m}} \cdots v_{i_{(r-1)m}}\,, \qquad v_{i_{(r+1)m}} \cdots v_{i_{km}} \qquad (r = 1 \cdots k)$$

the conditional probability $P\{v_{i_{rm}} = 1\}$ of the relation in braces satisfies the inequality

$$\left| \frac{g'(a_{\lambda_{rm}})}{\alpha + g'(a_{\lambda_{rm}})} - P\{v_{i_{rm}} = 1\} \right| < \psi(m), \tag{31}$$

where $\psi(m)$ depends only on m and is such that

$$\operatorname*{Lim}_{m\to 0} \psi(m) = 0\,. \tag{32}$$

For each m let

$$V'_m = v'_{i_{1m}}\,, v'_{i_{2m}} \cdots v'_{i_{(r-1)m}}\,, v'_{i_{(r+1)m}} \cdots v'_{i_{km}} \tag{33}$$

be a fixed sequence whose elements are either 0 or 1. We shall consider the conditional probability $P\{v_{i_{rm}} = s\}$, $(s = 0, 1)$ of the relation in braces subject to the condition that

$$v_{i_{jm}} = v'_{i_{jm}}\,, \qquad (j = 1, 2, \cdots (r-1), (r+1), (r+2), \cdots k). \tag{34}$$

Let a and b be two numbers such that $0 < a < b < 1$, and let m^* be a non-negative integer such that $m^* \leq m$, and $m^* \leq [\gamma(m+n)]$ where $[\gamma(m+n)]$ denotes the largest integer $\leq \gamma(m+n)$. Let $Q_m(a, b, m^*)$ denote the probability that, if m^* observations are made on X and $[\gamma(m+n)] - m^*$ observations are made on Y, the following conditions will be fulfilled:

(a) the total number of observations $< a$ is exactly $i_{rm} - 1$

(b) all observations are $< b$

(c) if the $[\gamma(m + n)]$ observations are arranged in ascending order and if $v_j^* = 0$ or 1 according as the jth element is an observation on X or on Y, then

$$v^*_{i_{jm}} = v'_{i_{jm}} \qquad (j = 1, 2, \cdots, r-1), \tag{35}$$

and

$$v^*_{i_{jm}-1} = v'_{i_{jm}} \qquad (j = r+1, r+2 \cdots k). \tag{36}$$

It is easy to see that the probability P_0 of the simultaneous fulfillment of the relations (34) and of $v_{i_{rm}} = 0$ is given by

$$P_0 = \int_0^1 \int_0^b \sum_{m^*} R_m(a, b, m^*) m'(1-b)^{m'-1}(1-g(b))^{n'}\, da\, db, \tag{37}$$

where

$$R_m(a, b, m^*) = {}^mC_{m^*}\, {}^nC_{[\gamma(m+n)]-m^*} \frac{dQ_m}{db}(a, b, m^*), \tag{38}$$

$$m' = m - m^*, \tag{39}$$

and

$$n' = n - [\gamma(m+n)] + m^*. \tag{40}$$

Similarly, the probability P_1 of the simultaneous fulfillment of the relations (34) and of $v_{i_{rm}} = 1$ is given by

$$P_1 = \int_0^1 \int_0^b \sum_{m^*} R_m(a, b, m^*)\, n' g'(a)(1-b)^{m'}(1-g(b))^{n'-1}\, da\, db. \tag{41}$$

Then

$$\frac{P\{v_{i_{rm}} = 0\}}{P\{v_{i_{rm}} = 1\}} = \frac{P_0}{P_1}. \tag{42}$$

Let $n_0 = \sum_{j > [\gamma(m+n)]} v_j$ and $m_0 = m + n - [\gamma(m+n)] - n_0$. The variables $(z_{i_{rm}} - a_{\lambda_{rm}})$, $(z_{[\gamma(m+n)]} - a_\gamma)$, $\left(\frac{m_0}{n_0} - \frac{\alpha(1-a_\gamma)}{(1-g(a_\gamma))}\right)$ all converge stochastically to zero.

Let $P_0(\epsilon)$ and $P_1(\epsilon)$ denote the values of the right members of (37) and (41), respectively, if the integration is restricted to the region where $a \leq b$, $|a - a_{\lambda_{rm}}| < \epsilon$, $|b - a_\gamma| < \epsilon$ and the summation is restricted to those values of m^* for which $\left| \frac{m'}{n'} - \frac{\alpha(1 - a_\gamma)}{(1 - g(a_\gamma))} \right| < \epsilon$. Hence, because of the aforementioned stochastic convergence, for all sufficiently large m

$$| P_s(\epsilon) - P_s | < \epsilon \qquad s = 1, 2. \tag{43}$$

Since $P_s > 0$, for sufficiently large m, also

$$\left| \frac{P_0(\epsilon)}{P_1(\epsilon)} - \frac{P_0}{P_1} \right| < \epsilon . \tag{44}$$

Since $g(x)$ and $g'(x)$ are continuous in the interval [0, 1] and hence uniformly continuous, it is clear that

$$\left| \frac{P_0(\epsilon)}{P_1(\epsilon)} - \frac{\alpha}{g'(a_{\lambda_{rm}})} \right| < c\epsilon , \tag{45}$$

where c is a fixed constant independent of m. From (44) and (45) it follows easily that, for any arbitrarily small ϵ',

$$\left| \frac{P_0}{P_1} - \frac{\alpha}{g'(a_{\lambda_{rm}})} \right| < \epsilon' \tag{46}$$

for sufficiently large m.

Since $P\{v_{i_{rm}} = 1\} = \frac{P_1}{P_0 + P_1}$, the required relation (31) follows. This completes the proof of Lemma 1.

Lemma 2. *If conditions a, b, and c of Lemma 1 are satisfied, then*

$$\lim_{m \to \infty} E\left(\frac{U}{m}; f; g\right) = 2 \int_0^1 \frac{g'(x)}{\alpha + g'(x)}\, dx \tag{47}$$

and

$$\lim_{m \to \infty} \sigma^2 \left(\frac{U}{m}; f; g\right) = 0. \tag{48}$$

Proof: Since

$$\begin{aligned} \frac{U}{m} &= \frac{1}{m} + \frac{1}{m} \sum_{j=2}^{m+n} (v_j - v_{j-1})^2 \\ &= \frac{1 + v_1 + v_{m+n}}{m} + \frac{2}{m} \sum_{j=2}^{m+n-1} v_j - \frac{2}{m} \sum_{j=2}^{m+n} v_{j-1} v_j , \end{aligned} \tag{49}$$

we have from Lemma 1,

$$\begin{aligned} E\left(\frac{U}{m}\right) &= \frac{2}{m} \left[\sum_j \frac{g'(a_{jm})}{\alpha + g'(a_{jm})} - \sum_j \left(\frac{g'(a_{jm})}{\alpha + g'(a_{jm})} \right)^2 \right] + \eta(m) + \eta^*(\gamma) \\ &= \frac{2}{m} \sum \left[\frac{\alpha g'(a_{jm})}{(\alpha + g'(a_{jm}))^2} \right] + \eta(m) + \eta^*(\gamma) , \end{aligned} \tag{50}$$

where

(51) $$\lim_{m\to\infty} \eta(m) = \lim_{\gamma\to 1} \eta^*(\gamma) = 0$$

and a_{jm} is the root of the equation

(52) $$ma_{jm} + ng(a_{jm}) = j \qquad (j = 2 \cdots m + n).$$

From equation (52) it follows that

(53) $$\lim_{m\to\infty} (a_{jm} - a_{(j-1)m})(m + ng'(a_{jm})) = 1$$

uniformly in j. Since γ may be chosen arbitrarily near to 1, the required result (47) follows easily from (50).

It remains to consider the variance of $\frac{U}{m}$. The expression

$$\frac{1 + v_1 + v_{m+n}}{m} + \frac{2}{m} \sum_{j=2}^{m+n-1} v_j$$

differs from $\frac{2}{\alpha}$ by at most $\frac{1}{m}$, so that its variance converges to zero with $m \to \infty$. In order to prove (48), it will be sufficient to show that the variance of

(54) $$W = \frac{1}{m} \sum_{j=2}^{m+n} v_{j-1} v_j$$

goes to zero with increasing m. From LEMMA 1 it follows that

(55) $$-z(m) < [E(v_i v_j v_k v_e) - E(v_i v_j)E(v_k v_e)] < z(m),$$

where $\lim_{m\to\infty} |z(m)| = 0$, provided only that the integers i, j, k, l are distinct and $< \gamma(m + n)$. The variance of mW is the sum of terms of the type occurring in (55). The number of terms for which i, j, k, l are distinct is of the order m^2. All other terms are of size at most 2 and their number is of the order m. Since the number γ may be chosen arbitrarily near to 1, the variance of W converges to zero with $m \to \infty$.

This proves LEMMA 2.

LEMMA 3. *If conditions a, b, and c of Lemma 1 are fulfilled, and if (24) holds, then*

(56) $$T = \int_0^1 \frac{g'(x)}{\alpha + g'(x)} dx < \frac{1}{1 + \alpha}.$$

Let $a_1 < a_3$ be any two real numbers and designate $\frac{a_1 + a_3}{2}$ by a_2. Let $F(x)$ be defined as follows:

(57) $$\begin{aligned} &F(a_1) = 0, \\ &F(x) = (x - a_i)b_i + F(a_i), \qquad (a_i \le x \le a_{i+1}\,;\, i = 1, 2). \end{aligned}$$

Let c be defined by

(58) $$F(a_3) = c(a_3 - a_1).$$

Then it is easy to verify that the maximum of

(59) $$T^* = \int_{a_1}^{a_3} \frac{F'(x)}{\alpha + F'(x)} dx$$

with respect to b_1 and b_2, subject to the restrictions that b_1 and b_2 be non-negative, and that a_1, a_3 and c be fixed ($c > 0$), occurs when and only when

(60) $$b_1 = b_2 = c.$$

Now define

(61) $$P_{ij} = \frac{i}{2^j}, \qquad P_{0j} = 0,$$
$$l_{ij} = \frac{g(P_{ij}) - g(P_{(i-1)j})}{2^j}$$

and

$$S_j = \frac{1}{2^j} \sum_{i=1}^{2^j} \frac{l_{ij}}{\alpha + l_{ij}}, \qquad (i = 1, 2, \cdots 2^j; j = 0, 1, 2 \cdots).$$

Repeated application of the result of the previous paragraph easily gives

(62) $$S_j \geq S_{j+1}.$$

From (24) it follows that there exists a positive integer j' such that $S_{j'} > S_{j'+1}$. Obviously

(63) $$S_0 = \frac{1}{1 + \alpha}$$

and

(64) $$\lim_{j \to \infty} S_j = T.$$

Hence LEMMA 3 is proved.

Proof of Theorem II: Let $\delta_1 > \delta_2 > \cdots > \delta_j > \cdots$ be an arbitrary but fixed sequence such that $\lim \delta_j = 0$. For $\delta = \delta_j$, let $I_1, \cdots, I_{k(j)}$ be a set of closed intervals such that no two intervals have an interior point in common and within which, by condition (A), $f'(x)$ and $g'(x)$ exist, are positive, and continuous. Let I_{0j} be the complementary set (with respect to the whole line). (It is easy to see that, if condition (A) is fulfilled, such a system can be constructed.) Let $U_i(i = 1, 2 \cdots k(j)$ and U_{0j} denote, respectively, the runs caused by the observations which fall in the intervals I_i, I_{0j}. Then

(65) $$\left| U - \sum_{i=1}^{k(j)} U_i - U_{0j} \right| \leq 2(k(j)).$$

From condition (A) it follows that, with a probability arbitrarily close to 1, for sufficiently large m,

$$U_{0j} < 3pm\delta_j, \tag{66}$$

where

$$p = \max\left[1, \frac{1}{\alpha}\right], \qquad (j = 1, 2 \cdots).$$

Let $[a_i \leq x < b_i]$, $i = 1, 2 \cdots$ denote the interval I_i, and let m_i and n_i denote the number of observations on X and Y, respectively, which fall in the interval I_i. Then $\frac{m_i}{m}$ and $\frac{n_i}{n}$ converge stochastically with increasing m to $[f(b_i) - f(a_i)]$ and $[g(b_i) - g(a_i)]$, respectively.

Within the interval $I_i (i = 1, 2 \cdots k)$ we perform the transformation

$$X^* = f(X), \qquad Y^* = f(Y), \tag{67}$$

which leaves U_i invariant. For fixed m_i, n_i the relative distribution of X^* is uniform and the relative distribution of Y^* fulfills condition (c) of LEMMA 1. Hence from LEMMA 2 we obtain that $\frac{U_i}{m}$ converges stochastically to

$$\operatorname*{Lim}_{m\to\infty} E\left(\frac{U_i}{m}; f; g\right) \leq \frac{2[f(b_i) - f(a_i)][g(b_i) - g(a_i)]}{[g(b_i) - g(a_i)] + \alpha[f(b_i) - f(a_i)]}. \tag{68}$$

It can be verified that the sum of the second members in (68) over all values i is less than or equal to $\frac{2}{1 + \alpha}$.

From (24) and condition (A) we get that, for sufficiently small δ_j, there exists at least one interval for which the first member of (68) is less than the second member. Hence

$$\Sigma < \frac{2}{1 + \alpha}, \tag{69}$$

where

$$\Sigma = \sum_{i=1}^{\infty} \operatorname*{Lim}_{m\to\infty} E\left(\frac{U_i}{m}; f; g\right). \tag{70}$$

Now take j so large that

$$3p\delta_j < \epsilon, \tag{71}$$

where

$$0 < 3\epsilon < \frac{2}{1 + \alpha} - \Sigma. \tag{72}$$

Since $\frac{U_i}{m}$ converges stochastically to its expected value, from (65), (66), (70), (71), and (72), it follows that, with a probability arbitrarily close to 1, for sufficiently large m,

$$\frac{U}{m} < \frac{2}{1 + \alpha} - \epsilon. \tag{73}$$

From (23) and THEOREM I we get

$$\lim_{m \to \infty} \frac{u_0(m)}{m} = \frac{2}{1 + \alpha}. \tag{74}$$

THEOREM II follows easily from (73) and (74).

8. Remarks on a proposed test. We have already remarked in Section 3 that the test proposed by W. R. Thompson is not consistent. To show this, we shall give two distribution functions $f(x)$ and $g(x)$ such that, although these functions will be very different, the probability of rejecting the hypothesis that they are the same will not approach one as the sample number approaches infinity.

Suppose, to simplify the notation, that the observations have been ordered according to size, i.e., that $x_1 < x_2 < \cdots < x_m$ and $y_1 < y_2 < \cdots < y_n$. Suppose further than $m = n$, and that the test is to be made on a level of significance $\beta > 0$. In the right member of (2) we need not exhibit n and shall replace k and k' by $k(m)$ and $k'(m)$ to show the dependence on m. We have, under the null hypothesis,

$$P\{x_{k(m)} < y_{k'(m)}\} = \psi(m, k(m), k'(m)) = \beta. \tag{75}$$

The sequence $\frac{k(m)}{m}$ is bounded, so that there exists a monotonically increasing subsequence $m_1, m_2 \cdots$ of the sequence of integers $1, 2 \cdots$ and a number h, $0 \le h \le 1$, such that

$$\lim_{i \to \infty} \frac{k(m_i)}{m_i} = h. \tag{76}$$

It is easy to see that then also

$$\lim_{i \to \infty} \frac{k'(m_i)}{m_i} = h. \tag{77}$$

We shall now assume that $0 < h < 1$. If $h = 0$ or 1 only a trivial alteration will be needed in the argument to follow. Let ϵ and δ be arbitrarily small positive numbers. We now consider two populations, A and B described as follows:

A) $$f(x) \equiv g(x) \equiv x \qquad (0 \le x \le 1),$$

B) $$f(x) \equiv x \qquad (0 \le x \le 1),$$

$$g(x) \equiv g(a_i) + \frac{(x - a_i)(g(a_{i+1}) - g(a_i))}{(a_{i+1} - a_i)} \qquad (a_i \le x \le a_{i+1}; i = 0, 1, \cdots, 4),$$

where

$$
\begin{aligned}
a_0 &= 0 & g(a_0) &= 0 \\
a_1 &= h - 2\delta > 0 & g(a_1) &= 0 \\
a_2 &= h - \delta & g(a_2) &= a_2 \\
a_3 &= h + \delta < 1 - \delta & g(a_3) &= a_3 \\
a_4 &= 1 - \delta & g(a_4) &= a_3 \\
a_5 &= 1 & g(a_5) &= 1
\end{aligned}
$$

The definition of $f(x)$ and $g(x)$ outside the interval $0 \leq x \leq 1$ is obvious. It will be shown that even for such different populations as A and B and for samples of size greater than that of any arbitrarily assigned number, the probability of rejecting the null hypothesis if B is true will be at most $\beta + \epsilon$.

Let h_1, h_2, h_3 denote the number of observations on X which fall in the intervals $0 < x \leq a_2$, $a_2 < x \leq a_3$, $a_3 < x \leq 1$, respectively (m fixed, of course). Let h'_1, h'_2, h'_3 be the corresponding numbers for Y. For a fixed m, the probability of a set h_1, h_2, h_3, h'_1, h'_2, h'_3 is the same whether the sample be drawn from the population A or B. From (76), (77), and multinomial law it follows that for all sufficiently large m_i the probability is at least $1 - \epsilon$ of the occurrence of a set h_1, h_2, h_3, h'_1, h'_2, h'_3 for which $x_{k(m_i)}$ and $y_{k'(m_i)}$ will both fall in the interval $a_2 < x \leq a_3$. Furthermore it is obvious that for all samples with fixed h_2, h'_2 the distribution within the interval $a_2 < x \leq a_3$ is the same whether the sample came from the population A or B. Hence even when the sample is drawn from the population B, the first member of (75) is $< \beta + \epsilon$. This completes the proof of the inconsistency of the test based on (75).

This test is consistent if the alternatives to the null hypothesis are limited, for example, to those where $g(x) \equiv f(x + c)$, c a constant.

REFERENCES

[1] William R. Thompson, *Annals of Math. Stat.*, Vol. 9, (1938), p. 281.
[2] Maurice Fréchet, *Généralités sur les Probabilités. Variables aléatoires*, Paris, (1937).
[3] J. Neyman and E. S. Pearson. *Statistical Research Memoirs.* University College, London. Vol. 1, (1936).

Columbia University,
New York, N. Y.

Reprinted from THE ANNALS OF MATHEMATICAL STATISTICS
Vol. XI, No. 3, September, 1940

THE FITTING OF STRAIGHT LINES IF BOTH VARIABLES ARE SUBJECT TO ERROR

BY ABRAHAM WALD

1. Introduction. The problem of fitting straight lines if both variables x and y are subject to error, has been treated by many authors. If we have $N > 2$ observed points (x_i, y_i) $(i = 1, \cdots, N)$, the usually employed method of least squares for determining the coefficients a, b, of the straight line $y = ax + b$ is that of choosing values of a and b which minimize the sum of the squares of the residuals of the y's, i.e. $\Sigma(ax_i + b - y_i)^2$ is a minimum. It is well known that treating y as an independent variable and minimizing the sum of the squares of the residuals of the x's, we get a different straight line as best fit. It has been pointed out[1] that if both variables are subject to error there is no reason to prefer one of the regression lines described above to the other. For obtaining the "best fit," which is not necessarily equal to one of the two lines mentioned, new criteria have to be found. This problem was treated by R. J. Adcock as early as 1877.[2]

He defines the line of best fit as the one for which the sum of the squares of the normal deviates of the N observed points from the line becomes a minimum. (Another early attempt to solve this problem by minimizing the sum of squares of the normal deviates was made by Karl Pearson.[3])

Many objections can be raised against this method. First, there is no justification for minimizing the sum of the squares of the *normal* deviates, and not the deviations in some other direction. Second, the straight line obtained by that method is not invariant under transformation of the coordinate system. It is clear that a satisfactory method should give results which do not depend on the choice of a particular coordinate system. This point has been emphasized by C. F. Roos. He gives[4] a good summary of the different methods and then proposes a general formula for fitting lines (and planes in case of more than two variables) which do not depend on the choice of the coordinate system.

---

[1] See for instance Henry Schultz' "The Statistical Law of Demand," *Jour. of Political Economy*, Vol. 33, Dec. (1925).

[2] *Analyst*, Vol. IV, p. 183 and Vol. V, p. 53.

[3] "On Lines and Planes of Closest Fit to Systems of Points in Space" *Phil. Mag.* 6th Ser. Vol. II (1901).

[4] "A General Invariant Criterion of Fit for Lines and Planes where all Variates are Subject to Error," *Metron*, February 1937. See also Oppenheim and Roos *Bulletin of the American Mathematical Society*, Vol. 34 (1928), pp. 140 141.

Roos' formula includes many previous solutions[5] as special cases. H. E. Jones[6] gives an interesting geometric interpretation of Roos' general formula.

It is a common feature of Roos' general formula and of all other methods proposed in recent years that the fitted straight line cannot be determined without *a priori* assumptions (independent of the observations) regarding the weights of the errors in the variables x and y. That is to say, either the standard deviations of the errors in x and in y are involved (or at least their ratio is included) in the formula of the fitted straight line and there is no method given by which those standard deviations can be estimated by means of the observed values of x and y.

R. Frisch[7] has developed a new general theory of linear regression analysis, when all variables are subject to error. His very interesting theory employs quite new methods and is not based on probability concepts. Also on the basis of Frisch's discussion it seems that there is no way of determining the "true" regression without *a priori* assumptions about the disturbing intensities.

T. Koopmans[8] combined Frisch's regression theory with the classical one in a new general theory based on probability concepts. Also, according to his theory, the regression line can be determined only if the ratio of the standard deviations of the errors is known.

In a recent paper R. G. D. Allen[9] gives a new interesting method for determining the fitted straight line in case of two variables x and y. Denoting by σ_ϵ the standard deviation of the errors in x, by σ_η the standard deviation of the errors in y and by ρ the correlation coefficient between the errors in the two variables, Allen emphasizes (p. 194)[9] that the fitted line can be determined only if the *values of two* of the three quantities σ_ϵ, σ_η, ρ are given *a priori*.

Finally I should like to mention a paper by C. Eisenhart,[10] which contains many interesting remarks related to the subject treated here.

In the present paper I shall deal with the case of two variables x and y in which the errors are uncorrelated. It will be shown that under certain conditions:

(1) The fitted straight line can be determined without making *a priori* assumptions (independent of the observed values x and y) regarding the standard deviations of the errors.

(2) The standard deviation of the errors can be well estimated by means of

[5] For instance also Corrado Gini's method described in his paper, "Sull' Interpolazione di una Retta Quando i Valori della Variable Independente sono Affecti da Errori Accidentalis," *Metron*, Vol. I, No. 3 (1921), pp. 63–82.

[6] "Some Geometrical Considerations in the General Theory of Fitting Lines and Planes," *Metron*, February 1937.

[7] *Statistical Confluence Analysis by Means of Complete Regression Systems*, Oslo, 1934.

[8] *Linear Regression Analysis of Economic Time Series*, Haarlem, 1937.

[9] "The Assumptions of Linear Regression," *Economica*, May 1939.

[10] "The interpretation of certain regression methods and their use in biological and industrial research," *Annals of Math. Stat.*, Vol. 10 (1939), pp. 162–186.

the observed values of x and y. The precision of the estimate increases with the number of the observations and would give the exact values if the number of observations were infinite. (See in this connection also condition V in section 3.)

2. Formulation of the Problem. Let us begin with a precise formulation of the problem. We consider two sets of random variables[11]

$$x_1, \cdots, x_N; \qquad y_1, \cdots, y_N.$$

Denote the expected value $E(x_i)$ of x_i by X_i and the expected value $E(y_i)$ of y_i by Y_i $(i = 1, \cdots, N)$. We shall call X_i the true value of x_i, Y_i the true value of y_i, $x_i - X_i = \epsilon_i$ the error in the i-th term of the x-set, and $y_i - Y_i = \eta_i$ the error in the i-th term of the y-set.

The following assumptions will be made:

I. *The random variables* $\epsilon_1, \cdots, \epsilon_N$ *each have the same distribution and they are uncorrelated, i.e.* $E(\epsilon_i \epsilon_j) = 0$ *for* $i \neq j$. *The variance of* ϵ_i *is finite.*

II. *The random variables* $\eta_1, \cdots, \eta_N$ *each have the same distribution and are uncorrelated, i.e.* $E(\eta_i \eta_j) = 0$ *for* $i \neq j$. *The variance of* η_i *is finite.*

III. *The random variables* ϵ_i *and* η_j $(i = 1, \cdots, N; j = 1, \cdots, N)$ *are uncorrelated, i.e.* $E(\epsilon_i \eta_j) = 0$.

IV. *A single linear relation holds between the true values* X *and* Y, *that is to say* $Y_i = \alpha X_i + \beta$ $(i = 1, \cdots, N)$.

Denote by ϵ a random variable having the same probability distribution as possessed by each of the random variables $\epsilon_1, \cdots, \epsilon_N$, and by η a random variable having the same distribution as $\eta_1, \cdots, \eta_N$.

The problem to be solved can be formulated as follows:

We know only two sets of observations: $x_1', \cdots, x_N'; y_1', \cdots, y_N'$, where x_i' denotes the observed value of x_i and y_i' denotes the observed value of y_i. We know neither the true values $X_1, \cdots, X_N; Y_1, \cdots, Y_N$, nor the coefficients α and β of the linear relation between them. We have to estimate by means of the observations $x_1', \cdots, x_N'; y_1', \cdots, y_N'$, (1) the values of α and β, (2) the standard deviation σ_ϵ of ϵ, and (3) the standard deviation σ_η of η.

Problems of this kind occur often in Economics, where we are dealing with time series. For example, denote by x_i the price of a certain good G in the period t_i, and by y_i the quantity of G demanded in t_i. In each time period t_i there exists a normal price X_i and a normal demand Y_i which would obtain if the influence of some accidental disturbances could be eliminated. If we have reason to assume that there exists between the normal price and the normal demand a linear relationship we have to deal with a problem of the kind de scribed above.

In the following discussions we shall use the notations x_i and y_i also for their

[11] A random or stochastic variable is a real variable associated with a probability distribution.

observed values x_i' and y_i' since it will be clear in which sense they are meant and no confusion can arise.

3. Consistent Estimates of the Parameters α, β, σ_ϵ, σ_η. For the sake of simplicity we assume that N is even. We consider the expression

$$(1)\qquad \begin{aligned} a_1 &= \frac{(x_1 + \cdots + x_m) - (x_{m+1} + \cdots + x_N)}{N}; \\ a_2 &= \frac{(y_1 + \cdots + y_m) - (y_{m+1} + \cdots + y_N)}{N}, \end{aligned}$$

where $m = N/2$. As an estimate of α we shall use the expression

$$(2)\qquad a = \frac{a_2}{a_1} = \frac{(y_1 + \cdots + y_m) - (y_{m+1} + \cdots + y_N)}{(x_1 + \cdots + x_m) - (x_{m+1} + \cdots + x_N)}.$$

We make the assumption

V. *The limit inferior of*

$$\left| \frac{(X_1 + \cdots + X_m) - (X_{m+1} + \cdots + X_N)}{N} \right| \qquad (N = 2, 3, \cdots \text{ad. inf.}$$

is positive.

We shall prove that a is a consistent estimate of α, i.e. a converges stochastically to α with $N \to \infty$, if the assumptions I–V hold. Denote the expected value of a_1 by $\bar{a}_1$ and the expected value of a_2 by $\bar{a}_2$. It is obvious that

$$(3)\qquad \begin{aligned} \bar{a}_1 &= \frac{(X_1 + \cdots + X_m) - (X_{m+1} + \cdots + X_N)}{N}, \\ \bar{a}_2 &= \frac{(Y_1 + \cdots + Y_m) - (Y_{m+1} + \cdots + Y_N)}{N}. \end{aligned}$$

On account of the condition IV we have

$$(4)\qquad \bar{a}_2 = \alpha \bar{a}_1, \quad \text{or} \quad \frac{\bar{a}_2}{\bar{a}_1} = \alpha.$$

The variance of $a_1 - \bar{a}_1$ is equal to σ_ϵ^2/N and the variance of $a_2 - \bar{a}_2$ is equal to σ_η^2/N. Hence a_1 and a_2 converge stochastically towards $\bar{a}_1$ and $\bar{a}_2$ respectively. From that and assumption V it follows that also $\frac{a_2}{a_1}$ converges stochastically towards $\frac{\bar{a}_2}{\bar{a}_1} = \alpha$. The intercept β of the regression line will be estimated by

$$(5)\qquad b = \bar{y} - a\bar{x}, \qquad \text{where } \bar{x} = \frac{x_1 + \cdots + x_N}{N} \quad \text{and} \quad \bar{y} = \frac{y_1 + \cdots + y_N}{N}.$$

Denote by $\bar{X}$ the arithmetic mean of $X_1, \cdots, X_N$ and by $\bar{Y}$ the arithmetic mean of $Y_1, \cdots, Y_N$. Since $\bar{y}$ converges stochastically towards $\bar{Y}$, $\bar{x}$ towards

$\bar{X}$, and a towards α, b converges stochastically towards $\bar{Y} - \alpha\bar{X}$. From condition IV it follows that $\bar{Y} - \alpha\bar{X} = \beta$. Hence b converges stochastically towards β.

Let us introduce the following notations:

$$s_x = \sqrt{\Sigma \frac{(x_i - \bar{x})^2}{N}} = \text{sample standard deviation of the } x\text{-observations,}$$

$$s_y = \sqrt{\Sigma \frac{(y_i - \bar{y})^2}{N}} = \text{sample standard deviation of the } y\text{-observations,}$$

$$s_{xy} = \Sigma \frac{(x_i - \bar{x})(y_i - \bar{y})}{N} = \text{sample covariance between the } x\text{-set and } y\text{-set.}$$

s_X, s_Y and s_{XY} denote the same expressions of the true values $X_1, \cdots, X_N$; $Y_1, \cdots, Y_N$.

It is obvious that

$$E(s_x^2) = s_X^2 + \sigma_\epsilon^2 \frac{N-1}{N}, \tag{6}$$

$$E(s_y^2) = s_Y^2 + \sigma_\eta^2 \frac{N-1}{N}, \tag{7}$$

$$E(s_{xy}) = s_{XY}, \tag{8}$$

where $E(s_x^2)$, $E(s_y^2)$, and $E(s_{xy})$ denote the expected values of s_x^2, s_y^2, and s_{xy}.[12]

Since $Y_i = \alpha X_i + \beta$, we have

$$s_Y = \alpha s_X, \tag{9}$$

$$s_{XY} = \alpha s_X^2. \tag{10}$$

From (8), (9) and (10) we get

$$s_X^2 = \frac{E(s_{xy})}{\alpha}, \tag{11}$$

$$s_Y^2 = \alpha E(s_{xy}). \tag{12}$$

If we substitute in (6) and (7) for s_X^2 and s_Y^2 their values in (11) and (12), we get

$$\sigma_\epsilon^2 = \left[E(s_x^2) - \frac{E(s_{xy})}{\alpha}\right] N/(N-1), \tag{13}$$

$$\sigma_\eta^2 = [E(s_y^2) - \alpha E(s_{xy})] N/(N-1). \tag{14}$$

---

[12] I observe that the equations (6), (7) and (8) are essentially the same as those investigated by R. Frisch, *Statistical Confluence Analysis* pp. 51–52. See also Allen's equations (4) l.c. p. 194.

Since s_x^2, s_y^2, s_{xy} converge stochastically towards their expected values and a converges stochastically towards α, the expressions

$$\left[s_x^2 - \frac{s_{xy}}{a}\right] N/(N-1) \tag{15}$$

and

$$[s_y^2 - a s_{xy}] N/(N-1) \tag{16}$$

are consistent estimates of σ_ϵ^2 and σ_η^2 respectively.

4. Confidence Interval for α. In this section, as well as in sections 5 and 6, only the assumptions I–IV are assumed to hold. In other words, all statements made in these sections are valid independently of Assumption V, except where the contrary is explicitly stated.

Let us introduce the following notation:

$$\bar{x}_1 = \frac{x_1 + \cdots + x_m}{m}; \qquad \bar{y}_1 = \frac{y_1 + \cdots + y_m}{m}$$

$$\bar{x}_2 = \frac{x_{m+1} + \cdots + x_N}{m}; \qquad \bar{y}_2 = \frac{y_{m+1} + \cdots + y_N}{m}$$

$$(s_x')^2 = \frac{\sum_{i=1}^{m} (x_i - \bar{x}_1)^2 + \sum_{j=m+1}^{N} (x_j - \bar{x}_2)^2}{N}$$

$$(s_y')^2 = \frac{\sum_{i=1}^{m} (y_i - \bar{y}_1)^2 + \sum_{j=m+1}^{N} (y_j - \bar{y}_2)^2}{N}$$

$$s_{xy}' = \frac{\sum_{i=1}^{m} (x_i - \bar{x}_1)(y_i - \bar{y}_1) + \sum_{j=m+1}^{N} (x_j - \bar{x}_2)(y_j - \bar{y}_2)}{N}.$$

$\bar{X}_1$, $\bar{X}_2$, $\bar{Y}_1$, $\bar{Y}_2$, $(s_X')^2$, $(s_Y')^2$ and s_{XY}' denote the same functions of the true values $X_1, \cdots, X_N$, $Y_1, \cdots, Y_N$. The expressions s_x', s_y', and s_{xy}' are slightly different from the corresponding expressions s_x, s_y, and s_{xy}. The reason for introducing these new expressions is that the distributions of s_x, s_y, and s_{xy} are not independent of the slope $a = \frac{a_2}{a_1}$ of the sample regression line, but s_x', s_y' and s_{xy}' are distributed independently from a (assuming that ϵ and η are normally distributed). The latter statement follows easily from the fact that according to (1) and (2) $a = \frac{\bar{y}_1 - \bar{y}_2}{\bar{x}_1 - \bar{x}_2}$ and s_x', s_y', s_{xy}' are distributed independently of $\bar{x}_1$, $\bar{x}_2$, $\bar{y}_1$ and $\bar{y}_2$.

In the same way as we derived (13) and (14), we get

$$\sigma_\epsilon^2 = \left[E(s_x')^2 - \frac{E(s_{xy}')}{\alpha}\right]N/(N-2), \tag{13'}$$

$$\sigma_\eta^2 = [E(s_y')^2 - \alpha E(s_{xy}')]N/(N-2). \tag{14'}$$

These formulae differ from the corresponding formulae (13) and (14) only in the denominator of the second factor, having there $N - 2$ instead of $N - 1$. This is due to the fact that the estimates s_x, s_y, s_{xy} are based on $N - 1$ degrees of freedom whereas s_x', s_y' and s_{xy}' are based only on $N - 2$ degrees of freedom. From (13′) and (14′) we get the following estimates[13] for σ_ϵ^2 and σ_η^2:

$$\left[(s_x')^2 - \frac{s_{xy}'}{\alpha}\right]N/(N-2), \tag{17}$$

$$[(s_y')^2 - \alpha s_{xy}']N/(N-2). \tag{18}$$

Hence we get as an estimate of $\sigma_\eta^2 + \alpha^2\sigma_\epsilon^2$ the expression:

$$\begin{aligned} s^2 &= [(s_y')^2 + \alpha^2(s_x')^2 - 2\alpha s_{xy}']N/(N-2) \\ &= \frac{N}{N-2}\left\{\frac{\sum_{i=1}^{m}[(y_i - \alpha x_i) - (\bar{y}_1 - \alpha\bar{x}_1)]^2 + \sum_{j=m+1}^{N}[(y_j - \alpha x_j) - (\bar{y}_2 - \alpha\bar{x}_2)]^2}{N}\right\}. \end{aligned} \tag{19}$$

Now we shall show that

$$\frac{(N-2)s^2}{\sigma_\eta^2 + \alpha^2\sigma_\epsilon^2} \tag{20}$$

has the χ^2-distribution with $N - 2$ degrees of freedom, provided that ϵ and η are normally distributed. In fact,

$$(y_i - \alpha x_i) - (\bar{y}_1 - \alpha\bar{x}_1) = \eta_i - \alpha\epsilon_i - (\bar{\eta}_1 - \alpha\bar{\epsilon}_1) \qquad (i = 1, \cdots, m)$$

and

$$(y_j - \alpha x_j) - (\bar{y}_2 - \alpha\bar{x}_2) = \eta_j - \alpha\epsilon_j - (\bar{\eta}_2 - \alpha\bar{\epsilon}_2) \quad (j = m+1, \cdots, N),$$

where

$$\bar{\epsilon}_1 = \frac{\epsilon_1 + \cdots + \epsilon_m}{m}, \qquad \bar{\epsilon}_2 = \frac{\epsilon_{m+1} + \cdots + \epsilon_N}{m},$$

$$\bar{\eta}_1 = \frac{\eta_1 + \cdots + \eta_m}{m}, \qquad \bar{\eta}_2 = \frac{\eta_{m+1} + \cdots + \eta_N}{m}.$$

Since the variance of $\eta_k - \alpha\epsilon_k$ is equal to $\sigma_\eta^2 + \alpha^2\sigma_\epsilon^2$ and since $\eta_k - \alpha\epsilon_k$ is uncorrelated with $\eta_l - \alpha\epsilon_l$ $(k \neq l)$ $(k, l = 1, \cdots, N)$, the expression (20) has the χ^2-distribution with $N - 2$ degrees of freedom.

---

[13] An "estimate" is usually a function of the observations not involving any unknown parameters. We designate here as estimates also some functions involving the parameter α.

Now we shall show that

$$\text{(21)} \qquad \frac{\sqrt{N}\, a_1(a - \alpha)}{\sqrt{\sigma_\eta^2 + \alpha^2 \sigma_\epsilon^2}}$$

is normally distributed with zero mean and unit variance. In fact from the equations (1)–(4) it follows that

$$a_1(a - \alpha) = \bar{a}_2 + \frac{\bar{\eta}_1 - \bar{\eta}_2}{2} - a_1\left(\frac{\bar{a}_2}{\bar{a}_1}\right)$$

$$= \bar{a}_2 + \frac{\bar{\eta}_1 - \bar{\eta}_2}{2} - \left(\bar{a}_1 + \frac{\bar{\epsilon}_1 - \bar{\epsilon}_2}{2}\right)\left(\frac{\bar{a}_2}{\bar{a}_1}\right)$$

$$= \frac{\bar{\eta}_1 - \bar{\eta}_2}{2} - \alpha\,\frac{\bar{\epsilon}_1 - \bar{\epsilon}_2}{2}.$$

Since the latter expression is normally distributed (provided that ϵ and η are normally distributed) with zero mean and variance $\frac{\sigma_\eta^2 + \alpha^2\sigma_\epsilon^2}{N}$, our statement about (21) is proved.

Obviously (20) and (21) are independently distributed, hence $\sqrt{N - 2}$ times the ratio of (21) to the square root of (20), namely,

$$\text{(22)} \qquad t = \sqrt{N-2}\,\frac{\sqrt{N}\, a_1(a-\alpha)}{\sqrt{N-2}\; s} = \frac{a_1(a-\alpha)\sqrt{N-2}}{\sqrt{(s_y')^2 + \alpha^2(s_x')^2 - 2\alpha s_{xy}'}}$$

has the Student distribution with $N - 2$ degrees of freedom. Denote by t_0 the critical value of t corresponding to a chosen probability level. The deviation of a from an assumed population value α is significant if

$$\left|\frac{a_1(a-\alpha)\sqrt{N-2}}{\sqrt{(s_y')^2 + \alpha^2(s_x')^2 - 2\alpha s_{xy}'}}\right| \geqq t_0.$$

The confidence interval for α can be obtained by solving the equation in α,

$$\text{(23)} \qquad a_1^2(a-\alpha)^2 = [(s_y')^2 + \alpha^2(s_x')^2 - 2\alpha s_{xy}']\,\frac{t_0^2}{N-2}.$$

Now we shall show that if the relation

$$\text{(24)} \qquad a_1^2 > \frac{(s_x')^2 t_0^2}{N-2},$$

holds, the roots α_1 and α_2 are real and a is contained in the interior of the interval $[\alpha_1\alpha_2]$. From (19) it follows that

$$(s_y')^2 + \alpha^2(s_x')^2 - 2\alpha s_{xy}' > 0$$

for all values of α. Hence, for $\alpha = a$ the left hand side of (23) is smaller than the right hand side. On account of (24) there exists a value $a' > a$ and a value $a'' < a$ such that the left hand side of (23) is greater than the right hand side for $\alpha = a'$ and $\alpha = a''$. Hence one root must lie between a and a' and the other root between a'' and a. This proves our statement. The relation (24) always holds for sufficiently large N if Assumption V is fulfilled. The confidence interval of α is the interval $[\alpha_1, \alpha_2]$. For very small N (24) may not hold.

Finally I should like to remark that no essentially better estimate of the variance of $\eta - \alpha\epsilon$ can be given than the expression s^2 in (19). In fact, we have $2N$ observations $x_1, \cdots, x_N; y_1, \cdots, y_N$. For the estimation of the variance of $\eta - \alpha\epsilon$ we must eliminate the unknowns $X_1, \cdots, X_N$ and β. (The unknowns $Y_1, \cdots, Y_N$ are determined by the relations $Y_i = \alpha X_i + \beta$ and α is involved in the expression whose variance is to be determined.) Hence we have at most $N - 1$ degrees of freedom and the estimate in (19) is based on $N - 2$ degrees of freedom.

5. Confidence Interval for β if α is Given. In this case the best estimate of β is given by the expression:

$$b_\alpha = \bar{y} - \alpha\bar{x} \text{ where } \bar{x} = \frac{x_1 + \cdots + x_N}{N} \text{ and } \bar{y} = \frac{y_1 + \cdots + y_N}{N}.$$

We have

$$b_\alpha - \beta = (\bar{y} - \bar{Y}) - \alpha(\bar{x} - \bar{X}) = \bar{\eta} - \alpha\bar{\epsilon}$$

where

$$\bar{\epsilon} = \frac{\epsilon_1 + \cdots + \epsilon_N}{N}, \text{ and } \bar{\eta} = \frac{\eta_1 + \cdots + \eta_N}{N}.$$

Hence,

$$\frac{\sqrt{N}\,(b_\alpha - \beta)}{\sqrt{\sigma_\eta^2 + \alpha^2\sigma_\epsilon^2}} \tag{25}$$

is normally distributed with zero mean and unit variance. It is obvious that the expressions (20) and (25) are independently distributed. Hence $\sqrt{N-2}$ times the ratio of (25) to the square root of (20), i.e.

$$t = \sqrt{N-2}\,\frac{\sqrt{N}\,(b_\alpha - \beta)}{\sqrt{N-2}\,s} = \frac{\sqrt{N-2}\,(b_\alpha - \beta)}{\sqrt{(s_y')^2 + \alpha^2(s_x')^2 - 2\alpha s_{xy}'}}$$

has the Student distribution with $N - 2$ degrees of freedom. Denoting by t_0 the critical value of t according to the chosen probability level, the confidence interval for β is given by the interval:

$$\left[b_\alpha + \frac{\sqrt{(s_y')^2 + \alpha^2(s_x')^2 - 2\alpha s_{xy}'}}{\sqrt{N-2}}\,t_0, \quad b_\alpha - \frac{\sqrt{(s_y')^2 + \alpha^2(s_x')^2 - 2\alpha s_{xy}'}}{\sqrt{N-2}}\,t_0\right].$$

6. Confidence Region for α and β Jointly. In most practical cases we want to know confidence limits for α and β jointly. A pair of values α, β can be represented in the plane by the point with the coordinates α, β. A region R of this plane is called confidence region of the true point (α, β) corresponding to the probability level P if the following two conditions are fulfilled.

(1) The region R is a function of the observations $x_1, \cdots, x_N; y_1, \cdots, y_N$, i.e. it is uniquely determined by the observations.

(2) Before performing the experiment the probability that we shall obtain observed values such that (α, β) will be contained in R, is exactly equal to P. P is usually chosen to be equal to .95 or .99.

We have shown that the expressions (21) and (25), i.e.

$$\frac{\sqrt{N}\, a_1(a - \alpha)}{\sqrt{\sigma_\eta^2 + \alpha^2 \sigma_\epsilon^2}}, \qquad \frac{\sqrt{N}\,(b_\alpha - \beta)}{\sqrt{\sigma_\eta^2 + \alpha^2 \sigma_\epsilon^2}}$$

are normally distributed with zero mean and unit variance. Now we shall show that these two quantities are independently distributed. For this purpose we have only to show that $\bar{x}$, $\bar{y}$, a_1 and a_2 are independently distributed (a_1 and a_2 are defined in (1)), but since

$$a_1 - E(a_1) = (\bar{\epsilon}_1 - \bar{\epsilon}_2)/2$$

$$a_2 - E(a_2) = (\bar{\eta}_1 - \bar{\eta}_2)/2$$

$$\bar{x} - E(\bar{x}) = \bar{\epsilon}$$

$$\bar{y} - E(\bar{y}) = \bar{\eta},$$

we have only to show that $\bar{\epsilon}$, $\bar{\eta}$, $\bar{\epsilon}_1 - \bar{\epsilon}_2$, $\bar{\eta}_1 - \bar{\eta}_2$ are independently distributed. We obviously have

$$\bar{\epsilon} = \frac{\bar{\epsilon}_1 + \bar{\epsilon}_2}{2}, \qquad \bar{\eta} = \frac{\bar{\eta}_1 + \bar{\eta}_2}{2}.$$

It is evident that $\bar{\epsilon}_1$, $\bar{\epsilon}_2$, $\bar{\eta}_1$ and $\bar{\eta}_2$ are independently distributed. Hence, $E[\bar{\epsilon}(\bar{\epsilon}_1 - \bar{\epsilon}_2)] = (E\bar{\epsilon}_1^2 - E\bar{\epsilon}_2^2)/2 = 0$ and also $E[\bar{\eta}(\bar{\eta}_1 - \bar{\eta}_2)] = (E\bar{\eta}_1^2 - E\bar{\eta}_2^2)/2 = 0$. Since $\bar{\epsilon}_1 - \bar{\epsilon}_2$, $\bar{\eta}_1 - \bar{\eta}_2$, and $\bar{\epsilon}$ and $\bar{\eta}$ are normally distributed, the independence of this set of variables is proved, and therefore also (21) and (25) are independently distributed. It is obvious that the expression (20) is distributed independently of (21) and (25). From this it follows that

$$(26) \qquad \frac{N-2}{2} \cdot \frac{N[a_1^2(a - \alpha)^2 + (\bar{y} - \alpha x - \beta)^2]}{(N-2)s^2} = \frac{(N-2)[a_1^2(a - \alpha)^2 + (\bar{y} - \alpha\bar{x} - \beta)^2]}{2[(s_y')^2 + \alpha^2(s_x')^2 - 2\alpha s_{xy}']}$$

has the F-distribution (analysis of variance distribution) with 2 and $N - 2$ degrees of freedom. The F-distribution is tabulated in Snedecor's book: *Calcu-*

lation and Interpretation of Analysis of Variance, Collegiate Press, Ames, Iowa, 1934. The distribution of $\frac{1}{2} \log F = z$ is tabulated in R. A. Fisher's book: *Statistical Methods for Research Workers*, London, 1936. Denote by F_0 the critical value of F corresponding to the chosen probability level P. Then the confidence region R is the set of points (α, β) which satisfy the inequality

$$\frac{N-2}{2} \cdot \frac{a_1^2(a-\alpha)^2 + (\bar{y} - \alpha\bar{x} - \beta)^2}{(s_y')^2 + \alpha^2(s_x')^2 - 2\alpha s_{xy}'} < F_0. \tag{27}$$

The boundary of the region is given by the equation

$$a_1^2(a-\alpha)^2 + (\bar{y} - \alpha\bar{x} - \beta)^2 = \frac{2F_0}{N-2}[(s_y')^2 + \alpha^2(s_x')^2 - 2\alpha s_{xy}']. \tag{28}$$

This is the equation of an ellipse. Hence the region R is the interior of the ellipse defined by the equation (28). If Assumption V holds, the length of the axes of the ellipse are of the order $1/\sqrt{N}$, hence with increasing N the ellipse reduces to a point.

7. The Grouping of the Observations. We have divided the observations in two equal groups G_1 and G_2, G_1 containing the first half $(x_1, y_1), \cdots, (x_m, y_m)$ and G_2 the second half $(x_{m+1}, y_{m+1}), \cdots, (x_N, y_N)$ of the observations. All the formulas and statements of the previous sections remain exactly valid for any arbitrary subdivision of the observations in two equal groups, provided that the subdivision is defined independently of the errors $\epsilon_1, \cdots, \epsilon_N$; $\eta_1, \cdots, \eta_N$. The question of which is the most advantageous grouping arises, i.e. for which grouping will a be the most efficient estimate of α (will lead to the shortest confidence interval for α). It is easy to see that the greater $|a_1|$ the more efficient is the estimate a of α. The expression $|a_1|$ becomes a maximum if we order the observations such that $x_1 \leq x_2 \leq \cdots \leq x_N$. That is to say $|a_1|$ becomes a maximum if we group the observations according to the following:

RULE I. *The point (x_i, y_i) belongs to the group G_1 if the number of elements x_j $(j \neq i)$ of the series $x_1, \cdots, x_N$ for which $x_j \leq x_i$ is less than $m = N/2$. The point (x_i, y_i) belongs to G_2 if the number of elements x_j $(j \neq i)$ for which $x_j \leq x_i$ is greater than or equal to m.*

This grouping, however, depends on the observed values $x_1, \cdots, x_N$ and is therefore in general not entirely independent of the errors $\epsilon_1, \cdots, \epsilon_N$. Let us now consider the grouping according to the following:

RULE II. *The point (x_i, y_i) belongs to the group G_1 if the number of elements X_j of the series $X_1, \cdots, X_N$ for which $X_j \leq X_i$ $(j \neq i)$ is less than m. The point (x_i, y_i) belongs to G_2 if the number of elements X_j for which $X_j \leq X_i$ $(j \neq i)$ is equal to or greater than m.*

The grouping according to Rule II is entirely independent of the errors $\epsilon_1, \cdots, \epsilon_N ; \eta_1, \cdots, \eta_N$. It is identical with the grouping according to Rule I in the following case: Denote by x the median of $x_1, \cdots, x_N$; assume that ϵ can take values only within the finite interval $[-c, +c]$ and that all the values $x_1, \cdots, x_N$ fall outside the interval $[x - c, x + c]$. It is easy to see that in this case $x_i \leq x$ $(i = 1, \cdots, N)$ holds if and only if $X_i \leq X$, where X denotes the median of $X_1, \cdots, X_N$. Hence the grouping according to Rule II is identical to that according to Rule I and therefore the grouping according to Rule I is independent of the errors $\epsilon_1, \cdots, \epsilon_N$. In such cases we get the best estimate of α by grouping the observations according to Rule I. Practically, we can use the grouping according to Rule I and regard it as independent of the errors $\epsilon_1, \cdots, \epsilon_N ; \eta_1, \cdots, \eta_N$ if there exists a positive value c for which the probability that $|\epsilon| \geq c$ is negligibly small and the number of observations contained in $[x - c, x + c]$ is also very small.

Denote by a' the value of a which we obtain by grouping the observations according to Rule I and by a'' the value of a if we group the observations according to Rule II. The value a'' is in general unknown, since the values $X_1, \cdots, X_N$ are unknown, except in the special case considered above, when we have $a'' = a'$. We will now show that an upper and a lower limit for a'' can always be given. First, we have to determine a positive value c such that the probability that $|\epsilon| \geq c$ is negligibly small. The value of c may often be determined before we make the observations having some *a priori* knowledge about the possible range of the errors. If this is not the case, we can estimate the value of c from the data. It is well known that if we have errors in both variables and fit a straight line by the method of least squares minimizing in the x-direction, the sum of the squared deviations divided by the number of degrees of freedom will overestimate σ_ϵ^2. Hence, if ϵ is normally distributed, we can consider the interval $[-3v, 3v]$ as the possible range of ϵ, i.e. $c = 3v$, where v^2 denotes the sum of the squared residuals divided by the number of degrees of freedom. If the distribution of ϵ is unknown, we shall have to take for c a somewhat larger value, for instance $c = 5v$. After having determined c, upper and lower limits for a'' can be given as follows: we consider the system S of all possible groupings satisfying the conditions:

(1) If $x_i \leq x - c$ the point (x_i, y_i) belongs to the group G_1.

(2) If $x_i \geq x + c$ the point (x_i, y_i) belongs to the group G_2.

We calculate the value of a according to each grouping of the system S and denote the minimum of these values by a^*, and the maximum by a^{**}. Since the grouping according to Rule II is contained in the system S, a^* is a lower and a^{**} an upper limit of a''.

Let g be a grouping contained in S and denote by I_g the confidence interval for α which we obtain from formula (23) using the grouping g. Denote further by I the smallest interval which contains the intervals I_g for all elements g of S. Then I contains also the confidence interval corresponding to the grouping according to Rule II. If we denote by P the chosen probability level (say

$P = .95$), then we can say: If we were to draw a sample consisting of N pairs of observations $(x_1, y_1), \cdots, (x_N, y_N)$, the probability is greater than or equal to P that we shall obtain a system of observations such that the interval I will include the true slope α.

The computing work for the determination of I may be considerable if the number of observations within the interval $[x - c, x + c]$ is not small. We can get a good approximation to I by less computation work as follows: First we calculate the slope a' using the grouping according to Rule I and determine the confidence interval $[a' - \delta, a' + \Delta]$ according to formula (23). Denote by $a(g)$ the value of the slope, i.e. the value of $\frac{\bar{y}_1 - \bar{y}_2}{\bar{x}_1 - \bar{x}_2}$, corresponding to a grouping g of the system S, and by $[a(g) - \delta_g, a(g) + \Delta_g]$ the corresponding confidence interval calculated from (23). Neglecting the differences $(\delta_g - \delta)$ and $(\Delta_g - \Delta)$, we obtain for I the interval $[a^* - \delta, a^{**} + \Delta]$.

If the difference $a^{**} - a^*$ is small, we can consider $I = [a^* - \delta, a^{**} + \Delta]$ as the correct confidence interval of α corresponding to the chosen probability level P. If, however, $a^{**} - a^*$ is large, the interval I is unnecessarily large. In such cases we may get a much shorter confidence interval by using some other grouping defined independently of the errors $\epsilon_1, \cdots, \epsilon_N; \eta_1, \cdots, \eta_N$. For instance if we see that the values $x_1, \cdots, x_N$ considered in the order as they have been observed, show a monotonically increasing (or decreasing) tendency, we shall define the group G_1 as the first half, and the group G_2 as the second half of the observations. Though we decide to make this grouping after having observed that the values $x_1, \cdots, x_N$ show a clear trend, the grouping can be considered as independent of the errors $\epsilon_1, \cdots, \epsilon_N$. In fact, if the range of the error ϵ is small in comparison to the true part X, the trend tendency of the value $x_1, \cdots, x_N$ will not be affected by the size of the errors $\epsilon_1, \cdots, \epsilon_N$. We may use for the grouping also any other property of the data which is independent of the errors.

The results of the preceding considerations can be summarized as follows: We use first the grouping according to Rule I, calculate the slope $a' = \frac{\bar{y}_1 - \bar{y}_2}{\bar{x}_1 - \bar{x}_2}$ and the corresponding confidence interval $[a' - \delta, a' + \Delta]$ (formula (23)). This confidence interval cannot be considered as exact since the grouping according to Rule I is not completely independent of the errors. In order to take account of this fact, we calculate a^* and a^{**}. If $a^{**} - a^*$ is small, we consider $I = [a^* - \delta, a^{**} + \Delta]$ with practical approximation as the correct confidence interval. If, however, $a^{**} - a^*$ is large, the interval I is unnecessarily large. We can only say that I is a confidence interval corresponding to a probability level greater than or equal to the chosen one. In such cases we should try to use some other grouping defined independently of the errors, which eventually will lead to a considerably shorter confidence interval.

Analogous considerations hold regarding the joint confidence region for α and β. We use the grouping according to Rule I and calculate from (27) the corresponding confidence region R. If $| a^{**} - a^* |$ and $| b^{**} - b^* |$ are small ($b^* = \bar{y} - a^*\bar{x}$ and $b^{**} = \bar{y} - a^{**}\bar{x}$) we enlarge R to a region $\bar{R}$ corresponding to the fact that a and b may take any values within the intervals $[a^{**}, a^*]$ and $[b^{**}, b^*]$ respectively. The region $\bar{R}$ can be considered with practical approximation as the correct confidence region. If $| a^{**} - a^* |$ or $| b^{**} - b^* |$ is large, we may try some other grouping defined independently of the errors, which may lead to a smaller confidence region. In any case $\bar{R}$ represents a confidence region corresponding to a probability level greater than or equal to the chosen one.

8. Some Remarks on the Consistency of the Estimates of $\alpha, \beta, \sigma_\epsilon, \sigma_\eta$. We have shown in section 3 that the given estimates of α, β, σ_ϵ and σ_η are consistent if condition V is satisfied.

If the values $x_1, \cdots, x_N$ are not obtained by random sampling, it will in general be possible to define a grouping which is independent of the errors and for which condition V is satisfied. We can sometimes arrange the experiments such that no values of the series $x_1, \cdots, x_N$ should be within the interval $[x - c, x + c]$ where x denotes the median of $x_1, \cdots, x_N$ and c the range of the error ϵ. In such cases, as we saw, the grouping according to Rule I is independent of the errors. Condition V is certainly satisfied if we group the data according to Rule I.

Let us now consider the case that $X_1, \cdots, X_N$ are random variables independently distributed, each having the same distribution. Denote by X a random variable having the same probability distribution as possessed by each of the random variables $X_1, \cdots, X_N$. Assuming that X has a finite second moment, the expression in condition V will approach zero stochastically with $N \to \infty$ for any grouping defined independently of the values $X_1, \cdots, X_N$. It is possible, however, to define a grouping independent of the errors (but not independent of $X_1, \cdots, X_N$) for which the expression in V does not approach zero, provided that X has the following property: There exists a real value λ such that the probability that X will lie within the interval $[\lambda - c, \lambda + c]$ (c denotes the range of the error ϵ) is zero, the probability that $X > \lambda + c$ is positive, and the probability that $X < \lambda - c$ is positive. The grouping can be defined, for instance, as follows:

The i-th observation (x_i, y_i) belongs to the group G_1 if $x_i \leq \lambda$ and to G_2 if $x_i > \lambda$. We continue the grouping according to this rule up to a value i for which one of the groups G_1, G_2 contains already $N/2$ elements. All further observations belong to the other group.

It is easy to see that the probability is equal to 1 that the relation $x_i \leq \lambda$ is equivalent to the relation $X_i < \lambda - c$ and the relation $x_i > \lambda$ is equivalent to the relation $X_i > \lambda + c$. Hence this grouping is independent of the errors. Since for this grouping condition V is satisfied, our statement is proved.

If X has not the property described above, it may happen that for every grouping defined independently of the errors, the expression in condition V converges always to zero stochastically. Such a case arises for instance if X, ϵ and η are normally distributed.[14] It can be shown that in this case no consistent estimates of the parameters α and β can be given, unless we have some additional information not contained in the data (for instance we know *a priori* the ratio $\sigma_\epsilon/\sigma_\eta$).

9. Structural Relationship and Prediction.[15] The problem discussed in this paper was the question as to how to estimate the relationship between the true parts X and Y. We shall call the relationship between the true parts the structural relationship. The problem of finding the structural relationship must not be confused with the problem of prediction of one variable by means of the other. The problem of prediction can be formulated as follows: We have observed N pairs of values $(x_1, y_1), \cdots, (x_N, y_N)$. A new observation on x is given and we have to estimate the corresponding value of y by means of our previous observations $(x_1, y_1), \cdots, (x_N, y_N)$. One might think that if we have estimated the structural relationship between X and Y, we may estimate y by the same relationship. That is to say, if the estimated structural relationship is given by $Y = aX + b$, we may estimate y from x by the same formula: $y = ax + b$. This procedure may lead, however, to a biased estimate of y. This is, for instance, the case if X, ϵ and η are normally distributed. It can easily be shown in this case that for any given x the conditional expectation of y is a linear function of x, that the slope of this function is different from the slope of the structural relationship, and that among all unbiased estimates of y which are linear functions of x, the estimate obtained by the method of least squares has the smallest variance. Hence in this case we have to use the least square estimate for purposes of prediction. Even if we would know exactly the structural relationship $Y = \alpha X + \beta$, we would get a biased estimate of y by putting $y = \alpha x + \beta$.

Let us consider now the following example: X is a random variable having a rectangular distribution with the range $[0, 1]$. The random variable ϵ has a rectangular distribution with the range $[-0.1, +0.1]$. For any given x let us denote the conditional expectation of y by $E(y \mid x)$ and the conditional expectation of X by $E(X \mid x)$. Then we obviously have

$$E(y \mid x) = \alpha E(X \mid x) + \beta.$$

Now let us calculate $E(X \mid x)$. It is obvious that the joint distribution of X and ϵ is given by the density function:

$$5\, dX\, d\epsilon,$$

[14] I wish to thank Professor Hotelling for drawing my attention to this case.

[15] I should like to express my thanks to Professor Hotelling for many interesting suggestions and remarks on this subject.

where X can take any value within the interval [0, 1] and ϵ can take any value within [−0.1, + 0.1]. From this we obtain easily that the joint distribution of x and X is given by the density function

$$5\,dx\,dX,$$

where x can take any value within the interval [−0.1, 1.1] and X can take any value lying in both intervals [0, 1] and $[x - 0.1, x + 0.1]$ simultaneously. Denote by I_x the common part of these two intervals. Then for any fixed x the relative distribution of X is given by the probability density

$$\frac{dX}{\int_{I_x} dX}.$$

Hence, we have

$$E(X \mid x) = \frac{\int_{I_x} X\,dX}{\int_{I_x} dX}.$$

We have to consider 3 cases:

(1) $$0.1 \leq x \leq 0.9.$$

In this case $I_x = [x - 0.1, x + 0.1]$ and

$$E(X \mid x) = \frac{\int_{x-0.1}^{x+0.1} X\,dX}{\int_{x-0.1}^{x+0.1} dX} = x.$$

(2) $-0.1 < x \leq 0.1$. Then $I_x = [0, x + 0.1]$ and

$$E(X \mid x) = \frac{\int_{0}^{x+0.1} X\,dX}{\int_{0}^{x+0.1} dX} = .5x + .05.$$

(3) $0.9 \leq x < 1.1$. Then $I_x = [x - 0.1, 1]$ and

$$E(X \mid x) = \frac{\int_{x-0.1}^{1} X\,dX}{\int_{x-0.1}^{1} dX} = .5x + .45.$$

Since

$$E(y \mid x) = \alpha E(X \mid x) + \beta,$$

we see that the structural relationship gives an unbiased prediction of y from x if $0.1 \leq x \leq 0.9$, but not in the other cases.

The problem of cases for which the structural relationship is appropriate also for purposes of prediction, needs further investigation. I should like to mention a class of cases where the structural relationship has to be used also for prediction. Assume that we have observed N values $(x_1, y_1), \cdots, (x_N, y_N)$ of the variables x and y for which the conditions I–IV of section 2 hold. Then we make a new observation on x obtaining the value x'. We assume that the last observation on x has been made under changed conditions such that we are sure that x' does not contain error, i.e. x' is equal to the true part X'. Such a situation may arise for instance if the error ϵ is due to errors of measurement and the last observation has been made with an instrument of great precision for which the error of measurement can be neglected. In such cases the prediction of the corresponding y' has to be made by means of the estimated structural relationship, i.e. we have to put $y' = ax' + b$.

The knowledge of the structural relationship is essential for constructing any theory in the empirical sciences. The laws of the empirical sciences mostly express relationships among a limited number of variables which would prevail exactly if the disturbing influence of a great number of other variables could be eliminated. In our experiments we never succeed in eliminating completely these disturbances. Hence in deducing laws from observations, we have the task of estimating structural relationships.

Columbia University,
New York, N. Y.

Made in the United States of America

Reprinted from The Annals of Mathematical Statistics
Vol. XII, No. 1, March, 1941

NOTE ON CONFIDENCE LIMITS FOR CONTINUOUS DISTRIBUTION FUNCTIONS

By A. Wald* and J. Wolfowitz

In a recent paper [1] we discussed the following problem: Let X be a stochastic variable with the cumulative distribution function $f(x)$, about which nothing is known except that it is continuous. Let $x_1, \cdots, x_n$ be n independent, random observations on X. The question is to give confidence limits for $f(x)$. We gave a theoretical solution when the confidence set is a particularly simple and important one, a "belt."

A particularly simple and expedient way from the practical point of view is to construct these belts of uniform thickness ([1], p. 115, equation 50). If the appropriate tables, as mentioned in our paper, were available, the construction of confidence limits, no matter how large the size of the sample, would be immediate.

Our formulas (11), (16), (19), (27) and (29) are not very practical for computation, particularly when the samples are large. We have recently learned that there exists a result by Kolmogoroff [2], generalized by Smirnoff [3],[1] which for large samples gives an easy method for constructing tables, i.e. of finding α when c and n are given (all notations as in [1]). The result of Kolmogoroff-Smirnoff is:

Let $c = \lambda/\sqrt{n}$. Then for any fixed $\lambda > 0$,

$$\lim_{n=\infty} \overline{P} = \lim_{n=\infty} \underline{P} = 1 - e^{-2\lambda^2}$$

$$\lim_{n=\infty} P = 1 - 2\sum_{m=1}^{\infty} (-1)^{(m-1)} e^{-2m^2\lambda^2}.$$

This series converges very rapidly.

REFERENCES

[1] Wald and Wolfowitz, "Confidence limits for continuous distribution functions," *Annals of Math. Stat.*, Vol. 10(1939), pp. 105–118.

[2] A. Kolmogoroff, "Sulla determinazione empirica di una leggi di distribuzione," *Giornale dell'Instituto Italiana degli Attuari*, Vol. 11(1933).

[3] N. Smirnoff, "Sur les ecarts de la courbe de distribution empirique," *Recueil Mathematique (Mathematicheski Sbornik)*, New series, Vol. 6(48)(1939), pp. 3–26.

---

* Columbia University, New York City.

[1] In the French résumé of Smirnoff's article, on page 26, due to a typographical error this formula is given with a factor $(-1)^m$ instead of the correct factor $(-1)^{m-1}$. The correct result follows from equation (112), page 23, of the Russian text when t is set equal to zero.

Reprinted from THE ANNALS OF MATHEMATICAL STATISTICS
Vol. XII, No. 1, March, 1941

ASYMPTOTICALLY MOST POWERFUL TESTS OF STATISTICAL HYPOTHESES[1]

BY ABRAHAM WALD[2]

Columbia University, New York City

1. Introduction. Let $f(x, \theta)$ be the probability density function of a variate x involving an unknown parameter θ. For testing the hypothesis $\theta = \theta_0$ by means of n independent observations $x_1, \cdots, x_n$ on x we have to choose a region of rejection W_n in the n-dimensional sample space. Denote by $P(W_n \mid \theta)$ the probability that the sample point $E = (x_1, \cdots, x_n)$ will fall in W_n under the assumption that θ is the true value of the parameter. For any region U_n of the n-dimensional sample space denote by $g(U_n)$ the greatest lower bound of $P(U_n \mid \theta)$. For any pair of regions U_n and T_n denote by $L(U_n, T_n)$ the least upper bound of

$$P(U_n \mid \theta) - P(T_n \mid \theta).$$

In all that follows we shall denote a region of the n-dimensional sample space by a capital letter with the subscript n.

Definition 1. A sequence $\{W_n\}$, $(n = 1, 2, \cdots, \text{ad inf.})$, of regions is said to be an asymptotically most powerful test of the hypothesis $\theta = \theta_0$ on the level of significance α if $P(W_n \mid \theta_0) = \alpha$ and if for any sequence $\{Z_n\}$ of regions for which $P(Z_n \mid \theta_0) = \alpha$, the inequality

$$\limsup_{n \to \infty} L(Z_n, W_n) \leq 0$$

holds.

Definition 2. A sequence $\{W_n\}$, $(n = 1, 2, \cdots, \text{ad inf.})$, of regions is said to be an asymptotically most powerful unbiased test of the hypothesis $\theta = \theta_0$ on the level of significance α if $P(W_n \mid \theta_0) = \lim_{n=\infty} g(W_n) = \alpha$, and if for any sequence $\{Z_n\}$ of regions for which $P(Z_n \mid \theta_0) = \lim_{n=\infty} g(Z_n) = \alpha$, the inequality

$$\limsup_{n \to \infty} L(Z_n, W_n) \leq 0$$

holds.

Let $\hat{\theta}_n(x_1, \cdots, x_n)$ be the maximum likelihood estimate of θ in the n-dimensional sample space. That is to say, $\hat{\theta}_n(x_1, \cdots, x_n)$ denotes the value of θ

---

[1] Presented to the American Mathematical Society at New York, February 24, 1940.

[2] Research under a grant-in-aid from the Carnegie Corporation of New York.

for which the product $\prod_{\nu=1}^{n} f(x_\nu, \theta)$ becomes a maximum. Let W'_n be the region defined by the inequality $\sqrt{n}(\hat{\theta}_n - \theta_0) \geq c'_n$, W''_n defined by the inequality $\sqrt{n}(\hat{\theta}_n - \theta_0) \leq c''_n$, and let W_n consists of all points for which at least one of the inequalities

$$\sqrt{n}(\hat{\theta}_n - \theta_0) \geq a_n, \qquad \sqrt{n}(\hat{\theta}_n - \theta_0) \leq -a_n$$

is satisfied. The constants a_n, c'_n, c''_n are chosen such that

$$P(W'_n \mid \theta_0) = P(W''_n \mid \theta_0) = P(W_n \mid \theta_0) = \alpha.$$

It will be shown in this paper that under certain restrictions on the probability density $f(x, \theta)$ the sequence $\{W'_n\}$ is an asymptotically most powerful test of the hypothesis $\theta = \theta_0$ if θ takes only values $\theta \geq \theta_0$. Similarly $\{W''_n\}$ is an asymptotically most powerful test if θ takes only values $\theta \leq \theta_0$. Finally $\{W_n\}$ is an asymptotically most powerful unbiased test if θ can take any real value.

2. Assumptions on the density function $f(x, \theta)$.

ASSUMPTION 1. *For any positive k*

$$\lim_{n=\infty} P(-k < \hat{\theta}_n - \theta < k \mid \theta) = 1$$

uniformly in θ, where $P(-k < \hat{\theta}_n - \theta < k \mid \theta)$ denotes the probability that $-k \leq \hat{\theta}_n - \theta \leq k$ under the assumption that θ is the true value of the parameter.

Assumption 1 implies somewhat more than consistency of the maximum likelihood estimate $\hat{\theta}_n$. In fact, consistency means only that for any positive k

$$\lim_{n=\infty} P(-k \leq \hat{\theta}_n - \theta \leq k \mid \theta) = 1,$$

without asking that the convergence should be uniform in θ. If $\hat{\theta}_n$ satisfies Assumption 1 we shall say that $\hat{\theta}_n$ is a uniformly consistent estimate of θ. A rigorous proof of the consistency of $\hat{\theta}_n$ (under certain restrictions on $f(x, \theta)$) was given by J. L. Doob.[3] In an appendix to this paper it will be shown that under certain conditions $\hat{\theta}_n$ is uniformly consistent.

Denote by $E_\theta[\psi(x)]$ the expected value of $\psi(x)$ under the assumption that θ is the true value of the parameter. That is to say,

$$E_\theta[\psi(x)] = \int_{-\infty}^{\infty} \psi(x) f(x, \theta)\, dx.$$

For any x, for any positive δ, and for any θ_1, denote by $\varphi_1(x, \theta_1, \delta)$ the greatest lower bound, and by $\varphi_2(x, \theta_1, \delta)$ the least upper bound of $\dfrac{\partial^2 \log f(x, \theta)}{\partial \theta^2}$ in the interval $\theta_1 - \delta \leq \theta \leq \theta_1 + \delta$.

ASSUMPTION 2. *There exists a positive value k_0 such that the expectations $E_\theta \varphi_1(x, \theta_1, \delta)$ and $E_\theta \varphi_2(x, \theta_1, \delta)$ exist and are continuous functions of θ, θ_1 and δ*

[3] J. L. Doob, "Probability and statistics," *Trans. Am. Math. Soc.*, Vol. 36 (1937).

in the domain D defined by the inequalities: $0 \leq \delta \leq \frac{1}{2}k_0$, $\theta_0 - \frac{1}{2}k_0 \leq \theta_1 \leq \theta_0 + \frac{1}{2}k_0$, $\theta_0 - k_0 \leq \theta \leq \theta_0 + k_0$. *Furthermore the expectations* $E_\theta[\varphi_1(x, \theta_1, \delta)]^2$ *and* $E_\theta[\varphi_2(x, \theta_1, \delta)]^2$ *exist in D and have a finite upper bound in D.*

ASSUMPTION 3. *There exists a positive value* k_0 *such that*

$$\int_{-\infty}^{\infty} \frac{\partial f(x, \theta)}{\partial \theta}\, dx = \int_{-\infty}^{\infty} \frac{\partial^2 f(x, \theta)}{\partial \theta^2}\, dx = 0 \quad \textit{for} \quad \theta_0 - k_0 \leq \theta \leq \theta_0 + k_0 .$$

Assumption 3 means simply that we may differentiate with respect to θ under the integral sign. In fact

$$\int_{-\infty}^{\infty} f(x, \theta)\, dx = 1$$

identically in θ. Hence

$$\frac{\partial}{\partial \theta} \int_{-\infty}^{\infty} f(x, \theta)\, dx = \frac{\partial^2}{\partial \theta^2} \int_{-\infty}^{\infty} f(x, \theta)\, dx = 0.$$

Differentiating under the integral sign, we obtain the relations in Assumption 3.

ASSUMPTION 4. *There exists a positive* η *and a positive* k_0 *such that*

$$E_\theta \left| \frac{\partial \log f(x, \theta)}{\partial \theta} \right|^{2+\eta}$$

exists and has a finite upper bound in the interval $\theta_0 - k_0 \leq \theta \leq \theta_0 + k_0$.

3. Some propositions. Denote $\sqrt{n}\,(\theta_n - \theta)$ by $z_n(\theta)$ and denote the probability $P[z_n(\theta) < t \mid \theta]$ by $\Phi_n(t, \theta)$.

PROPOSITION I. *Within the* θ*-interval* $[\theta_0 - \frac{1}{2}k_0, \theta_0 + \frac{1}{2}k_0]$ $\Phi_n(t, \theta)$ *converges with* $n \to \infty$ *uniformly in t and* θ *towards the cumulative normal distribution with zero mean and variance*

$$-1 \Big/ E_\theta \frac{\partial^2 \log f(x, \theta)}{\partial \theta^2}$$

PROOF: In all that follows we assume that θ takes only values in the interval $[\theta_0 - k_0, \theta_0 + k_0]$, except when the contrary is explicitly stated. Furthermore we introduce the variable θ_1 and assume that θ_1 takes only values in the interval $[\theta_0 - \frac{1}{2}k_0, \theta_0 + \frac{1}{2}k_0]$.

Because of Assumption 3 we have

$$(1) \qquad E_\theta \frac{\partial \log f(x, \theta)}{\partial \theta} = \int_{-\infty}^{\infty} \frac{\partial f(x, \theta)}{\partial \theta}\, dx = 0$$

Since

$$\frac{\partial^2 \log f(x, \theta)}{\partial \theta^2} = \frac{1}{f(x, \theta)} \cdot \frac{\partial^2 f(x, \theta)}{\partial \theta^2} - \frac{1}{[f(x, \theta)]^2} \left[\frac{\partial f(x, \theta)}{\partial \theta} \right]^2$$

we get from Assumption 3

$$(2) \qquad E_\theta \left[\frac{\partial \log f(x, \theta)}{\partial \theta} \right]^2 = -E_\theta \frac{\partial^2 \log f(x, \theta)}{\partial \theta^2}.$$

Hence

$$d(\theta) = -E_\theta \frac{\partial^2 \log f(x, \theta)}{\partial \theta^2} > 0. \tag{3}$$

Consider the Taylor expansion

$$\sum_{\alpha=1}^{n} \frac{\partial \log f(x_\alpha, \theta)}{\partial \theta} = \sum_{\alpha=1}^{n} \frac{\partial \log f(x_\alpha, \theta_1)}{\partial \theta} + (\theta - \theta_1) \sum_{\alpha=1}^{n} \frac{\partial^2 \log f(x_\alpha, \theta')}{\partial \theta^2} \tag{4}$$

where θ' lies in the interval $[\theta_1, \theta]$. Denote $\frac{1}{\sqrt{n}} \sum_\alpha \frac{\partial \log f(x_\alpha, \theta_1)}{\partial \theta}$ by $y_n(\theta_1)$. For $\theta = \hat{\theta}_n$ the left hand side of (4) is equal to zero. Hence we have

$$y_n(\theta_1) + [\sqrt{n}(\hat{\theta}_n - \theta_1)] \frac{1}{n} \sum_\alpha \frac{\partial^2 \log f(x_\alpha, \theta')}{\partial \theta^2} = 0, \tag{5}$$

or

$$y_n(\theta_1) + z_n(\theta_1) \frac{1}{n} \sum_\alpha \frac{\partial^2 \log f(x_\alpha, \theta')}{\partial \theta^2} = 0. \tag{6}$$

Let $Q_n(\theta_1)$ be the region defined by the inequality

$$\left| \frac{1}{n} \sum_\alpha \frac{\partial^2 \log f(x_\alpha, \theta')}{\partial \theta^2} + d(\theta_1) \right| < \nu \tag{7}$$

where ν denotes a positive number less than the greatest lower bound of $d(\theta_1)$. We shall prove that

$$\lim_{n=\infty} P[Q_n(\theta_1) \mid \theta_1] = 1 \tag{8}$$

uniformly in θ_1. Let τ_0 be a positive number such that

$$\left| E_{\theta_1} \varphi_i(x, \theta_1, \tau_0) - E_{\theta_1} \frac{\partial^2 \log f(x, \theta_1)}{\partial \theta^2} \right| < \frac{\nu}{2}, \qquad (i = 1, 2) \tag{9}$$

for all values of θ_1. Because of Assumption 2 such a τ_0 certainly exists. Denote by $R_n(\theta_1)$ the region defined by the inequality

$$| \hat{\theta}_n - \theta_1 | \leq \tau_0 . \tag{10}$$

On account of Assumption 1

$$\lim_{n=\infty} P[R_n(\theta_1) \mid \theta_1] = 1 \tag{11}$$

uniformly in θ_1. Since θ' lies in the interval $[\theta_1, \hat{\theta}_n]$, we have

$$| \theta' - \theta_1 | \leq \tau_0 \tag{12}$$

for all points in $R_n(\theta_1)$. Hence at any point in $R_n(\theta_1)$ the inequality

$$\sum_{\alpha=1}^{n} \varphi_1(x_\alpha, \theta_1, \tau_0) \leq \sum_{\alpha=1}^{n} \frac{\partial^2 \log f(x_\alpha, \theta')}{\partial \theta^2} \leq \sum_{\alpha=1}^{n} \varphi_2(x_\alpha, \theta_1, \tau_0) \tag{13}$$

holds.

Let $S_n(\theta_1)$ be defined by the inequality

$$\left|\frac{1}{n}\sum_{\alpha} \varphi_1(x_\alpha, \theta_1, \tau_0) - E_{\theta_1}\varphi_1(x, \theta_1, \tau_0)\right| < \frac{\nu}{2} \tag{14}$$

and $T_n(\theta_1)$ by the inequality

$$\left|\frac{1}{n}\sum_{\alpha} \varphi_2(x_\alpha, \theta_1, \tau_0) - E_{\theta_1}\varphi_2(x, \theta_1, \tau_0)\right| < \frac{\nu}{2}. \tag{15}$$

On account of Assumption 2 we have

$$\lim_{n=\infty} P[S_n(\theta_1) \mid \theta_1] = \lim_{n=\infty} P[T_n(\theta_1) \mid \theta_1] = 1 \tag{16}$$

uniformly in θ_1.

Denote by $U_n(\theta_1)$ the common part of the regions $R_n(\theta_1)$, $S_n(\theta_1)$ and $T_n(\theta_1)$. In $U_n(\theta_1)$ we have on account of (9), (14) and (15)

$$\left|\frac{1}{n}\sum_{\alpha} \varphi_i(x_\alpha, \theta_1, \tau_0) - E_{\theta_1}\frac{\partial^2 \log f(x, \theta_1)}{\partial\theta^2}\right| < \nu \qquad (i = 1, 2). \tag{17}$$

From this we obtain (7) because of (13). That is to say, the inequality (7) is valid everywhere in $U_n(\theta_1)$. Since

$$\lim_{n=\infty} P[U_n(\theta_1) \mid \theta_1] = 1$$

uniformly in θ_1, our statement about $Q_n(\theta_1)$ is proved. From (6) and (7) we get that everywhere in $Q_n(\theta_1)$ the inequalities hold:

$$\frac{y_n(\theta_1)}{d(\theta_1) + \nu} \leq z_n(\theta_1) \leq \frac{y_n(\theta_1)}{d(\theta_1) - \nu} \quad \text{if} \quad y_n(\theta_1) \geq 0; \tag{18}$$

$$\frac{y_n(\theta_1)}{d(\theta_1) + \nu} \geq z_n(\theta_1) \geq \frac{y_n(\theta_1)}{d(\theta_1) - \nu} \quad \text{if} \quad y_n(\theta_1) \leq 0. \tag{19}$$

Let $z_n^*(\theta_1)$ be defined as follows: $z_n^*(\theta_1) = z_n(\theta_1)$ at any point in $Q_n(\theta_1)$, and $z_n^*(\theta_1) = y_n(\theta_1)/d(\theta_1)$ at any point outside $Q_n(\theta_1)$.

On account of (8) we obviously have

$$\lim_{n=\infty} P[z_n^*(\theta_1) < t \mid \theta_1] - P[z_n(\theta_1) < t \mid \theta_1] = 0 \tag{20}$$

uniformly in t and θ_1.

From equation (1) it follows that $E_{\theta_1} y_n(\theta_1) = 0$. From Assumption 4 it follows on account of the general limit theorems

$$\lim_{n=\infty} P[y_n(\theta_1) < t \mid \theta_1] - \frac{1}{\sqrt{2\pi d(\theta_1)}}\int_{-\infty}^{t} e^{-\frac{1}{2}t^2/d(\theta_1)}\, dt = 0 \tag{21}$$

uniformly in t and θ_1. Hence

$$\lim_{n=\infty} P\left[\frac{y_n(\theta_1)}{d(\theta_1)} < t \mid \theta_1\right] - \sqrt{\frac{d(\theta_1)}{2\pi}}\int_{-\infty}^{t} e^{-\frac{1}{2}t^2 d(\theta_1)}\, dt = 0$$

uniformly in t and θ_1. Since ν can be chosen arbitrarily small, we get easily from (18), (19), (20) and (21)

$$\lim_{n=\infty} \left| P\left[\frac{y_n(\theta_1)}{d(\theta_1)} < t \mid \theta_1\right] - P[z_n(\theta_1) < t \mid \theta_1] \right| = 0 \tag{22}$$

uniformly in t and θ_1. Proposition 1 follows from (21) and (22).

PROPOSITION 2. *Let* $\{W_n\}$ *be a sequence of regions of size* α, *i.e.* $P(W_n \mid \theta_0) = \alpha$, *and let* $V_n(z)$ *be the region defined by the inequality*

$$(\hat{\theta}_n - \theta_0)\sqrt{n} < z.$$

Let $U_n(z)$ *be the intersection of* $V_n(z)$ *and* W_n, *and denote* $P[U_n(z) \mid \theta_0]$ *by* $F_n(z)$. *Denote furthermore* $P[W_n \mid \theta_0 + \mu/\sqrt{n}]$ *by* $G(\mu, n)$. *If* $F_n(z)$ *converges to* $F(z)$ *and if* $\lim_{n=\infty} \mu_n = \mu$, *then*

$$\lim_{n=\infty} G(\mu_n, n) = \int_{-\infty}^{\infty} e^{-\frac{1}{2}(\mu^2 - 2\mu z)/c}\, dF(z) \tag{23}$$

where

$$c = -1 \Big/ E_{\theta_0} \frac{\partial^2 \log f(x, \theta_0)}{\partial \theta^2}.$$

PROOF: First we show

$$\int_{-\infty}^{\infty} dF(z) = \alpha. \tag{24}$$

Denote $P[V_n(z) \mid \theta_0]$ by $\Phi_n(z)$. On account of Proposition 1 $\Phi_n(z)$ converges uniformly to the cumulative normal distribution $\psi(z)$ with zero mean and variance c. It is obvious that

$$F_n(z_2) - F_n(z_1) \le \Phi_n(z_2) - \Phi_n(z_1) \text{ for } z_2 > z_1. \tag{25}$$

Hence

$$F(z_2) - F(z_1) \le \psi(z_2) - \psi(z_1) \text{ for } z_2 > z_1. \tag{26}$$

From (25) we get

$$\left[\lim_{z=\infty} F_n(z)\right] - F_n(z) = \alpha - F_n(z) \le 1 - \Phi_n(z). \tag{27}$$

Hence

$$\alpha - F(z) \le 1 - \psi(z). \tag{28}$$

Since $F_n(z) \le \alpha$ and therefore also $F(z) \le \alpha$, we get from (28)

$$0 \le \alpha - F(z) \le 1 - \psi(z).$$

Hence

$$\lim_{z=\infty} F(z) = \alpha. \tag{29}$$

Since $F_n(z) \leq \Phi_n(z)$, we have $F(z) \leq \psi(z)$, and therefore

$$\lim_{z=-\infty} F(z) = 0. \tag{30}$$

The equation (24) follows from (29) and (30).

It follows easily from (26) that the integral on the right hand side of the equation (23) exists and is finite.

Let us denote $\theta_0 + \mu_n/\sqrt{n}$ by θ_n. Consider the Taylor expansions

$$\begin{aligned}\sum_\alpha \log f(x_\alpha, \theta_0) &= \sum_\alpha \log f(x_\alpha, \hat{\theta}_n) + (\theta_0 - \hat{\theta}_n) \sum_\alpha \frac{\partial}{\partial \theta} \log f(x_\alpha, \hat{\theta}_n) \\ &\quad + \tfrac{1}{2}(\theta_0 - \hat{\theta}_n)^2 \sum_\alpha \frac{\partial^2}{\partial \theta^2} \log f(x_\alpha, \theta_n')\end{aligned} \tag{31}$$

and

$$\begin{aligned}\sum_\alpha \log f(x_\alpha, \theta_n) &= \sum_\alpha \log f(x_\alpha, \hat{\theta}_n) + (\theta_n - \hat{\theta}_n) \sum_\alpha \frac{\partial}{\partial \theta} \log f(x_\alpha, \hat{\theta}_n) \\ &\quad + \tfrac{1}{2}(\theta_n - \hat{\theta}_n)^2 \sum_\alpha \frac{\partial^2}{\partial \theta^2} \log f(x_\alpha, \theta_n'')\end{aligned} \tag{32}$$

where θ_n' lies in the interval $[\theta_0, \hat{\theta}_n]$ and θ_n'' lies in the interval $[\theta_n, \hat{\theta}_n]$. Since $\hat{\theta}_n$ is the maximum likelihood estimate, we get from (31) and (32)

$$\sum_\alpha \log f(x_\alpha, \theta_0) = \sum_\alpha \log f(x_\alpha, \hat{\theta}_n) + \tfrac{1}{2}(\theta_0 - \hat{\theta}_n)^2 \sum_\alpha \frac{\partial^2}{\partial \theta^2} \log f(x_\alpha, \theta_n'), \tag{33}$$

$$\sum_\alpha \log f(x_\alpha, \theta_n) = \sum_\alpha \log f(x_\alpha, \hat{\theta}_n) + \tfrac{1}{2}(\theta_n - \hat{\theta}_n)^2 \sum_\alpha \frac{\partial^2}{\partial \theta^2} \log f(x_\alpha, \theta_n''). \tag{34}$$

Denote by β a real variable which can take any value between -2μ and $+2\mu$. Denote by R_n the region defined by the inequality

$$|\hat{\theta}_n - \theta_0| < n^{-\frac{1}{3}}. \tag{35}$$

From Proposition 1 it follows easily that

$$\lim_{n=\infty} P(R_n \mid \theta_0 + \beta/\sqrt{n}) = 1 \tag{36}$$

uniformly in β. Denote $2n^{-\frac{1}{3}}$ by τ_n. Then for almost all n the following inequalities hold at any point in R_n:

$$\sum_\alpha \varphi_1(x_\alpha, \theta_0, \tau_n) \leq \sum_\alpha \frac{\partial^2}{\partial \theta^2} \log f(x_\alpha, \theta_n') \leq \sum_\alpha \varphi_2(x_\alpha, \theta_0, \tau_n), \tag{37}$$

$$\sum_\alpha \varphi_1(x_\alpha, \theta_0, \tau_n) \leq \sum_\alpha \frac{\partial^2}{\partial \theta^2} \log f(x_\alpha, \theta_n'') \leq \sum_\alpha \varphi_2(x_\alpha, \theta_0, \tau_n). \tag{38}$$

Denote by S_n the region in which (35), (37) and (38) simultaneously hold. It is obvious that

$$\lim_{n=\infty} P(S_n \mid \theta_0 + \beta/\sqrt{n}) = 1$$

uniformly in β. Denote $\theta_0 + \beta/\sqrt{n}$ by $\theta_n(\beta)$. From Assumption 2 it follows easily that

$$\lim_{n=\infty} E_{\theta_n(\beta)} \left\{ \frac{\sum_\alpha \varphi_i(x_\alpha, \theta_0, \tau_n)}{n} \right\} = E_{\theta_0} \frac{\partial^2}{\partial\theta^2} \log f(x, \theta_0) = \frac{-1}{c} \qquad (i = 1, 2) \tag{39}$$

uniformly in β. Furthermore the variance of $\sum_\alpha \frac{\varphi_i(x_\alpha, \theta_0, \tau_n)}{n}$, if $\theta_n(\beta)$ is the true value of the parameter θ, converges to zero with $n \to \infty$ uniformly in β. Hence a sequence $\{\lambda_n\}$, ($n = 1, 2, \cdots$, ad inf.), of positive numbers can be given such that

$$\lim_{n=\infty} \lambda_n = 0 \tag{40}$$

and

$$\lim P[T_n \mid \theta_n(\beta)] = 1 \tag{41}$$

uniformly in β, where the region T_n is defined by the inequality

$$\left| \sum_\alpha \frac{\varphi_i(x_\alpha, \theta_0, \tau_n)}{n} + \frac{1}{c} \right| < \lambda_n n^{-\frac{1}{4}} \qquad (i = 1, 2). \tag{42}$$

From (37) and (38) it follows that in the intersection T'_n of T_n and S_n

$$\left| \frac{1}{n} \sum_\alpha \frac{\partial^2}{\partial\theta^2} \log f(x_\alpha, \theta'_n) + \frac{1}{c} \right| < \lambda_n n^{-\frac{1}{4}} \tag{43}$$

and

$$\left| \frac{1}{n} \sum_\alpha \frac{\partial^2}{\partial\theta^2} \log f(x_\alpha, \theta''_n) + \frac{1}{c} \right| < \lambda_n n^{-\frac{1}{4}}. \tag{44}$$

We get from (33), (34), (35), (43) and (44) that at any point in T'_n

$$\sum_\alpha \log f(x_\alpha, \theta_n) - \sum_\alpha \log f(x_\alpha, \theta_0) = \frac{n}{2c} [(\theta_0 - \hat{\theta}_n)^2 - (\theta_n - \hat{\theta}_n)^2] + \lambda'_n, \tag{45}$$

where $|\lambda'_n| \leq \rho\lambda_n$, and ρ denotes a constant not depending on n.
On account of (36) and (41) we have

$$\lim_{n=\infty} P[T'_n \mid \theta_n(\beta)] = 1 \tag{46}$$

uniformly in β.
Denote by $T''_n(z)$ the intersection of $U_n(z)$ (defined in Proposition 2) and T'_n. Denote furthermore $P[T''_n(z) \mid \theta_0]$ by $F^*_n(z)$.
Since

$$n[(\theta_0 - \hat{\theta}_n)^2 - (\theta_n - \hat{\theta}_n)^2] = n[(\theta_0 - \hat{\theta}_n)^2 - (\theta_0 - \hat{\theta}_n + \mu_n/\sqrt{n})^2]$$
$$= -\mu_n^2 + 2\sqrt{n}\,\mu_n(\hat{\theta}_n - \theta_0),$$

we get from (45) and (46)

$$\lim_{n=\infty} \left\{ P[T''_n(z) \mid \theta_n] - \int_{-\infty}^{z} e^{-\frac{1}{2}(\mu_n^2 - 2\mu_n t)/c} \, dF_n^*(t) \right\} = 0 \tag{47}$$

uniformly in z. It is obvious that

$$\lim_{n=\infty} \{ P[T''_n(z) \mid \theta_n] - P[U_n(z) \mid \theta_n] \} = 0 \tag{48}$$

uniformly in z. Hence we get from (47)

$$\lim_{n=\infty} \left\{ P[U_n(z) \mid \theta_n] - \int_{-\infty}^{z} e^{-\frac{1}{2}(\mu_n^2 - 2\mu_n t)/c} \, dF_n^*(t) \right\} = 0 \tag{49}$$

uniformly in z. It follows from (49) that for any positive L

$$\lim_{n=\infty} \left\{ P[U_n(L) \mid \theta_n] - P[U_n(-L) \mid \theta_n] - \int_{-L}^{L} e^{-\frac{1}{2}(\mu_n^2 - 2\mu_n t)/c} \, dF_n^*(t) \right\} = 0. \tag{50}$$

Since $\lim \mu_n = \mu$, $\lim_{n=\infty} [F_n^*(t) - F_n(t)] = 0$ uniformly in t, and since $\lim_{n=\infty} F_n(t) = F(t)$ uniformly in t, we get from (50)

$$\lim_{n=\infty} \{ P[U_n(L) \mid \theta_n] - P[U_n(-L) \mid \theta_n] \} = \int_{-L}^{L} e^{-\frac{1}{2}(\mu^2 - 2\mu t)/c} \, dF(t). \tag{51}$$

Now let us calculate the limit of $P[V_n(z) \mid \theta_n]$ if $n \to \infty$. The region $V_n(z)$ is defined by the inequality

$$(\hat{\theta}_n - \theta_0) \sqrt{n} < z. \tag{52}$$

This inequality can be written as follows:

$$(\hat{\theta}_n - \theta_n) \sqrt{n} < z - \mu_n. \tag{53}$$

Since $\lim \mu_n = \mu$, we get on account of Proposition 1

$$\begin{aligned} \lim_{n=\infty} P[(\hat{\theta}_n - \theta_n)\sqrt{n} < z - \mu_n \mid \theta_n] &= \frac{1}{\sqrt{2\pi c}} \int_{-\infty}^{z-\mu} e^{-\frac{1}{2}t^2/c} \, dt \\ &= \frac{1}{\sqrt{2\pi c}} \int_{-\infty}^{z} e^{-\frac{1}{2}(t-\mu)^2/c} \, dt \end{aligned} \tag{54}$$

Hence

$$\lim_{n=\infty} P[V_n(z) \mid \theta_n] = \frac{1}{\sqrt{2\pi c}} \int_{-\infty}^{z} e^{-\frac{1}{2}(t-\mu)^2/c} \, dt \tag{55}$$

uniformly in z.

For any positive ϵ let L_ϵ denote the positive number satisfying the condition:

$$\frac{1}{\sqrt{2\pi c}} \left[\int_{-\infty}^{-L_\epsilon} e^{-\frac{1}{2}(t-\mu)^2/c} \, dt + \int_{L_\epsilon}^{\infty} e^{-\frac{1}{2}(t-\mu)^2/c} \, dt \right] = \frac{\epsilon}{2}. \tag{56}$$

From (56) we easily get on account of (26)

$$(57)\qquad 0 \le \int_{-\infty}^{\infty} e^{-\frac{1}{2}(\mu^2 - 2\mu t)/c}\, dF(t) - \int_{-L_\epsilon}^{L_\epsilon} e^{-\frac{1}{2}(\mu^2 - 2\mu t)/c}\, dF(t) \le \frac{\epsilon}{2}.$$

Since the region $U_n(z_2) - U_n(z_1)$ is a subset of $V_n(z_2) - V_n(z_1)$ for $z_2 > z_1$, we have on account of (55) and (56)

$$(58)\quad \limsup_{n\to\infty} | \{P[U_n(\infty) \mid \theta_n] - P[U_n(L_\epsilon) \mid \theta_n] + P[U_n(-L_\epsilon) \mid \theta_n]\} | \le \frac{\epsilon}{2}.$$

Since

$$P[U_n(\infty) \mid \theta_n] = G(\mu_n, n),$$

we have

$$(59)\qquad \limsup_{n\to\infty} | G(\mu_n, n) - \{P[U_n(L_\epsilon) \mid \theta_n] - P[U_n(-L_\epsilon) \mid \theta_n]\} | \le \frac{\epsilon}{2}.$$

From (51), (57) and (59) we get

$$(60)\qquad \limsup_{n\to\infty} \left| G(\mu_n, n) - \int_{-\infty}^{\infty} e^{-\frac{1}{2}(\mu^2 - 2\mu t)/c}\, dF(t) \right| \le \epsilon.$$

Since ϵ can be chosen arbitrarily small, Proposition 2 is proved.

4. Theorems on asymptotically most powerful tests.

THEOREM 1: *Let M_n be the region defined by the inequality $\sqrt{n}\,(\hat{\theta}_n - \theta_0) \ge A_n$, where A_n is chosen such that $P(M_n \mid \theta_0) = \alpha$. Then $\{M_n\}$ is an asymptotically most powerful test of the hypothesis $\theta = \theta_0$, provided the parameter θ is restricted to values $\ge \theta_0$.*

PROOF: Assume that there exists a test $\{W_n\}$ of size α such that

$$(61)\qquad \limsup_{n\to\infty} L(W_n, M_n) = \delta > 0.$$

Then there exists a subsequence $\{n'\}$ of the sequence $\{n\}$ and a sequence $\{\theta_{n'}\}$ of parameter values $\ge \theta_0$ such that

$$(62)\qquad \lim_{n=\infty} \{P(W_{n'} \mid \theta_{n'}) - P(M_{n'} \mid \theta_{n'})\} = \delta$$

The expression

$$(63)\qquad (\theta_{n'} - \theta_0)\sqrt{n} = \mu_{n'} > 0$$

must be bounded. This can be proved as follows: Since under the assumption $\theta = \theta_0$ the distribution of $\sqrt{n}\,(\hat{\theta}_n - \theta_0)$ converges to a normal distribution with zero mean and finite variance, the sequence $\{A_n\}$ must be bounded. Hence M_n is defined by the inequality

$$(64)\qquad \hat{\theta}_n - \theta_0 \ge A_n/\sqrt{n} = \epsilon_n$$

where

$$\lim_{n=\infty} \epsilon_n = 0. \tag{65}$$

From Assumption 1, (64) and (65) it follows easily that if

$$\lim_{n=\infty} \theta_{n'} = \theta_1 > \theta_0, \qquad \lim_{n=\infty} P(M_{n'} \mid \theta_{n'}) = 1.$$

Hence on account of (62) we must have

$$\lim_{n=\infty} \theta_{n'} = \theta_0. \tag{66}$$

If there would exist a subsequence $\{n^*\}$ of $\{n'\}$ such that $\lim_{n=\infty} \mu_{n^*} = \infty$, then on account of (66) and Proposition 1 we would have $\lim_{n=\infty} P(M_{n^*} \mid \theta_{n^*}) = 1$, which is in contradiction to (62). Hence the expression (63) must be bounded. Let $\{n''\}$ be a subsequence of $\{n'\}$ such that

$$\lim_{n=\infty} \mu_{n''} = \mu > 0. \tag{67}$$

Denote by $F_n(z)$ the probability of the intersection of W_n and the region $(\hat{\theta}_n - \theta_0)\sqrt{n} < z$ under the hypothesis that $\theta = \theta_0$. Consider the subsequence $\{n'''\}$ of the sequence $\{n''\}$ such that $F_{n'''}(z)$ converges with $n \to \infty$ towards a function $F(z)$. The existence of such a subsequence $\{n'''\}$ can be proved as follows: Denote the probability $P[(\hat{\theta}_n - \theta_0)\sqrt{n} < z \mid \theta_0]$ by $\Phi_n(z)$. On account of Proposition 1, $\Phi_n(z)$ converges with $n \to \infty$ uniformly in z towards

$$\psi(z) = \frac{1}{\sqrt{2\pi c}} \int_{-\infty}^{z} e^{-\frac{1}{2}t^2/c}\, dt \tag{68}$$

where c has the same value in (23).
We obviously have

$$F_n(z_2) - F_n(z_1) \leq \Phi_n(z_2) - \Phi_n(z_1) \tag{69}$$

for any pair of values z_1, z_2 for which $z_2 > z_1$. Hence

$$\limsup_{n \to \infty} [F_n(z_2) - F_n(z_1)] \leq \psi(z_2) - \psi(z_1). \tag{70}$$

Since $F_n(z)$ is a monotonic function of z, our statement follows easily from (70) and the fact that $\psi(z)$ is uniformly continuous. Hence on account of Proposition 2 we have

$$\lim_{n=\infty} P(W_{n'''} \mid \theta_{n'''}) = \int_{-\infty}^{\infty} e^{-\frac{1}{2}(\mu^2 - 2\mu z)/c}\, dF(z) \tag{71}$$

and

$$\lim_{n=\infty} P(M_{n'''} \mid \theta_{n'''}) = \int_{-\infty}^{\infty} e^{-\frac{1}{2}(\mu^2 - 2\mu z)/c}\, d\Phi(z) \tag{72}$$

where

$$(73) \qquad \Phi(z) = 0 \text{ for } z \leq z_0 ,$$

$$(74) \qquad \Phi(z) = \psi(z) - \psi(z_0) \text{ for } z > z_0 ,$$

and z_0 is given by

$$(75) \qquad 1 - \psi(z_0) = \alpha.$$

From (62), (71) and (72) we get

$$(76) \qquad \int_{-\infty}^{\infty} e^{-\frac{1}{2}(\mu^2 - 2\mu z)/c} \, d[F(z) - \Phi(z)] = \delta > 0.$$

Consider a normally distributed variate y with mean ν and variance c. Let B be a critical region of size α for testing the hypothesis $\nu = 0$ by a single observation on y, i.e. B is a subset of the real axis $[-\infty, +\infty]$. Denote by $D(v)$ the intersection of B and the region $C(v)$ defined by the inequality $y < v$. Denote by $H(v)$ the probability of $D(v)$ under the hypothesis $\nu = 0$. Then the power of the test B with respect to the alternative $\nu = \mu$ is given by the following expression

$$(77) \qquad \int_{-\infty}^{\infty} e^{-\frac{1}{2}(\mu^2 - \mu v)/c} \, dH(v).$$

If the region B is given by the inequality $y \geq v_0$ where v_0 is chosen such that the size of B is equal to α, then $H(v) = \Phi(v)$ where the function Φ is defined by the equations (73), (74) and (75). Since the latter test is uniformly most powerful[4] with respect to all alternatives $\nu > 0$, for any positive μ the inequality

$$(78) \qquad \int_{-\infty}^{\infty} e^{-\frac{1}{2}(\mu^2 - \mu v)/c} \, d[H(v) - \Phi(v)] \leqq 0$$

holds. Let

$$\psi(v) = \frac{1}{\sqrt{2\pi c}} \int_{-\infty}^{v} e^{-\frac{1}{2}t^2/c} \, dt .$$

It is obvious that

$$(79) \qquad H(v_2) - H(v_1) \leq \psi(v_2) - \psi(v_1) \text{ for } v_2 > v_1$$

and

$$(80) \qquad \int_{-\infty}^{\infty} dH(v) = \alpha.$$

---

[4] See for instance J. Neyman and E. S. Pearson, "Contributions to the theory of testing statistical hypotheses," *Stat. Res. Memoirs*, Vol. 1 (1936).

On the other hand, if $K(v)$ is a monotonically non-decreasing non-negative function of v such that

$$K(v_2) - K(v_1) \leq \psi(v_2) - \psi(v_1) \text{ for } v_2 > v_1 \tag{79'}$$

and

$$\int_{-\infty}^{\infty} dK(v) = \alpha \tag{80'}$$

hold, then there exists a sequence $\{B^{(i)}\}$, $(i = 1, 2, \cdots$, ad inf.), of regions of size α such that

$$\lim_{i=\infty} H^{(i)}(v) = K(v)$$

uniformly in v. Since (78) holds for $H(v) = H^{(i)}(v)$, and since

$$H^{(i)}(v_2) - H^{(i)}(v_1) \leq \psi(v_2) - \psi(v_1) \text{ for } v_2 > v_1,$$

it is easy to see that (78) will hold also for $H(v) = K(v)$. Hence for any monotonically non-decreasing non-negative function $K(v)$ for which (79′) and (80′) are fulfilled, also (78) must hold. Since $F(v)$ is a distribution function which satisfies (79′) and (80′), we have a contradiction to (76). This proves Theorem 1.

THEOREM 2: *Let M_n be the region defined by the inequality $\sqrt{n}\,(\hat{\theta}_n - \theta_0) \leq A_n$, where A_n is chosen such that $P(M_n \mid \theta_0) = \alpha$. Then $\{M_n\}$ is an asymptotically most powerful test of the hypothesis $\theta = \theta_0$, provided that the parameter θ is restricted to values $\leq \theta_0$.*

We omit the proof since it is entirely analogous to that of Theorem 1.

THEOREM 3: *Let M_n be the region consisting of all points which satisfy at least one of the inequalities*

$$\sqrt{n}\,(\hat{\theta}_n - \theta_0) \leq -A_n, \qquad \sqrt{n}\,(\hat{\theta}_n - \theta_0) \geq A_n.$$

The constant $A_n > 0$ is chosen such that $P(M_n \mid \theta_0) = \alpha$. Then $\{M_n\}$ is an asymptotically most powerful unbiased test of the hypothesis $\theta = \theta_0$.

PROOF: Assume that there exists a sequence $\{W_n\}$ $(n = 1, 2, \cdots$, ad inf.) of regions such that

$$P(W_n \mid \theta_0) = \alpha \tag{81}$$

$$\lim_{n=\infty} g(W_n) = \alpha \tag{82}$$

and

$$\limsup_{n\to\infty} L(W_n, M_n) = \delta > 0. \tag{83}$$

We shall deduce a contradiction from this assumption. On account of (83) there exists a subsequence $\{n'\}$ of $\{n\}$ such that

$$\lim_{n=\infty} \{P(W_{n'} \mid \theta_{n'}) - P(M_{n'} \mid \theta_{n'})\} = \delta. \tag{84}$$

The expression

$$(85) \qquad (\theta_{n'} - \theta_0)\sqrt{n'} = \mu_{n'}$$

must be bounded. The proof of this statement is omitted, since it is analogous to the proof of the similar statement about (63). Hence there exists a subsequence $\{n''\}$ of $\{n'\}$ such that

$$(86) \qquad \lim_{n=\infty} \mu_{n''} = \mu.$$

Denote by $F_n(z)$ the probability of the intersection of W_n with the region $(\hat{\theta}_n - \theta_0)\sqrt{n} < z$ under the hypothesis $\theta = \theta_0$. Consider a subsequence $\{n'''\}$ of $\{n''\}$ such that $F_{n'''}(z)$ converges with $n \rightarrow \infty$ towards a function $F(z)$. The existence of such a sequence $\{n'''\}$ can be proved in the same way as the similar statement in the proof of Theorem 1. Hence on account of Proposition 2 and (86) we have

$$(87) \qquad \lim_{n=\infty} P(W_{n'''} \mid \theta_{n'''}) = \int_{-\infty}^{\infty} e^{-\frac{1}{2}(\mu^2 - 2\mu z)/c}\, dF(z)$$

and

$$(88) \qquad \lim_{n=\infty} P(M_{n'''} \mid \theta_{n'''}) = \int_{-\infty}^{\infty} e^{-\frac{1}{2}(\mu^2 - 2\mu z)/c}\, d\Phi(z)$$

where

$$(89) \qquad \Phi(z) = \frac{1}{\sqrt{2\pi c}} \int_{-\infty}^{z} e^{-\frac{1}{2}t^2/c}\, dt \quad \text{for} \quad z \leq -z_0,$$

$$(90) \qquad \Phi(z) = \Phi(-z_0) \quad \text{for} \quad -z_0 \leq z \leq z_0$$

$$(91) \qquad \Phi(z) = \Phi(-z_0) + \frac{1}{\sqrt{2\pi c}} \int_{z_0}^{z} e^{-\frac{1}{2}t^2/c}\, dt \quad \text{for} \quad z > z_0,$$

and

$$(92) \qquad \Phi(-z_0) = \tfrac{1}{2}\alpha.$$

From (84), (87) and (88) it follows that

$$(93) \qquad \int_{-\infty}^{\infty} e^{-\frac{1}{2}(\mu^2 - 2\mu z)/c}\, d[F(z) - \Phi(z)] = \delta.$$

Consider a normally distributed variate y with means ν and variance c. Let B an unbiased critical region of size α for testing the hypothesis $\nu = 0$ by a single observation on y, i.e. B is a subset of the real axis $[-\infty, +\infty]$. Denote by $D(v)$ the intersection of B with the region $C(v)$ defined by the inequality $y < v$. Denote by $H(v)$ the probability of $D(v)$ under the hypothesis $\nu = 0$. Then the power of the test B with respect to the alternative $\nu = \mu$ is given by

$$(94) \qquad \int_{-\infty}^{\infty} e^{-\frac{1}{2}(\mu^2 - 2\mu v)/c}\, dH(v).$$

If the region B consists of all points which satisfy at least one of the inequalities $y \leq -v_0$, $y \geq v_0$, and if $v_0 > 0$ is chosen such that the size of B is equal to α, then $H(v) = \Phi(v)$, where $\Phi(v)$ is defined by the equations (89)–(92). Since the latter test is a uniformly most powerful unbiased test,[5] for any μ the inequality

$$(95) \qquad \int_{-\infty}^{\infty} e^{-\frac{1}{2}(\mu^2 - 2\mu v)/c}\, d[H(v) - \Phi(v)] \leq 0$$

holds. Let

$$\psi(v) = \frac{1}{\sqrt{2\pi c}} \int_{-\infty}^{v} e^{-\frac{1}{2}t^2/c}\, dt.$$

It is obvious that

$$(96) \qquad H(v_2) - H(v_1) \leq \psi(v_2) - \psi(v_1) \quad \text{for } v_2 > v_1,$$

$$(97) \qquad \int_{-\infty}^{\infty} dH(v) = \alpha$$

and

$$(98) \qquad \int_{-\infty}^{\infty} e^{-\frac{1}{2}(\mu^2 - 2\mu v)/c}\, dH(v) \text{ has a minimum for } \mu = 0,$$

On the other hand, if $K(v)$ is a monotonically non-decreasing non-negative function of v such that

$$(96') \qquad K(v_2) - K(v_1) \leq \psi(v_2) - \psi(v_1) \text{ for } v_2 > v_1,$$

$$(97') \qquad \int_{-\infty}^{\infty} dK(v) = \alpha,$$

$$(98') \qquad \int_{-\infty}^{\infty} e^{-\frac{1}{2}(\mu^2 - 2\mu v)/c}\, dK(v) \text{ has a minimum for } \mu = 0,$$

then there exists a sequence $\{B^{(i)}\}$ ($i = 1, 2, \cdots$, ad inf.) of unbiased regions of size α such that

$$\lim_{i=\infty} H^{(i)}(v) = K(v)$$

uniformly in v. Since (95) holds for $H(v) = H^{(i)}(v)$ ($i = 1, 2, \cdots$, ad inf.), and since

$$H^{(i)}(v_2) - H^{(i)}(v_1) \leq \psi(v_2) - \psi(v_1) \text{ for } v_2 > v_1,$$

it is easy to see that (95) holds also for $H(v) = K(v)$. Hence for any monotonically non-decreasing non-negative function $K(v)$ for which (96′), (97′), and (98′) are fulfilled, also (95) must be fulfilled if we substitute $K(v)$ for $H(v)$.

[5] J. Neyman and E. S. Pearson, l. c., p. 29.

Since $F(v)$ is a distribution function which satisfies (96′), (97′) and (98′), we have a contradiction to (93). This proves Theorem 3.

5. Appendix. *Proof of the uniform consistency of* $\hat{\theta}_n$. It will be shown here that under certain conditions on the density function $f(x, \theta)$, Assumption 1, i.e. uniform consistency of $\hat{\theta}_n$, can be proved.

For any open subset ω of the θ-axis we denote by $\varphi(x, \omega)$ the least upper bound, and by $\psi(x, \omega)$ the greatest lower bound of $\frac{\partial^2 \log f(x, \theta)}{\partial \theta^2}$ with respect to θ in the set ω. For any function $\lambda(x)$ we denote by $E_\theta \lambda(x)$ the expected value of $\lambda(x)$ under the assumption that θ is the true value of the parameter, i.e.

$$E_\theta \lambda(x) = \int_{-\infty}^{\infty} \lambda(x) f(x, \theta)\, dx.$$

Denote furthermore by $P(\hat{\theta}_n \in \omega \mid \theta)$ the probability that $\hat{\theta}_n$ will fall in ω under the assumption that θ is the true value of the parameter. Finally denote by Ω the parameter space and assume that Ω is either the whole real axis or a subset of it.

PROPOSITION 3. *$\hat{\theta}_n$ is a uniformly consistent estimate of θ, i.e. for any positive k*

$$\lim_{n=\infty} P(-k < \hat{\theta}_n - \theta < k \mid \theta) = 1$$

uniformly for all θ in Ω, if the following two conditions are fulfilled:

Condition I. For all values θ in Ω

$$\int_{-\infty}^{\infty} \frac{\partial f(x, \theta)}{\partial \theta}\, dx = \int_{-\infty}^{\infty} \frac{\partial^2 f(x, \theta)}{\partial \theta^2}\, dx = 0.$$

Condition II. For any value θ in Ω there exists an open interval $\omega(\theta)$ containing θ and having the following three properties:

II_a. $$\lim_{n=\infty} P(\hat{\theta}_n \in \omega(\theta) \mid \theta] = 1$$

uniformly for all θ in Ω.

II_b. *$E_\theta \varphi^2[x, \omega(\theta)]$ is a bounded function of θ in Ω, and the least upper bound A of $E_\theta \varphi[x, \omega(\theta)]$ with respect to θ in Ω is negative.*

II_c. *$E_\theta \psi[x, \omega(\theta)]$ is a bounded function of θ in the set Ω.*

Condition I means simply that we may differentiate under the integral sign. In fact

$$\int_{-\infty}^{\infty} f(x, \theta) = 1$$

identically in θ. Hence

$$\frac{\partial}{\partial \theta} \int_{-\infty}^{\infty} f(x, \theta)\, dx = \frac{\partial^2}{\partial \theta^2} \int_{-\infty}^{\infty} f(x, \theta)\, dx = 0.$$

Differentiating under the integral sign, we obtain Condition I.

In case that $\omega(\theta)$ is the whole axis Condition II_a reduces to the condition that $\hat{\theta}_n$ exists.

In order to prove Proposition 3, we show first that for any positive η

$$\lim_{n=\infty} P\left[\left(-\eta < \frac{1}{n}\sum_{\alpha=1}^{n} \frac{\partial \log f(x_\alpha, \theta)}{\partial \theta} < \eta\right) \Big| \theta\right] = 1 \tag{99}$$

uniformly for all θ in Ω. We have on account of Condition I

$$E_\theta \frac{\partial \log f(x, \theta)}{\partial \theta} = E_\theta \frac{\partial f(x, \theta)}{\partial \theta} \Big/ f(x, \theta) = \int_{-\infty}^{\infty} \frac{\partial f(x, \theta)}{\partial \theta}\, dx = 0. \tag{100}$$

Since

$$\frac{\partial^2 \log f(x, \theta)}{\partial \theta^2} = \frac{\partial}{\partial \theta}\left[\frac{\partial f(x, \theta)}{\partial \theta} \Big/ f(x, \theta)\right] = \frac{\partial^2 f(x, \theta)}{\partial \theta^2} \Big/ f(x, \theta) - \left\{\frac{\partial f(x, \theta)}{\partial \theta} \Big/ [f(x, \theta]^2\right\}^2$$

we have on account of Condition I

$$E_\theta \left(\frac{\partial \log f(x, \theta)}{\partial \theta}\right)^2 = -E_\theta \frac{\partial^2 \log f(x, \theta)}{\partial \theta^2}. \tag{101}$$

According to Condition II $E_\theta \psi[x, \omega(\theta)] < 0$ and is a bounded function of θ. Since $E_\theta \dfrac{\partial^2 \log f(x, \theta)}{\partial \theta^2} < 0$ and $> E_\theta \psi[x, \omega(\theta)]$, the left hand side of (101), i.e. the variance of $\dfrac{\partial \log f(x, \theta)}{\partial \theta}$, is a bounded function of θ. From this and the equation (100) we obtain easily (99). Consider the Taylor expansion

$$\frac{1}{n}\sum_{\alpha} \frac{\partial \log f(x_\alpha, \theta)}{\partial \theta} = (\theta - \hat{\theta}_n)\frac{1}{n}\sum_{\alpha} \frac{\partial^2 \log f(x_\alpha, \theta'_n)}{\partial \theta^2}, \tag{102}$$

where θ'_n lies in the interval $[\theta, \hat{\theta}_n]$. Let ϵ be an arbitrary positive number and denote by $Q_n(\theta)$ the region defined by the inequality

$$\left|\frac{1}{n}\sum_{\alpha} \frac{\partial \log f(x_\alpha, \theta)}{\partial \theta}\right| \leq \epsilon. \tag{103}$$

On account of (99) we have

$$\lim_{n=\infty} P[Q_n(\theta) \mid \theta] = 1 \tag{104}$$

uniformly for all θ in Ω.

Denote by $R_n(\theta)$ the region defined by the inequality

$$\frac{1}{n}\sum_{\alpha} \varphi[x_\alpha, \omega(\theta)] < \tfrac{1}{2}A < 0. \tag{105}$$

On account of Condition II_b

$$\lim_{n=\infty} P[R_n(\theta) \mid \theta] = 1 \tag{106}$$

uniformly for all θ in Ω. Denote by $B_n(\theta)$ the region in which $\hat{\theta}_n \in \omega(\theta)$. Since in $B_n(\theta)$

$$\frac{1}{n} \sum \frac{\partial^2 \log f(x_\alpha, \theta_n')}{\partial \theta^2} \leq \frac{1}{n} \sum \varphi[x_\alpha, \omega(\theta)]$$

we have in the intersection $R_n'(\theta)$ of $R_n(\theta)$ and $B_n(\theta)$

$$\left| \frac{1}{n} \sum \frac{\partial^2 \log f(x_\alpha, \theta_n')}{\partial \theta^2} \right| > \left| \frac{A}{2} \right|. \tag{107}$$

Denote by $U_n(\theta)$ the intersection of $Q_n(\theta)$ and $R_n'(\theta)$. It is obvious that

$$\lim_{n=\infty} P[U_n(\theta) \mid \theta] = 1 \tag{108}$$

uniformly for all θ in Ω. From (102), (103) and (107) we get that in $U_n(\theta)$

$$|\theta - \hat{\theta}_n| \leq \frac{\epsilon}{|\frac{1}{2}A|} = \frac{2\epsilon}{|A|}. \tag{109}$$

Hence on account of (108)

$$\lim_{n=\infty} P\left(|\theta - \hat{\theta}_n| < \frac{2\epsilon}{|A|} \middle| \theta \right) = 1$$

uniformly for all θ in Ω. Since ϵ can be chosen arbitrarily, Proposition 3 is proved.

Conditions I and II are sufficient but not necessary for the uniform consistency of $\hat{\theta}_n$. For sufficiently small $\omega(\theta)$ the conditions II_b and II_c are rather weak. In fact, on account of (101) we have

$$E_\theta \frac{\partial^2 \log f(x, \theta)}{\partial \theta^2} < 0.$$

Hence for sufficiently small intervals $\omega(\theta)$, under certain continuity conditions, also $E_\theta \varphi[x, \omega(\theta)]$ will be negative. However, in some cases may be difficult to verify II_a for small $\omega(\theta)$. On the other hand, for sufficiently large $\omega(\theta)$ (certainly for $\omega(\theta) = [-\infty, +\infty]$) II_a can easily be verified, but the conditions II_b and II_c might be unnecessarily strong. In cases where II_b or II_c does not hold for $\omega(\theta) = [-\infty, +\infty]$ and the validity of II is not apparent, the following Lemma may be useful:

LEMMA: *Proposition* 3 *remains valid if we substitute for Condition II the conditions*

II′. *Denote by* T_n *the set of all points at which* $\hat{\theta}_n$ *exists and*

$$\sum_\alpha \frac{\partial}{\partial \theta} \log f(x_\alpha, \theta^*) = 0 \tag{110}$$

has at most one solution in θ^*. *Then* $\lim_{n=\infty} P[T_n \mid \theta] = 1$ *uniformly for all* θ *in* Ω, *and*

II″. *There exists a positive* k *such that for* $\omega(\theta) = I(\theta) = (\theta - k, \theta + k)$ *the following two conditions hold:*

II_b''. *$E_\theta\varphi^2[x, I(\theta)]$ is a bounded function of θ in Ω and the least upper bound A of $E_{\theta}\varphi[x, I(\theta)]$ with respect to θ in Ω is negative.*

II_c''. *$E_\theta\psi[x, I(\theta)]$ is a bounded function of θ in the set Ω.* In cases where II_b or II_c is not fulfilled for $\omega(\theta) = [-\infty, +\infty]$ the verification of II′ and II″ may be easier than that of II.

Our Lemma can be proved as follows: Consider the Taylor expansion

$$\frac{1}{n}\Sigma\frac{\partial}{\partial\theta}\log f(x_\alpha, \theta^*) = \frac{1}{n}\Sigma\frac{\partial}{\partial\theta}\log f(x_\alpha, \theta) + (\theta^* - \theta)\frac{1}{n}\Sigma\frac{\partial^2}{\partial\theta^2}\log f(x_\alpha, \theta') \tag{111}$$

where θ' lies in $[\theta, \theta^*]$. Denote by $V_n(\theta)$ the region defined by

$$\frac{1}{n}\Sigma\varphi[x_\alpha, I(\theta)] < \tfrac{1}{2}A < 0. \tag{112}$$

On account of II_b'' we have

$$\lim_{n=\infty} P[V_n(\theta) \mid \theta] = 1 \tag{113}$$

uniformly for all θ in Ω. Let $W_n(\theta)$ be the region defined by

$$\left|\frac{1}{n}\Sigma\frac{\partial}{\partial\theta}\log f(x_\alpha, \theta)\right| < \epsilon. \tag{114}$$

From Condition I and Condition II_c'' it follows easily that

$$\lim_{n=\infty} P[W_n(\theta) \mid \theta] = 1 \tag{115}$$

uniformly for all θ in Ω. For all values θ^* in the interval $I(\theta)$ we have

$$\frac{1}{n}\Sigma\varphi[x_\alpha, I(\theta)] \geq \frac{1}{n}\Sigma\frac{\partial^2}{\partial\theta^2}\log f(x_\alpha, \theta'). \tag{116}$$

Because of (112) and (116) we have in $V_n(\theta)$

$$\frac{1}{n}\Sigma\frac{\partial^2}{\partial\theta^2}\log f(x_\alpha, \theta') < \tfrac{1}{2}A < 0 \tag{117}$$

for all values θ^* in the interval $I(\theta)$. Let ϵ be less than $|\tfrac{1}{4}kA|$. Then in the intersection $W_n'(\theta)$ of the regions $V_n(\theta)$ and $W_n(\theta)$ we obviously have on account of (114) that the values of the left hand side of (111) for $\theta^* = \theta + k$ and $\theta^* = \theta - k$ will be of opposite sign. Hence at any point of $W_n'(\theta)$ the equation (110) has at least one root which lies in the interval $I(\theta)$. Since (110) has at most one root in T_n and since $\hat{\theta}_n$ is a root of (110), we get that at any point of the intersection $W_n''(\theta)$ of $W_n'(\theta)$ and T_n, $\hat{\theta}_n$ lies in $I(\theta)$. Since

$$\lim_{n=\infty} P[W''(\theta) \mid \theta] = 1 \quad \text{uniformly for all } \theta \text{ in } \Omega, \tag{118}$$

also

$$\lim_{n=\infty} P[\hat{\theta}_n \,\epsilon\, I(\theta) \mid \theta] = 1 \tag{119}$$

uniformly for all θ in Ω. The relation (119) combined with the conditions II_b'' and II_c'' is equivalent to Condition II. Hence our Lemma is proved.

Reprinted from THE ANNALS OF MATHEMATICAL STATISTICS
Vol. XII, No. 2, June, 1941

ON THE DISTRIBUTION OF WILKS' STATISTIC FOR TESTING THE INDEPENDENCE OF SEVERAL GROUPS OF VARIATES

BY A. WALD[1] AND R. J. BROOKNER[1]

Columbia University

1. **Introduction.** We consider p variates $x_1, x_2, \cdots, x_p$ which have a joint normal distribution. Let the variates be divided into k groups; group one containing $x_1, x_2, \cdots, x_{p_1}$, group two containing $x_{p_1+1}, x_{p_1+2}, \cdots, x_{p_2}$, etc. We are interested in testing the hypothesis that the set of all population correlation coefficients between any two variates which belong to different groups is zero.

Wilks[2] has derived, by using the Neyman-Pearson likelihood ratio criterion, a statistic based on N independent observations on each variate with which one may test this hypothesis. Let $\| r_{ij} \|$ be the matrix of sample correlation coefficients; Wilks' statistic, λ, is the ratio of the determinant of the p-rowed matrix of sample correlations to the product of the p_1-rowed determinant of correlations of the variates of group one, the $(p_2 - p_1)$-rowed determinant of correlations of the second group, etc. That is

$$\lambda = \frac{| r_{ij} |}{| r_{\alpha_1\beta_1} | \cdot | r_{\alpha_2\beta_2} | \cdots | r_{\alpha_k\beta_k} |}$$

where $| r_{\alpha_i\beta_i} |$ is the principal minor of $| r_{ij} |$ corresponding to the ith group.

In order to use the test, the distribution function of λ must be known. Wilks has shown that in certain cases the exact distribution is a simple elementary function; in other cases it is an elementary function, but one which is rather unwieldy and which does not lend itself readily to practical use. It is our purpose in this paper (1) to show a method by which the exact distribution can be explicitly given as an elementary function for a certain class of groupings of the variates, and (2) to give an expansion of the exact cumulative distribution function in an infinite series which is applicable to any grouping.

2. **The exact distribution of λ.** By the method to be described, the exact distribution of λ can be found when the numbers of variates in the groups are such that there are an odd number in at most one group. If the number of variates is small, say at most eight, the method will increase only slightly the list of distribution functions that Wilks gives in his paper.

---

[1] Research under a grant-in-aid of the Carnegie Corporation of New York.

[2] S. S. Wilks, "On the independence of k sets of normally distributed statistical variables," *Econometrica*, Vol. 3 (1935), pp. 309 326. Other references to Wilks in this paper except where otherwise noted are to this publication.

For purposes of deriving the distribution of λ we may assume that $E(x_u) = 0$, $(u = 1, 2, \cdots, p)$; that there are $n = N - 1$ independent observations $x_{u\alpha}$ $(\alpha = 1, 2, \cdots, n)$ on each variate x_u; and that the sample covariance between x_i and x_j is given by $s_{ij} = \sum_{\alpha=1}^{n} x_{i\alpha}x_{j\alpha}/n$. We define u' (a function of u) to be the total number of variables in all the groups which precede the group in which x_u lies. The complete theory is independent of the ordering of the groups and of the ordering of the variates within the groups; hence without loss of generality, we may assume that if any group contains an odd number of variates, it will be the last group, hence u' is always an even integer.

Wilks has shown that λ is a product $\prod_{u=p_1+1}^{p} z_u$ where each z_u is distributed independently of the others, and that the distribution of z_u is

$$\frac{z_u^{\frac{1}{2}(n-u-1)}(1 - z_u)^{\frac{1}{2}(u'-2)}}{\mathrm{B}[\frac{1}{2}(n - u + 1), u'/2]} dz_u . \tag{1}$$

Now let $y_u = \log z_u$, then the characteristic function of y_u is

$$\begin{aligned}\phi_u(t) &= \frac{1}{\mathrm{B}[\frac{1}{2}(n - u + 1), u'/2]} \int_0^1 e^{t \log z_u} z_u^{\frac{1}{2}(n-u-1)}(1 - z_u)^{\frac{1}{2}(u'-2)} dz_u \\ &= \frac{1}{\mathrm{B}[\frac{1}{2}(n - u + 1), u'/2]} \int_0^1 z_u^{\frac{1}{2}(n-u-1)+t}(1 - z_u)^{\frac{1}{2}(u'-2)} dz_u\end{aligned}$$

where t is a pure imaginary. It is known[3] that this integral, even with complex exponents, is the Beta-function so long as the real parts of both exponents are greater than minus one, so

$$\begin{aligned}\phi_u(t) &= \frac{\mathrm{B}[\frac{1}{2}(n - u + 1) + t, u'/2]}{\mathrm{B}[\frac{1}{2}(n - u + 1), u'/2]} \\ &= \frac{\Gamma[\frac{1}{2}(n - u + 1) + t\cdot]\Gamma[\frac{1}{2}(n - u + 1 + u')]}{\Gamma[\frac{1}{2}(n - u + 1 + u') + t]\cdot\Gamma[\frac{1}{2}(n - u + 1)]}.\end{aligned} \tag{2}$$

But here u' is always an even integer, hence by the well known recursion formula of the Gamma-function, which is valid for complex arguments excluding only negative integers

$$\phi_u(t) = c_u\{[\tfrac{1}{2}(n - u + 1) + t][\tfrac{1}{2}(n - u + 3) + t] \cdots [\tfrac{1}{2}(n - u + u' - 1) + t]\}^{-1}$$

where

$$c_u = [\tfrac{1}{2}(n - u + 1)][\tfrac{1}{2}(n - u + 3)] \cdots [\tfrac{1}{2}(n - u + u' - 1)].$$

[3] See Whittaker and Watson, *A Course in Modern Analysis*, Fourth edition 1927, Chap. 12.

Now set

$$y = \log \lambda = y_{p_1+1} + y_{p_1+2} + \cdots + y_p$$

and the characteristic function of y is

$$\phi(t) = \prod_{u=p_1+1}^{p} c_u\{[\tfrac{1}{2}(n - u + 1) + t][\tfrac{1}{2}(n - u + 3) + t] \cdots [\tfrac{1}{2}(n - u + u' - 1) + t]\}^{-1}.$$

From the characteristic function, we can obtain the distribution function, $g(y)$, of y by the relation

$$\begin{aligned} g(y) &= \frac{c_n}{2\pi i}\int_{-i\infty}^{i\infty} \frac{e^{-yt}\,dt}{\prod_{u=p_1+1}^{p}[\tfrac{1}{2}(n - u + 1) + t] \cdots [\tfrac{1}{2}(n - u + u' - 1) + t]} \\ &= \frac{c_n}{2\pi i}\int_{-i\infty}^{i\infty} \Phi(t)\,dt, \end{aligned}$$

where

$$c_n = \prod_{u=p_1+1}^{p} c_u.$$

The integration can be carried out by the method of residues; since y is always negative (the range of λ is from 0 to 1), on a half circle with center at the origin in the negative half of the complex t-plane, the integral of the function $\Phi(t)$ converges to zero as the radius of the circle becomes infinite. Since $\Phi(t)$ is analytic except for a finite number of poles on the negative real axis, $g(y)$ is c_n times the sum of the residues at these points.

Now $\Phi(t)$ is of the form $\dfrac{e^{-yt}}{P(t)}$ where $P(t)$ is a polynomial in t as follows: suppose that the groups contain $r_1, r_2, \cdots, r_k$ variables respectively, then let $(k_j + 1)$ be the number of these r's which are greater than or equal to j; then

$$P(t) = [\tfrac{1}{2}(n - 2) + t]^{k_1}[\tfrac{1}{2}(n - 3) + t]^{k_2}[\tfrac{1}{2}(n - 4) + t]^{k_3+k_1}[\tfrac{1}{2}(n - 5) + t]^{k_4+k_2} [\tfrac{1}{2}(n - 6) + t]^{k_5+k_3+k_1} \cdots [\tfrac{1}{2}(n - p + 1) + t]^{k_{p-2}+k_{p-4}+\cdots+k_{[\frac{1}{2}p]-[\frac{1}{2}(p-3)]}}.$$

where

$$[\sigma/2] = \begin{matrix} \sigma/2 \text{ if } \sigma \text{ is even} \\ (\sigma - 1)/2 \text{ if } \sigma \text{ is odd.} \end{matrix}$$

Then

$$g(y; r_1, r_2, \cdots, r_k) = c_n \sum_{\alpha=1}^{p-2} \frac{1}{\theta_\alpha!}\frac{d^{\theta_\alpha}}{dt^{\theta_\alpha}}[(t + \tfrac{1}{2}(n - \alpha - 1))^{\theta_\alpha+1}\Phi(t)]_{t=-\frac{1}{2}(n-\alpha-1)}$$

where

$$\theta_\alpha + 1 = k_\alpha + k_{\alpha-2} + \cdots + k_{[\frac{1}{2}(\alpha+2)]-[\frac{1}{2}(\alpha-1)]}.$$

It can be shown that θ_α is $\geqq 0$ for α between 1 and $p - 2$. Thus we have $g(y; r_1, r_2, \cdots, r_k)$ and from it we can calculate $f(\lambda; r_1, r_2, \cdots, r_k)$.

Suppose $p = 8$ and that the variables are divided into two groups of four each, then we will calculate the distribution function $f(\lambda; 4, 4)$. Now

$$g(y; 4, 4) = \frac{c_n}{2\pi i}\int_{-i\infty}^{i\infty} \frac{e^{-yt}\,dt}{[\frac{1}{2}(n-2)+t][\frac{1}{2}(n-3)+t][\frac{1}{2}(n-4)+t]^2 \cdot[\frac{1}{2}(n-5)+t]^2[\frac{1}{2}(n-6)+t][\frac{1}{2}(n-7)+t]}$$

and

$$c_n = \left(\frac{n-2}{2}\right)\left(\frac{n-3}{2}\right)\left(\frac{n-4}{2}\right)^2\left(\frac{n-5}{2}\right)^2\left(\frac{n-6}{2}\right)\left(\frac{n-7}{2}\right).$$

Then

$$g(y; 4, 4) = 16c_n\left[\frac{-e^{\frac{1}{2}(n-2)y}}{90} + e^{\frac{1}{2}(n-3)y} + \frac{8e^{\frac{1}{2}(n-4)y}}{9} - \frac{8e^{\frac{1}{2}(n-5)y}}{9} - e^{\frac{1}{2}(n-6)y} + \frac{e^{\frac{1}{2}(n-7)y}}{90} - \frac{ye^{\frac{1}{2}(n-4)y}}{3} + \frac{ye^{\frac{1}{2}(n-5)y}}{3}\right].$$

Since

$$y = \log\lambda, \qquad dy = \frac{d\lambda}{\lambda},$$

we have

$$f(\lambda; 4, 4,) = \frac{16c_n}{3}\left[-\frac{\lambda^{\frac{1}{2}(n-4)}}{30} + \frac{\lambda^{\frac{1}{2}(n-5)}}{2} - \frac{8\lambda^{\frac{1}{2}(n-6)}}{3} + \frac{8\lambda^{\frac{1}{2}(n-7)}}{3} - \frac{\lambda^{\frac{1}{2}(n-8)}}{2} + \frac{\lambda^{\frac{1}{2}(n-9)}}{30} - (\lambda^{\frac{1}{2}(n-7)} + \lambda^{\frac{1}{2}(n-6)})\log\lambda\right].$$

The cumulative distribution function is given by

$$J_w(4, 4) = \text{Prob}\,[\lambda \leq w; 4, 4]$$

$$= \frac{16c_n}{3}w^{\frac{1}{2}(n-7)}\left[\frac{1}{15(n-7)} - \frac{w^{\frac{1}{2}}}{n-6} - \frac{4(4n-23)w}{3(n-5)^2} + \frac{14(4n-13)w^{\frac{3}{2}}}{3(n-4)^2} + \frac{w^2}{n-3} - \frac{w^{\frac{5}{2}}}{15(n-2)} - \left(\frac{2w}{n-5} + \frac{2w^{\frac{3}{2}}}{n-4}\right)\log w\right].$$

Wilks' expression for the cumulative distribution function appears to be quite different, but if we substitute $n = N - 1$ and use the relation

$$\beta_{\sqrt{w}}(N-6; 4) = \frac{\Gamma(N-2)}{\Gamma(N-6)\cdot\Gamma(4)}\int_0^{\sqrt{w}} x^{N-7}(1-x)^3\,dx$$

$$= \tfrac{1}{6}(n-2)(n-3)(n-4)(n-5) \cdot\left[\frac{w^{\frac{1}{2}(n-5)}}{n-5} - \frac{3w^{\frac{1}{2}(n-4)}}{n-4} + \frac{3w^{\frac{1}{2}(n-3)}}{n-3} - \frac{w^{\frac{1}{2}(n-2)}}{n-2}\right]$$

it can be shown that the two formulas for the cumulative distribution are identical.

In cases where u' is not always an even integer, the exact distribution function of λ can still be obtained using this method. However, in such a case, the gamma functions do not cancel out and the integrand has an infinitude of poles, so the function is expressed by an infinite series. We will use a different method to obtain an infinite series expansion.

3. **A series expansion of the cumulative distribution function.** Let us put $v = -y$, and let the density function of v be $h(v)$, then from (2), we have

$$h(v)\,dv = dv\,\frac{c_n}{2\pi i}\int_{-i\infty}^{i\infty} e^{vt} \prod_{u=r_1+1}^{p} \frac{\Gamma[\frac{1}{2}(n-u+1)+t]\,dt}{\Gamma[\frac{1}{2}(n-u+1+u')+t]}.$$

Since v is a monotonic decreasing function of λ, and since the critical region for testing the null hypothesis is given by the inequality $\lambda < \lambda_0$, then the critical region will be defined by $v > v_0$, where v_0 is such that

$$\int_{v_0}^{\infty} h(v)\,dv$$

is equal to a chosen level of significance.

PROPOSITION 1.

$$h(v) = h_n(v)\bar{\psi}(v)$$

where $\bar{\psi}(v)$ *does not depend on* n, *and* $h_n(v) = c_n e^{-\frac{n}{2}v}$.

PROOF: Let

$$t' = t + \tfrac{1}{2}(n-p).$$

Then

$$h(v) = \frac{c_n}{2\pi i}\int_{-i\infty+\frac{1}{2}(n-p)}^{i\infty+\frac{1}{2}(n-p)} e^{v(t'-\frac{1}{2}(n-p))} \prod_{u} \frac{\Gamma[\frac{1}{2}(p-u+1)+t']\,dt'}{\Gamma[\frac{1}{2}(p-u+u'+1)+t']}.$$

Now the area in the complex plane bounded by the vertical line through $\frac{1}{2}(n-\text{p})$, by the vertical line through the origin, and by arcs of a circle with center at the origin of arbitrary radius is one in which the integrand is everywhere regular. Furthermore, the integral along the arcs approaches zero as the radius of the circle approaches infinity, hence the integrals along the vertical line through $\frac{1}{2}(n-p)$ and along the vertical axis are equal. Then we may write

$$\begin{aligned}\frac{e^{\frac{n}{2}v}}{c_n}h(v) &= \frac{1}{2\pi i}\int_{-i\infty}^{i\infty} e^{v(t'+p/2)} \prod_{u} \frac{\Gamma[\frac{1}{2}(p-u+1)+t']\,dt'}{\Gamma[\frac{1}{2}(p-u+u'+1)+t']}\\ &= \bar{\psi}(v).\end{aligned}$$

Therefore

$$h(v) = c_n e^{-\frac{n}{2}v}\bar{\psi}(v).$$

PROPOSITION 2.

$$I = \lim_{n\to\infty} \int_0^\infty \frac{c_n e^{-\frac{n}{2}v} v^{r-1}\, dv}{\Gamma(r)} = 1$$

where we define

$$r = \sum_{j=i+1}^{k} \sum_{i=1}^{k-1} \frac{r_i r_j}{2}$$

so that

$$\begin{aligned} r &= \tfrac{1}{2}[r_2 r_1 + r_3(r_1 + r_2) + \cdots + r_k(r_1 + r_2 + \cdots + r_{k-1})] \\ &= \tfrac{1}{2} \sum_u u'. \end{aligned}$$

PROOF: Let

$$\frac{n}{2} v = v^*$$

then

$$\begin{aligned} \int_0^\infty c_n e^{-\frac{n}{2}v} v^{r-1}\, dv &= \int_0^\infty c_n e^{-v^*} \left(\frac{2}{n}\right)^r (v^*)^{r-1}\, dv^* \\ &= c_n \left(\frac{2}{n}\right)^r \Gamma(r). \end{aligned}$$

Hence

$$I = \lim_{n\to\infty} c_n \left(\frac{2}{n}\right)^r$$

but

$$c_n = \prod_u \frac{\Gamma\frac{1}{2}(n - u + 1 + u')}{\Gamma\frac{1}{2}(n - u + 1)}$$

and therefore

$$I_u = \lim_{n\to\infty} \frac{\Gamma\frac{1}{2}(n - u + 1 + u')}{\Gamma\frac{1}{2}(n - u + 1)} \left(\frac{2}{n}\right)^{u'/2} = 1$$

by an application of the Stirling approximation. Therefore

$$I = \prod_u I_u = 1.$$

We then write

$$\psi(v) = \frac{\bar{\psi}(v)\Gamma(r)}{v^{r-1}}$$

hence

$$h(v) = \frac{c_n e^{-\frac{n}{2}v} v^{r-1} \psi(v)}{\Gamma(r)}. \tag{3}$$

PROPOSITION 3. *For any positive integer s,*

$$\lim_{n \to \infty} \left\{ n^s \cdot \text{Prob}\left(v > \frac{1}{\sqrt{n}} \right) \right\} = 0.$$

PROOF: Since $v = -\log \lambda$, the inequality $v > 1/\sqrt{n}$ is equivalent to the inequality $\lambda < e^{-1/\sqrt{n}}$. Since $\lambda = \prod_{u=p_1+1}^{p} z_u$, the inequality $\lambda < e^{-1/\sqrt{n}}$ implies that there exists at least one value of u for which

$$z_u < e^{-1/(p-p_1)\sqrt{n}}.$$

Hence

$$\sum_{u=p_1+1}^{p} P(z_u < e^{-1/(p-p_1)\sqrt{n}}) \geq P(\lambda < e^{-1/\sqrt{n}}) = P(v > 1/\sqrt{n}).$$

Hence in order to prove Proposition 3 we have only to show that for each u and any arbitrary positive integer s

$$\lim_{n \to \infty} \{ n^s \cdot P(z_u < e^{-1/(p-p_1)\sqrt{n}}) \} = 0.$$

From (1) we have

$$P(z_u < e^{-1/(p-p_1)\sqrt{n}})$$

$$= \frac{1}{B[\frac{1}{2}(n-u+1); u'/2]} \int_0^{e^{-1/(p-p_1)\sqrt{n}}} z_u^{\frac{1}{2}(n-u-1)} (1-z_u)^{\frac{1}{2}(u'-2)} \, dz_u .$$

Over the range of integration, we have $z_u \leq e^{-1/(p-p_1)\sqrt{n}}$ so

$$P(z_u < e^{-1/(p-p_1)\sqrt{n}}) \leq \frac{e^{\frac{1}{2}(n-u-1)/(p-p_1)\sqrt{n}}}{B[\frac{1}{2}(n-u+1); u'/2]} \int_0^{e^{-1/(p-p_1)\sqrt{n}}} (1-z_u)^{\frac{1}{2}(u'-2)} \, dz_u$$

$$= \frac{e^{-\frac{1}{2}(n-u-1)/(p-p_1)\sqrt{n}}}{B[\frac{1}{2}(n-u+1); u'/2]} \left[-\frac{2}{u'} (1-z_u)^{u'/2} \right]_0^{e^{-1/(p-p_1)\sqrt{n}}}$$

$$= \frac{2e^{-\frac{1}{2}(n-u-1)/(p-p_1)\sqrt{n}}}{u' \cdot B[\frac{1}{2}(n-u+1); u'/2]} [1 - (1 - e^{-1/(p-p_1)\sqrt{n}})^{u'/2}].$$

It follows from the Stirling formula that

$$\lim_{n \to \infty} \left(\frac{n}{2}\right)^{u'/2} B[\tfrac{1}{2}(n-u+1); u'/2] = \lim_{n \to \infty} \frac{\Gamma\frac{1}{2}(n-u+1)\Gamma(u'/2)}{\Gamma\frac{1}{2}(n-u+u'+1)} \left(\frac{n}{2}\right)^{u'/2}$$

$$= \Gamma(u'/2).$$

Since

$$\lim_{n\to\infty} n^{\frac{1}{2}u'+s} e^{-\sqrt{n}/2(p-p_1)} = 0$$

and

$$\lim_{n\to\infty} (1 - (1 - e^{-1/\sqrt{n}})) = 1,$$

the proposition follows.

Proposition 4. *The function $\psi(v)$ of formula (3) can be expanded in a power series, i.e.*

$$\psi(v) = \alpha_0 + \alpha_1 v + \alpha_2 v^2 + \cdots$$

with a finite radius of convergence.

Proof: Wilks[4] has considered the following integral equation:

$$\int_0^\beta w^t g(w)\, dw = CB^t \frac{\Gamma(b_1 + t)\cdot\Gamma(b_2 + t)\cdots\Gamma(b_q + t)}{\Gamma(c_1 + t)\cdot\Gamma(c_2 + t)\cdots\Gamma(c_q + t)},$$

where $C = \dfrac{\Gamma(c_1)\cdot\Gamma(c_2)\cdots\Gamma(c_q)}{\Gamma(b_1)\cdot\Gamma(b_2)\cdots\Gamma(b_q)}$, B and $g(w)$ are independent of t, and $b_i < c_i$ $(i = 1, 2, \cdots, q)$. Wilks has shown that the solution of the integral equation, $g(w)$, is given by the following expression:

$$\begin{aligned}
g(w) = {} & \frac{kw^{b_q-1}\left(1 - \frac{w}{B}\right)^{\gamma_q-\beta_q-1}}{B^{b_q}} \int_0^1\int_0^1 \cdots \int_0^1 v_1^{c_1-b_1-1} v_2^{c_2-b_2-1} \cdots v_{q-1}^{c_{q-1}-b_{q-1}-1} \\
& \times (1 - v_1)^{\gamma_{q-1}-\beta_{q-1}-1}(1 - v_2)^{\gamma_{q-2}-\beta_{q-2}-1} \cdots (1 - v_{q-1})^{\gamma_1-\beta_1-1} \\
& \times \left[1 - v_1\left(1 - \frac{w}{B}\right)\right]^{b_1-c_2}\left[1 - \{v_1 + v_2(1 - v_1)\}\left(1 - \frac{w}{B}\right)\right]^{b_2-c_3} \cdots \\
& \times \Big[1 - \{v_1 + v_2(1 - v_1) + \cdots \\
& \qquad + v_{q-1}(1 - v_1)(1 - v_2)\cdots(1 - v_{q-2})\}\left(1 - \frac{w}{B}\right)\Big]^{b_{q-1}-c_q} \\
& \times dv_1\, dv_2 \cdots dv_{q-1}
\end{aligned} \tag{4}$$

where

$$k = \prod_{i=1}^{q} \frac{\Gamma(c_i)}{\Gamma(b_i)\Gamma(c_i - b_i)}$$

and

$$\gamma_i = \sum_{j=0}^{i-1} c_{q-j} \qquad \beta_i = \sum_{j=0}^{i-1} b_{q-j}$$

---

[4] S. S. Wilks, "Certain generalizations in the analysis of variance," *Biometrika*, Vol. 24 (1932), pp. 474–5.

the range of w being $0 \leqq w \leqq B$. Wilks has furthermore shown that

$$(5) \quad \{v_1 + v_2(1 - v_1) + \cdots + v_i(1 - v_1)(1 - v_2) \cdots (1 - v_{i-1})\}\left(1 - \frac{w}{B}\right) < 1$$

for $w > 0$ and $0 \leqq v_i \leqq 1$ $(i = 1, 2, \cdots, q - 1)$.

We denote the left hand side of (5) by ζ_i. The factor $(1 - \zeta_i)^{b_i - c_{i+1}}$ can be expanded in a power series, i.e.

$$(6) \quad (1 - \zeta_i)^{b_i - c_{i+1}} = (1 - \zeta_i)^{-(c_{i+1} - b_i)}$$
$$= 1 + (c_{i+1} - b_i)\zeta_i + \tfrac{1}{2}(c_{i+1} - b_i)(c_{i+1} - b_i + 1)\zeta_i^2 + \cdots$$

with a radius of convergence equal to one. Since we will show shortly that for the choices we make for the b_i's and c_i's, $c_{i+1} \geq b_i$, then all coefficients in this last expansion are non-negative. Substituting this series expansion (6) in (4), and ordering it according to powers of $(1 - w/B)$, the expression under the integral sign (in 4) becomes

$$(7) \quad \begin{aligned} &\theta_0(v_1, v_2, \cdots v_{q-1}) \\ &+ \theta_1(v_1, \cdots v_{q-1})\left(1 - \frac{w}{B}\right) + \theta_2(v_1, \cdots, v_{q-1})\left(1 - \frac{w}{B}\right)^2 + \cdots. \end{aligned}$$

This series is uniformly convergent over the domain defined by the inequalities $0 \leqq v_i \leqq 1$ $(i = 1, 2, \cdots, q - 1)$ and $|1 - w/B| < 1$. We can even say that (7) is uniformly convergent for $|1 - w/B| < 1$ if we substitute for each θ_i the maximum of θ_i with respect to $v_1, v_2, \cdots, v_{q-1}$. Hence we may integrate the series (7) with respect to $v_1, v_2, \cdots v_{q-1}$ term by term, i.e.

$$(8) \quad \int_0^1 \int_0^1 \cdots \int_0^1 (7)\, dv_1\, dv_2 \cdots dv_{q-1} = \sigma_0 + \sigma_1\left(1 - \frac{w}{B}\right) + \sigma_2\left(1 - \frac{w}{B}\right)^2 + \cdots$$

and the series (8) is uniformly convergent for $|1 - w/B| < 1$. The coefficients $\sigma_0, \sigma_1, \cdots$ are non-negative.

The case of the λ statistic which we are considering is a special case of this integral equation which we obtain by making the following substitutions:

$$w = \lambda, \quad B = 1, \quad u = r + p_1, \quad q = p - p_1$$

$$b_r = \tfrac{1}{2}(n - u + 1), \quad c_r = \tfrac{1}{2}(n - u + u' + 1), \quad (r = 1, 2, \cdots, p - p_1)$$

Note that then

$$c_{r+1} - b_r = \tfrac{1}{2}[(u + 1)' - 1] \geqq 0.$$

Hence, according to (4)

$$g(\lambda)\, d\lambda = k \cdot \lambda^{\frac{1}{2}(n-p-1)}(1 - \lambda)^{\frac{1}{2}\Sigma u' - 1}\{\sigma_0 + \sigma_1(1 - \lambda) + \sigma_2(1 - \lambda^2) + \cdots\}\, d\lambda$$

where the infinite series converges for $|1 - \lambda| < 1$.

Now $v = -\log \lambda$, or $\lambda = e^{-v}$, hence

$$h(v)\, dv = k \cdot e^{-\frac{1}{2}(n-p+1)v} v^{r-1}\left(\frac{1 - e^{-v}}{v}\right)^{r-1}\{\epsilon_0 + \epsilon_1 v + \epsilon_2 v^2 + \cdots\}\, dv$$

where the series $\{\epsilon_0 + \epsilon_1 v + \epsilon_2 v^2 + \cdots\}$ is obtained from the series $\{\sigma_0 + \sigma_1(1 - \lambda) + \cdots\}$ by substituting for $(1 - \lambda)$ the Taylor expansion of $(1 - e^{-v})$. The series $\{\epsilon_0 + \epsilon_1 v + \epsilon_2 v^2 + \cdots\}$ has a finite radius of convergence.[5]

Hence the function $\psi(v)$ can be written as

$$\psi(v) = A \cdot e^{\frac{1}{2}(p-1)v} \left(\frac{1 - e^{-v}}{v}\right)^{r-1} \{\epsilon_0 + \epsilon_1 v + \epsilon_2 v^2 + \cdots\}$$

where A denotes a constant factor. Then since $e^{\frac{1}{2}(p-1)v}\left(\frac{1 - e^{-v}}{v}\right)^{r-1}$ can be expanded in a Taylor series around $v = 0$, Proposition 4 is proved.

4. **Evaluation of the coefficients in the expansion of $\psi(v)$.** Let the series expansion of $\psi(v)$ be

$$\psi(v) = \alpha_0 + \alpha_1 v + \alpha_2 v^2 + \cdots$$

Then we have

$$\int_0^\infty \frac{c_n e^{-\frac{1}{2}nv} v^{r-1}}{\Gamma(r)} (\alpha_0 + \alpha_1 v + \alpha_2 v^2 + \cdots)\, dv \equiv 1.$$

Now let $v^* = \frac{n}{2} v$, then

$$\int_0^\infty \left(\frac{2}{n}\right)^r \frac{c_n e^{-v^*} v^{*^{r-1}}}{\Gamma(r)} \left(\alpha_0 + \frac{2\alpha_1 v^*}{n} + \frac{4\alpha_2 v^{*^2}}{n^2} + \cdots\right) dv^* \equiv 1.$$

Suppose that the asymptotic expansion of $\left(\frac{n}{2}\right)^r \frac{1}{c_n}$ is given by

$$\beta_0 + \frac{\beta_1}{n} + \frac{\beta_2}{n^2} + \cdots.$$

On account of Proposition 3, we have that the asymptotic expansion in powers of $1/n$ of

$$(9) \qquad \int_0^{\sqrt{n}} \frac{e^{-v^*} v^{*^{r-1}}}{\Gamma(r)} \left(\alpha_0 + \frac{2\alpha_1}{n} v^* + \frac{4\alpha_2}{n^2} v^{*^2} + \cdots\right) dv^*$$

must be equal to the asymptotic expansion of $\left(\frac{n}{2}\right)^r \frac{1}{c_n}$. Since we may integrate in (9) term by term for sufficiently large n, we easily obtain

$$\alpha_0 = \beta_0, \qquad \alpha_1 = \frac{\beta_1}{2r}, \cdots \alpha_k = \frac{\beta_k}{2^k \cdot r(r+1) \cdots (r+k-1)}.$$

---

[5] See A. Gutzmer, *Theorie der Eindeutigen Analytischen Funktionen*, 1906, pp. 91–2.

The asymptotic expansion of $\left(\frac{n}{2}\right)^r \frac{1}{c_n}$ can be calculated in the following manner:

$$\left(\frac{n+2}{n}\right)^r \frac{c_n}{c_{n+2}} = \frac{\beta_0 + \frac{\beta_1}{n+2} + \frac{\beta_2}{(n+2)^2} + \cdots}{\beta_0 + \frac{\beta_1}{n} + \frac{\beta_2}{n^2} + \cdots}$$

and

$$\left(\frac{n+2}{n}\right)^r \frac{c_n}{c_{n+2}} = (1 + 2/n)^r \prod_u \frac{n - u + 1}{n - u + u' + 1}.$$

Equating the right hand members of these last two equations, and taking logs, we obtain

$$\log\left[\beta_0 + \frac{\beta_1}{n+2} + \frac{\beta_2}{(n+2)^2} + \cdots\right] = r \log(1 + 2/n) + \sum_u \log\left(1 - \frac{u-1}{n}\right)$$
$$- \sum_u \log\left(1 - \frac{u - u' - 1}{n}\right) + \log\left(\beta_0 + \frac{\beta_1}{n} + \frac{\beta_2}{n^2} + \cdots\right).$$

Then we expand each term in a series of powers of $1/n$ and equate coefficients of $1/n^i$ for each i. We obtain the following formulae for the first five β's:

$$\beta_0 = 1$$

$$\beta_1 = r + \tfrac{1}{4} \sum_u (u-1)^2 - \tfrac{1}{4} \sum_u (u - u' - 1)^2$$

$$\beta_2 = \beta_1 + \frac{\beta_1^2}{2} - \frac{2r}{3} + \frac{1}{12} \sum_u (u-1)^3 - \frac{1}{12} \sum_u (u - u' - 1)^3$$

$$\beta_3 = -\tfrac{4}{3}\beta_1 - \beta_1^2 - \tfrac{1}{3}\beta_1^3 + \beta_1\beta_2 + 2\beta_2 + \tfrac{2}{3}r$$
$$+ \tfrac{1}{24} \sum_u (u-1)^4 - \tfrac{1}{24} \sum_u (u - u' - 1)^4$$

$$\beta_4 = 2\beta_1 + 2\beta_1^2 + \beta_1^3 + \frac{\beta_1^4}{4} - 3\beta_1\beta_2 + \beta_1\beta_3 - \beta_1^2\beta_2 - 4\beta_2$$
$$+ \frac{\beta_2^2}{2} + 3\beta_3 - \frac{4}{5}r + \frac{1}{40} \sum_u (u-1)^5 - \frac{1}{40} \sum_u (u - u' - 1)^5.$$

5. **Practical use of the series.** In practical applications, the value of the statistic, say λ_0, is calculated, and it is desired that we determine whether or not this value of the statistic falls into the critical region. That is, for a particular grouping of the variates, for a particular number of degrees of freedom, and for a chosen level of significance α, there is determined from the distribution of λ, a value λ^* such that

$$\text{Prob}\,[\lambda < \lambda^*] = \alpha,$$

and if $\lambda_0 < \lambda^*$ we reject the hypothesis that in the population from which the sample is taken all the correlation coefficients between variates in different groups are zero.

Since v is a monotonic decreasing function of λ we make the test by computing $v_0 = -\log \lambda_0$ and we reject the hypothesis if $v_0 > v^*$ where $v^* = -\log \lambda^*$. But this is equivalent to computing Prob $[v > v_0]$ and if this value is less than α we reject the hypothesis. Now

$$\text{Prob}\,[v > v_0] = J_{v_0}(r_1, r_2, \cdots, r_k)$$
$$= \frac{C_n}{\Gamma(r)} \int_{v_0}^{\infty} e^{-\frac{1}{2}nv} v^{r-1}(1 + \alpha_1 v + \alpha_2 v^2 + \cdots)\, dv.$$

Setting $\frac{nv}{2} = z$

$$\text{Prob}\,[v > v_0] = \left(\frac{2}{n}\right)^r \frac{C_n}{\Gamma(r)} \int_{nv_0/2}^{\infty} e^{-z} z^{r-1} \left[1 + \alpha_1 \frac{2z}{n} + \alpha_2 \left(\frac{2}{n}\right)^2 z^2 + \cdots\right] dz.$$

On account of Proposition 3 we obtain an asymptotic expansion of Prob $[v > v_0]$ by integrating the right hand member of the above equation term by term. This can be expressed by means of the incomplete gamma function, which is tabulated[6] in the form

$$I(u, p) = \frac{\int_0^{u\sqrt{p+1}} v^p e^{-v}\, dv}{\Gamma(p + 1)}.$$

We obtain

$$\text{Prob}\,[v > v_0] = \left(\frac{2}{n}\right)^r c_n \left\{\left[1 - I\left(\frac{nv_0}{2\sqrt{r}}, r - 1\right)\right]\right.$$
$$\left. + \frac{\beta_1}{n}\left[1 - I\left(\frac{nv_0}{2\sqrt{r+1}}, r\right)\right] + \frac{\beta_2}{n^2}\left[1 - I\left(\frac{nv_0}{2\sqrt{r+2}}, r + 1\right)\right] + \cdots\right\}.$$

The values of the constant $K = \left(\frac{2}{n}\right)^r c_n$ and the values of $\beta_1, \beta_2, \beta_3, \beta_4$ are herein tabulated for any grouping which might be made on six or fewer variates. Some cases, such as groupings $(1, p - 1)$, in which case the distribution of λ is the distribution of the multiple correlation coefficient; and as the groupings $(2, p - 2)$, the exact distribution for which was given by Wilks as an incomplete Beta-function, are superfluous here. These cases are included only for the sake of completeness.

---

[6] K. Pearson (Editor), *Tables of the Incomplete Gamma Function*, Biometric Laboratory, London, 1922.

Table of the First Four β's

Grouping	r	β_1	β_2	β_3	β_4
2,1	1	2	4	8	16
1,1,1	1.5	2.75	6.28125	13.38281	27.57568
3,1	1.5	3.75	12.03125	36.91406	111.55225
2,2	2	5	19	65	211
2,1,1	2.5	5.75	23.53125	83.97656	279.50538
1,1,1,1	3	6.5	28.625	106.9375	366.39844
4,1	2	6	28	120	496
3,2	3	9	55	285	1351
3,1,1	3.5	9.75	62.53125	334.10156	1615.91163
2,2,1	4	11	77	439	2229
2,1,1,1	4.5	11.75	86.03125	506.16406	2628.23974
1,1,1,1,1	5	12.5	95.625	580.6875	3085.52344
5,1	2.5	8.75	55.78125	315.82031	1690.65282
4,2	4	14	125	910	5901
3,3	4.5	15.75	154.03125	1205.03906	8277.55226
4,1,1	4.5	14.75	136.28125	1015.50781	6693.45068
3,2,1	5.5	17.75	189.53125	1584.10156	11445.75538
2,2,2	6	19	214	1866	13947
3,1,1,1	6	18.5	203.625	1740.9375	12797.27344
2,2,1,1	6.5	19.75	229.03125	2042.16406	15530.08351
2,1,1,1,1	7	20.5	244.625	2230.1875	17257.64836
1,1,1,1,1,1	7.5	21.25	260.78125	2430.49219	19139.02892

Tables of the Constant $K = \left(\frac{2}{n}\right)^r C_n$

n	21	111	31	22	211	1111	41	311
10	.800	.738	.646	.560	.517	.477	.480	.310
11	.818	.761	.676	.595	.553	.515	.521	.352
12	.833	.780	.702	.625	.585	.548	.556	.390
13	.846	.796	.724	.651	.612	.576	.586	.424
14	.857	.810	.743	.674	.637	.602	.612	.455
15	.867	.822	.759	.693	.658	.624	.636	.482
16	.875	.833	.774	.711	.677	.645	.656	.508
17	.882	.843	.787	.727	.694	.663	.675	.531
18	.889	.851	.798	.741	.709	.679	.691	.552
19	.895	.859	.808	.754	.723	.694	.706	.571
20	.900	.866	.818	.765	.736	.708	.720	.589
22	.909	.878	.834	.785	.758	.732	.744	.620
24	.917	.888	.847	.802	.777	.752	.764	.647
26	.923	.896	.859	.817	.793	.770	.781	.671
28	.929	.903	.869	.829	.807	.785	.796	.691
30	.933	.910	.877	.840	.819	.798	.809	.710
35	.943	.922	.894	.862	.843	.825	.835	.747
40	.950	.932	.908	.879	.862	.846	.855	.776
45	.956	.940	.918	.892	.877	.862	.871	.799
50	.960	.946	.926	.902	.889	.875	.883	.818
55	.964	.950	.932	.911	.899	.886	.894	.833
60	.967	.954	.938	.918	.907	.895	.902	.846
65	.969	.958	.943	.924	.914	.903	.910	.858
70	.971	.961	.947	.930	.920	.910	.916	.867
80	.975	.966	.953	.938	.930	.921	.926	.883
90	.978	.970	.959	.945	.937	.929	.934	.896
100	.980	.973	.963	.951	.943	.936	.941	.906

Tables of the Constant K (ii)

n	221	2111	32	11111	51	42	33
10	.269	.248	.336	.229	.323	.168	.136
11	.310	.288	.379	.268	.369	.206	.171
12	.347	.325	.417	.304	.410	.243	.205
13	.381	.359	.451	.338	.445	.277	.237
14	.412	.390	.481	.368	.478	.309	.268
15	.441	.418	.508	.397	.506	.339	.297
16	.467	.444	.533	.423	.532	.367	.324
17	.490	.468	.556	.447	.555	.392	.350
18	.512	.490	.576	.470	.576	.416	.374
19	.532	.511	.595	.490	.596	.438	.396
20	.551	.530	.612	.510	.613	.459	.417
22	.584	.564	.642	.544	.644	.496	.455
24	.613	.593	.668	.575	.671	.529	.489
26	.638	.619	.691	.601	.694	.558	.519
28	.660	.642	.711	.625	.714	.584	.546
30	.680	.662	.728	.646	.731	.607	.570
35	.720	.704	.764	.689	.767	.654	.621
40	.751	.737	.791	.723	.794	.692	.661
45	.776	.763	.813	.751	.816	.722	.694
50	.797	.785	.830	.773	.833	.747	.721
55	.814	.803	.845	.792	.848	.768	.743
60	.828	.818	.857	.808	.860	.786	.762
65	.841	.831	.868	.822	.870	.801	.779
70	.852	.842	.877	.833	.879	.814	.793
80	.869	.861	.892	.853	.894	.836	.817
90	.883	.876	.903	.869	.905	.853	.836
100	.894	.888	.913	.881	.915	.867	.852

Tables of the Constant K (iii)

n	411	321	222	3111	2211	21111	111111
10	.155	.108	.094	.100	.087	.080	.076
11	.192	.140	.123	.130	.114	.106	.099
12	.228	.171	.152	.160	.142	.133	.125
13	.261	.201	.180	.189	.170	.160	.150
14	.292	.230	.208	.217	.197	.186	.176
15	.322	.257	.235	.244	.223	.212	.201
16	.349	.284	.261	.270	.248	.236	.225
17	.375	.309	.285	.295	.272	.260	.248
18	.398	.332	.308	.318	.295	.283	.271
19	.421	.354	.330	.340	.317	.304	.292
20	.442	.375	.351	.361	.338	.325	.313
22	.479	.414	.390	.400	.376	.363	.351
24	.512	.448	.424	.434	.411	.398	.385
26	.542	.479	.456	.465	.442	.430	.417
28	.568	.507	.484	.493	.471	.458	.446
30	.591	.532	.510	.519	.497	.484	.472
35	.640	.585	.564	.573	.552	.540	.528
40	.679	.628	.608	.616	.597	.585	.574
45	.710	.663	.644	.652	.633	.623	.612
50	.736	.692	.674	.681	.664	.654	.644
55	.758	.716	.700	.706	.690	.681	.671
60	.776	.737	.722	.728	.712	.704	.695
65	.792	.755	.740	.746	.732	.723	.715
70	.805	.771	.757	.762	.749	.741	.733
80	.828	.797	.784	.789	.777	.770	.762
90	.846	.818	.806	.811	.800	.793	.786
100	.860	.835	.824	.828	.818	.812	.806

Reprinted from THE ANNALS OF MATHEMATICAL STATISTICS
Vol. XII, No. 3, September, 1941

ON THE ANALYSIS OF VARIANCE IN CASE OF MULTIPLE CLASSIFICATIONS WITH UNEQUAL CLASS FREQUENCIES

BY ABRAHAM WALD[1]
Columbia University

In a previous paper[2] the author considered the case of a single criterion of classification with unequal class frequencies and derived confidence limits for σ'^2/σ^2 where σ'^2 denotes the variance associated with the classification, and σ^2 denotes the residual variance. The scope of the present paper is to extend those results to the case of multiple classifications with unequal class frequencies.

For the sake of simplicity of notations we will derive the required confidence limits in the case of a two-way classification, the extension to multiple classifications being obvious.

Consider a two-way classification with p rows and q columns. Let y be the observed variable, and let n_{ij} be the number of observations in the ith row and jth column. Denote by $y_{ij}^{(k)}$ the kth observation on y in the ith row and jth column ($k = 1, \cdots, n_{ij}$). Let the total number of observations be N. We order the N observations and let y_α be the αth observation on y in that order. Consider the variables:

$$t, t_1, \cdots, t_p, v_1, \cdots, v_q,$$

and denote by t_α the αth observation on t, by $t_{i\alpha}$ the αth observation on t_i and by $v_{j\alpha}$ the αth observation on v_j. The values of t_α, $t_{i\alpha}$ and $v_{j\alpha}$ are defined as follows:

$$\begin{aligned}
t_\alpha &= 1 \ (\alpha = 1, \cdots, N),\\
t_{i\alpha} &= 1 \text{ if } y_\alpha \text{ lies in the } i\text{th row},\\
t_{i\alpha} &= 0 \text{ if } y_\alpha \text{ does not lie in the } i\text{th row},\\
v_{j\alpha} &= 1 \text{ if } y_\alpha \text{ lies in the } j\text{th column},\\
v_{j\alpha} &= 0 \text{ if } y_\alpha \text{ does not lie in the } j\text{th column}.
\end{aligned}$$

We make the assumptions

$$y_{ij}^{(k)} = x_{ij}^{(k)} + \epsilon_i + \eta_j,$$

where the variates $x_{ij}^{(k)}$, ϵ_i, η_j ($i = 1, \cdots, p; j = 1, \cdots, q; k = 1, \cdots, n_{ij}$) are independently and normally distributed, the variance of $x_{ij}^{(k)}$ is σ^2, the variance of ϵ_i is σ'^2, the variance of η_j is σ''^2, and the mean values of ϵ_i and η_j are zero.

---

[1] Research under a grant-in-aid from the Carnegie Corporation of New York.

[2] "A note on the analysis of variance with unequal class frequencies," *Annals of Math. Stat.*, Vol. 11 (1940).

Let the sample regression of y on $t, t_1, \cdots, t_{p-1}, v_1, \cdots, v_{q-1}$ be

$$Y = at + b_1t_1 + \cdots + b_{p-1}t_{p-1} + d_1v_1 + \cdots + d_{q-1}v_{q-1}.$$

We want to derive confidence limits for

$$\sigma'^2/\sigma^2 = \lambda^2.$$

Let us introduce the notations:

$$\sum_\alpha t_\alpha t_{i\alpha} = a_{0i} \qquad (i = 1, \cdots, p-1),$$

$$\sum t_\alpha v_{j\alpha} = a_{0\,p-1+j} \qquad (j = 1, \cdots, q-1),$$

$$\sum t_{i\alpha} t_{j\alpha} = a_{ij} \qquad (i, j = 1, \cdots, p-1),$$

$$\sum t_{i\alpha} v_{j\alpha} = a_{i\,p-1+j} \qquad (i = 1, \cdots, p-1; j = 1, \cdots, q-1),$$

$$\sum v_{i\alpha} v_{j\alpha} = a_{p-1+i\;p-1+j} \qquad (i, j = 1, \cdots, q-1),$$

$$\| c_{ij} \| = \| a_{ij} \|^{-1} \qquad (i, j = 0, 1, \cdots, p+q-2).$$

Let the regression of $x_{ij}^{(k)}$ on $t, t_1, \cdots, t_{p-1}, v_1, \cdots, v_{q-1}$ be

$$X = a^*t + b_1^*t_1 + \cdots + b_{p-1}^*t_{p-1} + d_1^*v_1 + \cdots + d_{q-1}^*v_{q-1}.$$

The regression of $\epsilon_i + \eta_j$ on the same independent variables is evidently equal to

$$\begin{aligned}\epsilon_1t_1 + \cdots + \epsilon_pt_p + \eta_1v_1 + \cdots + \eta_qv_q \\ &= (\eta_q + \epsilon_p)t + (\epsilon_1 - \epsilon_p)t_1 + \cdots + (\epsilon_{p-1} - \epsilon_p)t_{p-1} \\ &\quad + (\eta_1 - \eta_q)v_1 + \cdots + (\eta_{q-1} - \eta_q)v_{q-1},\end{aligned}$$

since $t_p = t - t_1 - \cdots - t_{p-1}$ and $v_q = t - v_1 - \cdots - v_{q-1}$. Hence

$$b_i = b_i^* + (\epsilon_i - \epsilon_p), \qquad (i = 1, \cdots, p-1), \tag{1}$$

and therefore

$$\begin{aligned}\sigma_{b_ib_j} &= \sigma_{b_i^*b_j^*} + \sigma_{(\epsilon_i-\epsilon_p)(\epsilon_j-\epsilon_p)} = c_{ij}\sigma^2 + \sigma_{\epsilon_i\epsilon_j} + \sigma_{\epsilon_p\epsilon_p} \\ &= [c_{ij} + (1 + \delta_{ij})\lambda^2]\sigma^2, \quad (i, j = 1, \cdots, p-1),\end{aligned} \tag{2}$$

where δ_{ij} is the Kronecker delta, i.e. $\delta_{ij} = 0$ for $i \neq j$ and $\delta_{ii} = 1$. Denote $c_{ij} + (1 + \delta_{ij})\lambda^2$ by c'_{ij}. Since the expected value of b_i^* is equal to zero, on account of (1) also the expected value of b_i is equal to zero. Let

$$\| g_{ij} \| = \| c'_{ij} \|^{-1}, \qquad (i, j = 1, \cdots, p-1).$$

Then

$$\frac{1}{\sigma^2} \sum_{j=1}^{p-1} \sum_{i=1}^{p-1} g_{ij} b_i b_j \tag{3}$$

has the χ^2-distribution with $p - 1$ degrees of freedom. The expression

(4) $$\frac{1}{\sigma^2}\sum_{\alpha=1}^{N}(y_\alpha - Y_\alpha)^2,$$

has the χ^2-distribution $N - p - q + 1$ degrees of freedom. The expressions (3) and (4) are independently distributed. Hence

(5) $$\frac{N-p-q+1}{p-1}\,\frac{\Sigma\Sigma g_{ij}b_ib_j}{\Sigma(y_\alpha - Y_\alpha)^2},$$

has the F-distribution (analysis of variance distribution). We will now show that (5) is a monotonic function of λ^2. It is known that $\Sigma\Sigma g_{ij}b_ib_j$ is invariant under linear transformations, i.e.

$$\Sigma\Sigma g_{ij}b_ib_j = \Sigma\Sigma g'_{ij}b'_ib'_j,$$

where b'_i is an arbitrary linear function, say $\mu_{i1}b_1 + \cdots + \mu_{ip-1}b_{p-1}$ of $b_1, \cdots, b_{p-1}$ $(i = 1, \cdots, p-1)$ and

$$\| g'_{ij} \| = \| \sigma_{b'_ib'_j} \|^{-1}.$$

We can choose the matrix $\| \mu_{ij} \|$ such that

$$\epsilon'_i = \mu_{i1}(\epsilon_1 - \epsilon_p) + \cdots + \mu_{ip-1}(\epsilon_{p-1} - \epsilon_p), \qquad (i = 1, \cdots, p-1),$$

are independently distributed and $\sigma^2_{\epsilon'_i} = \sigma'^2$. The coefficients μ_{ij} of course do not depend on σ'. We have

$$\sigma_{b'_ib'_j} = \sigma_{b^{*\prime}_ib^{*\prime}_j} + \delta_{ij}\sigma'^2, \qquad (\delta_{ij} = \text{Kronecker delta}).$$

Now let

$$b''_i = \nu_{i1}b'_1 + \cdots + \nu_{ip-1}b'_{p-1}, \qquad (\nu = 1, \cdots, p-1),$$

where $\| \nu_{ij} \|$ is an orthogonal matrix and is chosen such that $b^{*\prime\prime}_1, \cdots, b^{*\prime\prime}_{p-1}$ are independently distributed. On account of the orthogonality of $\| \nu_{ij} \|$ we obviously have

$$\sigma^2_{b''_i} = \sigma^2_{b^{*\prime\prime}_i} + \sigma'^2; \qquad \sigma_{b''_ib''_j} = 0 \qquad \text{for } i \neq j.$$

Hence

(6) $$\sum\sum g_{ij}b_ib_j = \sum_{i=1}^{p-1}\frac{b''^2_i}{\sigma^2_{b^{*\prime\prime}_i} + \lambda^2\sigma^2}.$$

The right hand side of (6) is evidently a monotonic function of λ^2 which proves our statement. The endpoints of the confidence interval for λ^2 are the roots in λ^2 of the equations

(7) $$\frac{N-p-q+1}{p-1}\,\frac{\Sigma\Sigma g_{ij}b_ib_j}{\Sigma(y_\alpha - Y_\alpha)^2} = F_2; \qquad \frac{N-p-q+1}{p-1}\,\frac{\Sigma\Sigma g_{ij}b_ib_j}{\Sigma(y_\alpha - Y_\alpha)^2} = F_1,$$

where F_2 denotes the upper, and F_1 the lower critical value of F.

The derivation of the required confidence limits in case of classifications in more than two ways can be carried out in the same way and I shall merely state here the results.

Consider r criterions of classifications and denote by p_u the number of classes in the uth classification $(u = 1, \cdots, r)$. Denote by $n_{i_1 \cdots i_r}$ the number of observations which belong to the i_1th class of the first classification, i_2th class of the second classification, $\cdots$, and to the i_rth class of the rth classification. Let $y^{(k)}_{i_1 \cdots i_r}$ be the kth observation on y in the set of observations belonging to the classes mentioned above $(k = 1, \cdots, n_{i_1 \cdots i_r})$. We make the assumption

$$y^{(k)}_{i_1 \cdots i_r} = x^{(k)}_{i_1 \cdots i_r} + \epsilon^{(1)}_{i_1} + \cdots + \epsilon^{(r)}_{i_r},$$

where the variates

$$x^{(k)}_{i_1 \cdots i_r}, \ \epsilon^{(1)}_{i_1}, \cdots, \epsilon^{(r)}_{i_r} \quad (i_u = 1, \cdots, p_u;\ u = 1, \cdots, r;\ k = 1, \cdots, n_{i_1 \cdots i_r}),$$

are independently and normally distributed, the variance of $x^{(k)}_{i_1 \cdots i_r}$ is σ^2, the variance of $\epsilon^{(u)}_{i_u}$ is σ^2_u and the mean value of $\epsilon^{(u)}_{i_u}$ is zero $(i_u = 1, \cdots, p_u;$ $u = 1, \cdots, r)$.

Let N be the total number of observations. We order the observations in a certain order and denote by y_α the αth observation in that order $(\alpha = 1, \cdots, N)$. Consider the variables:

$$t, t^{(u)}_{i_u}, \qquad (u = 1, \cdots, r;\ i_u = 1, \cdots, p_u),$$

and denote by t_α the αth observation on t and by $t^{(u)}_{i_u}\alpha$ the αth observation on $t^{(u)}_{i_u}$. The values of t_α and $t^{(u)}_{i_u \alpha}$ are given as follows:

$$t_\alpha = 1 \ (\alpha = 1, \cdots, N),$$

$$t^{(u)}_{i_u \alpha} = 1 \text{ if } y_\alpha \text{ lies in the } i_u\text{th class of the } u\text{th classification},$$

$$t^{(u)}_{i_u \alpha} = 0 \text{ if } y_\alpha \text{ does not lie in the } i_u\text{th class of the } u\text{th classification}.$$

Let the sample regression of y on t, $t^{(u)}_{i_u}$ be given by

$$Y = at + \sum_{u=1}^{r} \sum_{i_u=1}^{p_u-1} b^{(u)}_{i_u} t^{(u)}_{i_u}.$$

Let the covariance of $b^{(u)}_{i_u}$ and $b^{(u)}_{j_u}$ be given by $C^{(u)}_{i_u j_u}\sigma^2$ under the assumption that $\sigma_1 = \sigma_2 = \cdots = \sigma_r = 0$. The matrix $\| C^{(u)}_{i_u j_u} \| \ (i_u, j_u = 1, \cdots, p_u - 1)$ can be calculated by known methods of the theory of least squares. Let

$$\| g^{(u)}_{i_u j_u} \| = \| C^{(u)}_{i_u j_u} + (1 + \delta_{i_u j_u}) \lambda^2_u \|^{-1} \qquad (i_u, j_u = 1, \cdots, p_u - 1),$$

where $\delta_{i_u j_u}$ is the Kronecker delta and $\lambda^2_u = \sigma^2_u / \sigma^2$. Then the lower and upper confidence limits for λ^2_u are given by the roots in λ^2_u of the equations

$$\text{(8)} \qquad \frac{N - \sum_{u=1}^{r} p_u + r - 1}{p_u - 1} \frac{\sum_{j_u=1}^{p_u-1} \sum_{i_u=1}^{p_u-1} g^{(u)}_{i_u j_u} b^{(u)}_{i_u} b^{(u)}_{j_u}}{\sum_{\alpha=1}^{N} (y_\alpha - Y_\alpha)^2} = F_i \qquad (i = 1, 2),$$

where F_2 is the upper and F_1 the lower critical value of the analysis of variance distribution with $p_u - 1$ and $N - \sum_{u=1}^{r} p_u + r - 1$ degrees of freedom. In case of a single criterion of classification the confidence limits (8) are identical with those given in my previous paper.

Made in the United States of America

Reprinted from THE ANNALS OF MATHEMATICAL STATISTICS
Vol. XII, No. 4, December, 1941

SOME EXAMPLES OF ASYMPTOTICALLY MOST POWERFUL TESTS

BY ABRAHAM WALD[1]

Columbia University

1. Introduction. In a previous paper[2] the author gave the definition of an asymptotically most powerful test and has shown that the commonly used tests, based on the maximum likelihood estimate, are asymptotically most powerful.

In this paper some further examples of asymptotically most powerful tests will be given. Let us first restate the definition of an asymptotically most powerful test. Let $f(x, \theta)$ be the probability density of a variate x involving an unknown parameter θ. For testing the hypothesis $\theta = \theta_0$ by means of n independent observations $x_1, \cdots, x_n$ on x we have to choose a region of rejection W_n in the n-dimensional sample space. Denote by $P(W_n \mid \theta)$ the probability that the sample point $E = (x_1, \cdots, x_n)$ will fall in W_n under the assumption that θ is the true value of the parameter. For any region U_n of the n-dimensional sample space denote by $g(U_n)$ the greatest lower bound of $P(U_n \mid \theta)$. For any pair of regions U_n and T_n denote by $L(U_n, T_n)$ the least upper bound of

$$P(U_n \mid \theta) - P(T_n \mid \theta).$$

In all that follows we shall denote a region of the n-dimensional sample space by a capital letter with the subscript n.

Definition 1: A sequence $\{W_n\}$ ($n = 1, 2, \cdots$, ad inf.) of regions is said to be an asymptotically most powerful test of the hypothesis $\theta = \theta_0$ on the level of significance α if $P(W_n \mid \theta_0) = \alpha$ and if for any sequence $\{Z_n\}$ of regions for which $P(Z_n \mid \theta_0) = \alpha$ the inequality

$$\limsup_{n \to \infty} L(Z_n, W_n) \leq 0$$

holds.

Definition 2: A sequence $\{W_n\}$ ($n = 1, 2, \cdots$, ad inf.) of regions is said to be an asymptotically most powerful unbiased test of the hypothesis $\theta = \theta_0$ on the level of significance α if $P(W_n \mid \theta_0) = \lim_{n=\infty} g(W_n) = \alpha$, and if for any sequence $\{Z_n\}$ of regions for which $P(Z_n \mid \theta_0) = \lim_{n=\infty} g(Z_n) = \alpha$, the inequality

$$\limsup_{n \to \infty} L(Z_n, W_n) \leq 0$$

holds.

---

[1] Research under a grant-in-aid of the Carnegie Corporation of New York.

[2] "Asymptotically most powerful tests of statistical hypotheses," *Annals of Math. Stat.* Vol. 12 (1941).

Consider the expression

$$(1) \qquad y_n(\theta) = \frac{1}{\sqrt{n}} \sum_{\alpha=1}^{n} \frac{\partial}{\partial\theta} \log f(x_\alpha, \theta).$$

Let W'_n be the region defined by the inequality $y_n(\theta_0) \geq c'_n$, W''_n defined by the inequality $y_n(\theta_0) \leq c''_n$, and W_n defined by the inequality $|y_n(\theta_0)| \geq c_n$, where the constants c'_n, c''_n and c_n are chosen such that

$$P(W'_n \mid \theta_0) = P(W''_n \mid \theta_0) = P(W_n \mid \theta_0) = \alpha.$$

It will be shown in this pàper that under certain restrictions on the probability density $f(x, \theta)$ the sequence $\{W'_n\}$ is an asymptotically most powerful test of the hypothesis $\theta = \theta_0$ if θ takes only values $\geq \theta_0$. Similarly $\{W''_n\}$ is an asymptotically most powerful test if θ takes only values $\leq \theta_0$. Finally $\{W_n\}$ is an asymptotically most powerful unbiased test if θ can take any real value.

Another example of an asymptotically most powerful unbiased test of the hypothesis $\theta = \theta_0$, as it will be shown, is the critical region of type A in the Neyman-Pearson theory of testing hypotheses. This fact gives a strong justification for the use of the critical region of type A.

2. Assumptions on the density function. Let ω be a subset of the real axis. Denote by θ^* a real variable which takes only values in ω and let θ be a variable which can take any real value. For any function $\psi(x)$ we denote by $E_\theta\psi(x)$ the expected value of $\psi(x)$ under the assumption that θ is the true value of the parameter, i.e.

$$E_\theta \psi(x) = \int_{-\infty}^{+\infty} \psi(x) f(x, \theta)\, dx.$$

For any x, for any positive δ and for any real value θ_1 denote by $\varphi_1(x, \theta_1, \delta)$ the greatest lower bound, and by $\varphi_2(x, \theta_1, \delta)$ the least upper bound of $\frac{\partial^2}{\partial\theta^2} \log f(x, \theta)$ in the interval $\theta_1 - \delta \leq \theta \leq \theta_1 + \delta$. In all that follows the symbol θ_i^*, for any integer i, will denote a value of θ^*, i.e., θ_i^* is a point of ω.

We say that a value θ lies in the ϵ-neighborhood of ω if there exists a value θ^* such that $|\theta - \theta^*| \leq \epsilon$.

Throughout the paper the following assumptions on $f(x, \theta)$ will be made:

Assumption 1: *For any pair of sequences $\{\theta_n\}$ and $\{\theta_n^*\}$ ($n = 1, 2, \cdots$, ad inf.) for which*

$$\lim_{n=\infty} E_{\theta_n} \frac{\partial}{\partial\theta} \log f(x, \theta_n^*) = 0$$

also

$$\lim_{n=\infty} (\theta_n - \theta_n^*) = 0.$$

Furthermore there exists a positive ϵ *such that* $E_\theta\left[\frac{\partial}{\partial\theta}\log f(x, \theta_1)\right]^2$ *is a bounded function of* θ *and* θ_1, $E_\theta \frac{\partial}{\partial\theta}\log f(x, \theta_1)$ *is a continuous function of* θ *and* θ_1 *and* $E_{\theta_1}\left[\frac{\partial}{\partial\theta}\log f(x, \theta_1)\right]^2 = d(\theta_1)$ *has a positive lower bound, where* θ_1 *can take any value in the* ϵ*-neighborhood of* ω.

ASSUMPTION 2: *There exists a positive* k_0 *such that* $E_{\theta_2}\varphi_1(x, \theta_1, \delta)$ *and* $E_{\theta_2}\varphi_2(x, \theta_1, \delta)$ *are uniformly continuous functions in the domain D defined as follows: the variables* θ_1 *and* θ_2 *may take any value in the* k_0*-neighborhood of* ω *and* δ *may take any value for which* $|\delta| \leq k_0$. *Furthermore it is assumed that*

$$E_{\theta_2}[\varphi_i(x, \theta_1, \delta)]^2, \qquad (i = 1, 2)$$

are bounded functions of θ_1, θ_2 *and* δ *in D.*

ASSUMPTION 3: *There exists a positive* k_0 *such that*

$$\int_{-\infty}^{+\infty}\frac{\partial}{\partial\theta}f(x, \theta)\,dx = \int_{-\infty}^{+\infty}\frac{\partial^2}{\partial\theta^2}f(x, \theta)\,dx = 0$$

for all θ *in the* k_0*-neighborhood of* ω.

Assumption 3 means simply that we may differentiate with respect to θ under the integral sign. In fact,

$$\int_{-\infty}^{+\infty} f(x, \theta)\,dx = 1,$$

identically in θ. Hence

$$\frac{\partial}{\partial\theta}\int_{-\infty}^{+\infty} f(x, \theta)\,dx = \frac{\partial^2}{\partial\theta^2}\int_{-\infty}^{+\infty} f(x, \theta)\,dx = 0.$$

Differentiating under the integral sign we obtain the relations in Assumption 3.

ASSUMPTION 4: *There exists a positive* k_0 *and a positive* η *such that*

$$E_\theta\left[\frac{\partial}{\partial\theta}\log f(x, \theta)\right]^{2+\eta}$$

is a bounded function of θ *in the* k_0*-neighborhood of* ω.

3. Some propositions. PROPOSITION 1: *To any positive* β *there exists a positive* γ *such that*

$$\lim_{n=\infty} P\left\{\frac{1}{\sqrt{n}}\,|y_n(\theta^*)| > \gamma \mid \theta\right\} = 1$$

uniformly in θ^* *and for all* θ *for which* $|\theta - \theta^*| \geq \beta$.

PROOF: From Assumption 1 it follows that $\left|E_\theta \frac{\partial}{\partial\theta}\log f(x, \theta^*)\right|$ has a positive lower bound in the domain $|\theta - \theta^*| \geq \beta$. Since according to Assumption 1 $E_\theta\left[\frac{\partial}{\partial\theta}\log f(x, \theta^*)\right]^2$ is a bounded function of θ and θ^*, Proposition 1 easily follows.

PROPOSITION 2: *There exists a positive ϵ such that*

$$\lim_{n=\infty} P[y_n(\theta) < t \mid \theta] = N(t \mid \theta)$$

uniformly in t and for all θ in the ϵ-neighborhood of ω where

$$d(\theta) = -E_\theta \frac{\partial^2}{\partial\theta^2}\log f(x, \theta) = E_\theta\left[\frac{\partial}{\partial\theta}\log f(x, \theta)\right]^2 \tag{2}$$

and

$$N(t \mid \theta) = \frac{1}{\sqrt{2\pi d(\theta)}}\int_{-\infty}^{t} e^{-\frac{1}{2}v^2/d(\theta)}\, dv. \tag{3}$$

Proposition 2 follows easily from Assumptions 3 and 4 and the general limit theorems.

PROPOSITION 3: *There exists a positive ϵ such that for any bounded sequence $\{\mu_n\}$*

$$\lim_{n=\infty}\left\{P\left[y_n(\theta) < t \mid \theta + \frac{\mu_n}{\sqrt{n}}\right] - \int_{-\infty}^{t} e^{\mu_n v - \frac{1}{2}\mu_n^2 d(\theta)}\, dN(v \mid \theta)\right\} = 0$$

uniformly in t and for all θ in the ϵ-neighborhood of ω.

PROOF: We have

$$y_n\left(\theta + \frac{\mu_n}{\sqrt{n}}\right) = y_n(\theta) + \frac{\mu_n}{\sqrt{n}}\frac{1}{\sqrt{n}}\sum_\alpha \frac{\partial^2}{\partial\theta^2}\log f(x_\alpha, \theta_n') \tag{4}$$

where θ_n' lies in the interval $\left[\theta, \theta + \frac{\mu_n}{\sqrt{n}}\right]$. From Assumption 2 and the above equation we easily obtain

$$\lim_{n=\infty}\left\{P\left[y_n\left(\theta + \frac{\mu_n}{\sqrt{n}}\right) < t \mid \theta + \frac{\mu_n}{\sqrt{n}}\right] - P\left[y_n(\theta) - \mu_n d(\theta) < t \mid \theta + \frac{\mu_n}{\sqrt{n}}\right]\right\} = 0 \tag{5}$$

uniformly in t and for all θ in the ϵ-neighborhood of ω. From Proposition 2 and (5) we get

$$\lim_{n=\infty}\left\{\int_{-\infty}^{t} dN(v \mid \theta) - P\left[y_n(\theta) < t + \mu_n d(\theta) \middle| \theta + \frac{\mu_n}{\sqrt{n}}\right]\right\} = 0$$

or

$$\lim_{n=\infty}\left\{\int_{-\infty}^{t-\mu_n d(\theta)} dN(v \mid \theta) - P\left[y_n(\theta) < t \mid \theta + \frac{\mu_n}{\sqrt{n}}\right]\right\} = 0 \tag{6}$$

uniformly in t and for all θ in the ϵ-neighbourhood of ω. This proves Proposition 3.

PROPOSITION 4: *There exists a positive* ϵ *such that for any positive* γ *and for any sequence* $\{\mu_n\}$ *for which* $\lim_{n=\infty} |\mu_n| = \infty$

$$\lim_{n=\infty} P\left\{| y_n(\theta^*) | > \gamma \mid \theta^* + \frac{\mu_n}{\sqrt{n}}\right\} = 1$$

uniformly in θ^*.

PROOF: If there exists a positive β such that $\left|\frac{\mu_n}{\sqrt{n}}\right| > \beta$ for almost all n, Proposition 4 follows from Proposition 1. Hence we have to consider only the case $\lim_{n=\infty} \frac{\mu_n}{\sqrt{n}} = 0$. Since

$$E_{\theta^*+(\mu_n/\sqrt{n})}\, y_n\left(\theta^* + \frac{\mu_n}{\sqrt{n}}\right) = 0,$$

we get from (4)

$$E_{\theta^*+(\mu_n/\sqrt{n})}[y_n(\theta^*)] + \mu_n E_{\theta^*+(\mu_n/\sqrt{n})} \frac{\sum_\alpha \frac{\partial^2}{\partial\theta^2} \log f(x_\alpha, \theta_n')}{n} = 0. \tag{7}$$

Since $\lim \frac{\mu_n}{\sqrt{n}} = 0$, we have on account of Assumption 2

$$\lim_{n=\infty} E_{\theta^*+(\mu_n/\sqrt{n})} \frac{\sum_\alpha \frac{\partial^2}{\partial\theta^2} \log f(x_\alpha, \theta_n')}{n} = E_{\theta^*} \frac{\partial^2}{\partial\theta^2} \log f(x, \theta^*)$$
$$= -E_{\theta^*}\left[\frac{\partial}{\partial\theta} \log f(x, \theta^*)\right]^2 = -d(\theta^*)$$

uniformly in θ^*. According to Assumption 1 $d(\theta^*)$ has a positive lower bound; hence on account of $\lim |\mu_n| = \infty$ we obtain from (7)

$$\lim_{n=\infty} | E_{\theta^*+(\mu_n/\sqrt{n})}\, y_n(\theta^*) | = \infty \tag{8}$$

uniformly in θ^*. The variance of $y_n(\theta^*)$ is equal to the variance of $\frac{\partial}{\partial\theta} \log f(x, \theta^*)$. On account of Assumption 1 the variance of $\frac{\partial}{\partial\theta} \log f(x, \theta^*)$ (under the assumption that $\theta^* + \frac{\mu_n}{\sqrt{n}}$ is the true value of the parameter) is a bounded function. Hence Proposition 4 is proved on account of (8).

PROPOSITION 5: *Let* $\{W_n(\theta^*)\}$ *be a sequence of regions of size* α, *i.e.* $P[W_n(\theta^*) \mid \theta^*] = \alpha$, *and let* $V_n(\theta^*, y)$ *be the region defined by the inequality*

$y_n(\theta^*) < y$. *Let* $U_n(\theta^*, y)$ *be the intersection of* $V_n(\theta^*, y)$ *and* $W_n(\theta^*)$ *and denote* $P[U_n(\theta^*, y) \mid \theta^*]$ *by* $F_n(y \mid \theta^*)$. *Denote furthermore* $P\left[W_n(\theta^*) \mid \theta^* + \frac{\mu}{\sqrt{n}}\right]$ *by* $G(\theta^*, \mu, n)$. *If* $\{\theta_n^*\}$ *and* $\{\mu_n\}$ *are two sequences such that* $\lim_{n=\infty} d(\theta_n^*) = d$; $\lim_{n=\infty} F_n(y \mid \theta_n^*) = F(y)$ *and* $\lim \mu_n = \mu$ *then*

$$\lim_{n=\infty} G(\theta_n^*, \mu_n, n) = \int_{\infty}^{+\infty} e^{\mu y - \frac{1}{2}\mu^2 d}\, dF(y).$$

PROOF: Let $\lim \mu_n = \mu$ and consider the Taylor expansion

$$\sum_\alpha \log f\left(x_\alpha, \theta^* + \frac{\mu_n}{\sqrt{n}}\right) = \sum_\alpha \log f(x_\alpha, \theta^*) + \frac{\mu_n}{\sqrt{n}} \sum_\alpha \frac{\partial}{\partial\theta} \log f(x_\alpha, \theta^*) + \frac{1}{2}\frac{\mu_n^2}{n}\frac{\partial^2}{\partial\theta^2}\sum_\alpha \log f(x_\alpha, \theta_n') \tag{9}$$

where θ_n' lies in the interval $\left[\theta^*, \theta^* + \frac{\mu_n}{\sqrt{n}}\right]$. From this we easily get on account of Assumption 2 and the fact that $\{\mu_n\}$ is bounded

$$\log \prod_{\alpha=1}^{n} \frac{f\left(x_\alpha, \theta^* + \frac{\mu_n}{\sqrt{n}}\right)}{f(x_\alpha, \theta^*)} = \mu_n y_n(\theta^*) - \tfrac{1}{2}\mu_n^2 d(\theta^*) + \epsilon(\theta^*, n) \tag{10}$$

where for arbitrary positive η

$$\lim_{n=\infty} P\left\{\mid \epsilon(\theta^*, n) \mid < \eta \mid \theta^* + \frac{\mu_n}{\sqrt{n}}\right\} = 1 \tag{11}$$

uniformly in θ^*. Denote by $R_n(\theta^*)$ the region defined by

$$\mid \epsilon(\theta^*, n) \mid < \eta > 0. \tag{12}$$

On account of (11) we have

$$\lim_{n=\infty} P\left[R_n(\theta^*) \mid \theta^* + \frac{\mu_n}{\sqrt{n}}\right] = 1, \tag{13}$$

uniformly in θ^*. Denote the intersection of $R_n(\theta^*)$ and $W_n(\theta^*)$ by $Q_n(\theta^*)$, and the intersection of $R_n(\theta^*)$ and $U_n(\theta^*, y)$ by $T_n(\theta^*, y)$. Furthermore denote $P[T_n(\theta^*, y) \mid \theta^*]$ by $\bar{F}_n(y \mid \theta^*)$. Then we have

$$e^{-\eta} \int_{-\infty}^{t} e^{\mu_n y - \frac{1}{2}\mu_n^2 d(\theta^*)}\, d\bar{F}_n(y \mid \theta^*) \le P\left[T_n(\theta^*, t) \mid \theta^* + \frac{\mu_n}{\sqrt{n}}\right] \le e^{\eta} \int_{-\infty}^{t} e^{\mu_n y - \frac{1}{2}\mu_n^2 d(\theta^*)}\, d\bar{F}_n(y \mid \theta^*) \tag{14}$$

for all values of t and θ^*. Furthermore we obviously have

$$\lim_{n=\infty} \left\{ G(\theta^*, \mu_n, n) - P\left[Q_n(\theta^*) \mid \theta^* + \frac{\mu_n}{\sqrt{n}} \right] \right\} = 0 \tag{15}$$

uniformly in θ^*, and

$$\lim_{n=\infty} [\bar{F}_n(t \mid \theta^*) - F_n(t \mid \theta^*)] = 0 \tag{16}$$

uniformly in θ^* and t. Since η may be chosen arbitrarily small, it follows from (14) and (15) that to any $\epsilon > 0$, η may be chosen such that

$$\limsup_{n=\infty} \left| G(\theta_n^*, \mu_n, n) - \int_{-\infty}^{+\infty} e^{\mu_n t - \frac{1}{2}\mu_n^2 d(\theta_n^*)} \, d\bar{F}_n(t \mid \theta_n^*) \right| \leqq \frac{\epsilon}{2} \tag{17}$$

for any sequence $\{\theta_n^*\}$.

To each ϵ let L_ϵ be a positive number such that L_ϵ depends only on ϵ and

$$\int_{-\infty}^{-L_\epsilon} e^{\mu_n t - \frac{1}{2}\mu_n^2 d(\theta^*)} \, dN(t \mid \theta^*) + \int_{L_\epsilon}^{\infty} e^{\mu_n t - \frac{1}{2}\mu_n^2 d(\theta^*)} \, dN(t \mid \theta^*) \leq \frac{\epsilon}{2} \tag{18}$$

for all n and for all values of θ^*. Since $d(\theta^*)$ has a positive lower and a finite upper bound, it is easy to verify that such a L_ϵ exists. From (18) and Proposition 3 it follows

$$\begin{aligned} \limsup_{n\to\infty} \Bigg\{ & P\left[y_n(\theta_n^*) < -L_\epsilon \mid \theta_n^* + \frac{\mu_n}{\sqrt{n}} \right] \\ & + P\left[y_n(\theta_n^*) > L_\epsilon \mid \theta_n^* + \frac{\mu_n}{\sqrt{n}} \right] \Bigg\} \leq \frac{\epsilon}{2} \end{aligned} \tag{19}$$

for any arbitrary sequence $\{\theta_n^*\}$. Since the difference $U_n(\theta^*, t_2) - U_n(\theta^*, t_1)$ is a subset of the difference $V_n(\theta^*, t_2) - V_n(\theta^*, t_1)$ and since $T_n(\theta^*, t_2) - T_n(\theta^*, t_1)$ is a subset of $U_n(\theta^*, t_2) - U_n(\theta^*, t_1)$ for $t_2 > t_1$, we get from (18) and (19)

$$\begin{aligned} \limsup_{n\to\infty} \Bigg\{ & P\left[U_n(\theta_n^*, -L_\epsilon) \mid \theta_n^* + \frac{\mu_n}{\sqrt{n}} \right] + P\left[W_n(\theta_n^*) \mid \theta_n^* + \frac{\mu_n}{\sqrt{n}} \right] \\ & - P\left[U_n(\theta_n^*, L_\epsilon) \mid \theta_n^* + \frac{\mu_n}{\sqrt{n}} \right] \Bigg\} \leq \frac{\epsilon}{2} \end{aligned} \tag{20}$$

and

$$\begin{aligned} \limsup_{n\to\infty} \Bigg\{ & P\left[T_n(\theta_n^*, -L_\epsilon) \mid \theta_n^* + \frac{\mu_n}{\sqrt{n}} \right] + P\left[Q_n(\theta_n^*) \mid \theta_n^* + \frac{\mu_n}{\sqrt{n}} \right] \\ & - P\left[T_n(\theta_n^*, L_\epsilon) \mid \theta_n^* + \frac{\mu_n}{\sqrt{n}} \right] \Bigg\} \leq \frac{\epsilon}{2} \end{aligned} \tag{21}$$

for any sequence $\{\theta_n^*\}$. On account of (14) we get from (21)

$$e^{-\eta} \limsup_{n\to\infty} \left\{ \int_{-\infty}^{-L_\epsilon} e^{\mu_n t - \frac{1}{2}\mu_n^2 d(\theta_n^*)} \, d\bar{F}_n(t \mid \theta_n^*) + \int_{L_\epsilon}^{\infty} e^{\mu_n t - \frac{1}{2}\mu_n^2 d(\theta_n^*)} \, d\bar{F}_n(t \mid \theta_n^*) \right\} \leq \frac{\epsilon}{2}. \tag{22}$$

From (17) and (22) we obtain

$$(23)\qquad \limsup_{n\to\infty} \left| G(\theta_n^*, \mu_n, n) - \int_{-L_\epsilon}^{L_\epsilon} e^{\mu_n t - \frac{1}{2}\mu_n^2 d(\theta_n^*)}\, d\bar{F}_n(t \mid \theta_n^*) \right| \le \epsilon\left(\frac{1+e^\eta}{2}\right)$$

for any sequence $\{\theta_n^*\}$. Consider now the sequence $\{\theta_n^*\}$ which satisfies the conditions of Proposition 5. Since $F_n(t \mid \theta_n^*)$ converges to $F(t)$ uniformly in t, on account of (16) also $\bar{F}_n(t \mid \theta_n^*)$ converges to $F(t)$ uniformly in t. Hence we obtain from (23)

$$(24)\qquad \limsup_{n\to\infty} \left| G(\theta_n^*, \mu_n, n) - \int_{-L_\epsilon}^{L_\epsilon} e^{\mu t - \frac{1}{2}\mu^2 d}\, dF(t) \right| \le \epsilon\left(\frac{1+e^\eta}{2}\right).$$

Since ϵ and η may be chosen arbitrarily small, Proposition 5 follows from (24).

4. Some theorems and corollaries. THEOREM 1. *Denote by $S_n(\theta^*)$ the region defined by the inequality $y_n(\theta^*) \ge A_n(\theta^*)$ where $A_n(\theta^*)$ is chosen such that $P[S_n(\theta^*) \mid \theta^*] = \alpha$. For any region $W_n(\theta^*)$ denote by $L_n[W_n(\theta^*)]$ the least upper bound of $P[W_n(\theta^*) \mid \theta] - P[S_n(\theta^*) \mid \theta]$ with respect to θ^* and θ, where θ is restricted to values $\ge \theta^*$. Then for any sequence $\{W_n(\theta^*)\}$ for which $P[W_n(\theta^*) \mid \theta^*] = \alpha$,*

$$\limsup_{n\to\infty} L_n[W_n(\theta^*)] \le 0.$$

PROOF: Assume that Theorem 1 is not true. Then there exists a sequence of integers $\{n'\}$, a sequence $\{\theta_{n'}^*\}$ and a sequence $\{\theta_{n'}\}$ $(\theta_{n'} \ge \theta_{n'}^*)$ such that

$$(25)\qquad \lim_{n=\infty} \{P[W_{n'}(\theta_{n'}^*) \mid \theta_{n'}] - P[S_{n'}(\theta_{n'}^*) \mid \theta_{n'}]\} = \delta > 0.$$

On account of Proposition 2 and Assumption 2 the sequence $\{A_{n'}(\theta_{n'}^*)\}$ is bounded. Then it follows easily from (25) and Proposition 4 (taking in account that $E_\theta \frac{\partial}{\partial\theta} \log f(x, \theta^*) > 0$ for $\theta > \theta^*$

$$(26)\qquad (\theta_{n'} - \theta_{n'}^*)\sqrt{n'} = \mu_{n'} > 0$$

must be bounded. Denote by $\{n''\}$ a subsequence of $\{n'\}$ such that

$$(27)\qquad \lim d(\theta_{n''}^*) = d$$

$$(28)\qquad \lim \mu_{n''} = \mu, \quad \text{and}$$

$$(29)\qquad \lim F_{n''}(t \mid \theta_{n''}^*) = F(t)$$

uniformly in t where

$$F_n(t \mid \theta^*) = P[U_n(\theta^*, t) \mid \theta^*]$$

and $U_n(\theta^*, t)$ is the intersection of $W_n(\theta^*)$ and the region $y_n(\theta^*) < t$. The existence of a subsequence $\{n''\}$ such that (29) holds follows from the fact that

$$(30)\qquad F_n(t_2 \mid \theta^*) - F_n(t_1 \mid \theta^*) \le \Phi_n(t_2 \mid \theta^*) - \Phi_n(t_1 \mid \theta^*) \quad \text{for} \quad t_2 > t_1,$$

and

$$\lim_{n=\infty} \Phi_n(t \mid \theta^*_{n''}) = \frac{1}{\sqrt{2\pi d}} \int_{-\infty}^{t} e^{-\frac{1}{2}v^2/d}\, dv = N(t), \tag{31}$$

where $\Phi_n(t \mid \theta^*)$ denotes the probability $P[y_n(\theta^*) < t \mid \theta^*]$. Furthermore it can easily be shown that

$$\int_{-\infty}^{+\infty} dF(t) = \alpha. \tag{32}$$

On account of Proposition 5 we get from (25), (27), (28), (29), (30) and (31)

$$\int_{-\infty}^{+\infty} e^{\mu t - \frac{1}{2}\mu^2 d}\, dF(t) - \int_{A}^{\infty} e^{\mu t - \frac{1}{2}\mu^2 d}\, dN(t) = \delta, \tag{33}$$

where A denotes a value such that

$$\int_{A}^{\infty} dN(t) = \alpha.$$

It has been shown in a previous paper[3] that (33) leads to a contradiction. Hence Theorem 1 is proved.

THEOREM 2: *Denote by* $S_n(\theta^*)$ *the region defined by the inequality* $y_n(\theta^*) \leq A_n(\theta^*)$ *where* $A_n(\theta^*)$ *is chosen such that* $P[S_n(\theta^*) \mid \theta^*] = \alpha$. *For any region* $W_n(\theta^*)$ *denote by* $L_n[W_n(\theta^*)]$ *the least upper bound of*

$$P[W_n(\theta^*) \mid \theta] - P[S_n(\theta^*) \mid \theta]$$

with respect to θ^* *and* θ, *where* θ *is restricted to values* $\leq \theta^*$. *Then for any sequence* $\{W_n(\theta^*)\}$ *for which* $P[W_n(\theta^*) \mid \theta^*) = \alpha$,

$$\limsup_{n \to \infty} L_n[W_n(\theta^*)] \leq 0.$$

The proof is omitted, since it is analogous to that of Theorem 1.

THEOREM 3: *Let* $\{W_n(\theta^*)\}$ *be for each* θ^* *a sequence of regions for which* $P[W_n(\theta^*) \mid \theta^*] = \alpha$ *and* $\lim_{n=\infty} g[W_n(\theta^*)] = \alpha$ *uniformly in* θ^*. *Denote by* $L_n[W_n(\theta^*)]$ *the least upper bound of*

$$P[W_n(\theta^*) \mid \theta] - P[|\, y_n(\theta^*) \,| \geq A_n(\theta^*) \mid \theta]$$

with respect to θ *and* θ^*, *where* $A_n(\theta^*)$ *is chosen such that*

$$P[|\, y_n(\theta^*) \,| \geq A_n(\theta^*) \mid \theta^*] = \alpha.$$

Then

$$\limsup_{n \to \infty} L_n[W_n(\theta^*)] \leq 0.$$

[3] See p. 12 of the paper cited in [2].

PROOF: Denote $P[y_n(\theta^*) < t \mid \theta^*]$ by $\Phi_n(t \mid \theta^*)$ and denote by $F_n(t \mid \theta^*)$ the probability (under the hypothesis $\theta = \theta^*$) of the intersection of $W_n(\theta^*)$ with the region $y_n(\theta^*) < t$. Assume that Theorem 3 is not true. Then there exists a subsequence $\{n''\}$, a sequence $\{\theta^*_{n''}\}$ and a sequence $\{\theta_{n''}\}$ such that

$$\lim_{n=\infty} d(\theta^*_{n''}) = d; \qquad \lim\, (\theta_{n''} - \theta^*_{n''})\sqrt{n''} = \lim \mu_{n''} = \mu;$$

$$\lim F_{n''}(t \mid \theta^*_{n''}) = F(t)$$

uniformly in t, and

$$\int_{-\infty}^{+\infty} e^{\mu t - \frac{1}{2}\mu^2 d}\, dF(t) - \int_{-\infty}^{-A} e^{\mu t - \frac{1}{2}\mu^2 d}\, dN(t) - \int_{A}^{\infty} e^{\mu t - \frac{1}{2}\mu^2 d}\, dN(t) = \delta \tag{34}$$

where A is a positive number such that

$$\int_{-\infty}^{-A} dN(t) = \frac{\alpha}{2}, \qquad \text{and} \qquad N(t) = \frac{1}{\sqrt{2\pi d}} \int_{-\infty}^{t} e^{-\frac{1}{2}v^2/d}\, dv.$$

This can be proved in the same way as (33) has been proved. The author has shown in a previous paper[4] that (34) leads to a contradiction. Hence Theorem 3 is proved.

THEOREM 4: *Denote by $A_n(\theta^*)$ the region of type[5] A of size α for testing the hypothesis $\theta = \theta^*$. Denote by $B_n(\theta^*)$ the region $|y_n(\theta^*)| \geq C_n(\theta^*)$ where $C_n(\theta^*)$ is determined such that*

$$P[|y_n(\theta^*)| \geq C_n(\theta^*) \mid \theta^*] = \alpha.$$

Then, under the assumption that $E_\theta \left[\frac{\partial^2}{\partial \theta^2} \log f(x, \theta^*) \right]^2$ *is bounded,*

$$\lim_{n=\infty} \{P[A_n(\theta^*) \mid \theta] - P[B_n(\theta^*) \mid \theta]\} = 0$$

uniformly in θ and θ^*.

PROOF: The region $A_n(\theta^*)$ is given by the inequality[6]

$$\left[\sum_\alpha \frac{\partial}{\partial \theta} \log f(x_\alpha, \theta^*) \right]^2 + \sum_\alpha \frac{\partial^2}{\partial \theta^2} \log f(x_\alpha, \theta^*) \geq k'_n(\theta^*) \left[\sum_\alpha \frac{\partial}{\partial \theta} \log f(x_\alpha, \theta^*) \right] + k''_n(\theta^*), \tag{35}$$

where $k'_n(\theta^*)$ and $k''_n(\theta^*)$ are chosen such that $A_n(\theta^*)$ should be unbiased and of size α. . The inequality (35) can be written also in the form

$$[y_n(\theta^*)]^2 + \frac{1}{n} \sum_\alpha \frac{\partial^2}{\partial \theta^2} \log f(x_\alpha, \theta^*) \geq l'_n(\theta^*) y_n(\theta^*) + l''_n(\theta^*). \tag{36}$$

---

[4] See p. 14 of the paper cited in [2].

[5] Neyman, J. and Pearson, E. S., "Contributions to the theory of testing statistical hypotheses," *Stat. Res. Mem.*, Vol. 1.

[6] See the paper cited in [5].

Let $\{\mu_n\}$ be a bounded sequence. From Assumption 2 it follows that for any positive ϵ

$$(37) \qquad P\left\{\left|\frac{1}{n}\sum_{\alpha}\frac{\partial^2}{\partial\theta^2}\log f(x_\alpha, \theta^*) + d(\theta^*)\right| < \epsilon \mid \theta^* + \frac{\mu_n}{\sqrt{n}}\right\} = 1$$

uniformly in θ^*. Since (37) holds for arbitrarily small ϵ, we get easily on account of Proposition 3

$$(38) \qquad \lim_{n=\infty}\left\{P\left[A_n(\theta^*) \mid \theta^* + \frac{\mu_n}{\sqrt{n}}\right] - P\left[A_n'(\theta^*) \mid \theta^* + \frac{\mu_n}{\sqrt{n}}\right]\right\} = 0$$

uniformly in θ^*, where $A_n'(\theta^*)$ is defined by

$$(39) \qquad [y_n(\theta^*)]^2 \geq l_n'(\theta^*)y_n(\theta^*) + l_n''(\theta^*) + d(\theta^*).$$

Since $A_n(\theta^*)$ is unbiased and of size α, we have on account of (38) and (39)

$$(40) \qquad \lim l_n'(\theta^*) = 0 \quad \text{and}$$

$$(41) \qquad \lim l_n''(\theta^*) + d(\theta^*) = \lambda(\theta^*) > 0$$

uniformly in θ^*, where $\lambda(\theta^*)$ is given by the condition

$$(42) \qquad \frac{1}{\sqrt{2\pi d(\theta^*)}}\int_{-\sqrt{\lambda(\theta^*)}}^{+\sqrt{\lambda(\theta^*)}} e^{-\frac{1}{2}t^2/d(\theta^*)}\,dt = \alpha.$$

Inequality (39) is obviously equivalent to the simultaneous inequalities:

$$y_n(\theta^*) \leq c_n'(\theta^*) \qquad \text{and} \qquad y_n(\theta^*) \geq c_n''(\theta^*)$$

where $c_n'(\theta^*)$ and $c_n''(\theta^*)$ are the roots of the equation in $y_n(\theta^*)$

$$[y_n(\theta^*)]^2 = l_n'(\theta^*)y_n(\theta^*) + l_n''(\theta^*) + d(\theta^*).$$

Since

$$\lim c_n'(\theta^*) = -\sqrt{\lambda(\theta^*)} \qquad \text{and} \qquad \lim c_n''(\theta^*) = +\sqrt{\lambda(\theta^*)}$$

uniformly in θ^*, from Proposition 3 it follows that

$$(43) \qquad \lim_{n=\infty}\left\{P\left[A_n(\theta^*) \mid \theta^* + \frac{\mu_n}{\sqrt{n}}\right] - \int_{-\infty}^{-\sqrt{\lambda(\theta^*)}} e^{\mu_n t - \frac{1}{2}\mu_n^2 d(\theta^*)}\,dN(t \mid \theta^*) - \int_{+\sqrt{\lambda(\theta^*)}}^{\infty} e^{\mu_n t - \frac{1}{2}\mu_n^2 d(\theta^*)}\,dN(t \mid \theta^*)\right\} = 0$$

uniformly in θ^*.

Now let us consider a sequence $\{\nu_n\}$ such that $\lim |\nu_n| = \infty$ and $\lim \frac{\nu_n}{\sqrt{n}} = 0$.

We shall prove that

$$P\left[A_n(\theta^*) \mid \theta^* + \frac{\nu_n}{\sqrt{n}}\right] = 1 \tag{44}$$

uniformly in θ^*. Since $E_\theta\left[\frac{\partial^2}{\partial\theta^2}\log f(x, \theta^*)\right]^2$ is assumed to be bounded,

$$E_{\theta^*+(\nu_n/\sqrt{n})}\left[\frac{\partial^2}{\partial\theta^2}\log f(x, \theta^*)\right] \tag{45}$$

and

$$E_{\theta^*+(\nu_n/\sqrt{n})}\left[\frac{\partial^2}{\partial\theta^2}\log f(x, \theta^*)\right]^2 \tag{46}$$

are bounded functions of θ^* and n. We get by Taylor expansion

$$\begin{aligned}\sum_\alpha \frac{\partial}{\partial\theta}\log f(x_\alpha, \theta^*) &= \sum_\alpha \frac{\partial}{\partial\theta}\log f\left(x_\alpha, \theta^* + \frac{\nu_n}{\sqrt{n}}\right) \\ &\quad - \frac{\nu_n}{\sqrt{n}}\sum_\alpha \frac{\partial^2}{\partial\theta^2}\log f(x_\alpha, \bar\theta_n^*)\end{aligned} \tag{47}$$

where $\bar\theta_n^*$ lies in $\left[\theta^*, \theta^* + \frac{\nu_n}{\sqrt{n}}\right]$. Hence

$$E_{\theta^*+(\nu_n/\sqrt{n})}[y_n(\theta^*)] = -\nu_n E_{\theta^*+(\nu_n/\sqrt{n})}\left[\frac{1}{n}\frac{\partial^2}{\partial\theta^2}\sum_\alpha \log f(x_\alpha, \bar\theta_n^*)\right]. \tag{48}$$

From Assumption 2 and $\lim |\nu_n| = \infty$ it follows that the absolute value of the right hand side of (48) converges to ∞. Hence

$$\lim | E_{\theta^*+\nu_n/\sqrt{n}}[y_n(\theta^*)] | = \infty.$$

Since on account of Assumption 1

$$E_{\theta^*+(\nu_n/\sqrt{n})}\left[\frac{\partial}{\partial\theta}\log f(x_\alpha, \theta^*)\right]^2$$

is a bounded function of n and θ^*, also the variance of $y_n(\theta^*)$ (under the assumption that $\theta = \theta^* + \nu_n/\sqrt{n}$ is the true value of the parameter) is a bounded function of n and θ^*. Hence for any arbitrary large constant C

$$\lim P\left[\,| y_n(\theta^*) | \geq C \mid \theta^* + \frac{\nu_n}{\sqrt{n}}\right] = 1, \tag{49}$$

uniformly in θ^*. The equation (44) follows easily from (36), (40), (41), (45), (46) and (49).

Consider a sequence $\{\rho_n\}$ such that $\left|\frac{\rho_n}{\sqrt{n}}\right| > \beta > 0$ for all n. Then it follows easily from Proposition 1 that for any arbitrary C

$$\lim P\left[\, | y_n(\theta^*) | \geq C \,|\, \theta^* + \frac{\rho_n}{\sqrt{n}}\right] = 1 \tag{50}$$

uniformly in θ^*. Since $E_\theta\left[\frac{\partial^2}{\partial\theta^2} \log f(x_\alpha, \theta^*)\right]^2$ is assumed to be bounded, and therefore also $E_\theta \frac{\partial^2}{\partial\theta^2} \log f(x, \theta^*)$ is bounded, there exists a finite g such that

$$\lim P\left\{\left|\frac{1}{n} \sum_\alpha \frac{\partial^2}{\partial\theta^2} \log f(x_\alpha, \theta^*)\right| < g \,|\, \theta^* + \frac{\rho_n}{\sqrt{n}}\right\} = 1 \tag{51}$$

uniformly in θ^*. From (36), (40), (41), (50) and (51) it follows

$$\lim P\left[A_n(\theta^*) \,|\, \theta^* + \frac{\rho_n}{\sqrt{n}}\right] = 1 \tag{52}$$

uniformly in θ^*. Since on account of Propositions 3 and 4, the relations (43), (44) and (52) hold if we substitute $B_n(\theta^*)$ for $A_n(\theta^*)$, Theorem 4 is proved.

If Assumptions 1–4 are fulfilled for the set ω consisting of the single point $\theta = \theta_0$, then we get from Theorems 1–4 the following corollaries:

COROLLARY 1: *Let W_n' be the region defined by the inequality $y_n(\theta_0) \geq c_n'$, W_n'' defined by the inequality $y_n(\theta_0) \leq c_n''$, and W_n defined by the inequality $| y_n(\theta_0) | \geq c_n$, where the constants c_n', c_n'' and c_n are chosen such that*

$$P(W_n' \mid \theta_0) = P(W_n'' \mid \theta_0) = P(W_n \mid \theta_0) = \alpha.$$

Then $\{W_n'\}$ is an asymptotically most powerful test of the hypothesis $\theta = \theta_0$ if θ takes only values $\geq \theta_0$. Similarly $\{W_n''\}$ is an asymptotically most powerful test if θ takes only values $\leq \theta_0$. Finally $\{W_n\}$ is an asymptotically most powerful unbiased test if θ can take any real value.

COROLLARY 2: *The sequence $\{A_n(\theta_0)\}$ is an asymptotically most powerful unbiased test of the hypothesis $\theta = \theta_0$, where $A_n(\theta_0)$ denotes the critical region of type A for testing $\theta = \theta_0$.*

Reprinted from THE ANNALS OF MATHEMATICAL STATISTICS
Vol. XIII, No. 2, June, 1942

ASYMPTOTICALLY SHORTEST CONFIDENCE INTERVALS[1]

BY ABRAHAM WALD[2]
Columbia University

The theory of confidence intervals, based on the classical theory of probability, has been treated by J. Neyman.[3] While Neyman considers the case of small samples, we shall deal here with the limit properties of the confidence intervals if the number of observations approaches infinity.

1. Definitions. We will start with some of Neyman's definitions. Let $f(x, \theta)$ be the probability density function of a variate x involving an unknown parameter θ. Denote by E_n a point of the n-dimensional sample space of n independent observations on x. If $\rho(E_n)$ denotes for each E_n a subset of the real axis, the symbol $P[\rho(E_n)c\theta' \mid \theta'']$ will denote the probability that $\rho(E_n)$ contains θ' under the hypothesis that θ'' is the true value of the parameter. Let $\underline{\theta}(E_n)$ and $\bar{\theta}(E_n)$ be two real functions defined over the whole sample space such that $\underline{\theta}(E_n) \leq \bar{\theta}(E_n)$. The interval $\delta(E_n) = [\underline{\theta}(E_n, \bar{\theta}(E_n)]$ is called a confidence interval of θ corresponding to the confidence coefficient α $(0 < \alpha < 1)$ if $P[\delta(E_n)c\theta \mid \theta] = \alpha$ for all values of θ.

The interval function $\delta(E_n)$ is called a shortest confidence interval of θ corresponding to the confidence coefficient α if

(a) $P[\delta(E_n)c\theta \mid \theta] = \alpha$ for all values of θ, and

(b) for any interval function $\delta'(E_n)$ which satisfies the condition (a) we have

$$P[\delta(E_n)c\theta' \mid \theta''] \leq P[\delta'(E_n)c\theta' \mid \theta''],$$

for arbitrary values θ' and θ''.

The interval function $\delta(E_n)$ is called a shortest unbiased confidence interval of θ if the following three conditions are fulfilled:

(a) $P[\delta(E_n)c\theta \mid \theta] = \alpha$ for all values of θ.

(b) $P[\delta(E_n)c\theta' \mid \theta''] \leq \alpha$ for all values of θ' and θ''.

(c) For any interval function $\delta'(E_n)$ for which the conditions (a) and (b) are satisfied, we have

$$P[\delta(E_n)c\theta' \mid \theta''] \leq P[\delta'(E_n)c\theta' \mid \theta''],$$

for all values of θ' and θ''.

For any relation R we shall denote by $P(R \mid \theta)$ the probability that R holds under the hypothesis that θ is the true value of the parameter. Similarly for

[1] Presented at a joint meeting of the Institute of Mathematical Statistics and the American Mathematical Society in Hanover, September, 1940.

[2] Research under a grant-in-aid from the Carnegie Corporation of New York.

[3] J. NEYMAN, "Outline of a theory of statistical estimation based on the classical theory of probability," *Phil. Trans. Roy. Soc. London*, Vol. 236 (1937), pp. 333–380.

any region Q_n of the n-dimensional sample space the symbol $P(Q_n \mid \theta)$ will denote the probability that the sample point E_n falls in Q_n under the hypothesis that θ is the true value of the parameter.

In all that follows we shall denote a region of the n-dimensional sample space by a capital letter with the subscript n.

A real function $\bar{\theta}(E_n)$ is called a best upper estimate of θ if the following two conditions are fulfilled:

(a) $P[\theta \leq \bar{\theta}(E_n) \mid \theta] = \alpha$ for all values of θ.

(b) For any function $\bar{\theta}'(E_n)$ which satisfies the condition (a) we have

$$P[\theta' \leq \bar{\theta}(E_n) \mid \theta''] \leq P[\theta' \leq \bar{\theta}'(E_n) \mid \theta'']$$

for all values θ' and θ'' for which $\theta' \geq \theta''$.

A real function $\underline{\theta}(E_n)$ is called a best lower estimate of θ if the following two conditions are fulfilled:

(a) $P[\theta \geq \underline{\theta}(E_n) \mid \theta] = \alpha$ for all values of θ.

(b) For any function $\underline{\theta}'(E_n)$ which satisfies the condition (a) we have

$$P[\theta' \geq \underline{\theta}(E_n) \mid \theta''] \leq P[\theta' \geq \underline{\theta}'(E_n) \mid \theta'']$$

for all values of θ' and θ'' for which $\theta' \leq \theta''$.

We will extend the above definitions of Neyman to the limit case when n approaches infinity.

DEFINITION I: *A sequence of interval functions* $\{\delta_n(E_n)\}$ $(n = 1, 2, \cdots)$ *is called an asymptotically shortest confidence interval of* θ *if the following two conditions are fulfilled:*

(a) $P[\delta_n(E_n) c \theta \mid \theta] = \alpha$ *for all values of* θ.

(b) *For any sequence of interval functions* $\{\delta'_n(E_n)\}$ $(n = 1, 2, \cdots, ad\ inf.)$ *which satisfies* (a), *the least upper bound of*

$$P[\delta_n(E_n) c \theta' \mid \theta''] - P[\delta'_n(E_n) c \theta' \mid \theta'']$$

with respect to θ' *and* θ'' *converges to zero as* $n \to \infty$.

DEFINITION II: *A sequence of interval functions* $\{\delta_n(E_n)\}$ *is called an asymptotically shortest unbiased confidence interval of* θ *if the following three conditions are fulfilled:*

(a) $P[\delta_n(E_n) c \theta \mid \theta] = \alpha$ *for all values of* θ.

(b) *The least upper bound of* $P[\delta_n(E_n) c \theta' \mid \theta'']$ *with respect to* θ' *and* θ'' *converges to* α *with* $n \to \infty$.

(c) *For any sequence of interval functions* $\{\delta'_n(E_n)\}$ *which satisfies the conditions* (a) *and* (b), *the least upper bound of*

$$P[\delta_n(E_n) c \theta' \mid \theta''] - P[\delta'_n(E_n) c \theta' \mid \theta'']$$

with respect to θ' *and* θ'' *converges to zero with* $n \to \infty$.

DEFINITION III: *A sequence of real functions* $\{\bar{\theta}_n(E_n)\}$ $(n = 1, 2, \cdots, ad\ inf.)$ *is called an asymptotically best upper estimate of* θ *if the following two conditions are fulfilled:*

(a) $P[\theta \leq \bar{\theta}_n(E_n) \mid \theta] = \alpha$ *for all values of* θ.

(b) *For any sequence of functions* $\{\bar{\theta}'_n(E_n)\}$ *which satisfies* (a) *the least upper bound of*

$$P[\theta' \leq \bar{\theta}_n(E_n) \mid \theta''] - P[\theta' \leq \bar{\theta}'_n(E_n) \mid \theta'']$$

in the domain $\theta' \geq \theta''$ *converges to zero with* $n \to \infty$.

DEFINITION IV: *A sequence of real functions* $\{\underline{\theta}_n(E_n)\}$ *is called an asymptotically best lower estimate of* θ *if the following two conditions are fulfilled:*

(a) $P[\theta \geq \underline{\theta}_n(E_n) \mid \theta] = \alpha$ *for all values of* θ.

(b) *For any sequence of functions* $\{\underline{\theta}'_n(E_n)\}$ *which satisfies* (a) *the least upper bound of*

$$P[\theta' \geq \underline{\theta}_n(E_n) \mid \theta''] - P[\theta' \geq \underline{\theta}'_n(E_n) \mid \theta'']$$

in the domain $\theta' \leq \theta''$ *converges to zero with* $n \to \infty$.

2. Two Propositions. PROPOSITION I: *Let* $\{W_n(\theta)\}$ $(n = 1, 2, \cdots, ad\ inf.)$ *be for each* θ *a sequence of regions such that the following two conditions are fulfilled:*

(a) $P[W_n(\theta) \mid \theta] = 1 - \alpha$ *for all values of* θ.

(b) *For any sequence of regions* $\{Z_n(\theta)\}$ *which satisfies* (a) *the least upper bound of*

$$P[Z_n(\theta') \mid \theta''] - P[W_n(\theta') \mid \theta'']$$

in the domain $\theta' \geq \theta''(\theta' \leq \theta'')$ *converges to zero with* $n \to \infty$.

Denote by $\rho_n(E_n)$ *the set of all values of* θ *for which* E_n *does not lie in* $W_n(\theta)$. *Then we have*

(c) $P[\rho_n(E_n)c\theta \mid \theta] = \alpha$ *for all values of* θ.

(d) *For any sequence of set functions* $\{\rho'_n(E_n)\}$ *which satisfies* (c), *the least upper bound of*

$$P[\rho_n(E_n)c\theta' \mid \theta''] - P[\rho'_n(E_n)c\theta' \mid \theta'']$$

in the domain $\theta' \geq \theta''(\theta' \leq \theta'')$ *converges to zero with* $n \to \infty$.

PROPOSITION II: *Let* $\{W_n(\theta)\}$ *be for each* θ *a sequence of regions such that the following three conditions are fulfilled:*

(a) $P(W_n(\theta) \mid \theta] = 1 - \alpha$ *for all values of* θ.

(b) *The greatest lower bound of* $P[W_n(\theta') \mid \theta'']$ *converges to* $1 - \alpha$ *with* $n \to \infty$.

(c) *For any sequence* $\{W'_n(\theta)\}$ *which satisfies* (a) *and* (b), *the least upper bound of*

$$P[W'_n(\theta') \mid \theta''] - P[W_n(\theta') \mid \theta'']$$

with respect to θ' *and* θ'' *converges to* 0 *with* $n \to \infty$.

Denote by $\rho_n(E_n)$ *the set of all values of* θ *for which* E_n *does not lie in* $W_n(\theta)$. *Then we have*

(d) $P[\rho_n(E_n)c\theta \mid \theta] = \alpha$ *for all values of* θ.

(e) *The least upper bound of* $P[\rho_n(E_n)c\theta' \mid \theta'']$ *converges to* α *with* $n \to \infty$.

(f) *For any sequence of setfunctions* $\{\rho'_n(E_n)\}$ *which satisfies* (d) *and* (e), *the least upper bound of*

$$P[\rho_n(E_n)c\theta' \mid \theta''] - P[\rho'_n(E_n)c\theta' \mid \theta'']$$

with respect to θ' *and* θ'' *converges to* 0 *with* $n \to \infty$.

The validity of the above propositions follows easily from the identity

$$P[\rho_n(E_n)c\theta' \mid \theta''] = 1 - P[W_n(\theta') \mid \theta''].$$

3. Assumptions on the probability density function. For any function $\psi(x)$ denote by $E_\theta\psi(x)$ the expected value of $\psi(x)$ under the assumption that θ is the true value of the parameter, i.e.

$$E_\theta \psi(x) = \int_{-\infty}^{+\infty} \psi(x) f(x, \theta)\, dx.$$

For any x, for any positive δ, and for any real value θ' denote by $\varphi_1(x, \theta', \delta)$ the greatest lower bound, and by $\varphi_2(x, \theta', \delta)$ the least upper bound of $\frac{\partial^2}{\partial\theta^2} \log f(x, \theta)$ in the interval $\theta' - \delta \leq \theta \leq \theta' + \delta$.

Throughout this paper the following assumptions on $f(x, \theta)$ will be made:

ASSUMPTION I: *The expectation* $E_{\theta'} \frac{\partial}{\partial\theta} \log f(x, \theta'')$ *is a continuous function of* θ' *and* θ'', *and for any pair of sequences* $\{\theta'_n\}$ *and* $\{\theta''_n\}$ $(n = 1, 2, \cdots, \textit{ad inf.})$ *for which*

$$\lim_{n=\infty} E_{\theta'_n} \frac{\partial}{\partial\theta} \log f(x, \theta''_n) = 0$$

also

$$\lim_{n=\infty} (\theta'_n - \theta''_n) = 0.$$

Furthermore

$$E_{\theta'} \left[\frac{\partial}{\partial\theta} \log f(x, \theta'')\right]^2$$

is a bounded function of θ' *and* θ'', *and* $E_\theta \left[\frac{\partial}{\partial\theta} \log f(x, \theta)\right]^2 = d(\theta)$ *has a positive lower bound.*

ASSUMPTION II: *There exists a positive value* k_0 *such that the expectations* $E_{\theta'}\varphi_1(x, \theta'', \delta)$ *and* $E_{\theta'}\varphi_2(x, \theta'', \delta)$ *are uniformly continuous functions of* θ', θ'' *and* δ *where* δ *takes only values for which* $|\delta| \leq k_0$. *Furthermore it is assumed that* $E_{\theta'}[\varphi_i(x, \theta'', \delta)]^2$ $(i = 1, 2)$ *are bounded functions of* θ', θ'' *and* δ $(|\delta| \leq k_0)$.

ASSUMPTION III: *The relations*

$$\int_{-\infty}^{+\infty} \frac{\partial}{\partial\theta} f(x, \theta)\, dx = \int_{-\infty}^{+\infty} \frac{\partial^2}{\partial\theta^2} f(x, \theta)\, dx = 0$$

hold.

The above assumption means simply that we may differentiate with respect to θ under the integral sign. In fact

$$\int_{-\infty}^{+\infty} f(x, \theta)\, dx = 1$$

identically in θ. Hence

$$\frac{\partial}{\partial\theta}\int_{-\infty}^{+\infty} f(x, \theta)\, dx = \frac{\partial^2}{\partial\theta^2}\int_{-\infty}^{+\infty} f(x, \theta)\, dx = 0.$$

Differentiating under the integral sign, we obtain the relations in Assumption III.

ASSUMPTION IV: *There exists a positive* η *such that*

$$E_\theta\left[\frac{\partial}{\partial\theta}\log f(x, \theta)\right]^{2+\eta}$$

is a bounded function of θ.

4. Some theorems. The assumptions on $f(x, \theta)$ made in this paper become identical with the assumptions I–IV formulated in a previous paper[4] if a certain set ω involved in those assumptions is put equal to the whole real axis $(-\infty, +\infty)$. Hence we can make use of all results obtained in that paper putting $\omega = (-\infty, +\infty)$. Among others, the following statements have been proved there:

(A) Denote $\sum_{\alpha=1}^{n} \frac{1}{\sqrt{n}} \frac{\partial}{\partial\theta} \log f(x_\alpha, \theta)$ by $y_n(\theta, E_n)$ and let $R_n(\theta)$ be the region defined by the inequality $y_n(\theta, E_n) \geq A_n(\theta)$ where $A_n(\theta)$ is chosen such that $P[R_n(\theta) \mid \theta] = 1 - \alpha$. Then for any sequence of regions $\{Z_n(\theta)\}$ for which $P[Z_n(\theta) \mid \theta] = 1 - \alpha$, the least upper bound of

$$P[Z_n(\theta') \mid \theta''] - P[R_n(\theta') \mid \theta'']$$

in the set $\theta'' \geq \theta'$ converges to 0 with $n \to \infty$.

(B) Let $S_n(\theta)$ be the region defined by the inequality $y_n(\theta, E_n) \leq B_n(\theta)$ where $B_n(\theta)$ is defined such that $P[S_n(\theta) \mid \theta] = 1 - \alpha$. Then for any sequence of regions $\{Z_n(\theta)\}$ for which $P[Z_n(\theta) \mid \theta] = 1 - \alpha$, the least upper bound of

$$P[Z_n(\theta') \mid \theta''] - P[S_n(\theta') \mid \theta'']$$

in the set $\theta'' \leq \theta'$ converges to 0 with $n \to \infty$.

(C) Denote by $T_n(\theta)$ the region defined by $\mid y_n(\theta, E_n) \mid \geq C_n(\theta)$ where $C_n(\theta)$ is chosen such that

(a) $P[T_n(\theta) \mid \theta] = 1 - \alpha$.

Then $T_n(\theta)$ satisfies also the following two conditions:

(b) The greatest lower bound of $P[T_n(\theta')\theta'']$ converges to $1 - \alpha$ with $n \to \infty$.

(c) For any sequence of regions $\{Z_n(\theta)\}$ which satisfies (a) and (b), the least upper bound of

$$P[Z_n(\theta') \mid \theta''] - P[T_n(\theta') \mid \theta'']$$

converges to 0 with $n \to \infty$.

---

[4] A. WALD, "Some examples of asymptotically most powerful tests," *Annals of Math. Stat.*, Vol. 12 (1941), pp. 396–408.

On account of Propositions I and II we easily get the following theorems:

THEOREM I: *Denote by* $\xi_n(E_n)$ *the set of all values of* θ *for which* $y_n(\theta, E_n) \leq A_n(\theta)$ *and* $A_n(\theta)$ *is defined such that* $P[y_n(\theta, E_n) > A_n(\theta) \mid \theta] = 1 - \alpha$. *Then* $\xi_n(E_n)$ *satisfies the following two conditions:*

(a) $P[\xi_n(E_n)c\theta \mid \theta] = \alpha$ *for all values of* θ.

(b) *For any sequence of setfunctions* $\{\xi_n'(E_n)\}$ *which satisfies the condition* (a), *the least upper bound of*

$$P[\xi_n(E_n)c\theta' \mid \theta''] - P[\xi_n'(E_n)c\theta' \mid \theta'']$$

in the set $\theta'' \geq \theta'$ *converges to* 0 *with* $n \to \infty$.

THEOREM II: *Denote by* $\zeta_n(E_n)$ *the set of all values of* θ *for which* $y_n(\theta, E_n) \geq B_n(\theta)$ *and* $B_n(\theta)$ *is defined such that* $P[y_n(\theta, E_n) < B_n(\theta) \mid \theta] = 1 - \alpha$. *Then* $\zeta_n(E_n)$ *satisfies the following two conditions:*

(a) $P[\zeta_n(E_n)c\theta \mid \theta] = \alpha$ *for all values of* θ.

(b) *For any sequence of setfunctions* $\{\zeta_n'(E_n)\}$ *which satisfies the condition* (a), *the least upper bound of*

$$P[\zeta_n(E_n)c\theta' \mid \theta''] - P[\zeta_n'(E_n)c\theta' \mid \theta'']$$

in the set $\theta'' \leq \theta'$ *converges to* 0 *with* $n \to \infty$.

THEOREM III: *Denote by* $\rho_n(E_n)$ *the set of all values of* θ *for which* $| y_n(\theta, E_n) | \leq C_n(\theta)$ *and* $C_n(\theta)$ *is chosen such that* $P[| y_n(\theta, E_n) | > C_n(\theta) \mid \theta] = 1 - \alpha$. *Then* $\rho_n(E_n)$ *satisfies the following three conditions:*

(a) $P[\rho_n(E_n)c\theta \mid \theta] = \alpha$ *for all values of* θ.

(b) *The least upper bound of* $P[\rho_n(E_n)c\theta' \mid \theta'']$ *converges to* α *with* $n \to \infty$.

(c) *For any sequence of setfunctions* $\{\rho_n'(E_n)\}$ *which satisfies the conditions* (a) *and* (b), *the least upper bound of*

$$P[\rho_n(E_n)c\theta' \mid \theta''] - P[\rho_n'(E_n)c\theta' \mid \theta'']$$

converges to zero with $n \to \infty$.

Now we shall investigate the question whether the sets $\xi_n(E_n)$, $\zeta_n(E_n)$ and $\rho_n(E_n)$ are intervals. For this purpose we will prove some propositions.

PROPOSITION III: *Let* ϵ *and* D *be two positive numbers such that* $\epsilon < D$. *Denote by* $Q_n(\theta, \epsilon, D)$ *the region which consists of all points* E_n *for which*

$$y_n(\theta + \epsilon', E_n) \leq -n^{\frac{1}{3}}, \qquad \text{and} \qquad y_n(\theta - \epsilon', E_n) \geq n^{\frac{1}{3}}$$

for all values ϵ' *in the interval* $[\epsilon, D]$. *Then we have*

$$\lim_{n=\infty} P[Q_n(\theta, \epsilon, D) \mid \theta] = 1 \tag{1}$$

uniformly in θ.

PROOF: Let $\epsilon_1, \epsilon_2, \cdots, \epsilon_r$ be a sequence of points in the interval $[\epsilon, D]$ such that $\epsilon_1 - \epsilon = \epsilon_2 - \epsilon_1 = \cdots = \epsilon_r - \epsilon_{r-1} = D - \epsilon_r = k_0$ (say), where r is chosen sufficiently large such that Assumption II holds for $| \delta | \leq k_0$. Denote by $R_n(\theta, \epsilon_i)$ the region in which

$$y_n(\theta + \epsilon_i, E_n) \leq -n^{\frac{1}{3}}. \tag{2}$$

We will show that

(3) $$\lim_{n=\infty} P[R_n(\theta, \epsilon_i) \mid \theta] = 1$$

uniformly in θ.

From Assumption I it follows that the greatest lower bound of

$$\left| E_\theta \frac{\partial}{\partial\theta} \log f(x, \theta + \epsilon') \right|$$

with regard to ϵ' in the interval $[\epsilon, D]$ is positive. Let this greatest lower bound be $A > 0$. Since on account of Assumption I $E_\theta \frac{\partial}{\partial\theta} \log f(x, \theta + \epsilon')$ is a continuous function of ϵ', it does not change sign in the interval $\epsilon \leq \epsilon' \leq D$. Since this is true for arbitrarily small ϵ and since $E_\theta \left[\frac{\partial}{\partial\theta} \log f(x,\theta)\right]^2 = -E_\theta \frac{\partial^2}{\partial\theta^2} \log f(x, \theta)$ has a positive lower bound (Assumption I), it follows easily on account of Assumption II that

$$E_\theta \frac{\partial}{\partial\theta} \log f(x, \theta + \epsilon') < 0.$$

Hence

(4) $$E_\theta \frac{\partial}{\partial\theta} \log f(x, \theta + \epsilon') \leq -A < 0 \quad \text{for} \quad \epsilon \leq \epsilon' \leq D,$$

and therefore

(5) $$E_\theta y_n(\theta + \epsilon', E_n) \leq -A\sqrt{n} \qquad \text{for } \epsilon \leq \epsilon' \leq D.$$

From Assumption II it follows that the variance of $y_n(\theta + \epsilon', E_n)$ is a bounded function of θ and ϵ'. Hence

(6) $$\lim_{n=\infty} P[y_n(\theta + \epsilon_i, E_n) \leq -\tfrac{1}{2}A\sqrt{n} \mid \theta] = 1$$

uniformly in θ. The equation (3) is a consequence of (6).

Denote by $S_n(\theta, \epsilon_i)$ the region in which

$$\left| \frac{1}{n} \sum_\alpha \varphi_i(x_\alpha, \theta + \epsilon_i, k_0) \right| < C \qquad (i = 1, 2)$$

where C is greater than the least upper bound of $| E_\theta \varphi_i(x, \theta', k_0) |$ with respect to θ and θ'. Then we have on account of Assumption II:

(7) $$\lim_{n=\infty} P[S_n(\theta, \epsilon_i) \mid \theta] = 1 \qquad (i = 1, 2, \cdots, r)$$

uniformly in θ. In the region $S_n(\theta, \epsilon_i)$ we obviously have

(8) $$y_n(\theta + \epsilon_i', E_n) \leq y_n(\theta + \epsilon_i, E_n) + 2k_0\sqrt{n}C$$

for all values ϵ_i' in the interval $[\epsilon_i - k_0, \epsilon_i + k_0]$. By choosing r sufficiently large we can always achieve that

$$2k_0 C \leq \frac{A}{4}.$$

Denote by $T_n(\theta, \epsilon_i)$ the region in which

$$(9) \qquad y_n(\theta + \epsilon_i', E_n) \leq -\frac{A}{4}\sqrt{n} \quad \text{for} \quad \epsilon_i - k_0 \leq \epsilon_i' \leq \epsilon_i + k_0.$$

From (6), (7) and (8) we get

$$(10) \qquad \lim_{n=\infty} P[T_n(\theta, \epsilon_i) \mid \theta] = 1$$

uniformly in θ. Let $Q_n'(\theta, \epsilon, D)$ be the common part of the r regions $T_n(\theta, \epsilon_1), \cdots, T_n(\theta, \epsilon_r)$, i.e. $Q_n'(\theta, \epsilon, D)$ is the set of all points E_n for which

$$y_n(\theta + \epsilon', E_n) \leq -\frac{A}{4}\sqrt{n}$$

for all ϵ' in the interval $[\epsilon, D]$. Since r is a fixed positive integer not depending on n, we get from (10)

$$(11) \qquad \lim_{n=\infty} P[Q_n'(\theta, \epsilon, D) \mid \theta] = 1$$

uniformly in θ.

In the same way we can prove that

$$(12) \qquad \lim_{n=\infty} P[Q_n''(\theta, \epsilon, D) \mid \theta] = 1$$

uniformly in θ, where $Q_n''(\theta, \epsilon, D)$ denotes the region in which

$$y_n(\theta - \epsilon', E_n) \geq \frac{A}{4}\sqrt{n} \quad \text{for all} \quad \epsilon' \text{ in } [\epsilon, D].$$

Proposition III follows from (11) and (12).

PROPOSITION IV: *Denote by $V_n(\theta, \epsilon)$ the region in which*

$$\frac{\partial}{\partial\theta} y_n(\theta', E_n) < -n^{\frac{1}{3}}$$

for all values θ' in the interval $[\theta - \epsilon, \theta + \epsilon]$. There exists a positive ϵ such that

$$\lim_{n=\infty} P[V_n(\theta, \epsilon) \mid \theta] = 1$$

uniformly in θ.

PROOF: Since the least upper bound of $E_\theta\varphi_2(x, \theta, 0)$ is <0, we get from Assumption II that the least upper bound of $E_\theta\varphi_2(x, \theta, \epsilon)$ is <0 for sufficiently small $\epsilon > 0$. Denote the least upper bound of $E_\theta \varphi_2(x, \theta, \epsilon)$ by $-B$ and let the region in which

$$\frac{1}{n} \sum_\alpha \varphi_2(x_\alpha, \theta, \epsilon) < -\tfrac{1}{2}B$$

be denoted by $W_n(\theta, \epsilon)$. From Assumption II it follows that

$$\lim_{n=\infty} P[W_n(\theta, \epsilon) \mid \theta] = 1$$

uniformly in θ. Since for almost all n $W_n(\theta, \epsilon)$ is a subset of $V_n(\theta, \epsilon)$, Proposition IV is proved.

PROPOSITION V: *Let $A_n(\theta)$, $B_n(\theta)$, $C_n(\theta)$ be the functions as defined in Theorems I–III. There exists a finite value G such that*

$$| A_n(\theta) | < G, \qquad | B_n(\theta) | < G \quad \text{and} \quad | C_n(\theta) | < G$$

for all θ and all n.

Proposition V follows easily from the fact that the variance of $y_n(\theta, E_n)$ is a bounded function of n and θ.

Let D be an arbitrary positive number and denote by $W_n(\theta, D)$ the region consisting of all points E_n for which the following conditions are fulfilled:

(a) The equation $y_n(\theta', E_n) = A_n(\theta')$ has exactly one root in θ' which lies in the interval $[\theta - D, \theta + D]$.

(b) The equation $y_n(\theta', E_n) = B_n(\theta')$ has exactly one root in θ' which lies in the interval $[\theta - D, \theta + D]$.

(c) The equation $y_n(\theta', E_n) = C_n(\theta')$ has exactly one root in θ' which lies in the interval $[\theta - D, \theta + D]$.

(d) The equation $y_n(\theta', E_n) = -C_n(\theta')$ has exactly one root in θ' which lies in the interval $[\theta - D, \theta + D]$.

(e) The common part of $[\theta - D, \theta + D]$ and $\xi_n(E_n)$ is the interval $[\theta_n'(E_n), D]$ where $\theta_n'(E_n)$ denotes the root of the equation in (a).

(f) The common part of $\zeta_n(E_n)$ and $[\theta - D, \theta + D]$ is the interval $[-D, \theta_n''(E_n)]$ where $\theta_n''(E_n)$ denotes the root of the equation in (b).

(g) The common part of $\rho_n(E_n)$ and $[\theta - D, \theta + D]$ is the interval $[\underline{\theta}_n(E_n), \bar{\theta}_n(E_n)]$ where $\underline{\theta}_n(E_n)$ denotes the root of the equation in (c) and $\bar{\theta}_n(E_n)$ denotes the root of the equation in (d).

From Propositions III–V follows easily the following

PROPOSITION VI: *For any positive value D*

$$\lim_{n=\infty} P[W_n(\theta, D) \mid \theta] = 1,$$

uniformly in θ, provided that the functions $A_n(\theta)$, $B_n(\theta)$ and $C_n(\theta)$ are continuous and of bounded variation in any finite interval.

We will show that Proposition VI remains valid for $D = +\infty$, if we make the following

ASSUMPTION V: *Denote by* $\psi(x, \theta, D)$ *the least upper bound of* $\frac{\partial}{\partial\theta} \log f(x, \theta')$ *with respect to* θ' *where* $\theta' \geq \theta + D$. *Denote furthermore by* $\psi^*(x, \theta, D)$ *the greatest lower bound of* $\frac{\partial}{\partial\theta} \log f(x,\theta')$ *with respect to* θ' *where* $\theta' \leq \theta - D$. *There exists a positive D such that the least upper bound of* $E_\theta\psi(x, \theta, D)$ *with respect to* θ *is negative, the greatest lower bound of* $E_\theta\psi^*(x, \theta, D)$ *with respect to* θ *is positive, and the variances of* $\psi(x, \theta, D)$ *and* $\psi^*(x, \theta, D)$ *are bounded functions of* θ. *(The variances are calculated under the assumption that* θ *is the true value of the parameter.)*

It follows easily from Assumption V that

$$\lim_{n=\infty} P\left[\frac{1}{\sqrt{n}}\sum_\alpha \psi(x_\alpha, \theta, D) < -n^{\frac{1}{3}} \mid \theta\right] = \lim_{n=\infty} P\left[\frac{1}{\sqrt{n}}\sum_\alpha \psi^*(x_\alpha, \theta, D) > n^{\frac{1}{3}} \mid \theta\right] = 1$$

uniformly in θ.

Since

$$\frac{1}{\sqrt{n}}\sum_\alpha \psi(x_\alpha, \theta, D) \geq y_n(\theta', E_n) \quad \text{for} \quad \theta' \geq \theta + D$$

and

$$\frac{1}{\sqrt{n}}\sum_\alpha \psi^*(x_\alpha, \theta, D) \leq y_n(\theta', E_n) \quad \text{for} \quad \theta' \leq \theta - D,$$

Proposition VI remains valid if we substitute $+\infty$ for D.

Hence we obtain the following

COROLLARY: *If the assumptions* I–V *are fulfilled and if* $A_n(\theta)$, $B_n(\theta)$ *and* $C_n(\theta)$ *are continuous and of bounded variation in any finite interval, then*

(a) *The root* $\theta_n'(E_n)$ *of the equation* $y_n(\theta, E_n) = A_n(\theta)$ *in* θ *is an asymptotically best lower estimate of* θ.

(b) *The root* $\theta_n''(E_n)$ *of the equation* $y_n(\theta, E_n) = B_n(\theta)$ *in* θ *is an asymptotically best upper estimate of* θ.

(c) *The interval* $[\underline{\theta}_n(E_n), \bar{\theta}_n(E_n)$ *is an asymptotically shortest unbiased confidence interval of* θ, *where* $\underline{\theta}_n(E_n)$ *denotes the root of the equation* $y_n(\theta, E_n) = +C_n(\theta)$, *and* $\bar{\theta}_n(E_n)$ *denotes the root of the equation* $y_n(\theta, E_n) = -C_n(\theta)$.

5. Some Remarks. 1. I should like to make a few remarks about the relationship of these results to those obtained by S. S. Wilks.[5] The definition of a shortest confidence interval underlying Wilks' investigations is somewhat different from that of Neyman's which has been used in this paper. According to Wilks, a confidence interval $\delta(E_n)$ is called shortest in the average if the expected

[5] S. S. WILKS, "Shortest average confidence intervals from large samples," *Annals of Math. Stat.*, Vol. 9 (1938), pp. 166–175.

value of the length of $\delta(E_n)$ is a minimum. The main result obtained by Wilks can be formulated as follows: The confidence interval $[\underline{\theta}_n(E_n), \bar{\theta}_n(E_n)]$ given in our Corollary is asymptotically shortest in the average compared with all confidence intervals computed on the basis of functions belonging to a certain class C. In the present paper no restriction to a certain class of functions has been made.

2. If the parameter space Ω is not the whole real axis, but an open subset of it, and if the assumptions I–V are fulfilled when θ can take only values in Ω, the previously proved Corollary remains valid. If Ω is a bounded set, Assumption V is a consequence of Assumptions I–IV.

Made in the United States of America

Reprinted from THE ANNALS OF MATHEMATICAL STATISTICS
Vol. XIII, No. 3, September, 1942

ON THE CHOICE OF THE NUMBER OF CLASS INTERVALS IN THE APPLICATION OF THE CHI SQUARE TEST

BY H. B. MANN AND A. WALD[1]

Columbia University

Introduction. To test whether a sample has been drawn from a population with a specified probability distribution, the range of the variable is divided into a number of class intervals and the statistic,

$$\sum_{i=1}^{i=k} \frac{(\alpha_i - Np_i)^2}{Np_i} = \chi^2, \tag{1}$$

computed. In (1) k is the number of class intervals, α_i the number of observations in the ith class, p_i the probability that an observation falls into the ith class (calculated under the hypothesis to be tested). It is known that under the null hypothesis (hypothesis to be tested) the statistic (1) has asymptotically the chi-square distribution with $k - 1$ degrees of freedom, when each Np_i is large. To test the null hypothesis the upper tail of the chi-square distribution is used as a critical region.

In the literature only rules of thumb are found as to the choice of the number and lengths of the class intervals. It is the purpose of this paper to formulate principles for this choice and to determine the number and lengths of the class intervals according to these principles.

If a choice is made as to the number of class intervals it is always possible to find alternative hypotheses with class probabilities equal to the class probabilities under the null hypothesis. The least upper bound of the "distances" of such alternative distributions from the null hypothesis distribution can evidently be minimized by making the class probabilities under the null hypothesis equal to each other. By the distance of two distribution functions we mean the least upper bound of the absolute value of the difference of the two cumulative distribution functions. We have therefore based this paper on a procedure by which the lengths of the class intervals are determined so that the probability of each class under the null hypothesis is equal to $1/k$ where k is the number of class intervals.[2]

Let $C(\Delta)$ be the class of alternative distributions with a distance $\geqslant \Delta$ from the null hypothesis. Let $f(N, k, \Delta)$ be the greatest lower bound of the power of the chi-square test with sample size N and number of class intervals k with respect to alternatives in $C(\Delta)$. The maximum of $f(N, k, \Delta)$ with respect to k is a function $\Phi(N, \Delta)$ of N and Δ. It is most desirable to maximize $f(N, k, \Delta)$ for

---

[1] Research under a grant in aid from the Carnegie Corporation of New York.

[2] This procedure was first used by H. Hotelling. "The consistency and ultimate distribution of optimum statistics," *Trans. Am. Math. Soc.*, Vol. 32, pp. 851.) It has been advocated by E. J. Gumbel in a paper which will appear shortly.

such values of Δ for which $\Phi(N, \Delta)$ is neither too large nor too small and in this paper we propose to determine Δ so that $\Phi(N, \Delta)$ is equal to $\frac{1}{2}$.

Hence we introduce the following definitions:

DEFINITION 1. *A positive integer k is called best with respect to the number of observations N if there exists a Δ such that $f(N, k, \Delta) = \frac{1}{2}$ and $f(N, k', \Delta) \leqslant \frac{1}{2}$ for any positive integer k'.*

DEFINITION 2. *A positive integer k is called ϵ-best ($0 \leqslant \epsilon \leqslant 1$) with respect to the number of observations N if ϵ is the smallest number in the interval [0, 1] for which the following condition is fulfilled: There exists a Δ such that $f(N, k, \Delta) \geqslant \frac{1}{2} - \epsilon$ and $f(N, k', \Delta) \leqslant \frac{1}{2} + \epsilon$ for any positive integer k'.*

It is obvious that an ϵ-best k is a best k if $\epsilon = 0$. If ϵ is very small an ϵ-best k is for all practical purposes equivalent to a best k.

Since $f(N, k, \Delta)$ is a continuous function of Δ it is easy to see that for any pair of positive integers k and N there exists exactly one value ϵ such that k is ϵ-best with respect to the number of observations N. Since the value of this ϵ is a function of k and N we will denote it by $\epsilon(k, N)$.

DEFINITION 3. *A sequence $\{k_N\}$ of positive integers is called best in the limit if* $\lim_{N=\infty} \epsilon(k_N, N) = 0$.

In this paper the following theorem is proved:

THEOREM 1. *Let* $k_N = 4\sqrt[5]{\dfrac{2(N-1)^2}{c^2}}$ *where c is determined so that* $\dfrac{1}{\sqrt{2\pi}}\displaystyle\int_c^\infty e^{-x^2/2}\,dx$ *is equal to the size of the critical region (probability of the critical region under the null hypothesis) then the sequence $\{k_N\}$ is best in the limit. Furthermore* $\lim_{N=\infty} f(N, k_N, \Delta_N) = \frac{1}{2}$ *for* $\Delta_N = \dfrac{5}{k_N} - \dfrac{4}{k_N^2}$.

It is further shown that for $N \geqslant 450$, if the 5% level of significance is used, and for $N \geqslant 300$, if the 1% level of significance is used, the value of $\epsilon(k_N, N)$ is small so that for practical purposes k_N can be considered as a best k. The authors are convinced although no rigorous proof has been given that $\epsilon(k_N, N)$ is quite small for $N \geqslant 200$ and is very likely to be small even for considerably lower values of N.

1. Mean value and standard deviation of the statistic under alternative hypotheses. It is well known that every continuous distribution can by a simple transformation be transformed into a rectangular distribution with range [0, 1]. We may therefore for convenience assume that the hypothesis to be tested is that of a rectangular distribution with the range [0, 1]. Moreover as mentioned earlier we assume that a procedure is chosen by which the class probabilities under the null hypothesis are equal to each other.

The statistic whose mean value and standard deviation is to be determined is

$$\sum_{i=1}^{i=k} x_i^2 = \chi'^2 \quad \text{where} \quad x_i = \sqrt{\frac{k}{N}}\left(\alpha_i - \frac{N}{k}\right).$$

Let p_i be the probability under the alternative hypothesis that one observation will fall into the ith class. The probability of obtaining certain specified values $\alpha_1, \alpha_2, \cdots, \alpha_k$ is given by

$$f(\alpha_1, \alpha_2, \cdots \alpha_k) = \frac{N!}{\alpha_1! \alpha_2! \cdots \alpha_k!} p_1^{\alpha_1} p_2^{\alpha_2} \cdots p_k^{\alpha_k}.$$

Since $\sum_{i=1}^{i=k} \alpha_i = N$ we have

$$\sum_{i=1}^{i=k} x_i^2 = \frac{k}{N} \sum_{i=1}^{i=k} \alpha_i^2 - N.$$

We consider the function

$$(p_1 e^{t_1} + p_2 e^{t_2} + \cdots p_k e^{t_k})^N = \Sigma f(\alpha_1, \alpha_2, \cdots \alpha_k) e^{\alpha_1 t_1 + \alpha_2 t_2 + \cdots \alpha_k t_k}.$$

Differentiating twice and then setting $t_i = 0$ for $i = 1, 2, \cdots k$ we obtain

(2) $\quad N(N-1)p_i^2 + Np_i = E(\alpha_i^2), \quad N(N-1)p_i p_j = E(\alpha_i \alpha_j)$ for $i \neq j$.

Hence

$$E\left(\sum_{i=1}^{i=k} a_i^2\right) = N(N-1) \sum_{i=1}^{i=k} p_i^2 + N,$$

and

(3) $$E(\chi'^2) = k(N-1) \sum_{i=1}^{i=k} p_i^2 + k - N.$$

To compute the standard deviation of χ'^2 we put

$$\mu_i = \left(Np_i - \frac{N}{k}\right)\sqrt{\frac{k}{N}} = \sqrt{Nk}\left(p_i - \frac{1}{k}\right),$$

$$y_i = (\alpha_i - Np_i)\sqrt{\frac{k}{N}} \quad \text{hence} \quad y_i = x_i - \mu_i, \quad E(y_i) = 0.$$

We have

$$\sigma_{\chi'^2}^2 = E\left[\sum_{i=1}^{i=k}(y_i + \mu_i)^2 - E\left(\sum_{i=1}^{i=k}(y_i + \mu_i)^2\right)\right]^2$$

$$= E\left(\sum_{i=1}^{i=k} y_i^2 + 2\sum_{i=1}^{i=k} y_i \mu_i - E\left(\sum_{i=1}^{i=k} y_i^2\right)\right)^2.$$

Let

$$\sqrt{\frac{N}{k}}\, y_i = z_i, \qquad \sqrt{\frac{N}{k}}\, \mu_i = \nu_i;$$

then

$$\nu_i = N\left(p_i - \frac{1}{k}\right), \qquad z_i = \alpha_i - Np_i.$$

We now assume that N is so large that the joint distribution of the z_i is sufficiently well approximated by a multivariate normal distribution. Then

$E(z_i^2 z_j) = 0, \quad E(z_i^4) = 3[E(z_i^2)]^2, \quad E(z_i^2 z_j^2) = E(z_i^2)E(z_j^2) + 2[E(z_i z_j)]^2$ for $i \neq j$.

We have the well known relations

$$E(z_i^2) = E(\alpha_i^2) - N^2 p_i^2 = Np_i(1 - p_i),$$

$$E(z_i z_j) = E(\alpha_i \alpha_j) - N^2 p_i p_j = -Np_i p_j .$$

Using the above equations we obtain

$$\sigma_{\chi'^2}^2 = \frac{k^2}{N^2}\left\{E\left(\sum_{i=1}^{i=k} z_i^2\right)^2 - \left(E\sum_{i=1}^{i=k} z_i^2\right)^2 + 4E\left(\sum_{i=1}^{i=k} z_i \nu_i\right)^2\right\},$$

$$E\left(\sum_{i=1}^{i=k} z_i^2\right)^2 - \left(E\sum_{i=1}^{i=k} z_i^2\right)^2$$

$$= N^2\left\{3\sum_{i=1}^{i=k} p_i^2(1 - p_i)^2 + \sum_{i \neq j}[p_i p_j(1 - p_i)(1 - p_j) + 2p_i^2 p_j^2] - \left[\sum_{i=1}^{i=k} p_i(1 - p_i)\right]^2\right\}$$

$$= 2N^2\left[\sum_{i=1}^{i=k} p_i^2(1 - p_i)^2 + \sum_{i \neq j} p_i^2 p_j^2\right]$$

$$= 2N^2\left[\sum_{i=1}^{i=k} p_i^2 - 2\sum_{i=1}^{i=k} p_i^3 + \left(\sum_{i=1}^{i=k} p_i^2\right)^2\right].$$

Further

$$E\left(\sum_{i=1}^{i=k} z_i \nu_i\right)^2 = E\left(\sum_{i=1}^{i=k} z_i^2 \nu_i^2\right) + E\left(\sum_{i \neq j} z_i z_j \nu_i \nu_j\right)$$

$$= N^3\left[\sum_{i=1}^{i=k} p_i(1 - p_i)\left(p_i - \frac{1}{k}\right)^2 - \sum_{i \neq j} p_i p_j\left(p_i - \frac{1}{k}\right)\left(p_j - \frac{1}{k}\right)\right]$$

$$= N^3\left[\sum_{i=1}^{i=k} p_i\left(p_i - \frac{1}{k}\right)^2 - \left[\sum_{i=1}^{i=k} p_i\left(p_i - \frac{1}{k}\right)\right]^2\right]$$

$$= N^3\left[\sum_{i=1}^{i=k} p_i^3 - \frac{2}{k}\sum_{i=1}^{i=k} p_i^2 + \frac{1}{k^2} - \left[\sum_{i=1}^{i=k} p_i^2 - \frac{1}{k}\right]^2\right]$$

$$= N^3\left[\sum_{i=1}^{i=k} p_i^3 - \left(\sum p_i^2\right)^2\right].$$

Substituting this into the formula for $\sigma_{\chi'^2}^2$ we finally obtain

$$\sigma_{\chi'^2}^2 = 2k^2\left\{\sum_{i=1}^{i=k} p_i^2 + 2(N - 1)\sum_{i=1}^{i=k} p_i^3 - (2N - 1)\left(\sum_{i=1}^{i=k} p_i^2\right)^2\right\}. \tag{4}$$

2. The Taylor expansion of the power. Let C be determined so that the probability under the null hypothesis that $\sum_{i=1}^{i=k} x_i^2 \geqslant C$ is equal to the size λ_0 of the critical region. Let $P\left(\sum_{i=1}^{i=k} x_i^2 \geqslant C\right)$ be the probability under the alternative hypothesis that $\sum_{i=1}^{i=k} x_i^2 \geqslant C$. Then the power P is given by

$$P\left(\sum_{i=1}^{i=k} x_i^2 \geqslant C\right), \tag{5}$$

where

$$x_i = \frac{\alpha_i - \dfrac{N}{k}}{\sqrt{\dfrac{N}{k}}}.$$

Hence

$$\sum_{i=1}^{i=k} x_i^2 = \frac{k}{N}\left(\sum_{i=1}^{i=k} \alpha_i^2 - \frac{N^2}{k}\right),$$

and (5) can be written in the form

$$P\left(\sum_{i=1}^{i=k} \alpha_i^2 \geqslant C'\right) \tag{6}$$

where C' is a certain function of N and k. Let $\delta_i = p_i - \dfrac{1}{k}$, where p_i is the probability of the ith class interval under the alternative hypothesis.

Expanding P into a power series we obtain (in this and the following derivations, we take all partial differential quotients at the point $\delta_1 = \delta_2 = \cdots = \delta_k = 0$)

$$P = \lambda_0 + \sum_{i=1}^{i=k} \delta_i \frac{\partial P}{\partial \delta_i} + \frac{1}{2}\left\{\sum_{i=1}^{i=k} \delta_i^2 \frac{\partial^2 P}{\partial \delta_i^2} + \sum_{i \neq j} \delta_i \delta_j \frac{\partial^2 P}{\partial \delta_i \partial \delta_j}\right\} + \cdots.$$

Since P is a symmetric function of the δ_i we have for $\delta_1 = \delta_2 = \cdots = \delta_k = 0$

$$\frac{\partial^2 P}{\partial \delta_i^2} = \frac{\partial^2 P}{\partial \delta_1^2}, \qquad \frac{\partial^2 P}{\partial \delta_i \partial \delta_j} = \frac{\partial^2 P}{\partial \delta_1 \partial \delta_2} \qquad \text{for } i \neq j.$$

Furthermore $\sum_{i=1}^{i=k} \delta_i = 0$. Therefore

$$P = \lambda_0 + \frac{1}{2}\left\{\frac{\partial^2 P}{\partial \delta_1^2} \sum_{i=1}^{i=k} \delta_i^2 + \frac{\partial^2 P}{\partial \delta_1 \partial \delta_2} \sum_{i \neq j} \delta_i \delta_j\right\} + \cdots.$$

We shall first show that the terms of second order are always positive. This shows that the test is unbiased and justifies again the choice of equal class probabilities under the null hypothesis since this assures unbiasedness and mini-

mizes among all unbiased tests the g.l.b. of the distances of such alternatives whose power is equal to the size of the critical region.
The power is given by

$$P = \sum_{\alpha_1^2+\alpha_2^2+\cdots \alpha_k^2 \geq C'} \frac{N!}{\alpha_1!\alpha_2! \cdots \alpha_k!} p_1^{\alpha_1} p_2^{\alpha_2} \cdots p_k^{\alpha_k}.$$

Since $\sum_{i=1}^{i=k} \delta_i^2 = -\sum_{i \neq j} \delta_i \delta_j$ we obtain for the second order terms

$$\frac{\partial^2 P}{\partial \delta_1^2} \sum_{i=1}^{i=k} \delta_i^2 + \frac{\partial^2 P}{\partial \delta_1 \partial \delta_2} \sum_{i \neq j} \delta_i \delta_j = \left(\frac{\partial^2 P}{\partial \delta_1^2} - \frac{\partial^2 P}{\partial \delta_1 \partial \delta_2}\right) \sum_{i=1}^{i=k} \delta_i^2$$

(7)

$$= \sum_{\alpha_1^2+\alpha_2^2+\cdots+\alpha_k^2 \geq C'} (\alpha_1^2 - \alpha_1 - \alpha_1 \alpha_2) p(\alpha_1, \alpha_2 \cdots \alpha_k) \sum_{i=1}^{i=k} \delta_i^2$$

where

$$p(\alpha_1, \cdots \alpha_k) = \frac{N!}{\alpha_1!\alpha_2! \cdots \alpha_k!} \frac{1}{k^N}.$$

In the following derivation extend all sums if not otherwise stated over all terms for which $\sum_{i=1}^{i=k} \alpha_i^2 \geqslant C'$ and use the relation $\sum_{i=1}^{i=k} \alpha_i = N$. We have because of the symmetry

$$\sum \alpha_1 p(\alpha_1, \alpha_2, \cdots \alpha_k) = \frac{N}{k} \sum p(\alpha_1, \alpha_2 \cdots \alpha_k) = \frac{N}{k} \lambda_0,$$

$$\sum \alpha_1 \alpha_2 p(\alpha_1, \alpha_2, \cdots \alpha_k) = \frac{1}{k(k-1)} \sum \left(N^2 - \sum_{i=1}^{i=k} \alpha_i^2\right) p(\alpha_1, \alpha_2, \cdots \alpha_k)$$

$$= \frac{N^2 \lambda_0}{k(k-1)} - \frac{1}{k-1} \sum \alpha_1^2 p(\alpha_1, \alpha_2, \cdots \alpha_k).$$

Hence the coefficient of the second order term becomes

$$\frac{k}{k-1} \sum \alpha_1^2 p(\alpha_1, \alpha_2, \cdots \alpha_k) - \frac{N}{k} \lambda_0 - \frac{N^2}{k(k-1)} \lambda_0$$

$$= \frac{1}{k-1} \sum \sum_{i=1}^{i=k} \alpha_i^2 p(\alpha_1, \alpha_2, \cdots \alpha_k) - \frac{N}{k} \lambda_0 - \frac{N^2}{k(k-1)} \lambda_0.$$

But

$$\frac{\sum \sum_{i=1}^{i=k} \alpha_i^2 p(\alpha_1, \alpha_2, \cdots \alpha_k)}{\lambda_0} > E\left(\sum_{i=1}^{i=k} \alpha_i^2\right),$$

since the conditional mean for values of $\sum_{i=1}^{i=k} \alpha_i^2 \geqslant C'$ must be larger than the mean of all values of $\sum_{i=1}^{i=k} \alpha_i^2$. Since $E\left(\sum_{i=1}^{i=k} \alpha_i^2\right) = \frac{N^2}{k} - \frac{N}{k} + N$, we obtain

$$\frac{1}{k-1} \sum \sum_{i=1}^{i=k} \alpha_i^2 p(\alpha_1, \alpha_2 \cdots \alpha_k)$$

$$> \frac{\lambda_0}{k-1}\left(\frac{N^2}{k} + \frac{N(k-1)}{k}\right) = \lambda_0\left(\frac{N^2}{k(k-1)} + \frac{N}{k}\right)$$

and hence the coefficient of $\sum_{i=1}^{i=k} \delta_i^2$ is larger than 0.

To prove Theorem 1, we will have to determine the alternative distribution for which $\sum_{i=1}^{i=k} \delta_i^2$ becomes a minimum subject to the condition that the distance from the null hypothesis should be greater than or equal to a given Δ.

Hence we have to find a distribution function $F(x)$ such that $| F(x) - x | \geqslant \Delta$ for at least one value x and $\sum_{i=1}^{i=k} \delta_i^2 = \sum_{i=1}^{i=k} \left(p_i - \frac{1}{k}\right)^2 = \sum_{i=1}^{i=k} p_i^2 - \frac{1}{k}$ is a minimum where $p_i = F\left(\frac{i}{k}\right) - F\left(\frac{i-1}{k}\right)$. Instead of minimizing $\sum_{i=1}^{i=k} \delta_i^2$ we may minimize $\sum_{i=1}^{i=k} p_i^2$, since the two expressions differ merely by a constant. There will be two different solutions for $F(x)$ depending on whether $F(x) - x \geqslant \Delta$ or $F(x) - x \leqslant -\Delta$ for at least one value x. Because of symmetry we restrict ourselves to the case in which $F(x) - x \geqslant \Delta$ for at least one value of x.

Let a be a value for which $F(a) - a \geqslant \Delta$ and suppose that

$$\frac{l-1}{k} < a \leqslant \frac{l}{k}$$

then

$$F(a) \geqslant a + \Delta,$$

$$F\left(\frac{l}{k}\right) = \frac{l}{k} + \epsilon.$$

We prove first

$$\epsilon \geqslant \Delta - \frac{1}{k}.$$

Proof: Since $F\left(\frac{l}{k}\right) - F(a) \geqslant 0$ we have

$$F\left(\frac{l}{k}\right) = F(a) + F\left(\frac{l}{k}\right) - F(a) \geqslant a + \Delta$$

and

$$\epsilon = F\left(\frac{l}{k}\right) - \frac{l}{k} \geqslant a + \Delta - \frac{l}{k} \geqslant \frac{l-1}{k} + \Delta - \frac{l}{k} = \Delta - \frac{1}{k}.$$

If $\Delta \leqslant \frac{1}{k}$ we can always find a distribution function in $C(\Delta)$ for which $p_i = \frac{1}{k}$. Hence we consider only the case $k > \frac{1}{\Delta}$. We must minimize $\sum_{i=1}^{i=k} p_i^2$ under the condition $\sum_{i=1}^{i=l} p_i = \frac{l}{k} + \epsilon, \sum_{i=l+1}^{i=k} p_i = \frac{k-l}{k} - \epsilon.$ We therefore minimize

$$\Phi = \sum_{i=1}^{i=k} p_i^2 - 2\lambda_1 \sum_{i=1}^{i=l} p_i - 2\lambda_2 \sum_{i=l+1}^{i=k} p_i .$$

This leads to

$$p_i = \begin{cases} \frac{1}{k} + \frac{\epsilon}{l} & \text{for } i = 1, \cdots l \\ \frac{1}{k} - \frac{\epsilon}{k-l} & \text{for } i = (l+1), \cdots k. \end{cases}$$

We then have

$$\sum_{i=1}^{i=k} p_i^2 = l\left(\frac{1}{k} + \frac{\epsilon}{l}\right)^2 + (k-l)\left(\frac{1}{k} - \frac{\epsilon}{k-l}\right)^2 = \frac{1}{k} + \frac{\epsilon^2 k}{l(k-l)}.$$

This is smallest if $\epsilon = \Delta - \frac{1}{k}$ and $l = \frac{k}{2}$. The following discontinuous distribution function gives these values for ϵ, l and p_i and has the distance Δ from the rectangular distribution.

$$\begin{aligned} &F(x) = x\left[1 + 2\left(\Delta - \frac{1}{k}\right)\right] && \text{for } 0 \leqslant x \leqslant \frac{1}{2} - \frac{1}{k}, \\ &F(x) = \frac{1}{2} + \Delta - \frac{1}{k} && \text{for } \frac{1}{2} - \frac{1}{k} < x \leqslant \frac{1}{2}, \\ (8) \quad &F(x) = x\left[1 - 2\left(\Delta - \frac{1}{k}\right)\right] + 2\left(\Delta - \frac{1}{k}\right) && \text{for } \frac{1}{2} \leqslant x \leqslant 1, \\ &F(x) = 0 && \text{for } 0 \leqslant x, \\ &F(x) = 1 && \text{for } x \geqslant 1. \end{aligned}$$

3. Solution for large N**.** Denote by $F(\Delta, k)$ the distribution function (8) of $C(\Delta)$ which makes $\sum_{i=1}^{i=k} \delta_i^2$ a minimum if the test is made with k class intervals.

Assume that k is large enough that χ'^2 can be taken as normally distributed. The power of the test is then given by

$$(9)\qquad \frac{1}{\sqrt{2\pi}\,\sigma'}\int_{(k-1)+c\sqrt{2(k-1)}}^{\infty} e^{-\frac{1}{2\sigma'^2}\left(\sum_{i=1}^{i=k} x_i^2 - E\left(\sum_{i=1}^{i=k} x_i^2\right)\right)^2} d\left(\sum_{i=1}^{i=k} x_i^2\right) = \frac{1}{\sqrt{2\pi}}\int_{\frac{k-1-E\left(\sum_{i=1}^{i=k} x_i^2\right)+c\sqrt{2(k-1)}}{\sigma'}}^{\infty} e^{-\frac{1}{2}y^2}\,dy,$$

where σ' is the standard deviation of $\sum_{i=1}^{i=k} x_i^2$ and c is determined so that $\frac{1}{\sqrt{2\pi}}\int_c^{\infty} e^{-\frac{1}{2}y^2}\,dy$ is equal to the size of the critical region. Hence to maximize the power with respect to k is equivalent to maximizing

$$\psi(k) = \frac{E\left(\sum_{i=1}^{i=k} x_i^2\right) - (k-1) - c\sqrt{2(k-1)}}{\sigma'}$$

with respect to k.

Under the alternative $F(\Delta, k)$ we obtain

$$E\left(\sum_{i=1}^{i=k} x_i^2\right) - (k-1) = k(N-1)\sum_{i=1}^{i=k} p_i^2 + k - N - k + 1 = 4(N-1)\left(\Delta - \frac{1}{k}\right)^2$$

Hence

$$\psi(k) = \frac{4(N-1)\left(\Delta - \frac{1}{k}\right)^2 - c\sqrt{2(k-1)}}{\sigma'}.$$

We choose Δ so that this maximum power is exactly $\frac{1}{2}$, that is, so that $\psi(k) = 0$ for that k which maximizes $\psi(k)$. Denote this value of Δ by Δ_N and let k_N be the value of k which maximizes $\psi(k)$. The differential-quotient of the numerator of $\psi(k)$ with respect to k is then equal to 0 for $k = k_N$. Hence

$$(10)\qquad 8(N-1)\left(\Delta_N - \frac{1}{k_N}\right)\frac{1}{k_N^2} = \frac{c}{\sqrt{2(k_N-1)}}.$$

Furthermore since $\psi(k_N) = 0$ we have

$$(11)\qquad 4(N-1)\left(\Delta_N - \frac{1}{k_N}\right)^2 = c\sqrt{2(k_N-1)}.$$

Solving equations (10) and (11) we obtain

$$(12)\qquad \Delta_N = \frac{5}{k_N} - \frac{4}{k_N^2}$$

and

$$\sqrt[5]{\frac{k_N^8}{(k_N-1)^3}} = 4\sqrt[5]{\frac{2(N-1)^2}{c^2}}$$

or since $k_N > 3$,

$$k_N < 4\sqrt[5]{\frac{2(N-1)^2}{c^2}} < k_N + 1.$$

Hence

$$\text{(13)} \quad \text{either } k_N = \left[4\sqrt[5]{\frac{2(N-1)^2}{c^2}}\right] \quad \text{or} \quad k_N = \left[4\sqrt[5]{\frac{2(N-1)^2}{c^2}}\right] + 1,$$

is the value of k for which the power with respect to $F(\Delta_N, k)$ becomes a maximum. We have merely to show that $\psi''(k)$ is negative for $k = k_N$.

Using the fact that $\psi(k_N) = \psi'(k_N) = 0$ we obtain

$$\sigma'\psi''(k_N) = \frac{-16(N-1)}{k_N^3}\Delta_N + \frac{24(N-1)}{k_N^4} + \frac{c}{(\sqrt{2(k_N-1)})^3}.$$

Substituting for Δ_N the right hand side of (12) we obtain on account of (10)

$$\sigma'\psi''(k_N) = \frac{-56(N-1)}{k_N^4} + \frac{64(N-1)}{k_N^5} + \frac{8(N-1)}{2(k-1)}\left(\frac{4}{k_N^3} - \frac{4}{k_N^4}\right).$$

Using $2(k-1) > k$ we obtain

$$\psi''(k_N) < \frac{1}{k^4\sigma'}\left(-24(N-1) + \frac{32}{k}(N-1)\right)$$

which is negative. σ' can be shown to be of order $k_N^{\frac{1}{2}}$; $\psi''(k_N)$ is, therefore, of order $\frac{N}{k_N^{4+\frac{1}{2}}} = 0\left(\frac{1}{N^{\frac{4}{5}}}\right)$. The maximum is, therefore, rather flat for large values of N.

We shall now show that if k is large enough to assume χ'^2 to be normally distributed then $F(\Delta, k)$ is the alternative which gives the smallest power compared with all alternatives in the class $C(\Delta)$ provided the power for the alternative $F(\Delta, k)$ equals $\frac{1}{2}$.

We know that $E\left(\sum_{i=1}^{i=k} x_i^2\right)$ is smallest for $F(\Delta, k)$. Since the power with respect to $F(\Delta, k)$ equals $\frac{1}{2}$ we have

$$E\left(\sum_{i=1}^{i=k} x_i^2\right) - (k - 1 - c\sqrt{2(k-1)} = 0.$$

Thus the lower limit of the integral in (9) becomes negative for every other alternative and the power will be larger than $\frac{1}{2}$.

The power with respect to $F(\Delta_N, k_N)$ is equal to $\frac{1}{2}$, hence if we choose $k = k_N$ the power of the test will be $\geqslant \frac{1}{2}$ for all alternatives in the class $C(\Delta_N)$. On the other hand if we choose $k \neq k_N$ then there will be at least one alternative in

[3] Cantelli's formula and its proof are given by Fréchet in his book *Recherches Théoriques Modernes sur la Théorie de Probabilités*, Paris (1937), pp. 123–126.

$C(\Delta_N)$ for which the power is $<\frac{1}{2}$. (For instance $F(\Delta_N, k)$ is such an alternative.)

The above statements have been derived under the assumption that χ'^2 is normally distributed. Hence if the distribution of χ'^2 were exactly normal $k_N = 4\sqrt[5]{\frac{2(N-1)^2}{c^2}}$ would be a best k and for this k_N and $\Delta_N = \frac{5}{k_N} - \frac{4}{k_N^2}$ the greatest lower bound of the power in the class $C(\Delta_N)$ would be exactly $\frac{1}{2}$. Since the distribution of χ'^2 approaches the normal distribution with $k \to \infty$ the sequence $\{k_N\}$ is best in the limit and Theorem 1 stated in the introduction is proved.

For the purposes of practical applications, it is not enough to know that $\{k_N\}$ is best in the limit. We have to know for what values of N k_N can be considered practically as a best k, i.e. for what values of N the quantity $\epsilon(k_N, N)$ defined in the introduction is sufficiently small. The quantity $\epsilon(k_N, N)$ is certainly small if for the number of class intervals k_N the distribution of χ'^2 is near to normal and if the power with respect to at least one alternative of the class $C(\Delta_N)$ is smaller than $\frac{1}{2}$ also in the case when the number of class intervals is too small to assume a normal distribution for χ'^2.

We shall in the following assume that for $k > 13$ the normal distribution is a sufficiently good approximation. Actually we need not assume a normal distribution but only that the probability is close to $\frac{1}{2}$ that the statistic will exceed its mean value.

Cantelli[3] gave the following formula. Let M_r be the rth moment of a distribution about x_0. Let d be any arbitrary positive number. Let $P(|x - x_0| \leqslant d)$ be the probability that $|x - x_0| \leqslant d$ then the following inequalities hold:

$$\text{If} \quad \frac{M_r}{d^r} \leqslant \frac{M_{2r}}{d^{2r}} \quad \text{then} \quad P(|x - x_0| \leqslant d) \geqslant 1 - \frac{M_r}{d^r}.$$

$$\text{If} \quad \frac{M_r}{d^r} \geqslant \frac{M_{2r}}{d^{2r}} \quad \text{then} \quad P(|x - x_0| \leqslant d) \geqslant 1 - \frac{M_{2r} - M_r^2}{(d^r - M_r)^2 + M_{2r} - M_r^2}.$$

Since χ'^2 can only take positive values we have

$$\text{(14)} \quad \text{If} \quad \frac{E(\chi'^2)}{c_k} \leqslant \frac{\sigma_{\chi'^2}^2 + [E(\chi'^2)]^2}{c_k^2} \quad \text{then} \quad P(\chi'^2 \leqslant c_k) \geqslant 1 - \frac{E(\chi'^2)}{c_k}.$$

$$\text{(15)} \quad \text{If} \quad \frac{E(\chi'^2)}{c_k} \geqslant \frac{\sigma_{\chi'^2}^2 + [E(\chi'^2)]^2}{c_k^2}$$

$$\text{then} \quad P(\chi'^2 \leqslant c_k) \geqslant 1 - \frac{\sigma_{\chi'^2}^2}{(c_k - E(\chi'^2))^2 + \sigma_{\chi'^2}^2}.$$

Where c_k is determined so that $P(\chi^2 \geqslant c_k)$ equals the size of the critical region if the null hypothesis is true and the number of class intervals equals k. c_k can be obtained from a table of the chi-square distribution.

For $F(\Delta_N, k)$ we obtain with $\Delta_N' = \frac{5}{k_N} - \frac{4}{k_N^2} - \frac{1}{k}$ from (3) and (4)

$$E(\chi'^2) = (k - 1) + 4(N - 1)\Delta_N'^2,$$
$$\sigma^2_{\chi'^2} = 2(k - 1) + 8\Delta_N'^2(k + 2N - 4) - 32(2N - 1)\Delta_N'^4.$$

By numerically calculating $E(\chi'^2)$ and $\sigma_{\chi'^2}$ for $N = 450$ and a 5% level of significance, for $N = 300$ and a 1% level of significance, and for $k = 13, 12 \cdots \left[\frac{1}{\Delta_N}\right] + 1$ it can be shown that for these values of N and k

$$(16) \qquad \frac{E(\chi'^2)}{c_k} \geqslant \frac{\sigma^2_{\chi'^2} + [E(\chi'^2)]^2}{c_k^2}.$$

Hence we have to use (15). From (16) it follows that $c_k > E(\chi'^2)$. If $P(\chi'^2 \leqslant c_k \leqslant \frac{1}{2}$ we obtain on account of (15) and (16)

$$\frac{\sigma^2_{\chi'^2}}{(c_k - E(\chi'^2))^2 + \sigma^2_{\chi'^2}} \geqslant \frac{1}{2}, \qquad \sigma_{\chi'^2} + E(\chi'^2) \geqslant c_k.$$

Numerical calculation shows that for the values of N and k and the significance levels considered

$$(17) \qquad \sigma_{\chi'^2} + E(\chi'^2) < c_k.$$

It can then be shown that for $N \geqslant 450$ and $N \geqslant 300$ respectively $N\Delta_N'$ decreases with N. A simple argument then shows that (16) and (17) are also true for all values $N \geqslant 450$ and $N \geqslant 300$ respectively. Hence the power with respect to $F(\Delta_N, k)$ is $<\frac{1}{2}$ for these values of N. Thus we see: For $N \geqslant 450$ if the 5% level is used, and for $N \geqslant 300$ if the 1% level is used, the value $k_N = 4\sqrt[5]{\frac{2(N-1)^2}{c^2}}$ can be considered for practical purposes as a best k. The value c is determined so that $\frac{1}{\sqrt{2\pi}}\int_c^\infty e^{-\frac{1}{2}t^2}\,dt$ is equal to the size of the critical region.

Reprinted from THE ANNALS OF MATHEMATICAL STATISTICS
Vol. XIII, No. 4, December, 1942

SETTING OF TOLERANCE LIMITS WHEN THE SAMPLE IS LARGE *

BY ABRAHAM WALD

Columbia University

1. Introduction. Let $f(x_1, \cdots, x_p, \theta_1, \cdots, \theta_k)$ be the joint probability density function of the variates $x_1, \cdots, x_p$ involving k unknown parameters $\theta_1, \cdots, \theta_k$. A sample of size n is drawn from this population. Denote by $x_{i\alpha}(i = 1, \cdots, p; \alpha = 1, \cdots, n)$ the α-th observation on x_i. We will deal here with the following two problems of setting tolerance limits, which are of importance in the mass production of a product:

Problem 1. *For any two positive numbers* $\beta < 1$ *and* $\gamma < 1$ *we have to construct* p *pairs of functions of the observations* $L_i(x_{11}, \cdots, x_{pn})$ *and* $U_i(x_{11}, \cdots, x_{pn})$ $(i = 1, \cdots, p)$ *such that*

$$(1) \quad P\left\{\int_{L_p}^{U_p} \cdots \int_{L_1}^{U_1} f(x_1, \cdots, x_p, \theta_1, \cdots, \theta_k)\, dx_1 \cdots dx_p \geq \gamma \mid \theta_1, \cdots, \theta_k\right\} = \beta,$$

where for any relation R, $P(R \mid \theta_1, \cdots, \theta_k)$ *denotes the probability that* R *holds, calculated under the assumption that* $\theta_1, \cdots, \theta_k$ *are the true values of the parameters.*

Problem 2. *For any positive numbers* $\beta < 1, \lambda < 1$ *and for any positive integer* N *we have to construct* p *pairs of functions of the observations* $L_i(x_{11}, \cdots, x_{pn})$ *and* $U_i(x_{11}, \cdots, x_{pn})$ *with the following property: Let* $y_{i\alpha}(i = 1, \cdots, p; \alpha = 1, \cdots, N)$ *be the* α*-th observation on the variate* x_i *in a second sample of size* N *drawn from the same population as the first sample has been drawn. Denote by* M *the number of different values of* α *for which the* p *inequalities*

$$L_i(x_{11}, \cdots, x_{pn}) \leq y_{i\alpha} \leq U_i(x_{11}, \cdots, x_{pn}) \quad (i = 1, \cdots, p),$$

are fulfilled. Then

$$(2) \qquad P(M \geq \lambda N \mid \theta_1, \cdots, \theta_k) = \beta,$$

where $\theta_1, \cdots, \theta_k$ *denote the unknown parameter values of the population from which the observations* $x_{i\alpha}$ *and* $y_{i\alpha}$ *have been drawn.*

The functions L_i and U_i are called the tolerance limits for the variate x_i. We will say that L_i is the lower, and U_i the upper tolerance limit of x_i. In general, there exist infinitely many tolerance limits L_i and U_i which are solutions of Problem 1 or Problem 2. It is clear that the tolerance limits L_i and U_i are the more favorable the smaller the difference $U_i - L_i$. Hence if there exist several solutions for the tolerance limits L_i and U_i we should select that one for which the difference $U_i - L_i$ becomes a minimum in some sense.

S. S. Wilks[1] gave a solution of Problems 1 and 2 in the univariate case, i.e.

[1] S. S. Wilks, "Determination of sample sizes for setting tolerance limits," *Annals of Math. Stat.*, Vol. 12 (1941). See also his paper on the same subject presented at the meeting of the Institute of Mathematical Statistics in Poughkeepsie, September, 1942.

* ϵ in (5) and in the following unnumbered expression should be changed to ξ. The minus signs in (16), (21), (29), (30) and the equations for ξ and ξ^* at the end of Sec. 5 should be changed to plus signs.

if $p = 1$. It seems that Wilks' solution is the best possible one if nothing is known about the probability density function except that it is continuous. However, if it is known a priori that the unknown density function is an element of a k-parameter family of functions, it will in general be possible to derive tolerance limits which are considerably better than those proposed by Wilks.

Wilks' results can easily be extended to the multivariate case provided the variates $x_1, \cdots, x_p$ are known to be independently distributed.[2] This is a serious restriction, since in many practical cases the independence of the variates $x_1, \cdots, x_p$ cannot be assumed. The case of dependent variates has not been treated by Wilks.

In this paper we give a solution of problems 1 and 2 when the size n of the sample is large. In the next section a lemma is proved which will be used in the derivation of tolerance limits. In section 3 the univariate case is treated and in section 4 the results are extended to the multivariate case.

2. A lemma. We will prove the following

LEMMA: *Let* $\{x_{1n}\}, \cdots, \{x_{rn}\}$ $(n = 1, 2, \cdots, \text{ad inf.})$ *be r sequences of random variables and let $a_1, \cdots, a_r$ be r constants such that the joint distribution of $\sqrt{n}(x_{1n} - a_1), \cdots, \sqrt{n}(x_{rn} - a_r)$ converges with $n \to \infty$ towards the r-variate normal distribution with zero means and finite non-singular covariance matrix $\|\sigma_{ij}\|$ $(i, j = 1, \cdots, r)$. Furthermore, let $g(u_1, \cdots, u_r)$ be a function of r variables $u_1, \cdots, u_r$ which admits continuous first derivatives in the neighborhood of the point $u_1 = a_1, \cdots, u_r = a_r$. Assume that at least one of the first partial derivatives of $g(u_1, \cdots, u_r)$ is not zero at the point $u_1 = a_1, \cdots, u_r = a_r$. Then the distribution of $\sqrt{n}[g(x_{1n}, \cdots, x_{rn}) - g(a_1, \cdots, a_r)]$ converges with $n \to \infty$ towards the normal distribution with zero mean and variance $\sigma_g^2 = \sum_j \sum_i \sigma_{ij} g_i g_j$ where g_i denotes the partial derivative of $g(u_1, \cdots, u_r)$ with respect to u_i taken at $u_1 = a_1, \cdots, u_r = a_r$.*

Proof: Since the joint distribution of $\sqrt{n}(x_{1n} - a_1), \cdots, \sqrt{n}(x_{rn} - a_r)$ approaches an r-variate normal distribution with zero means and finite non-singular covariance matrix, the probability that

$$a_i - \frac{1}{\sqrt[3]{n}} \leq x_{in} \leq a_i + \frac{1}{\sqrt[3]{n}} \qquad (i = 1, \cdots, r) \tag{3}$$

holds, converges to 1 with $n \to \infty$. From (3) and the continuity of the first derivatives of $g(u_1, \cdots, u_r)$ it follows easily that for any positive ϵ the probability that

$$\begin{aligned} &\sum_{i=1}^{r} \sqrt{n}\,(x_{in} - a_i)g_i - \epsilon \\ &\leq \sqrt{n}\,[g(x_{1n}, \cdots, x_{rn}) - g(a_1, \cdots, a_r)] \leq \sum_i \sqrt{n}\,(x_{in} - a_i)g_i + \epsilon \end{aligned} \tag{4}$$

[2] This was mentioned by Wilks in his paper presented at the meeting of the Institute of Mathematical Statistics in Poughkeepsie, N. Y., September, 1942.

holds, converges to 1 with $n \to \infty$. Since the limit distribution of $\sum_i \sqrt{n}(x_{in} - a_i)g_i$ is normal with zero mean and variance equal to $\Sigma\Sigma\sigma_{ij}g_ig_j$, our Lemma follows easily from the fact that the quantity ϵ in (4) can be chosen arbitrarily small.

3. The univariate case. In this section we assume that $p = 1$. Hence the probability density function $f(x_1, \cdots, x_p, \theta_1, \cdots, \theta_k)$ is replaced by the univariate density function $f(x, \theta_1, \cdots, \theta_k)$. In order to simplify the notations, the letter θ without any subscript will be used to denote the set of parameter values $\theta_1, \cdots, \theta_k$.

For any positive $\xi < 1$ let $\varphi(\theta, \xi)$ and $\psi(\theta, \xi)$ be two functions of θ such that

$$\int_{\varphi(\theta, \xi)}^{\psi(\theta, \epsilon)} f(x, \theta)\, dx = \xi. \tag{5}$$

If $f(x, \theta)$ is a continuous function of x, functions $\varphi(\theta, \xi)$ and $\psi(\theta, \xi)$ satisfying (5) exist. It is clear that for any function $\varphi(\theta, \xi)$ subject to the condition

$$\int_{-\infty}^{\varphi(\theta, \epsilon)} f(x, \theta)\, dx < 1 - \xi$$

there exists a function $\psi(\theta, \xi)$ such that (5) holds. We will choose $\varphi(\theta, \xi)$ and $\psi(\theta, \xi)$ so that (5) is satisfied and

$$\psi(\theta, \xi) - \varphi(\theta, \xi) \leq \bar{\psi}(\theta, \xi) - \bar{\varphi}(\theta, \xi) \tag{6}$$

for any value of θ and for any functions $\bar{\varphi}(\theta, \xi)$ and $\bar{\psi}(\theta, \xi)$ which satisfy (5).

Let $\hat{\theta}_i$ $(i = 1, \cdots, k)$ be the maximum likelihood estimate of θ_i calculated from the observations $x_{11}, \cdots, x_{pn}$. We propose the use of the tolerance limits

$$L = \varphi(\hat{\theta}, \xi) \quad \text{and} \quad U = \psi(\hat{\theta}, \xi) \tag{7}$$

where the value of the constant ξ has to be properly determined. Problem 1 is solved if we can determine ξ as a function of β and γ such that

$$P\left\{\int_{\varphi(\hat{\theta}, \xi)}^{\psi(\hat{\theta}, \xi)} f(x, \theta)\, dx \geq \gamma \mid \theta\right\} = \beta. \tag{8}$$

Problem 2 is solved if we determine ξ as a function of β, λ and N such that

$$P(M \geq \lambda N \mid \theta) = \beta \tag{9}$$

where M denotes the number of observation in the second sample which lie between the tolerance limits $\varphi(\hat{\theta}, \xi)$ and $\psi(\hat{\theta}, \xi)$. The use of tolerance limits of the form (7) seems to be well justified by the fact that the functions $\varphi(\theta, \xi)$ and $\psi(\theta, \xi)$ satisfy (5) and (6) and that $\hat{\theta}_i$ is an optimum estimate of θ_i $(i = 1, \cdots, k)$.

Now we will derive the large sample distribution of

$$I(\hat{\theta}, \theta, \xi) = \int_{\varphi(\dot{\theta}, \xi)}^{\psi(\hat{\theta}, \xi)} f(x, \theta)\, dx. \tag{10}$$

We obviously have

$$I(\theta, \theta, \xi) = \xi. \tag{11}$$

We will assume that the limit joint distribution of $\sqrt{n}(\hat{\theta}_1 - \theta_1), \cdots, \sqrt{n}(\hat{\theta}_k - \theta_k)$ is normal with mean values 0 and non-singular covariance matrix $\| \sigma_{ij}(\theta) \| = \| c_{ij}(\theta) \|^{-1}$ where $c_{ij}(\theta)$ denotes the expected value of $-\dfrac{\partial^2 \log f(x, \theta)}{\partial\theta_i\, \partial\theta_j}$ $(i, j = 1, \cdots, k)$. This is known to be true if $f(x, \theta)$ satisfies some regularity conditions.[3] Furthermore we assume that $\varphi(\theta, \xi)$ and $\psi(\theta, \xi)$ admit continuous first partial derivatives with respect to $\theta_1, \cdots, \theta_k$ and that $f(x, \theta)$ is a continuous function of x in the neighborhood of $x = \varphi(\theta, \xi)$ and $x = \psi(\theta, \xi)$. We have

$$\left.\frac{\partial I(\hat{\theta}, \theta, \xi)}{\partial\hat{\theta}_i}\right|_{\hat{\theta}=\theta} = \frac{\partial\psi(\theta, \xi)}{\partial\theta_i} f[\psi(\theta, \xi), \theta] - \frac{\partial\varphi(\theta, \xi)}{\partial\theta_i} f[\varphi(\theta, \xi), \theta] \tag{12}$$

Assuming that at least one of the derivatives $\left.\dfrac{\partial I(\hat{\theta}, \theta, \xi)}{\partial\hat{\theta}_i}\right|_{\hat{\theta}=\theta}$ is not zero, it follows from our Lemma that
$\sqrt{n}[I(\hat{\theta}, \theta, \xi) - I(\theta, \theta, \xi)] = \sqrt{n}[I(\hat{\theta}, \theta, \xi) - \xi]$ is in the limit normally distributed with zero mean and variance

$$\begin{aligned}\sigma^2(\theta, \xi) = {} & \{f[\psi(\theta, \xi), \theta]\}^2 \sum_j \sum_i \frac{\partial\psi(\theta, \xi)}{\partial\theta_i} \frac{\partial\psi(\theta, \xi)}{\partial\theta_j} \sigma_{ij}(\theta) \\ & - 2f[\psi(\theta, \xi), \theta] f[\varphi(\theta, \xi), \theta] \sum_j \sum_i \frac{\partial\psi(\theta, \xi)}{\partial\theta_i} \frac{\partial\varphi(\theta, \xi)}{\partial\theta_j} \sigma_{ij}(\theta) \\ & + \{f[\varphi(\theta, \xi), \theta]\}^2 \sum_j \sum_i \frac{\partial\varphi(\theta, \xi)}{\partial\theta_i} \frac{\partial\varphi(\theta, \xi)}{\partial\theta_j} \sigma_{ij}(\theta).\end{aligned} \tag{13}$$

For any positive $\beta < 1$ denote by λ_β the value for which

$$\frac{1}{\sqrt{2\pi}} \int_{\lambda_\beta}^{\infty} \theta^{-\frac{1}{2}t^2}\, dt = \beta. \tag{14}$$

Then the probability that

$$I(\hat{\theta}, \theta, \xi) \geq \xi + \lambda_\beta \frac{\sigma(\theta, \xi)}{\sqrt{n}}, \tag{15}$$

converges with $n \to \infty$ towards β.

Let

$$\bar{\xi}(\beta, \gamma, \hat{\theta}) = \gamma - \lambda_\beta \frac{\sigma(\hat{\theta}, \gamma)}{\sqrt{n}} \tag{16}$$

---

[3] See for instance J. L. Doob, "Probability and statistics," *Trans. Amer. Math. Soc.*, October, 1934.

If $\sigma(\theta, \xi)$ is continuous in θ and ξ, it follows easily from (15) that the probability that

$$I[\hat{\theta},\ \theta,\ \bar{\xi}(\beta,\ \gamma,\ \hat{\theta})]\ \geq\ \gamma \tag{17}$$

holds, converges to β with $n \to \infty$. Hence we can summarize our results in the following

THEOREM 1: *Let $\varphi(\theta, \xi)$ and $\psi(\theta, \xi)$ be two functions satisfying* (5) *and* (6). *Furthermore, let the functions $I(\hat{\theta}, \theta, \xi)$, $\sigma^2(\theta, \xi)$ and $\bar{\xi}(\beta, \gamma, \hat{\theta})$ be defined by* (10), (13) *and* (16) *respectively. Denote by $\theta_1^0, \cdots, \theta_k^0$ the true values of the parameters. It is assumed that there exist two positive numbers ϵ and δ such that the following three conditions are fulfilled:*

(a) *For any point θ for which $\sum_{i=1}^{k} (\theta_i - \theta_i^0)^2 \leq \epsilon$ the limit joint distribution of $\sqrt{n}(\hat{\theta}_1 - \theta_1), \cdots, \sqrt{n}(\hat{\theta}_k - \theta_k)$, calculated under the assumption that θ is the true parameter point, is normal with zero means and a finite non-singular covariance matrix $\| \sigma_{ij}(\theta) \|$ where $\sigma_{ij}(\theta)$ is a continuous function of θ in the domain $\sum_i (\theta_i - \theta_i^0)^2 \leq \epsilon$.*

(b) *The partial derivatives $\left.\dfrac{\partial I(\hat{\theta}, \theta, \xi)}{\partial \hat{\theta}_i}\right|_{\hat{\theta}=\theta}$ $(i = 1, \cdots, k)$ are continuous functions of θ and ξ in the domain*

$$\sum_{i=1}^{k} (\theta_i - \theta_i^0)^2 \leq \epsilon \quad \textit{and} \quad | \xi - \gamma | \leq \delta.$$

(c) *At least one of the partial derivatives $\left.\dfrac{\partial I(\hat{\theta}, \theta^0, \gamma)}{\partial \hat{\theta}_i}\right|_{\hat{\theta}=\theta^0}$ $(i = 1, \cdots, k)$ is not equal to zero.*

Then the probability that

$$I[\hat{\theta},\ \theta^0,\ \bar{\xi}(\beta,\ \gamma,\ \hat{\theta})]\ \geq\ \gamma,$$

holds, converges to β with $n \to \infty$.

From Theorem 1 we obtain the following

LARGE SAMPLE SOLUTION OF PROBLEM 1. *For large n we can approximate the lower and upper tolerance limits by $\varphi[\hat{\theta}, \bar{\xi}(\beta, \gamma, \hat{\theta})]$ and $\psi[\hat{\theta}, \bar{\xi}(\beta, \gamma, \hat{\theta})]$ respectively, where $\bar{\xi}(\beta, \gamma, \hat{\theta})$ is given by* (16).

Now we will deal with Problem 2. We distinguish two cases

$$\lim_{n \to \infty} \frac{N}{n} = \infty. \tag{a}$$

It is easy to see that in this case the solution of Problem 2 is obtained from that of Problem 1 by substituting λ for γ. Hence for large n the tolerance limits can be approximated by $\varphi[\hat{\theta},\ \bar{\xi}(\beta,\ \lambda,\ \hat{\theta})]$ and $\psi[\hat{\theta},\ \bar{\xi}(\beta,\ \lambda,\ \hat{\theta})]$ respectively.

For these tolerance limits condition 2 is fulfilled in the limit, i.e. $\lim_{n=\infty} P(M \geq \lambda N \mid \theta_1, \cdots, \theta_k) = \beta$

(b) The integers n and N approach infinity while $\frac{N}{n}$ remains bounded.

Denote $\sqrt{n}\,[I(\hat{\theta}, \theta, \xi) - \xi]$ by u and $\sqrt{N}\left(\frac{M(\xi)}{N} - \xi\right)$ by v, where $M(\xi)$ denotes the number of observations in the second sample which fall between the limits $\varphi(\hat{\theta}, \xi)$ and $\psi(\hat{\theta}, \xi)$. For any fixed value of u the conditional expected value of $\frac{M(\xi)}{N}$ is given by $\xi + \frac{u}{\sqrt{n}}$ and the conditional variance of $\frac{M(\xi)}{N}$ is given by $\frac{1}{N}\left(\xi + \frac{u}{\sqrt{n}}\right)\left(1 - \xi - \frac{u}{\sqrt{n}}\right)$. Hence the conditional expected value of v is equal to $u\sqrt{\frac{N}{n}}$ and the conditional variance of v is equal to $\left(\xi + \frac{u}{\sqrt{n}}\right)\left(1 - \xi - \frac{u}{\sqrt{n}}\right)$. Since the limit distribution of u is normal with zero mean and standard deviation $\sigma(\theta, \xi)$ given in (13), we find that the limit bivariate distribution of u and v is given by

$$(18) \qquad \frac{1}{2\pi\sigma(\theta, \xi)\sqrt{\xi(1 - \xi)}} \exp\left[-\frac{u^2}{2\sigma^2(\theta, \xi)} - \frac{\left(v - \sqrt{\frac{N}{n}}\,u\right)^2}{2\xi(1 - \xi)}\right] du\, dv.$$

From (18) it follows that the limit distribution of v is normal with zero mean and variance

$$(19) \qquad \begin{aligned} \sigma_v^2 &= \sigma^2(\theta, \xi)\left(\frac{1}{\sigma^2(\theta, \xi)} + \frac{N}{n\xi(1 - \xi)}\right)\xi(1 - \xi) \\ &= \frac{n\xi(1 - \xi) + N\sigma^2(\theta, \xi)}{n}. \end{aligned}$$

From (19) it follows easily that the probability that

$$(20) \qquad \frac{M(\xi)}{N} \geq \xi + \frac{\lambda_\beta \sigma_v}{\sqrt{N}}$$

converges to β with $n \to \infty$. Let

$$(21) \qquad \xi^*(\beta, \lambda, \hat{\theta}) = \lambda - \frac{\lambda_\beta}{\sqrt{N}}\sqrt{\frac{n\lambda(1 - \lambda) + N\sigma^2(\hat{\theta}, \lambda)}{n}}.$$

From (20) it follows that the probability that

$$\frac{M}{N} \geq \lambda,$$

converges to β with $n \to \infty$. The letter M denotes the number of observations in the second sample which lie between the limits $\varphi[\theta, \xi^*(\beta, \lambda, \hat{\theta})]$ and $\psi[\hat{\theta}, \xi^*(\beta, \lambda, \hat{\theta})]$.

We can summarize our results in the following

THEOREM 2. *Let $\varphi(\theta, \xi)$ and $\psi(\theta, \xi)$ be two functions satisfying* (5) *and* (6). *Two samples of size n and N respectively are drawn and the maximum likelihood estimate $\hat{\theta}$ is calculated from the first sample only. Assume that conditions* (a), (b) *and* (c) *of Theorem* 1 *are satisfied. Let $\bar{\xi}(\beta, \gamma, \hat{\theta})$ and $\xi^*(\beta, \lambda, \hat{\theta})$ be defined by* (16) *and* (21) *respectively.*

If n and $\frac{N}{n}$ both approach infinity, the probability that $\frac{M}{N} \geq \lambda$ holds, converges to β, where M denotes the number of observations in the second sample which lie between the limits $\varphi[\hat{\theta}, \bar{\xi}(\beta, \lambda, \hat{\theta})]$ and $\psi[\hat{\theta}, \bar{\xi}(\beta, \lambda, \hat{\theta})]$.

If n and N approach infinity while $\frac{N}{n}$ remains bounded, the probability that $\frac{M}{N} \geq \lambda$ holds, converges to β, where M denotes the number of observations in the second sample which lie between the limits $\varphi[\hat{\theta}, \xi^*(\beta, \lambda, \hat{\theta})]$ and $\psi[\hat{\theta}, \xi^*(\beta, \lambda, \hat{\theta})]$.

From Theorem 2 we obtain the following

LARGE SAMPLE SOLUTION OF PROBLEM 2. *If n and $\frac{N}{n}$ both approach infinity the lower and upper tolerance limits can be approximated by $\varphi[\hat{\theta}, \bar{\xi}(\beta, \lambda, \hat{\theta})]$ and $\psi[\hat{\theta}, \bar{\xi}(\beta, \lambda, \hat{\theta})]$ respectively. If n and N both approach infinity while $\frac{N}{n}$ remains bounded, the tolerance limits can be approximated by $\varphi[\hat{\theta}, \xi^*(\beta, \lambda, \hat{\theta})]$ and $\psi[\hat{\theta}, \xi^*(\beta, \lambda, \hat{\theta})]$ respectively. The expressions $\bar{\xi}(\beta, \lambda, \hat{\theta})$ and $\xi^*(\beta, \lambda, \hat{\theta})$ are given by* (16) *and* (21) *respectively.*

4. The multivariate case. For any positive $\xi < 1$ let $\varphi_i(\theta, \xi)$ and $\psi_i(\theta, \xi)$ $(i = 1, \cdots, p)$ be p pairs of functions of θ such that

$$\int_{\varphi_p(\theta,\xi)}^{\psi_p(\theta,\xi)} \cdots \int_{\varphi_1(\theta,\xi)}^{\psi_1(\theta,\xi)} f(x_1, \cdots, x_p, \theta)\, dx_1 \cdots dx_p = \xi. \tag{22}$$

If $f(x_1, \cdots, x_p, \theta)$ is a continuous function of $x_1, \cdots, x_p$, functions $\varphi_i(\theta, \xi)$ and $\psi_i(\theta, \xi)$ $(i = 1, \cdots, p)$ satisfying (22) certainly exist. As in the univariate case, there will be infinitely many sets of p pairs of functions $\varphi_i(\theta, \xi)$ and $\psi_i(\theta, \xi)$ which satisfy (22). Since we wish to have tolerance limits as narrow as possible, we will try to choose the functions $\varphi_i(\theta, \xi)$ and $\psi_i(\theta, \xi)$ so that $\psi_i(\theta, \xi) - \varphi_i(\theta, \xi)$ should be as small as possible. Since it is impossible to minimize all p differences $\psi_1(\theta, \xi) - \varphi_1(\theta, \xi), \cdots, \psi_p(\theta, \xi) - \varphi_p(\theta, \xi)$ simultaneously, we will have to be satisfied with some compromise solution. For example, we could minimize the product $\prod [\psi_i(\theta, \xi) - \varphi_i(\theta, \xi)]$ or some other function of the p differences $\psi_i(\theta, \xi) - \varphi_i(\theta, \xi)$. Another reasonable procedure would be to minimize

$\prod_i [\psi_i(\theta, \xi) - \varphi_i(\theta, \xi)]$ subject to (22) and the condition that for any i and j, $\dfrac{\psi_i(\theta, \xi) - \varphi_i(\theta, \xi)}{\psi_j(\theta, \xi) - \varphi_j(\theta, \xi)}$ is equal to the ratio of the standard deviation of x_i to that of x_j.

Here we will deal with the problem of deriving tolerance limits for the variates $x_1, \cdots, x_p$ after the functions $\varphi_i(\theta, \xi)$ and $\psi_i(\theta, \xi)$ have been chosen. Since the theory of the multivariate case is very similar to that of the univariate case, we will merely outline it briefly.

As tolerance limits for x_i we will use the functions $\varphi_i(\hat\theta, \xi)$ and $\psi_i(\hat\theta, \xi)$ where the value of ξ has to be properly determined. Problem 1 is solved if we can determine ξ as a function of β and γ so that

$$P\left\{\int_{\varphi_p(\hat\theta,\xi)}^{\psi_p(\hat\theta,\xi)} \cdots \int_{\varphi_1(\hat\theta,\xi)}^{\psi_1(\hat\theta,\xi)} f(x_1, \cdots, x_p, \theta)\, dx_1 \cdots dx_p \geq \gamma \mid \theta\right\} = \beta. \tag{23}$$

Problem 2 is solved if we determine ξ as a function of β, λ and N such that condition 2 is fulfilled. Let

$$I(\hat\theta, \theta, \xi) = \int_{\varphi_p(\hat\theta,\xi)}^{\psi_p(\hat\theta,\xi)} \cdots \int_{\varphi_1(\hat\theta,\xi)}^{\psi_1(\hat\theta,\xi)} f(x_1, \cdots, x_p, \theta)\, dx_1 \cdots dx_p \tag{24}$$

and let

$$\begin{aligned} I_i(\hat\theta, \theta, \xi, x_i) = \int_{\varphi_p(\hat\theta,\xi)}^{\psi_p(\hat\theta,\xi)} \cdots \int_{\varphi_{i+1}(\hat\theta,\xi)}^{\psi_{i+1}(\hat\theta,\xi)} \int_{\varphi_{i-1}(\hat\theta,\xi)}^{\psi_{i-1}(\hat\theta,\xi)} \\ \cdots \int_{\varphi_1(\hat\theta,\xi)}^{\psi_1(\hat\theta,\xi)} f(x_1, \cdots, x_p, \theta)\, dx_1 \cdots dx_{i-1}\, dx_{i+1} \cdots dx_p. \end{aligned} \tag{25}$$

We have

$$\begin{aligned} \left.\frac{\partial I(\hat\theta, \theta, \xi)}{\partial \hat\theta_i}\right|_{\hat\theta=\theta} &= \sum_{s=1}^{p} \frac{\partial \psi_s(\theta, \xi)}{\partial \theta_i} I_s[\theta, \theta, \xi, \psi_s(\theta, \xi)] \\ &\quad - \sum_{s=1}^{p} \frac{\partial \varphi_s(\theta, \xi)}{\partial \theta_i} I_s[\theta, \theta, \xi, \varphi_s(\theta, \xi)]. \end{aligned} \tag{26}$$

Assuming that the partial derivatives $\left.\dfrac{\partial I(\hat\theta, \theta, \xi)}{\partial \hat\theta_i}\right|_{\hat\theta=\theta}$ $(i = 1, \cdots, k)$ are continuous functions and that $\left.\dfrac{\partial I(\hat\theta, \theta, \xi)}{\partial \hat\theta_i}\right|_{\hat\theta=\theta}$ is not zero for at least one value of i, it follows from our Lemma that $\sqrt{n}\,[I(\hat\theta, \theta, \xi) - I(\theta, \theta, \xi)] = \sqrt{n}\,[I(\hat\theta, \theta, \xi) - \xi]$ is in the limit normally distributed with mean value zero and variance

$$\begin{aligned} \bar\sigma^2(\theta, \xi) = \sum_{q=1}^{p}\sum_{s=1}^{p}\sum_{j=1}^{k}\sum_{i=1}^{k} & \frac{\partial \psi_s(\theta, \xi)}{\partial \theta_i}\frac{\partial \psi_q(\theta, \xi)}{\partial \theta_j} I_s[\theta, \theta, \xi, \psi_s(\theta, \xi)] I_q[\theta, \theta, \xi, \psi_s(\theta, \xi)]\sigma_{ij}(\theta) \\ & - 2\sum_s\sum_q\sum_i\sum_j \frac{\partial \psi_s(\theta, \xi)}{\partial \theta_i}\frac{\partial \varphi_q(\theta, \xi)}{\partial \theta_j} \\ & \cdot I_s[\theta, \theta, \xi, \psi_s(\theta, \xi)] I_q[\theta, \theta, \xi, \varphi_q(\theta, \xi)]\sigma_{ij}(\theta) \\ & + \sum_s\sum_q\sum_j\sum_i \frac{\partial \varphi_s(\theta, \xi)}{\partial \theta_i}\frac{\partial \varphi_q(\theta, \xi)}{\partial \theta_j} \\ & \cdot I_s[\theta, \theta, \xi, \varphi_s(\theta, \xi)] I_q[\theta, \theta, \xi, \varphi_q(\theta, \xi)]\sigma_{ij}(\theta) \end{aligned} \tag{27}$$

where $\| \sigma_{ij}(\theta) \|$ is the limit covariance matrix of $\sqrt{n}(\hat{\theta}_1 - \theta_1), \cdots, \sqrt{n}(\hat{\theta}_k - \theta_k)$.

For any positive $\beta > 1$, let λ_β be the real value defined by the equation

$$\frac{1}{\sqrt{2\pi}} \int_{\lambda_\beta}^{\infty} e^{-\frac{1}{2}t^2} \, dt = \beta. \tag{28}$$

Let

$$\bar{\xi}(\beta, \gamma, \hat{\theta}) = \gamma - \lambda_\beta \frac{\bar{\sigma}(\hat{\theta}, \gamma)}{\sqrt{n}} \tag{29}$$

and

$$\zeta^*(\beta, \lambda, \hat{\theta}) = \lambda - \frac{\lambda_\beta}{\sqrt{N}} \sqrt{\frac{n\lambda(1 - \lambda) + N\bar{\sigma}^2(\hat{\theta}, \lambda)}{n}}. \tag{30}$$

We can easily prove the following two theorems:

THEOREM 3. *Let $\varphi_i(\theta, \xi)$ and $\psi_i(\theta, \xi)$ $(i = 1, \cdots, p)$ be p pairs of functions which satisfy* (22). *Let the functions $I(\hat{\theta}, \theta, \xi)$, $\bar{\sigma}^2(\theta, \xi)$ and $\bar{\xi}(\beta, \gamma, \hat{\theta})$ be defined by* (24), (27) *and* (29) *respectively. Denote by $\theta_1^0, \cdots, \theta_k^0$ the true values of the parameters $\theta_1, \cdots, \theta_k$. It is assumed that there exist two positive numbers ϵ and δ such that the following three conditions are fulfilled:*

(a) *For any point θ for which $\sum_{i=1}^{k} (\theta_i - \theta_i^0)^2 \leq \epsilon$ the limit joint distribution of $\sqrt{n}(\hat{\theta}_1 - \theta_1), \cdots, \sqrt{n}(\hat{\theta}_k - \theta_k)$, calculated under the assumption that θ is the true parameter point, is normal with zero means and a finite non-singular covariance matrix $\| \sigma_{ij}(\theta) \|$ where $\sigma_{ij}(\theta)$ is a continuous function of θ in the domain $\sum_{i} (\theta_i - \theta_i^0)^2 \leq \epsilon$.*

(b) *The partial derivatives $\left. \frac{\partial I(\hat{\theta}, \theta, \xi)}{\partial \hat{\theta}_i} \right|_{\hat{\theta}=\theta}$ $(i = 1, \cdots, k)$ are continuous functions of θ and ξ in the domain $\sum_{i=1}^{k} (\theta_i - \theta_i^0)^2 \leq \epsilon$ and $| \xi - \gamma | \leq \delta$.*

(c) *At least one of the partial derivatives $\left. \frac{\partial I(\hat{\theta}, \theta^0, \gamma)}{\partial \hat{\theta}_i} \right|_{\hat{\theta}=\theta^0}$ $(i = 1, \cdots, k)$ is not equal to zero.*

Then the probability that

$$I[\hat{\theta}, \theta^0, \bar{\xi}(\beta, \gamma, \hat{\theta})] \geq \gamma$$

holds, converges to β with $n \to \infty$.

THEOREM 4. *Let $\varphi_i(\theta, \xi)$ and $\psi_i(\theta, \xi)$ $(i = 1, \cdots, p)$ be p pairs of functions which satisfy* (22). *Two samples of size n and N respectively are drawn and the maximum likelihood estimate $\hat{\theta}$ is calculated from the first sample only. Assume that conditions* (a), (b) *and* (c) *of Theorem* 3 *are fulfilled and let $\bar{\xi}(\beta, \gamma, \hat{\theta})$ and $\zeta^*(\beta, \lambda, \hat{\theta})$ be defined by* (29) *and* (30) *respectively. Denote by $y_{i\alpha}$ the outcome of the α-th observation on the i-th variate in the second sample.*

If n and $\frac{N}{n}$ both approach infinity, the probability that $M \geq \lambda N$ holds converges to β, where M denotes the number of different values of α for which

$$\varphi_i[\hat{\theta}, \bar{\zeta}(\beta, \lambda, \hat{\theta})] \leq y_{i\alpha} \leq \psi_i[\hat{\theta}, \bar{\zeta}(\beta, \lambda, \hat{\theta})] \qquad (i = 1, \cdots, p).$$

If n and N approach infinity while $\frac{N}{n}$ remains bounded, the probability that $M \geq \lambda N$ holds converges to β where M denotes the number of different values of α for which

$$\varphi_i[\hat{\theta}, \zeta^*(\beta, \lambda, \hat{\theta})] \leq y_{i\alpha} \leq \psi_i[\hat{\theta}, \zeta^*(\beta, \lambda, \hat{\theta})] \qquad (i = 1, \cdots, p).$$

The proofs of Theorems 3 and 4 are omitted since they are similar to the proofs of Theorems 1 and 2.

From Theorem 3 we obtain the following

LARGE SAMPLE SOLUTION OF PROBLEM 1. *For large n we can approximate the lower and upper tolerance limits for x_i by $\varphi_i[\hat{\theta}, \bar{\zeta}(\beta, \gamma, \hat{\theta})]$ and $\psi_i[\hat{\theta}, \bar{\zeta}(\beta, \gamma, \hat{\theta})]$ respectively where $\bar{\zeta}(\beta, \gamma, \theta)$ is given by* (29).

From Theorem 4 we obtain the following

LARGE SAMPLE SOLUTION OF PROBLEM 2. *If n and $\frac{N}{n}$ approach infinity, the lower and upper tolerance limits for x_i can be approximated by $\varphi_i[\hat{\theta}, \bar{\zeta}(\beta, \lambda, \hat{\theta})]$ and $\psi_i[\hat{\theta}, \bar{\zeta}(\beta, \lambda, \hat{\theta})]$ respectively. If n and N both approach infinity while $\frac{N}{n}$ remains bounded, the tolerance limits for x_i can be approximated by $\varphi_i[\hat{\theta}, \zeta^*(\beta, \lambda, \hat{\theta})]$ and $\psi_i[\hat{\theta}, \zeta^*(\beta, \lambda, \hat{\theta})]$ respectively. The expressions $\zeta(\beta, \lambda, \hat{\theta})$ and $\zeta^*(\beta, \lambda, \hat{\theta})$ are defined in* (29) *and* (30) *respectively.*

5. An example. Let x be a normally distributed variate with mean value θ_1 and standard deviation θ_2, i.e. the probability density function of x is given by

$$f(x, \theta_1, \theta_2) = \frac{1}{\sqrt{2\pi}\,\theta_2} e^{-\frac{1}{2}(x-\theta_1)^2/\theta_2^2}.$$

For any positive $\xi < 1$ let $\rho(\xi)$ be the value for which

$$\frac{1}{\sqrt{2\pi}} \int_{-\rho(\xi)}^{\rho(\xi)} e^{-\frac{1}{2}t^2}\, dt = \xi.$$

Then the functions

$$\varphi(\theta, \xi) = \theta_1 - \rho(\xi)\theta_2$$

and

$$\psi(\theta, \xi) = \theta_1 + \rho(\xi)\theta_2$$

satisfy conditions (5) and (6).

We have

$$\hat{\theta}_1 = \frac{x_1 + \cdots + x_n}{n} = \bar{x} \quad \text{and} \quad \hat{\theta}_2 = \sqrt{\frac{\sum_{\alpha=1}^{n} (x_\alpha - \bar{x})^2}{n}}.$$

The variance of $\sqrt{n}(\hat{\theta}_1 - \theta_1)$ is equal to θ_2^2 and the limit variance of $\sqrt{n}(\hat{\theta}_2 - \theta_2)$ is equal to $\frac{1}{2}\theta_2^2$. Since the covariance of $\hat{\theta}_1$ and $\hat{\theta}_2$ is equal to zero, we obtain from (13)

$$\sigma^2(\theta, \xi) = 2\left\{\frac{1}{\sqrt{2\pi}\,\theta_2} e^{-\frac{1}{2}[\rho(\xi)]^2}\right\}\{\theta_2^2 + \tfrac{1}{2}\theta_2^2[\rho(\xi)]^2\}$$

$$- 2\left\{\frac{1}{\sqrt{2\pi}\,\theta_2} e^{-\frac{1}{2}[\rho(\xi)]^2}\right\}\{\theta_2^2 - \tfrac{1}{2}\theta_2^2[\rho(\xi)]^2\}$$

$$= \frac{1}{\pi}[\rho(\xi)]^2 e^{-[\rho(\xi)]^2}.$$

Hence for large n the tolerance limits satisfying (1) can be approximated by $\hat{\theta}_1 - \rho(\bar{\xi})\hat{\theta}_2$ and $\hat{\theta}_1 + \rho(\bar{\xi})\hat{\theta}_2$ respectively where

$$\bar{\xi} = \gamma - \lambda_\beta \frac{\rho(\gamma)}{\sqrt{n\pi}} e^{-\frac{1}{2}\rho[(\gamma)]^2}$$

and λ_β is the value determined by the equation

$$\frac{1}{\sqrt{2\pi}} \int_{\lambda_\beta}^{\infty} e^{-\frac{1}{2}t^2}\, dt = \beta.$$

If n and N are large, the tolerance limits satisfying (2) can be approximated by $\hat{\theta}_1 - \rho(\xi^*)\hat{\theta}_2$ and $\hat{\theta}_1 + \rho(\xi^*)\hat{\theta}_2$ respectively where

$$\xi^* = \lambda - \lambda_\beta \sqrt{\frac{\lambda(1-\lambda)}{N} + \frac{[\rho(\lambda)]^2}{n\pi} e^{-[\rho(\lambda)]^2}}.$$

Made in United States of America

Reprinted from THE ANNALS OF MATHEMATICAL STATISTICS
Vol. XIII, No. 4, December, 1942

ON THE POWER FUNCTION OF THE ANALYSIS OF VARIANCE TEST[1] *

BY ABRAHAM WALD

Columbia University

It is known[2] that the general problem of the analysis of variance can be reduced by an orthogonal transformation to the following canonical form: Let the variates $y_1, \cdots, y_p, z_1, \cdots, z_n$ be independently and normally distributed with a common unknown variance σ^2. The mean values of $z_1, \cdots, z_n$ are known to be zero, and the mean values $\eta_1, \cdots, \eta_p$ of the variates $y_1, \cdots, y_p$ are unknown. The canonical form of the analysis of variance test is the test of the hypothesis that

$$\eta_1 = \eta_2 = \cdots = \eta_r = 0 \qquad (r \leq p) \tag{1}$$

where a single observation is made on each of the variates $y_1, \cdots, y_p, z_1, \cdots, z_n$.

In the theory of the analysis of variance the test of the hypothesis (1) is based on the critical region

$$\frac{y_1^2 + \cdots + y_r^2}{z_1^2 + \cdots + z_n^2} \geq c \tag{2}$$

where the constant c is chosen so that the size of the critical region is equal to the level of significance α we wish to have. The critical region (2) is identical with the critical region

$$\frac{y_1^2 + \cdots + y_r^2}{y_1^2 + \cdots + y_r^2 + z_1^2 + \cdots + z_n^2} \geq c' = \frac{c}{c+1}. \tag{3}$$

It is known that the power function of the critical region (3) depends only on the single parameter

$$\lambda = \frac{1}{\sigma^2} \sum_{i=1}^{r} \eta_i^2. \tag{4}$$

Denote the power function of the critical region (3) by $\beta_0(\lambda)$. P. L. Hsu has proved[3] the following optimum property of the region (3): *Let W be a critical region which satisfies the following two conditions*:

(a) *The size of W is equal to the size of the region* (3).

---

[1] Presented at a joint meeting of the Institute of Mathematical Statistics and the American Mathematical Society in New York, December, 1941.

[2] See for instance P. C. TANG, "The power function of the analysis of variance tests," *Stat. Res. Mem.*, Vol. 2, 1938.

[3] P. L. HSU, "Analysis of variance from the power function standpoint," *Biometrika*, January, 1941.

* The proof of Lemma 2 is incorrect; a correct proof is given in "Note on a lemma" {61}.

(b) *The power function of W depends on the single parameter* λ.
Then $\beta(\lambda) \leq \beta_0(\lambda)$ *where* $\beta(\lambda)$ *denotes the power function of* W.

Condition (b) is a serious restriction in Hsu's result. In this paper we shall prove an optimum property of $\beta_0(\lambda)$ where $\beta_0(\lambda)$ is compared with the power function of any other critical region of size equal to that of (3).

For any given values $\eta'_{r+1}, \cdots, \eta'_p, \sigma'$ and λ denote by $S(\eta'_{r+1}, \cdots, \eta'_p, \sigma', \lambda)$ the sphere defined by the equations

$$(5)\qquad \eta_1^2 + \cdots + \eta_r^2 = \lambda\sigma'^2; \qquad \eta_i = \eta'_i (i = r+1, \cdots, p); \qquad \sigma = \sigma'.$$

For any region W denote by $\beta_W(\eta_1, \cdots, \eta_p, \sigma)$ the power function of W, i.e. $\beta_W(\eta_1, \cdots, \eta_p, \sigma)$ denotes the probability that the sample point will fall within W calculated under the assumption that $\eta_1, \cdots, \eta_p$ and σ are the true values of the parameters. We will denote by $\gamma_W(\eta'_{r+1}, \cdots, \eta'_p, \sigma', \lambda)$ the integral of the power function $\beta_W(\eta'_1, \cdots, \eta'_p, \sigma')$ over the surface $S(\eta'_{r+1}, \cdots, \eta'_p, \sigma', \lambda)$ divided by the area of $S(\eta'_{r+1}, \cdots, \eta'_p, \sigma', \lambda)$, i.e.

$$(6)\qquad \gamma_W(\eta'_{r+1}, \cdots, \eta'_p, \sigma', \lambda) = \left[\int_{S(\eta'_{r+1},\cdots,\eta'_p,\sigma',\lambda)} dA\right]^{-1} \int_{S(\eta'_{r+1},\cdots,\eta'_p,\sigma',\lambda)} \beta_W(\eta'_1, \cdots, \eta'_p, \sigma')\, dA.$$

We will prove the following

THEOREM: *If* W *is a critical region of size equal to that of* (3), *i.e.* $\beta_W(0, \cdots, 0, \eta_{r+1}, \cdots, \eta_p, \sigma) = \beta_0(0)$, *then*

$$(7)\qquad \gamma_W(\eta'_{r+1}, \cdots, \eta'_p, \sigma', \lambda) \leq \beta_0(\lambda)$$

for arbitrary values $\eta'_{r+1}, \cdots, \eta'_p, \sigma'$ *and* λ.

If W satisfies Hsu's condition (*b*) then the power function $\beta_W(\eta_1, \cdots, \eta_p, \sigma)$ is constant on the surface $S(\eta_{r+1}, \cdots, \eta_p, \sigma, \lambda)$ and therefore $\gamma_W(\eta_{r+1}, \cdots, \eta_p, \sigma, \lambda) = \beta_W(\eta_1, \cdots, \eta_p, \sigma)$. Hence Hsu's result is an immediate consequence of our Theorem.

Denote $|\sqrt{y_1^2 + \cdots + y_r^2 + z_1^2 + \cdots + z_n^2}|$ by t and for any values $a_{r+1}, \cdots, a_p, b$ let $R(a_{r+1}, \cdots, a_p, b)$ be the set of all sample points for which

$$y_i = a_i (i = r+1, \cdots, p) \quad \text{and} \quad t = b.$$

For any region W of the sample space we denote by $W(y_{r+1}, \cdots, y_p, t)$ the common part of W and $R(y_{r+1}, \cdots, y_p, t)$.

In order to prove our Theorem we first show the validity of the following

LEMMA 1: *For any critical region* Z *there exists a function* $\varphi_Z(y_{r+1}, \cdots, y_p, t)$ *of the variables* $y_{r+1}, \cdots, y_p, t$ *such that the critical region* Z^* *defined by the inequality*

$$y_1^2 + \cdots + y_r^2 \geq \varphi_Z(y_{r+1}, \cdots, y_p, t)$$

satisfies the following two conditions:

$$(a)\qquad \beta_Z(0, \cdots, 0, \eta_{r+1}, \cdots, \eta_p, \sigma) = \beta_{Z^*}(0, \cdots, 0, \eta_{r+1}, \cdots, \eta_p, \sigma);$$

(b) $$\gamma_Z(\eta_{r+1}, \cdots, \eta_p, \sigma, \lambda) \leq \gamma_{Z^*}(\eta_{r+1}, \cdots, \eta_p, \sigma, \lambda).$$

PROOF: Denote by $P_Z(y_{r+1}, \cdots, y_p, t)$ the conditional probability of $Z(y_{r+1}, \cdots, y_p, t)$ calculated under the condition that the sample point lies in $R(y_{r+1}, \cdots, y_p, t)$ and under the assumption that $\eta_1 = \cdots = \eta_r = 0$. Denote by $F(d, t)$ the conditional probability that

$$y_1^2 + \cdots + y_r^2 \geq d$$

calculated under the condition that the sample point lies in $R(y_{r+1}, \cdots, y_p, t)$ and under the assumption that $\eta_1 = \cdots = \eta_r = 0$. It is easy to verify that the values of $F(d, t)$ and $P_Z(y_{r+1}, \cdots, y_p, t)$ do not depend on the unknown parameters $\eta_{r+1}, \cdots, \eta_p, \sigma$. Since $F(d, t)$ is a continuous function of d and since $F(t^2, t) = 0$, there exists a function $\varphi_Z(y_{r+1}, \cdots, y_p, t)$ such that

$$F[\varphi_Z(y_{r+1}, \cdots, y_p, t), t] = P_Z(y_{r+1}, \cdots, y_p, t).$$

For this function $\varphi_Z(y_{r+1}, \cdots, y_p, t)$ the region Z^* certainly satisfies condition (a) of Lemma 1. We will show that condition (b) is also satisfied. Consider the ratio

$$\frac{\int_{S(\eta_{r+1},\cdots,\eta_p,\sigma,\lambda)} \exp\left[-\frac{1}{2\sigma^2}\sum_{i=1}^{p}(y_i-\eta_i)^2 - \frac{1}{2\sigma^2}\sum_{\alpha=1}^{n} z_\alpha^2\right] dA}{\exp\left[-\frac{1}{2\sigma^2}\left(\sum_{i=1}^{r} y_i^2 + \sum_{i=r+1}^{p}(y_i-\eta_i)^2 + \sum_{\alpha=1}^{n} z_\alpha^2\right)\right]} \tag{8}$$

$$= e^{-\frac{1}{2}\lambda} \int_{S(\eta_{r+1},\cdots,\eta_p,\sigma,\lambda)} e^{\sum_{i=1}^{r} y_i\eta_i/\sigma^2}\, dA.$$

Denote $\left|\sqrt{\sum_{i=1}^{r} y_i^2}\right|$ by r_y. Then we have

$$\int_{S(\eta_{r+1},\cdots,\eta_p,\sigma,\lambda)} e^{\sum_{i=1}^{r} y_i\eta_i/\sigma^2}\, dA = \int_{(\eta_{r+1},\cdots,\eta_p,\sigma,\lambda)} e^{\sqrt{\lambda}\, r_y \cos[\alpha(\eta)]/\sigma}\, dA, \tag{9}$$

where $\alpha(\eta)$ denotes the angle $(0 \leq \alpha(\eta) \leq \pi)$ between the vector y with the components $y_1, \cdots, y_r$ and the vector η with the components $\eta_1, \cdots, \eta_r$. Because of the symmetry of the sphere, the value of the right hand side of (9) is not changed if we substitute $\beta(\eta)$ for $\alpha(\eta)$ where $\beta(\eta)$ denotes the angle $(0 \leq \beta(\eta) \leq \pi)$ between the vector η and an arbitrarily chosen fixed vector u. Hence the value of the right hand side of (9) depends only on r_y, i.e.

$$\int_{S(\eta_{r+1},\cdots,\eta_p,\sigma,\lambda)} e^{\sqrt{\lambda}\, r_y \cos[\alpha(\eta)]/\sigma}\, dA \tag{10}$$

$$= \int_{S(\eta_{r+1},\cdots,\eta_p,\sigma,\lambda)} e^{\sqrt{\lambda}\, r_y \cos[\beta(\eta)]/\sigma}\, dA = I(r_y).$$

Now we will show that $I(r_y)$ is a monotonically increasing function of r_y. We have

$$\frac{dI(r_y)}{dr_y} = \frac{\sqrt{\lambda}}{\sigma} \int_{S(\eta_{r+1}, \cdots, \eta_p, \sigma, \lambda)} \cos [\beta(\eta)] \, e^{\sqrt{\lambda}\, r_y \cos [\beta(\eta)]/\sigma} \, dA. \tag{11}$$

Denote by ω_1 the subset of $S(\eta_{r+1}, \cdots, \eta_p, \sigma, \lambda)$ in which $0 \leq \beta(\eta) \leq \frac{\pi}{2}$ and by ω_2 the subset in which $\frac{\pi}{2} \leq \beta(\eta) \leq \pi$. Because of the symmetry of the sphere we obviously have

$$\begin{aligned} \int_{\omega_2} \cos [\beta(\eta)] e^{\sqrt{\lambda}\, r_y \cos [\beta(\eta)]/\sigma} \, dA &= \int_{\omega_1} \cos [\pi - \beta(\eta)] e^{\sqrt{\lambda}\, r_y \cos [\pi - \beta(\eta)]/\sigma} \, dA \\ &= -\int_{\omega_1} \cos [\beta(\eta)] e^{-\sqrt{\lambda}\, r_y \cos [\beta(\eta)]/\sigma} \, dA. \end{aligned} \tag{12}$$

Hence

$$\frac{dI(r_y)}{dr_y} = \frac{\sqrt{\lambda}}{\sigma} \int_{\omega_1} \cos [\beta(\eta)] \{ e^{\sqrt{\lambda}\, r_y \cos [\beta(\eta)]/\sigma} - e^{-\sqrt{\lambda}\, r_y \cos [\beta(\eta)]/\sigma} \} \, dA. \tag{13}$$

The right hand side of (13) is positive. Hence $I(r_y)$, and therefore also the left hand side of (8), is a monotonically increasing function of r_y.

Let $P_1(y'_{r+1}, \cdots, y'_p, t', \eta_1, \cdots, \eta_p, \sigma) \, dy_{r+1} \cdots dy_p \, dt$ be the probability that the sample point will fall in the intersection of Z and the set

$$y'_i - \tfrac{1}{2} dy_i \leq y_i \leq y'_i + \tfrac{1}{2} dy_i (i = r + 1, \cdots, p), \quad t' - \tfrac{1}{2} dt \leq t \leq t' + \tfrac{1}{2} dt$$

Similarly let $P_2(y'_{r+1}, \cdots, y'_p, t', \eta_1, \cdots, \eta_p, \sigma) \, dy_{r+1} \cdots dy_p \, dt$ be the unconditional probability that the sample point will fall in the intersection of Z^* and the set

$$y'_i - \tfrac{1}{2} dy_i \leq y_i \leq y'_i + \tfrac{1}{2} dy_i (i = r + 1, \cdots, p), \quad t' - \tfrac{1}{2} dt \leq t \leq t' + \tfrac{1}{2} dt.$$

Since the function $\varphi_Z(y_{r+1}, \cdots, y_p, t)$ has been defined so that

$$P_Z(y_{r+1}, \cdots, y_p, t) = F[\varphi(y_{r+1}, \cdots, y_p, t), t],$$

we obviously have

$$\begin{aligned} &P_1(y_{r+1}, \cdots, y_p, t, 0, \cdots, 0, \eta_{r+1}, \cdots, \eta_p, \sigma) \\ &\qquad = P_2(y_{r+1}, \cdots, y_p, t, 0, \cdots, 0, \eta_{r+1}, \cdots, \eta_p, \sigma). \end{aligned} \tag{14}$$

Using a lemma[4] by Neyman and Pearson, we easily obtain

$$\begin{aligned} &\int_{S(\eta_{r+1}, \cdots, \eta_p, \sigma, \lambda)} P_2(y_{r+1}, \cdots, y_p, t, \eta_1, \cdots, \eta_p, \sigma) \, dA \\ &\qquad \geq \int_{S(\eta_{r+1}, \cdots, \eta_p, \sigma, \lambda)} P_1(y_{r+1}, \cdots, y_p, t, \eta_1, \cdots, \eta_p, \sigma) \, dA \end{aligned} \tag{15}$$

[4] J. Neyman and E. S. Pearson, "Contributions to the theory of testing statistical hypotheses," *Stat. Res. Mem.*, Vol. 1, London, 1936.

from (14) and the fact that the left hand side of (8) is a monotonically increasing function of $r_y^2 = y_1^2 + \cdots + y_r^2$. Condition (b) is an immediate consequence of (15). Hence Lemma 1 is proved.

For the proof of our theorem we will also need the following

LEMMA 2: *Let $v_1, \cdots, v_k$ be k normally and independently distributed variates with a common variance σ^2. Denote the mean value of v_i by $\alpha_i (i = 1, \cdots, k)$ and let $f(v_1, \cdots, v_k, \sigma)$ be a function of the variables $v_1, \cdots, v_k$ and σ which does not involve the mean values $\alpha_1, \cdots, \alpha_k$. Then, if the expected value of $f(v_1, \cdots, v_k, \sigma)$ is equal to zero, $f(v_1, \cdots, v_k, \sigma)$ is identically equal to zero, except perhaps on a set of measure zero.*

PROOF: Lemma 2 is obviously proved for all values of σ if we prove it for $\sigma = 1$. Hence we will assume that $\sigma = 1$. It is known that a k-variate distribution which has moments equal to those of the joint distribution of $v_1, \cdots, v_k$, must be identical with the joint distribution of $v_1, \cdots, v_k$. That is to say, the joint distribution of $v_1, \cdots, v_k$ is uniquely determined by its moments. Hence if

$$\text{(16)} \quad \int_{-\infty}^{+\infty} \cdots \int_{-\infty}^{+\infty} v_1^{r_1} v_2^{r_2} \cdots v_k^{r_k} g(v_1, \cdots, v_k) e^{-\frac{1}{2}\sum_{i=1}^{k} (v_i - \alpha_i)^2} dv_1 \cdots dv_k = 0$$

for any set $(r_1, \cdots, r_k)$ of non-negative integers, then $g(v_1, \cdots, v_k)$ must be equal to zero except perhaps on a set of measure zero. Now let $f(v_1, \cdots, v_k)$ be a function whose expected value is zero, i.e.

$$\text{(17)} \quad \int_{-\infty}^{+\infty} \cdots \int_{-\infty}^{+\infty} f(v_1, \cdots, v_k) e^{-\frac{1}{2}\sum_{i=1}^{k} (v_i - \alpha_i)^2} dv_1 \cdots dv_k = 0$$

identically in $\alpha_1, \cdots, \alpha_k$. From (17) it follows that

$$\text{(18)} \quad \int_{-\infty}^{+\infty} \cdots \int_{-\infty}^{+\infty} f(v_1, \cdots, v_k) e^{-\frac{1}{2}\sum_{i=1}^{k} v_i^2 + \sum_{i=1}^{k} \alpha_i v_i} dv_1 \cdots dv_k = 0$$

identically in $\alpha_1, \cdots, \alpha_k$. Differentiating the left hand side of (18) r_1 times with respect to α_1, r_2 times with respect to α_2, $\cdots$, and r_k times with respect to α_k, we obtain

$$\text{(19)} \quad \int_{-\infty}^{+\infty} \cdots \int_{-\infty}^{+\infty} v_1^{r_1} \cdots v_k^{r_k} f(v_1, \cdots, v_k) e^{-\frac{1}{2}\Sigma (v_i - \alpha_i)^2} dv_1 \cdots dv_k = 0.$$

From (16) and (19) it follows that $f(v_1, \cdots, v_k) = 0$. Hence Lemma 2 is proved.

Using Lemmas 1 and 2 we can easily prove our theorem. Because of Lemma 1 we can restrict ourselves to critical regions W which are given by an inequality of the following type

$$y_1^2 + \cdots + y_r^2 \geq \varphi(y_{r+1}, \cdots, y_p, t)$$

where $\varphi(y_{r+1}, \cdots, y_p, t)$ is some function of $y_{r+1}, \cdots, y_p$ and t. The above inequality can be written as

$$\frac{y_1^2 + \cdots + y_r^2}{t^2} \geq \psi(y_{r+1}, \cdots, y_p, t). \tag{20}$$

For any given values of $y_{r+1}, \cdots, y_p, t$ denote by $P(y_{r+1}, \cdots, y_p, t)$ the conditional probability that (20) holds calculated under the assumption that $\eta_1 = \cdots = \eta_r = 0$. It is obvious that $P(y_{r+1}, \cdots, y_p, t)$ does not depend on the unknown parameters $\eta_{r+1}, \cdots, \eta_p, \sigma$. If we denote by W the critical region defined by the inequality (20), we have

$$\begin{aligned} &\beta_W(0, \cdots, 0, \eta_{r+1}, \cdots, \eta_p, \sigma) \\ &= \int_{-\infty}^{+\infty} \cdots \int_{-\infty}^{+\infty} \int_0^{\infty} P(y_{r+1}, \cdots, y_p, t)\rho_1(y_{r+1}, \cdots, y_p, \eta_{r+1}, \cdots, \eta_p, \sigma) \\ &\qquad \times \rho_2(t, \sigma)\, dy_{r+1} \cdots dy_p\, dt \end{aligned} \tag{21}$$

where $\rho_1(y_{r+1}, \cdots, y_p, \eta_{r+1}, \cdots, \eta_p, \sigma)$ denotes the joint probability density function of $y_{r+1}, \cdots, y_p$ and $\rho_2(t, \sigma)$ denotes the probability density function of t calculated under the assumption that $\eta_1 = \cdots = \eta_r = 0$. In order to satisfy the condition of our Theorem, the function ψ in (20) must be chosen so that

$$\begin{aligned} &\int_{-\infty}^{+\infty} \cdots \int_{-\infty}^{+\infty} \int_0^{\infty} P(y_{r+1}, \cdots, y_p, t)\rho_1(y_{r+1}, \cdots, y_p, \eta_{r+1}, \cdots, \eta_p, \sigma) \\ &\qquad \times \rho_2(t, \sigma)\, dy_{r+1} \cdots dy_p\, dt = \beta_0(0). \end{aligned} \tag{22}$$

Let

$$\int_0^{\infty} P(y_{r+1}, \cdots, y_p, t)\rho_2(t, \sigma)\, dt = Q(y_{r+1}, \cdots, y_p, \sigma). \tag{23}$$

Then we obtain from (22)

$$\int_{-\infty}^{+\infty} \cdots \int_{-\infty}^{+\infty} Q(y_{r+1}, \cdots, y_p, \sigma)\rho_1\, dy_{r+1} \cdots dy_p = \beta_0(0). \tag{24}$$

From (24) and Lemma 2 it follows that

$$Q(y_{r+1}, \cdots, y_p, \sigma) = \beta_0(0) \tag{25}$$

except perhaps on a set of measure zero. From (23), (25) and a result[5] by P. L. Hsu we obtain

$$P(y_{r+1}, \cdots, y_p, t) = \beta_0(0) \tag{26}$$

except perhaps on a set of measure zero.

It follows easily from (26) that $\psi(y_{r+1}, \cdots, y_p, t)$ is equal to a fixed constant except perhaps on a set of measure zero. This proves our Theorem.

---

[5] P. L. Hsu, "Notes on Hotelling's generalized T," *Annals of Math. Stat.*, Vol. 9, p. 237.

Reprinted from THE ANNALS OF MATHEMATICAL STATISTICS
Vol. XIV, No. 1, March, 1943

AN EXTENSION OF WILKS' METHOD FOR SETTING TOLERANCE LIMITS

BY ABRAHAM WALD

Columbia University

1. Introduction. Let x be a random variable and let $f(x)$ be its probability density function. Suppose that nothing is known about $f(x)$ except that it is continuous. Let $x_1, \cdots, x_n$ be n independent observations on x. The problem of setting tolerance limits can be formulated as follows: *For some given positive values* $\beta < 1$ and $\gamma < 1$ *we have to construct two functions* $L(x_1, \cdots, x_n)$ *and* $M(x_1, \cdots, x_n)$, *called tolerance limits, such that the probability that*

$$\int_L^M f(t)\,dt \geq \gamma, \tag{1}$$

holds, is equal to β. This problem has recently been solved by S. S. Wilks[1] in a very satisfactory way when nothing is known about $f(x)$ except that it is continuous. Wilks proposes the following solution: Let $x_1, \cdots, x_n$ be the observed values of x arranged in order of increasing magnitude. Then $L = x_r$ and $M = x_{n-r+1}$ where r denotes a positive integer. The exact sampling distribution of the statistic $\int_{x_r}^{x_{n-r+1}} f(t)\,dt$ is derived by Wilks and this provides the solution for the problem of setting tolerance limits. A very important feature of Wilks' solution is the fact that the distribution of $\int_{x_r}^{x_{n-r+1}} f(t)\,dt$ is entirely independent of the unknown density function $f(x)$, i.e. the distribution of $\int_{x_r}^{x_{n-r+1}} f(t)\,dt$ is the same for any arbitrary continuous density function $f(x)$.

In this paper we shall give an extension of Wilks' method to the multivariate case. Let $x_1, \cdots, x_p$ be a set of p random variables with the joint probability density function $f(x_1, \cdots, x_p)$. Suppose that nothing is known about $f(x_1, \cdots, x_p)$ except that it is a continuous function of $x_1, \cdots, x_p$. A sample of n independent observations is drawn and the α-th observation on x_i is denoted by $x_{i\alpha}$ $(i = 1, \cdots, p; \alpha = 1, \cdots, n)$. The problem of setting tolerance limits for $x_1, \cdots, x_p$ can be formulated as follows: *For some given positive values* $\beta < 1$ *and* $\gamma < 1$ *we have to construct* p *pairs of functions of the observations* $L_i(x_{11}, \cdots, x_{pn})$ *and* $M_i(x_{11}, \cdots, x_{pn})$ $(i = 1, \cdots, p)$ *such that the probability that*

$$\int_{L_p}^{M_p} \cdots \int_{L_1}^{M_1} f(t_1, \cdots, t_p)\,dt_1 \cdots dt_p \geq \gamma, \tag{2}$$

[1] S. S. Wilks, "Determination of sample sizes for setting tolerance limits," *Annals of Math. Stat.*, Vol. 12 (1941).

holds, is equal to β. The functions L_i and M_i are called the lower and upper tolerance limits of x_i. A natural extension of Wilks' procedure would seem to be the following: Let $x_{i1}, \cdots, x_{in}$ be the observations on x_i arranged in order of increasing magnitude and let $L_i = x_{ir_i}$ and $M_i = x_{is_i}$ $(i = 1, \cdots, p)$ where r_i and s_i denote some integers. However, this choice of the tolerance limits does not provide a satisfactory solution of our problem, since the distribution of (2) is *not* independent of the unknown density function $f(x_1, \cdots, x_p)$. It will be shown in this paper that by a slight modification of the above procedure the distribution of (2) becomes entirely independent of the unknown density function $f(x_1, \cdots, x_p)$. In section 2 we will treat the bivariate case and in section 3 we will extend the results to multivariate distributions.

2. The bivariate case. In this section we deal with the case when $p = 2$. Let $x_{11}, \cdots, x_{1n}$ be the observations on x_1 arranged in order of increasing magnitude. We may assume that $x_{11} < x_{12} < \cdots < x_{1n}$ since the probability of an equality sign is equal to zero. We define

$$L_1 = x_{1r_1} \quad \text{and} \quad M_1 = x_{1s_1}, \tag{3}$$

where r_1 and s_1 denote some positive integers and $r_1 < s_1 \leq n$. Consider only those sample points $(x_{1\alpha}, x_{2\alpha})$ for which $x_{1r_1} < x_{1\alpha} < x_{1s_1}$, i.e. consider the sample points $(x_{1,r_1+1}, x_{2,r_1+1}), \cdots, (x_{1,s_1-1}, x_{2,s_1-1})$. Denote by $x'_{2,r_1+1}, \cdots, x'_{2,s_1-1}$ the values $x_{2,r_1+1}, \cdots, x_{2,s_1-1}$ arranged in order of increasing magnitude. We define

$$L_2 = x'_{2r_2} \quad \text{and} \quad M_2 = x'_{2s_2}, \tag{4}$$

where r_2 and s_2 denote some positive integers for which $r_2 < s_2 \leq s_1 - r_1 - 1$.

We will show that the distribution of the statistic

$$Q = \int_{L_2}^{M_2} \int_{L_1}^{M_1} f(t_1, t_2)\, dt_1\, dt_2, \tag{5}$$

is entirely independent of the unknown density function $f(x_1, x_2)$. Denote by $\varphi(x_1)$ the marginal distribution of x_1, i.e.

$$\varphi(x_1) = \int_{-\infty}^{+\infty} f(x_1, x_2)\, dx_2. \tag{6}$$

Furthermore denote by $\psi(x_2, L_1, M_1)$ the conditional distribution of x_2 calculated under the condition that $L_1 < x_1 < M_1$. Hence

$$\psi(x_2, L_1, M_1) = \frac{\displaystyle\int_{L_1}^{M_1} f(x_1, x_2)\, dx_1}{\displaystyle\int_{-\infty}^{+\infty} \int_{L_1}^{M_1} f(x_1, x_2)\, dx_1\, dx_2}. \tag{7}$$

Let

$$P = \int_{L_1}^{M_1} \varphi(t)\, dt \tag{8}$$

and

$$(9) \qquad \bar{P} = \int_{L_2}^{M_2} \psi(t, L_1, M_1)\, dt$$

From (5), (8) and (9) it follows that

$$(10) \qquad Q = P\bar{P}.$$

It is obvious that the distribution of P is given by Wilks' formula. Since Wilks derived the distribution only when $s_1 = n - r_1 + 1$, we will briefly give here the derivation for any integers r_1 and s_1.

Let $\int_{-\infty}^{x_{1r_1}} \varphi(t)\, dt = u$ and $\int_{x_{1s_1}}^{\infty} \varphi(t)\, dt = v$. Then the joint probability density function of u and v is given by

$$(11) \qquad cu^{r_1-1}(1 - u - v)^{s_1-r_1-1}\, v^{n-s_1}\, du\, dv,$$

where c is a constant. We obviously have $P = 1 - u - v$. The joint density function of P and u is given by

$$(12) \qquad cu^{r_1-1}P^{s_1-r_1-1}(1 - u - P)^{n-s_1}\, du\, dP,$$

where u is restricted to the interval $[0, 1 - P]$. Hence the distribution of P is given by

$$\begin{aligned} cP^{s_1-r_1-1} &\int_0^{1-P} u^{r_1-1}(1 - u - P)^{n-s_1}\, du \\ &= cP^{s_1-r_1-1}(1-P)^{n-s_1+r_1-1} \int_0^{1-P} \left(\frac{u}{1-P}\right)^{r_1-1} \left(1 - \frac{u}{1-P}\right)^{n-s_1} du \\ &= cP^{s_1-r_1-1}(1-P)^{n-s_1+r_1} \int_0^1 T^{r_1-1}(1-T)^{n-s_1}\, dT \\ &= c'P^{s_1-r_1-1}(1-P)^{n-s_1+r_1}. \end{aligned}$$

Since the integral of the density function of P over the range of P must be equal to 1, we find that

$$c' = \Gamma(n + 1)/\Gamma(s_1 - r_1)\Gamma(n - s_1 + r_1 + 1).$$

Hence the probability density function of P is given by

$$(13) \qquad \frac{\Gamma(n+1)}{\Gamma(s_1 - r_1)\Gamma(n - s_1 + r_1 + 1)} P^{s_1-r_1-1}(1 - P)^{n-s_1+r_1}\, dP.$$

Since $x_{2,r_1+1}, \cdots, x_{2,s_1-1}$ can be considered as $s_1 - r_1 - 1$ independent observations on a random variable t the distribution of which is given by $\psi(t, L_1, M_1)\, dt$, for any given values L_1, M_1 the conditional distribution of $\bar{P}$ is given by the expression we obtain from (13) by substituting r_2 for r_1, s_2 for s_1 and $s_1 - r_1 - 1$ for n. Hence the conditional distribution of $\bar{P}$ is given by

$$(14) \qquad \frac{\Gamma(s_1 - r_1)}{\Gamma(s_2 - r_2)\Gamma(s_1 - r_1 - s_2 + r_2)} (\bar{P})^{s_2 - r_2 - 1}(1 - \bar{P})^{s_1 - r_1 - 1 - s_2 + r_2}\, d\bar{P}.$$

Since the expression (14) does not involve the quantities L_1 and M_1, $\bar{P}$ is distributed independently of L_1 and M_1. Hence the joint density function of P and $\bar{P}$ is given by the product of (13) and (14), i.e. by

$$(15) \qquad AP^{s_1 - r_1 - 1}(1 - P)^{n - s_1 + r_1}(\bar{P})^{s_2 - r_2 - 1}(1 - \bar{P})^{s_1 - r_1 - 1 - s_2 + r_2}\, dP\, d\bar{P},$$

where A denotes the product of the constant coefficients in (13) and (14). From (15) it follows that the joint distribution of P and $Q = P\bar{P}$ is given by

$$(16) \qquad A(1 - P)^{n - s_1 + r_1}Q^{s_2 - r_2 - 1}(P - Q)^{s_1 - r_1 - 1 - s_2 + r_2}\, dP\, dQ.$$

Since the range of P is the interval $[Q, 1]$, the distribution of Q is given by

$$(17) \qquad AQ^{s_2 - r_2 - 1} \int_Q^1 (1 - P)^{n - s_1 + r_1}(P - Q)^{s_1 - r_1 - 1 - s_2 + r_2}\, dP.$$

Let $R = P - Q$. Then we have

$$\int_Q^1 (1 - P)^{n - s_1 + r_1}(P - Q)^{s_1 - r_1 - 1 - s_2 + r_2}\, dP$$

$$(18) \qquad = \int_0^{1-Q} (1 - Q - R)^{n - s_1 + r_1} R^{s_1 - r_1 - 1 - s_2 + r_2}\, dR$$

$$= (1 - Q)^{n - 1 - s_2 + r_2}(1 - Q) \int_0^1 (1 - T)^{n - s_1 + r_1}\, T^{s_1 - r_1 - 1 - s_2 + r_2}\, dT.$$

From (17) and (18) it follows that the probability density function of Q is given by

$$(19) \qquad \frac{\Gamma(n + 1)}{\Gamma(s_2 - r_2)\Gamma(n - s_2 + r_2 + 1)} Q^{s_2 - r_2 - 1}(1 - Q)^{n - s_2 + r_2}\, dQ.$$

3. The multivariate case. We may assume that no two elements of the matrix $\| x_{i\alpha} \|$ $(i = 1, \cdots, p;\ \alpha = 1, \cdots, n)$ are equal, since the probability of this event is equal to 1. For each α let $t_\alpha (\alpha = 1, \cdots, n)$ be the point with the coordinates $x_{1\alpha}, \cdots, x_{p\alpha}$. Let $x_{11}, \cdots, x_{1n}$ be the observations on x_1 arranged in order of increasing magnitude. Then $L_1 = x_{1r_1}$ and $M_1 = x_{1s_1}$. The quantities L_i and M_i $(i = 2, \cdots, p)$ are defined in the following manner: Let S be the set of all points t_α for which

$$L_j < x_{j\alpha} < M_j \qquad (j = 1, \cdots, i - 1).$$

Arrange the i-th coordinates of the points in S in order of increasing magnitude. Then L_i is equal to the r_i-th element and M_i is equal to the s_i-th element of this ordered sequence. We will derive the distribution of

$$(20) \qquad Q_p = \int_{L_p}^{M_p} \cdots \int_{L_1}^{M_1} f(x_1, \cdots, x_p)\, dx_1 \cdots dx_p.$$

Let

$$Q_i = \int_{-\infty}^{+\infty} \cdots \int_{-\infty}^{+\infty} \int_{L_i}^{M_i} \cdots \int_{L_1}^{M_1} f(x_1, \cdots, x_p)\, dx_1 \cdots dx_p \tag{21}$$

$$(i = 1, \cdots, p - 1).$$

Denote by $\varphi_i(x_i, L_1, M_1, \cdots, L_{i-1}, M_{i-1})$ $(i = 2, \cdots, p)$ the conditional probability density function of x_i calculated under the condition that $L_j \leq x_j \leq M_j$ $(j = 1, \cdots, i - 1)$. Let

$$\bar{P}_i = \int_{L_i}^{M_i} \varphi_i(x_i, L_1, M_1, \cdots, L_{i-1}, M_{i-1})\, dx_i. \tag{22}$$

We obviously have

$$Q_{i+1} = Q_i \bar{P}_{i+1} \qquad (i = 1, \cdots, p - 1). \tag{23}$$

We will prove that the probability density function of Q_i is given by

$$\frac{\Gamma(n+1)}{\Gamma(s_i - r_i)\Gamma(n - s_i + r_i + 1)} Q_i^{s_i - r_i - 1}(1 - Q_i)^{n - s_i + r_i}\, dQ_i, \quad (i = 1, \cdots, p). \tag{24}$$

This is certainly true for $i = 1, 2$. We will assume that it is true for $i = j$ and we will prove it for $i = j + 1$. It is easy to see that Q_j and $\bar{P}_{j+1}$ are independently distributed and that the probability density function of $\bar{P}_{j+1}$ is given by

$$\frac{\Gamma(s_j - r_j)}{\Gamma(s_{j+1} - r_{j+1})\Gamma(s_j - r_j - s_{j+1} + r_{j+1})} \cdot (\bar{P}_{j+1})^{s_{j+1} - r_{j+1} - 1}(1 - \bar{P}_{j+1})^{s_j - r_j - 1 - s_{j+1} + r_{j+1}}\, d\bar{P}_{j+1}. \tag{25}$$

The joint distribution of Q_j and $\bar{P}_{j+1}$ is of the same form as the joint distribution of P and $\bar{P}$ in section 2. Hence the distribution of $Q_j\bar{P}_{j+1}$ can be obtained from the distribution of $Q = P\bar{P}$ by substituting r_{j+1} for r_2 and s_{j+1} for s_2. The distribution of Q is given in (19). Making the above substitution in formula (19) we obtain formula (24) for $i = j + 1$. Hence the validity of (24) is proved for $i = 1, 2, \cdots, p$. In particular, the distribution of Q_p is given by

$$\frac{\Gamma(n+1)}{\Gamma(s_p - r_p)\Gamma(n - s_p + r_p + 1)} Q_p^{s_p - r_p - 1}(1 - Q_p)^{n - s_p + r_p}\, dQ_p. \tag{26}$$

It is interesting to note that the distribution of Q_p does not depend on the integers $r_1, s_1, \cdots, r_{p-1}, s_{p-1}$. The construction of the tolerance limits L_i, M_i $(i = 1, \cdots, p)$, as proposed here, is somewhat asymmetric, since it depends on the order of the variates $x_1, \cdots, x_p$. In practical applications the asymmetry of the construction will be very slight, since in most practical cases the integers r_p and s_p will be chosen so that $(s_p - r_p - 1)/n$ will be near to 1. If, for example, $(s_p - r_p - 1)/n \geq .95$, the tolerance limits will be affected only very slightly by a permutation of the variates $x_1, \cdots, x_p$. However, it would be desirable to find a construction which is entirely independent of the order of the variates $x_1, \cdots, x_p$.

4. Tolerance regions composed of several rectangles. For the sake of simplicity we will consider here the bivariate case. All results obtained in this section can be extended without any difficulty to the multivariate case.

In section 2 the tolerance region has been a single rectangle in the plane (x_1, x_2) determined by the four lines $x_1 = L_1$, $x_1 = M_1$; $x_2 = L_2$ and $x_2 = M_2$. If the variates x_1 and x_2 are strongly correlated, a tolerance region of rectangle shape seems to be unfavorable, since it will cover an unnecessarily large area in the (x_1, x_2) plane. The situation is illustrated in figure 1 where the scatter of a

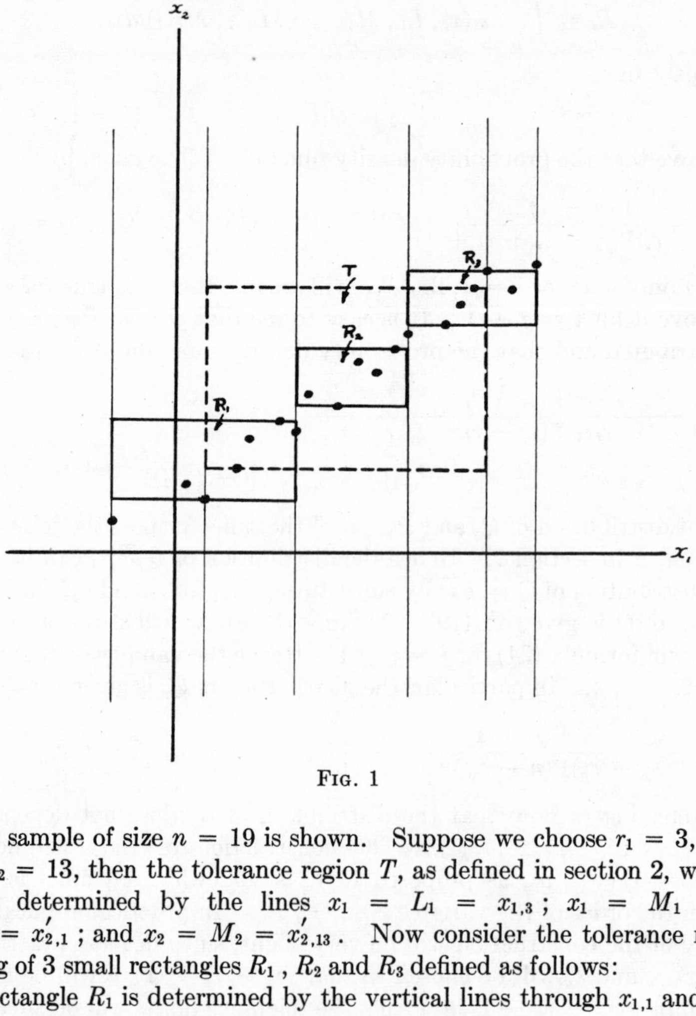

FIG. 1

bivariate sample of size $n = 19$ is shown. Suppose we choose $r_1 = 3$, $s_1 = 17$; $r_2 = 1$, $s_2 = 13$, then the tolerance region T, as defined in section 2, will be the rectangle determined by the lines $x_1 = L_1 = x_{1,3}$; $x_1 = M_1 = x_{1,17}$; $x_2 = L_2 = x'_{2,1}$; and $x_2 = M_2 = x'_{2,13}$. Now consider the tolerance region T' consisting of 3 small rectangles R_1, R_2 and R_3 defined as follows:

The rectangle R_1 is determined by the vertical lines through $x_{1,1}$ and $x_{1,7}$ and the horizontal lines through the sample points with smallest and largest ordinate, restricting ourselves to points which have abscissa values in the interior of the interval $[x_{1,1}, x_{1,7}]$. Similarly R_2 is determined by the vertical lines through

$x_{1,7}$ and $x_{1,13}$ and the horizontal lines through the sample points with largest and smallest ordinate, restricting ourselves to points with abscissa values in the interior of $[x_{1,7}, x_{1,13}]$. Finally R_3 is determined by the vertical lines through $x_{1,13}$ and $x_{1,19}$ and the horizontal lines through the sample points with largest and smallest ordinate, restricting ourselves to points whose abscissa values lie in the interior of $[x_{1,13}, x_{1,19}]$. The region T' consisting of the rectangles R_1, R_2 and R_3 has a much smaller area than the region T. As we will see later, the probability distribution of the statistic $\iint_{T'} f(x_1, x_2)dx_1\,dx_2$ is exactly the same as that of $\iint_{T} f(x_1, x_2)dx_1\,dx_2$. Thus the use of T' may be preferred to that of T.

We will consider tolerance regions T^* of the following general shape: Let $m_1, \cdots, m_k$ be k positive integers such that $1 \leq m_1, m_k \leq n$ and $m_{i+1} - m_i \geq 3$ where n is the size of the bivariate sample. Let V_i be the vertical line in the (x_1, x_2) plane given by the equation $x_1 = x_{1,m_i}$ $(i = 1, \cdots, k)$. The number of sample points which lie between the vertical lines V_i and V_{i+1} is obviously equal to $m_{i+1} - m_i - 1$. Through each point which lies between the vertical lines V_i and V_{i+1} we draw a horizontal line. In this way we obtain $m_{i+1} - m_i - 1$ horizontal lines $W_{i,1}, \cdots, W_{i,m_{i+1}-m_i-1}$ where the line $W_{i,j+1}$ is above the line $W_{i,j}$. Denote by R_{ij} $(i = 1, \cdots, k - 1; j = 1, \cdots, m_{i+1} - m_i - 2)$ the rectangle determined by the lines V_i, V_{i+1}, $W_{i,j}$, $W_{i,j+1}$. Let T^* be a region composed of s different rectangles R_{ij}. The regions T and T' in the example illustrated in figure 1 are special cases of the type of regions T^* as described above. For the region T we have $k = 2$, $m_1 = 3$, $m_2 = 17$, $s = 12$, and for the region T' we have $k = 4$, $m_1 = 1$, $m_2 = 7$, $m_3 = 13$, $m_4 = 19$ and $s = 12$.

Let Q^* be given by $\iint_{T^*} f(x_1, x_2)dx_1\,dx_2$. We will prove that the probability density function of Q^* is given by

$$\frac{\Gamma(n+1)}{\Gamma(s)\Gamma(n-s+1)} Q^{*s-1}(1 - Q^*)^{n-s}\,dQ^*. \tag{27}$$

Let $f_i(x_2)\,dx_2$ be the conditional distribution of x_2 under the restriction that $x_{1,m_i} < x_1 < x_{1,m_{i+1}}$. Thus, we have

$$f_i(x_2) = \frac{\displaystyle\int_{x_{1,m_i}}^{x_{1,m_{i+1}}} f(x_1, x_2)\,dx_1}{\displaystyle\int_{-\infty}^{+\infty}\int_{x_{1,m_i}}^{x_{1,m_{i+1}}} f(x_1, x_2)\,dx_1\,dx_2}. \tag{28}$$

Denote by $\varphi(x_1)\,dx_1$ the marginal distribution of x_1, i.e.

$$\varphi(x_1) = \int_{-\infty}^{+\infty} f(x_1, x_2)\,dx_2$$

Let

$$P_i^* = \int_{x_{1,m_i}}^{x_{1,m_{i+1}}} \varphi(x_1)\,dx_1 \qquad (i = 1, \cdots, k-1) \tag{29}$$

and

$$\cdot\ \bar{P}_i^* = \sum_j \int_{a_{ij}}^{b_{ij}} f_i(x_2)\,dx_2 \qquad (i = 1, \cdots, k-1) \tag{30}$$

where a_{ij} is the ordinate of the lower corners and b_{ij} is the ordinate of the upper corners of the rectangle R_{ij} and the summation is to be taken over all values of j for which R_{ij} is included in T^*. It is clear that

$$Q^* = P_1^* \bar{P}_1^* + \cdots + P_{k-1}^* \bar{P}_{k-1}^* . \tag{31}$$

Let y be any random variable which has a continuous probability density function, say $\psi(y)\,dy$. Furthermore let $y_1, \cdots, y_n$ be n independent observations on y. Let $\psi_i(y)\,dy$ be the conditional density function of y under the condition that y is restricted to the interval $[y_{m_i}, y_{m_{i+1}}]$. Let

$$P' = \sum_{i,j} \int_{y_{m_i+j}}^{y_{m_i+j+1}} \psi(y)\,dy \tag{32}$$

where the summation is taken over all pairs i, j for which R_{ij} is contained in T^*. Let

$$P_i' = \int_{y_{m_i}}^{y_{m_{i+1}}} \psi(y)\,dy, \qquad \text{and}$$

$$\bar{P}_i' = \sum_j \int_{y_{m_i+j}}^{y_{m_i+j+1}} \psi_i(y)\,dy,$$

where the summation is to be taken over all values j for which R_{ij} is contained in T^*. We obviously have

$$P' = P_1' \bar{P}_1' + \cdots + P_{k-1}' \bar{P}_{k-1}' . \tag{33}$$

It is easy to verify that (*i*) the joint distribution of $P_1', \cdots, P_{k-1}'$ is the same as the joint distribution of $P_1^*, \cdots, P_{k-1}^*$; (ii) the distribution of $\bar{P}_i'$ is the same as that of $\bar{P}_i^*$ $(i = 1, \cdots, k-1)$; (iii) the variates $\bar{P}_1', \cdots, \bar{P}_{k-1}'$ are independent of each other and also of $P_1', \cdots, P_{k-1}'$; (iv) the variates $\bar{P}_1^*, \cdots, \bar{P}_{k-1}^*$ are independent of each other and also of $P_1^*, \cdots, P_{k-1}^*$. Hence it follows from (31) and (33) that the distribution of Q^* is the same as that of P'. Now we will derive the distribution of P'. The expression P' can be written in the following form:

$$P' = \sum_{i=1}^{l} \int_{y_{r_i}}^{y_{s_i}} \psi(y)\,dy, \tag{34}$$

where $r_1, s_1, \cdots, r_l, s_l$ are some positive integers for which $1 \leq r_1 < s_1 < r_2 < s_2 < \cdots < r_l < s_l \leq n$. Let

$$(35)\quad \begin{aligned} P'' &= \sum_{i=1}^{l-1} \int_{y_{r_i}}^{y_{s_i}} \psi(y)\,dy + \int_{y_{s_{l-1}}}^{y_{s_{l-1}+s_l-r_l}} \psi(y)\,dy \\ &= \sum_{i=1}^{l-2} \int_{y_{r_i}}^{y_{s_i}} \psi(y)\,dy + \int_{y_{r_{l-1}}}^{y_{s_{l-1}+s_l-r_l}} \psi(y)\,dy. \end{aligned}$$

For any fixed value $y_{s_{l-1}}$ denote by $\psi_1(y)$ the conditional probability density of y under the restriction that $y < y_{s_{l-1}}$ and by $\psi_2(y)$ the conditional distribution of y under the restriction that $y > y_{s_{l-1}}$. Let

$$P = \int_{-\infty}^{y_{s_{l-1}}} \psi(y)\,dy \qquad P_1 = \sum_{i=1}^{l-1} \int_{y_{r_i}}^{y_{s_i}} \psi_1(y)\,dy;$$

$$P_2 = \int_{y_{r_l}}^{y_{s_l}} \psi_2(y)\,dy \quad \text{and} \quad P_3 = \int_{y_{s_{l-1}}}^{y_{s_{l-1}+s_l-r_l}} \psi_2(y)\,dy.$$

Then it follows from (34) and (35) that

$$(36)\quad \begin{aligned} P' &= PP_1 + (1 - P)P_2, \\ P'' &= PP_1 + (1 - P)P_3. \end{aligned}$$

For calculating the distributions of P_2 and P_3 we may consider the variates $y_{s_{l-1}+1}, \cdots, y_n$ as $n - s_{l-1}$ independent observations drawn from a population which has the distribution $\psi_2(y)\,dy$. Hence, the distribution of P_2 can be derived from (13) and it is easy to verify that the distribution of P_3 is the same as that of P_2. It is clear that P_2 is independent of P and P_1. Similarly P_3 is independent of P and P_1. Hence, because of (36) the distribution of P' must be the same as that of P''.

In the same way we find that the distribution of P'' is the same as the distribution of

$$P''' = \sum_{i=1}^{l-3} \int_{y_{r_i}}^{y_{s_i}} \psi(y)\,dy + \int_{y_{r_{l-2}}}^{y_{s_{l-2}+s_{l-1}+s_l-r_l-r_{l-1}}} \psi(y)\,dy$$

Thus, by induction we see that the distribution of P' is the same as the distribution of the statistic $P_0 = \int_{y_{r_1}}^{y_{r_1+s}} \psi(y)\,dy$ where $s = \sum_{i=1}^{l} (s_i - r_i)$. From (13) it follows that the distribution of P_0 is given by

$$\frac{\Gamma(n + 1)}{\Gamma(s)\Gamma(n - s + 1)} P_0^{s-1}(1 - P_0)^{n-s}\,dP_0.$$

Hence, we have proved that the distribution of Q^* is given by (27).

5. Summary of the results and numerical illustrations. I shall give here a summary of the results obtained and a few illustrative examples. The multivariate case being a straightforward extension of the bivariate case, I shall discuss merely the latter. Consider a pair of random variables x and y. Denote by $f(x, y)dx\,dy$ the joint probability density function of x and y and suppose that nothing is known about $f(x, y)$ except that it is continuous. A sample of n pairs of independent observations $(x_1, y_1), \cdots, (x_n, y_n)$ is drawn from this bivariate population. The sample can be represented by n points $p_1, \cdots, p_n$ in the plane (x, y), p_i being the point with the coordinates x_i and y_i. In section 2 we have dealt with the problem of finding a rectangle T in the plane (x, y), called tolerance region, such that we can state with high probability, say with probability .98 or .99, that the proportion Q of the bivariate universe included in the rectangle T is not less than a given number b, say not less than .98 or .99. The rectangle T is constructed as follows: Suppose that the points $p_1, \cdots, p_n$ are arranged in order of increasing magnitude of their abscissa values, i.e. $x_1 < x_2 < \cdots < x_n$. We draw a vertical line V_{r_1} through the point p_{r_1} and a vertical line V_{s_1} through p_{s_1} where r_1 and s_1 are positive integers such that $1 \leq r_1, r_1 \leq s_1 - 3$ and $s_1 \leq n$. We consider the set S consisting of the points $p_{r_1+1}, \cdots, p_{s_1-1}$ which lie between the vertical lines V_{r_1} and V_{s_1}. We draw a horizontal line H_{r_2} through the point of S which has the r_2-th smallest ordinate in S. Finally a horizontal line H_{s_2} is drawn through the point of S which has the s_2-th smallest ordinate in S. The values r_2 and s_2 are positive integers for which $r_2 < s_2$. The tolerance region T is the rectangle determined by the lines V_{r_1}, V_{s_1}, H_{r_2} and H_{s_2}. The probability p that at least the porportion $b(0 < b < 1)$ of the universe is included in T is given by

$$(37) \qquad p = \int_b^1 \frac{\Gamma(n+1)}{\Gamma(s_2 - r_2)\Gamma(n - s_2 + r_2 + 1)} Q^{s_2-r_2-1}(1-Q)^{n-s_2+r_2}\,dP.$$

It is known that if a random variable $v(0 \leq v \leq 1)$ has the distribution

$$\frac{\Gamma(c+d)}{\Gamma(c)\Gamma(d)} v^{c-1}(1-v)^{d-1}\,dv,$$

and $2c$ and $2d$ are positive integers, then $\frac{2c}{2d}\frac{1-v}{v}$ has the F-distribution (analysis of variance distribution) with $2d$ and $2c$ degrees of freedom. Thus,

$$(39) \qquad \frac{2(s_2 - r_2)}{2(n - s_2 + r_2 + 1)} \frac{1-Q}{Q} = F$$

has the F-distribution with $2(n - s_2 + r_2 + 1)$ and $2(s_2 - r_2)$ degrees of freedom. From (37) it follows that p is equal to the probability that

$$F \leq \frac{2(s_2 - r_2)}{2(n - s_2 + r_2 + 1)} \frac{1-b}{b}$$

where F has the analysis of variance distribution with $2(n - s_2 + r_2 + 1)$ and $2(s_2 - r_2)$ degrees of freedom. For the case $r_1 = 1$, $s_1 = n$, $r_2 = 1$ and $s_2 =$

$n - 2$, the following table gives the value of the sample size n which is necessary for having the probability p that at least the proportion b of the universe is included in the tolerance rectangle T.

	$b = .97$	$b = .975$	$b = .98$	$b = .985$	$b = .99$
$p = .99$	332	398	499	668	1001
$p = .95$	256	309	385	515	771

Thus, if we want the probability to be .99 that the tolerance region will include at least 98 per cent of the universe, the sample size must be 499.

In section 4 tolerance regions are considered which are composed of several rectangles. Such a tolerance region may be more favorable than a single rectangle if x and y are highly correlated. As an illustration we consider tolerance regions T^* constructed as follows: Suppose that n is divisible by 4 and the sample points $p_1, \ldots, p_n$ are arranged in order of increasing magnitude of their abscissa values. We draw the vertical lines V_0, V_1, V_2, V_3 and V_4 through the points p_1, $p_{n/4}$, $p_{n/2}$, $p_{3n/4}$ and p_n. Let $R_i (i = 1, 2, 3, 4)$ be the rectangle determined by the vertical lines V_{i-1} and V_i and the horizontal lines H_i and H_i' where H_i and H_i' are defined as follows: consider only the points which lie between the two vertical lines V_{i-1} and V_i (points on the vertical lines are excluded). From these select the point with the smallest and the point with the largest ordinate. The lines H_i and H_i' are the horizontal lines which go through these two points respectively. The tolerance region T^* is composed of the four rectangles R_1, R_2, R_3 and R_4. The number of rectangles R_{ij} (defined in section 4) included in T^* is equal to $s = n - 9$. Thus, according to the results of section 4 the probability distribution of the proportion Q^* of the universe included in the region T^* is given by

$$\frac{\Gamma(n+1)}{\Gamma(n-9)\Gamma(10)} (Q^*)^{n-8} (1 - Q^*)^9 \, dQ^*.$$

Numerical calculations show that for $n = 1000$ the probability is .99 that at least 98.1 per cent of the universe will be included in the tolerance region T^*.

Made in the United States of America

Reprinted from THE ANNALS OF MATHEMATICAL STATISTICS
Vol. XIV, No. 2, June, 1943

ON THE EFFICIENT DESIGN OF STATISTICAL INVESTIGATIONS

BY ABRAHAM WALD

Columbia University

1. Introduction. A theory of efficient design of statistical investigations has been developed by R. A. Fisher[1] and his followers mainly in connection with agricultural experimentation. However, the same methods can be applied to other fields also. All statistical designs treated in the aforementioned theory refer to problems of testing linear hypotheses. By testing a linear hypothesis we mean the following problem: Let $y_1, \cdots, y_N$ be N independently and normally distributed variates with a common variance σ^2. It is assumed that the expected value of y_α is given by

$$E(y_\alpha) = \beta_1 x_{1\alpha} + \beta_2 x_{2\alpha} + \cdots + \beta_p x_{p\alpha} \qquad (\alpha = 1, \cdots, N) \tag{1}$$

where the quantities $x_{i\alpha} (i = 1, \cdots, p; \alpha = 1, \cdots, N)$ are known constants and $\beta_1, \cdots, \beta_p$ are unknown constants. The coefficients $\beta_1, \cdots, \beta_p$ are called the population regression coefficients of y on $x_1, x_2, \cdots,$ and x_p, respectively. The hypothesis that the unknown regression coefficients $\beta_1, \cdots, \beta_p$ satisfy a set of linear equations

$$g_{i1}\beta_1 + \cdots + g_{ip}\beta_p = g_i \qquad (i = 1, \cdots, r; r \leqq p), \tag{2}$$

is called a linear hypothesis. The problem under consideration is that of testing the hypothesis (2) on the basis of the observed values $y_1, \cdots, y_N$.

In many cases the experimenter has a certain amount of freedom in the choice of the values $x_{i\alpha}$. The efficiency of the test is greatly affected by the values of $x_{i\alpha}$. The statistical investigation is efficiently designed if the values $x_{i\alpha}$ are chosen so that the sensitivity of the test is maximized. Let us illustrate this by a simple example. Suppose that x and y have a bivariate normal distribution and we want to test the hypothesis that the regression coefficient β of y on x has a particular value β_0. Suppose, furthermore, that the test has to be carried out on the basis of N pairs of observations $(x_1, y_1), \cdots, (x_N, y_N)$, where the experiments are performed in such a way that $x_1, \cdots, x_N$ are not random variables but have predetermined fixed values. It is known that the variance of the least square estimate b of β is inversely proportional to $\sum_{\alpha=1}^{N} (x_\alpha - \bar{x})^2$ where $\bar{x} = (x_1 + \cdots + x_N)/N$. Hence, if we can freely choose the values $x_1, \cdots, x_N$ in a certain domain D, the greatest sensitivity of the test will be achieved by choosing $x_1, \cdots, x_N$ so that $\Sigma(x_\alpha - \bar{x})^2$ becomes a maximum.

In the next section we will introduce a measure of the efficiency of the design

[1] See for instance R. A. FISHER, *The Design of Experiments*, Oliver and Boyd, London, 1935.

of a statistical investigation for testing a linear hypothesis. In sections 3 and 4 it will be shown that some well known experimental designs, used widely in agricultural experimentation, are most efficient in the sense of the definition given in section 2.

2. A measure of the efficiency of the design of a statistical investigation for testing a linear hypothesis. The hypothesis (2) can be reduced by a suitable linear transformation to the canonical form

$$\beta_1 = \beta_2 = \cdots = \beta_r = 0, \qquad (r \leq p). \tag{3}$$

Hence, we can restrict ourselves without loss of generality to the consideration of the hypothesis (3).

Denote $\sum_{\alpha=1}^{N} x_{i\alpha} x_{j\alpha}$ by a_{ij} and let the matrix $\| c_{ij} \|$ be the inverse of the matrix $\| a_{ij} \|$ $(i, j = 1, \cdots, p)$. Denote by b_i the least square estimate of β_i $(i = 1, \cdots, p)$. It is known that the estimates $b_1, \cdots, b_p$ have a joint normal distribution with mean values $\beta_1, \cdots, \beta_p$, respectively. It is furthermore known that the covariance of b_i and b_j is equal to $c_{ij}\sigma^2$. The statistic used for testing the hypothesis (3) is given by

$$F = \frac{N-p}{r} \frac{\sum_{l=1}^{r} \sum_{m=1}^{r} a^*_{lm} b_l b_m}{\sum_{\alpha=1}^{N} (y_\alpha - b_1 x_{1\alpha} \cdots - b_p x_{p\alpha})^2} \tag{4}$$

where $\| a^*_{lm} \|$ is the inverse of $\| c_{lm} \|$ $((l, m = 1, \cdots, r)$. The statistic F has the F-distribution with r and $N - p$ degrees of freedom. The critical region for testing the hypothesis (3) is given by the inequality

$$F \geqq F_0, \tag{5}$$

where the constant F_0 is determined so that the probability that $F \geq F_0$ (calculated under the assumption that (3) holds) is equal to the level of significance we wish to have.

It is known that the power function[2] of the critical region (5) depends only on the single parameter

$$\lambda = \frac{1}{\sigma^2} \sum_{l=1}^{r} \sum_{m=1}^{r} a^*_{lm} \beta_l \beta_m. \tag{6}$$

Furthermore this power function is a monotonically increasing function of λ. The coefficients a^*_{lm} are functions of the quantities $x_{i\alpha}$ $(i = 1, \cdots, p;\ \alpha = 1, \cdots, N)$. The choice of the values $x_{i\alpha}$ $(i = 1, \cdots, p;\ \alpha = 1, \cdots, N)$ is the better the greater the corresponding value of λ. If $r = 1$, the expression λ

---

[2] See for instance P. C. Tang, "The power function of the analysis of variance tests," *Stat. Res. Mem.*, Vol. II, 1938.

reduces to $\frac{1}{\sigma^2} a_{11}^* \beta_1^2$. Hence, if $r = 1$, we maximize λ by maximizing a_{11}^*. Since $a_{11}^* = 1/c_{11}$, we maximize λ by minimizing c_{11}. Thus, if $r = 1$, we can say that we obtain the most powerful test by minimizing c_{11}, i.e. by minimizing the variance of b_1. If $r > 1$, the difficulty arises that no set of values $x_{i\alpha}$ $(i = 1, \cdots, p; \alpha = 1, \cdots, N)$ can be found for which λ becomes a maximum irrespective of the values of the unknown parameters $\beta_1, \cdots, \beta_r$. Hence, if $r > 1$, we have to be satisfied with some compromise solution. For this purpose let us consider the unit sphere

$$\beta_1^2 + \cdots + \beta_r^2 = 1, \tag{7}$$

in the space of the parameters $\beta_1, \cdots, \beta_r$. It is known that the smallest root in ρ of the determinantal equation

$$\begin{vmatrix} a_{11}^* - \rho & a_{12}^* & \cdots & a_{1r}^* \\ a_{21}^* & a_{22}^* - \rho & \cdots & a_{2r}^* \\ & & \cdots & \\ a_{r1}^* & a_{r2}^* & \cdots & a_{rr}^* - \rho \end{vmatrix} = 0, \tag{8}$$

is equal to the minimum value of $\sigma^2\lambda$ on the unit sphere (7). Similarly the greatest root of (8) is equal to the maximum value of $\sigma^2\lambda$ on the sphere (7). The compromise solution of maximizing the smallest root of (8) seems to be a very reasonable one. However, for the sake of certain mathematical simplifications, we propose to maximize the product of the r roots of (8). Since the product of the roots of (8) is equal to the determinant

$$\begin{vmatrix} a_{11}^* & \cdots & a_{1r}^* \\ \cdot & \cdots & \cdot \\ a_{r1}^* & \cdots & a_{rr}^* \end{vmatrix}, \tag{9}$$

we have to maximize the determinant (9). The value of the determinant $| c_{lm} |$ $(l, m = 1, \cdots, r)$ is the reciprocal of that of (9). Hence we maximize (9) by minimizing the determinant $| c_{lm} |$. The generalized variance of the set of variates $b_1, \cdots, b_r$ is equal to the product of σ^{2r} and the determinant $| c_{lm} |$. Thus, our result can be expressed as follows: The optimum choice of the values of $x_{i\alpha}$ is that for which the generalized variance of the variates $b_1, \cdots, b_r$ becomes a minimum.

Any set of pN values $x_{i\alpha}$ $(i = 1, \cdots, p; \alpha = 1, \cdots, N)$ can be represented by a point in the pN-dimensional Cartesian space. Denote by D the set of all points in the pN-dimensional space which we are free to choose. If N is fixed and if any point of D can be equally well chosen, the following two definitions seem to be appropriate:

DEFINITION 1. *Denote by c the minimum value of the determinant $| c_{lm} |$ $(l, m = 1, \cdots, r)$ in the domain D. Then the ratio $c/| c_{lm} |$ is called the efficiency of the design of the statistical investigation for testing the hypothesis* (3).

DEFINITION 2. *The design of the statistical investigation for testing the hypothesis* (3) *is said to be most efficient if its efficiency is equal to* 1.

3. Efficiency of the Latin square design. A widely used and important design in agricultural experimentation is the so-called Latin square. Suppose we wish to find out by experimentation whether there is any significant difference among the yields of m different varieties $v_1, \cdots, v_m$. For this purpose the experimental area is subdivided into m^2 plots lying in m rows and m columns and each plot is assigned to one of the varieties $v_1, \cdots, v_m$. If each variety appears exactly once in each row and exactly once in each column, we have a Latin square arrangement. Denote by y_{ijk} the yield of the variety v_k on the plot which lies in the i-th row and j-th column. The subscript k is, of course, a single valued function of the subscripts i and j, since to each plot only one variety is assigned. The following assumptions are made: the variates y_{ijk} are independently and normally distributed with a common variance σ^2 and the expected value of y_{ijk} is given by

$$E(y_{ijk}) = \mu_i + \nu_j + \rho_k. \tag{10}$$

The parameters σ^2, μ_i, ν_j and ρ_k are unknown. The hypothesis to be tested is the hypothesis that variety has no effect on yield, i.e.

$$\rho_1 = \rho_2 = \cdots = \rho_k. \tag{11}$$

We associate the positive integer $\alpha(i, j) = (i - 1)m + j$ with the plot which lies in the i-th row and j-th column. $(i, j = 1, \cdots, m)$. It is clear that for any positive integer $\alpha \leq m^2$ there exists exactly one plot, i.e. exactly one pair of values i and j, such that $\alpha = \alpha(i, j)$. In the following discussions the symbol y_α $(\alpha = 1, \cdots, m^2)$ will denote the yield y_{ijk} where the indices i and j are determined so that $\alpha(i, j) = \alpha$. The plot in the i-th row and j-th column will be called the α-th plot where $\alpha = \alpha(i, j)$.

We define the symbols $t_{i\alpha}$, $u_{j\alpha}$, $z_{k\alpha}$ $(i, j, k = 1, \cdots, m; \alpha = 1, \cdots, m^2)$, as follows: $t_{i\alpha} = 1$ if the α-th plot lies in the i-th row, and $t_{i\alpha} = 0$ otherwise. Similarly $u_{j\alpha} = 1$ if the α-th plot lies in the j-th column, and $u_{j\alpha} = 0$ otherwise. Finally $z_{k\alpha} = 1$ if the k-th variety is assigned to the α-th plot, and $z_{k\alpha} = 0$ otherwise. Then equation (10) can be written as

$$\begin{aligned} E(y_\alpha) = \mu_1 t_{1\alpha} + \cdots + \mu_m t_{m\alpha} + \nu_1 u_{1\alpha} + \cdots \\ + \nu_m u_{m\alpha} + \rho_1 z_{1\alpha} + \cdots + \rho_m z_{m\alpha}. \end{aligned} \tag{12}$$

Denote the arithmetic means $\frac{1}{m^2}\sum_{\alpha=1}^{m^2} t_{i\alpha}$, $\frac{1}{m^2}\sum_{\alpha=1}^{m^2} u_{i\alpha}$, and $\frac{1}{m^2}\sum_{\alpha=1}^{m^2} z_{i\alpha}$ by $\bar{t}_i$, $\bar{u}_i$ and $\bar{z}_i$ respectively. Let $t'_{i\alpha} = t_{i\alpha} - \bar{t}_i$, $u'_{i\alpha} = u_{i\alpha} - \bar{u}_i$, $z'_{i\alpha} = z_{i\alpha} - \bar{z}_i$, $\mu'_i = \mu_i - \mu_m$, $\nu'_i = \nu_i - \nu_m$ and $\rho'_i = \rho_i - \rho_m$ for $i = 1, \cdots, m - 1$. Let furthermore $w_\alpha = 1$ for $\alpha = 1, \cdots, m^2$. Then we have

$$\begin{cases} t_{i\alpha} = t'_{i\alpha} + \bar{t}_i w_\alpha; \quad u_{i\alpha} = u'_{i\alpha} + \bar{u}_i w_\alpha; \quad z_{i\alpha} = z'_{i\alpha} + \bar{z}_i w_\alpha; \\ \qquad\qquad\qquad\qquad\qquad\qquad (i = 1, \cdots, m-1) \\ t_{m\alpha} = (1 - \bar{t}_1 - \cdots - \bar{t}_{m-1})w_\alpha - t'_{1\alpha} - \cdots - t'_{m-1,\alpha}, \\ u_{m\alpha} = (1 - \bar{u}_1 - \cdots - \bar{u}_{m-1})w_\alpha - u'_{1\alpha} - \cdots - u'_{m-1,\alpha}, \\ z_{m\alpha} = (1 - \bar{z}_1 - \cdots - \bar{z}_{m-1})w_\alpha - z'_{1\alpha} - \cdots - z'_{m-1,\alpha}. \end{cases} \tag{13}$$

From (12) and (13) we obtain

$$E(y_\alpha) = \xi w_\alpha + \sum_{i=1}^{m-1} \mu_i' t_{i\alpha}' + \sum_{i=1}^{m-1} \nu_i' u_{i\alpha}' + \sum_{i=1}^{m-1} \rho_i' z_{i\alpha}' \tag{14}$$

where

$$\xi = \sum_{i=1}^{m-1} \mu_i' \bar{t}_i + \sum_{i=1}^{m-1} \nu_i' \bar{u}_i + \sum_{i=1}^{m-1} \rho_i' \bar{z}_i + \mu_m + \nu_m + \rho_m.$$

The hypothesis (11) can be written as

$$\rho_1' = \rho_2' = \cdots = \rho_{m-1}' = 0. \tag{15}$$

This is a linear hypothesis in canonical form as given in (3). The values $z_{i\alpha}'$ $(i = 1, \cdots, m - 1; \alpha = 1, \cdots, m^2)$ depend on the way in which the varieties $v_1, \cdots, v_m$ are assigned to the m^2 plots. We will show that we obtain a most efficient design if we distribute the varieties over the m^2 plots in a Latin square arrangement, i.e. if each variety appears exactly once in each row and exactly once in each column.

Let $q_{1\alpha} = w_\alpha$, $q_{i+1,\alpha} = t_{i\alpha}'$ $(i = 1, \cdots, m - 1)$, $q_{m+j,\alpha} = u_{j\alpha}'$ $(j = 1, \cdots, m - 1)$ and $q_{2m-1+k,\alpha} = z_{k\alpha}'$ $(k = 1, \cdots, m - 1)$. Denote $\sum_{\alpha=1}^{m^2} q_{i\alpha} q_{j\alpha}$ by a_{ij} $(i, j = 1, 2, \cdots, 3m - 2)$ and let the matrix $\| c_{ij} \|$ be the inverse of the matrix $\| a_{ij} \|$ $(i, j = 1, \cdots, 3m - 2)$. Let us denote by Δ the determinant $| a_{ij} |$ $(i, j = 1, \cdots, 3m - 2)$, by Δ_1 the determinant $| a_{ij} |$ $(i, j = 1, \cdots, 2m - 1)$, by Δ_2 the determinant $| a_{ij} |$ $(i, j = 2m, \cdots, 3m - 2)$ and Δ_2' the determinant $| c_{ij} |$ $(i, j = 2m, \cdots, 3m - 2)$. We have to show that for the Latin square arrangement Δ_2' becomes a minimum. From a known theorem[3] about determinants it follows that

$$\Delta_2' = \Delta_1/\Delta. \tag{16}$$

Hence, we have merely to show that Δ/Δ_1 becomes a maximum for the Latin square arrangement. Denote by $\bar{\Delta}$, $\bar{\Delta}_1$ and $\bar{\Delta}_2$ the values taken by Δ, Δ_1 and Δ_2, respectively, in the case of a Latin square arrangement. Since, for the Latin square arrangement, as is known,

$$\sum_{\alpha=1}^{m^2} z_{k\alpha}' u_{j\alpha}' = \sum_{\alpha=1}^{m^2} z_{k\alpha}' t_{i\alpha}' = \sum_{\alpha=1}^{m^2} z_{k\alpha}' w_\alpha = 0 \qquad (i, j, k = 1, \cdots, m - 1)$$

we have

$$\frac{\bar{\Delta}}{\bar{\Delta}_1} = \bar{\Delta}_2. \tag{17}$$

Since the matrix $\| a_{ij} \|$ $(i, j = 1, \cdots, 3m - 2)$ is positive definite we have

$$\frac{\Delta}{\Delta_1} \leqq \Delta_2. \tag{18}$$

---

[3] See M. Bôcher, *Introduction to Higher Algebra*, 1931, pp. 31.

Because of (17) and (18) the Latin square design is proved to be most efficient if we show that $\Delta_2 \leq \bar{\Delta}_2$.

Denote by Δ_2^* the m-rowed determinant $| a_{ij} |$ $(i, j = 1, 2m, 2m + 1, \cdots, 3m - 2)$. Since $a_{1j} = 0$ for $j \neq 1$, we have

$$\Delta_2^* = a_{11}\Delta_2 = m^2\Delta_2. \tag{19}$$

Denote $\sum_{\alpha=1}^{m^2} z_{i\alpha}z_{j\alpha}$ by b_{ij} $(i, j = 1, \cdots, m)$. Then

$$\begin{cases} b_{ij} = 0, & \text{for } i \neq j \\ \text{and } b_{ii} = N_i, \end{cases} \tag{20}$$

where N_i denotes the number of plots to which the variety v_i has been assigned. Because of (20) we have

$$\begin{vmatrix} b_{11} & \cdots & b_{1m} \\ \cdot & \cdots & \cdot \\ \cdot & \cdots & \cdot \\ b_{m1} & & b_{mm} \end{vmatrix} = N_1 N_2 \cdots N_m. \tag{21}$$

According to (13) we have

$$\begin{aligned} z'_{i\alpha} + \bar{z}_i w_\alpha &= z_{i\alpha}, \qquad (i = 1, \cdots, m - 1) \\ -z'_{1\alpha} - \cdots - z'_{m-1,\alpha} + w_\alpha(1 - \bar{z}_1 - \cdots - \bar{z}_{m-1}) &= z_{m\alpha}. \end{aligned} \tag{22}$$

The determinant of these equations is given by

$$\lambda = \begin{vmatrix} 1 & 0 & 0 & \cdots & 0 & 0 & \bar{z}_1 \\ 0 & 1 & 0 & \cdots & 0 & 0 & \bar{z}_2 \\ \cdot & \cdot & \cdot & \cdots & \cdot & \cdot & \cdot \\ 0 & 0 & 0 & \cdots & 0 & 1 & \bar{z}_{m-1} \\ -1 & -1 & -1 & \cdots & -1 & -1 & \delta \end{vmatrix} \tag{23}$$

where $\delta = 1 - \bar{z}_1 - \bar{z}_2 - \cdots - \bar{z}_{m-1}$. It is easy to verify that

$$\lambda = 1. \tag{24}$$

From (21), (22) and (24) it follows that

$$\Delta_2^* = N_1 N_2 \cdots N_m. \tag{25}$$

Hence, from (19) we obtain

$$\Delta_2 = N_1 N_2 \cdots N_m / m^2. \tag{26}$$

In the case of a Latin square design we have $N_1 = N_2 = \cdots = N_m = m$. Hence

$$\bar{\Delta}_2 = m^{m-2}. \tag{27}$$

Because of the condition $N_1 + N_2 + \cdots + N_m = m^2$, the right hand side of (26) becomes a maximum when $N_1 = N_2 = \cdots = N_m = m$. Thus $\Delta_2 \leq \bar{\Delta}_2$ and consequently the Latin square design is proved to be most efficient.

4. Efficiency of Graeco-Latin and higher squares. Consider m varieties $v_1, \cdots, v_m$ and m treatments $q_1, \cdots, q_m$ Suppose that we wish to find out by experimentation whether the yield is affected by varieties or treatments. For this purpose the experimental area is subdivided into m^2 plots lying in m rows and m columns and to each plot one of the varieties and one of the treatments is assigned. We call this arrangement a Graeco-Latin square if the following conditions are fulfilled: 1) each variety appears exactly once in each row and exactly once in each column; 2) each treatment appears exactly once in each row and exactly once in each column; 3) each variety is combined with each of the treatments exactly once.

The following general abstract scheme includes the Latin square and Graeco-Latin square as special cases: Consider an r-way classification with m classes in each classification. Denote by $y_{a_1 a_2 \cdots a_r}$ the value of a certain characteristic of an individual who is classified in the a_1-class of the first classification, in the a_2-class of the second classification, $\cdots$, and in the a_r-class of the r-th classification. Suppose that m^2 observations are made for the purpose of investigating the effect of the classes on the value of the characteristic under consideration. We will say that we have a generalized Latin square design if the following condition is fulfilled: *Let i, j, m' and m'' be an arbitrary set of four positive integers for which $i \neq j$, $i \leq r$, $j \leq r$, $m' \leq m$ and $m'' \leq m$. Then among the m^2 individuals observed there exists exactly one individual who belongs to the m'-class of the i-th classification and m''-class of the j-th classification.*

It is clear that if $r = 3$ the above scheme is a Latin square. If $r = 4$ we have a Graeco-Latin square.

Assume that the observations $y_{a_1 \cdots a_r} (a_1, a_2, \cdots, a_r = 1, \cdots, m)$ are normally and independently distributed with a common variance σ^2. Assume furthermore that the expected value of $y_{a_1 \cdots a_r}$ is given by

$$E(y_{a_1 a_2 \cdots a_r}) = \gamma_{1a_1} + \cdots + \gamma_{ra_r} .$$

The parameters σ^2 and γ_{ia} $(i = 1, \cdots, r;\ a = 1, \cdots, m)$ are unknown constants. Suppose that we wish to test the hypothesis that

$$\gamma_{i1} = \gamma_{i2} = \cdots = \gamma_{im} . \tag{28}$$

It can be shown that if the number of observations is limited to m^2, we obtain a most efficient design by constructing a generalized Latin square. The proof of this statement is similar to that of the efficiency of the Latin square and is therefore omitted.

Reprinted from THE ANNALS OF MATHEMATICAL STATISTICS
Vol. XIV, No. 3, September, 1943

ON STOCHASTIC LIMIT AND ORDER RELATIONSHIPS

BY H. B. MANN AND A. WALD[1]
Columbia University

1. Introduction. The concept of a stochastic limit is frequently used in statistical literature. Writers of papers on problems in statistics and probability usually prove only those special cases of more general theorems which are necessary for the solution of their particular problems. Thus readers of statistical papers are confronted with the necessity of laboriously ploughing through details, a task which is made more difficult by the fact that no uniform notation has as yet been introduced. It is therefore the purpose of the present paper to outline a systematic theory of stochastic limit and order relationships and at the same time to propose a convenient notation analogous to the notation of ordinary limit and order relationships. The theorems derived in this paper are of a more general nature and seem to contain to the authors' knowledge all previous results in the literature. For instance the so-called δ-method for the derivation of asymptotic standard deviations and limit distributions, also two lemmas by J. L. Doob [1] on products, sums and quotients of random variables and a theorem derived by W. G. Madow [2] are special cases of our results. It is hoped that such a general theory together with a convenient notation will considerably facilitate the derivation of theorems concerning stochastic limits and limit distributions. In section 2 we define the notion of convergence in probability and that of stochastic order and derive 5 theorems of a very general nature. Section 2 contains 2 corollaries of these general theorems which have so far been most important in applications.

We shall frequently need the concept of a vector. A vector $a = (a^1, \cdots, a^r)$ is an ordered set of r numbers $a^1, \cdots, a^r$. The numbers $a^1, \cdots, a^r$ are called the components of a. If the components are random variables then the vector is called a random vector.

We shall generally denote by a, b constant vectors by x, y random vectors and by $a^1, \cdots, a^r, x^1, \cdots, x^r$ their components. Differing from the usual practice we shall put $|a| = (|a^1|, \cdots, |a^r|)$ and we shall write $a < b$ or $a \leq b$ if $a^i < b^i$ or $a^i \leq b^i$ for every i. This notation saves a great amount of writing, since all our theorems except theorem 4 are valid for sequences of any number of jointly distributed variates.

We shall review here the ordinary order notation. In all that follows let $f(N)$ be a positive function defined for all positive integers N.

[1] Research under a grant-in-aid from the Carnegie Corporation of New York.

We write

$a_N = o[f(N)]$ if $\lim_{N\to\infty} a_N/f(N) = 0$.

$a_N = O[f(N)]$ if $|a_N| \leq Mf(N)$ for all N and a fixed $M > 0$.

$a_N = \Omega[f(N)]$ if $0 < M'f(N) \leq |a_N| \leq Mf(N)$ for almost all N and for two fixed numbers $M > M' > 0$.

$a_N = \omega[f(N)]$ if $0 < Mf(N) \leq |a_N|$ for almost all N and a fixed $M > 0$.

For instance, $\log N = o(N^\epsilon)$ for every $\epsilon > 0$, or $\sin N/N = O(1/N)$, $3 + 4\cdot N/(4 + 8\sqrt{N}) = \Omega(\sqrt{N})$ $5/\sin N = \omega(1)$.

For any statement V we shall denote by $P(V)$ the probability that V holds.

2. General theorems on stochastic limit and order relationships.

DEFINITION 1. *We write* $\operatorname{plim}_{N\to\infty} x_N = 0$. (*In words x_N converges in probability to* 0 *with increasing* N) *if for every* $\epsilon > 0$ $\lim P(|x_N| \leq \epsilon) = 1$. *Further* $\operatorname{plim}_{N\to\infty} x_N = x$ *if* $\operatorname{plim}_{N\to\infty} (x_N - x) = 0$.

DEFINITION 2. *We write* $x_N = o_p[f(N)]$ (x_N *is of probability order* $o[f(N)]$) *if* $\operatorname{plim}_{N\to\infty} x_N/f(N) = 0$.

DEFINITION 3. *We write* $x_N = O_p[f(N)]$ (x_N *is of probability order* $O[f(N)]$) *if for each* $\epsilon > 0$ *there exists an* $A_\epsilon > 0$ *such that* $P(|x_N| \leq A_\epsilon f(N)) \geq 1 - \epsilon$ *for all values of* N.

DEFINITION 4. $x_N = \Omega_p[f(N)]$ *if for each* $\epsilon > 0$ *there exist two numbers* $A_\epsilon > 0$ *and* $B_\epsilon > 0$ *and an integer* N_ϵ *such that* $P[A_\epsilon f(N) \leq |x_N| \leq B_\epsilon f(N)] \geq 1 - \epsilon$ *for all* $N \geq N_\epsilon$.

DEFINITION 5. $x_N = \omega_p[f(N)]$ *if for every* $\epsilon > 0$ *there exists an* $A_\epsilon > 0$ *and an integer* N_ϵ *such that* $P[A_\epsilon f(N) < |x_N|] \geq 1 - \epsilon$ *for all* $N \geq N_\epsilon$.

Let E denote a vector space. For any subset E' of E the symbol $a \subset E'$ will mean that a is an element of E'.

Since $P(x \subset E_1 \,\&\, x \subset E_2) \geq P(x \subset E_1) - P(x \not\subset E_2)$ we evidently have

LEMMA 1. *If* $P(x \subset E_1) \geq 1 - \epsilon$, $P(x \subset E_2) \geq 1 - \epsilon'$, *then* $P(x \subset E_1;\ x \subset E_2) \geq 1 - \epsilon - \epsilon'$.

We now put $O^1 = o$, $O^2 = O$, $O^3 = \Omega$, $O^4 = \omega$.

THEOREM 1. *For every* $\epsilon > 0$ *let* $\{R_N(\epsilon)\}$ *be a sequence of subsets of the r-dimensional Cartesian space such that* $P(x_N \subset R_N(\epsilon)) \geq 1 - \epsilon$ *for all* N *greater than a certain integer* N_ϵ. *Let* $\{g_N(x)\}$ *be a sequence of functions of* $x = (x^1, x^2, \cdots x^r)$ *such that* $g_N(a_N) = O^i[f(N)]$ *for any* $\epsilon > 0$ *and for any sequence* $\{a_N\}$ *for which* $a_N \subset R_N(\epsilon)$. *Then we have* $g_N(x_N) = O_p^i[f(N)]$.

PROOF: For $i = 1, 2, 3$, there exists a positive integer $\bar{N}_\epsilon$ such that $|g_N(a)|$ is a bounded function of a in $R_N(\epsilon)$ for $N > \bar{N}_\epsilon$. For otherwise we could construct a sequence $\{a_N\}$ with a_N in $R_N(\epsilon)$ such that $|g_N(a_N)| > Mf(N)$ for any M and for infinitely many values of N which contradicts the hypothesis of our theorem. Hence there exists an $\bar{N}_\epsilon$ such that for $N > \bar{N}_\epsilon$ the function $|g_N(a)|$ is bounded in $R_N(\epsilon)$. Let $M_N(\epsilon)$ be the l.u.b. of $|g_N(a)|/f(N)$ in $R_N(\epsilon)$. We can construct a sequence $\{a_N\}$ with $a_N \subset R_N(\epsilon)$ such that $|g_N(a_N)|/f(N) \geq M_N(\epsilon)/2$ for all $N > \bar{N}_\epsilon$. Hence for $i = 2, 3$ the sequence $M_N(\epsilon)$ must be bounded and for

$i = 1$ we must have $\lim_{N\to\infty} M_N(\epsilon) = 0$. Let $M(\epsilon)$ be the l.u.b. of $M_N(\epsilon)$. For $i = 3, 4$ one shows in exactly the same manner the existence of a g.l.b. $\bar{M}(\epsilon)$ of $|g_N(a)|/f(N)$ if $a \subset R_N(\epsilon)$ and for $N > N'_\epsilon$. Hence for sufficiently large N we have

$$P[|g_N(x_N)| \leq M_N(\epsilon)f(N)] \geq 1 - \epsilon \text{ with } \lim_{N\to\infty} M_N(\epsilon) = 0 \text{ for } i = 1,$$

$$P[|g_N(x_N)| \leq M(\epsilon)f(N)] > 1 - \epsilon \qquad \text{for } i = 2,$$

$$P[\bar{M}(\epsilon)f(N) \leq |g_N(x_N)| \leq M(\epsilon)f(N)] \geq 1 - \epsilon \qquad \text{for } i = 3,$$

$$P[\bar{M}(\epsilon)f(N) \leq |g_N(x_N)|] \geq 1 - \epsilon \qquad \text{for } i = 4.$$

For $i = 2$ the existence of an $M'(\epsilon)$ such that $P[|g_N(x_N)| \leq M'(\epsilon)f(N)] \geq 1 - \epsilon$ for all N follows easily from this result. Hence our theorem is proved.

COROLLARY 1. *If $x_N^j = O_p^{ij}[f_j(N)]$ for $j = 1, 2, \cdots, r$ and $\{\bar{R}_N(\epsilon)\}$ is a sequence of subsets of the k-dimensional space $y^1, y^2, \cdots y^k$ such that $P[y_N \subset \bar{R}_N(\epsilon)] \geq 1 - \epsilon$ for sufficiently large N, and if $\{g_N(x^1, x^2, \cdots x^r, y^1, y^2, \cdots y^k,)\}$ is a sequence of functions of $x^1, x^2, \cdots x^r, y^1, y^2, \cdots y^k$ such that for any $\epsilon > 0$ we have $g_N(a_N, b_N) = O^i[f(N)]$ for every sequence $\{a_N, b_N\}$ with $a_N^j = O^{ij}[f_j(N)]$ $(j = 1, 2, \cdots, r)$ and $b_N \subset \bar{R}_N(\epsilon)$, then $g_N(x_N, y_N) = O_p^i[f(N)]$.*

PROOF: It follows from Lemma 1, the definition of the relation $x_N^j = O_p^{ij}[f_j(N)]$ and the hypothesis of our corollary that for any $\epsilon > 0$ there exists a sequence of subsets $\{R_N(\epsilon)\}$ of the space $x^1, \cdots, x^r, y^1, \cdots, y^k$ which satisfies the conditions of Theorem 1 with respect to the sequence of functions $\{g_N\}$. Hence Corollary 1 is an immediate consequence of Theorem 1.

Corollary 1 implies *inter alia* that all operational rules for the ordinary order and limit relations are also applicable to stochastic limit and order relations. For instance $o[f(N)]/\Omega[g(N)] = o[f(N)/g(N)]$. Hence also $o_p[f(N)]/\Omega_p[g(N)] = o_p[f(N)/g(N)]$.

DEFINITION 6. *For any N let R_N be a region, $f_N(a)$ a function defined on R_N. The sequence $\{f_N(a)\}$ will be said to be uniformly continuous with respect to $\{R_N\}$ if the following condition is fulfilled. For every $\epsilon > 0$ there exists a vector $\delta > 0$ such that for almost all N*

$$|f_N(a + \bar{\delta}) - f_N(a)| \leq \epsilon \qquad \textit{for any } |\bar{\delta}| < \delta, \textit{ and for any } a \subset R_N$$

THEOREM 2. *Let $\operatorname{plim}_{N\to\infty} (x_N - y_N) = 0$. For every $\epsilon > 0$ let $\{R_N(\epsilon)\}$ be a sequence of subsets of the r-dimensional vector space such that for almost all N we have $P[y_N \subset R_N(\epsilon)] \geq 1 - \epsilon$. If the sequence of functions $\{f_N(a)\}$ is uniformly continuous with respect to $\{R_N(\epsilon)\}$ for every $\epsilon > 0$, then $\operatorname{plim}_{N\to\infty} [f_N(x_N) - f_N(y_N)] = 0$.*

PROOF: We have $f_N(x_N) - f_N(y_N) = f_N(y_N + z_N) - f_N(y_N)$ where $z_N^j = o(1)$ for $j = 1, \cdots, r$. Because of the uniform continuity of $f_N(a)$ with respect to $R_N(\epsilon)$ we see that for every sequence $\{a_N, b_N\}$ with $a_N \subset R_N(\epsilon)$ and $b_N^j = o(1)$ $(j = 1, 2, \cdots, r)$.

$$f_N(a_N + b_N) - f_N(a_N) = o(1).$$

Hence Theorem 2 follows from Corollary 1.

In the following we shall abbreviate "cumulative distribution function" by d.f.

DEFINITION 7. *Let $\{x_N\}$ be a sequence of random variables. Let F_N be the d.f. of x_N. Let x have the distribution F. We shall write $d\infty(x_N) = d(x)$ if $\lim_{N\to\infty} F_N = F$ in every continuity point of F.*

THEOREM 3. *Let $\operatorname{plim}_{N\to\infty} (x_N - y_N) = 0$ and $d\infty(y_N) = d(y)$; then $d\infty(x_N) = d(y)$.*

PROOF: Let G_N, F_N be the d.f.'s of x_N, y_N resp. For any $\delta > 0$ we have

$$P(y_N \leq a + \delta) \geq P(x_N \leq a; y_N \leq a + \delta) \geq P(x_N \leq a; | y_N - x_N | \leq \delta)$$
$$\geq P(x_N \leq a) - P(| y_N - x_N | > \delta),$$
$$P(x_N \leq a) \geq P(x_N \leq a; y_N \leq a - \delta) \geq P(y_N \leq a - \delta)$$
$$- P(| x_N - y_N | > \delta).$$

Hence since $P(y_N \leq a) = F_N(a)$, $P(x_N \leq a) = G_N(a)$, $\lim_{N\to\infty} P(| x_N - y_N | > \delta) = 0$ we have lim. sup. $F_N(a + \delta) \geq$ lim. sup. $G_N(a) \geq$ lim. inf. $G_N(a) \geq$ lim. inf. $F_N(a - \delta)$.

If $a + \delta$ and $a - \delta$ are continuity points of F we have

$$F(a + \delta) \geq \text{lim. sup. } G_N(a) \geq \text{lim. inf. } G_N(a) \geq F(a - \delta) .$$

For any $\delta_0 > 0$ there exists a positive $\delta < \delta_0$ such that $a - \delta$ and $a + \delta$ are continuity points of F. Hence we can choose δ arbitrarily small and if a is a continuity point of F we must have

$$\text{lim. } G_N(a) = F(a).$$

THEOREM 4. *Let x_N, y_N be two sequences of one-dimensional vectors and let $\operatorname{plim}_{N\to\infty} (x_N - y_N) = 0$. Let F_N, G_N be the cumulative distribution functions of x_N and y_N respectively. Let $R_N(\epsilon)$ be the set of points a for which $| F_N(a) - G_N(a) | > \epsilon$. Let $M_N(\epsilon)$ be the Lebesgue measure of this set. Then $\lim_{N\to\infty} M_N(\epsilon) = 0$ for every $\epsilon > 0$.*

We first prove the following lemma.

LEMMA 2. *Let δ, ϵ be any arbitrary positive numbers and let f be a distribution function. The set of points a for which $f(a + \delta) - f(a) \geq \epsilon$ has at most the Lebesgue measure δ/ϵ.*

PROOF: The points a for which $f(a + \delta) - f(a) \geq \epsilon$ must have a lower bound $\bar{a}$. Otherwise we could find infinitely many such points whose distance from each other is more than δ. But this contradicts the requirement that $f(\infty) = 1$. Let a_1 be the g.l.b. of the a's. Then for any $\eta > 0$ in the interval $(a_1 \leq x \leq a_1 + \delta + \eta)$ the value of F increases at least by the amount ϵ. Let now a_2 be the g.l.b. of the a's outside of this interval. We continue our construction by constructing the interval $(a_2 \leq x \leq a_2 + \delta + \eta)$ and so forth. But after at most

$1/\epsilon$ such steps the construction must stop. Hence all points a for which $f(a+\delta) - f(a) \geq \epsilon$ are contained in at most $1/\epsilon$ intervals of length $\delta + \eta$. Hence since η was arbitrary the Lebesgue measure of this set is at most δ/ϵ.

We come now to the proof of our theorem. We have

$$P(x_N \leq a) \geq P(x_N \leq a; y_N \leq a + \delta) \geq P(x_N \leq a) - P(|x_N - y_N| > \delta),$$

$$P(y_N \leq a + \delta) \geq P(x_N \leq a; y_N \leq a + \delta) \geq P(y_N \leq a + \delta)$$
$$-P(|x_N - y_N| > \delta) - P(a \leq x_N \leq a + 2\delta).$$

Therefore

$$P(x_N \leq a; y_N \leq a + \delta) = P(x_N \leq a) - \bar{\theta}_N P(|x_N - y_N| > \delta)$$
$$= P(y_N \leq a + \delta) - \bar{\theta}'_N P(|x_N - y_N| > \delta) - \bar{\theta}'_N P(a \leq x_N \leq a + 2\delta),$$

where $0 \leq \bar{\theta}_N \leq 1$, $0 \leq \bar{\theta}'_N \leq 1$. Hence

$$P(y_N \leq a + \delta) = P(x_N \leq a) + \theta_N P(|x_N - y_N| > \delta)$$
$$+ \theta'_N[F_N(a + 2\delta) - F_N(a)]$$

where $|\theta_N|, |\theta'_N| \leq 1$.

By hypothesis we have $P(|x_N - y_N| \geq 1/m) \leq 1/m$ for almost all N and every integer m. Hence we can choose a sequence $\{\delta_N\}$ with $\delta_N > 0$ in such a way that $\lim_{N\to\infty} \delta_N = 0$, $\lim_{N\to\infty} P(|x_N - y_N| \geq \delta_N) = 0$. We can then choose N_ϵ so that $P(|x_N - y_N| \geq \delta_N) \leq \epsilon/3$ for $N \geq N_\epsilon$. Applying Lemma 2 we see that except for a set of measure at most $6\,\delta_N/\epsilon$ we have $F_N(a + 2\delta_N) - F_N(a) \leq \epsilon/3$. Similarly the set of points for which $g_N(a + \delta_N) - g_N(a) \geq \epsilon/3$ has at most the Lebesgue measure $3\,\delta_N/\epsilon$. Hence, except in a set of points whose measure is at most $9\,\delta_N/\epsilon$, we have

$$|G_N(a) - F_N(a)| \leq \epsilon,$$

and this completes the proof of Theorem 4.

THEOREM 4a. *Let* $\operatorname{plim}_{N\to\infty} (x_N - y_N) = 0$. *Let* F_N, G_N *be the distribution functions of* x_N, y_N *respectively. Furthermore, let* $R_N(\epsilon)$ *be the set of points inside an r-dimensional cube where* $|F_N - G_N| \geq \epsilon$ *and let* $M_N(\epsilon)$ *be the Lebesgue measure of* $R_N(\epsilon)$, *then* $\lim_{N\to\infty} M_N(\epsilon) = 0$.

We prove first

LEMMA 2a. *Let* $\delta = (\delta^1, \delta^2, \cdots, \delta^r) > 0$ *and max.* $\delta^i = d$. *Let I be the cube defined by* $(-A \leq x^i \leq A, i = 1, 2, \cdots r)$. *Let furthermore f be a d.f. Then the Lebesgue measure of the points a in I for which* $f(a+\delta) - f(a) \geq \epsilon$ *is at most* $dr^2 A^{r-1}/\epsilon$.

PROOF: Let $f_1(x^1), f_2(x^2), \cdots f_r(x^r)$ be the marginal distributions of $x^1, x^2, \cdots x^r$ respectively. It follows from Lemma 2 that the linear Lebesgue measure of those numbers a^i for which $f_i(a^i + \delta^i) - f_i(a^i) \geq \epsilon/r$ is smaller than rd/ϵ. We form the set $(x^i = a^i \;\&\; x \subset I)$ for every such a^i and for $i = 1, 2, \cdots r$. The

Lebesgue measure of the sum $R(\epsilon)$ of all these sets is at most $r^2 dA^{r-1}/\epsilon$. We shall show that $R(\epsilon)$ contains all points a inside I for which $f(a + \delta) - f(a) \geq \epsilon$. We have

$$f(a^1 + \delta^1, a^2 + \delta^2, \cdots, a^r + \delta^r) - f(a^1, a^2, \cdots a^r) = \Delta_1 + \Delta_2 + \cdots + \Delta_r,$$

where $\Delta_i = f(a^1, a^2, \cdots a^{i-1}, a^i + \delta^i, \cdots a^r + \delta^r) - f(a^1, \cdots a^i, a^{i+1} + \delta^{i+1}, \cdots a^r + \delta^r)$. If $f(a + \delta) - f(a) \geq \epsilon$ then we must have for at least one i

$$\Delta_i \geq \epsilon/r.$$

But Δ_i is the probability of a subset of the set $T = (a^i \leq x^i \leq a^i + \delta^i)$ and $f_i(a^i + \delta^i) - f_i(a^i)$ is the probability of T itself. Hence

$$\epsilon/r \leq \Delta_i \leq f_i(a^i + \delta^i) - f_i(a^i),$$

and if $(a^1, a^2, \cdots a^r)$ is in I then it is contained in $R(\epsilon)$. Hence Lemma 2a is proved.

The proof of Theorem 4a using Lemma 2a is similar to that of Theorem 4 and therefore it is omitted.

The Jordan measure of a set R with respect to the distribution function F is defined as follows. We consider only intervals whose boundary points are continuity points of F. We cover R with the sum I of a finite number of intervals. (The intervals themselves may also be infinite. For instance the sets $a \leq x < \infty$, $a < x < \infty$ are also considered intervals.) We consider $M(I) = \int_I dF$ for every I covering R. The g.l.b. of all such $M(I)$ is called the exterior Jordan measure $M(R)$ of R. Similarly we consider all sums $\bar{I}$ of a finite number of intervals which are contained in R. The l.u.b. of $\int_{\bar{I}} dF$ is called the interior Jordan measure $\bar{M}(R)$ of R. If $M(R) = \bar{M}(R)$ then $M(R)$ is called the Jordan measure of R.

Lemma 3. *Let $F_N(x)$ be a sequence of d.f.'s such that $\lim_{N\to\infty} F_N(x) = F(x)$ in every continuity point of $F(x)$. Let $h(x)$ be a bounded function such that the discontinuity points of $h(x)$ have the Jordan measure 0 with respect to F and such that $\int_{-\infty}^{+\infty} h(x)\, dF_N(x)$ and $\int_{-\infty}^{+\infty} h(x)\, dF(x)$ exist. Then $\lim_{N\to\infty} \int_{-\infty}^{+\infty} h(x)\, dF_N(x) = \int_{-\infty}^{+\infty} h(x)\, dF(x)$.*

Proof: There is only an enumerable set of hyperplanes parallel to the plane $x^i = 0$ which have positive probability with respect to F. Hence we can find for every δ an interval net whose cells have a diameter at most δ and such that the boundary points of every cell are continuity points of F.

We first determine a closed finite interval I such that $\int_I dF(x) \geq 1 - \frac{\epsilon}{2}$ and such that the boundary points of I are continuity points of F. We further determine a sum I' of a finite number of open intervals such that I' contains all discontinuity points of h, $\int_{I'} dF(x) \leq \frac{\epsilon}{2}$, and such that the boundary of I' does not contain any discontinuity points of F. All this is possible by hypothesis and because the set of hyperplanes with positive probability is enumerable. Let R be the subset of I consisting of all points of I which are not contained in I'. R is a closed set and can be decomposed into a finite number of intervals. The function h is continuous in R and therefore uniformly continuous. We can therefore cover R by a finite set of intervals such that the variation of h in every interval is less than ϵ and such that the boundary points of each interval are continuity points of F. Let $I_1, I_2, \cdots I_k$ be such a finite set of intervals. Let x_j be any point in I_j. We have

$$\begin{aligned}
|H_N| = \left| \int_{-\infty}^{+\infty} h(x)\, dF_N(x) - \int_{-\infty}^{+\infty} h(x)\, dF(x) \right| &= \left| \sum_{j=1}^{k} \int_{I_j} [h(x) - h(x_j)]\, dF_N(x) \right. \\
&\quad - \sum_{j=1}^{k} \int_{I_j} [h(x) - h(x_j)]\, dF(x) + \sum_{j=1}^{k} h(x_j) \left[\int_{I_j} dF_N(x) - \int_{I_j} dF(x) \right] \\
&\quad \left. + \int_{x \notin R} h(x)\, dF_N(x) - \int_{x \notin R} h(x)\, dF(x) \right| \\
&\leq \epsilon + \epsilon + \sum_{j=1}^{k} h(x_j) \left[\int_{I_j} dF_N(x) - \int_{I_j} dF(x) \right] \\
&\quad + \max.\, h(x) \left[\int_{x \notin R} dF_N(x) + \epsilon \right].
\end{aligned}$$

But $\lim_{N \to \infty} \int_R dF_N(x) \geq 1 - \epsilon$. Hence

$$\text{lim. sup. } H_N \leq 2\epsilon + 2\epsilon \max.\, h(x)\,.$$

Since ϵ was arbitrary, we must have $\lim_{N \to \infty} H_N = 0$.

We are now prepared to prove

THEOREM 5. Let $d\infty(x_N) = d(x)$. *Let* $g(x)$ *be a Borel measurable function such that the set* R *of discontinuity points of* $g(x)$ *is closed and* $P(x \subset R) = 0$. *Then* $d\infty[g(x_N)] = d[g(x)]$.

PROOF: Let F_N be the d.f. of x_N, F the d.f. of x, F_{Ng}, F_g the d.f.'s of $g(x_N)$, $g(x)$ resp. Then $\lim_{N \to \infty} F_N = F$ in every cont. point of F. Let $h(x)$ be defined as follows:

$$h(x) = 1 \text{ if } g(x) \leq a\,,$$

$$h(x) = 0 \text{ if } g(x) > a\,.$$

The discontinuities of h are contained in the set M of all points where $g(x) = a$ and is continuous or where $g(x)$ is discontinuous. The set R of discontinuity points of $g(x)$ is closed and of measure 0 with respect to F. We can therefore subtract from M a sum R^* of a finite number of open intervals of arbitrarily small measure with respect to F which contains all discontinuity points of $g(x)$. This difference set M' is closed and contains only points where $g(x) = a$ and

$x \notin R$. If a is a continuity point of F_g then the Borel measure of M' with respect to F is 0. Since M' is closed, its Jordan measure is also 0. Hence the Jordan measure of the discontinuity points of $h(x)$ is 0 if a is a continuity point of F_g. Since $g(x)$ is Borel measurable, $\int_{-\infty}^{+\infty} h(x)\, dF_N(x) = F_{Ng}(a)$ and $\int_{-\infty}^{+\infty} h(x)\, dF(x) = F_g(a)$ exist for every a. Hence by Lemma 3 $\lim_{N\to\infty} F_{Ng}(a) = F_g(a)$ in every continuity point of F_g and this proves our theorem.

3. Corollaries and applications. Corollary 2.* *If* $\text{plim}_{N\to\infty} (x_N - y_N) = 0$, $d\infty(y_N) = d(y)$ *and if* f *is continuous except in a set* R *for which* $\lim_{N\to\infty} P(y_N \subset R) = 0$ *then* $\text{plim}_{N\to\infty} f(x_N) - f(y_N) = 0$.

Proof: Let I be a closed interval such that $P(y_N \subset I) \geq 1 - \epsilon/2$. Let I' be a sum of open intervals containing all discontinuity points of $f(x)$ in I and such that $P(y_N \subset I') \leq \epsilon/2$ for sufficiently large N. The set J of points of I which are not points of I' is a closed set. Hence f is uniformly continuous in J and $P(y_N \subset J) \geq 1 - \epsilon$ for sufficiently large N. In Theorem 2 we put $R_N(\epsilon) = J$, $f_N = f$. Then all conditions of Theorem 2 are satisfied and it follows that $\text{plim}_{N\to\infty} [f(x_N) - f(y_N)] = 0$.

If, moreover, the set of discontinuity points of f is closed then by Theorems 3 and 5 $d\infty[f(x_N)] = d\infty[f(y_N)] = d[f(y)]$.

Special cases of Corollary 2 have been proved by J. L. Doob and W. G. Madow (2).

Theorem 5 is very useful in deriving limit distributions.

It follows for instance from Theorem 5 that if $d\infty(x_N) = d(x)$, $d\infty(y_N) = d(y)$, where x, y are independently and normally distributed with mean 0 and equal variances, then $d\infty(x_N/y_N) = d(x/y)$. That is to say the distribution of x_N/y_N converges to a Cauchy distribution.

It also follows from Theorem 5 that under very general conditions the limit distribution of $t = \sqrt{N}(\bar{x} - \mu)/s$ is normal. ($\bar{x}$ = sample mean, μ = population mean, s^2 = sample variance.) For we have under very general conditions $d\infty \sqrt{N}(\bar{x} - \mu) = d(\xi)$, plim $s = \sigma$, where ξ is normally distributed with variance σ^2.

Applying Theorem 5 it can also easily be shown that under very general conditions the limit distribution of T^2 is a chi-square distribution if the means of all variates are 0. Hotelling's T^2 (the generalized Student ratio) for a p-variate distribution is defined as follows:

$$T^2 = N \sum_{i=1}^{p} \sum_{j=1}^{p} A_{ij} \xi_i \xi_j \quad \text{where} \quad \| A_{ij} \| = \| s_{ij} \|^{-1}, \qquad \xi_i = \bar{x}^i,$$

where s_{ij} is the sample covariance between x^i and x^j.

We have $d\infty(A_{ij}) = d(\sigma^{ij})$, where $\| \sigma_{ij} \|^{-1} = \| \sigma^{ij} \|$. If $E(x^i) = 0$ for $i = 1, 2, \cdots p$ then $d\infty(\sqrt{N}\, \xi_i) = d(\eta_i)$ where the η_i have a joint normal distribution

* This corollary is incorrect as it stands. The condition on f should read: "If for every $\epsilon > 0$ the discontinuity points of f can be embedded into an open set $R(\epsilon)$ such that $\lim P(Y_N \subset R(\epsilon)) \leq \epsilon$."

with covariance matrix $\| \sigma_{ij} \|$. Hence

$$d\infty(T^2) = d\left[\sum_{i=1}^{p}\sum_{j=1}^{p}\sigma^{ij}\eta_i\eta_j\right] = d\left(\sum_{i=1}^{p}\eta_i'^2\right),$$

where the η_i' are normally and independently distributed with variance 1. Hence the distribution of T^2 converges to a chi-square distribution with p degrees of freedom.

If the samples are drawn from a sequence of populations $\{\pi_N\}$ all with the same covariance matrix and such that $\lim_{N\to\infty} \sqrt{N}\mu_{iN} = \mu_i$ where μ_{iN} is the mean value of the ith variate in the Nth population, then one sees in exactly the same way that the limit distribution of T^2 is a non-central square distribution with p degrees of freedom.

The limit distribution of T^2 has been derived by W. G. Madow (2).

COROLLARY 3. *Let x_N, y_N be r-dimensional vectors $d\infty(y_N) = d(y)$ and $x_N - y_N = O_p[f(N)]$ with $\lim_{N\to\infty} f(N) = 0$. Let $g(x)$ be a function admitting continuous jth derivatives except in a set R with $\lim_{N\to\infty} P(y_N \subset R) = 0$. Let*

$$T_j(x, a) = \sum_{i=1}^{r}\left(\frac{\partial g}{\partial x^i}\right)_{x=a}(x^i - a^i) + \cdots + \left[\sum_{i=1}^{r}(x^i - a^i)\left(\frac{\partial}{\partial x^i}\right)_{x=a}\right]^j g,$$

then

$$g(x_N) - g(y_N) - T_j(x_N, y_N) = o_p\{[f(N)]^j\}.$$

Since the jth derivatives are continuous except in a set of limit measure 0 we can determine a closed set $R(\epsilon)$ on which they are uniformly continuous and so that $P(y_N \subset R(\epsilon)) \geq 1 - \epsilon$ for sufficiently large N. Then for every sequence with $a_N - b_N = O(f(N))$, $b_N \subset R(\epsilon)$ we have

$$g(a_N) - g(b_N) - T_j(a_N, b_N) = o[f(N)^j].$$

Hence Corollary 3 follows from Theorem 1.

Corollary 3 was first proved by W. G. Madow [2] and J. L. Doob [1] for the important case that y_N is a constant.

The following example will illustrate Corollary 3. Let x, y be normally and independently distributed random variables with mean 0 and variance 1; $\{z_N\}$, $\{z_N'\}$ sequences of random variables with $\operatorname{plim}_{N\to\infty} \sqrt{N}\, z_N = \operatorname{plim}_{N\to\infty} \sqrt{N}\, z_N' = 1$. Let $x_N = x + z_N$, $y_N = y + z_N'$. We consider the function $g(x, y) = x^3/3 + y^3/3 + 2x - 2y + 5$. Applying Corollary 1 it is easy to verify that $g(x_N, y_N) - g(x, y) = \Omega_p[1/\sqrt{N}]$, $z_N = O_p(1/\sqrt{N})$, $z_N' = O_p(1/\sqrt{N})$. Hence applying Corollary 3 for $j = 1$ we have

$$g(x_N, y_N) - g(x, y) - (x^2 + 2)z_N - (y^2 - 2)z_N' = o_p(1/\sqrt{N}).$$

Multiplying by $\sqrt{N}$ we have

$$[g(x_N, y_N) - g(x, y)]\sqrt{N} - [(x^2 + 2)z_N + (y^2 - 2)z_N']\sqrt{N} = o_p(1).$$

This is equivalent to

$$\operatorname*{plim}_{N\to\infty}[\sqrt{N}(g(x_N, y_N) - g(x, y))] = x^2 + y^2.$$

Hence the distribution of $\sqrt{N}(g(x_N, y_N) - g(x, y))$ converges to the chi-square distribution with 2 degrees of freedom.

If $\operatorname*{plim}_{N\to\infty} x_N = a$ and $\{\sigma_N\}$ is a sequence of numbers with $\lim_{N\to\infty} \sigma_N = 0$ such that $d\infty[(x_N^i - a^i)/\sigma_N] = d(\xi_i)$ where the ξ_i are constants or random variables and if g admits continuous first derivatives at $x = a$ at least one of which is different from 0, then putting $\left(\frac{\partial g}{\partial x^i}\right)_{x=a} = g_i$, we have

$$g(x_N) - g(a) = g_1(x_N^1 - a^1) + \cdots + g_r(x_N^r - a^r) + o_p(\sigma_N).$$

Hence applying Theorems 3 and 5 we have

$$\text{(i)} \qquad d\infty\left[\frac{g(x_N) - g(a)}{\sigma_N}\right] = d(g_1\xi_1 + \cdots + g_r\xi_r).$$

That is to say the distribution of $[g(x_N) - g(a)]/\sigma_N$ converges to the distribution of $\sum_{i=1}^{p} g_i\xi_i$ in all continuity points of the latter. A corresponding result can be obtained from Corollary 3 if all first derivatives are 0 at $x = a$ and at least one second derivative is different from 0 and so forth.

A method of deriving limiting distributions and limit standard deviations based on (i) is known as the δ-method and has been extensively applied in statistical literature.

REFERENCES

[1] J. L. Doob, "The limiting distribution of certain statistics." *Annals of Math. Stat.*, Vol. 6 (1935).

[2] W. G. Madow, "Limiting distribution of quadratic and bilinear forms." *Annals of Math. Stat.*, Vol. 11 (1940).

Reprinted from
ECONOMETRICA, Journal of the Econometric Society
Vol. 11, Nos. 3 and 4, July-October, 1943
The University of Chicago, Chicago, Illinois, U.S.A.

ON THE STATISTICAL TREATMENT OF LINEAR STOCHASTIC DIFFERENCE EQUATIONS

By H. B. MANN and A. WALD

INTRODUCTION

THE VALUES OF a great many variables important in the study of economics depend on the values previously assumed by these variables. For example present prices of a certain set of commodities may depend on the prices of these commodities in previous time periods, etc.

Such a relationship is usually described by a set of stochastic difference equations. Thus if $x_{1t}, \cdots, x_{rt}$ are the values of the variables $x_1, \cdots, x_r$ at the time t we should have equations of the following type:

$$\phi_i(x_{1t}, \cdots, x_{rt}, x_{1,t-1}, \cdots, x_{r,t-1}, \cdots, x_{1,t-p}, \cdots, x_{r,t-p}, \epsilon_{1t}, \cdots, \epsilon_{rt}) = 0 \qquad (i = 1, 2 \cdots, r),$$

where $\epsilon_{1t}, \cdots, \epsilon_{rt}$ are random variables representing the effect of errors and disturbing factors. In practice mostly linear functions ϕ_i are considered.

In the case $r=1$ we have a single equation of the form

$$x_t + \alpha_1 x_{t-1} + \cdots + \alpha_p x_{t-p} + \alpha_0 = \epsilon_t. \tag{1}$$

The ϵ_t's are usually assumed to be independently distributed random variables with mean 0 and to have the same distribution for each value of t.

Ordinarily some or all of the coefficients $\alpha_1, \cdots, \alpha_p, \alpha_0$ in equation (1) are unknown. Also the distribution of ϵ_t will usually involve a finite number of unknown parameters. Thus we have to deal with the following statistical problem: On the basis of the observed values of x_t for a finite number of time points t we want to set up statistical estimates for the unknown α's and for some of the unknown parameters of the distribution of ϵ_t. In addition to the problem of estimation we may also have the problem of predicting future values of x_t on the basis of previously observed values. The problems of estimation and prediction are closely connected. In fact, if a_i are functions of the observations which are used as statistical estimates of α_i and if the distributions of a_i and ϵ_t are known, then the future values $x_{t+1}, x_{t+2}, \cdots, x_{t+j}$ may be predicted on the basis of equation (1) by substituting a_i for α_1 and zero for ϵ_t. The distribution of these predicted values can be derived from the distribution of the a_i and ϵ_t. Hence the fundamental problem is to derive estimates of the α_i and then to study the joint distribution of these estimates.

In estimating the values of unknown parameters the method of maximum likelihood is widely used and has been shown to have certain optimum properties at least in the case when the observations are independent of each other. This method consists of the following procedure: Let $f(x, \theta_1, \cdots, \theta_p)$ be the elementary probability law of the random variable x. Assume that a sample of N independent observations $x_1, x_2, \cdots, x_N$ is drawn. Then the joint elementary probability law of $x_1, x_2, \cdots, x_N$ is given by $f(x_1, \theta_1, \cdots, \theta_p)f(x_2, \theta_1, \cdots, \theta_p) \cdots f(x_N, \theta_1, \cdots, \theta_p) = F(x_1, x_2, \cdots, x_N, \theta_1, \cdots, \theta_p)$. The values $\bar{\theta}_1, \cdots, \bar{\theta}_p$ of the parameters $\theta_1, \cdots, \theta_p$ for which F is maximized are called the maximum-likelihood estimates of $\theta_1, \cdots, \theta_p$. J. L. Doob [1][1] proved that under certain restrictive conditions on the elementary probability law $f(x, \theta_1, \cdots, \theta_p)$ the maximum-likelihood estimate is consistent and its limit distribution normal.

However, in the case of a linear stochastic difference equation we cannot use these results of the general theory of the maximum-likelihood estimate, since the observations are not independent of each other. Nevertheless, by formally applying the maximum-likelihood method we shall obtain certain functions of the observations which will be shown to be consistent estimates of the α's. We shall also show that these estimates are in the limit normally distributed.

The authors believe that the maximum-likelihood estimates of the α's have the same optimum properties that they are known to possess in the case of independent observations. However, no rigorous proof of this statement has as yet been given.

In the case of a single equation we shall be led, as we shall see later, to the ordinary least-squares procedure. However, the least-squares procedure is not generally applicable to the case of a system of equations with several variables; it may lead to inconsistent estimates. T. Haavelmo [2] pointed out this fact in his extensive discussion of the statistical problems connected with systems of stochastic difference equations. The problem of the exact distribution of the maximum-likelihood estimate of the coefficient of a single linear stochastic difference equation with $p=1$ under the assumption that the population value of this coefficient is zero has been treated by R. L. Anderson [3], T. Koopmans [4], and others. It appears from T. Koopmans' paper that this distribution is difficult to derive and cumbersome to handle.

[1] Numbers in brackets refer to references listed at end of this article.

In this paper we deal with the general problem of estimating the coefficients of linear stochastic difference equations but confine ourselves to the distribution of the maximum-likelihood estimates obtained from large samples. It is shown that the maximum-likelihood estimates are consistent and their limit distributions normal.

We shall give now some definitions and notations which will be used throughout this paper.

For any statement A we shall denote by $P(A)$ the probability that A is true.

Let $\{x_N\}$ be a sequence of random variables. We shall write $\text{plim}_{N\to\infty} x_N = a$ to mean that the probability limit of x_N equals a, i.e., that $\lim_{N\to\infty} P(|x_N - a| \leqq \epsilon) = 1$ for every positive ϵ.

Let $t_N(x_1, \cdots, x_N)$ be an estimate of the parameter θ computed from N observations $x_1, \cdots, x_N$. We call t_N a consistent estimate of θ if $\text{plim}_{N\to\infty} t_N = \theta$.

For any random variable x the symbol $E(x)$ will denote the expected value of x, and $\sigma(x)$ the standard deviation of x.

PART I

THE CASE OF A SINGLE EQUATION

1. *Assumptions Underlying the Statistical Analysis*

Let x_t be a variable depending on t where t can take only integral values.

Assumption I. x_t *satisfies the equation*

$$x_t = \alpha_1 x_{t-1} + \alpha_2 x_{t-2} + \cdots + \alpha_p x_{t-p} + \alpha_0 + \epsilon_t, \tag{1'}$$

where the α*'s are known or unknown constants, the random variables* $\epsilon_1, \epsilon_2, \cdots,$ *ad inf. are independently distributed, each having the same distribution, and* $E(\epsilon_t) = 0$.

Assumption II. All roots of the equation in ρ

$$\rho^p - \alpha_1 \rho^{p-1} \cdots - \alpha_p = 0 \tag{2}$$

are <1 *in absolute value.*

Since by assumption 1 is not a root of (2), we must have

$$1 - \alpha_1 - \cdots - \alpha_p \neq 0. \tag{3}$$

Assumption II implies, as we shall show, that $\lim_{t\to\infty} E(x_t)$ and $\lim_{t\to\infty} E(x_t^2)$ exist and are finite. On the other hand if $\lim_{t\to\infty} E(x_t)$ and $\lim_{t\to\infty} E(x_t^2)$ exist, then the roots of equation (2) are smaller than 1 in absolute value.

Assumption III. All moments of the ϵ_t *exist and are finite.*

2. *The Maximum-Likelihood Estimates of the α's*

If the variables ϵ_t are normally distributed with mean 0 and variance σ^2 then the joint probability distribution of $\epsilon_1, \cdots, \epsilon_N$ is given by

$$\frac{1}{(2\pi)^{N/2}\sigma^N}\exp\left(-\frac{1}{2\sigma^2}\sum_{t=1}^{N}\epsilon_t^2\right)d\epsilon_1\cdots d\epsilon_N$$

$$(4)\qquad = \frac{1}{(2\pi)^{N/2}\sigma^N}$$

$$\exp\left(-\frac{1}{2\sigma^2}\sum_{t=1}^{N}(x_t-\alpha_1x_{t-1}-\cdots-\alpha_px_{t-p}-\alpha_0)^2\right)d\epsilon_1\cdots d\epsilon_N.$$

We have $dx_t/d\epsilon_t=1$, $dx_{t-i}/d\epsilon_t=0$ for $i>0$. Hence it follows from (4) that the joint probability density of $x_1, \cdots, x_N$ is given by

$$(4')\qquad \frac{1}{(2\pi)^{N/2}\sigma^N}\exp\left[-\frac{1}{2\sigma^2}\sum_{t=1}^{N}(x_t-\alpha_1x_{t-1}-\cdots-\alpha_px_{t-p}-\alpha_0)^2\right]dx_1\cdots dx_N.$$

To obtain the maximum-likelihood estimates of $\alpha_1, \cdots, \alpha_p, \alpha_0, \sigma^2$ we have to maximize (4′) or its logarithm with respect to $\alpha_1, \cdots, \alpha_p, \alpha_0, \sigma^2$. This leads to the following equations:

$$\sum_{t=1}^{N}x_tx_{t-j}-\sum_{i=1}^{p}a_i\sum_{t=1}^{N}x_{t-i}x_{t-j}-a_0\sum_{t=1}^{N}x_{t-j}=0\quad(j=1,\cdots,p),$$

$$(5)\qquad \sum_{t=1}^{N}x_t-\sum_{i=1}^{p}a_i\sum_{t=1}^{N}x_{t-i}-Na_0=0,$$

$$s^2=\frac{1}{N}\sum_{t=1}^{N}(x_t-a_1x_{t-1}-\cdots-a_px_{t-p}-a_0)^2,$$

where $a_1, \cdots, a_p, a_0, s^2$ are the maximum-likelihood estimates of $\alpha_1, \cdots, \alpha_p, \alpha_0, \sigma^2$ respectively.[2]

In deriving the large-sample distributions of these maximum-likelihood estimates we shall not assume normality of the distribution of ϵ_t. If the quantities $x_1, \cdots, x_N$ were independently distributed, then it would follow from known results that our estimates of the $\alpha_i, \alpha_0, \sigma^2$ are consistent. Since $x_1, \cdots, x_N$ are obviously not independ-

[2] We shall refer to the solutions of (5) as maximum-likelihood estimates also in the case when the ϵ_t are not normally distributed. The phrase maximum-likelihood estimate is then to be considered only as an abbreviation and the reader should not be misled by it into thinking that we assume a normal distribution of the ϵ_t.

ent, we have to prove this property of our estimates. We shall prove even more. We shall prove that $\sqrt{N}(a_i-\alpha_i)$, $(i=0, 1, \cdots, p)$, are in the limit normally distributed with 0 mean and finite covariance matrix. The essential point in this proof is the fact that under our assumptions the stochastic limits $\text{plim}_{N\to\infty} \sum_{t=1}^{N} x_{t-i}x_{t-j}/N$ and $\text{plim}_{N\to\infty} \sum_{t=1}^{N} x_{t-i}/N$ exist and are finite. We shall prove this point in the next section.

3. *Proof of the Existence of* $\text{plim}_{N\to\infty} \sum_{t=1}^{N} x_{t-i}x_{t-j}/N$ *and* $\text{plim}_{N\to\infty}\sum_{t=1}^{N} x_{t-i}/N$

From (1) it follows that x_t is a linear function of $\epsilon_\tau(\tau=1, \cdots, t)$. Hence we may write

$$x_t = \phi_0(t) + \phi_1(t)\epsilon_1 + \cdots + \phi_t(t)\epsilon_t. \tag{6}$$

But then we have from (1)

$$\begin{aligned}\phi_0(t) + \phi_1(t)\epsilon_1 + \cdots + \phi_t(t)\epsilon_t \\ = \alpha_0 + \alpha_1[\phi_0(t-1) + \phi_1(t-1)\epsilon_1 + \cdots + \phi_{t-1}(t-1)\epsilon_{t-1}] \\ + \alpha_2[\phi_0(t-2) + \cdots + \phi_{t-2}(t-2)\epsilon_{t-2}] \\ + \alpha_p[\phi_0(t-p) + \cdots + \phi_{t-p}(t-p)\epsilon_{t-p}] \\ + \epsilon_t.\end{aligned}$$

This is an identity in $\epsilon_\tau(\tau=1, \cdots, t)$ and therefore we must have

$$\begin{aligned}\phi_0(t) &= \alpha_0 + \alpha_1\phi_0(t-1) + \cdots + \alpha_p\phi_0(t-p), \\ \phi_\tau(t) &= \alpha_1\phi_\tau(t-1) + \cdots + \alpha_p\phi_\tau(t-p) \quad (\tau = 1, \cdots, t-p), \\ \phi_\tau(t) &= \alpha_1\phi_\tau(t-1) + \cdots + \alpha_i\phi_\tau(t-i) \\ &\qquad (\tau = t-i; i = 1, \cdots, p-1), \\ \phi_t(t) &= 1, \\ \phi_t(t-h) &= 0 \qquad (h > 0).\end{aligned}$$

This gives a system of difference equations for $\phi_\tau(t)$ and $\phi_0(t)$:

$$\begin{aligned}\phi_0(t) &= \alpha_0 + \alpha_1\phi_0(t-1) + \cdots + \alpha_p\phi_0(t-p) \qquad (t > 0), \\ \phi_\tau(t) &= \alpha_1\phi_\tau(t-1) + \cdots + \alpha_p\phi_\tau(t-p) \qquad (t > \tau > 0),\end{aligned}$$

satisfying the initial conditions

$$\begin{aligned}&\phi_\tau(\tau) = 1,\ \phi_\tau(\tau-1) = 0, \cdots, \phi_\tau(\tau-p+1) = 0, \\ &\phi_0(0) = x_0,\ \phi_0(-1) = x_{-1}, \cdots, \phi_0(-p+1) = x_{-p+1}.\end{aligned} \tag{7}$$

Solving first the homogeneous difference equations we obtain

$$\phi_\tau(t) = P_{1\tau}(t)\rho_1{}^t + \cdots + P_{l\tau}(t)\rho_l{}^t \qquad (\tau > 0), \tag{8}$$

where $\rho_1, \cdots, \rho_l$ are the roots of the equation

$$\rho^p = \alpha_1 \rho^{p-1} + \cdots + \alpha_{p-1}\rho + \alpha_p$$

and $P_{i\tau}(t)$ is a polynomial in t whose degree is one less than the multiplicity of the root ρ_i. Since according to (3) $1-\alpha_1- \cdots -\alpha_p \neq 0$, the solution of the nonhomogeneous equation is given by

$$\phi_0(t) = P_1(t)\rho_1^t + \cdots + P_l(t)\rho_l^t + c, \tag{9}$$

where

$$c = \alpha/(1 - \alpha_1 - \cdots - \alpha_p).$$

The expressions $P_{1\tau}(t), \cdots, P_{l\tau}(t)$ can easily be determined from the initial conditions (7). In fact we have the following initial conditions:

$$1 = \sum_{i=1}^{l} P_{i\tau}(\tau)\rho_i^\tau,$$

$$0 = \sum_{i=1}^{l} P_{i\tau}(\tau - j)\rho_i^{\tau-j} \quad (j = 1, \cdots, p-1; \tau = 1, 2, \cdots, \text{ad inf.}).$$

It is known that the polynomials $P_{i\tau}(t)$ are uniquely determined by the initial conditions and by the condition that the degree of $P_{i\tau}(t)$ is one less than the multiplicity of the root ρ_i. Let

$$P_{i\tau}^*(t) = P_{i1}(t - \tau + 1)/\rho_i^{\tau-1} \qquad (\tau = 2, 3, \cdots).$$

Obviously the degree of $P_{i\tau}^*(t)$ is the same as that of $P_{i1}(t)$ and consequently is one less than the multiplicity of the root ρ_i. By substituting $P_{i\tau}^*(t)$ for $P_{i\tau}(t)$ in the initial conditions, it can easily be verified that the initial conditions are satisfied. Since the initial conditions determine uniquely the polynomials $P_{i\tau}(t)$ we must have

$$P_{i\tau}(t) = P_{i\tau}^*(t) = P_{i1}(t - \tau + 1)/\rho_i^{\tau-1} \qquad (\tau = 2, 3, \cdots).$$

For the sake of simplicity we shall carry our proofs out only for the case when all roots are different. The proof in the more general case is completely analogous.

We then have

$$\phi_0(t) = \sum_{i=1}^{p} \mu_i \rho_i^t + c, \tag{10}$$

$$\phi_\tau(t) = \sum_{i=1}^{p} \lambda_i \rho_i^{t-\tau+1}, \tag{11}$$

where μ_i, λ_i are constants.

Let δ_{ij} denote the Kronecker δ; i.e., $\delta_{ij}=0$ if $i \neq j$ and $\delta_{ij}=1$ if $i=j$. Since the covariance $\sigma_{\epsilon_\tau \epsilon_{\tau'}}$ is equal to $\delta_{\tau\tau'}\,\sigma^2$ where σ^2 is the variance of ϵ_τ we obtain from (6), (10), and (11)

(12)
$$\begin{aligned} E\left(\sum_{t=1}^{N} x_t x_{t-i}/N\right) &= \frac{\sigma^2}{N}\sum_{t=1}^{N}\sum_{\tau=1}^{t}\phi_\tau(t)\phi_\tau(t-i) \\ &\quad + \frac{1}{N}\sum_{t=1}^{N}\phi_0(t)\phi_0(t-i) \\ &= \frac{\sigma^2}{N}\sum_{t=1}^{N}\sum_{\tau=1}^{t-i}\sum_{\alpha=1}^{p}\sum_{\beta=1}^{p}\lambda_\alpha\lambda_\beta\rho_\alpha{}^{t-\tau+1}\rho_\beta{}^{t-\tau-i+1} \\ &\quad + \frac{1}{N}\sum_{t=1}^{N}\sum_{\alpha=1}^{p}\sum_{\beta=1}^{p}\mu_\alpha\mu_\beta\rho_\alpha{}^t\rho_\beta{}^{t-i} \\ &\quad + \frac{c}{N}\left(\sum_{t=1}^{N}\sum_{\alpha=1}^{p}\mu_\alpha\rho_\alpha{}^t + \sum_{t=1}^{N}\sum_{\alpha=1}^{p}\mu_\alpha\rho_\alpha{}^{t-i}\right) \\ &\quad + c^2. \end{aligned}$$

Since $|\rho_\alpha| < 1$ $(\alpha = 1, \cdots, p)$ we certainly have

(13)
$$\begin{aligned} &\lim_{N\to\infty}\frac{c}{N}\left[\sum_{t=1}^{N}\sum_{\alpha=1}^{p}(\mu_\alpha\rho_\alpha{}^t + \mu_\alpha\rho_\alpha{}^{t-i})\right] = 0, \\ &\lim_{N\to\infty}\frac{1}{N}\sum_{t=1}^{N}\sum_{\alpha=1}^{p}\sum_{\beta=1}^{p}\mu_\alpha\mu_\beta\rho_\alpha{}^t\rho_\beta{}^{t-i} = 0. \end{aligned}$$

Furthermore it is clear that

(14)
$$\begin{aligned} \lim_{t\to\infty}\sum_{\tau=1}^{t-i}\lambda_\alpha\lambda_\beta\rho_\alpha{}^{t-\tau+1}\rho_\beta{}^{t-\tau-i+1} &= \lambda_\alpha\lambda_\beta\rho_\alpha\rho_\beta{}^{-i+1}\sum_{m=i}^{\infty}(\rho_\alpha\rho_\beta)^m \\ &= \lambda_\alpha\lambda_\beta\rho_\alpha{}^{i+1}\rho_\beta/(1-\rho_\alpha\rho_\beta). \end{aligned}$$

From (14) it follows that

(15)
$$\begin{aligned} \lim_{N\to\infty}\frac{\sigma^2}{N}\sum_{t=1}^{N}\sum_{\tau=1}^{t-i}\sum_{\alpha=1}^{p}\sum_{\beta=1}^{p}\lambda_\alpha\lambda_\beta\rho_\alpha{}^{t-\tau+1}\rho_\beta{}^{t-\tau-i+1} \\ = \sigma^2\sum_{\alpha=1}^{p}\sum_{\beta=1}^{p}\lambda_\alpha\lambda_\beta\rho_\alpha{}^{i+1}\rho_\beta/(1-\rho_\alpha\rho_\beta). \end{aligned}$$

From (12), (13), and (15) we obtain

(16)
$$\lim_{N\to\infty}\sum_{t=1}^{N}x_t x_{t-i}/N = \sigma^2\sum_{\alpha=1}^{p}\sum_{\beta=1}^{p}\lambda_\alpha\lambda_\beta\rho_\alpha{}^{i+1}\rho_\beta/(1-\rho_\alpha\rho_\beta) + c^2.$$

In a similar way it can be shown that

(16′)
$$\lim_{t\to\infty}E(x_t x_{t-i}) = \sigma^2\sum_{\alpha=1}^{p}\sum_{\beta=1}^{p}\lambda_\alpha\lambda_\beta\rho_\alpha{}^{i+1}\rho_\beta/(1-\rho_\alpha\rho_\beta) + c^2.$$

Now we shall show that the variance of $\sum_{t=1}^{N} x_t x_{t-i}/N$ is of order $1/N$. We have

$$\sum_{t=1}^{N} x_t x_{t-i} = \sum_{t=1}^{N} [x_t - \phi_0(t)][x_{t-i} - \phi_0(t-i)] + \sum_{t=1}^{N} x_t \phi_0(t-i) + \sum_{t=1}^{N} \phi_0(t) x_{t-i} - \sum_{t=1}^{N} \phi_0(t)\phi_0(t-i). \tag{17}$$

We shall first show that the variance of $\sum_{t=1}^{N} [x_t - \phi_0(t)][x_{t-1} - \phi_0(t-i)]/N$ is of order $1/N$.

From (6) it follows that

$$E\left[\sum_{t=1}^{N} [x_t - \phi_0(t)][x_{t-i} - \phi_0(t-i)]\right]^2 / N^2$$

$$= \frac{1}{N^2} E\left[\sum_{t=1}^{N} \sum_{\tau=1}^{t} \sum_{\tau'=1}^{t-i} \phi_\tau(t)\phi_{\tau'}(t-i)\epsilon_\tau\epsilon_{\tau'}\right]^2$$

$$= \frac{1}{N^2} \sum_{t=1}^{N} \sum_{\tau=1}^{t} \sum_{\tau'=1}^{t-i} \sum_{t'=1}^{N} \sum_{\tau''=1}^{t'} \sum_{\tau'''=1}^{t'-i} \phi_\tau(t)\phi_{\tau'}(t-i)\phi_{\tau''}(t')\phi_{\tau'''}(t'-i) \cdot E(\epsilon_\tau\epsilon_{\tau'}\epsilon_{\tau''}\epsilon_{\tau'''}).$$

But $E(\epsilon_\tau\epsilon_{\tau'}\epsilon_{\tau''}\epsilon_{\tau'''})$ is different from 0 only if either $\tau=\tau'$ and $\tau''=\tau'''$ or $\tau=\tau''$ and $\tau'=\tau'''$, or $\tau=\tau'''$ and $\tau'=\tau''$. Moreover $E(\epsilon_\tau\epsilon_{\tau'}\epsilon_{\tau''}\epsilon_{\tau'''})$ is equal to the fourth moment μ_4 of ϵ_t if $\tau=\tau'=\tau''=\tau'''$ and is equal to σ^4 or 0 if two of the ϵ's are different. Hence we obtain

$$E\left\{\sum_{t=1}^{N} [x_t - \phi_0(t)][x_{t-i} - \phi_0(t-i)]/N\right\}^2$$

$$= \frac{1}{N^2}\Bigg[\sigma^4 \sum_{t=1}^{N} \sum_{\tau=1}^{t-i} \sum_{t'=1}^{N} \sum_{\tau'=1}^{t'-i} \phi_\tau(t)\phi_\tau(t-i)\phi_{\tau'}(t')\phi_{\tau'}(t'-i)$$

$$+ \sigma^4 \sum_{t=1}^{N} \sum_{t'=1}^{N} \sum_{\tau=1}^{\min(t,t')} \sum_{\tau'=1}^{\min(t,t'-i)} \phi_\tau(t)\phi_{\tau'}(t-i)\phi_\tau(t')\phi_{\tau'}(t'-i)$$

$$+ \sigma^4 \sum_{t=1}^{N} \sum_{t'=1}^{N} \sum_{\tau=1}^{\min(t,t'-i)} \sum_{\tau'=1}^{\min(t-i,t')} \phi_\tau(t)\phi_{\tau'}(t-i)\phi_{\tau'}(t')\phi_\tau(t'-i)$$

$$+ (\mu_4 - 3\sigma^4) \sum_{t=1}^{N} \sum_{t'=1}^{N} \sum_{\tau=1}^{\min(t-i,t'-i)} \phi_\tau(t)\phi_\tau(t-i)\phi_\tau(t')\phi_\tau(t'-i)\Bigg].$$

Since

$$\left[E\left\{\sum_{t=1}^{N}[x_t - \phi_0(t)][x_{t-i} - \phi_0(t-i)/N]\right\}\right]^2$$
$$= \frac{\sigma^4}{N^2}\sum_{t=1}^{N}\sum_{t'=1}^{N}\sum_{\tau=1}^{t-i}\sum_{\tau'=1}^{t'-i}\phi_\tau(t)\phi_\tau(t-i)\phi_{\tau'}(t')\phi_{\tau'}(t'-i),$$

we have

$$\sigma^2\left\{\sum_{t=1}^{N}[x_t - \phi_0(t)][x_{t-i} - \phi_0(t-i)]/N\right\} \tag{18}$$
$$= \frac{1}{N^2}[(\mu_4 - 3\sigma^4)d + \sigma^4(b+c)],$$

where

$$d = \sum_{t=1}^{N}\sum_{t'=1}^{N}\sum_{\tau=1}^{\min(t-i,t'-i)}\phi_\tau(t)\phi_\tau(t-i)\phi_\tau(t')\phi_\tau(t'-i), \tag{19}$$

$$b = \sum_{t=1}^{N}\sum_{t'=1}^{N}\sum_{\tau=1}^{\min(t,t')}\sum_{\tau'=1}^{\min(t,t'-i)}\phi_\tau(t)\phi_{\tau'}(t-i)\phi_\tau(t')\phi_{\tau'}(t'-i), \tag{20}$$

$$c = \sum_{t=1}^{N}\sum_{\tau'=1}^{N}\sum_{\tau=1}^{\min(t,t'-i)}\sum_{\tau'=1}^{\min(t',t-i)}\phi_\tau(t)\phi_{\tau'}(t-i)\phi_{\tau'}(t')\phi_\tau(t'-i). \tag{21}$$

We have

$$|\phi_\tau(t)| = \left|\sum_{\alpha=1}^{p}\lambda_\alpha\rho_\alpha^{t-\tau+1}\right| \leqq M\rho^{t-\tau+1}$$

where

$$M = \sum_{\alpha=1}^{p}|\lambda_\alpha|, \qquad \rho = \max|\rho_\alpha|.$$

Hence

$$|d|, |b|, |c| \leqq \sum_{t=1}^{N}\sum_{t'=1}^{N}\sum_{\tau=1}^{\min(t,t')}\sum_{\tau'=1}^{\min(t,t')}\rho^{-2i}M^4\rho^{t+t'-2\tau+2}\rho^{t+t'-2\tau'+2}$$
$$\leqq \sum_{t=1}^{N}\sum_{t'=1}^{N}\frac{\rho^{-2i}}{(1-\rho^2)^2}M^4(\rho^2)^{t-\min(t,t')}(\rho^2)^{t'-\min(t,t')}$$
$$\leqq M'\sum_{t=1}^{N}\left[\sum_{t'=1}^{t}(\rho^2)^{t-t'} + \sum_{t'=t+1}^{N}(\rho^2)^{t'-t}\right]$$
$$\leqq M'N(1+\rho^2)/(1-\rho^2) \leqq M''N,$$

where M'' is independent of N. Hence

$$\sigma^2\left\{\sum_{t=1}^{N}[x_t - \phi_0(t)][x_{t-i} - \phi_0(t-i)]/N\right\} \leqq M'''N/N^2 = M'''/N,$$

where M''' is independent of N.

In analogous manner we can show that the variances of $\sum_{t=1}^{N} x_t\phi_0(t-i)/N$, $\sum_{t=1}^{N}\phi_0(t)x_{t-i}/N$ are of the order $1/N$. Hence from (17) we see that also the variance of $\sum_{t=1}^{N} x_t x_{t-i}/N$ is of the order $1/N$.

Since we have shown that $\lim E[\sum_{t=1}^{N} x_{t-i}x_{t-j}/N]$ exists and that the variance of $\sum_{t=1}^{N} x_{t-i}x_{t-j}/N$ is of order $1/N$, it follows from Tchebycheff's inequality that

$$P(|\textstyle\sum_{t=1}^{N} x_{t-i}x_{t-j}/N - E\sum_{t=1}^{N} x_{t-i}x_{t-j}/N| > \epsilon) \leqq M'''/\epsilon^2 N.$$

Hence if we put $\lim E[\sum_{t=1}^{N} x_{t-j}x_{t-j}/N] = D_{ij}$, we must have

$$\operatorname*{plim}_{N\to\infty} \sum_{t=1}^{N} x_{t-i}x_{t-j}/N = D_{ij}.$$

Now we shall prove that

$$\operatorname*{plim}_{N\to\infty} \sum_{t=1}^{N} x_t/N = \lim_{N\to\infty} E\left(\sum_{t=1}^{N} x_t/N\right) = \lim_{t\to\infty} E(x_t) = c. \tag{22}$$

Since

$$x_t = \phi_0(t) + \phi_1(t)\epsilon_1 + \cdots + \phi_t(t)\epsilon_t$$

we have that

$$\lim_{t\to\infty} E(x_t) = \lim_{t\to\infty} \phi_0(t) = c.$$

Furthermore we see that

$$\lim_{N\to\infty} E\left(\sum_{t=1}^{N} x_t/N\right) = \lim_{N\to\infty} \sum_{t=1}^{N} \phi_0(t)/N = c.$$

Hence in order to prove (22) we have merely to show that $\sigma^2(\sum x_t/N)$ is of order $1/N$. In fact we have

$$\sum_{t=1}^{N} x_t/N = \sum_{t=1}^{N} \phi_0(t)/N + \sum_{t=1}^{N} [\phi_1(t)/N]\epsilon_1 + \cdots + \sum_{t=1}^{N} [\phi_t(t)/N]\epsilon_t.$$

Hence

$$\sigma^2\left(\sum_{t=1}^{N} x_t/N\right) = \frac{\sigma^2}{N^2}\left\{\left[\sum_{t=1}^{N}\phi_1(t)\right]^2 + \cdots + \left[\sum_{t=1}^{N}\phi_t(t)\right]^2\right\}.$$

We have proved that

$$|\phi_\tau(t)| \leqq M\rho^{t-\tau+1}$$

where M is a fixed positive constant. Hence $\sum_{t=1}^{N}\phi_\tau(t)$ is a bounded function of τ and N. From this it follows that $\sigma^2(\sum_{t=1}^{N} x_t/N)$ is of order $1/N$. Hence (22) is proved.

4. *Proof of the Consistency of the Maximum-likelihood Estimates a_i, a_0, s^2*

We put

$$y_i = \frac{1}{\sqrt{N}} \sum_{t=1}^{N} (x_t - \alpha_1 x_{t-1} - \cdots - \alpha_p x_{t-p} - \alpha_0) x_{t-i},$$

$$y_0 = \frac{1}{\sqrt{N}} \sum_{t=1}^{N} (x_t - \alpha_1 x_{t-1} - \cdots - \alpha_p x_{t-p} - \alpha_0).$$

Then because of (5) we have

$$(23)\quad y_i = \sum_{j=1}^{p} \sqrt{N}(a_j - \alpha_j) \sum_{t=1}^{N} x_{t-i}x_{t-j}/N + \sqrt{N}(a_0 - \alpha_0) \sum_{t=1}^{N} x_{t-i}/N \qquad (i = 1, \cdots, p),$$

$$y_0 = \sum_{j=1}^{p} \sqrt{N}(a_j - \alpha_j) \sum_{t=1}^{N} x_{t-j}/N + \sqrt{N}(a_0 - \alpha_0).$$

But

$$(24)\quad y_i = \frac{1}{\sqrt{N}} \sum_{t=1}^{N} \epsilon_t x_{t-i}, \qquad y_0 = \frac{1}{\sqrt{N}} \sum_{t=1}^{N} \epsilon_t.$$

Since ϵ_τ is independent of x_{t-i} we can easily show that

$$E(y_i) = 0 \qquad (i = 0, 1, \cdots, p),$$

$$(25)\quad E(y_i y_j) = \sigma^2 E\left(\sum_{t=1}^{N} x_{t-i}x_{t-j}/N\right) \qquad (i, j = 1, \cdots, p),$$

$$E(y_i y_0) = \sigma^2 E\left(\sum_{t=1}^{N} x_{t-i}/N\right) \qquad (i = 1, \cdots, p),$$

$$E(y_0^2) = \sigma^2.$$

We put $\sum_{t=1}^{N} x_{t-i}x_{t-j}/N = D_{ijN}$, $\sum_{t=1}^{N} x_{t-i}/N = D_{i0N} = D_{0iN}$. Let furthermore $D_{00N} = 1$. Then equations (23) can be written

$$(26)\quad y_i = \sum_{j=0}^{p} \sqrt{N}(a_j - \alpha_j) D_{ijN} \qquad (i = 0, 1, \cdots, p).$$

Denoting the constant c by $D_{0i} = D_{i0}$ $(i = 1, \cdots, p)$ and putting D_{00} equal to 1, we have

$$(27)\quad \operatorname*{plim}_{N \to \infty} D_{ijN} = D_{ij} \qquad (i, j = 0, 1, \cdots, p).$$

Now we shall show that the determinant value D of the matrix $\|D_{ij}\|(i, j=0, 1, \cdots, p)$ is different from 0. Consider the expected value

$$E[(b_1x_{t-1} + \cdots + b_px_{t-p} + b_0)^2], \tag{28}$$

where $b_1, \cdots, b_p, b_0$ are arbitrary real values subject to the restriction $\sum_{i=0}^{p} b_i^2 \neq 0$. Since for any t' the conditional variance of $x_{t'}$, given $x_{t'-1}, \cdots, x_{t'-p}$, is equal to $\sigma^2 > 0$, we see that the expected value (28), as well as the limit value of (28) when $t \to \infty$, must be positive for arbitrary values $b_0, \cdots, b_p$ subject to the restriction $\sum_{i=0}^{p} b_i^2 \neq 0$. But this is possible only if the matrix $\|D_{ij}\|(i, j, = 0, 1, \cdots, p)$ is positive definite. Hence we have proved that $D > 0$.

From (25) we see that $E(y_i) = 0$ and that the covariance matrix of the random variables $y_0, y_1 \cdots, y_p$ is a bounded function of N. Hence from (26), (27), and the fact that $D > 0$ it follows easily that lim $E\sqrt{N}(a_j - \alpha_j) = 0$, $(j = 0, 1, \cdots, p)$ and that the covariance matrix of the variates $\sqrt{N}(a_0 - \alpha_0)$, $\sqrt{N}(a_1 - \alpha_1), \cdots, \sqrt{N}(a_p - \alpha_p)$ is a bounded function of N. Thus we must have

$$\text{plim}\ a_j = \alpha_j \qquad (j = 0, 1, \cdots, p).$$

We now proceed to prove the consistency of the estimate s^2 of σ^2. We have

$$\begin{aligned} s^2 &= \frac{1}{N} \sum_{t=1}^{N} (x_t - a_1x_{t-1} - \cdots - a_px_{t-p} - a_0)^2 \\ &= \frac{1}{N}\left(\sum_{t=1}^{N} x_t^2 - a_1 \sum_{t=1}^{N} x_tx_{t-1} - \cdots - a_p \sum_{t=1}^{N} x_tx_{t-p} - a_0 \sum_{t=1}^{N} x_t \right) \\ &\quad - \frac{1}{N} \sum_{i=1}^{p} a_i \left(\sum_{t=1}^{N} x_tx_{t-i} - a_1 \sum_{t=1}^{N} x_{t-1}x_{t-i} - \cdots \right. \\ &\quad \left. - a_p \sum_{t=1}^{N} x_{t-p}x_{t-i} - a_0 \sum_{t=1}^{N} x_{t-i} \right) \\ &\quad - \frac{1}{N} a_0 \left(\sum_{t=1}^{N} x_t - a_1 \sum_{t=1}^{N} x_{t-1} - \cdots - a_p \sum_{t=1}^{N} x_{t-p} - Na_0 \right). \end{aligned} \tag{29}$$

Hence on account of (5) we have

$$\begin{aligned} s^2 = \sum_{t=1}^{N} x_t^2/N - a_1 \sum_{t=1}^{N} x_tx_{t-1}/N - \cdots \\ - a_p \sum_{t=1}^{N} x_tx_{t-p}/N - a_0 \sum_{t=1}^{N} x_t/N. \end{aligned} \tag{30}$$

Since the expressions $\sum_{t=1}^{N} x_t x_{t-j}/N$ and $\sum_{t=1}^{N} x_t/N$ converge stochastically to their limit expectations and since $\text{plim}_{N\to\infty} a_i = \alpha_i$, we see that the stochastic limit of the right-hand side of (30) is not changed if we replace a_i by $\alpha_i (i=0, 1, \cdots, p)$. Since the right-hand side of (30) is identically equal to

$$(31) \qquad \frac{1}{N} \sum_{t=1}^{N} (x_t - a_1 x_{t-1} - \cdots - a_p x_{t-p} - a_0)^2,$$

the stochastic limit of (31) is not changed if we replace a_i by $\alpha_i (i=0, 1, \cdots, p)$. Then from (29) it follows that

$$\begin{aligned} \text{plim}_{N\to\infty} s^2 &= \text{plim}_{N\to\infty} \frac{1}{N} \sum_{t=1}^{N} (x_t - \alpha_1 x_{t-1} - \cdots - \alpha_p x_{t-p} - \alpha_0)^2 \\ &= \text{plim}_{N\to\infty} \frac{1}{N} \sum_{t=1}^{N} \epsilon_t^2 = \sigma^2. \end{aligned}$$

5. *Large-Sample Distribution of the Maximum-Likelihood Estimates*

Now we shall derive the limit distribution of the maximum-likelihood estimates $a_i (i=0, 1, \cdots, p)$. For this purpose we shall show first that the variates $y_0, y_1, \cdots, y_p$ have a joint normal distribution in the limit. For proving the latter statement we need the following lemma.

Lemma 1. *Let* $\{x_n\}$ *be a sequence of random variables such that* $\lim_{n\to\infty} P(x_n \leqq t) = F(t)$. *Let* $\{z_n^i\}$ $(i=1, 2, \cdots, ad\ inf.)$ *be a sequence of sequences of random variables such that* $\lim_{n\to\infty} E(z_n^i)^2 = \epsilon_i$ *and* $\lim_{i\to\infty} \epsilon_i = 0$. *Then for every* $\delta > 0$ *there exists an integer* $\bar{i}$ *such that*

$$F(t - \delta) - \delta \leqq \lim_{n\to\infty} P(x_n + z_n^i \leqq t) \leqq F(t + \delta) + \delta \qquad (i > \bar{i}).$$

Proof: We have

$$P(x_n + z_n^i \leqq t) \geqq P(x_n \leqq t - \delta \ \&\ z_n^i \leqq \delta) \geqq P(x_n \leqq t - \delta) - P(z_n^i \geqq \delta)$$

and

$$P(x_n \leqq t + \delta) \geqq P(x_n + z_n^i \leqq t \ \&\ z_n^i \geqq -\delta) \geqq P(x_n + z_n^i \leqq t) - P(z_n^i \leqq -\delta).$$

But by choosing i and n sufficiently large we can make the probabilities $P(z_n^i \geqq \delta)$ and $P(z_n^i \leqq -\delta)$ less than δ and obtain our lemma by letting n approach infinity.

We have by (24)

$$y_i = \sum_{t=1}^{N} \epsilon_t x_{t-i}/\sqrt{N} \quad (i = 1, \cdots, p), \qquad y_0 = \sum_{t=1}^{N} \epsilon_t/\sqrt{N}.$$

We put

$$x_t = x_t' + x_t'',$$

where

$$\begin{aligned} x_t' &= \epsilon_t + \phi_{t-1}(t)\epsilon_{t-1} + \cdots + \phi_{t'+1}(t)\epsilon_{t'+1}, \\ x_t'' &= \phi_{t'}(t)\epsilon_{t'} + \cdots + \phi_1(t)\epsilon_1, \qquad (32) \\ t' &= t - d \qquad (d > 0). \end{aligned}$$

We have

$$(33) \qquad |\phi_j(t)| = |\lambda_1\rho_1^{t-j+1} + \cdots + \lambda_p\rho_p^{t-j+1}| \leqq pM\rho^{t-j+1},$$

where $M = \max(|\lambda_1|, |\lambda_2|, \cdots, |\lambda_p|)$ and $\rho = \max(|\rho_1|, \cdots, |\rho_p|)$. Hence

$$\begin{aligned} E(x_t'')^2 &= \sigma^2 \sum_{i=1}^{t-d} \phi_i^2(t) \leqq p^2M^2\sigma^2(\rho^{2t} + \rho^{2(t-1)} + \cdots + \rho^{2(d+1)}) \qquad (34) \\ &\leqq \frac{p^2M^2\sigma^2\rho^{2(d+1)}}{1 - \rho^2}. \end{aligned}$$

We put

$$y_i = y_i' + y_i'',$$

where

$$y_i' = \sum_{t=1}^{N} x_{t-i}'\epsilon_t/\sqrt{N}, \quad y_i'' = \sum_{t=1}^{N} x_{t-i}''\epsilon_t/\sqrt{N} \qquad (i = 1, \cdots, p).$$

Then

$$\sigma^2(y_i'') = \sum_{t=1}^{N} E(x_{t-i}'')^2/N\sigma^2 \leqq p^2M^2\sigma^2(\rho^2)^{d+1}/(1 - \rho^2).$$

By choosing d sufficiently large we can make $\sigma^2(y_i'') = E(y_i'')^2$ as small as we wish. But then, by Lemma 1, the distribution of y_i' differs arbitrarily little from the distribution of y_i. Hence if we can prove that y_i' is in the limit normally distributed for any d, then also y_i must be in the limit normally distributed.

We shall show now that y_i' is in the limit normally distributed.

We put $d+i+1=d^*$ and let n be an integer $>2d^*$.

Suppose that $N=kn+m$ where k is a positive integer and $0\leqq m<n$. Denote $\sum_{t=jn+1}^{(j+1)n-d^*-1} x_t{}'_{-i}\epsilon_t$ by b_{j+1} and $\sum_{(j+1)n-d^*}^{(j+1)n} x_t{}'_{-i}\epsilon_t$ by a_{j+1} $(j=0, \cdots, k-1)$. Then we have

$$\sqrt{N}\, y_i{}' = \sum_{j=1}^{k} (b_j + a_j) + \sum_{kn+1}^{kn+m} x_t{}'_{-i}\epsilon_t.$$

Evidently for any fixed n the limit distribution of $(1/\sqrt{N}) \sum_{j=1}^{k}(b_j+a_j)$ when $N\rightarrow\infty$ is the same as the distribution of y_i. Hence we have merely to show that $(1/N)\sum_{j=1}^{k}(b_j+a_j)$ is in the limit normally distributed. It can be seen that for any value τ the variate ϵ_τ does not occur in both b_i and b_j if $i\neq j$. Hence $b_1, \cdots, b_k$ are independently distributed. Similarly we see that $a_1, \cdots, a_k$ are independent. We have

$$(35) \qquad \sigma^2(a_j) = \sigma^2 \sum_{jn-d^*}^{jn} E(x'_{t-i})^2, \qquad \sigma^2(b_j) = \sigma^2 \sum_{(j-1)n+1}^{jn-d^*-1} E(x'_{t-i})^2.$$

As we have shown $\lim E(x_{t-i})^2$ exists. Since according to (34) $E(x_{t-i})^2$ differs from $E(x'_{t-i})^2$ by an arbitrarily small quantity we have that

$$(36) \qquad \sigma^2(a_j)/\sigma^2(b_j) \leqq M(d^* + 1)/n,$$

where M is a fixed constant independent of j, d^*, and n. By choosing n sufficiently large we can make this ratio arbitrarily small. Since $a_1, \cdots, a_k$ are independent of each other and also $b_1, \cdots, b_k$ are independent of each other, it follows from (36) that

$$(37) \qquad \frac{\sigma^2\left(\sum_{j=1}^{k} a_j\right)}{\sigma^2\left(\sum_{j=1}^{k} b_j\right)} = \sum_{j=1}^{k} \sigma^2(a_j)/\sum_{j=1}^{k} \sigma^2(b_j) \leqq M(d^* + 1)/n.$$

It can be seen from (35) that $\sum_{j=1}^{k}\sigma^2(b_j)/N$ is a bounded function of N. Hence it follows from Lemma 1 that the limit distribution of $(1/\sqrt{N})\sum_{j=1}^{k}(a_j+b_j)$ is the same as the limit distribution of $(1/\sqrt{N}) \sum_{j=1}^{k}b_j$. Thus the limit distribution of y_i is the same as the limit distribution of $(1/\sqrt{N})\sum b_j$. From the definition of b_j it follows that b_j is a quadratic function of the ϵ_t's, the number of terms and the coefficients being bounded functions of j. Hence since ϵ_t has finite moments by assumption, the third absolute moment of b_j is a bounded function of j. Furthermore $\sigma^2(b_j)$ converges to a finite positive limit with $j\rightarrow\infty$. The b_j's therefore fulfil the Liapounoff conditions and the distribution of $\sum_{j=1}^{k}b_j/\sqrt{N}$ approaches a normal distribution. Hence we have proved that y_i is in the limit normally distributed.

The proof carried through for each y_i can be carried through in exactly the same manner for every linear combination.

$$y = \gamma_1 y_1 + \cdots + \gamma_k y_k + \gamma_0 y_0.$$

Since the moments of ϵ_t are finite by assumption, we see that the moments of any linear combination $\sum_{i=0}^{p} \gamma_i y_i$ converge with $N \to \infty$ to the moments of a normal distribution. The fact that $y_0, y_1, \cdots, y_p$ have in the limit a joint normal distribution follows from the following lemma.

Lemma 2. Let $Y_1^{(n)}, \cdots, Y_k^{(n)}$ be k sequences of random variables such that for any set of coefficients $\gamma_1, \cdots, \gamma_k$ for which $\sum \gamma_i^2 \neq 0$ the moments of $\gamma_1 Y_1^{(n)} + \cdots + \gamma_k Y_k^{(n)}$ converge with $n \to \infty$ to the moments of a normal distribution with 0 mean and finite variance. Then the joint limiting distribution of the variates $Y_1^{(n)}, \cdots, Y_k^{(n)}$ is a multivariate normal distribution.

This lemma can be proved as follows: There exists a linear transformation

$$\bar{Y}_i^{(n)} = \sum_{j=1}^{k} \gamma_{ij} Y_j^{(n)} \qquad (i = 1, \cdots, k)$$

such that the covariance matrix of $\bar{Y}_1^{(n)}, \cdots, \bar{Y}_k^{(n)}$ converges with $n \to \infty$ to the unit matrix. Then it follows from the assumptions of our lemma that the moments of

$$\gamma_1 \bar{Y}_1^{(n)} + \cdots + \gamma_k \bar{Y}_k^{(n)} / \sqrt{\gamma_1^2 + \cdots + \gamma_k^2}$$

converge to the moments of a normal distribution with zero mean and unit variance for arbitrary values $\gamma_1, \cdots, \gamma_k$ subject to the restrictions $\sum \gamma_j^2 \neq 0$. Hence for any positive integer r we have

$$\lim_{n \to \infty} E\left(\sum_{i=1}^{k} \gamma_i \bar{Y}_i^{(n)} / \sqrt{\gamma_1^2 + \cdots + \gamma_k^2} \right)^r$$

$$(38) \qquad = \lim_{n \to \infty} \left[\frac{1}{\sqrt{\sum_{i=1}^{k} \gamma_i^2}} \right]^r \sum_{l_1 + \cdots + l_k = r} r!/l_1! l_2! \cdots l_k! \gamma_1^{l_1} \cdots \gamma_k^{l_k} \cdot E[(\bar{Y}_1^{(n)})^{l_1} \cdots (\bar{Y}_k^{(n)})^{l_k}] = \mu_r$$

where μ_r denotes the rth moment of the normal distribution with 0 mean and unit variance.

Since (38) must be fulfilled identically in $\gamma_1, \cdots, \gamma_k$ we must have

$$\lim_{n \to \infty} E\{(\bar{Y}_1^{(n)})^{l_1} \cdots (\bar{Y}_k^{(n)})^{l_k}\} = \mu_{l_1} \mu_{l_2} \cdots \mu_{l_k}.$$

Hence the variates $\bar{Y}_1^{(n)}, \cdots, \bar{Y}_k^{(n)}$ are in the limit normally and independently distributed. From this it follows that the joint limiting distribution of $Y_1^{(n)}, \cdots, Y_k^{(n)}$ is a multivariate normal distribution and Lemma 2 is proved.

Thus we have arrived at the important result that $y_0, \cdots, y_p$ have in the limit a joint normal distribution with zero means and finite covariance matrix.

From (25) we see that the covariance matrix of the variates $y_0, y_1, \cdots, y_p$ is given by

$$\|\sigma(y_i y_j)\| = \sigma^2 \|E(D_{ijN})\| \qquad (i, j = 0, 1, \cdots, p). \tag{39}$$

Since $\operatorname{plim}_{N\to\infty} s^2 = \sigma^2$ and since $\operatorname{plim}_{N\to\infty}[D_{ijN} - E(D_{ijN})] = 0$ we have that

$$\operatorname*{plim}_{N\to\infty} [\sigma(y_i y_j) - s^2 D_{ijN}] = 0. \tag{40}$$

Let $\|c_{ijN}\|$ be equal to $\|D_{ijN}\|^{-1}$. Then we obtain from (26)

$$\sqrt{N}(a_i - \alpha_i) = \sum_{j=0}^{p} c_{ijN} y_j \qquad (i = 0, \cdots, p). \tag{41}$$

Let $\|c_{ij}\| = \|D_{ij}\|^{-1}$. Since $\operatorname{plim}_{N\to\infty} D_{ijN} = D_{ij}$ and since the matrix $\|D_{ij}\|$ is positive definite, we see that the limit distribution of $\sum_{j=0}^{p} c_{ijN} y_j$ is the same as the limit distribution of $\sum_{j=0}^{p} c_{ij} y_j$. From this and (41) it follows that the joint limiting distribution of $\sqrt{N}(a_j - \alpha_j)(j = 0, 1, \cdots, p)$ is a multivariate normal distribution with 0 means and finite covariance matrix.

Denote $\sqrt{N}(a_i - \alpha_i)$ by $\xi_i (i = 0, \cdots, p)$. We shall calculate the limiting covariance matrix of the variates $\xi_0, \cdots, \xi_p$.

Since $\operatorname{plim}_{N\to\infty} c_{ijN} = c_{ij}$, it follows from (41) that

$$\begin{aligned} \operatorname*{plim}_{N\to\infty} \sigma^2(\xi_i \xi_j) &= \operatorname*{plim}_{N\to\infty} \sum_{l=0}^{p} \sum_{k=0}^{p} c_{ik} c_{jl} \sigma(y_i y_j) \\ &= \sum_{l=0}^{p} \sum_{k=0}^{p} c_{ik} c_{jl} D_{kl} \sigma^2. \end{aligned} \tag{42}$$

Since $\|c_{ij}\| = \|D_{ij}\|^{-1}$ and since $D_{ij} = D_{ji}$ we have that

$$\sum_{l=0}^{p} \sum_{k=0}^{p} c_{ik} c_{jl} D_{kl} = c_{ij}.$$

Hence we obtain from (42)

$$\operatorname*{plim}_{N\to\infty} \sigma(\xi_i \xi_j) = c_{ij}\sigma^2 = \operatorname*{plim}_{N\to\infty} s^2 c_{ijN}. \tag{43}$$

Our results can be summarized in the following

Theorem. Let $\xi_i = \sqrt{N}(a_i - \alpha_i)$ $(i = 0, \cdots, p)$ *where* a_i *is the maximum-likelihood estimate of* α_i. *Furthermore let* $D_{ijN} = \sum_{t=1}^{N} x_{t-i}x_{t-j}/N \cdot (i, j = 1, \cdots, p)$, $D_{i0N} = D_{0iN} = \sum_{t=1}^{N} x_{t-i}/N$ $(i = 1, \cdots, p)$, $D_{00N} = 1$, *and* $\|c_{ijN}\| = \|D_{ijN}\|^{-1}(i, j = 0, \cdots, p)$. *Finally let* s^2 *be the maximum-likelihood estimate of* σ^2. *The joint limiting distribution of* $\xi_0, \xi_1, \cdots, \xi_p$ *is a multivariate normal distribution with* 0 *means and finite covariance matrix. The limiting value of the covariance between* ξ_i *and* ξ_j *is equal to the stochastic limit of* $s^2 c_{ijN}$.

For large N we can replace the covariance between ξ_i and ξ_j by the quantity $s^2 c_{ijN}$ which can be calculated from the observations. Thus we see that the joint limiting distribution of the estimates $a_i (i = 0, \cdots, p)$ is the same as the distribution we should obtain from the classical regression theory if x_t is considered as the dependent variable and $x_{t-1}, \cdots, x_{t-p}$ as the independent variables.

In other words a single stochastic difference equation in one variable can be treated for large N as a classical regression problem where x_t is the dependent variable and $x_{t-1}, \cdots, x_{t-p}$ are the independent variables.

6. *Estimation and Prediction*

On the basis of the joint limiting distribution of the maximum-likelihood estimates $a_0, a_1, \cdots, a_p$ we can carry out tests of significance or derive confidence regions for the unknown coefficients $\alpha_0, \cdots, \alpha_p$ provided that N is large. If we want to test the hypothesis that, for a given i, $\alpha_i = \alpha_i^0$ where α_i^0 is some specified value, we proceed as follows: We form the ratio

$$\sqrt{N}\,(a_i - \alpha_i^0)/s\sqrt{c_{iiN}}. \tag{44}$$

This ratio is in the limit normally distributed with zero mean and unit variance if the hypothesis to be tested is true. We decide to reject the hypothesis under consideration if the ratio (44) has an absolute value which exceeds the critical value d_0 corresponding to the chosen level of significance. If we choose 5 per cent as the level of significance then $d_0 = 1.96$.

A confidence interval for the unknown coefficient α_i is obtained as follows: Consider the inequality

$$\sqrt{N}\,(a_i - \alpha_i)/s\sqrt{c_{iiN}} \leqq d_0. \tag{45}$$

The set of all values α_i which are not rejected on the basis of the observations, i.e., the set of all values α_i which satisfy (45), forms a confidence interval for the unknown coefficient α_i.

Now suppose that we want to test the values of several coefficients α_i simultaneously. More precisely, we want to test the hypothesis that $\alpha_{1'} = \alpha_{1'}^0, \cdots, \alpha_{q'} = \alpha_{q'}^0$, where $1' < \cdots < q'$ are nonnegative integers

$\leqq p$ and $\alpha_{1'}^0, \cdots, \alpha_{q'}^0$ are some given specified values. The proper large-sample test procedure can be carried out as follows: We form the expression

$$T^2 = \frac{1}{s^2} \sum_{j=1}^{q} \sum_{i=1}^{q} N(a_{i'} - \alpha_{i'})(a_{j'} - \alpha_{j'}) \overset{*}{c}_{i'j'N}, \tag{46}$$

where

$$\|\overset{*}{c}_{i'j'N}\| = \|c_{i'j'N}\|^{-1} \qquad (i, j = 1, \cdots, q). \tag{47}$$

The expression (46) has in the limit the χ^2-distribution with q degrees of freedom. We reject the hypothesis under consideration if the expression (45) exceeds the critical value χ_0^2 of the χ^2-distribution corresponding to the chosen level of significance.

A confidence region for the unknown parameters $\alpha_{1'}, \cdots, \alpha_{q'}$ is given by the set of all points $(\alpha_{1'}^0, \cdots, \alpha_{q'}^0)$ which satisfy the inequality

$$T^2 \leqq \chi_0^2. \tag{48}$$

The region defined by the inequality (48) is an ellipsoid with the center at the point $(a_{1'}, \cdots, a_{q'})$.

If $q=p+1$ then $1'=0$, $2'=1, \cdots, q'=p$ and the expression (46) goes over into

$$\frac{1}{s^2} \sum_{j=0}^{p} \sum_{i=0}^{p} N(a_i - \alpha_i^0)(a_j - \alpha_j^0) D_{ijN}. \tag{49}$$

As a problem of prediction let us consider the following simple case: Suppose we want to predict the future value x_{t+1} on the basis of the past observed values $x_1, \cdots, x_t$. It is clear that the conditional expected value of x_{t+1} given $x_1, \cdots, x_t$ is equal to

$$E(x_{t+1}/x_1, \cdots, x_t) = \alpha_1 x_t + \alpha_2 x_{t-1} + \cdots + \alpha_p x_{t-p+1} + \alpha_0. \tag{50}$$

Hence the expression

$$X_{t+1} = a_1 x_t + a_2 x_{t-1} + \cdots + a_p x_{t-p+1} + a_0 \tag{51}$$

can be used as a point estimate of x_{t+1}. If N is large, the sampling variance of a_i can be neglected as compared with the variance σ^2 of ϵ_t. Hence for large N we can say that the distribution of

$$x_{t+1} - (a_1 x_t + a_2 x_{t-1} + \cdots + a_p x_{t-p+1} + a_0)$$

is the same distribution as that of ϵ_{t+1}. Thus we obtain confidence limits for x_{t+1} by setting confidence limits for ϵ_{t+1}. The problem of predicting $x_{t+h}(h>1)$ on the basis of $x_1, \cdots, x_t$ can be treated in a similar way provided that h is small. If h is large the sampling variation of the maximum-likelihood estimates $a_0, a_1 \cdots, a_p$ cannot be neglected in setting up confidence limits for x_{t+h}.

PART II

THE CASE OF SEVERAL EQUATIONS IN SEVERAL VARIABLES

1. *Assumptions Underlying the Statistical Analysis*

Let $x_{1t}, \cdots, x_{rt}$ be a set of r random variables depending on t where t can take only integral values.

Assumption I_2. The set of variables $x_{1t}, \cdots, x_{rt}$ satisfies the system of linear stochastic difference equations

$$\sum_{j=1}^{r} \sum_{k=0}^{p_{ij}} \alpha_{ijk} x_{j,t-k} + \alpha_i = \epsilon_{it} \qquad (i = 1, \cdots, r), \tag{52}$$

where p_{ij} is the highest lag in the variable x_{jt} occurring in the ith equation, the α's are known or unknown constants, the random vectors $(\epsilon_{11}, \cdots, \epsilon_{r1})$, $(\epsilon_{12}, \cdots, \epsilon_{r2}), \cdots$ ad inf. are independently distributed each having the same distribution and $E(\epsilon_{it}) = 0$ for all values of i and t.

Since the joint distribution of the variates $\epsilon_{1t}, \cdots, \epsilon_{rt}$ does not depend on t the covariance between ϵ_{it} and ϵ_{jt} also does not depend on t. We shall denote this covariance by σ_{ij}.

Assumption II_2. The determinant value $|\sigma_{ij}|$ of the matrix $\|\sigma_{ij}\|$ $(j, i = 1, \cdots, r)$ is not equal to 0.

Assumption III_2. All moments of ϵ_{it} are finite.

Assumption IV_2. Denote $\sum_{k=0}^{p_{ij}} \alpha_{ijk}\rho^{-k}$ by $b_{ij}(\rho)$ and let $|b_{ij}(\rho)|$ be the determinant value of the matrix $\|b_{ij}(\rho)\|$ $(i, j = 1, \cdots, r)$. It is assumed that all roots of the equation

$$|b_{ij}(\rho)| = 0 \tag{53}$$

are smaller than 1 in absolute value.

2. *Application of the Maximum-Likelihood Method*

If the variables $\epsilon_{1t}, \cdots, \epsilon_{rt}$ have a joint normal distribution then the joint distribution of the $\epsilon_{it} (i = 1, \cdots, r; t = 1, \cdots, N)$ is given by

$$\frac{C}{|\sigma_{ij}|^{N/2}} \exp\left(-\frac{1}{2} \sum_{t=1}^{N} \sum_{i=1}^{r} \sum_{j=1}^{r} \sigma^{ij} L_{it} L_{jt}\right) d\epsilon_{11} \cdots d\epsilon_{1t} \cdots d\epsilon_{r1} \cdots d\epsilon_{rt},$$

where C is a constant,

$$L_{it} = \epsilon_{it} = \sum_{j=1}^{r} \sum_{k=0}^{p_{ij}} \alpha_{ijk} x_{j,t-k} + \alpha_i,$$

and

$$\|\sigma^{ij}\| = \|\sigma_{ij}\|^{-1} \qquad (i, j = 1, \cdots, r).$$

We have $\partial\epsilon_{i\alpha}/\partial x_{j\alpha} = \alpha_{ij0}$, $\partial\epsilon_{i\alpha}/\partial x_{j\beta} = 0$ for $\beta > \alpha$. Hence the joint distribution of the $x_{it}(i=1, \cdots, r; t=1, \cdots, N)$ is given by

$$P = (C/|\sigma_{ij}|^{N/2})J^N \exp\left(-\frac{1}{2}\sum_{t=1}^{N}\sum_{i=1}^{r}\sum_{j=1}^{r}\sigma^{ij}L_{it}L_{jt}\right) \tag{54}$$
$$\cdot dx_{11}\cdots dx_{1N}\cdots dx_{r1}\cdots dx_{rN},$$

where $J = |\alpha_{ij0}|$.

Put $\sum_{t=1}^{N} L_{it}L_{jt} = Q_{ij}$, then

$$\log P = N \log J + \log C + (N/2) \log |\sigma^{ij}| - (1/2)\sum_{i=1}^{r}\sum_{j=1}^{r}\sigma^{ij}Q_{ij}.$$

Maximizing $\log P$ with respect to α_{ijk} and σ^{ij} we obtain the following system of equations:

$$(N/J)(\partial J/\partial\alpha_{ijk}) - \frac{1}{2}\sum_{l=1}^{r}\sigma^{il}\partial Q_{il}/\partial\alpha_{ijk} = 0 \tag{55}$$
$$(i = 1, \cdots, r; k = 1, \cdots, p_{ij}),$$
$$\|\sigma_{ij}\| = \|Q_{ij}/N\|.$$

Let us denote $(1/N)(\sum_{t=1}^{N} x_{i, t-k}x_{j,t-l})$ by D_{ikjlN}. Just as in the case of a single equation the essential point in our considerations will be the fact that

$$\operatorname*{plim}_{N\to\infty} (D_{ijklN} - E(D_{ikjlN})) = 0$$

and that $\lim_{N\to\infty} E(D_{ikjlN})$ exists and is finite. In the next section we shall outline a proof of this fact. This proof is very similar to the proof given in the case of one equation and we shall discuss only those parts of it in which the generalization from the case of a single equation is not immediately obvious.

3. *Proof of the Existence of* $\operatorname{plim}_{N\to\infty} \sum_{t=1}^{N} x_{i,t-k}x_{j,t-l}/N$

It follows from (52) that x_{it} is a linear function of $\epsilon_{j\tau}(j=1, \cdots, r; \tau=1, \cdots, t)$. Hence

$$x_{it} = \phi_i(t) + \sum_{\tau=1}^{t}\sum_{j=1}^{r}\phi_{ij\tau}(t)\epsilon_{j\tau}. \tag{56}$$

Substituting this in (52) we obtain

$$\sum_{j=1}^{r}\sum_{k=0}^{p_{ij}} \alpha_{ijk}\phi_j(t-k) + \sum_{j=1}^{r}\sum_{k=0}^{p_{ij}}\sum_{l=1}^{r}\sum_{\tau=1}^{t} \alpha_{ijk}\phi_{jl\tau}(t-k)\epsilon_{l\tau} + \alpha_i = \epsilon_{it}.$$

This is an identity in the ϵ_{jt}'s. Hence we must have

$$(57)\quad \begin{aligned} &\sum_{j=1}^{r}\sum_{k=0}^{p_{ij}} \alpha_{ijk}\phi_j(t-k) + \alpha_i = 0 && (i = 1, \cdots, r),\\ &\sum_{j=1}^{r}\sum_{k=0}^{p_{ij}} \alpha_{ijk}\phi_{jl\tau}(t-k) = 0 && (t > \tau;\ i,l = 1, \cdots, r),\\ &\sum_{j=1}^{r}\sum_{k=0}^{p_{ij}} \alpha_{ijk}\phi_{jit}(t-k) = 1, &&\\ &\sum_{j=1}^{r}\sum_{k=0}^{p_{ij}} \alpha_{ijk}\phi_{jlt}(t-k) = 0 && (l \neq i). \end{aligned}$$

Since according to (56)

$$\phi_{ij\tau}(t) = 0 \qquad (\tau > t;\ i, j = 1, \cdots, r),$$

we have that

$$(58)\qquad \phi_{jlt}(t-k) = 0 \qquad (k = 1, \cdots, p_{ij}).$$

From (58) and the last two equations in (57) we obtain

$$(59)\quad \begin{aligned} &\sum_{j=1}^{r} \alpha_{ij0}\phi_{jit}(t) = 1, &&\\ &\sum_{j=1}^{r} \alpha_{ij0}\phi_{jlt}(t) = 0 && (l \neq i). \end{aligned}$$

It is clear that functions $\phi_{ij\tau}(t)$ satisfying the equations (57) and (58) exist. For the sake of simplicity we shall assume that all roots of the equation (53) are distinct. It is known from general mathematical theory that the solution of the system of linear difference equations (57) must then be of the form:

$$\phi_{ij\tau}(t) = \lambda_{1j\tau}A_{1i}\rho_1{}^t + \cdots + \lambda_{\pi j\tau}A_{\pi i}\rho_\pi{}^t,$$
$$\phi_i(t) = \lambda_1 A_{1i}\rho_1{}^t + \cdots + \lambda_\pi A_{\pi i}\rho_\pi{}^t + c_i,$$

where the ρ_i are the solutions in ρ of the equation (53), the $A_{\beta i}$ are not all equal to zero and satisfy the systems of equations

$$\sum_{j=1}^{r} A_{\beta j}b_{ij}(\rho_\beta) = 0 \qquad (i = 1, \cdots, r),$$

the constants c_i are the solutions of the system of equations

$$\sum_{j=1}^{r} \sum_{k=0}^{p_{ij}} \alpha_{ijk} c_j = -\alpha_i \qquad (i = 1, \cdots, r).$$

(The determinant of the latter system is equal to $|b_{ij}(1)|$ which is unequal to 0 by Assumption IV_2) and the $\lambda_{\beta j\tau}$ and λ_β are constants. From the equations (57), (58), (59) it follows in a similar way to that for the case of a single equation that

$$\lambda_{\beta j\tau} = \lambda_{\beta j1}/\rho_\beta^{\tau-1}.$$

Writing $\lambda_{\beta j}$ instead of $\lambda_{\beta j1}$ we therefore have

$$\lambda_{\beta j\tau} = \lambda_{\beta j}/\rho_\beta^{\tau-1}.$$

Hence we have

$$\phi_{ij\tau}(t) = \lambda_{1j} A_{1i} \rho_1^{t-\tau+1} + \cdots + \lambda_{\pi j} A_{\pi i} \rho_\pi^{t-\tau+1}. \tag{60}$$

Using (60) we can show in complete analogy to the case of a single equation that $\text{plim}_{N\to\infty}[\sum_{t=1}^N x_{i,t-k}x_{j,t-l}/N - E(\sum_{t=1}^N x_{i,t-k}x_{j,t-l}/N)] = 0$ and that $\lim_{N\to\infty} E(\sum_{t=1}^N x_{i,t-k}x_{j,t-l}/N) = \lim_{t\to\infty} E(x_{i,t-k}x_{j,t-l})$. Denote this limit expected value by D_{ikjl}. Then we have

$$\underset{N\to\infty}{\text{plim}}\, D_{ikjlN} = D_{ikjl} \qquad (i, j = 1, \cdots, r). \tag{61}$$

Similarly it can be seen that

$$\underset{N\to\infty}{\text{plim}} \sum_{t=1}^{N} x_{it}/N = \lim_{N\to\infty} E\left(\sum_{t=1}^{N} x_{it}/N\right) = \lim_{t\to\infty} E(x_{it}) = c_i \qquad (i = 1, \cdots, r). \tag{62}$$

4. *Treatment of the Equation* (52) *when* $\|\alpha_{ij0}\|$ *Is Equal to the Unit Matrix and* $\sigma_{ij}=0$ *for* $i\neq j$

First we shall treat the equations (52) in the special case when $\|\alpha_{ij0}\|(i, j=1, \cdots, r)$ is equal to the unit matrix, i.e., $\alpha_{ij0}=0$ if $i\neq j$ and $\alpha_{ii0}=1$, and $\sigma_{ij}=0$ for $i\neq j$. The reason for discussing this special case separately is that it is an important one in applications and the results are very simple. The statistical treatment of such a system of equations can be reduced to the case of a single equation. In the special case under consideration the Jacobian is equal to 1 and the system of equations (55) reduces to

$$\partial Q_{ii}/\partial \alpha_{ijk} = 0, \qquad \sigma_{ii} = Q_{ii}/N \qquad (i = 1, \cdots, r). \tag{63}$$

Hence we can proceed exactly as in the case of a single equation and prove the consistency of the maximum-likelihood estimates a_{ijk}, a_i, and s_{ii} of α_{ijk}, α_i, and σ_{ii} respectively. Furthermore we can prove in a way analogous to the case of a single equation that the variates $\sqrt{N}(a_{ijk}-\alpha_{ijk})$ and $\sqrt{N}(a_i-\alpha_i)(i, j=1, \cdots, r)(k=1, \cdots, p_{ij})$ have in the limit a joint normal distribution with zero means and finite covariance matrix.

Now we shall derive the limit covariance matrix of the variates $\sqrt{N}(a_{ijk}-\alpha_{ijk})$, $\sqrt{N}(a_i-\alpha_i)(i, j=1, \cdots, r;\ k=1, \cdots, p_{ij})$. For the sake of convenience we introduce the following notations: The variable x_{0t} is identically equal to 1, i.e., for all values of t, x_{0t} is equal to 1. The symbols α_{i01} and a_{i01} will stand for α_i and a_i respectively. The value of p_{i0} is defined to be equal to 1. The symbols D_{ikjlN} and D_{ikjl} have been previously defined only for positive values of i and j. Now we shall extend this definition to include the cases when i or j or both are equal to zero. We put

$$(64)\qquad \begin{aligned} &\sum_{t=1}^{N} x_{i,t-k}x_{j,t-l}/N = D_{ikjlN}, \\ &\lim_{N\to\infty} E\left(\sum_{t=1}^{N} x_{i,t-k}x_{j,t-l}/N\right) = D_{ikjl} \qquad (i, j = 0, \cdots, r). \end{aligned}$$

It is clear that

$$(65)\qquad D_{0k0lN} = D_{0k0l} = 1.$$

From (62) it follows that

$$(66)\qquad D_{ik0l} = D_{0lik} = c_i \qquad (i = 1, \cdots, r).$$

We put

$$(67)\qquad \sqrt{N}\, y_{ijk} = \tfrac{1}{2}\partial Q_{ii}/\partial\alpha_{ijk} = \sum_{t=1}^{N} \epsilon_{it}x_{j,t-k}$$
$$(i = 1, \cdots, r;\ j = 0, \cdots, r;\ k = 1, \cdots, p_{ij}).$$

According to (63) we have

$$(68)\qquad \sum_{t=1}^{N} x_{j,t-k}\left(x_{it} + \sum_{l=1}^{p_{im}}\sum_{m=0}^{r} a_{iml}x_{m,t-l}\right) = 0$$
$$(i = 1, \cdots, r;\ j = 0, \cdots, r).$$

From (67) and (68) we obtain

$$(69)\qquad \sqrt{N}\, y_{ijk} = \sum_{t=1}^{N} x_{j,t-k}\left[\sum_{l=1}^{p_{im}}\sum_{m=0}^{r} (a_{iml} - \alpha_{iml})x_{m,t-l}\right].$$

Let ξ_{ijk} be equal to $\sqrt{N}(a_{ijk}-\alpha_{ijk})$. Then we obtain from (69)

(70)
$$\begin{aligned} y_{ijk} &= \sum_{l=1}^{p_{im}} \sum_{m=0}^{r} \xi_{iml} \left(\sum_{t=1}^{N} x_{j,t-k} x_{m,t-l}/N \right) \\ &= \sum_{l=1}^{p_{im}} \sum_{m=0}^{r} D_{jkmlN} \xi_{iml} \qquad (i = 1, \cdots, r; j = 0, \cdots, r). \end{aligned}$$

In order to obtain the limit covariance matrix of the variates ξ_{ijk} we need to know the limit covariance matrix of the variates y_{ijk}. First we show that the covariance between y_{ijk} and $y_{i'j'k'}$ is zero if $i \neq i'$. In fact, since $E(y_{ijk}) = 0$, we have

(71)
$$\begin{aligned} \sigma(y_{ijk} y_{i'j'k'}) &= E(y_{ijk} y_{i'j'k'}) = \frac{1}{N} \sum_{t=1}^{N} E(\epsilon_{it} \epsilon_{i't}) E(x_{j,t-k} x_{j',t-k'}) \\ &= \sigma_{ii'} E\left(\sum_{t=1}^{N} x_{j,t-k} x_{j',t-k'}/N \right). \end{aligned}$$

Since $\sigma_{ii'} = 0$ for $i \neq i'$, we obtain from (71)

(72)
$$(y_{ijk} y_{i'j'k'}) = 0 \quad \text{for} \quad i \neq i'.$$

If $i = i'$ it follows from (71) that

(73)
$$\lim_{N \to \infty} \sigma(y_{ijk} y_{ij'k'}) = \sigma_{ii} D_{jkj'k'}.$$

Denote by m_i the number of different pairs (j, k) for which y_{ijk} is defined. For a given value i we can arrange the m_i pairs (j, k) in an ordered sequence. We shall denote y_{ijk} by y_{iu} where (j, k) is the uth pair in the ordered sequence. Similarly, we shall denote $D_{jkj'k'N}$ by $D_{uu'N}$ and $D_{jkj'k'}$ by $D_{uu'}$ respectively where (j, k) is the uth element and (j', k') the u'th element in the ordered sequence. Then equations (70) and (73) can be written

(74)
$$y_{iu} = \sum_{v=1}^{m_i} D_{uvN} \xi_{iv}$$

and

(75)
$$\lim_{N \to \infty} \sigma(y_{iu} y_{iu'}) = \sigma_{ii} D_{uu'}.$$

Let

$$\|C_{uu'N}\| = \|D_{uu'N}\|^{-1}, \quad \|C_{uu'}\| = \|D_{uu'}\|^{-1} \quad (u, u' = 1, \cdots, m_i).$$

Then it can be shown on the basis of (74) and (75) and the relation $\text{plim}_{N \to \infty} D_{uvN} = D_{uv}$ that

(76)
$$\lim_{N \to \infty} \sigma(\xi_{iu} \xi_{iv}) = \sigma_{ii} C_{uv} = \underset{N \to \infty}{\text{plim}}\, s_{ii} C_{uvN}.$$

The proof of (76) is entirely analogous to that of (43) in the case of a single equation and is therefore omitted. Since $\sigma(y_{ijk}y_{i'j'k'})=0$ for $i\neq i'$ we obtain

$$(77) \qquad \lim_{N\to\infty} \sigma(\xi_{iu}\xi_{i'v}) = 0 \quad \text{for} \quad i \neq i'.$$

Thus (76) and (77) furnish the complete limit covariance matrix of the variables ξ_{ijk}.

Knowing that the joint limiting distribution of the variates ξ_{ijk} is a multivariate normal distribution with zero means and covariance matrix given by (76) and (77), tests of significance, derivation of confidence regions, and prediction of future values can be carried out in the same way as has been shown in the case of a single equation.

Frequently some of the coefficients α_{ijk} are known a priori to have some particular values. All of our results in this section can be extended without any difficulty to this case. Suppose that the coefficients α_{ijk} are classified into two groups. The first group contains all the coefficients α_{ijk} which are unknown and the second group contains those which have values known a priori. For any α_{ijk} in the second group we put $a_{ijk}=\alpha_{ijk}$. The variates y_{ijk} are defined only for those triples (i, j, k) for which α_{ijk} lies in the first group. Then all our results remain valid.

5. *Treatment of the Equations* (52) *when* $\|\alpha_{ij0}\|$ *Is Equal to the Unit Matrix and Nothing Is Known about the Covariance Matrix* $\|\sigma_{ij}\|$

We shall consider the following special case: The matrix $\|\alpha_{ij0}\|$ is equal to the unit matrix and p_{ij} has the same value for all pairs (i, j), i.e., $p_{ij}=p$ (say). It is furthermore assumed that all coefficients $\alpha_{ijk}(k>0)$ and all covariances σ_{ij} are unknown. Then the Jacobian J is equal to 1 and the system of equations (55) reduces to

$$(78) \qquad \sum_{l=1}^{r} \sigma^{il}\partial Q_{il}/\partial\alpha_{ijk} = 0; \qquad \sigma_{ij} = Q_{ij}/N.$$

Denote $\sum_{t=1}^{N} x_{j,t-k}L_l$ by M_{jkl}. Then it follows from (78) that

$$(79) \quad \sum_{l=1}^{r} \sigma^{il}M_{jkl} = 0 \quad (i = 1, \cdots, r; j = 1, \cdots, r; k = 1, \cdots, p).$$

For fixed values of j and k the equations (79) form a system of r homogeneous linear equations in $M_{jk1}, \cdots, M_{jkr}$. Since the determinant $|\sigma^{il}|$ of these equations is different from 0, the equations (79) are equivalent to

$$M_{jkl} = 0 \qquad\qquad (l = 1, \cdots, r).$$

Thus (78) can be written as

$$\partial Q_{il}/\partial \alpha_{ijk} = M_{jkl} = 0; \qquad \sigma_{ij} = Q_{ij}/N. \tag{80}$$

In deriving the equations (80) we made use of the assumptions that $p_{ij}=p$ and that all coefficients α_{ijk} are unknown. Otherwise it would not have been true that for any fixed values j, k the equations (79) hold for all values of i, i.e., for $i=1, 2, \cdots, r$.

Since the maximum-likelihood estimates are solutions of the equations (80), the joint limiting distribution of the maximum-likelihood estimates can be derived in the same way as it has been done in the special case discussed in the previous section. Thus we shall merely give the results.

The symbols D_{ikjlN} and D_{ikjl} are defined as in the previous section. We arrange the pairs (m, n) $(m=1, \cdots, r; n=1, \cdots, p)$ in an ordered sequence. We shall denote a_{ijk} and α_{ijk} by a_{iu} and α_{iu} respectively if (j, k) is the uth pair in the ordered sequence. Similarly we shall denote D_{ikjlN} and D_{ikjl} by D_{uvN} and D_{uv} respectively if (i, k) is the uth and (j, l) the vth pair in the ordered sequence. Let

$$||C_{uvN}|| = ||D_{uvN}||^{-1} \quad \text{and} \quad ||C_{uv}|| = ||D_{uv}||^{-1}.$$

The joint limiting distribution of the variates $\sqrt{N}(a_{iu}-\alpha_{iu})=\xi_{iu}$ $(u=1, \cdots, rp; i=1, \cdots, r)$ is a multivariate normal distribution with zero means and finite covariance matrix. The limit covariance of ξ_{iu} and ξ_{jv} is equal to $\sigma_{ij}C_{uv}$. Since $\text{plim}_{N\to\infty} s_{ij}C_{uvN}=\sigma_{ij}C_{uv}$, for large N we can replace the unknown value $\sigma_{ij}C_{uv}$ by the observed value $s_{ij}C_{uvN}$.

On the basis of the joint normal distribution of the maximum-likelihood estimates ξ_{ijk} we can carry out tests of significance or derive confidence regions for the unknown coefficients α_{ijk} in the usual manner.

We want to make a few remarks concerning the process of prediction, since this differs in some respects from the case when $\sigma_{ij}=0$ for $i\neq j$. If we want to predict the value of x_{it} on the basis of the values $x_{j\tau}$ $(j=1, \cdots, r; \tau=1, \cdots, t-1)$ then the process of prediction can be carried out in the same way as in the case when $\sigma_{ij}=0$ for $i\neq j$. A somewhat different situation arises, however, if we want to predict $x_{1t}, \cdots,$ $x_{qt}(q<r)$ on the basis of $x_{q+1,t}, \cdots, x_{rt}$ and all previously observed values $x_{j\tau}(j=1, \cdots, r; \tau<t)$. For this purpose we need the conditional distribution of $x_{1t}, \cdots, x_{qt}$ given $x_{q+1,t}, \cdots, x_{rt}$ and $x_{j\tau}(j=1, \cdots, r;$ $\tau<t)$. Let $M_{it}=x_{it}+\sum_{j=1}^{r}\sum_{k=1}^{p}a_{ijk}x_{jt-k}+a_i$. Since in the subpopulation considered $x_{it}-M_{it}$ is a fixed constant, our problem is equivalent to the problem of predicting $M_{1t}, \cdots, M_{qt}$ when the values of $M_{q+1,t}, \cdots, M_{rt}$ and $x_{j\tau}(\tau<t)$ are given. Since for large N the sampling variance of the maximum-likelihood estimates a_{ijk} can be neglected as

compared to the variance σ_{ii} of x_{it} the joint distribution of the variates $\epsilon_{1t}, \cdots, \epsilon_{rt}$ is the same as the joint limiting distribution of $M_{1t}, \cdots, M_{rt}$. Thus the problem of predicting $M_{1t}, \cdots, M_{qt}$ can be reduced to the problem of predicting $\epsilon_{1t}, \cdots, \epsilon_{qt}$ when $\epsilon_{q+1,t}, \cdots, \epsilon_{rt}$ are given. We assume that $\epsilon_{1t}, \cdots, \epsilon_{rt}$ have a joint normal distribution. Let

$$\|\sigma_{ij}^*\| = \|\sigma_{ij}\|^{-1} \qquad (i, j = q+1, \cdots, r); \tag{81}$$

then for any index $m \leqq q$ the conditional expected value of ϵ_{mt} given $\epsilon_{q+1,t}, \cdots, \epsilon_{rt}$ is known to be equal to

$$E(\epsilon_{mt} \mid \epsilon_{q+1,t}, \cdots, \epsilon_{rt}) = \sum_{v=q+1}^{r} \sum_{u=q+1}^{r} \sigma_{uv}^* \sigma_{mv} \epsilon_{ut}. \tag{82}$$

Let

$$\|\bar{\sigma}_{ij}\| = \|\sigma^{ij}\|^{-1} \qquad (i, j = 1, \cdots, q). \tag{83}$$

It is known that the conditional covariance matrix of $\epsilon_{1t}, \cdots, \epsilon_{qt}$ given $\epsilon_{q+1,t}, \cdots, \epsilon_{rt}$ is equal to $\|\bar{\sigma}_{ij}\| (i, j=1, \cdots, q)$. Thus

$$E_{mt} = \sum_{v=q+1}^{r} \sum_{u=q+1}^{r} \sigma_{uv}^* \sigma_{mv} \epsilon_{ut}$$

can be used as a point estimate of ϵ_{mt}. A confidence interval for ϵ_{mt} is given by $E_{mt} \pm \lambda\sqrt{\bar{\sigma}_{mm}}$ where λ is a constant corresponding to the confidence coefficient we wish to have. For large N we can replace σ_{ij} by its estimate s_{ij} and ϵ_{it} can be replaced by M_{it}. Thus we arrive at the following results: For large N we can use

$$E_{mt}' = \sum_{v=q+1}^{r} \sum_{u=q+1}^{r} s_{uv}^* s_{mv} M_{ut}$$

as a point estimate for M_{mt}. A confidence interval for M_{mt} is given by $E_{mt}' \pm \lambda\sqrt{\bar{s}_{mm}}$ where λ is a constant determined so that the confidence coefficient should have the required magnitude.

6. *The General Case*

In Sections 4 and 5 we have considered the special case when $\|\alpha_{ij0}\|$ is equal to the unit matrix. Here we shall deal with the general case.

The treatment of the general case presents a number of difficulties which do not arise in the special cases discussed before. Suppose that in the equations (52) all coefficients $\alpha_{ijk}(i, j=1, \cdots, r; k=1, \cdots, p_{ij})$, $\alpha_i(i=1, \cdots, r)$ and all covariances σ_{ij} are unknown. Assume furthermore that p_{ij} has the same value for all indices (i, j), i.e., $p_{ij}=p$ (say). Then it is not possible to have consistent estimates for all the parameters $\alpha_{ijk}, \alpha_i, \sigma_{ij}$. This can be seen as follows: Let $\|\lambda_{uv}\|$ $(u, v=1, \cdots,$

r) be an r-rowed square matrix with nonvanishing determinant. Let furthermore

$$\bar{\alpha}_{ujk} = \sum_{v=1}^{r} \lambda_{uv}\alpha_{vjk}, \qquad \bar{\alpha}_u = \sum_{v=1}^{r} \lambda_{uv}\alpha_v,$$

$$\bar{\epsilon}_{ut} = \sum_{v=1}^{r} \lambda_{uv}\epsilon_{vt}.$$

Then the system of equations

$$\sum_{j=1}^{r}\sum_{k=0}^{p} \bar{\alpha}_{ujk}x_{j,t-k} + \bar{\alpha}_u = \bar{\epsilon}_{ut} \qquad (u = 1, \cdots, r) \tag{84}$$

is equivalent to the system (52). The random variables $\bar{\epsilon}_{ut}$ satisfy all the assumptions made concerning the random variables ϵ_{ut}. Thus, on the basis of the observed values $x_{1\tau}, \cdots, r_{r\tau}(\tau = 1, \cdots, N)$ it is not possible to distinguish the system (84) from (52). On the other hand the coefficients $\bar{\alpha}_{ijk}$, $\bar{\alpha}_i$ in (84) are different from the coefficients α_{ijk}, α_i in (52). Hence it is impossible to have consistent estimates for all the unknown parameters α_{ijk}, α_i, and σ_{ij}.

In all that follows we shall assume that the determinant value $|\alpha_{ij0}|$ of the matrix $\|\alpha_{ij0}\|(i, j=1, \cdots, r)$ is different from zero. Then the system of equations (52) can be brought to the following reduced form:

$$x_{it} + \sum_{j=1}^{r}\sum_{k=1}^{p_{ij}} \alpha_{ijk}{}^* x_{j,t-k} + \alpha_i{}^* = \epsilon_{it}{}^* \qquad (i = 1, \cdots, r), \tag{85}$$

where $\alpha_{ijk}{}^*$ is a linear function of $\alpha_{1jk}, \cdots, \alpha_{rjk}$; $a_i{}^*$ is a linear function of $\alpha_1, \cdots, \alpha_r$, and $\epsilon_{it}{}^*$ is a linear function of $\epsilon_{1t}, \cdots, \epsilon_{rt}$. Denote the covariance of $\epsilon_{it}{}^*$ and $\epsilon_{jt}{}^*$ by $\sigma_{ij}{}^*$. Of course, the parameters $\alpha_{ijk}{}^*$, $\alpha_i{}^*$, $\sigma_{ij}{}^*$ depend also on the coefficients α_{ij0}. It follows from the result obtained in Section 5 that the parameters $\alpha_{ijk}{}^*$, $\alpha_i{}^*$, and $\sigma_{ij}{}^*$ can be estimated from the observations. Since the system (85) is equivalent to (52), it is clear that the observations cannot yield any additional information as to the values of the parameters α_{ij0}, α_{ijk} $(k=1, \cdots, p_{ij})$, α_i, and σ_{ij} beyond that contained in the estimates of the parameters $\alpha_{ijk}{}^*$, $\alpha_i{}^*$, and $\sigma_{ij}{}^*$. While the parameters α_{ijk}, α_i, and σ_{ij} $(i, j=1, \cdots, r$; $k=0, 1, \cdots, p_{ij})$ determine uniquely the values of $\alpha_{ijk}{}^*$ $(k>0)$, $\alpha_i{}^*$, and $\sigma_{ij}{}^*$, the converse is not true. Thus it is impossible to have estimates of all the parameters α_{ijk}, α_i, and σ_{ij} on the basis of the observations alone. However, if we have some a priori knowledge as to the values of α_{ijk}, α_i, and σ_{ij}, for instance, if we know that some of the coefficients α_{ijk}, α_i are equal to 0, then it may be possible to estimate all the unknown parameters in the original system (52). In order that we should be able to estimate all the unknown parameters in this system, it is necessary and sufficient that the values of the parameters α_{ijk}^*, α_i^*, and σ_{ij}^*, together with the available a priori knowledge, should determine uniquely the values of all the parameters α_{ijk} $(i, j=1, \cdots, r; k=0, 1 \cdots, p_{ij})$, α_i, and σ_{ij}.

Since (85) is equivalent to (52), we can discard the system (52) and treat the system (85): A definite advantage is gained by this procedure if there is no a priori knowledge whatsoever as to the values of α_{ijk}, α_i, and σ_{ij}, or if the a priori knowledge available does not imply any restrictions as to the values of α_{ijk}^* $(i, j=1, \cdots, r; k=1, \cdots, p_{ij}=p)$, α_i^*, σ_{ij}^*. For in this case we have to deal with the special situation discussed in Section 5 and the statistical treatment of (85) does not present any difficulties. If, however, the available a priori knowledge implies certain restrictions as to the values of the parameters α_{ijk}^*, α_i^*, and σ_{ij}^*, then the statistical treatment of (85) may be just as involved as that of (52) and no advantage is gained. In such cases it may even be preferable to treat the equations in their original form (52), since usually the coefficients in (52) have immediate meaning in economic theory, while the coefficients in (85) derive their significance merely from the fact that they are certain functions of the coefficients in (52).

In some practical cases one may find it perhaps worth while to discard the available a priori knowledge and to treat the equations (85) as if the parameters α_{ijk}^*, α_i^*, σ_{ij}^* were not subject to any restrictions. By doing so we decrease the efficiency of our statistical procedure but we gain in simplicity. In all that follows we shall assume that the available a priori knowledge implies restrictions as to the values of α_{ijk}^*, α_i^*, and σ_{ij}^* and that we are not willing to discard this information in order to gain in simplicity. Furthermore we assume that this a priori knowledge and the values of α_{ijk}^*, α_i^*, and σ_{ij}^* uniquely determine all the parameters α_{ijk}, α_i, and σ_{ij}. Since in such cases no advantage is gained by treating (85) instead of (52), we shall treat the equations in the original form (52). A frequent form of a priori knowledge is the following: It is known that ϵ_{it} is independent of ϵ_{jt} for $i \neq j$ and the values of some of the coefficients α_{ijk}, α_i are known to be zero. We shall restrict ourselves to the case when the a priori knowledge is of the type described above. There is no loss of generality in assuming that $\sigma_{ii}=1$, since this can always be achieved by multiplying the equations (52) by a constant factor. Then of course we cannot assume that $\alpha_{ii0}=1$. On the basis of these assumptions we shall procede to prove the consistency of the maximum-likelihood estimates and to derive their joint limiting distribution.

A. Consistency of the maximum-likelihood estimates. Let a_{ijk}, a_i be the maximum-likelihood estimates of the unknown coefficients α_{ijk} and α_i respectively. For triples (i, j, k) for which α_{ijk} is known to be zero we shall put $a_{ijk}=\alpha_{ijk}=0$. Similarly $a_i=\alpha_i=0$ if the value of α_i is a priori known to be zero. We bring the equations (52) into the reduced form

$$x_{it} + \sum_{j=1}^{r}\sum_{k=1}^{P} \alpha_{ijk}^* x_{j,t-k} + \alpha_i^* = \epsilon_{it}^* \qquad (i = 1, \cdots, r), \tag{86}$$

where P is the maximum of the integers p_{ij}. Denote the covariance of ϵ_{it}^* and ϵ_{jt}^* by σ_{ij}^*. Our a priori knowledge imposes certain restrictions on the parameters α_{ijk}^*, α_i^*, and σ_{ij}^*. Denote by a_{ijk}^*, a_i^*, and s_{ij}^* the maximum-likelihood estimates of α_{ijk}^*, α_i^*, and σ_{ij}^* respectively calculated under the restrictions imposed by the a priori knowledge. Since according to our assumption the parameters α_{ijk}^*, α_i^*, and σ_{ij}^* together with the a priori knowledge determine uniquely the coefficients α_{ijk} and α_i the consistency of the estimates a_{ijk} and a_i is proved if we prove the consistency of the estimates a_{ijk}^*, a_i^*, and s_{ij}^*. Denote by a_{ijk}', a_i', and s_{ij}' the maximum-likelihood estimates we should obtain if the parameters α_{ijk}^*, α_i^*, and σ_{ij}^* were not subject to any restrictions. Let

$$p = \frac{1}{(2\pi)^{rN/2}\left|\sigma_{ij}^*\right|^{N/2}} \exp\left[-\frac{1}{2}\sum_{j=1}^{r}\sum_{i=1}^{r}\sum_{t=1}^{N}\sigma^{*ij}L_{it}L_{jt}\right],$$

where $L_{it}=x_{it}+\sum_{j=1}^{r}\sum_{k=1}^{P}\alpha_{ijk}^*x_{j,t-k}+\alpha_i^*$. Denote by p_0 the value of p at the true values of the parameters α_{ijk}^*, α_i^*, and σ_{ij}^*. Let p^* be the value of p if we substitute a_{ijk}^*, a_i^*, and s_{ij}^* for α_{ijk}^*, α_i^*, and σ_{ij}^* respectively. Finally denote by p' the value of p we obtain if we replace α_{ijk}^*, α_i^*, and σ_{ij}^* by a_{ijk}', a_i', and s_{ij}' respectively. Then we obviously have

$$\log p_0 \leqq \log p^* \leqq \log p'. \tag{87}$$

From our results in Section 5 it follows that a_{ijk}', a_i', and s_{ij}' converge stochastically to α_{ijk}^*, α_i^*, and σ_{ij}^* respectively. From this and the fact that $(1/N)\sum_{t=1}^{N}x_{i,t-k}x_{j,t-l}$ converges stochastically to a fixed constant D_{ikjl} we easily obtain

$$\operatorname{plim}\left[(\log p_0)/N - (\log p')/N\right] = 0. \tag{88}$$

From (87) and (88) it follows that

$$\operatorname{plim}\left[(\log p^*)/N - (\log p')/N\right] = 0. \tag{89}$$

Let $\bar{p}$ be the expression we obtain from p if we replace $(1/N)\sum_{t=1}^{N}x_{i,t-k}x_{j,t-l}$ by its stochastic limit D_{ikjl}. Let $\bar{a}_{ijk}^*$, $\bar{a}_i^*$, and $\bar{s}_{ij}^*$

be the maximum-likelihood estimates we obtain instead of a_{ijk}^*, a_i^* and s_{ij}^* if we replace p by $\bar{p}$. The symbols $\bar{a}_{ijk}'$, $\bar{a}_i'$, and $\bar{s}_{ij}'$ are similarly defined. Then we obviously have

$$\text{(90)} \quad \text{plim } (\bar{a}_{ijk}' - \alpha_{ijk}^*) = \text{plim } (\bar{a}_i' - \alpha_i^*) = \text{plim } (\bar{s}_{ij}' - \sigma_{ij}^*) = 0,$$

$$\text{(91)} \quad \text{plim } (\bar{a}_{ijk}^* - a_{ijk}^*) = \text{plim } (\bar{a}_i^* - a_i^*) = \text{plim } (\bar{s}_{ij}^* - s_{ij}^*) = 0,$$

and, on account of (89),

$$\text{(92)} \qquad \text{plim } (\log \bar{p}^* - \log \bar{p}')/N = 0.$$

Since $\bar{a}_{ijk}'$, $\bar{a}_i'$, and s_{ij}' are unique solutions of the maximum-likelihood equations, the value of $\bar{p}$ is less than $\bar{p}'$ for any set of parameter values α_{ijk}^*, α_i^*, σ_{ij}^* which are not equal to the maximum-likelihood estimates. From this, (92), and the fact that $(\log \bar{p})/N$ does not depend on N, it follows that

$$\text{(93)} \quad \text{plim } (\bar{a}_{ijk}^* - \bar{a}_{ijk}') = \text{plim } (\bar{a}_i^* - \bar{a}_i') = \text{plim } (\bar{s}_{ij} - \bar{s}_{ij}') = 0.$$

The consistency of a_{ijk}^*, a_i^*, and s_{ij}^* follows from (90), (91), and (93). Hence the consistency of the estimates a_{ijk} and a_i is proved.

B. Limiting distribution of the maximum-likelihood estimates. It will be convenient to denote α_i by α_{i00} and a_i by a_{i00}. Furthermore we put $x_{0t} = 1$ for all values of t. Then equations (52) can be written

$$\text{(94)} \qquad \sum_{j=0}^{r} \sum_{k=0}^{p_{ij}} \alpha_{ijk} x_{j,t-k} = \epsilon_{it} \qquad (i = 1, \cdots, r),$$

where $p_{i0} = 0$. Denote by J the determinant value of the matrix $|\alpha_{ij0}|$ $(i, j = 1, \cdots, r)$. Then the maximum-likelihood estimates a_{ijk} are roots of the equation

$$\text{(95)} \qquad (N\partial J/\partial \alpha_{ijk})/J - \left(\sum_{j=0}^{r} \sum_{l=0}^{p_{ij}} \sum_{t=1}^{N} \alpha_{ijl} x_{j,t-l} \right) x_{j,t-k} = 0.$$

The equations (95) are formed for all triples (i, j, k) for which a_{ijk} is unknown. For $k > 0$ we have $\partial J/\partial \alpha_{ijk} = 0$. Also for $j = k = 0$ the derivative $\partial J/\partial \alpha_{ijk}$ is zero.

We put

$$\text{(96)} \quad y_{ijk} = (\sqrt{N}\, \partial J/\partial \alpha_{ijk})/J - \left(\sum_{j=0}^{r} \sum_{l=0}^{p_{ij}} \sum_{t=1}^{N} \alpha_{ijl} x_{j,t-l} \right) x_{j,t-k}/\sqrt{N},$$

where y_{ijk} is defined only for triples (i, j, k) for which α_{ijk} is unknown. In order to derive the joint limiting distribution of the maximum-likelihood estimates a_{ijk} we need to know the joint limiting distribution of the variates y_{ijk}. It can be shown that the joint limiting distribution of the variates y_{ijk} is a multivariate normal distribution. The proof of this statement is entirely analogous to that given in the case of a single equation and is therefore omitted. Thus for specifying completely the limiting distribution of the variates y_{ijk}, we merely need to know the limiting mean values and covariances of the variates y_{ijk}. First we show that

$$E(y_{ijk}) = 0. \tag{97}$$

From (96) it follows that

$$y_{ijk} = \sqrt{N}\,\partial J/\partial\alpha_{ijk} - \frac{1}{\sqrt{N}}\sum_{t=1}^{N}\epsilon_{it}x_{j,t-k}. \tag{98}$$

Since $\partial J/\partial\alpha_{ijk}=0$ for $k>0$ and since $E(\epsilon_{it}x_{j,t-k})=0$ for $k>0$, equation (97) is proved for $k>0$. For $j=k=0$ we have $\partial J/\partial\alpha_{ijk}=0$ and $E(\epsilon_{it}x_{0t})=E(\epsilon_{it})=0$. Hence we see that (97) holds also when $j=k=0$. Thus equation (97) is completely proved if we show that it holds when $j>0$ and $k=0$. Let $\|\alpha^{ij}\|=\|\alpha_{ij0}\|^{-1}(i,\ j=1,\cdots,r)$. Then $(1/J)(\partial J/\partial\alpha_{ij0})=\alpha^{ji}$ and from (98) we obtain

$$y_{ij0} = \sqrt{N}\,\alpha^{ji} - \frac{1}{\sqrt{N}}\sum_{t=1}^{N}E(\epsilon_{it}x_{jt}). \tag{98'}$$

Hence

$$E(y_{ij0}) = \sqrt{N}\,\alpha^{ji} - \frac{1}{\sqrt{N}}\sum_{t=1}^{N}E(\epsilon_{it}x_{jt}). \tag{99}$$

Since the covariance matrix of the variates $\epsilon_{1t},\cdots,\epsilon_{rt}$ is the unit matrix it can easily be shown that

$$E(\epsilon_{it}x_{jt}) = \alpha^{ji}. \tag{100}$$

From (99) and (100) we obtain $E(y_{ij0})=0$. Hence (97) is proved in all cases.

Now we shall derive the limiting covariance matrix of the variates y_{ijk}. We have

$$\begin{aligned}\sigma(y_{ijk}y_{lmn}) = \frac{1}{N}E\Big\{&\Big[(N/J)(\partial J/\partial\alpha_{ijk}) - \sum_{t=1}^{N}\epsilon_{it}x_{j,t-k}\Big]\\ &\cdot\Big[(N/J)(\partial J/\partial\alpha_{lmn}) - \sum_{t=1}^{N}\epsilon_{lt}x_{m,t-n}\Big]\Big\}.\end{aligned} \tag{101}$$

From (97) it follows that

$$\begin{aligned}&E\Big[(N/J)(\partial J/\partial\alpha_{ijk}) - \sum_{t=1}^{N}\epsilon_{it}x_{j,t-k}\Big]\\ &\quad = E\Big[(N/J)(\partial J/\partial\alpha_{lmn}) - \sum_{t=1}^{N}\epsilon_{lt}x_{m,t-n}\Big] = 0.\end{aligned} \tag{102}$$

From (101) and (102) we easily obtain

$$\sigma(y_{ijk}y_{lmn}) = \frac{1}{N}\left[E\left(\sum_{t=1}^{N} \epsilon_{it}x_{j,t-k}\right)\left(\sum_{t=1}^{N} \epsilon_{lt}x_{m,t-n}\right) - (N^2/J^2)(\partial J/\partial \alpha_{ijk})(\partial J/\partial \alpha_{lmn})\right]. \tag{103}$$

We shall denote $(1/N)\sum_{t=1}^{N} x_{j,t-k}x_{m,t-n}$ by D_{jkmnN} and plim D_{jkmnN} by D_{jkmn}. Furthermore we denote $E(D_{jkmnN})$ by B_{jkmnN}. We obviously have

$$\text{plim } B_{jkmnN} = D_{jkmn}. \tag{104}$$

In order to evaluate the right-hand side of (103) we shall distinguish different cases.

Case I. $k>0$ and $n>0$. In this case we have

$$\sigma(y_{ijk}y_{lmn}) = \frac{1}{N}E\left[\sum_{t=1}^{N}\sum_{t'=1}^{N} \epsilon_{it}x_{j,t-k}\epsilon_{lt'}x_{m,t'-n}\right]. \tag{105}$$

Since

$$E(\epsilon_{it}x_{j,t-k}\epsilon_{lt'}x_{mt'-n}) = 0 \quad \text{for} \quad t \neq t',$$

we obtain

$$\begin{aligned} \sigma(y_{ijk}y_{lmn}) &= \frac{1}{N}E\left[\sum_{t=1}^{N} \epsilon_{it}\epsilon_{lt}x_{j,t-k}x_{m,t-n}\right] \\ &= E(\epsilon_{it}\epsilon_{lt})E\left(\frac{1}{N}\sum_{t=1}^{N} x_{j,t-k}x_{m,t-n}\right) = \delta_{il}B_{jkmnN}, \end{aligned} \tag{106}$$

where δ_{il} is the Kronecker δ, i.e., $\delta_{il}=0$ for $i\neq l$ and $\delta_{ii}=1$.

Case II. $k=0$ and $n>0$. In this case we have

$$\sigma(y_{ij0}y_{lmn}) = \frac{1}{N}E\left[\sum_{t'=1}^{N}\sum_{t=1}^{N} \epsilon_{it}\epsilon_{lt'}x_{jt}x_{m,t'-n}\right]. \tag{107}$$

We have

$$\begin{aligned} E(\epsilon_{it}\epsilon_{lt'}x_{jt}x_{m,t'-n}) &= E[\epsilon_{it}\epsilon_{lt'}(x_{jt} - \alpha^{ji}\epsilon_{it})x_{m,t'-n}] \\ &\quad + \alpha^{ji}E(\epsilon_{it}^2\epsilon_{lt'}x_{m,t'-n}). \end{aligned} \tag{108}$$

We shall prove that the right-hand side of (108) is equal to 0 if $t\neq t'$. This is clearly true when $t'>t$. If $t>t'$ then ϵ_{it} is independent of $\epsilon_{lt'}$ and of $x_{m,t'-n}$. Since ϵ_{it} is independent of $(x_{jt}-\alpha^{ji}\epsilon_{it})$, we see that the right-hand side of (108) is equal to zero if $t>t'$. Hence we have

$$
(109) \quad \sigma(y_{ij0}y_{lmn}) = \frac{1}{N} E\left\{ \sum_{t=1}^{N} \epsilon_{it}\epsilon_{lt}(x_{jt} \quad \alpha^{ji}\epsilon_{it})x_{m,t-n} + \alpha^{ji} \sum_{t=1}^{N} E(\epsilon_{it}^2 \epsilon_{lt} x_{m,t-n}) \right\}.
$$

If $l \neq i$ then ϵ_{it} is independent of the set of variates ϵ_{lt}, $(x_{jt} - \alpha^{ji}\epsilon_{it})$, and $x_{m,t-n}$. From this it follows easily that the right-hand side of (109) is zero for $l \neq i$. We have

$$
(110) \quad \begin{aligned} E[\epsilon_{it}^2(x_{jt} - \alpha^{ji}\epsilon_{it})x_{m,t-n}] &= E(\epsilon_{it}^2)E[(x_{jt} - \alpha^{ji}\epsilon_{it})x_{m,t-n}] \\ &= E(x_{jt}x_{m,t-n}). \end{aligned}
$$

Furthermore we have

$$
(111) \quad E(\epsilon_{it}^3 x_{m,t-n}) = E(\epsilon_{it}^3)E(x_{m,t-n}) = \mu_3^{(i)}E(x_{m,t-n}),
$$

where $\mu_3^{(i)}$ denotes the third moment of ϵ_{it}. From (109), (110), and (111) we obtain

$$
(112) \quad \sigma(y_{ij0}y_{lmn}) = B_{j0mnN} + \mu_3^{(i)}\alpha^{ji}B_{00mnN}.
$$

Thus our final result can be summarized in the equation

$$
(113) \quad \sigma(y_{ij0}y_{lmn}) = \delta_{il}(B_{j0mnN} + \alpha^{ji}\mu_3^{(i)}B_{00mnN}),
$$

where δ_{il} is the Kronecker delta.

Case III. $k>0$ *and* $n=0$. The results in this case can be obtained from those in Case II by interchanging the pairs (j, k) and (m, n). Thus we have

$$
(114) \quad \sigma(y_{ijk}y_{lm0}) = \delta_{il}(B_{m0jkN} + \alpha^{mi}\mu_3^{(i)}B_{00jkN}).
$$

Case IV. $k=n=0$. For treating this case we shall break it down into four subcases IV_1–IV_4

Subcase IV_1. $j>0$, $m>0$, $k=n=0$. In this case we have on account of (97)

$$
E(y_{ij0}y_{lm0}) = E[(y_{ij0} - \sqrt{N}\,\alpha^{ji})(y_{lm0} - \sqrt{N}\,\alpha^{ml})] - N\alpha^{ji}\alpha^{ml}.
$$

Hence on account of (98′) we obtain

$$
(114') \quad \sigma(y_{ij0}y_{lm0}) = E\left[\frac{1}{N}\sum_{t=1}^{N}\sum_{t'=1}^{N} \epsilon_{it}\epsilon_{lt'}x_{jt}x_{mt'}\right] - N\alpha^{ji}\alpha^{ml}.
$$

We have the identity

$$
(115) \quad \begin{aligned} E(\epsilon_{it}\epsilon_{lt'}x_{jt}x_{mt'}) &= E[\epsilon_{it}\epsilon_{lt'}(x_{jt} - \alpha^{ji}\epsilon_{it})(x_{mt'} - \alpha^{ml}\epsilon_{lt'})] \\ &\quad + \alpha^{ji}E[\epsilon_{lt'}\epsilon_{it}^2(x_{mt'} - \alpha^{ml}\epsilon_{lt'})] \\ &\quad + \alpha^{ml}E[\epsilon_{lt'}^2\epsilon_{it}(x_{jt} - \alpha^{ji}\epsilon_{it})] \\ &\quad + \alpha^{ml}\alpha^{ji}E(\epsilon_{lt'}^2\epsilon_{it}^2). \end{aligned}
$$

From (59) it follows that $\phi_{jit}(t)=\alpha^{ji}$. Hence according to (56) $x_{jt}-\alpha^{ji}\epsilon_{it}$ is independent of ϵ_{it}. We shall frequently use this fact in the following derivations.

If t is different from t', say $t>t'$, then ϵ_{it} is independent of the set of variables $\epsilon_{lt'}$, $(x_{jt}-\alpha^{ji}\epsilon_{it})$ and $(x_{mt'}-\alpha^{ml}\epsilon_{lt'})$. Hence since $E(\epsilon_{it})=0$ we have

$$E\left[\epsilon_{it}\epsilon_{lt'}(x_{jt} - \alpha^{ji}\epsilon_{it})(x_{mt'} - \alpha^{ml}\epsilon_{lt'})\right] = E(\epsilon_{it})E\left[\epsilon_{lt'}(x_{jt} - \alpha^{ji}\epsilon_{it})(x_{mt'} - \alpha^{ml}\epsilon_{lt'})\right] = 0.$$

Similarly for $t'>t$

$$E\left[\epsilon_{it}\epsilon_{lt'}(x_{jt} - \alpha^{ji}\epsilon_{it})(x_{mt'} - \alpha^{ml}\epsilon_{lt'})\right] = E(\epsilon_{lt'})E\left[\epsilon_{it}(x_{jt} - \alpha^{ji}\epsilon_{it})(x_{mt'} - \alpha^{ml}\epsilon_{lt'})\right] = 0.$$

Hence

$$\text{(116)}\qquad E\left[\epsilon_{it}\epsilon_{lt'}(x_{jt} - \alpha^{ji}\epsilon_{it})(x_{mt'} - \alpha^{ml}\epsilon_{lt'})\right] = 0 \qquad (t \neq t').$$

On account of the independence of ϵ_{ut} and $(x_{vt}-\alpha^{vu}\epsilon_{ut})$ $(v=1, \cdots, r;$ $u=1, \cdots, r)$ we obtain

$$\text{(117)}\quad E\left[\epsilon_{lt'}\epsilon_{it}^2(x_{mt'} - \alpha^{ml}\epsilon_{lt'})\right] = E\left[\epsilon_{lt'}^2\epsilon_{it}(x_{jt} - \alpha^{ji}\epsilon_{it})\right] = 0 \quad (t \neq t').$$

Furthermore we have, because of $E(\epsilon_{lt'}^2)=1$,

$$E(\epsilon_{lt'}^2\epsilon_{it}^2) = E(\epsilon_{lt'}^2)E(\epsilon_{it}^2) = 1.$$

Hence on account of (115), (116), (117) we have

$$\text{(118)}\qquad E(\epsilon_{it}\epsilon_{lt'}x_{jt}x_{mt'}) = \alpha^{ml}\alpha^{ji} \qquad (t \neq t').$$

For $t=t'$ we use the identity

$$\text{(119)}\qquad \begin{aligned} E(\epsilon_{it}\epsilon_{lt}x_{jt}x_{mt}) = {} & E\left[\epsilon_{it}\epsilon_{lt}(x_{jt} - \alpha^{ji}\epsilon_{it})(x_{mt} - \alpha^{mi}\epsilon_{it})\right] \\ & + \alpha^{ji}E\left[\epsilon_{it}^2\epsilon_{lt}(x_{mt} - \alpha^{mi}\epsilon_{it})\right] \\ & + \alpha^{mi}E\left[\epsilon_{it}^2\epsilon_{lt}(x_{jt} - \alpha^{ji}\epsilon_{it})\right] \\ & + \alpha^{ji}\alpha^{mi}E(\epsilon_{it}^3\epsilon_{lt}). \end{aligned}$$

Since ϵ_{it} is independent of $(x_{jt}-\alpha^{ji}\epsilon_{it})$, $(x_{mt}-\alpha^{mi}\epsilon_{it})$, and of ϵ_{lt} if $i\neq l$, we have, for $i\neq l$,

$$\text{(120)}\qquad \begin{aligned} E\left[\epsilon_{it}\epsilon_{lt}(x_{jt} - \alpha^{ji}\epsilon_{it})(x_{mt} - \alpha^{mi}\epsilon_{it})\right] & = E(\epsilon_{it})E\left[\epsilon_{lt}(x_{jt} - \alpha^{ji}\epsilon_{it})(x_{mt} - \alpha^{mi}\epsilon_{it})\right] \\ & = 0, \\ E\left[\epsilon_{it}^2\epsilon_{lt}(x_{mt} - \alpha^{mi}\epsilon_{it})\right] & = E(\epsilon_{it}^2)E\left[\epsilon_{lt}(x_{mt} - \alpha^{mi}\epsilon_{it})\right] \\ & = E\left[\epsilon_{lt}(x_{mt} - \alpha^{mi}\epsilon_{it})\right], \\ E\left[\epsilon_{it}^2\epsilon_{lt}(x_{jt} - \alpha^{ji}\epsilon_{it})\right] & = E(\epsilon_{it}^2)E\left[\epsilon_{lt}(x_{jt} - \alpha^{ji}\epsilon_{it})\right] \\ & = E\left[\epsilon_{lt}(x_{jt} - \alpha^{ji}\epsilon_{it})\right], \\ E(\epsilon_{it}^3\epsilon_{lt}) & = 0 \qquad (i \neq l). \end{aligned}$$

Hence we have on account of (100), (119), and (120)

$$
\begin{aligned}
E(\epsilon_{it}\epsilon_{lt}x_{jt}x_{mt}) &= \alpha^{ji}E[\epsilon_{lt}(x_{mt} - \alpha^{mi}\epsilon_{it})] + \alpha^{mi}E[\epsilon_{lt}(x_{jt} - \alpha^{ji}\epsilon_{it}] \\
&= \alpha^{ji}E(\epsilon_{lt}x_{mt}) + \alpha^{mi}E(\epsilon_{lt}x_{jt}) \qquad (121)\\
&= \alpha^{ji}\alpha^{ml} + \alpha^{mi}\alpha^{jl} \qquad (i \neq 1).
\end{aligned}
$$

From (121) and (118) we see that for $i \neq l$ in the double sum on the right-hand side of (114′) the term $\alpha^{ji}\alpha^{ml}$ occurs N^2 times, the term $\alpha^{mi}\alpha^{jl}$ N times. Hence we obtain

$$
(122) \qquad \sigma(y_{ij0}y_{lm0}) = \alpha^{mi}\alpha^{jl} \quad \text{for} \quad i \neq 1.
$$

For $i = l$ we have

$$
\begin{aligned}
&E[\epsilon_{it}^2(x_{jt} - \alpha^{ji}\epsilon_{it})(x_{mt} - \alpha^{mi}\epsilon_{it})] \\
&\quad = E(\epsilon_{it}^2)E[(x_{jt} - \alpha^{ji}\epsilon_{it})(x_{mt} - \alpha^{mi}\epsilon_{it})] \\
(123) &\quad = E[(x_{jt} - \alpha^{ji}\epsilon_{it})(x_{mt} - \alpha^{mi}\epsilon_{it})] \\
&\quad = E[x_{jt}(x_{mt} - \alpha^{mi}\epsilon_{it})] = E(x_{jt}x_{mt}) - \alpha^{mi}E(\epsilon_{it}x_{jt}) \\
&\quad = E(x_{jt}x_{mt}) - \alpha^{ji}\alpha^{mi}.
\end{aligned}
$$

Furthermore we have

$$
(124) \quad E[\epsilon_{it}^3(x_{mt} - \alpha^{mi}\epsilon_{it})] = E(\epsilon_{it}^3)E[(x_{mt} - \alpha^{mi}\epsilon_{it})] = \mu_3^{(i)}E(x_{mt}),
$$

where $\mu_3^{(i)} = E(\epsilon_{it}^3)$. Similarly we obtain

$$
(125) \quad E[\epsilon_{it}^3(x_{jt} - \alpha^{ji}\epsilon_{it})] = E(\epsilon_{it}^3)E(x_{jt} - \alpha^{mi}\epsilon_{it}) = \mu_3^{(i)}E(x_{jt}).
$$

Putting $E(\epsilon_{it}^4) = \mu_4^{(i)}$ we finally obtain from (119), (123), (124), and (125)

$$
\begin{aligned}
(126) \quad E(\epsilon_{it}^2x_{jt}x_{mt}) &= E(x_{jt}x_{mt}) + \mu_3^{(i)}[\alpha^{ji}E(x_{mt}) + \alpha^{mi}E(x_{jt})] \\
&\quad + (\mu_4^{(i)} - 1)\alpha^{ji}\alpha^{mi}.
\end{aligned}
$$

Hence from (114), (118), and (126) we obtain

$$
\begin{aligned}
\sigma(y_{ij0}y_{im0}) &= \frac{1}{N}\sum_{t=1}^{N} E(x_{jt}x_{mt}) \\
&\quad + \mu_3^{(i)}\left[\alpha^{ji}\sum_{t=1}^{N} E(x_{mt})/N + \alpha^{mi}\sum_{t=1}^{N} E(x_{jt})/N\right] \qquad (127)\\
&\quad + (\mu_4^{(i)} - 2)\alpha^{ji}\alpha^{mi} \\
&= B_{j0m0N} + \mu_3^{(i)}(\alpha^{ji}B_{m000N} + \alpha^{mi}B_{j000N}) \\
&\quad + (\mu_4^{(i)} - 2)\alpha^{ji}\alpha^{mi}.
\end{aligned}
$$

We can summarize our result as follows:

$$\begin{aligned}(128)\qquad \sigma(y_{ij0}y_{lm0}) = \alpha^{il}\alpha^{mi} + \delta_{il}[B_{j0m0N} + \mu_3^{(i)}\alpha^{il}B_{m00N} \\ + \mu_3^{(l)}\alpha^{mi}B_{j000N} + (\mu_4^{(i)} - 3)\alpha^{il}\alpha^{mi}] \\ (j > 0,\ m > 0),\end{aligned}$$

where δ_{ij} denotes the Kronecker delta.

Subcase IV_2. $j>0$, $m=0$. In this case we have from (98)

$$\begin{aligned}(129)\qquad \sigma(y_{ij0}y_{l00}) &= E\left[\frac{1}{N}\sum_{t=1}^{N}\sum_{t'=1}^{N}\epsilon_{it}\epsilon_{lt'}x_{jt}\right] - \alpha^{ji}E\left[\sum_{t=1}^{N}\epsilon_{lt}\right] \\ &= E\left[\frac{1}{N}\sum_{t=1}^{N}\sum_{t'=1}^{N}\epsilon_{it}\epsilon_{lt'}x_{jt}\right].\end{aligned}$$

We have the identity

$$(130)\qquad E(\epsilon_{lt'}\epsilon_{it}x_{jt}) = E[\epsilon_{lt'}\epsilon_{it}(x_{jt} - \alpha^{ji}\epsilon_{it})] + \alpha^{ji}E(\epsilon_{it}^2\epsilon_{lt'}).$$

We see from (130) that $E(\epsilon_{lt'}\epsilon_{it}x_{jt})=0$ if $i\neq l$ or $t\neq t'$, since then the set of variables $\epsilon_{lt'}$, ϵ_{it}, $(x_{jt}-\alpha^{ji}\epsilon_{it})$ are independent of each other. We furthermore have from (130) and the fact that $E(\epsilon_{it}^2)=1$

$$(131)\qquad E(\epsilon_{it}^2x_{jt}) = E(x_{jt}) + \alpha^{ji}\mu_3^{(i)},$$

where $\mu_3^{(i)}$ is the third moment of ϵ_{it}. Hence we have

$$\begin{aligned}(132)\qquad \sigma(y_{ij0}y_{l00}) &= \delta_{il}\left(\sum_{l=1}^{N}E(x_{jt})/N + \alpha^{ji}\mu_3^{(i)}\right) \\ &= \delta_{il}(B_{j000N} + \alpha^{ji}\mu_3^{(i)}).\end{aligned}$$

Subcase IV_3. $j=0$, $m>0$. In this case we have

$$(133)\qquad \sigma(y_{i00}y_{lm0}) = \sigma(y_{lm0}y_{i00}).$$

Hence from (132) and (133) we see that

$$(134)\qquad \sigma(y_{i00}y_{lm0}) = \delta_{il}(B_{m000N} + \alpha^{ml}\mu_3^{(l)}).$$

Subcase IV_4. $j=m=0$. In this case we obtain from (98)

$$(135)\qquad \sigma(y_{i00}y_{l00}) = E\left(\frac{1}{N}\sum_{t=1}^{N}\sum_{t'=1}^{N}\epsilon_{it}\epsilon_{lt'}\right).$$

Because of the independence of ϵ_{it}, $\epsilon_{lt'}$ for $i\neq l$ or $t\neq t'$ and because of $E(\epsilon_{it}^2)=1$ we have

$$(136)\qquad E(\epsilon_{it}\epsilon_{lt'}) = \delta_{il}\delta_{tt'}.$$

From (135) and (136) we obtain

$$(137) \qquad \sigma(y_{i00}y_{l00}) = \delta_{il}.$$

Summarizing our results we have

$$(138) \quad \sigma(y_{ijk}y_{lmn}) = \begin{cases} \delta_{il}B_{jkmnN} & \text{if } k > 0, n > 0, \\ \delta_{il}(B_{j0mnN} + \alpha^{ji}\mu_3^{(i)}B_{00mnN}) & \text{if } k = 0, n > 0, \\ \delta_{il}(B_{m0jkN} + \alpha^{mi}\mu_3^{(i)}B_{00jkN}) & \text{if } k > 0, n = 0, \\ \alpha^{jl}\alpha^{mi} + \delta_{il}[B_{j0m0N} + \mu_3^{(i)}\alpha^{jl}B_{m000N} & \\ \quad + \mu_3^{(l)}\alpha^{mi}B_{j000N} + (\mu_4^{(i)} - 3)\alpha^{jl}\alpha^{mi}] & \\ \qquad\qquad \text{if } k = 0, n = 0, j > 0, k > 0, & \\ \delta_{il}(B_{j000N} + \alpha^{ji}\mu_3^{(i)}) & \text{if } j > 0, m = k = n = 0, \\ \delta_{il}(B_{m000N} + \alpha^{mi}\mu_3^{(i)}) & \text{if } m > 0, j = k = n = 0, \\ \delta_{il} & \text{if } j = k = m = n = 0. \end{cases}$$

Since according to (104) $\text{plim}_{N\to\infty}B_{jkmnN} = D_{jkmn} = \text{plim}_{N\to\infty}D_{jkmnN}$, we have

$$(139) \qquad \operatorname*{plim}_{N\to\infty} (B_{jkmnN} - D_{jkmnN}) = 0.$$

Hence the limit covariance matrix of the variates y_{ijk} is not changed if we replace in (138) B_{jkmnN} by D_{jkmnN}. Furthermore because of $\text{plim}_{N\to\infty}a_{ijk} = \alpha_{ijk}$, we can in (138) replace the parameters α_{ij0} by their consistent estimates a_{ij0}. Finally the moments $\mu_3^{(i)}$, $\mu_4^{(i)}$ can also be replaced by consistent sample estimates if they are not a priori known. Hence all quantities occuring on the right-hand side of (138) can be replaced by sample estimates without changing the limiting values of the covariances of the variates y_{ijk}. The limit covariance matrix of the y_{ijk} can be obtained from (138) by replacing B_{jkmnN} by its stochastic limit D_{jkmn}.

From the limit covariance matrix of the $y_{ijk}(i=1, \cdots, r; j=0, \cdots, r; k=0, \cdots, p_{ij})$ we shall derive the limit covariance matrix of the quantities $\sqrt{N}(a_{ijk}-\alpha_{ijk})(i=1, \cdots, r;\ j=0, \cdots, r;\ k=0, \cdots, p_{ij})$.

We shall denote by a^{ji} the value we obtain if we replace in the expression α^{ji} the parameters α_{ij0} by their estimates a_{ij0}. We furthermore put $\alpha^{ij0} = \alpha^{ij}, a^{ij0} = a^{ij}(i,\ j=1, \cdots,\ r)$, $\alpha^{i0k} = \alpha^{0ik} = a^{i0k} = a^{0ik} = 0$ and $\alpha^{ijk} = a^{ijk} = 0$ for $k>0$. Then from (55) and (96) we obtain

$$(140) \qquad y_{ijk} = \sum_{l=0}^{r}\sum_{m=0}^{p_{il}} \sqrt{N}\,(a_{ilm} - \alpha_{ilm})D_{jklmN} - \sqrt{N}\,(a^{ijk} - \alpha^{ijk}).$$

Since plim $a_{ijk} = \alpha_{ijk}$, we can for large N replace $a^{ji0} - \alpha^{ji0}$ by the linear term of its Taylor expansion. Thus we obtain

$$\tag{140'} \begin{aligned} y_{ijk} = & \sum_{l=1}^{r} \sum_{m=1}^{p_{il}} \sqrt{N}\,(a_{ilm} - \alpha_{ilm}) D_{jklmN} \\ & + \sum_{u=1}^{r} \sum_{v=1}^{r} (\partial \alpha^{jik} / \partial \alpha_{uvk}) \sqrt{N}\,(a_{uvk} - \alpha_{uvk}). \end{aligned}$$

For large N we can replace $\partial \alpha^{jik} / \partial \alpha_{uvk}$ by $\partial a^{jik} / \partial a_{uvk}$. Thus finally we can write

$$\tag{140''} \begin{aligned} y_{ijk} = & \sum_{l=1}^{r} \sum_{m=1}^{p_{il}} \sqrt{N}\,(a_{ilm} - \alpha_{ilm}) D_{jklmN} \\ & + \sum_{u=1}^{r} \sum_{v=1}^{r} (\partial a^{jik} / \partial a_{uvk}) \sqrt{N}\,(a_{uvk} - \alpha_{uvk}). \end{aligned}$$

We shall now enumerate the triples (i, j, k) for which α_{ijk} is unknown and write $y_u = y_{ijk}$ if (i, j, k) is the uth triple in our enumeration. We furthermore put $\xi_u = \sqrt{N}(a_{ijk} - \alpha_{ijk})$ if (i, j, k) is the uth triple in our enumeration. Let there be h triples (i, j, k). Then we have from (140″)

$$\tag{141} y_u = \sum_{u'=1}^{h} b_{uu'N} \xi_{u'},$$

where $b_{uu'N}$ may be determined from (140″). We shall now further assume that the distribution of the ϵ_{it} is such that $\operatorname{plim}_{N\to\infty} |b_{uu'N}|$ is different from 0. As we shall show later this is always the case if the ϵ_{it} are normally and independently distributed.

We put $\|b_{uu'N}\|^{-1} = \|c_{uu'N}\|$. Then

$$\xi_u = \sum_{u'=1}^{r} c_{uu'N} y_{u'}.$$

Hence for large N

$$\sigma(\xi_u \xi_v) = E\left[\sum_{u'=1}^{h} \sum_{v'=1}^{h} c_{uu'N} c_{vv'N} y_{u'} y_{v'}\right] = \sum_{u'=1}^{h} \sum_{v'=1}^{h} c_{uu'N} \sigma(y_{u'} y_{v'}) c_{vv'N}.$$

The above equation can be written as the following matrix equation

$$\tag{142} \|\sigma(\xi_u \xi_v)\| = \|c_{uvN}\|\,\|\sigma(y_u y_v)\|\,\|c_{uvN}\|^t,$$

where $\|c_{uvN}\|^t$ is the transposed matrix of $\|c_{uvN}\|$. The coefficients c_{uvN} do not involve any unknown parameters; they can be calculated from the observations. As mentioned before the formula (142) is valid only for large N. More precisely, the difference between the right-hand side expression in (142) and the exact value of the covariance between ξ_u and ξ_v converges stochastically to zero.

If we know a priori that the ϵ_{it} are normally and independently distributed with means zero and variances 1, the limit covariance matrix of the variables ξ_u can be considerably simplified. According to (54) the probability density p of the variables $x_{1t}, \cdots, x_{rt}$ is given by

$$p = \frac{J^N}{(2\pi)^{rN/2}} \exp\left(-\frac{1}{2}\sum_{i=1}^{r} L_{it}^2\right), \tag{143}$$

where

$$L_{it} = \sum_{i=1}^{r}\sum_{j=0}^{r}\sum_{k=0}^{p_{ij}} \alpha_{ijk}x_{j,t-k}.$$

We have

$$\sqrt{N}\, y_u = \partial \log p/\partial\alpha_u. \tag{144}$$

Hence, writing dx for $dx_{11}, \cdots, dx_{r1}, \cdots, dx_{1N}, \cdots, dx_{rN}$, we have

$$\begin{aligned} E(y_u) &= \frac{1}{\sqrt{N}}\int_R (\partial \log p/\partial\alpha_u)p\,dx \\ &= \frac{1}{\sqrt{N}}\int_R [(\partial p/\partial\alpha_u)/p]p\,dx = \int_R (\partial p/\partial\alpha_u)dx, \end{aligned} \tag{145}$$

where the symbol $\int_R$ denotes the integral over the whole space. Since $\int_R p\,dx = 1$ identically in $\alpha_u (u=1, \cdots, h)$ we must have

$$\int_R (\partial p/\partial\alpha_u)dx = 0. \tag{146}$$

From (145) and (146) we obtain

$$E(y_u) = 0. \tag{147}$$

Furthermore we have from (144)

$$N\sigma(y_u y_v) = E[(\partial \log p/\partial\alpha_u)(\partial \log p/\partial\alpha_v)]. \tag{148}$$

We shall prove that

$$E[(\partial \log p/\partial\alpha_u)(\partial \log p/\partial\alpha_v)] = -E(\partial^2 \log p/\partial\alpha_u\partial\alpha_v). \tag{149}$$

We start from the equation

$$\int_R (\partial \log p/\partial\alpha_u)p\,dx = 0. \tag{150}$$

Differentiating (150) with respect to α_v we obtain

$$\int_R [(\partial^2 \log p/\partial\alpha_u\partial\alpha_v)p + (\partial \log p/\partial\alpha_u)(\partial p/\partial\alpha_v)]dx = 0. \tag{151}$$

Since

$$\partial p/\partial\alpha_v = [(\partial p/\partial\alpha_v)/p]p = (\partial \log p/\partial\alpha_v)p, \tag{152}$$

(149) follows from (151) and (152). Hence we obtain

$$N\sigma(y_u y_v) = -E(\partial^2 \log p/\partial\alpha_u\partial\alpha_v). \tag{153}$$

It can be shown that the results in (153) check with those in (138) if we put $\mu_3^{(i)} = (\mu_4^{(i)} - 3) = 0$ and use the identity $\partial\alpha^{ji}/\partial\alpha_{lm0} = -\alpha^{mi}\alpha^{jl}$.

Expanding $\partial \log p/\partial\alpha_u$ into a Taylor series around $a_u (u = 1, \cdots, h)$ and putting $\xi_u = \sqrt{N}(a_u - \alpha_u)$ we obtain

$$\begin{aligned}\sqrt{N}\, y_u = \partial \log p/\partial\alpha_u &= (\partial \log p/\partial\alpha_u)_{\alpha_x = a_x (x=1,\cdots,h)} \\ &\quad + \sum_{v=1}^{h} \frac{1}{\sqrt{N}} (\partial^2 \log p/\partial\alpha_u\partial\alpha_v)_{\alpha_x = \bar{a}_y (x=1,\cdots,h)} \xi_v.\end{aligned} \tag{154}$$

where $\bar{a}_x$ is a value between a_x and $\alpha_x (x = 1, \cdots, h)$. Since the quantities a_x have been determined so that

$$(\partial \log p/\partial\alpha_u)_{\alpha_x = \bar{a}_x (x=1,\cdots,h)} = 0, \tag{155}$$

we obtain from (154) and (155)

$$y_u = -\sum_{v=1}^{h} \frac{1}{\sqrt{N}} (\partial^2 \log p/\partial\alpha_u\partial\alpha_v)_{\alpha_x = \bar{a}_x (x=1,\cdots,h)} \xi_v. \tag{156}$$

Since plim $a_u = \alpha_u$ the limit covariance matrix of the variates ξ_u is not changed if in (156) we replace $(1/N)(\partial^2 \log p/\partial\alpha_u\partial\alpha_v)_{\alpha_x = \bar{a}_x (x=1,\ldots,h)}$ by $(1/N)(\partial^2 \log p/\alpha_u\partial\alpha_v)$ taken at the point α_x $(x = 1, \cdots, h)$. Moreover, since $(1/N)(\partial^2 \log p/\partial\alpha_u\partial\alpha_v)$ can be shown to converge to its limit expectation, we may replace $(1/N)(\partial^2 \log p/\partial\alpha_u\partial\alpha_v)$ by its expectation. Hence on account of (153)

$$y_u = -\sum_{v=1}^{h} \frac{1}{N} E(\partial^2 \log p/\partial\alpha_u\partial\alpha_v)\xi_v = \sum_{v=1}^{h} \sigma(y_u y_v)\xi_v. \tag{157}$$

Equation (157) implies that the covariance matrix of the variates ξ_u is equal to the inverse of the covariance matrix of the variates y_u. Since (157) is valid only in the limit, we can merely state that the limit covariance matrix of the variates ξ_u is equal to the inverse of the limit covariance matrix of the variates y_u.

7. *Examples*

In his paper "The Statistical Implications of a System of Simultaneous Equations," [2] T. Haavelmo discussed the following system of equations:

$$Y = aX + \epsilon_1, \qquad X = bY + \epsilon_2, \tag{158}$$

where ϵ_1, ϵ_2 are assumed to be normally and independently distributed with mean 0. Bringing this system to the form (85), which we shall call the reduced form, Haavelmo obtains

$$Y = \epsilon_1^*, \qquad X = \epsilon_2^*, \tag{159}$$

where

$$\epsilon_1^* = (\epsilon_1 + a\epsilon_2)/(1 - ab), \qquad \epsilon_2^* = (b\epsilon_1 + \epsilon_2)/(1 - ab).$$

In the form (158) 4 parameters are unknown, namely a, b, σ_1^2, σ_2^2, where σ_1^2 and σ_2^2 are the variances of ϵ_1 and ϵ_2 respectively. From the reduced form (159) we can see that we can estimate not more than three parameters namely $E(X^2)$, $E(Y^2)$, $E(XY)$. Hence it is impossible to estimate *all* unknown parameters in the system (158) with any method, whatever the number of observations.

We can see this easily directly as follows: The system (158) is equivalent to

$$\begin{aligned} Y &= X(1 - a)/(1 - b) + (\epsilon_1 + \epsilon_2)/(1 - b) = a^*X + \epsilon_1', \\ X &= Y(1 + b)/(1 + a) + (\epsilon_2 - \epsilon_1)/(1 + a) = b^*Y + \epsilon_2', \end{aligned} \tag{160}$$

where again ϵ_1', ϵ_2' are normally and independently distributed. The coefficients in (160), $a^* = (1-a)/(1-b)$, $b^* = (1+b)/(1+a)$, are certainly different from the coefficients in (158). Moreover the variance of ϵ_1' and ϵ_2' are different from the variances of ϵ_1 and ϵ_2 respectively. Hence it is clear that a and b as well as σ_1^2 and σ_2^2 cannot be estimated from any sample and from the conditions (158) alone.

If we should have the additional knowledge that $\sigma_1^2 = \sigma_2^2 = \sigma^2$ then (158) would uniquely determine the parameters a, b, and σ^2. The number of unknown parameters is then three, that is to say, the same as the number of unknown parameters in the reduced form. The unreduced form,however, does not offer any advantage over the reduced form. We should therefore also in this case prefer to solve the equations in the reduced form.

Haavelmo then discussed an economic model which leads to the following system of difference equations:

$$\begin{aligned} u_t &= \alpha r_t + \beta + \epsilon_{1t}, \\ v_t &= k(u_t - u_{t-1}) + \epsilon_{2t}, \\ r_t &= u_t + v_t, \end{aligned} \tag{161}$$

where ϵ_{1t}, ϵ_{2t} are assumed to be independently and normally distributed with mean 0. Eliminating the third of the equations (161) we obtain

$$(162)\qquad \begin{aligned} (1-\alpha)u_t - \alpha v_t - \beta &= \epsilon_{1t}, \\ v_t - ku_t + ku_{t-1} &= \epsilon_{2t}. \end{aligned}$$

The reduced form of the system (162) is given by

$$(163)\qquad \begin{aligned} u_t + \alpha_{11}u_{t-1} + \alpha_1 &= \epsilon_{1t}^*, \\ v_t + \alpha_{21}u_{t-1} + \alpha_2 &= \epsilon_{2t}^*, \end{aligned}$$

where α_{11}, α_{21}, α_1, α_2, and the parameters σ_{11}^*, σ_{22}^*, σ_{12}^* can be expressed in terms of the five parameters of the system (161), i.e.,

$$\alpha_{11}=\alpha k/(1-\alpha-\alpha k),\quad \alpha_{21}=(1-\alpha)k/(1-\alpha-\alpha k),\quad \alpha_1=-\beta/(1-\alpha-\alpha k),$$

$$\alpha_2=-\alpha k/(1-\alpha-\alpha k),\qquad \sigma_{11}^*=(\sigma_1^2+\alpha^2\sigma_2^2)/(1-\alpha-\alpha k)^2,$$

$$\sigma_{12}^*=[k\sigma_1^2+\alpha(1-\alpha)\sigma_2^2]/(1-\alpha-\alpha k)^2,$$

$$\sigma_{22}^*=[k^2\sigma_1^2+(1-\alpha)^2\sigma_2^2]/(1-\alpha-\alpha k)^2.$$

In this case we can estimate the 7 parameters α_{11}, α_{21}, α_1, α_2, σ_{11}^*, σ_{12}^*, σ_{22}^* from the sample. The number of unknown parameters in the original form is 5, namely α, k, β, σ_1^2, σ_2^2. Hence by using the reduced form and estimating the 7 parameters as if they were not subject to any restrictions we should throw away some information and lose in efficiency.

SUMMARY

In the first part of this paper the statistical treatment of a single stochastic difference equation

$$\text{(A)}\qquad x_t = \alpha_1x_{t-1} + \alpha_2x_{t-2} + \cdots + \alpha_px_{t-p} + \epsilon_t$$

is discussed where $\alpha_1, \cdots, \alpha_p$, and α_0 are constant coefficients and $\epsilon_1, \epsilon_2, \cdots$, are random variables. It is assumed that $\epsilon_1, \epsilon_2, \cdots$, are independently distributed each having the same distribution and $E(\epsilon_t)=0$. Furthermore it is assumed that all moments of ϵ_t are finite and that the roots of the equation in ρ

$$\rho^p - \alpha_1\rho^{p-1} - \cdots - \alpha_p = 0$$

are less than 1 in absolute value. Suppose that the value of x_t has been observed for $t=1-p, 1-p+1, \cdots, N$. As statistical estimates of the coefficients $\alpha_1, \cdots, \alpha_p, \alpha_0$ the values $a_1, \cdots, a_p, a_0$ are used for which

$$\sum_{t=1}^{N} (x_t - \alpha_1x_{t-1} - \cdots - \alpha_px_{t-p} - \alpha_0)^2$$

becomes a minimum. The estimates $a_1, \cdots, a_p, a_0$ are the maximum-likelihood estimates of $\alpha_1, \cdots, \alpha_p, \alpha_0$, respectively, if ϵ_t is normally distributed. Although it is not assumed that the distribution of ϵ_t is normal, the estimates $a_1, \cdots, a_p, a_0$ are referred to as maximum-likelihood estimates and the joint limiting distribution of these estimates is derived.

Our main result in Part I can be summarized as follows: For large N the stochastic difference equation (A) can be treated in exactly the same way as a classical regression problem where x_t is the dependent variable and $x_{t-1}, \cdots, x_{t-p}$ are the independent variables. That is to say, the estimates of the coefficients $\alpha_1, \cdots, \alpha_p, \alpha_0$, as well as the joint limiting distribution of these estimates, are the same as if (A) were treated as a classical regression problem. Hence, the joint limiting distribution of $\sqrt{N}(a_1-\alpha_1), \cdots, \sqrt{N}(a_p-\alpha_p)$, and $\sqrt{N}(a_0-\alpha_0)$ is a multivariate normal distribution with zero means and a finite covariance matrix. The covariance between $\xi_i=\sqrt{N}(a_i-\alpha_i)$ and $\xi_j=\sqrt{N}(a_j-\alpha_j)$ $(i, j=0, 1, \cdots, p)$ can be obtained as follows: Denote $(1/N)\sum_{t=1}^{N} x_{t-i}x_{t-j}$ by $D_{ijN}(i, j=1, \cdots, p)$, $(1/N)\sum_{t=1}^{N} x_{t-i}$ by $D_{i0N}=D_{0iN}$ and let $D_{00N}=1$. Furthermore let $\|c_{ijN}\|=\|D_{ijN}\|^{-1}(i, j=0, 1, \cdots, p)$ and $s^2=(1/N)\sum_{t=1}^{N}(x_t-a_1x_{t-1}- \cdots -a_px_{t-p}-a_0)^2$. Then the limit covariance between ξ_i and ξ_j is equal to the stochastic limit of s^2c_{ijN}. Thus, for large N the covariance of ξ_i and ξ_j can be replaced by the quantity s^2c_{ijN} which can be calculated from the observations.

The second part of the paper deals with a system of r difference equations

$$\text{(B)} \qquad \sum_{j=1}^{r}\sum_{k=0}^{p_{ij}} \alpha_{ijk}x_{j,t-k} + \alpha_i = \epsilon_{it},$$

where p_{ij} is the highest lag in the variable x_{jt} occurring in the ith equation, the α's are known or unknown constants, the random vectors $(\epsilon_{11}, \cdots, \epsilon_{r1}), (\epsilon_{12}, \cdots, \epsilon_{r2}), \cdots$, are independently distributed each having the same distribution, and $E(\epsilon_{it})=0$ for all values of i and t. It is assumed that ϵ_{it} has finite moments and that the roots of the equation in ρ

$$\begin{vmatrix} b_{11}(\rho) & \cdots & b_{1r}(\rho) \\ \cdot & \cdots & \cdot \\ b_{r1}(\rho) & \cdots & b_{rr}(\rho) \end{vmatrix} = 0$$

are smaller than 1 in absolute value. The expression $b_{ij}(\rho)$ is given by $\sum_{k=1}^{p_{ij}} \alpha_{ijk}\rho^{-k}$.

The maximum-likelihood estimates of α_{ijk} and α_i are derived under the assumption that the joint distribution of $\epsilon_{1t}, \cdots, \epsilon_{rt}$ is normal.

Denote these maximum-likelihood estimates by a_{ijk} and a_i respectively. The joint limiting distribution of these estimates is derived without assuming normality of the distribution of ϵ_{it}. To summarize the results obtained, we distinguish 3 different cases.

Case A. The matrix $\|\alpha_{ij0}\|$ *is equal to the unit matrix and* $\sigma(\epsilon_{it}\epsilon_{jt})=0$ *for* $i\neq j$. In this case the maximum-likelihood estimates a_{ijk} and a_i are those values of α_{ijk} and α_i for which the sum of squares

$$\sum_{t=1}^{N}\sum_{i=1}^{r}\left[\sum_{j=1}^{r}\sum_{k=0}^{p_{ij}}(\alpha_{ijk}x_{j,t-k}+\alpha_i)\right]^2$$

becomes a minimum, i.e., the maximum-likelihood estimates are equal to the least-squares estimates. The joint limiting distributing of $\xi_{ijk}=\sqrt{N}(a_{ijk}-\alpha_{ijk})$ and $\xi_i=\sqrt{N}(a_i-\alpha_i)$ is a multivariate normal distribution with zero means and finite covariance matrix.

To give the explicit formulas for the covariances, the following notations are introduced: The variable x_{0t} is identically equal to 1. The symbols a_{i00}, α_{i00}, ξ_{i00} will stand for a_i, α_i, and ξ_i respectively. Furthermore, we put

$$p_{i0}=0, \qquad D_{ikjlN}=\frac{1}{N}\sum_{t=1}^{N}x_{it-k}x_{jt-l},$$

$$Q_{il}=\left(\sum_{j=0}^{r}\sum_{k=0}^{p_{ij}}\alpha_{ijk}x_{j,t-k}\right)\left(\sum_{j=0}^{r}\sum_{k=0}^{p_{lj}}\alpha_{ljk}x_{j,t-k}\right),$$

and $\bar{Q}_{il}$ is obtained from Q_{il} by substituting a_{ijk} for α_{ijk}. For any given value i ($i=1,\cdots,r$) denote by m_i the number of different pairs (j,k) for which a_{ijk} is defined, i.e., for which α_{ijk} is unknown. Then we denote ξ_{ijk} by ξ_{iu} if (j,k) is the uth element in this sequence. Similarly we denote D_{iklmN} by $D_{uvN}^{(i)}$ if (j,k) is the uth element and (l,m) the vth element in this sequence (the superscript i is used because the ordered sequence of pairs (j,k) may depend on i). Let $\|c_{uvN}^{(i)}\|=\|D_{uvN}^{(i)}\|^{-1}$ ($u,v=1,\cdots,m_i$). The limit covariance between ξ_{iu} and ξ_{lv} is zero if $i\neq l$ The limit covariance between ξ_{iu} and ξ_{iv} is equal to the stochastic limit of $(\bar{Q}_{ii}/N)C_{uvN}^{(i)}$. Thus, for large N the covariance of ξ_{iu} and ξ_{iv} can be replaced by $(\bar{Q}_{ii}/N)C_{uvN}^{(i)}$ which can be calculated from the observations.

Case B. Here we consider the case in which $\|\alpha_{ij0}\|$ is the unit matrix and nothing is known about the covariance matrix $\|\sigma_{ij}\|$ of the variates $\epsilon_{it},\cdots,\epsilon_{rt}$. It is furthermore assumed that p_{ij} has the same value p for all pairs (i,j) ($i,j=1,\cdots,r$) and that all coefficients $\alpha_{ijk}(k>0)$ and α_i are unknown. We shall use the notations introduced in Case A. The maximum-likelihood estimates $a_{ijk}(i=1,\cdots,r;\ j=0,\cdots,r)$ are again equal to the least-squares estimates. The joint limiting distribution of the variates $\xi_{ijk}=\sqrt{N}(a_{ijk}-\alpha_{ijk})$ is a multivariate normal distribution with zero means and finite covariance matrix. To express the covariances, we arrange the pairs (0, 0) and $(j, k)(j=1, \cdots, r; k=1, \cdots, p)$ in an ordered sequence. Denote D_{ijklN} by D_{uvN} if (i, k) is the uth and (j, l) the vth element in the ordered sequence. Similarly we denote ξ_{ijk} by ξ_{iu} if (j, k) is the uth element of the ordered sequence. Then the limit covariance between ξ_{iu} and ξ_{jv} is equal to the stochastic limit of $(\overline{Q}_{ij}/N)C_{uvN}$ where $\|C_{uvN}\|=\|D_{uvN}\|^{-1}$.

Case C. Here we deal with the general case in which $\|\alpha_{ij0}\|$ is not equal to the unit matrix. The treatment of the general case presents a number of difficulties which do not arise in the special cases A and B. Suppose that all coefficients α_{ijk} and α_i in equation (B) are unknown and that nothing is known about the covariance matrix $\|\sigma_{ij}\|$ of the variates $\epsilon_{1t}, \cdots, \epsilon_{rt}$. Suppose furthermore that p_{ij} has the same value for all pairs $(i, j)(i, j=1, \cdots, r)$. Then it is not possible to have consistent estimates for all parameters α_{ijk}, α_i, and σ_{ij}. Let

$$\text{(C)} \qquad x_{it}+\sum_{j=1}^{r}\sum_{k=1}^{p_{ij}}\overset{*}{\alpha}_{ijk}x_{j,t-k}+\alpha_i^*=\epsilon_{it}^*$$

be the reduced form of the equations (B). Then the coefficient α_{ijk}^* is a linear function of $\alpha_{ijk}, \cdots, \alpha_{rjk}$; α_i^* is a linear function of $\alpha_1, \cdots, \alpha_r$, and ϵ_{it}^* is a linear function of $\epsilon_{1t}, \cdots, \epsilon_{rt}$. Denote the covariance of ϵ_{it}^* and ϵ_{jt}^* by σ_{ij}^*. Of course, α_{ijk}^*, α_i^*, and σ_{ij}^* depend also on α_{ij0}. It has been shown that the observations x_{it} cannot yield any additional information as to the values α_{ijk}, α_i, and σ_{ij} beyond that contained in the estimates of the parameters α_{ijk}^*, α_i^*, and σ_{ij}^*.

If there is no a priori knowledge whatsoever as to the values α_{ijk}, α_i, and σ_{ij}, or if the a priori knowledge available does not imply any restrictions on the values of α_{ijk}^*, α_i^*, and σ_{ij}^*, then it is best to treat the difference equations in the reduced form (C), i.e., to estimate α_{ijk}^*, α_i^*, and σ_{ij}^*. Since the reduced equations (C) satisfy all conditions of Case B, the procedure described in Case B can be applied. If, however, the available a priori knowledge implies certain restrictions on the values of the parameters α_{ijk}^*, α_i^*, and σ_{ij}^*, then the statistical treatment of the reduced equations (C) may be just as involved as that of the original system (B) and no advantage is gained. In some practical cases one may find it worth while to discard the available a priori knowledge and to treat the reduced equations (C) as if the parameters α_{ijk}^*, α_i^*, and σ_{ij}^* were not subject to any restrictions. By doing so we decrease the efficiency of our statistical procedure, but we gain in simplicity. In some cases the gain in simplicity may compensate for the loss in efficiency.

If the a priori knowledge implies restrictions on the parameters α_{ijk}^*, α_i^*, and σ_{ij}^*, and if we are not willing to discard the available information, then it seems preferable to treat the equations in the original form (B). A frequent form of the a priori knowledge is the following: It is known that $\epsilon_{1t}, \cdots, \epsilon_{rt}$ are independently distributed and the values of some of the coefficients α_{ijk}, α_i are known to be zero. The treatment of the equations (B) is carried out in the case when the a priori knowledge is of the type described above. Furthermore it is assumed that this a priori knowledge and the parameters α_{ijk}^*, α_i^*, and σ_{ij}^* uniquely determine all parameters α_{ijk}, α_i, and σ_{ij}. Finally it is assumed that $\sigma^2(\epsilon_{it}) = 1$, since this can always be achieved by multiplying the equations (B) by a constant factor. Under these conditions the joint limiting distribution of the maximum-likelihood estimates of α_{ijk} and α_i is derived. While the distribution of the maximum-likelihood estimates is derived without assuming normality of the distribution of ϵ_{it}, the results become much simpler if the distribution of ϵ_{it} is normal. We shall summarize here merely the results obtained under the assumption of normality of the distribution of ϵ_{it}. Denote α_i by α_{i00}, a_i by a_{i00} and let $\xi_{ijk} = \sqrt{N}(a_{ijk} - \alpha_{ijk})$. Arrange the triples (i, j, k) for which α_{ijk} is unknown in an ordered sequence. Denote α_{ijk}, a_{ijk}, and ξ_{ijk} by α_u, a_u, and ξ_u, respectively, if (i, j, k) is the uth triple in the ordered sequence. Let p be the joint probability density of the variates $x_{11}, \cdots, x_{rN}$ and let $\bar{p}$ be the expression we obtain from p if we replace α_u by a_u. Then our results can be summarized as follows: The joint limiting distribution of the variates ξ_u is a multivariate normal distribution with zero means and finite covariance matrix. The limit value of the covariance matrix $\|\sigma(\xi_u\xi_v)\|$ is equal to the stochastic limit of $\left\| -\frac{1}{N} \frac{\partial^2 \log \bar{p}}{\partial a_u \partial a_v} \right\|^{-1}$. Hence, for large N the matrix $\|\sigma(\xi_u\xi_v)\|$ can be replaced by $\left\| -\frac{1}{N} \frac{\partial^2 \log \bar{p}}{\partial a_u \partial a_v} \right\|^{-1}$. The quantity $\frac{\partial^2 \log \bar{p}}{\partial a_u \partial a_v}$ can be calculated from the observations.

Columbia University

REFERENCES

[1] J. L. Doob, "Probability and Statistics," *Transactions of the American Mathematical Society*, Vol. 36, 1934, pp. 759–775.

[2] T. Haavelmo, "The Statistical Implications of a System of Simultaneous Equations," ECONOMETRICA, Vol. 11, January, 1943, pp. 1–12.

[3] R. L. Anderson, "Distribution of the Serial Correlation Coefficient," *Annals of Mathematical Statistics*, Vol. 13, March, 1942, pp. 1–13.

[4] Tjalling Koopmans, "Serial Correlation and Quadratic Forms in Normal Variables," *Annals of Mathematics Statistics*, Vol. 13, March, 1942, pp. 14–33.

Reprinted from the
TRANSACTIONS OF THE AMERICAN MATHEMATICAL SOCIETY
Vol. 54, No. 3, pp. 426-482
November, 1943

TESTS OF STATISTICAL HYPOTHESES CONCERNING SEVERAL PARAMETERS WHEN THE NUMBER OF OBSERVATIONS IS LARGE(1)

BY

ABRAHAM WALD

TABLE OF CONTENTS

1. **Introduction.** In this paper we shall deal with the following general problem: Let $f(x^1, x^2, \cdots, x^r, \theta^1, \cdots, \theta^k)$ be the joint probability density function of the variates (chance variables) $x^1, \cdots, x^r$ involving k unknown parameters $\theta^1, \cdots, \theta^k$. Any set of k values $\theta^1, \cdots, \theta^k$ can be represented by a point θ in the k-dimensional Cartesian space with the coordinates $\theta^1, \cdots, \theta^k$. We shall denote the set of all possible parameter points by Ω. The set Ω is called parameter space. The parameter space Ω may be the whole k-dimensional Cartesian space, or a subset of it. For any subset ω of Ω, we shall denote by H_ω the hypothesis that the parameter point lies in ω. If ω consists of a single point, H_ω is called a simple hypothesis, otherwise H_ω is called a composite hypothesis. In this paper we shall discuss the question of an appropriate test of the hypothesis H_ω based on a large number of independent observations on $x^1, \cdots, x^r$.

For simplicity we shall introduce the following notations: The letter θ or θ_i for any subscript i will denote a point in the parameter space Ω. The letter x

Some of the results contained in this paper were presented to the Society, February 22, 1941 and September 2, 1941; received by the editors March 31, 1943.

(1) Research under a grant-in-aid from the Carnegie Corporation of New York.

will denote the random vector with the components $x^1, \cdots, x^r$, and x_α will denote the random vector with the components $x_\alpha^1, \cdots, x_\alpha^r$ where x_α^i $(i=1, 2, \cdots, r)$ denotes the αth observation on x^i. In general, the components of a vector v in the s-dimensional space will be denoted by $v^1, \cdots, v^s$, that is the components of a vector will always be indicated by superscripts. Throughout this paper all vectors have their initial points at the origin. We denote by E_n a sample point in the rn-dimensional sample space of n independent observations on the random vector x. For any relation R we denote by $P(R|\theta)$ the probability that R holds under the assumption that θ is the true parameter point. A region (subset) of the rn-dimensional sample space will always be denoted by a capital letter with the subscript n. For any region W_n the symbol $P(W_n|\theta)$ will denote the probability that E_n falls within W_n under the assumption that θ is the true parameter point. Throughout this paper the word "region" will be used synonymously with "subset," since in the theory of testing statistical hypotheses it is customary to call the subsets which are used as criterions of rejection, critical regions.

By maximum likelihood estimates $\hat{\theta}_n^1, \cdots, \hat{\theta}_n^k$ of $\theta^1, \cdots, \theta^k$ we mean values of $\theta^1, \cdots, \theta^k$ for which $\prod_{\alpha=1}^n f(x_\alpha^1, \cdots, x_\alpha^r, \theta^1, \cdots, \theta^k)$ becomes a maximum. The subscript n in the symbol $\hat{\theta}_n^i$ will indicate that the maximum likelihood estimate is based on n independent observations on $x', \cdots, x^r$. A region W_n in the rn-dimensional sample space is called a critical region for testing the hypothesis H_ω if we decide to reject H_ω when and only when the observed sample point falls within W_n. For any θ not in ω the value of $P(W_n|\theta)$ is called the power of the critical region W_n with respect to the alternative hypothesis θ. The least upper bound of $P(W_n|\theta)$ with respect to θ, restricting θ to ω, will be called the size of the critical region W_n. A critical region is considered the better, the smaller its size and the greater its power.

In several previous publications(2) the author has considered the case of a single unknown parameter θ and the problem of testing a simple hypothesis $\theta=\theta_0$. It was shown, among other things, that under certain conditions the critical region given by the inequality $|n^{1/2}(\hat{\theta}_n-\theta_0)| \geq A_n$ has certain optimum properties. Here the symbol A_n denotes some properly chosen constant. In this paper the general case of several unknown parameters is treated and simple as well as composite hypotheses are considered.

By an equality or inequality among vectors we shall mean that the equality or inequality holds for all components. For example, if $\hat{\theta}_n$ denotes the vector with the components $\hat{\theta}_n^1, \cdots, \hat{\theta}_n^k$, where $\hat{\theta}_n^i$ is the maximum likelihood estimate of θ^i, and if A is a real number, then the inequality

$$n^{1/2}(\hat{\theta}_n - \theta) < A$$

(2) *Asymptotically most powerful tests of statistical hypotheses*, Ann. Math. Statist. vol. 12 (1941). *Some examples of asymptotically most powerful tests*, Ann. Math. Statist. ibid.

denotes the set of inequalities

$$n^{1/2}(\hat{\theta}_n^i - \theta^i) < A \qquad (i = 1, \cdots, k).$$

Or, if t is a vector with the components $t^1, \cdots, t^k$, then

$$n^{1/2}(\hat{\theta}_n - \theta) < t$$

denotes the set of inequalities

$$n^{1/2}(\hat{\theta}_n^i - \theta^i) < t^i \qquad (i = 1, \cdots, k).$$

2. **Assumptions on the density function $f(x,\theta)$.** For any function $\psi(x)$ we shall use the symbol $\int_{-\infty}^{+\infty}\psi(x)dx$ as an abbreviation for $\int_{-\infty}^{+\infty} \cdots \int_{-\infty}^{+\infty}\psi(x)\,dx^1 \cdots dx^r$. Denote by $E_\theta\psi(x)$ the expected value of $\psi(x)$ under the assumption that θ is the true parameter point, that is

$$E_\theta\psi(x) = \int_{-\infty}^{+\infty} \psi(x)f(x, \theta)dx.$$

For any x, any positive value δ, and any θ_1 denote by $\psi_{ij}(x, \theta_1, \delta)$ the greatest lower bound, and by $\phi_{ij}(x, \theta_1, \delta)$ the least upper bound of $\partial^2 \log f(x, \theta)/\partial\theta^i\partial\theta^j$ with respect to θ in the θ-interval $|\theta - \theta_1| \leq \delta$.

Throughout this paper the following assumptions on $f(x, \theta)$ are made:

ASSUMPTION I. *Denote by D_n the set of all sample points E_n for which the maximum likelihood estimate $\hat{\theta}_n = (\hat{\theta}_n^1, \cdots, \hat{\theta}_n^k)$ exists and the second order partial derivatives $\partial^2 f(x_\alpha, \theta)/\partial\theta^i\partial\theta^j$ $(\alpha = 1, \cdots, n; i, j = 1, \cdots, k)$ are continuous functions of θ. It is assumed that*

$$\lim_{n\to\infty} P(D_n \mid \theta) = 1 \qquad \textit{uniformly in } \theta.$$

If for a sample point E_n there exist several maximum likelihood estimates, we can select one of them by some given rule. Hence we shall consider $\hat{\theta}_n$ as a single-valued function of E_n defined for all points of D_n.

ASSUMPTION II. *For any positive ϵ*

$$\lim_{n=\infty} P[\,|\hat{\theta}_n - \theta| < \epsilon \mid \theta] = 1$$

uniformly in θ, where $\hat{\theta}_n$ denotes the vector with the components $\hat{\theta}_n^1, \cdots, \hat{\theta}_n^k$ and $\hat{\theta}_n^i$ is the maximum likelihood estimate of θ^i.

Assumption II is somewhat more than consistency of the maximum likelihood estimate $\hat{\theta}_n$. In fact consistency means only that for any positive ϵ

$$\lim_{n=\infty} P[\,|\hat{\theta}_n - \theta| < \epsilon \mid \theta] = 1,$$

without requiring that the convergence be uniform in θ. If $\hat{\theta}_n$ satisfies Assumption II we shall say that $\dot{\theta}_n$ is a uniformly consistent estimate of θ. A rigorous proof of the consistency of $\dot{\theta}_n$ (under certain restrictions on $f(x, \theta)$) was given by J. L. Doob(3). The uniform consistency of $\dot{\theta}_n$ together with the uniform consistency of the likelihood ratio test will be proved on the basis of some weak assumptions on $f(x, \theta)$ in a forthcoming paper.

ASSUMPTION III. *The following three conditions are fulfilled:*

(a) *For any sequences* $\{\theta_{1n}\}, \{\theta_{2n}\}$, *and* $\{\delta_n\}$ *for which* $\lim_{n=\infty} \theta_{1n} = \lim_{n=\infty} \theta_{2n} = \theta$ *and* $\lim \delta_n = 0$ *we have*

$$\lim_{n=\infty} E_{\theta_{1n}} \psi_{ij}(x, \theta_{2n}, \delta_n) = \lim_{n=\infty} E_{\theta_{1n}} \phi_{ij}(x, \theta_{2n}, \delta_n) = E_\theta \frac{\partial^2 \log f(x, \theta)}{\partial\theta^i \partial\theta^j}$$

uniformly in θ.

(b) *There exists a positive* ϵ *such that the expectations* $E_{\theta_1}[\psi_{ij}(x, \theta_2, \delta)]^2$ *and* $E_{\theta_1}[\phi_{ij}(x, \theta_2, \delta)]^2$ *are bounded functions of* θ_1, θ_2 *and* δ *in the domain* D_ϵ *defined by the inequalities* $|\theta_1 - \theta_2| \leqq \epsilon$ *and* $|\delta| \leqq \epsilon$.

(c) *The greatest lower bound with respect to* θ *of the absolute value of the determinant of the matrix* $\| -E_\theta(\partial^2 \log f(x, \theta))/\partial\theta^i\partial\theta^j \|$ *is positive.*

ASSUMPTION IV. $\int_{-\infty}^{+\infty} \partial f(x, \theta)/\partial\theta^i \, dx = \int_{-\infty}^{+\infty} \partial^2 f(x, \theta)/\partial\theta^i\partial\theta^j \, dx = 0$.

Assumption IV simply means that we may differentiate with respect to θ under the integral sign. In fact

$$\int_{-\infty}^{+\infty} f(x, \theta) dx = 1$$

identically in θ. Hence

$$\frac{\partial}{\partial\theta^i} \int_{-\infty}^{+\infty} f(x, \theta) dx = \frac{\partial^2}{\partial\theta^i\partial\theta^j} \int_{-\infty}^{+\infty} f(x, \theta) dx = 0.$$

Differentiating under the integral sign, we obtain the relations in Assumption IV.

ASSUMPTION V. *There exists a positive* η *such that*

$$E_\theta \left| \frac{\partial}{\partial\theta^i} \log f(x, \theta) \right|^{2+\eta} \qquad (i = 1, \cdots, k)$$

are bounded functions of θ.

3. **The joint limit distribution of** $\dot{\theta}_n$. Denote $n^{1/2}(\dot{\theta}_n^i - \theta^i)$ by $z_n^i(\theta^i)$ and let $z_n(\theta)$ be the vector with the components $z_n^1(\theta^1), \cdots, z_n^k(\theta^k)$. For any constant vector t denote the probability $P[z_n(\theta) < t | \theta]$ by $\Phi_n(t | \theta)$. We shall prove the following proposition.

---

(3) J. L. Doob, *Probability and statistics*, Trans. Amer. Math. Soc. vol. 36 (1934) pp. 759–775.

PROPOSITION I. *The distribution function $\Phi_n(t\mid\theta)$ converges with $n\to\infty$, uniformly in t and θ, towards the cumulative multivariate normal distribution with zero means and covariance matrix*

$$\|\sigma_{ij}(\theta)\| = \|c_{ij}(\theta)\|^{-1}$$

where

$$c_{ij}(\theta) = -E_\theta \partial^2 \log f(x,\theta)/\partial\theta^i\partial\theta^j.$$

Proof. Because of Assumption IV, we have

$$E_\theta \frac{\partial \log f(x,\theta)}{\partial\theta^i} = \int_{-\infty}^{+\infty} \frac{\partial f(x,\theta)}{\partial\theta^i}\, dx = 0. \tag{1}$$

From

$$\frac{\partial^2 \log f(x,\theta)}{\partial\theta^i\partial\theta^j} = \frac{1}{f(x,\theta)}\,\frac{\partial^2 f(x,\theta)}{\partial\theta^i\partial\theta^j} - \frac{1}{f^2}\left[\frac{\partial f}{\partial\theta^i}\,\frac{\partial f}{\partial\theta^j}\right]$$

we obtain because of Assumption IV

$$E_\theta\left\{\left[\frac{\partial \log f}{\partial\theta^i}\,\frac{\partial \log f}{\partial\theta^j}\right]\right\} = -E_\theta \frac{\partial^2 \log f(x,\theta)}{\partial\theta^i\partial\theta^j} = c_{ij}(\theta). \tag{2}$$

From (2) it follows that the matrix $\|c_{ij}(\theta)\|$ is positive definite or semidefinite. Because of condition (c) of Assumption III the matrix $\|c_{ij}(\theta)\|$ must be positive definite. For any point E_n of the set D_n defined in Assumption I consider the Taylor expansion

$$\begin{aligned}\sum_\alpha \frac{\partial \log f(x_\alpha,\theta_1)}{\partial\theta^i} &= \sum_\alpha \frac{\partial \log f(x_\alpha,\theta)}{\partial\theta^i}\\ &\quad + \sum_j (\theta_1^j - \theta^j)\left[\sum_\alpha \frac{\partial^2 \log f(x_\alpha,\bar\theta)}{\partial\theta^i\partial\theta^j}\right],\end{aligned} \tag{3}$$

where $\bar\theta$ lies on the segment connecting the points θ and θ_1. Denote $n^{-1/2}\sum_\alpha \partial \log f(x_\alpha,\theta)/\partial\theta^i$ by $y_n^i(\theta)$ and let $y_n(\theta)$ be the vector with the components $y_n^1(\theta),\cdots,y_n^k(\theta)$. Substituting $\dot\theta_n$ for θ_1 in (3), the left-hand side of (3) becomes equal to zero and we obtain

$$y_n^i(\theta) + \sum_j \left\{\left[n^{1/2}(\dot\theta_n^j - \theta^j)\right]\frac{1}{n}\left[\sum_\alpha \frac{\partial^2 \log f(x_\alpha,\bar\theta)}{\partial\theta^i\partial\theta^j}\right]\right\} = 0 \tag{4}$$

or

$$y_n^i(\theta) + \sum_j z_n^j(\theta)\,\frac{1}{n}\left[\sum_\alpha \frac{\partial^2 \log f(x_\alpha,\bar\theta)}{\partial\theta^i\partial\theta^j}\right] = 0. \tag{5}$$

Let ν be an arbitrary positive number and let $Q_n(\theta)$ be the subset of D_n for which the inequality

$$\left| \frac{1}{n} \sum_\alpha \frac{\partial^2 \log f(x_\alpha, \bar{\theta})}{\partial\theta^i \partial\theta^j} + c_{ij}(\theta) \right| < \nu \tag{6}$$

holds. We will prove that

$$\lim_{n=\infty} P[Q_n(\theta) \mid \theta] = 1 \tag{7}$$

uniformly in θ.

Let τ_0 be a positive number such that

$$\begin{aligned} \left| E_\theta \phi_{ij}(x, \theta, \tau_0) - E_\theta \frac{\partial^2 \log f(x, \theta)}{\partial\theta^i \partial\theta^j} \right| &< \frac{\nu}{2}, \\ \left| E_\theta \psi_{ij}(x, \theta, \tau_0) - E_\theta \frac{\partial^2 \log f(x, \theta)}{\partial\theta^i \partial\theta^j} \right| &< \frac{\nu}{2} \end{aligned} \tag{8}$$

for all values of θ. Because of condition (a) of Assumption III, such a τ_0 certainly exists. Denote by $R_n(\theta)$ the subset of D_n consisting of all points E_n for which the inequality

$$| \dot{\theta}_n - \theta | \leq \tau_0 \tag{9}$$

holds. Because of Assumption II

$$\lim_{n=\infty} P[R_n(\theta) \mid \theta] = 1 \tag{10}$$

uniformly in θ. Since $\bar{\theta}$ lies in the interval $[\dot{\theta}_n, \theta]$ we have for all points of $R_n(\theta)$

$$| \bar{\theta}^i - \theta^i | \leq \tau_0 \qquad (i = 1, \cdots, k). \tag{11}$$

Hence at any point of $R_n(\theta)$ the inequality

$$\sum_\alpha \psi_{ij}(x_\alpha, \theta, \tau_0) \leq \sum_\alpha \frac{\partial^2 \log f(x_\alpha, \bar{\theta})}{\partial\theta^i \partial\theta^j} \leq \sum_\alpha \phi_{ij}(x_\alpha, \theta, \tau_0) \tag{12}$$

holds. Let the region $S_n(\theta)$ be defined by the inequality

$$\left| \frac{1}{n} \sum_\alpha \phi_{ij}(x_\alpha, \theta, \tau_0) - E_\theta \phi_{ij}(x, \theta, \tau_0) \right| < \frac{\nu}{2} \tag{13}$$

and $T_n(\theta)$ be defined by the inequality

$$\left| \frac{1}{n} \sum_\alpha \psi_{ij}(x_\alpha, \theta, \tau_0) - E_\theta \psi_{ij}(x, \theta, \tau_0) \right| < \frac{\nu}{2}. \tag{14}$$

It follows from (b) of Assumption III and Tshebysheff's inequality that

$$\lim_{n=\infty} P[S_n(\theta) \mid \theta] = \lim_{n=\infty} P[T_n(\theta) \mid \theta] = 1 \tag{15}$$

uniformly in θ. Denote by $U_n(\theta)$ the common part of the regions $R_n(\theta)$, $S_n(\theta)$ and $T_n(\theta)$. In $U_n(\theta)$ we have because of (8), (13) and (14)

$$\begin{aligned} &\left| \frac{1}{n} \sum_\alpha \phi_{ij}(x_\alpha, \theta, \tau_0) - E_\theta \frac{\partial^2 \log f(x, \theta)}{\partial\theta^i \partial\theta^j} \right| < \nu, \\ &\left| \frac{1}{n} \sum_\alpha \psi_{ij}(x_\alpha, \theta, \tau_0) - E_\theta \frac{\partial^2 \log f(x, \theta)}{\partial\theta^i \partial\theta^j} \right| < \nu. \end{aligned} \tag{16}$$

From this we obtain (6) because of (12). That is to say, the inequality (6) is valid everywhere in $U_n(\theta)$. Since $\lim_{n=\infty} P[U_n(\theta) | \theta] = 1$ uniformly in θ and $U_n(\theta)$ is a subset of D_n, our statement about $Q_n(\theta)$ is proved.

Since the determinant $|c_{ij}(\theta)|$ has a positive lower bound, we obtain easily from (5) and (6)

$$\overset{i}{z}_n(\theta) = \sum_j \overset{j}{y}_n(\theta) [\sigma_{ij}(\theta) + \nu\epsilon_{ijn}(E_n, \theta, \nu)], \tag{17}$$

where $\epsilon_{ijn}(E_n, \theta, \nu)$ is a bounded function of E_n, θ, and ν, provided that, for each θ, E_n is restricted to points of $Q_n(\theta)$ and $|\nu|$ is less than a certain positive number ν_0.

Let $\bar{z}_n(\theta)$ be defined as follows: $\bar{z}_n(\theta) = z_n(\theta)$ at any point of $Q_n(\theta)$, and

$$\overset{i}{\bar{z}}_n(\theta) = \sum_j \overset{j}{y}_n(\theta)\sigma_{ij}(\theta)$$

at any point outside $Q_n(\theta)$. It follows from (7) that

$$\lim_{n=\infty} \{P[\bar{z}_n(\theta) < t \mid \theta] - P[z_n(\theta) < t \mid \theta]\} = 0 \tag{18}$$

uniformly in t and θ.

Denote $\sum_j y_n^j(\theta)\sigma_{ij}(\theta)$ by $\dot{z}_n^i(\theta)$ and let $\dot{z}_n(\theta)$ be the vector with the components $\dot{z}_n^1(\theta), \cdots, \dot{z}_n^k(\theta)$. From (1), (2), Assumption V and the general limit theorems it follows that $P[y_n(\theta) < t | \theta]$ converges with $n \to \infty$, uniformly in t and θ, towards the k-variate cumulative normal distribution with zero means and covariance matrix $\|c_{ij}(\theta)\|$. From this we easily obtain that $P[\dot{z}_n(\theta) < t | \theta]$ converges with $n \to \infty$, uniformly in t and θ, towards the cumulative joint normal distribution with zero means and covariance matrix $\|\sigma_{ij}(\theta)\|$. Since ν can be chosen arbitrarily small, we obviously have because of (17)

$$\lim_{n=\infty} \{P[\dot{z}_n(\theta) < t \mid \theta] - P[\bar{z}_n(\theta) < t \mid \theta]\} = 0 \tag{19}$$

uniformly in t and θ. Proposition I follows from (18) and (19).

4. **Reduction of the general problem to the case of a multivariate normal distribution.** In this section we shall prove two lemmas which will enable us to reduce the general problem of large sample inference to the case where the variates under consideration have a joint normal distribution.

LEMMA 1. *For each positive integer n there exists a set-function* $W_n^*(W_n)$ *defined over all Borel measurable subsets* W_n *of the rn-dimensional sample space such that the following two conditions are fulfilled:*

(a) *For each* W_n, $W_n^*(W_n)$ *is a Borel measurable subset of the rn-dimensional sample space with the following property: For each point of* $W_n^*(W_n)$ *the maximum likelihood estimate exists and if a sample point* E_n *lies in* $W_n^*(W_n)$ *then also all those points* E_n' *lie in* $W_n^*(W_n)$ *for which* $\dot{\theta}_n(E_n')=\dot{\theta}_n(E_n)$.

(b) $\text{Lim}_{n=\infty}\{P[W_n^*(W_n)|\theta]-P[W_n|\theta]\}=0$ *uniformly in* θ *and* W_n.

Proof. Let λ be a real variable which takes only non-negative values and consider the region $W_n(\theta, \lambda)$ defined as the common part of the region W_n and the region $|n^{1/2}(\dot{\theta}_n-\theta)|\leqq\lambda$. Similarly let $W_n^*(W_n, \theta, \lambda)$ be the intersection of $W_n^*(W_n)$ and the region $|n^{1/2}(\dot{\theta}_n-\theta)|\leqq\lambda$.

For any function $\Phi(v)$ we will denote by $\text{g.l.b.}_v\,\Phi(v)$ and $\text{l.u.b.}_v\,\Phi(v)$ the greatest lower bound and the least upper bound of $\Phi(v)$ respectively.

Since, on account of Proposition I, for any sequence $\{\lambda_n\}$ for which $\lim_{n=\infty}\lambda_n=\infty$

$$\lim_{n=\infty}\{\underset{\theta}{\text{g.l.b.}}\,[P[\,|\,n^{1/2}(\dot{\theta}_n-\theta)\,|\leqq\lambda_n\,|\,\theta]]\}=1,$$

Lemma I is proved if we show that there exists a sequence $\{\lambda_n\}$ not depending on θ and W_n such that $\lim_{n=\infty}\lambda_n=\infty$ and

$$\lim_{n=\infty}\{P[W_n(\theta,\lambda_n)\,|\,\theta]-P[W_n^*(W_n,\theta,\lambda_n)\,|\,\theta]\}=0 \tag{20}$$

uniformly in W_n and θ.

Let q be a real variable restricted to values greater than 1. For any set of k integers $(r_1, \cdots, r_k)$ and for any value q denote by $I_n(r_1, \cdots, r_k, q)$ the region defined by the inequalities:

$$\frac{r_1-1/2}{qn^{1/2}}<\dot{\theta}_n^1<\frac{r_1+1/2}{qn^{1/2}},\ \cdots,\ \frac{r_k-1/2}{qn^{1/2}}<\dot{\theta}_n^k<\frac{r_k+1/2}{qn^{1/2}}. \tag{21}$$

Furthermore denote by $\theta_n(r_1, \cdots, r_k, q)$ the parameter point with the coordinates $r_1/qn^{1/2}, \cdots, r_k/qn^{1/2}$. We order the system of all sets of k integers $(r_1, \cdots, r_k)$ in a sequence and we shall denote by $I_{ns}(q)$ the interval $I_n(r_1, \cdots, r_k, q)$ where $(r_1, \cdots, r_k)$ is the sth element in the ordered sequence $(s=1, 2, \cdots, \text{ad inf.})$. Similarly $\theta_{ns}(q)$ denotes the parameter point $\theta_n(r_1, \cdots, r_k, q)$ where $(r_1, \cdots, r_k)$ is the sth element in the ordered sequence.

Let $J_{ns}(W_n, q)$ be the common part of the three regions $I_{ns}(q)$, W_n and $Q_n[\theta_{ns}(q)]$ where for any θ, $Q_n(\theta)$ is defined by the inequalities

$$(22)\qquad \begin{aligned} \left| \frac{1}{n} \sum_\alpha \phi_{ij}\left(x_\alpha, \theta, \frac{1}{n^{1/3}}\right) + c_{ij}(\theta) \right| &\leqq \frac{1}{n^{1/3}} + \nu(n) + \bar{\nu}(n), \\ \left| \frac{1}{n} \sum_\alpha \psi_{ij}\left(x_\alpha, \theta, \frac{1}{n^{1/3}}\right) + c_{ij}(\theta) \right| &\leqq \frac{1}{n^{1/3}} + \nu(n) + \bar{\nu}(n). \end{aligned}$$

The expression $\nu(n)$ is equal to $\text{l.u.b.}_\theta\ [E_\theta\phi_{ij}(x, \theta, 2/n^{1/3}) - E_\theta\psi_{ij}(x, \theta, 2/n^{1/3})]$ and the expression $\bar{\nu}(n)$ is defined as follows: Denote by $c_{ij}(\theta, n)$ the least upper bound of $|c_{ij}(\theta) - c_{ij}(\bar{\theta})|$ with respect to $\bar{\theta}$ where $\bar{\theta}$ can take only values in the interval $[\theta - 1/n^{1/3}, \theta + 1/n^{1/3}]$. Then $\bar{\nu}(n)$ is defined as the least upper bound of $c_{ij}(\theta, n)$ with respect to θ, i and j. Because of condition (a) of Assumption III, we obtain

$$(23)\qquad \lim_{n=\infty} \nu(n) = \lim_{n=\infty} \bar{\nu}(n) = 0.$$

Let $J^*_{ns}(W_n, q)$ be a subset of $I_{ns}(q)$ such that the following two conditions are fulfilled:

(24a) If E_n is an element of $J^*_{ns}(W_n, q)$ then also all those points E_n' for which $\dot{\theta}_n(E_n') = \dot{\theta}_n(E_n)$ are elements of $J^*_{ns}(W_n, q)$, that is $J^*_{ns}(W_n, q)$ can be represented as a subset of the space of the maximum likelihood estimates. Furthermore $J^*_{ns}(W_n, q)$ is an interval in this space.

(24b) $\text{Lim}_{n=\infty}\{\text{l.u.b.}_{s,W_n}|P[J_{ns}(W_n, \ q)|\theta_{ns}(q)] - \mathfrak{P}[J^*_{ns}(W_n, \ q)|\theta_{ns}(q)]|\}$ $=0$ where $\mathfrak{P}(V_n|\theta)$ denotes the probability of V_n calculated on the basis that the joint distribution of $n^{1/2}(\dot{\theta}^1_n - \theta^1), \cdots, n^{1/2}(\dot{\theta}^k_n - \theta^k)$ is normal with zero means and covariance matrix $\|\sigma_{ij}(\theta)\| = \|c_{ij}(\theta)\|^{-1}$.

The existence of such a set $J^*_{ns}(W_n, q)$ can be proved as follows: Obviously there exists a subset $J^*_{ns}(W_r, q)$ of $I_{ns}(q)$ such that (24a) is fulfilled and

$$\mathfrak{P}[J^*_{ns}(W_n, q) \mid \theta_{ns}(q)] = \min \{P[J_{ns}(W_n, q) \mid \theta_{ns}(q)], \mathfrak{P}[I_{ns}(q) \mid \theta_{ns}(q)]\}.$$

Since $J_{ns}(W_n, q)$ is a subset of $I_{ns}(q)$ and since $\lim_{n=\infty} \{P[I_{ns}(q)|\theta_{ns}(q)] - \mathfrak{P}[I_{ns}(q)|\theta_{ns}(q)]\} = 0$ uniformly in s, the above defined subset $J^*_{ns}(W_n, q)$ satisfies also the condition (24b). We define

$$(25)\qquad W^*_n(W_n, q) = \sum_{s=1}^{\infty} J^*_{ns}(W_n, q).$$

Furthermore we define the regions $J_n(W_n, \theta, \lambda, q)$ and $J^*_n(W_n, \theta, \lambda, q)$ as follows:

$$(26)\qquad J_n(W_n, \theta, \lambda, q) = \sum_s J_{ns}(W_n, q),$$

$$(27)\qquad J^*_n(W_n, \theta, \lambda, q) = \sum_s J^*_{ns}(W_n, q).$$

where the summation is to be taken over all values of s for which $|\theta-\theta_{ns}(q)| \leqq \lambda/n^{1/2}$.

Let $\bar{J}_{ns}^*(W_n, q)$ be the intersection of $J_{ns}^*(W_n, q)$ and $Q_n[\theta_{ns}(q)]$, and let $\bar{J}_n^*(W_n, \theta, \lambda, q) = \sum_s \bar{J}_{ns}^*(W_n, q)$ where the summation is to be taken over all values of s for which $|\theta-\theta_{ns}(q)| \leqq \lambda/n^{1/2}$. Furthermore let $W_n^*(W_n, \theta, \lambda, q)$ be the intersection of $W_n^*(W_n, q)$ with the region $|n^{1/2}(\dot{\theta}_n-\theta)| \leqq \lambda$.

If for a value of s we have $n^{1/2}|\theta-\theta_{ns}(q)| \leqq \lambda$ then for all points of $I_{ns}(q)$ we have $n^{1/2}|\theta-\dot{\theta}_n| \leqq \lambda+1/q$. If there exists a point in $I_{ns}(q)$ for which $n^{1/2}|\theta-\dot{\theta}_n| \leqq \lambda-1/q$ then $n^{1/2}|\theta-\theta_{ns}(q)| \leqq \lambda$. Hence $W_n^*(W_n, \theta, \lambda-1/q, q)$ is a subset of $J_n^*(W_n, \theta, \lambda, q)$, and the latter is a subset of $W_n^*(W_n, \theta, \lambda+1/q, q)$. Thus

$$\begin{aligned} &| P[W_n^*(W_n, \theta, \lambda, q) \mid \theta] - P[J_n^*(W_n, \theta, \lambda, q) \mid \theta] | \\ (28) \qquad &\leqq P[W_n^*(W_n, \theta, \lambda + 1/q, q) \mid \theta] - P[W_n^*(W_n, \theta, \lambda - 1/q, q) \mid \theta] \\ &\leqq P[\lambda - 1/q \leqq n^{1/2} | \dot{\theta}_n - \theta | \leqq \lambda + 1/q \mid \theta]. \end{aligned}$$

According to condition (b) of Assumption III $E_\theta[\phi_{ij}(x, \theta, \delta)]^2$ and $E_\theta[\psi_{ij}(x, \theta, \delta)]^2$ are bounded functions of θ. Hence also $E_\theta\phi_{ij}(x, \theta, \delta)$ and $E_\theta\psi_{ij}(x, \theta, \delta)$ are bounded functions of θ. Substituting $\delta=0$ we find that $c_{ij}(\theta)$ is a bounded function of θ. From the boundedness of $c_{ij}(\theta)$ and from the fact that the determinant $|c_{ij}(\theta)|$ has a positive lower bound (condition (c) of Assumption III) it follows because of Proposition I that

$$(29) \qquad \lim_{q=\infty} \{\underset{\theta}{\text{l.u.b.}}\, P[\lambda - 1/q \leqq n^{1/2} | \dot{\theta}_n - \theta | \leqq \lambda + 1/q \mid \theta]\} = 0$$

uniformly in λ. From (28) and (29) we obtain

$$(30) \lim_{n=\infty} \sup \{\underset{\theta, W_n}{\text{l.u.b.}} | P[W_n^*(W_n, \theta, \lambda, q) \mid \theta] - P[J_n^*(W_n, \theta, \lambda, q) \mid \theta] |\} = \epsilon_1(\lambda, q)$$

where $\lim_{q=\infty} \epsilon_1(\lambda, q) = 0$ uniformly in λ.

Denote by $R_n(\theta, \lambda, q)$ the common part of the regions $Q_n[\theta_{ns}(q)]$ formed for all values s for which $|\theta-\theta_{ns}(q)| \leqq \lambda/n^{1/2}$. Then for almost all n the region $R_n(\theta, \lambda, q)$ contains the region $T_n(\theta)$ as a subset, where $T_n(\theta)$ is defined by the inequalities

$$(31) \qquad \begin{aligned} &\left| \frac{1}{n} \sum_\alpha \phi_{ij}\left(x_\alpha, \theta, \frac{2}{n^{1/3}}\right) + c_{ij}(\theta) \right| < \frac{1}{n^{1/3}} + \nu(n), \\ &\left| \frac{1}{n} \sum_\alpha \psi_{ij}\left(x_\alpha, \theta, \frac{2}{n^{1/3}}\right) + c_{ij}(\theta) \right| < \frac{1}{n^{1/3}} + \nu(n). \end{aligned}$$

In fact, from (31) it follows that

$$
(32)\quad
\begin{aligned}
c_{ij}(\theta) - \frac{1}{n^{1/3}} - \nu(n) &\leqq \frac{1}{n}\sum_\alpha \phi_{ij}\left(x_\alpha, \theta, \frac{2}{n^{1/3}}\right) \leqq c_{ij}(\theta) + \frac{1}{n^{1/3}} + \nu(n),\\
c_{ij}(\theta) - \frac{1}{n^{1/3}} - \nu(n) &\leqq \frac{1}{n}\sum_\alpha \psi_{ij}\left(x_\alpha, \theta, \frac{2}{n^{1/3}}\right) \leqq c_{ij}(\theta) + \frac{1}{n^{1/3}} + \nu(n).
\end{aligned}
$$

Since

$$
\begin{aligned}
\frac{1}{n}\sum_\alpha \psi_{ij}\left(x_\alpha, \theta, \frac{2}{n^{1/3}}\right) &\leqq \frac{1}{n}\sum_\alpha \phi_{ij}\left(x_\alpha, \theta_{ns}(q), \frac{1}{n^{1/3}}\right)\\
&\leqq \frac{1}{n}\sum_\alpha \phi_{ij}\left(x_\alpha, \theta, \frac{2}{n^{1/3}}\right)
\end{aligned}
$$

and

$$
\begin{aligned}
\frac{1}{n}\sum_\alpha \psi_{ij}\left(x_\alpha, \theta, \frac{2}{n^{1/3}}\right) &\leqq \frac{1}{n}\sum_\alpha \psi_{ij}\left(x_\alpha, \theta_{ns}(q), \frac{1}{n^{1/3}}\right)\\
&\leqq \frac{1}{n}\sum_\alpha \phi_{ij}\left(x_\alpha, \theta, \frac{2}{n^{1/3}}\right)
\end{aligned}
$$

for almost all n and for all s for which $n^{1/2}|\theta-\theta_{ns}(q)| \leqq \lambda$, we obtain from (32)

$$
(33)\quad
\begin{aligned}
c_{ij}(\theta) - \frac{1}{n^{1/3}} - \nu(n) &\leqq \frac{1}{n}\sum_\alpha \phi_{ij}\left(x_\alpha, \theta_{ns}(q), \frac{1}{n^{1/3}}\right) \leqq c_{ij}(\theta) + \frac{1}{n^{1/3}} + \nu(n),\\
c_{ij}(\theta) - \frac{1}{n^{1/3}} - \nu(n) &\leqq \frac{1}{n}\sum_\alpha \psi_{ij}\left(x_\alpha, \theta_{ns}(q), \frac{1}{n^{1/3}}\right) \leqq c_{ij}(\theta) + \frac{1}{n^{1/3}} + \nu(n),
\end{aligned}
$$

for almost all n and for all s for which $n^{1/2}|\theta-\theta_{ns}(q)| \leqq \lambda$. Since $|c_{ij}[\theta_{ns}(q)] - c_{ij}(\theta)| \leqq \bar{\nu}(n)$ for almost all n and for all s for which $n^{1/2}|\theta-\theta_{ns}(q)| \leqq \lambda$, we obtain from (33)

$$
(34)\quad
\begin{aligned}
c_{ij}[\theta_{ns}(q)] - \frac{1}{n^{1/3}} - \nu(n) - \bar{\nu}(n) &\leqq \frac{1}{n}\sum_\alpha \phi_{ij}\left(x_\alpha, \theta_{ns}(q), \frac{1}{n^{1/3}}\right)\\
&\leqq c_{ij}[\theta_{ns}(q)] + \frac{1}{n^{1/3}} + \nu(n) + \bar{\nu}(n),\\
c_{ij}[\theta_{ns}(q)] - \frac{1}{n^{1/3}} - \nu(n) - \bar{\nu}(n) &\leqq \frac{1}{n}\sum_\alpha \psi_{ij}\left(x_\alpha, \theta_{ns}(q), \frac{1}{n^{1/3}}\right)\\
&\leqq c_{ij}[\theta_{ns}(q)] + \frac{1}{n^{1/3}} + \nu(n) + \bar{\nu}(n)
\end{aligned}
$$

for almost all n and for all s for which $n^{1/2}|\theta-\theta_{ns}(q)| \leqq \lambda$. The inequalities (34) are equivalent to the inequalities (22) if in (22) $\theta_{ns}(q)$ is substituted for θ. Hence our statement about the region $R_n(\theta, \lambda, q)$ is proved.

Consider the region $U_n(\theta)$ defined by the inequalities

$$\left|\frac{1}{n}\sum_\alpha \phi_{ij}\left(x_\alpha, \theta, \frac{2}{n^{1/3}}\right) - E_\theta\phi_{ij}\left(x, \theta, \frac{2}{n^{1/3}}\right)\right| < \frac{1}{n^{1/3}},$$
$$\left|\frac{1}{n}\sum_\alpha \psi_{ij}\left(x_\alpha, \theta, \frac{2}{n^{1/3}}\right) - E_\theta\psi_{ij}\left(x, \theta, \frac{2}{n^{1/3}}\right)\right| < \frac{1}{n^{1/3}}. \tag{35}$$

Since

$$\nu(n) \geqq \left| E_\theta\phi_{ij}\left(x, \theta, \frac{2}{n^{1/3}}\right) + c_{ij}(\theta)\right|$$

and

$$\nu(n) \geqq \left| c_{ij}(\theta) + E_\theta\psi_{ij}\left(x, \theta, \frac{2}{n^{1/3}}\right)\right|,$$

the validity of (35) implies the validity of (31). Hence $U_n(\theta)$ is a subset of $T_n(\theta)$. From condition (b) of Assumption III and Tshebysheff's inequality it follows that $\lim_{n=\infty} P[U_n(\theta)|\theta]=1$ uniformly in θ. Hence $\lim_{n=\infty} P[T_n(\theta)|\theta]=1$ uniformly in θ. Thus, as can easily be seen,

$$\lim_{n=\infty} P[R_n(\theta, \lambda_n, q) \mid \theta] = 1 \tag{36}$$

for any bounded sequence $\{\lambda_n\}$ uniformly in θ and q.

Let $\dot{J}_{ns}(W_n, q)$ be the intersection of the regions W_n and $I_{ns}(q)$. Furthermore let $\dot{J}_n(W_n, \theta, \lambda, q)$ be equal to $\sum_s \dot{J}_{ns}(W_n, q)$ where the summation is to be taken over all values of s for which $n^{1/2}|\theta-\theta_{ns}(q)| \leqq \lambda$. Then the common part of $\sum_{s=1}^\infty I_{ns}(q)$ and $W_n(\theta, \lambda-1/q)$ is a subset of $\dot{J}_n(W_n, \theta, \lambda, q)$, and the last is a subset of the common part of $\sum_{s=1}^\infty I_{ns}(q)$ and $W_n(\theta, \lambda+1/q)$. Hence, since $P[\sum_{s=1}^\infty I_{ns}(q)|\theta]=1$, we have

$$\begin{aligned} &| P[W_n(\theta, \lambda) \mid \theta] - P[\dot{J}_n(W_n, \theta, \lambda, q) \mid \theta]| \\ &\qquad \leqq P[W_n(\theta, \lambda + 1/q) \mid \theta] - P[W_n(\theta, \lambda - 1/q) \mid \theta] \\ &\qquad \leqq P[\lambda - 1/q \leqq n^{1/2} | \dot{\theta}_n - \theta | \leqq \lambda + 1/q \mid \theta]. \end{aligned} \tag{37}$$

From (37) and (29) it follows that for any sequence $\{q_n\}$ for which $\lim q_n = \infty$, we have

$$\lim_{n=\infty} | P[W_n(\theta, \lambda) \mid \theta] - P[\dot{J}_n(W_n, \theta, \lambda, q_n) \mid \theta]| = 0 \tag{38}$$

uniformly in θ, W_n and λ.

Since the common part of the regions $\dot{J}_n(W_n, \theta, \lambda, q)$ and $R_n(\theta, \lambda, q)$ is contained as a subset in $J_n(W_n, \theta, \lambda, q)$ and since the last is a subset of $\dot{J}_n(W_n, \theta, \lambda, q)$, we obtain from (38) and (36) that for any bounded sequence

$\{\lambda_n\}$ and for any sequence $\{q_n\}$ for which $\lim_{n=\infty} q_n = \infty$ we have

$$\lim_{n=\infty} \{P[W_n(\theta, \lambda_n) \mid \theta] - P[J_n(W_n, \theta, \lambda_n, q_n) \mid \theta]\} = 0 \tag{39}$$

uniformly in θ and W_n.

Since the common part of the regions $J_n^*(W_n, \theta, \lambda, q)$ and $R_n(\theta, \lambda, q)$ is a subset of $\bar{J}_n^*(W_n, \theta, \lambda, q)$ and the last is a subset of $J_n^*(W_n, \theta, \lambda, q)$ we obtain from (36) that for any bounded sequence $\{\lambda_n\}$ and for any sequence $\{q_n\}$ for which $\lim_{n=\infty} q_n = \infty$, we have

$$\lim_{n=\infty} \{P[\bar{J}_n^*(W_n, \theta, \lambda_n, q_n) \mid \theta] - P[J_n^*(W_n, \theta, \lambda_n, q_n) \mid \theta]\} = 0 \tag{40}$$

uniformly in θ and W_n.

Now we shall evaluate the limit values of $P[J_n(W_n, \theta, \lambda, q) | \theta]$ and $P[\bar{J}_n^*(W_n, \theta, \lambda, q) | \theta]$. Denote by $A_n(\lambda, q)$ the domain in the space of the variables θ, W_n, θ' and E_n defined as follows: θ and W_n can take arbitrary values, θ' is restricted to values for which $|\theta' - \theta| \leqq \lambda / n^{1/2}$; and for any θ and W_n, E_n is restricted to points which lie in the sum of the sets $J_n(W_n, \theta, \lambda, q)$ and $\bar{J}_n^*(W_n, \theta, \lambda, q)$. Denote furthermore by $\rho_n(\theta', \theta, \dot{\theta}_n)$ the function

$$\rho_n(\theta', \theta, \dot{\theta}_n) = -\frac{1}{2} \sum_j \sum_i n(\theta'^i - \dot{\theta}_n^i)(\theta'^j - \dot{\theta}_n^j) c_{ij}(\theta). \tag{41}$$

Consider the Taylor expansion

$$\begin{aligned} &\sum_\alpha \log f(x_\alpha, \theta') \\ &= \sum_\alpha \log f(x_\alpha, \dot{\theta}_n) + \frac{n}{2} \sum_j \sum_i (\theta'^i - \dot{\theta}_n^i)(\theta'^j - \dot{\theta}_n^j) \frac{1}{n} \sum_\alpha \frac{\partial^2 \log f(x_\alpha, \bar{\theta})}{\partial \theta^i \partial \theta^j} \end{aligned} \tag{42}$$

where $\bar{\theta}$ lies in the interval $[\theta', \dot{\theta}_n]$.

Since in the domain $A_n(\lambda, q)$ any point E_n lies in the sum of the sets $J_n(W_n, \theta, \lambda, q)$ and $\bar{J}_n^*(W_n, \theta, \lambda, q)$, it follows from the definitions of these sets that E_n lies in the set $\sum_s Q_n[\theta_{ns}(q)]$ where the summation is to be taken over all values of s for which $n^{1/2}|\theta - \theta_{ns}(q)| \leqq \lambda$ (the set $Q_n(\theta)$ is defined in (22)). Hence for any E_n in the domain $A_n(\lambda, q)$ we have

$$\begin{aligned} &\left| \frac{1}{n} \sum_\alpha \phi_{ij}\left(x_\alpha, \theta_{ns}(q), \frac{1}{n^{1/3}}\right) + c_{ij}[\theta_{ns}(q)] \right| \leqq \frac{1}{n^{1/3}} + \nu(n) + \bar{\nu}(n), \\ &\left| \frac{1}{n} \sum_\alpha \psi_{ij}\left(x_\alpha, \theta_{ns}(q), \frac{1}{n^{1/3}}\right) + c_{ij}[\theta_{ns}(q)] \right| \leqq \frac{1}{n^{1/3}} + \nu(n) + \bar{\nu}(n) \end{aligned} \tag{43}$$

for that value of s for which E_n lies in $I_{ns}(q)$. In all that follows, with any point of the domain $A_n(\lambda, q)$ we shall associate the integer s for which E_n lies in $I_{ns}(q)$. Since $n^{1/2}|\theta-\theta_{ns}(q)| \leqq \lambda$ in the domain $A_n(\lambda, q)$, we have in the domain $A_n(\lambda, q)$

(44)
$$\frac{1}{n}\sum_\alpha \psi_{ij}\left[x_\alpha, \theta_{ns}(q), \frac{1}{n^{1/3}}\right] \leqq \frac{1}{n}\sum_\alpha \psi_{ij}\left(x_\alpha, \theta, \frac{1}{2n^{1/3}}\right) \leqq \frac{1}{n}\sum_\alpha \phi_{ij}\left[x_\alpha, \theta_{ns}(q), \frac{1}{n^{1/3}}\right],$$
$$\frac{1}{n}\sum_\alpha \psi_{ij}\left[x_\alpha, \theta_{ns}(q), \frac{1}{n^{1/3}}\right] \leqq \frac{1}{n}\sum_\alpha \phi_{ij}\left(x_\alpha, \theta, \frac{1}{2n^{1/3}}\right) \leqq \frac{1}{n}\sum_\alpha \phi_{ij}\left[x_\alpha, \theta_{ns}(q), \frac{1}{n^{1/3}}\right]$$

for almost all values of n. From the definition of $\bar{\nu}(n)$ and from the validity of the inequality $n^{1/2}|\theta-\theta_{ns}(q)| \leqq \lambda$ in $A_n(\lambda, q)$ we find that in the domain $A_n(\lambda, q)$

(45)
$$\left| c_{ij}[\theta_{ns}(q)] - c_{ij}(\theta) \right| \leqq \bar{\nu}(n)$$

for almost all values of n. From (43), (44) and (45) it follows that in the domain $A_n(\lambda, q)$

(46)
$$\left|\frac{1}{n}\sum_\alpha \phi_{ij}\left(x_\alpha, \theta, \frac{1}{2n^{1/3}}\right) + c_{ij}(\theta)\right| \leqq \frac{1}{n^{1/3}} + \nu(n) + 2\bar{\nu}(n),$$
$$\left|\frac{1}{n}\sum_\alpha \psi_{ij}\left(x_\alpha, \theta, \frac{1}{2n^{1/3}}\right) + c_{ij}(\theta)\right| \leqq \frac{1}{n^{1/3}} + \nu(n) + 2\bar{\nu}(n)$$

for almost all values of n.

Since $n^{1/2}|\theta-\theta_{ns}(q)| \leqq \lambda$ in $A_n(\lambda, q)$, we have

(47)
$$n^{1/2}|\theta - \dot{\theta}_n| \leqq \lambda + 1/q \quad \text{in} \quad A_n(\lambda, q)$$

and therefore

(48)
$$n^{1/2}|\theta' - \dot{\theta}_n| \leqq 2\lambda + 1/q \quad \text{in} \quad A_n(\lambda, q).$$

Since $\bar{\theta}$ lies in the interval $[\theta', \dot{\theta}_n]$, from (48) we obtain

(49)
$$n^{1/2}|\bar{\theta} - \dot{\theta}_n| \leqq 2\lambda + 1/q \quad \text{in} \quad A_n(\lambda, q).$$

From (47) and (49) it follows that in $A_n(\lambda, q)$

(50)
$$\sum_\alpha \psi_{ij}\left(x_\alpha, \theta, \frac{1}{2n^{1/3}}\right) \leqq \sum_\alpha \frac{\partial^2 \log f(x_\alpha, \bar{\theta})}{\partial\theta^i\partial\theta^j} \leqq \sum_\alpha \phi_{ij}\left(x_\alpha, \theta, \frac{1}{2n^{1/3}}\right)$$

for almost all n. Because of (23), (46) and (50) we obtain from (41) and (42)

$$\lim_{n=\infty}\left\{\underset{A_n(\lambda,q)}{\text{l.u.b.}}\left|\sum_\alpha \log f(x_\alpha,\theta') - \sum_\alpha \log f(x_\alpha,\dot\theta_n) - \rho_n(\theta',\theta,\dot\theta_n)\right|\right\} = 0, \tag{51}$$

and

$$\begin{aligned}\lim_{n=\infty}\Bigg\{\underset{A_n(\lambda,q)}{\text{l.u.b.}}\Bigg|&\left[\sum_\alpha \log f(x_\alpha,\theta) - \sum_\alpha \log f(x_\alpha,\theta')\right]\\ &- \left[\rho_n(\theta,\theta,\dot\theta_n) - \rho_n(\theta',\theta,\dot\theta_n)\right]\Bigg|\Bigg\} = 0.\end{aligned} \tag{52}$$

Denote $P[J_{ns}(W_n, q)|\theta_{ns}(q)]$ by $P_{ns}(W_n, q)$ and $P[\bar{J}^*_{ns}(W_n, q)|\theta_{ns}(q)]$ by $\bar{P}^*_{ns}(W_n, q)$. Substituting $\theta_{ns}(q)$ for θ' we easily obtain from (52)

$$\begin{aligned}\lim_{n=\infty}\Bigg\{\underset{\theta,W_n}{\text{l.u.b.}}\Bigg|&P[J_n(W_n,\theta,\lambda,q)\,|\,\theta]\\ &-\sum_s P_{ns}(W_n,q)\exp\left(\rho_n[\theta,\theta,\overset{*}{\theta}_{ns}(q)] - \rho_n[\theta_{ns}(q),\theta,\overset{*}{\theta}_{ns}(q)]\right)\Bigg|\Bigg\} = 0\end{aligned} \tag{53}$$

where the summation with respect to s is to be taken over all values for which $n^{1/2}|\theta-\theta_{ns}(q)| \leqq \lambda$ and $\theta^*_{ns}(q)$ is a parameter point for which

$$n^{1/2}|\,\theta_{ns}(q) - \overset{*}{\theta}_{ns}(q)\,| \leqq 1/q. \tag{54}$$

Since $c_{ij}(\theta)$ is a uniformly continuous and bounded function of θ, it follows from (41) and (54) that

$$\begin{aligned}|\,\rho(\theta,\theta,\overset{*}{\theta}_{ns}(q)) - \rho[\theta,\theta,\theta_{ns}(q)]\,| &= \phi_1[\theta,\overset{*}{\theta}_{ns}(q),\theta_{ns}(q)]/q,\\ |\,\rho[\theta_{ns}(q),\theta,\overset{*}{\theta}_{ns}(q)] - \rho[\theta_{ns}(q),\theta,\theta_{ns}(q)]\,| &= \phi_2[\theta,\overset{*}{\theta}_{ns}(q),\theta_{ns}(q)]/q,\end{aligned} \tag{55}$$

where $\phi_1[\theta, \theta^*_{ns}(q), \theta_{ns}(q)]$ and $\phi_2[\theta, \theta^*_{ns}(q), \theta_{ns}(q)]$ are bounded functions of θ, $\theta^*_{ns}(q)$ and $\theta_{ns}(q)$ in the domain $n^{1/2}|\theta-\theta_{ns}(q)| \leqq \lambda$. From (53) and (55) it follows that

$$\begin{aligned}\limsup_{n=\infty}\Bigg\{\underset{\theta,W_n}{\text{l.u.b.}}\Bigg|&P[J_n(W_n,\theta,\lambda,q)\,|\,\theta]\\ &-\sum_s P_{ns}(W_n,q)\exp\left(\rho_n[\theta,\theta,\theta_{ns}(q)] - \rho_n[\theta_{ns}(q),\theta,\theta_{ns}(q)]\right)\Bigg|\Bigg\} = \epsilon(\lambda,q)\end{aligned} \tag{56}$$

where

$$\lim_{q=\infty}\epsilon(\lambda,q) = 0 \tag{57}$$

uniformly in λ over any finite positive interval. Similarly we obtain

$$\limsup_{n=\infty} \left\{ \underset{\theta, W_n}{\text{l.u.b.}} \left| P[\bar{J}_n^*(W_n, \theta, \lambda, q) \mid \theta] - \sum_s \bar{P}_{ns}^*(W_n, q) \exp(\rho_n[\theta, \theta, \theta_{ns}(q)] - \rho_n[\theta_{ns}(q), \theta, \theta_{ns}(q)]) \right| \right\} = \eta(\lambda, q) \tag{58}$$

where

$$\lim_{q=\infty} \eta(\lambda, q) = 0 \tag{59}$$

uniformly in λ over any finite positive interval. In the formulas (56) and (58) the summation with respect to s is to be taken for all values for which $|\theta - \theta_{ns}(q)| \leqq \lambda/n^{1/2}$. The expression $\rho_n[\theta_{ns}(q), \theta, \theta_{ns}(q)]$ is obviously equal to zero, hence (56) and (58) can be simplified by substituting zero for this expression. Denote $P[J_{ns}^*(W_n, q) \mid \theta_{ns}(q)]$ by $P_{ns}^*(W_n, q)$. Because of Proposition I we have

$$\lim_{n=\infty} \{\mathfrak{P}[J_{ns}^*(W_n, q) \mid \theta_{ns}(q)] - P_{ns}^*(W_n, q)\} = 0$$

uniformly in s and W_n. Hence we obtain from condition (b) of (24)

$$\lim_{n=\infty} \{P_{ns}(W_n, q) - P_{ns}^*(W_n, q)\} = 0 \tag{60}$$

uniformly in s and W_n. Since $\bar{J}_{ns}^*(W_n, q)$ is the intersection of $Q_n[\theta_{ns}(q)]$ with $J_{ns}^*(W_n, q)$ and since

$$\lim_{n=\infty} P[Q_n[\theta_{ns}(q)] \mid \theta_{ns}(q)] = 1$$

uniformly in s, we have

$$\lim_{n=\infty} \{\bar{P}_{ns}^*(W_n, q) - P_{ns}^*(W_n, q)\} = 0 \tag{61}$$

uniformly in s and W_n. Since for any given λ and q the number of different values of s satisfying the inequality $n^{1/2}|\theta - \theta_{ns}(q)| \leqq \lambda$ is a bounded function of θ, from (56), (58), (60) and (61) we obtain

$$\limsup_{n=\infty} \{ \underset{\theta, W_n}{\text{l.u.b.}} \mid P[J_n(W_n, \theta, \lambda, q) \mid \theta] - P[\bar{J}_n^*(W_n, \theta, \lambda, q) \mid \theta] \mid \} = \zeta(\lambda, q) \leqq \epsilon(\lambda, q) + \eta(\lambda, q). \tag{62}$$

Hence

$$\lim_{q=\infty} \zeta(\lambda, q) = 0 \tag{63}$$

uniformly in λ over any finite positive interval.

For any positive λ' the sets $J_n(W_n, \theta, \lambda+\lambda', q)-J_n(W_n, \theta, \lambda, q)$, $\bar{J}_n^*(W_n, \theta, \lambda+\lambda', q)-\bar{J}_n^*(W_n, \theta, \lambda, q)$, $J_n^*(W_n, \theta, \lambda+\lambda', q)-J^*(W_n, \theta, \lambda, q)$, $W_n^*(W_n, \theta, \lambda+\lambda', q)-W_n^*(W_n, \theta, \lambda, q)$ and $W_n(\theta, \lambda+\lambda')-W_n(\theta, \lambda)$ are subsets of the set defined by the inequality

$$\lambda - 1/q \leqq n^{1/2}|\theta - \dot{\theta}_n| \leqq \lambda + \lambda' + 1/q.$$

Since for any sequence $\{\lambda_n\}$ for which $\lim_{n=\infty} \lambda_n = \infty$ we have

$$\lim_{n=\infty} P[\lambda_n - 1/q \leqq n^{1/2}|\theta - \dot{\theta}_n| \leqq \lambda_n + \lambda' + 1/q \mid \theta] = 0$$

uniformly in θ, q and λ', (39) and (40) hold for any arbitrary sequence $\{\lambda_n\}$ and (63) holds uniformly in λ where λ can take any positive value. Thus from (30), (39), (40), (62) and (63) we obtain

$$(64)\quad \limsup_{n=\infty} \{\underset{\theta, W_n}{\text{l.u.b.}} \mid P[W_n(\theta, \lambda) \mid \theta] - P[W_n^*(W_n, \theta, \lambda, q) \mid \theta] \mid \} = \epsilon_3(\lambda, q),$$

where

$$(65)\quad \lim_{q=\infty} \epsilon_3(\lambda, q) = 0$$

uniformly in λ. Let $\{q_i\}$ ($i=1, 2, \cdots$, ad inf.) be a sequence of positive integers such that $\lim_{i=\infty} q_i = +\infty$. Furthermore let $\{\eta_i\}$ be a sequence of positive numbers such that $\lim_{i=\infty} \eta_i = 0$. We define $W_n^*(W_n)$ as follows:

$$(66)\quad W_n^*(W_n) = W_n^*(W_n, q_{i+1}) \quad \text{for} \quad n_i < n \leqq n_{i+1} \quad (i = 0, 1, \cdots, \text{ad inf.}).$$

The sequence $\{n_i\}$ ($i=0, 1, 2, \cdots$, ad inf.) of integers is chosen as follows: Denote by $F_n(\lambda, q)$ the expression

$$\underset{\theta, W_n}{\text{l.u.b.}} \mid P[W_n(\theta, \lambda) \mid \theta] - P[W_n^*(W_n, \theta, \lambda, q) \mid \theta] \mid .$$

The integer n_0 is put equal to 0 and n_i is chosen such that

$$(67)\quad \begin{gathered} n_i > n_{i-1}, \\ F_n(\lambda_i, q_{i+1}) < \epsilon_3(\lambda_i, q_{i+1}) + \eta_i \end{gathered}$$

for all $n > n_i$, and where $\{\lambda_i\}$ denotes a sequence of numbers such that $\lim_{i=\infty} \lambda_i = +\infty$.

Let $\lambda_n' = \lambda_i$, $\eta_n' = \eta_i$, and $q_n' = q_{i+1}$ for $n_i < n \leqq n_{i+1}$ ($i=0, 1, 2, \cdots$, ad inf.). Then from (64), (65) and (67) we obtain

$$(68)\quad \lim_{n=\infty} \{\underset{\theta, W_n}{\text{l.u.b.}} \mid P[W_n(\theta, \lambda_n') \mid \theta] - P[W_n^*(W_n, \theta, \lambda_n', q_n') \mid \theta] \mid \} = 0.$$

Denote by $W_n^*(W_n, \theta, \lambda)$ the intersection of $W_n^*(W_n)$ and the set defined by the inequality $n^{1/2}|\theta-\dot\theta_n| \leqq \lambda$. Since $W_n^*(W_n, \theta, \lambda, q)$ is the intersection of $W_n^*(W_n, q)$ and the region $n^{1/2}|\theta-\dot\theta_n| \leqq \lambda$, it follows from (66) that

$$W_n^*(W_n, \theta, \lambda_n', q_n') = W_n^*(W_n, \theta, \lambda_n'). \tag{69}$$

Equation (20) follows from (68) and (69). Hence Lemma 1 is proved.

We shall say that a region V_n^* lies in the space of the maximum likelihood estimates if it has the following property: If E_n is an element of V_n^* then also all those points E_n' for which $\dot\theta_n(E_n') = \dot\theta_n(E_n)$ are elements of V_n^*. In all the following considerations the symbol * as a superscript in the notation of a region will indicate that the region lies in the space of the maximum likelihood estimates, except if a statement to the contrary is explicitly made. For any region V_n^* we shall denote by $\mathfrak{P}(V_n^*|\theta)$ the probability that the sample point will fall in V_n^* calculated under the assumption that $n^{1/2}(\dot\theta_n^1-\theta^1), \cdots, n^{1/2}(\dot\theta_n^k-\theta^k)$ have a joint normal distribution with zero means and covariance matrix $\|\sigma_{ij}(\theta)\| = \|c_{ij}(\theta)\|^{-1}$.

LEMMA 2. *There exists a function $W_n^*(R_n^*)$ defined over all Borel measurable subsets R_n^* such that*

$$\lim_{n=\infty} \{P[R_n^* | \theta] - \mathfrak{P}[W_n^*(R_n^*) | \theta]\} = 0$$

uniformly in θ and R_n^*.

Proof. Since we assumed that the set $J_{ns}^*(W_n, q)$ defined in (24) is an interval in the space of the maximum likelihood estimates, it follows from Proposition I that

$$\lim_{n=\infty} \{P[J_{ns}^*(W_n, q) | \theta] - \mathfrak{P}[J_{ns}^*(W_n, q) | \theta]\} = 0 \tag{70}$$

uniformly in θ, W_n, and s. Let $W_n^*(W_n, q)$ be the set defined in (25) and let $W_n^*(W_n, \theta, \lambda, q)$ be the intersection of $W_n^*(W_n, q)$ and the region $n^{1/2}|\theta-\dot\theta_n| \leqq \lambda$. For given values of λ and q the number of different values of s, for which $J_{ns}^*(W_n, q)$ has at least a point common with the region $n^{1/2}|\theta-\dot\theta_n| \leqq \lambda$, is a bounded function of θ. Hence it follows from (70) that

$$\lim_{n=\infty} \{P[W_n^*(W_n, \theta, \lambda, q) | \theta] - \mathfrak{P}[W_n^*(W_n, \theta, \lambda, q) | \theta]\} = 0 \tag{71}$$

uniformly in θ and W_n. From (64), (65) and (71) we obtain

$$\limsup_{n=\infty} \{\underset{\theta, W_n}{\text{l.u.b.}} | P[W_n(\theta, \lambda) | \theta] - \mathfrak{P}[W_n^*(W_n, \theta, \lambda, q) | \theta] |\} = \epsilon(\lambda, q) \tag{72}$$

where

$$(73) \qquad \lim_{q=\infty} \epsilon(\lambda, q) = 0$$

uniformly in λ. The set $W_n(\theta, \lambda)$ denotes the intersection of W_n and the region $n^{1/2}|\theta - \hat{\theta}_n| \leqq \lambda$.

Let $\{q_i\}$ ($i=1, 2, \cdots$, ad inf.) be a sequence of positive integers such that $\lim_{i=\infty} q_i = \infty$. Furthermore let $\{\eta_i\}$ be a sequence of positive numbers such that $\lim_{i=\infty} \eta_i = 0$. We define $W_n^*(W_n)$ as follows

$$(74) \quad W_n^*(W_n) = W_n^*(W_n, q_{i+1}) \text{ for } n_i < n \leqq n_{i+1} \quad (i = 0, 1, 2, \cdots, \text{ad inf.}).$$

The sequence $\{n_i\}$ ($i=0, 1, 2, \cdots$, ad inf.) of integers is chosen as follows: Denote by $F_n(\lambda, q)$ the expression

$$\underset{\theta, W_n}{\text{l.u.b.}} \left| P[W_n(\theta, \lambda) \mid \theta] - \mathfrak{P}[W_n^*(W_n, \theta, \lambda, q) \mid \theta] \right|.$$

The integer n_0 is put equal to zero and n_i is chosen so that

$$(75) \qquad \begin{gathered} n_i > n_{i-1}, \\ F_n(\lambda_i, q_{i+1}) < \epsilon(\lambda_i, q_{i+1}) + \eta_i \end{gathered}$$

for all $n > n_i$, and $\{\lambda_i\}$ denotes a sequence of numbers such that $\lim \lambda_i = \infty$. Let $\lambda_n' = \lambda_i$, $\eta_n' = \eta_i$ and $q_n' = q_{i+1}$ for $n_i < n \leqq n_{i+1}$ ($i=0, 1, \cdots$, ad inf.). From (72), (73) and (75) we obtain

$$(76) \quad \lim_{n=\infty} \left\{ \underset{\theta, W_n}{\text{l.u.b.}} \left| P[W_n(\theta, \lambda_n') \mid \theta] - \mathfrak{P}[W_n^*(W_n, \theta, \lambda_n', q_n') \mid \theta] \right| \right\} = 0.$$

Denote by $W_n^*(W_n, \theta, \lambda)$ the intersection of $W_n^*(W_n)$ and the region $n^{1/2}|\theta - \hat{\theta}_n| \leqq \lambda$. Because of (74) we obviously have

$$W_n^*(W_n, \theta, \lambda_n') = W_n^*(W_n, \theta, \lambda_n', q_n').$$

Hence from (76) we obtain

$$(77) \quad \lim_{n=\infty} \left\{ \underset{\theta, W_n}{\text{l.u.b.}} \left| P[W_n(\theta, \lambda_n') \mid \theta] - \mathfrak{P}[W_n^*(W_n, \theta, \lambda_n') \mid \theta] \right| \right\} = 0.$$

Since $\lim_{n=\infty} \lambda_n' = \infty$, it follows from (77) that

$$(78) \quad \lim_{n=\infty} \left\{ \underset{\theta, W_n}{\text{l.u.b.}} \left| P[W_n \mid \theta] - \mathfrak{P}[W_n^*(W_n) \mid \theta] \right| \right\} = 0.$$

The region W_n may be any Borel measurable subset of the rn-dimensional sample space. In particular, W_n may be any Borel measurable subset R_n^* in the space of the maximum likelihood estimates. Hence Lemma 2 follows from (78).

On the basis of Lemmas 1 and 2 we can restrict ourselves in case of large samples to subsets of the space of the maximum likelihood estimates and we can substitute $\mathfrak{P}[V_n^*|\theta]$ for $P[V_n^*|\theta]$. Hence, if the sample is sufficiently large, the problem of statistical inference concerning the unknown parameter θ can be reduced to the case where the variates involved have a joint normal distribution.

5. **Tests of simple hypotheses which have uniformly best average power over a family of surfaces.** For any value c let K_c denote a surface in the parameter space. For instance K_c may be defined by the equation $\phi(\theta)=c$ where $\phi(\theta)$ denotes some analytic function of θ. Consider furthermore a non-negative function $w(\theta)$ of θ, called a weight function. For any function $\psi(\theta)$ of θ the symbol $\int_{K_c}\psi(\theta)dA$ will denote the surface integral of the function $\psi(\theta)$ over the surface K_c.

DEFINITION I. *A critical region W_n is said to have uniformly best average power with respect to the surfaces K_c and the weight function $w(\theta)$ if for any region Z_n of size equal to that of W_n we have $\int_{K_c}P(W_n|\theta)w(\theta)dA \geqq \int_{K_c}P(Z_n|\theta)w(\theta)dA$ for all values c for which K_c is defined.*

Let $y^1, \cdots, y^k$ be k variates which have a joint normal distribution. The mean values $\theta^1, \cdots, \theta^k$ of the variates $y^1, \cdots, y^k$ are unknown, but the covariance matrix $\|\sigma_{ij}\|$ $(i, j=1, \cdots, k)$ is known and is nonsingular. Suppose that we wish to test the simple hypothesis that $\theta=\theta_0$. Consider the family of ellipsoids given by

$$\sum_j \sum_i \lambda_{ij}[\theta^i - \theta_0^i][\theta^j - \theta_0^j] = c, \tag{79}$$

where $\|\lambda_{ij}\| = \|\sigma_{ij}\|^{-1}$. For any c denote by S_c the ellipsoid given by (79). Consider a nonsingular linear transformation of the parameter space

$$\theta'^i - \theta_0^i = \beta_{i1}(\theta^1 - \theta_0^1) + \cdots + \beta_{ik}(\theta^k - \theta_0^k) \tag{80}$$

such that the family of ellipsoids S_c is transformed into a family of concentric spheres with the center at θ_0. Denote by S_c' the image of S_c. For any point θ and for any positive ρ consider the set $\omega(\theta, \rho)$ consisting of all points θ_1 which lie on the same S_c as θ and for which $|\theta_1-\theta| \leqq \rho$. Let

$$\xi(\theta) = \lim_{\rho=0} \frac{A[\omega'(\theta, \rho)]}{A[\omega(\theta, \rho)]}, \tag{81}$$

where $\omega'(\theta, \rho)$ is the image of $\omega(\theta, \rho)$ and for any set ω, $A(\omega)$ denotes the area of ω.

PROPOSITION II. *If the variates $y^1, \cdots, y^k$ have a joint normal distribution with unknown mean values $\theta^1, \cdots, \theta^k$ and a known covariance matrix $\|\sigma_{ij}\|$, then for testing the hypothesis $\theta=\theta_0$ on the basis of a single observation on each of the variates $y^1, \cdots, y^k$, the critical region given by the inequality*

$$(82) \qquad \sum_j \sum_i \lambda_{ij}(y^i - \theta_0^i)(y^j - \theta_0^j) \geqq d \qquad (\|\lambda_{ij}\| = \|\sigma_{ij}\|^{-1})$$

has uniformly best average power with respect to the surfaces S_c defined in (79) *and the weight function given in* (81).

Proof. Consider the linear transformation

$$(83) \qquad y'^i - \theta_0^i = \beta_{i1}(y^1 - \theta_0^1) + \cdots + \beta_{ik}(y^k - \theta_0^k),$$

where the matrix $\|\beta_{ij}\|$ is the same as in (80). The variates y'^i $(i=1, \cdots, k)$ are normally and independently distributed with mean values $\theta_0^1, \cdots, \theta_0^k$ (under the hypothesis $\theta=\theta_0$) and have a common variance σ^2. We will assume $\sigma^2=1$, since this can always be achieved by multiplying the matrix $\|\beta_{ij}\|$ by a proportionality factor. The critical region W given in (82) will be transformed into the region W' given by

$$(84) \qquad (y'^1 - \theta_0^1)^2 + \cdots + (y'^k - \theta_0^k)^2 \geqq d.$$

Because of (81) we obviously have

$$(85) \qquad \int_{S_c} P(Z \mid \theta)\xi(\theta)dA = \int_{S'_c} P(Z' \mid \theta')dA,$$

where Z denotes an arbitrary region in the space of $y^1, \cdots, y^k$ and Z' is the image of Z in the space of $y'^1, \cdots, y'^k$. Hence in order to prove Proposition II we have merely to show that

$$(86) \qquad \int_{S'_c} P(W' \mid \theta')dA \geqq \int_{S'_c} P(Z' \mid \theta')dA$$

for any region Z' in the space of $y'^1, \cdots, y'^k$ which has a size equal to that of W'.

By a lemma of Neyman and Pearson(4) we see easily that (86) is proved, if we can show that there exists a function $d(c)$ of c such that

$$(87) \qquad \int_{S'_c} p(y' \mid \theta')dA / p(y' \mid \theta_0) \geqq d(c) \text{ within } W'$$

and

$$(88) \qquad \int_{S'_c} p(y' \mid \theta')dA / p(y' \mid \theta_0) \leqq d(c) \text{ outside } W'$$

for all positive values of c, where $p(y' \mid \theta')$ denotes the joint density function of $y'^1, \cdots, y'^k$ under the assumption that the true means are $\theta'^1, \cdots, \theta'^k$.

(4) J. Neyman and E. S. Pearson, *Contributions to the theory of testing statistical hypotheses*, Statistical Research Memoirs vol. 1 (1936).

If we denote $y'^i-\theta_0^i$ by v^i and $\theta'^i-\theta_0^i$ by θ^{*i}, we have

$$p(y' \mid \theta') = \frac{1}{(2\pi)^{k/2}} \exp\left(-2^{-1}\sum (v^i - \theta^{*i})^2\right)$$

and

$$p(y' \mid \theta_0) = \frac{1}{(2\pi)^{k/2}} \exp\left(-2^{-1}\sum (v^i)^2\right).$$

Hence

$$\int_{S_c'} p(y' \mid \theta')dA = \left(\frac{1}{2\pi}\right)^{k/2} \int_{S_c'} \exp\left(-2^{-1}\sum (v^i-\theta^{*i})^2\right)dA$$

$$= \left(\frac{1}{2\pi}\right)^{k/2} \exp\left(-2^{-1}\sum (v^i)^2\right)$$

$$\int_{S_c'} \exp\left(\sum v^i\theta^{*i} - 2^{-1}\sum (\theta^{*i})^2\right)dA$$

$$= p(y' \mid \theta_0) \exp\left(-2^{-1}\sum (\theta^{*i})^2\right) \int_{S_c'} \exp\left(\sum v^i\theta^{*i}\right)dA,$$

since $\sum(\theta^{*i})^2$ is constant on the surface S_c'. Hence (87) and (88) can be written

$$(89) \qquad I(v^1, \cdots, v^k) = \int_{S_c'} \exp \sum v^i\theta^{*i}dA \geqq d^*(c) \text{ within } W',$$

$$(90) \qquad I(v^1, \cdots, v^k) \leqq d^*(c) \text{ outside } W'.$$

Denote $\left|\left(\sum(v^i)^2\right)^{1/2}\right|$ by r_v and $\left|\left(\sum(\theta^{*i})^2\right)^{1/2}\right|$ by r^*. On the surface S_c' we have $r^*=c$. Denote by $\alpha(\theta^*)$ the angle $(0 \leqq \alpha \leqq \pi)$ between the vector v and the vector θ^*. Then we have

$$I(v^1, \cdots, v^k) = \int_{S_c'} \exp\left(cr_v \cos\left[\alpha(\theta^*)\right]\right)dA.$$

Because of the symmetry of the sphere, the value of this integral will not be changed if we substitute $\beta(\theta^*)$ for $\alpha(\theta^*)$ where $\beta(\theta^*)$ denotes the angle $(0 \leqq \beta(\theta^*) \leqq \pi)$ between the vector θ^* and an arbitrarily chosen fixed vector u. Hence $I(v^1, \cdots, v^k)$ depends only on r_v, that is $I(v^1, \cdots, v^k) = I(r_v)$. The inequalities (89) and (90) are obviously proved if we can show that $I(r_v)$ is a monotonically increasing function of r_v. We have

$$(91) \qquad \frac{dI(r_v)}{dr_v} = \int_{S_c'} c \cos\left[\beta(\theta^*)\right] \exp\left(cr_v \cos\left[\beta(\theta^*)\right]\right)dA.$$

Denote by ω_1 the subset of S_c' in which $0 \leqq \beta(\theta^*) \leqq \pi/2$, and by ω_2 the subset in which $\pi/2 \leqq \beta(\theta^*) \leqq \pi$. Because of the symmetry of the sphere we obviously have

$$\int_{\omega_2} c \cos [\beta(\theta^*)] \exp (cr_v \cos [\beta(\theta^*)])dA$$

(92) $$= \int_{\omega_1} c \cos [\pi - \beta(\theta^*)] \exp (cr_v \cos [\pi - \beta(\theta^*)])dA$$

$$= - \int_{\omega_1} c \cos [\beta(\theta^*)] \exp (- cr_v \cos [\beta(\theta^*)])dA.$$

Hence

(93) $$\frac{dI(r_v)}{dr_v} = c \int_{\omega_1} \cos [\beta(\theta^*)] \{\exp (cr_v \cos [\beta(\theta^*)]) - \exp (-cr_v \cos [\beta(\theta^*)])\} dA.$$

The right-hand side of (93) is positive. Hence Proposition II is proved.

Now let us turn back to the general problem of r variates $x^1, \cdots, x^r$ whose joint probability density function $f(x^1, \cdots, x^r, \theta^1, \cdots, \theta^k) = f(x, \theta)$ involves k unknown parameters, as considered in the previous sections.

DEFINITION II. *A sequence* $\{W_n\}$ $(n=1, \cdots$, ad inf.) *of critical regions of size* α *for testing the simple hypothesis* $\theta = \theta_0$ *is said to have asymptotically best average power with respect to the family of surfaces* K_c *and the weight function* $w(\theta)$ *if for any sequence* $\{Z_n\}$ *for which* $P(Z_n | \theta_0) = \alpha$ *we have*

$$\limsup_{n=\infty} \left\{ \operatorname{l.u.b.}_{c} \left[\int_{K_c} P(Z_n | \theta) \frac{w(\theta)}{A(K_c)} dA - \int_{K_c} P(W_n | \theta) \frac{w(\theta)}{A(K_c)} dA \right] \right\} \leqq 0,$$

where

$$A(K_c) = \int_{K_c} w(\theta) dA.$$

We shall prove the following theorem.

THEOREM I. *Let* W_n^* *be a critical region for testing* $\theta = \theta_0$ *defined by the inequality*

$$n \sum_j \sum_i (\dot{\theta}_n^i - \theta_0^i)(\dot{\theta}_n^j - \theta_0^j) c_{ij}(\dot{\theta}_n) \geqq d_n,$$

where the real number d_n *is chosen so that* $P(W_n^* | \theta_0) = \alpha$. *Denote by* S_c *the surface in the parameter space defined by the equation*

$$\sum \sum (\theta^i - \theta_0^i)(\theta^j - \theta_0^j) c_{ij}(\theta_0) = c.$$

Furthermore let $\xi(\theta)$ *be the weight function as defined in* (81) *where* $\|c_{ij}(\theta_0)\|$ *is substituted for* $\|\lambda_{ij}\|$. *Then the test* $\{W_n^*\}$ *has asymptotically best average power with respect to the family of surfaces* S_c *and the weight function* $\xi(\theta)$.

Proof. Because of Lemma 1 we can restrict ourselves to subsets of the space of the maximum likelihood estimates. Let us assume that Theorem I is not true. Then it follows from Lemma 2 that a sequence of values $\{c_n\}$ and a sequence of regions $\{Z_n^*\}$ exist such that $P(Z_n^*|\theta_0)=\alpha$ and

$$\limsup_{n=\infty}\left\{\int_{S_{c_n}}\mathfrak{P}(Z_n^*\,|\,\theta)\zeta_n(\theta)dA-\int_{S_{c_n}}\mathfrak{P}(W_n^*\,|\,\theta)\zeta_n(\theta)dA\right\}=\delta>0, \tag{94}$$

where

$$\zeta_n(\theta)=\xi(\theta)\bigg/\int_{S_{c_n}}\xi(\theta)dA.$$

From (94) it follows that there exists a subsequence $\{n'\}$ of the sequence $\{n\}$ such that

$$\lim_{n=\infty}\left\{\int_{S_{c_{n'}}}\mathfrak{P}(Z_{n'}^*\,|\,\theta)\zeta_{n'}(\theta)dA-\int_{S_{c_{n'}}}\mathfrak{P}(W_{n'}^*\,|\,\theta)\zeta_{n'}(\theta)dA\right\}=\delta>0. \tag{95}$$

It is easy to verify that

$$\lim_{n=\infty}\mathfrak{P}(W_{n'}^*\,|\,\theta_{n'})=1$$

if $\theta_{n'}$ is a point of $S_{c_{n'}}$ and if $\lim_{n=\infty}n'c_{n'}=+\infty$. Under the latter condition also

$$\lim_{n=\infty}\int_{S_{c_{n'}}}\mathfrak{P}(W_{n'}^*\,|\,\theta)\zeta_{n'}(\theta)dA=1.$$

Thus (95) can hold only if the sequence $\{n'c_{n'}\}$ is bounded. If $\{n'c_{n'}\}$ is bounded, we obviously have for any sequence of regions $\{V_n^*\}$

$$\lim_{n=\infty}\left\{\int_{S_{c_{n'}}}\mathfrak{P}(V_{n'}^*\,|\,\theta)\zeta_{n'}(\theta)dA-\int_{S_{c_{n'}}}\bar{P}(V_{n'}^*\,|\,\theta)\zeta_{n'}(\theta)dA\right\}=0,$$

where $\bar{P}(V_n^*|\theta)$ denotes the probability of V_n^* calculated under the assumption that $n^{1/2}(\dot{\theta}_n^1-\theta^1),\cdots,n^{1/2}(\dot{\theta}_n^k-\theta^k)$ have a joint normal distribution with zero means and covariance matrix equal to $\|c_{ij}(\theta_0)\|^{-1}$. Hence from (95) we obtain

$$\lim_{n=\infty}\left\{\int_{S_{c_{n'}}}\bar{P}(Z_{n'}^*\,|\,\theta)\zeta_{n'}(\theta)dA-\int_{S_{c_{n'}}}\bar{P}(W_{n'}^*\,|\,\theta)\zeta_{n'}(\theta)dA\right\}=\delta>0. \tag{96}$$

Denote by $\overline{W}_n^*$ the region defined by the inequality

$$n\sum\sum(\dot{\theta}_n^i-\theta_0^i)(\dot{\theta}_n^j-\theta_0^j)c_{ij}(\theta_0)\geqq\bar{d}_n,$$

where $\bar{d}_n$ is chosen so that $\bar{P}(\overline{W}_n^*|\theta_0)=\alpha$. Furthermore denote by $\bar{Z}_n^*$ the sum of Z_n^* and the region $n^{1/2}|\theta_0-\dot{\theta}_n|\leqq\lambda_n$, where λ_n is chosen so that

$\bar{P}(\bar{Z}_n^*|\theta_0)=\alpha$. Since, as can easily be seen,

$$\lim_{n=\infty}\left\{\int_{S_{c_{n'}}}\bar{P}(\bar{W}^*_{n'}\mid\theta)\zeta_{n'}(\theta)dA-\int_{S_{c_{n'}}}\bar{P}(W^*_{n'}\mid\theta)\zeta_{n'}(\theta)dA\right\}=0$$

and

$$\lim_{n=\infty}\left\{\int_{S_{c_{n'}}}\bar{P}(\overset{*}{\bar{Z}}_{n'}\mid\theta)\zeta_{n'}(\theta)dA-\int_{S_{c_{n'}}}\bar{P}(\overset{*}{Z}_{n'}\mid\theta)\zeta_{n'}(\theta)dA\right\}=0,$$

we obtain from (96) a contradiction to Proposition II. Hence Theorem I is proved.

6. Tests of simple hypotheses which have best constant power on a family of surfaces.

DEFINITION III. *A critical region W_n for testing $\theta=\theta_0$ is said to have uniformly best constant power on the family of surfaces $\{K_c\}$ if the following two conditions are fulfilled:*

(a) *$P(W_n|\theta_1)=P(W_n|\theta_2)$ for any pair of points θ_1, θ_2 which lie on the same surface K_c.*

(b) *$P(W_n|\theta)\geqq P(Z_n/\theta)$ for any Z_n which satisfies condition* (a) *and for which $P(Z_n|\theta_0)=P(W_n|\theta_0)$.*

From Proposition II we obtain the following:

PROPOSITION III. *Let $y^1, \cdots, y^k$ be k variates which have a joint normal distribution with unknown mean values $\theta^1, \cdots, \theta^k$ and a known covariance matrix $\|\sigma_{ij}\|=\|\lambda_{ij}\|^{-1}$. Then for testing $\theta=\theta_0$, the region defined in* (82) *has uniformly best constant power on the surfaces S_c defined by the equation*

$$\sum\sum(\theta^i-\theta_0^i)(\theta^j-\theta_0^j)\lambda_{ij}=c. \tag{97}$$

Since the critical region defined in (82) satisfies condition (a) of Definition III, Proposition III is an immediate consequence of Proposition II.

DEFINITION IV. *A sequence of critical regions $\{W_n\}$ for testing $\theta=\theta_0$ is said to be of size α and to have asymptotically best constant power on the surfaces K_c if the following three conditions are fulfilled:*

(a) $P(W_n|\theta_0)=\alpha$ $(n=1, 2, \cdots$, ad inf.).

(b) $\lim_{n=\infty}\{\text{l.u.b.}_c\,[\text{l.u.b.}_{\theta\in K_c}P(W_n|\theta)-\text{g.l.b.}_{\theta\in K_c}P(W_n|\theta)]\}=0$, *where the symbol* $\text{l.u.b.}_{\theta\in K_c}$ *means that the least upper bound is to be taken with respect to θ restricting θ to points of K_c.*

(c) *For any sequence $\{Z_n\}$ which satisfies* (a) *and* (b) *we have*

$$\lim_{n=\infty}\{\underset{\theta}{\text{l.u.b.}}\,[P(Z_n|\theta)-P(W_n|\theta)]\}=0.$$

It is easy to verify that the sequence $\{W_n^*\}$ defined in Theorem I satisfies the conditions (a) and (b) for $K_c = S_c$, where S_c denotes the surface defined in Theorem I. Thus from Theorem I we obtain the following theorem.

THEOREM II. *Let* $\{W_n^*\}$ *and* S_c *be defined as in Theorem* I. *For testing* $\theta = \theta_0$, *the sequence* $\{W_n^*\}$ *has asymptotically best constant power on the surfaces* S_c.

7. **Most stringent tests of simple hypotheses.** Let θ and θ_0 be two parameter points and let α denote a positive number less than 1. We denote by $P_n(\theta, \theta_0, \alpha)$ the least upper bound of $P(W_n | \theta)$ with respect to W_n, where W_n is restricted to regions for which $P(W_n | \theta_0) = \alpha$. It is clear that if W_n is a critical region of size α for testing $\theta = \theta_0$, its power function can nowhere exceed the value of $P_n(\theta, \theta_0, \alpha)$, that is $P(W_n | \theta) \leqq P_n(\theta, \theta_0, \alpha)$ for all values of θ.

DEFINITION V. *A critical region* W_n *is said to be a most stringent test of the hypothesis* $\theta = \theta_0$ *on the level of significance* α *if* $P(W_n | \theta_0) = \alpha$ *and if*

$$\underset{\theta}{\text{l.u.b.}} \left[P_n(\theta, \theta_0, \alpha) - P(W_n \mid \theta)\right] \leqq \underset{\theta}{\text{l.u.b.}} \left[P_n(\theta, \theta_0, \alpha) - P(Z_n \mid \theta)\right]$$

for all regions Z_n *for which* $P(Z_n | \theta_0) = \alpha$.

We shall prove the following proposition.

PROPOSITION IV. *Let* $y^1, \cdots, y^k$ *be* k *variates which have a joint normal distribution with unknown mean values* $\theta^1, \cdots, \theta^k$ *and a known covariance matrix* $\|\sigma_{ij}\| = \|\lambda_{ij}\|^{-1}$. *Then for testing* $\theta = \theta_0$ *the region* W *defined in* (82) *is a most stringent test.*

Proof. We shall assume that Proposition IV does not hold and we shall arrive at a contradiction. If Proposition IV is not true, then there exists a region Z in the space of $y^1, \cdots, y^k$ such that $P(Z | \theta_0) = \alpha$ and

$$(98) \qquad \underset{\theta}{\text{l.u.b.}} \left[P(\theta, \theta_0, \alpha) - P(W \mid \theta)\right] > \underset{\theta}{\text{l.u.b.}} \left[P(\theta, \theta_0, \alpha) - P(Z \mid \theta)\right].$$

Let S_c be the surface defined by the equation

$$\sum_j \sum_i (\theta^i - \theta_0^i)(\theta^j - \theta_0^j)\lambda_{ij} = c.$$

The functions $P_1(\theta, \theta_0, \alpha)$ and $P(W | \theta)$ are constant on the surface S_c. Hence, on account of (98), there exists a value c_0 such that

$$P(Z \mid \theta) > P(W \mid \theta)$$

for all points θ on S_{c_0}. But this is a contradiction to Proposition II. Hence Proposition IV is proved.

DEFINITION VI. *A sequence of critical regions* $\{W_n\}$ *is said to be an asymptotically most stringent test of the hypothesis* $\theta = \theta_0$ *on the level of significance* α

if $P(W_n|\theta_0)=\alpha$ *and if for any* $\{Z_n\}$ *for which* $P(Z_n|\theta_0)=\alpha$ *we have*

$$\limsup_{n=\infty} \{\underset{\theta}{\text{l.u.b.}} [P_n(\theta,\theta_0,\alpha)-P(W_n|\theta)] - \underset{\theta}{\text{l.u.b.}} [P_n(\theta,\theta_0,\alpha)-P(Z_n|\theta)]\} \leqq 0.$$

We shall prove the following theorem.

THEOREM III. *Let* W_n^* *be the region defined in Theorem* I. *Then the sequence* $\{W_n^*\}$ *is an asymptotically most stringent test of the hypothesis* $\theta=\theta_0$.

Proof. Denote by $\mathfrak{P}_n(\theta,\theta_0,\alpha)$ the least upper bound of $\mathfrak{P}(Z_n^*|\theta)$ with respect to Z_n^*, where Z_n^* is restricted to regions in the space of the maximum likelihood estimates for which $\mathfrak{P}(Z_n^*|\theta_0)=\alpha$. Because of Lemma 1 we have

$$\lim_{n=\infty} [P_n(\theta,\theta_0,\alpha)-\mathfrak{P}_n(\theta,\theta_0,\alpha)]=0 \tag{99}$$

uniformly in θ. Denote by $\bar{P}_n(\theta,\theta_0,\alpha)$ the least upper bound of $\bar{P}(Z_n^*|\theta)$ with respect to Z_n^*, where Z_n^* is restricted to regions in the space of the maximum likelihood estimates for which $\bar{P}(Z_n^*|\theta_0)=\alpha$. The symbol $\bar{P}(V_n^*|\theta)$ denotes the probability of V_n^* calculated under the assumption that the joint distribution of $n^{1/2}(\hat{\theta}_n^1-\theta^1),\cdots,n^{1/2}(\hat{\theta}_n^k-\theta^k)$ is normal with zero means and covariance matrix $\|c_{ij}(\theta_0)\|^{-1}$. For any positive λ we have

$$\lim_{n=\infty} [\mathfrak{P}_n(\theta,\theta_0,\alpha)-\bar{P}_n(\theta,\theta_0,\alpha)]=0 \tag{100}$$

uniformly in θ in the domain $|n^{1/2}(\theta-\theta_0)|\leqq\lambda$. Since for any sequence $\{\theta_n\}$ for which $\lim|n^{1/2}(\theta_n-\theta_0)|=+\infty$, we have

$$\lim_{n=\infty}\mathfrak{P}_n(\theta_n,\theta_0,\alpha)=\lim_{n=\infty}\bar{P}_n(\theta_n,\theta_0,\alpha)=1,$$

we obtain from (100)

$$\lim_{n=\infty} [\mathfrak{P}_n(\theta,\theta_0,\alpha)-\bar{P}_n(\theta,\theta_0,\alpha)]=0 \tag{101}$$

uniformly in θ. For any c let S_c be the surface defined by

$$\sum_{j=1}^{k}\sum_{i=1}^{k}(\theta^i-\theta_0^i)(\theta^j-\theta_0^j)c_{ij}(\theta_0)=c.$$

Obviously $\bar{P}_n(\theta,\theta_0,\alpha)$ is constant along the surface S_c. From (99) and (101) we obtain

$$\lim_{n=\infty}\{\underset{\theta\in S_c}{\text{l.u.b.}}\, P_n(\theta,\theta_0,\alpha)-\underset{\theta\in S_c}{\text{g.l.b.}}\, P_n(\theta,\theta_0,\alpha)\}=0 \tag{102}$$

uniformly in θ. We shall derive a contradiction from the assumption that Theorem III is not true. If Theorem III is not true, there exists a sequence $\{Z_n\}$ of regions such that $P(Z_n|\theta_0)=\alpha$ and

$$\text{(103)}\quad \limsup_{n=\infty} \left\{ \underset{\theta}{\text{l.u.b.}} \left[P_n(\theta, \theta_0, \alpha) - P(W_n^* \mid \theta) \right] - \underset{\theta}{\text{l.u.b.}} \left[P_n(\theta, \theta_0, \alpha) - P(Z_n \mid \theta) \right] \right\} = \delta > 0.$$

On account of (102) and since

$$\lim_{n=\infty} \left[\underset{\theta \in S_c}{\text{l.u.b.}}\, P(W_n^* \mid \theta) - \underset{\theta \in S_c}{\text{g.l.b.}}\, P(W_n^* \mid \theta) \right] = 0$$

uniformly in c (see Theorem II), we obtain from (103) that there exists a sequence $\{c_n\}$ and a subsequence $\{n'\}$ of $\{n\}$ such that for all points $\theta_{n'}$ of $S_{c_{n'}}$

$$P(Z_{n'} \mid \theta_{n'}) > P(W_{n'}^* \mid \theta_{n'}) + \delta/2$$

for all n greater than a certain n_0. But this contradicts Theorem I. Hence Theorem III is proved.

8. **Definitions of "best" tests of composite hypotheses.** In this section we shall extend the definitions given in the previous sections to the case of composite hypotheses. Let ω be a subset of the parameter space and denote by H_ω the hypothesis that the true parameter point is contained in ω. In all that follows the letter θ printed in boldface will indicate that the parameter point lies in ω. For example, the symbol l.u.b.$_{\boldsymbol{\theta}}$ $f(\boldsymbol{\theta})$ denotes the least upper bound of the function $f(\boldsymbol{\theta})$ with respect to $\boldsymbol{\theta}$ where $\boldsymbol{\theta}$ is restricted to points of ω. For any point $\boldsymbol{\theta}$ and for any real value c let $K_c(\boldsymbol{\theta})$ denote a surface in the parameter space. For instance $K_c(\boldsymbol{\theta})$ may be given by r equations in θ

$$\phi_1(\theta, \boldsymbol{\theta}) = \cdots = \phi_r(\theta, \boldsymbol{\theta}) = 0,$$

where $\phi_1(\theta, \boldsymbol{\theta}), \cdots, \phi_r(\theta, \boldsymbol{\theta})$ are some analytic functions of θ and $\boldsymbol{\theta}$.

DEFINITION VII. *A critical region W_n for testing H_ω is said to have uniformly best average power with respect to a family of surfaces $K_c(\boldsymbol{\theta})$ and a weight function $w(\theta)$ if for any Z_n for which*

$$\underset{\boldsymbol{\theta}}{\text{l.u.b.}}\, P(Z_n \mid \boldsymbol{\theta}) = \underset{\boldsymbol{\theta}}{\text{l.u.b.}}\, P(W_n \mid \boldsymbol{\theta})$$

we have

$$\int_{K_c(\boldsymbol{\theta})} P(W_n \mid \theta) w(\theta) dA \geqq \int_{K_c(\boldsymbol{\theta})} P(Z_n \mid \theta) w(\theta) dA$$

for any $\boldsymbol{\theta}$ and for any c for which $K_c(\boldsymbol{\theta})$ is defined.

DEFINITION VIII. *A sequence* $\{W_n\}$ ($n=1, 2, \cdots$, ad inf.) *of critical regions for testing the hypothesis H_ω is said to have asymptotically best average power with respect to a family of surfaces $K_c(\boldsymbol{\theta})$ and a weight function $w(\theta)$ if the following two conditions are fulfilled:*

(a) *There exists a fixed* α *such that*

$$\underset{\boldsymbol{\theta}}{\text{l.u.b.}}\ P(W_n \mid \boldsymbol{\theta}) = \alpha \qquad (n = 1, 2, \cdots, \text{ad inf.})$$

(b) *For any sequence* $\{Z_n\}$ *for which* l.u.b.$_{\boldsymbol{\theta}}$ $P(Z_n \mid \boldsymbol{\theta}) = \alpha$, *we have*

$$\limsup_{n=\infty} \left\{ \underset{c,\boldsymbol{\theta}}{\text{l.u.b.}} \left[\int_{K_c(\boldsymbol{\theta})} P(Z_n \mid \theta) \frac{w(\theta)}{A[K_c(\boldsymbol{\theta})]} dA - \int_{K_c(\boldsymbol{\theta})} P(W_n \mid \theta) \frac{w(\theta)}{A[K_c(\boldsymbol{\theta})]} dA \right] \right\} \leqq 0,$$

where

$$A[K_c(\boldsymbol{\theta})] = \int_{K_c(\boldsymbol{\theta})} w(\theta) dA.$$

DEFINITION IX. *A critical region* W_n *for testing* H_ω *is said to have uniformly best constant power on the family of surfaces* $K_c(\boldsymbol{\theta})$ *if the following two conditions are fulfilled:*

(a) $P(W_n \mid \theta') = P(W_n \mid \theta'')$ *for all pairs of points* θ' *and* θ'' *which lie on the same surface* $K_c(\boldsymbol{\theta})$.

(b) $P(W_n \mid \theta) \geqq P(Z_n \mid \theta)$ *for any* θ *not in* ω *and for any* Z_n *which satisfies* (a) *and the condition*

$$\underset{\boldsymbol{\theta}}{\text{l.u.b.}}\ P(Z_n \mid \boldsymbol{\theta}) = \underset{\boldsymbol{\theta}}{\text{l.u.b.}}\ P(W_n \mid \boldsymbol{\theta}).$$

DEFINITION X. *A sequence of critical regions* $\{W_n\}$ *for testing* H_ω *is said to have asymptotically best constant power on the surfaces* $K_c(\boldsymbol{\theta})$ *if the following three conditions are fulfilled:*

(a) l.u.b.$_{\boldsymbol{\theta}}$ $P(W_n \mid \boldsymbol{\theta}) = \alpha$ $(n = 1, 2, \cdots, \text{ad inf.})$.

(b) $\lim_{n=\infty} \{\text{l.u.b.}_{c,\boldsymbol{\theta}} [\text{l.u.b.}_{\theta \in K_c(\boldsymbol{\theta})} P(W_n \mid \theta) - \text{g.l.b.}_{\theta \in K_c(\boldsymbol{\theta})} P(W_n \mid \theta)]\} = 0$.

(c) *For any sequence* $\{Z_n\}$ *which satisfies* (a) *and* (b) *we have*

$$\lim_{n=\infty} \{ \underset{\theta \in \bar{\omega}}{\text{l.u.b.}}\ [P(Z_n \mid \theta) - P(W_n \mid \theta]\} = 0,$$

where $\bar{\omega}$ *is the complement of* ω.

DEFINITION XI. *Denote by* $P_n(\theta, \omega, \alpha)$ *the least upper bound of* $P(Z_n \mid \theta)$ *with respect to* Z_n *subject to the condition* l.u.b.$_{\boldsymbol{\theta}}$ $P(Z_n \mid \boldsymbol{\theta}) = \alpha$. *A critical region* W_n *is said to be a most stringent test of the hypothesis* H_ω *if for some positive* α

$$\underset{\boldsymbol{\theta}}{\text{l.u.b.}}\ P(W_n \mid \theta) = \alpha$$

and

$$\underset{\theta}{\text{l.u.b.}}\ [P_n(\theta, \omega, \alpha) - P(W_n \mid \theta)] \leqq \underset{\theta}{\text{l.u.b.}}\ [P_n(\theta, \omega, \alpha) - P(Z_n \mid \theta)]$$

for all regions Z_n *for which* l.u.b.$_{\boldsymbol{\theta}}$ $P(Z_n \mid \boldsymbol{\theta}) = \alpha$.

DEFINITION XII. *A sequence of critical regions* $\{W_n\}$ *is said to be an asymptotically most stringent test of the hypothesis* H_ω *if the following two conditions are fulfilled:*

(a) *There exists a positive* α *such that*

$$\underset{\boldsymbol{\theta}}{\text{l.u.b.}}\ P(W_n \mid \boldsymbol{\theta}) = \alpha \qquad (n = 1, 2, \cdots, \text{ad inf.}).$$

(b) *For any sequence* $\{Z_n\}$ *which satisfies* (a) *for the same* α *we have*

$$\limsup_{n=\infty} \{\underset{\theta}{\text{l.u.b.}}\ [P_n(\theta, \omega, \alpha) - P(W_n \mid \theta)] - \underset{\theta}{\text{l.u.b.}}\ [P_n(\theta, \omega, \alpha) - P(Z_n \mid \theta)]\} \leqq 0.$$

In Definitions VII–XII we have formulated the condition

$$\underset{\boldsymbol{\theta}}{\text{l.u.b.}}\ P(W_n \mid \boldsymbol{\theta}) = \alpha.$$

The question can be raised whether, in place of this condition, the requirement that

$$(*) \qquad P(W_n \mid \boldsymbol{\theta}) = \alpha$$

for all points $\boldsymbol{\theta}$ should be made; or whether the weaker condition that

$$(**) \qquad \lim_{n=\infty} P(W_n \mid \boldsymbol{\theta}) = \alpha$$

uniformly in $\boldsymbol{\theta}$ should be required. Condition (*) has the serious drawback that regions satisfying it do not always exist. Even in cases where (*) can be fulfilled, it imposes too strong a restriction on the possible choice of W_n, which does not seem to be quite justified. It is conceivable that in some cases a region W_n' may exist which does not satisfy (*) but has such an advantageous power function that we prefer it to any region W_n which satisfies(*).

As to the condition (**), we shall see that it is satisfied for the sequence $\{W_n\}$ which is shown in this paper to be asymptotically best according to all three definitions VIII, X and XII. Hence the same sequence $\{W_n\}$ remains asymptotically best if we replace the condition l.u.b.$_{\boldsymbol{\theta}}$ $P(W_n \mid \boldsymbol{\theta}) = \alpha$ by (**) in the definitions VIII, X and XII.

In the following §§9–11 we shall discuss a linear hypothesis of the following type: $\theta^1 = \theta_0^1, \cdots, \theta^r = \theta_0^r$ $(r < k)$, where $\theta_0^1, \cdots, \theta_0^r$ are some specified values. That is to say, the set ω is the set of all points θ for which the above equations hold. In §12 the general composite hypothesis will be discussed.

9. **Tests of linear composite hypotheses which have uniformly best average power over a family of surfaces.** Let H_ω be the hypothesis that $\theta^1 = \theta_n^1, \cdots, \theta^r = \theta_0^r$ $(r < k)$. We shall introduce the following notation: For any parameter point $\theta = (\theta^1, \cdots, \theta^k)$ the symbol ${}_1\theta$ will denote the vector in the r-dimensional space with the components $\theta^1, \cdots, \theta^r$, and ${}_2\theta$ will denote the vector in the $k-r$ dimensional space with the components $\theta^{r+1}, \cdots, \theta^k$. For any function $\psi(\theta)$ of θ we shall use the synonymous nota-

tion $\psi({}_1\theta, {}_2\theta)$. For instance $P(W|{}_1\theta, {}_2\theta)$ is synonymous with $P(W|\theta)$.

Let $y^1, \cdots, y^k$ be k variates which have a joint normal distribution with unknown mean values $\theta^1, \cdots, \theta^k$ and known covariance matrix $\|\sigma_{ij}\| = \|\lambda_{ij}\|^{-1}$ $(i, j = 1, \cdots, k)$, which is nonsingular. Denote by W the region in the space of $y^1, \cdots, y^k$ given by the inequality

$$\sum_{q=1}^{r}\sum_{p=1}^{r} \bar{\lambda}_{pq}(y^p - \theta_0^p)(y^q - \theta_0^q) \geqq d, \tag{104}$$

where $\|\bar{\lambda}_{pq}\| = \|\sigma_{pq}\|^{-1}$ $(p, q = 1, \cdots, r)$. Consider the nonsingular linear transformation of the variates $y^1, \cdots, y^k$ given by the equations

$$\begin{aligned} y'^p - \theta_0^p &= \beta_{p1}(y^1 - \theta_0^1) + \cdots + \beta_{pr}(y^r - \theta_0^r) \qquad (p = 1, \cdots, r), \\ y'^t &= \gamma_{t1}y^1 + \cdots + \gamma_{tk}y^k \qquad (t = r+1, \cdots, k), \end{aligned} \tag{105}$$

such that $y'^1, \cdots, y'^k$ are independently distributed with unit variances. Denote by $S_c(\boldsymbol{\theta})$ the surface given by the equations

$$\begin{aligned} &\sum_{q=1}^{r}\sum_{p=1}^{r} \bar{\lambda}_{pq}(\theta^p - \theta_0^p)(\theta^q - \theta_0^q) = c, \\ &\gamma_{t1}\theta^1 + \cdots + \gamma_{tk}\theta^k = \sum_{i=1}^{k} \gamma_{ti}\boldsymbol{\theta}^i \qquad (t = r+1, \cdots, k). \end{aligned} \tag{106}$$

Consider the transformation of the parameter space given by

$$\begin{aligned} \theta'^p - \theta_0^p &= \beta_{p1}(\theta^1 - \theta_0^1) + \cdots + \beta_{pr}(\theta^r - \theta_0^r) \qquad (p = 1, \cdots, r), \\ \theta'^t &= \gamma_{t1}\theta^1 + \cdots + \gamma_{tk}\theta^k \qquad (t = r+1, \cdots, k), \end{aligned} \tag{107}$$

where the coefficients β_{pq} and γ_{tj} are the same as in (105). The transformation (107) transforms the surface $S_c(\boldsymbol{\theta})$ into a sphere $S_c'(\boldsymbol{\theta})$ given by

$$\sum_{p=1}^{r} (\theta'^p - \theta_0^p)^2 = c, \qquad \theta'^t = \sum_{i=1}^{k} \gamma_{ti}\boldsymbol{\theta}^i = \boldsymbol{\theta}'^t. \tag{108}$$

For any point θ and for any positive ρ consider the set $\omega(\theta, \rho)$ consisting of all points ${}^*\theta$ which lie on the same $S_c(\boldsymbol{\theta})$ as θ, and for which $|{}^*\theta - \theta| \leqq \rho$. Let

$$\xi(\theta) = \lim_{\rho=0} \frac{A[\omega'(\theta, \rho)]}{A[\omega(\theta, \rho)]}, \tag{109}$$

where $\omega'(\theta, \rho)$ is the image of $\omega(\theta, \rho)$ (by transformation (107)) and, for any set ω, $A(\omega)$ denotes the $(r-1)$-dimensional area of ω. We shall prove the following proposition.

PROPOSITION V. *Let $y^1, \cdots, y^k$ be k variates which have a joint normal distribution with unknown mean values $\theta^1, \cdots, \theta^k$ and known covariance matrix $\|\sigma_{ij}\| = \|\lambda_{ij}\|^{-1}$. For testing the hypothesis ${}_1\theta = {}_1\theta_0$ on the basis of a single* *observation on each of the variates* $y^1, \cdots, y^k$, *the critical region* W *given in* (104) *has uniformly best average power with respect to the family of surfaces* $S_c(\boldsymbol{\theta})$ *defined in* (106) *and the weight function* $\xi(\theta)$ *given in* (109).

Proof. Because of (109) we have

$$\int_{S_c(\boldsymbol{\theta})} P(Z \mid \theta)\xi(\theta)dA = \int_{S'_c(\boldsymbol{\theta})} P(Z' \mid \theta')dA, \tag{110}$$

where Z denotes an arbitrary region in the space of $y^1, \cdots, y^k$ and Z' is the image of Z by transformation (105). The region W is transformed into W' given by

$$(y'^1 - \theta_0^1)^2 + \cdots + (y'^r - \theta_0^r)^2 \geqq d. \tag{111}$$

In order to prove Proposition V we have merely to show that

$$\int_{S'_c(\boldsymbol{\theta})} P(W' \mid \theta')dA \geqq \int_{S'_c(\boldsymbol{\theta})} P(Z' \mid \theta')dA \tag{112}$$

for any $c>0$, for any $\boldsymbol{\theta}$, and for any region Z' in the space of $y'^1, \cdots, y'^k$ for which $\text{l.u.b.}_{{}_2\theta'} P(Z' \mid {}_1\theta_0, {}_2\theta') = \text{l.u.b.}_{{}_2\theta'} P(W' \mid {}_1\theta_0, {}_2\theta')$. For any point θ' of $S'_c(\boldsymbol{\theta})$ we have ${}_2\theta' = {}_2\boldsymbol{\theta}'$. By a lemma of Neyman and Pearson, (112) is proved if we can show that there exists a function $d(c)$ such that

$$\frac{\int_{S'_c(\boldsymbol{\theta})} p(y'^1, \cdots, y'^k \mid {}_1\theta', {}_2\boldsymbol{\theta}')dA}{p(y'^1, \cdots, y'^k \mid {}_1\theta_0, {}_2\boldsymbol{\theta}')} \left\{ \begin{array}{l} \geqq d(c) \text{ within } W', \\ \leqq d(c) \text{ outside } W' \end{array} \right. \tag{113}$$

for all values of c and $\boldsymbol{\theta}$ where $p(y'^1, \cdots, y'^k \mid \theta')$ denotes the joint probability density of $y'^1, \cdots, y'^k$ under the assumption that θ' is the true parameter point. Obviously

$$\frac{p(y'^1, \cdots, y'^k \mid {}_1\theta', {}_2\boldsymbol{\theta}')}{p(y'^1, \cdots, y'^k \mid {}_1\theta_0, {}_2\boldsymbol{\theta}')} = \frac{p(y'^1, \cdots, y'^r \mid {}_1\theta')}{p(y'^1, \cdots, y'^r \mid {}_1\theta_0)}.$$

Hence (113) is equivalent to

$$\frac{\int_{S'_c(\boldsymbol{\theta})} p(y'^1, \cdots, y'^r \mid {}_1\theta')dA}{p(y'^1, \cdots, y'^r \mid {}_1\theta_0)} \left\{ \begin{array}{l} \geqq d(c) \text{ within } W', \\ \leqq d(c) \text{ outside } W'. \end{array} \right. \tag{114}$$

The proof of (114) is omitted, since it is the same as that of the inequalities (87) and (88). Hence Proposition V is proved.

Let the critical region W_n^* be defined by the following inequality

$$n \sum_{q=1}^{r} \sum_{p=1}^{r} (\dot{\theta}_n^p - \theta_0^p)(\dot{\theta}_n^q - \theta_0^q)\bar{c}_{pq}(\dot{\theta}_n) \geqq d_n, \tag{115}$$

where $\|\bar{c}_{pq}(\theta)\| = \|\sigma_{pq}(\theta)\|^{-1}$ $(p, q=1, \cdots, r)$ and $\|\sigma_{ij}(\theta)\| = \|c_{ij}(\theta)\|^{-1}$ $(i, j$

$=1, \cdots, k$). The constant d_n is chosen so that

$$\underset{\theta}{\text{l.u.b.}}\ P(W^* \mid \theta) = \alpha.$$

Let $Z_n^i(\theta) = n^{1/2}(\theta_n^i - \theta^i)$ $(i=1, \cdots, k)$ and consider the nonsingular linear transformation

$$\begin{aligned} \bar{Z}_n^p(\theta) &= \beta_{p1}(\theta)Z_n^1(\theta) + \cdots + \beta_{pr}(\theta)Z_n^r(\theta) \qquad (p = 1, \cdots, r), \\ \bar{Z}_n^t(\theta) &= \gamma_{t1}(\theta)Z_n^1(\theta) + \cdots + \gamma_{tk}(\theta)Z_n^k(\theta) \qquad (t = r+1, \cdots, k), \end{aligned} \tag{116}$$

such that $\bar{Z}_n^1(\theta), \cdots, \bar{Z}_n^k(\theta)$ would be independently distributed with unit variances if the covariance matrix of $Z_m^1(\theta), \cdots, Z_n^k(\theta)$ were given by $\|\sigma_{ij}(\theta)\|$. Denote by $S_c(\boldsymbol{\theta})$ the surface defined by the equations

$$\begin{aligned} &\sum_{q=1}^{r}\sum_{p=1}^{r} (\theta^p - \theta_0^p)(\theta^q - \theta_0^q)\bar{c}_{pq}(\boldsymbol{\theta}) = c, \\ &\gamma_{t1}(\boldsymbol{\theta})\theta^1 + \cdots + \gamma_{tk}(\boldsymbol{\theta})\theta^k = \sum \gamma_{ti}(\boldsymbol{\theta})\boldsymbol{\theta}^i \quad (t = r+1, \cdots, k). \end{aligned} \tag{117}$$

For any positive δ denote by S_δ the set of all points θ for which $|\theta^p - \theta_0^p| \leq \delta$ $(p=1, \cdots, r)$. We shall prove the existence of a positive δ such that for any point θ in S_δ there exists exactly one surface $S_c(\boldsymbol{\theta})$, that is exactly one value of c and exactly one point $\boldsymbol{\theta}$, such that θ lies on the surface $S_c(\boldsymbol{\theta})$. This statement is obviously proved if we show that for any point θ in S_δ the set of $k-r$ equations

$$\gamma_{t1}(\boldsymbol{\theta})(\theta^1 - \boldsymbol{\theta}^1) + \cdots + \gamma_{tk}(\boldsymbol{\theta})(\theta^k - \boldsymbol{\theta}^k) = 0 \qquad (t = r+1, \cdots, k) \tag{118}$$

has a unique solution in the unknowns $\boldsymbol{\theta}^{r+1}, \cdots, \boldsymbol{\theta}^k$. From the definition of the quantities $\beta_{pq}(\boldsymbol{\theta})$ and $\gamma_{ti}(\boldsymbol{\theta})$ it follows that

$$A(\boldsymbol{\theta})\|\sigma_{ij}(\boldsymbol{\theta})\|\bar{A}(\boldsymbol{\theta}) = I, \tag{119}$$

where $A(\boldsymbol{\theta})$ denotes the matrix

$$\left\| \begin{matrix} \beta_{11}(\boldsymbol{\theta}) & \beta_{12}(\boldsymbol{\theta}) & \cdots & \beta_{1r}(\boldsymbol{\theta}) & 0 & \cdots & 0 \\ \beta_{12}(\boldsymbol{\theta}) & \beta_{22}(\boldsymbol{\theta}) & \cdots & \beta_{2r}(\boldsymbol{\theta}) & 0 & \cdots & 0 \\ \cdot & \cdot & \cdots & \cdot & \cdot & \cdots & \cdot \\ \beta_{r1}(\boldsymbol{\theta}) & \beta_{r2}(\boldsymbol{\theta}) & \cdots & \beta_{rr}(\boldsymbol{\theta}) & 0 & \cdots & 0 \\ \gamma_{r+1\,1}(\boldsymbol{\theta}) & \gamma_{r+1\,2}(\boldsymbol{\theta}) & \cdots & \gamma_{r+1\,r}(\boldsymbol{\theta}) & \gamma_{r+1\,r+1}(\boldsymbol{\theta}) & \cdots & \gamma_{r+1\,k}(\boldsymbol{\theta}) \\ \cdot & \cdot & \cdots & \cdot & \cdot & \cdots & \cdot \\ \gamma_{k1}(\boldsymbol{\theta}) & \gamma_{k2}(\boldsymbol{\theta}) & \cdots & \gamma_{kr}(\boldsymbol{\theta}) & \gamma_{k\,r+1}(\boldsymbol{\theta}) & \cdots & \gamma_{kk}(\boldsymbol{\theta}) \end{matrix} \right\|, \tag{120}$$

$\bar{A}(\boldsymbol{\theta})$ is the transposed of $A(\boldsymbol{\theta})$ and I denotes the unit matrix. Since $\sigma_{ij}(\boldsymbol{\theta})$ is a continuous and bounded function of $\boldsymbol{\theta}$ and since the determinant $|\sigma_{ij}(\boldsymbol{\theta})|$ has a positive lower bound, we find that $\beta_{pq}(\boldsymbol{\theta})$ and $\gamma_{ti}(\boldsymbol{\theta})$ are continuous and bounded functions of $\boldsymbol{\theta}$ and that the absolute value of the determinant $A(\boldsymbol{\theta})$ has a positive lower bound. Hence also the absolute value of the determinant

$$\gamma(\boldsymbol{\theta}) = \begin{vmatrix} \gamma_{r+1\, r+1}(\boldsymbol{\theta}) & \cdots & \gamma_{r+1\, k}(\boldsymbol{\theta}) \\ \cdot & \cdots & \cdot \\ \gamma_{k\, r+1}(\boldsymbol{\theta}) & \cdots & \gamma_{kk}(\boldsymbol{\theta}) \end{vmatrix} \tag{121}$$

has a positive lower bound.

Let $\theta = \boldsymbol{\theta}^*$ where $\boldsymbol{\theta}^*$ denotes an arbitrary point of ω. Then, since the determinant in (121) has a positive lower bound, the equations (118) have a unique solution in the unknowns $\boldsymbol{\theta}^{r+1}, \cdots, \boldsymbol{\theta}^k$, namely the solution $\boldsymbol{\theta} = \boldsymbol{\theta}^*$. Furthermore we see that the Jacobian of the equations (118), taken at the point $\boldsymbol{\theta} = \boldsymbol{\theta}^*$, is equal to $\gamma(\boldsymbol{\theta}^*)$. Since the absolute value of $\gamma(\boldsymbol{\theta}^*)$ has a positive lower bound, there exists a positive δ such that the equations (118) have a unique solution in $\boldsymbol{\theta}$ if $|\theta - \boldsymbol{\theta}^*| \leqq \delta$. This proves the existence of a positive δ such that for any point θ in S_δ there exists exactly one surface $S_c(\boldsymbol{\theta})$ such that θ lies on $S_c(\boldsymbol{\theta})$.

Since for the critical region W_n^* defined in (115) we obviously have $\lim_{n=\infty} P(W_n^* | \theta) = 1$ uniformly over the domain $|\theta^p - \theta_0^p| \geqq \delta$, we shall restrict ourselves to the consideration of points θ for which $|\theta^p - \theta_0^p| \leqq \delta$ $(p = 1, \cdots, r)$.

Consider the transformation of the parameter space given by

$$\begin{aligned} \theta'^p &= \theta_0^p = \beta_{p1}(\boldsymbol{\theta})(\theta^1 - \theta_0^1) + \cdots + \beta_{pr}(\boldsymbol{\theta})(\theta^r - \theta_0^r) \qquad (p = 1, \cdots, r), \\ \theta'^t &= \gamma_{t1}(\boldsymbol{\theta})\theta^1 + \cdots + \gamma_{tk}(\boldsymbol{\theta})\theta^k \qquad (t = r+1, \cdots, k), \end{aligned} \tag{122}$$

where $\boldsymbol{\theta}$ denotes the point for which θ lies on $S_c(\boldsymbol{\theta})$. The transformation (122) transforms $S_c(\boldsymbol{\theta})$ into the sphere $S_c'(\boldsymbol{\theta})$ given by

$$\sum_{p=1}^{r} (\theta'^p - \theta_0^p)^2 = c; \quad \theta'^t = \sum \gamma_{ti}(\boldsymbol{\theta})\boldsymbol{\theta}^i = \boldsymbol{\theta}'^t \qquad (t = r+1, \cdots, k). \tag{123}$$

We define a weight function $\xi(\theta)$ as follows:

$$\xi(\theta) = \lim_{\rho=0} \frac{A[\omega'(\theta, \rho)]}{A[\omega(\theta, \rho)]}, \tag{124}$$

where the symbols on the right-hand side have the same meaning as in (109).

THEOREM IV. *Let the critical region* W_n^* *for testing* ${}_1\theta = {}_1\theta_0$ *be the region defined in* (115). *Furthermore, let* $S_c(\boldsymbol{\theta})$ *be the surface defined in* (117) *and let* $\xi(\theta)$ *be the weight function defined in* (124). *Then* $\{W_n^*\}$ *has asymptotically best average power with respect to the family of surfaces* $S_c(\boldsymbol{\theta})$ *and the weight function* $\xi(\theta)$.

Proof. Because of Lemma 1, we can restrict ourselves to subsets of the space of the maximum likelihood estimates. Because of Lemma 2, Theorem IV is proved if we show that for any $\{Z_n^*\}$ for which l.u.b.$_\theta$ $P(Z_n^*|\theta)=\alpha$ we have

$$\limsup_{n=\infty}\left\{\underset{c,\theta}{\text{l.u.b.}}\left[\int_{S_c(\theta)}\mathfrak{P}(Z_n^*\mid\theta)\zeta_n(\theta)dA-\int_{S_c(\theta)}\mathfrak{P}(W_n^*\mid\theta)\zeta_n(\theta)dA\right]\right\}\leqq 0,$$

where

$$\zeta_n(\theta)=\xi(\theta)\bigg/\int_{S_c(\theta)}\xi(\theta)dA.$$

If Theorem IV were not true, there would exist a sequence $\{c_n\}$, a sequence $\{\theta_n\}$, a sequence $\{Z_n^*\}$, and a subsequence $\{n'\}$ of $\{n\}$ such that

(125)
$$\underset{\theta}{\text{l.u.b.}}\ P(Z_n^*\mid\theta)=\alpha,$$
$$\lim_{n=\infty}\left\{\int_{S_{c_{n'}}(\theta_{n'})}\mathfrak{P}(Z_{n'}^*\mid\theta)\zeta_{n'}(\theta)dA-\int_{S_{c_{n'}}(\theta_{n'})}\mathfrak{P}(W_{n'}^*\mid\theta)\zeta_{n'}(\theta)dA\right\}=\delta>0.$$

It is easy to verify that for any sequence $\{c_n\}$ for which $\lim nc_n=+\infty$ we have $\int_{S_{c_n}(\theta)}\mathfrak{P}(W_n^*|\theta)\zeta_n(\theta)dA=1$ uniformly in θ. Hence (125) can hold only if the sequence $\{n'c_{n'}\}$ is bounded. If $\{n'c_{n'}\}$ is bounded, for any sequence of regions $\{V_n^*\}$ we obviously have

$$(126)\quad \lim_{n=\infty}\left\{\int_{S_{c_{n'}}(\theta_{n'})}\mathfrak{P}(V_{n'}^*\mid\theta)\zeta_{n'}(\theta)dA-\int_{S_{c_{n'}}(\theta_{n'})}P_{\theta_{n'}}(V_{n'}^*\mid\theta)\zeta_{n'}(\theta)dA\right\}=0,$$

where $P_\theta(V_n^*|\theta)$ denotes the probability of V_n^* calculated under the assumption that $n^{1/2}(\dot\theta_n^1-\theta^1),\cdots,n^{1/2}(\dot\theta_n^k-\theta^k)$ have a joint normal distribution with zero means and covariance matrix $\sigma_{ij}(\theta)$. Let $W_n^*(\theta)$ be the region defined by

$$n\sum_{q=1}^{r}\sum_{p=1}^{r}(\dot\theta_n^p-\theta_0^p)(\dot\theta_n^q-\theta_0^q)\bar c_{pq}(\theta)\geqq d_n.$$

It is clear that if $\{n'c_{n'}\}$ is bounded, we have

$$(127)\quad \lim_{n=\infty}\{P_\theta(W_{n'}^*\mid\theta)-P_\theta[W_{n'}^*(\theta)\mid\theta]\}=0$$

uniformly in θ and $\boldsymbol\theta$ over the domain in which θ is a point of $S_{c_{n'}}(\theta)$. From (125), (126) and (127) we obtain

(128)
$$\lim_{n=\infty}\left\{\int_{S_{c_{n'}}(\theta_{n'})}P_{\theta_{n'}}(Z_{n'}^*\mid\theta)\zeta_{n'}(\theta)dA\right.$$
$$\left.-\int_{S_{c_{n'}}(\theta_{n'})}P_{\theta_{n'}}[W_{n'}^*(\theta_{n'})\mid\theta]\zeta_{n'}(\theta)dA\right\}=\delta>0.$$

The surface $S_c(\theta)$ defined in (106) is identical with the surface $S_c(\theta)$ defined in (117) if in (106) we substitute $c_{ij}(\boldsymbol{\theta})$ for λ_{ij}, that is we substitute $\bar{c}_{pq}(\boldsymbol{\theta})$ for $\bar{\lambda}_{pq}$. Similarly, if we substitute $c_{ij}(\boldsymbol{\theta})$ for λ_{ij} for any point of the surface $S_c(\boldsymbol{\theta})$, the value of the weight function $\xi(\theta)$ defined in (109) is the same as the value of $\xi(\theta)$ defined in (124). Hence, since

$$\lim_{n=\infty} \left[\underset{\boldsymbol{\theta}}{\text{l.u.b.}}\ P_{\boldsymbol{\theta}}(Z^*_{n'} \mid \boldsymbol{\theta})\right] = \alpha,$$

equation (128) is in contradiction to Proposition V. Thus Theorem IV is proved.

10. **Tests of linear composite hypotheses which have best constant power on a family of surfaces.** The critical region W defined in (104) satisfies condition (a) of Definition IX if $K_c(\boldsymbol{\theta})$ is equal to $S_c(\boldsymbol{\theta})$ given in (106). Hence from Proposition V we obtain the following proposition.

PROPOSITION VI. *The region W given in* (104) *has uniformly best constant power on the surfaces $S_c(\boldsymbol{\theta})$ defined in* (106).

If W_n^* is the region defined in (115) and if $K_c(\boldsymbol{\theta})$ is equal to the surface $S_c(\boldsymbol{\theta})$ defined in (117), then $\{W_n^*\}$ satisfies conditions (a) and (b) of Definition X. Hence, from Theorem IV we easily obtain the following theorem.

THEOREM V. *Let W_n^* be the region defined in* (115) *and let $S_c(\boldsymbol{\theta})$ be the surface defined in* (117), *then for testing ${}_1\theta = {}_1\theta_0$, $\{W_n^*\}$ has asymptotically best constant power along the surfaces $S_c(\boldsymbol{\theta})$.*

11. **Most stringent tests of linear composite hypotheses.** We shall prove the following proposition.

PROPOSITION VII. *Let $y^1, \cdots, y^k$ be k variates which have a joint normal distribution with unknown mean values $\theta^1, \cdots, \theta^k$ and known covariance matrix $\|\sigma_{ij}\| = \|\lambda_{ij}\|^{-1}$. For testing the hypothesis ${}_1\theta = {}_1\theta_0$ on the basis of a single observation on each of the variates $y^1, \cdots, y^k$, the region W given in* (104) *is a most stringent test.*

Proof. First we shall show that $P(\theta, \omega, \alpha)$ is constant along $S_c(\boldsymbol{\theta})$ where $S_c(\boldsymbol{\theta})$ is defined in (106). Consider a linear transformation of $y^1, \cdots, y^k$ as defined in (105). Then the transformed variates $y'^1, \cdots, y'^k$ are independently distributed with unit variances. Denote by θ' the image of θ obtained by the transformation (107) and let $P'(\theta', \omega, \alpha)$ be equal to l.u.b. $P(Z' \mid \theta')$ with respect to Z', where Z' may be any region in the space of $y'^1, \cdots, y'^k$ subject to the condition that

$$\underset{{}_2\theta'}{\text{l.u.b.}}\ P(Z' \mid {}_1\theta_0, {}_2\theta') = \alpha.$$

Obviously $P'(\theta', \omega, \alpha) = P(\theta, \omega, \alpha)$.

Hence we have merely to show that $P'(\theta', \omega, \alpha)$ is constant along $S'_c(\boldsymbol{\theta})$ where $S'_c(\boldsymbol{\theta})$ is the image of $S_c(\boldsymbol{\theta})$ and is given by

$$\sum_{p=1}^{r} (\theta'^p - \theta_0^p)^2 = c; \qquad \theta'^t = \boldsymbol{\theta}'^t \qquad (t = r+1, \cdots, k).$$

Let $P^*(\theta', {}_1\theta_0, \alpha)$ be equal to the least upper bound of $P(Z' \mid \theta')$ with respect to all regions Z' in the space of $y'^1, \cdots, y'^k$ for which $P(Z' \mid {}_1\theta_0, {}_2\theta') = \alpha$. Obviously $P^*(\theta', {}_1\theta_0, \alpha) \geqq P'(\theta', \omega, \alpha)$ It is easy to verify that the region V' for which $P(V' \mid {}_1\theta', {}_2\theta') = P^*(\theta', {}_1\theta_0, \alpha)$ is a subset in the space of $y'^1, \cdots, y'^r$. Hence

$$P(V' \mid {}_1\theta_0, {}_2\theta_1') = P(V' \mid {}_1\theta_0, {}_2\theta_2')$$

for any pair of points θ_1' and θ_2', and therefore

$$P^*(\theta', {}_1\theta_0, \alpha) = P'(\theta', \omega, \alpha) = P(\theta, \omega, \alpha).$$

Since $P^*(\theta', {}_1\theta_0, \alpha)$ is constant along $S_c(\boldsymbol{\theta})$, our statement is proved. From this and Proposition V, Proposition VII easily follows.

THEOREM VI. *Let W_n^* be the region defined in* (115). *Then* $\{W_n^*\}$ *is an asymptotically most stringent test of the hypothesis* ${}_1\theta = {}_1\theta_0$.

Proof. Denote by $\mathfrak{P}_n(\theta, \omega, \alpha)$ the least upper bound of $\mathfrak{P}(Z_n^* \mid \theta)$ with respect to Z_n^*, where Z_n^* is restricted to regions in the space of the maximum likelihood estimates for which

$$\underset{\boldsymbol{\theta}}{\text{l.u.b.}}\ \mathfrak{P}(Z_n^* \mid \boldsymbol{\theta}) = \alpha.$$

On account of Lemmas 1 and 2 we have

$$(129) \qquad \lim_{n=\infty} \{P_n(\theta, \omega, \alpha) - \mathfrak{P}_n(\theta, \omega, \alpha)\} = 0$$

uniformly in θ.

Denote by $\overline{P}_n(\theta, {}_1\theta_0, \alpha)$ the least upper bound of $\overline{P}_n(Z_n^* \mid \theta)$ with respect to Z_n^* where Z_n^* is restricted to regions in the space of the maximum likelihood estimates for which

$$\underset{{}_2\theta}{\text{l.u.b.}}\ \overline{P}(Z_n^* \mid {}_1\theta_0, {}_2\theta) = \alpha.$$

The symbol $\overline{P}(V_n^* \mid \theta)$ denotes the probability of V_n^* calculated under the assumption that the joint distribution of $n^{1/2}(\hat\theta_n^1 - \theta^1), \cdots, n^{1/2}(\hat\theta_n^k - \theta^k)$ is normal with zero means and covariance matrix $\|c_{ij}(\boldsymbol{\theta})\|^{-1}$, where $\boldsymbol{\theta}$ denotes that point for which θ lies on the surface $S_c(\boldsymbol{\theta})$ defined in (117). It can be shown that for any positive λ we have

$$(130) \qquad \lim_{n=\infty} \{\mathfrak{P}_n(\theta, \omega, \alpha) - \overline{P}_n(\theta, {}_1\theta_0, \alpha)\} = 0$$

uniformly in θ in the domain $|{}_1\theta - {}_1\theta_0| \leqq \lambda/n^{1/2}$. Since for any sequence $\{\theta_n\}$ for which

$$\lim_{n=\infty} n^{1/2}\left(\sum_{p=1}^{r}(\theta^p - \theta_0^p)^2\right)^{1/2} = +\infty,$$

we have

$$\lim_{n=\infty} \mathfrak{P}(\theta_n, \omega, \alpha) = \lim_{n=\infty} \overline{P}_n(\theta_n, {}_1\theta_0, \alpha) = 1,$$

we obtain from (130)

$$\lim_{n=\infty} \{\mathfrak{P}_n(\theta, \omega, \alpha) - \overline{P}_n(\theta, {}_1\theta_0, \alpha)\} = 0 \tag{131}$$

uniformly in θ. The function $\overline{P}_n(\theta, {}_1\theta_0, \alpha)$ is constant along the surface $S_c(\boldsymbol{\theta})$ defined in (117). This can be proved in the same way as the constancy of $P(\theta, \omega, \alpha)$ on $S_c(\boldsymbol{\theta})$ defined in (106). Hence from (129) and (131) we obtain

$$\lim_{n=\infty} \{\underset{\theta \in S_c(\boldsymbol{\theta})}{\text{l.u.b.}}\ P_n(\theta, \omega, \alpha) - \underset{\theta \in S_c(\boldsymbol{\theta})}{\text{g.l.b.}}\ P_n(\theta, \omega, \alpha)\} = 0 \tag{132}$$

uniformly in $\boldsymbol{\theta}$ and c. According to Theorem V we have

$$\lim_{n=\infty} \{\underset{\theta \in S_c(\boldsymbol{\theta})}{\text{l.u.b.}}\ P(W_n^* \mid \theta) - \underset{\theta \in S_c(\boldsymbol{\theta})}{\text{g.l.b.}}\ P(W_n^* \mid \theta)\} = 0 \tag{133}$$

uniformly in c and $\boldsymbol{\theta}$. Theorem VI follows from (132), (133) and Theorem IV.

12. **The general composite hypothesis.** In §§9–11 we have considered the linear composite hypothesis ${}_1\theta = {}_1\theta_0$. Now we shall discuss a general composite hypothesis H_ω where ω denotes a subset of the parameter space given by r equations

$$\xi^1(\theta) = \xi^2(\theta) = \cdots = \xi^r(\theta) = 0 \qquad (r < k), \tag{134}$$

that is, ω is the set of all points θ which satisfy equations (134). We make the following assumption.

Assumption VI. *There exist $k-r$ functions $\xi^{r+1}(\theta), \cdots, \xi^k(\theta)$ such that the following three conditions are fulfilled:*

(a) *The transformation which transforms the point θ into the point ξ with the coordinates $\xi^1(\theta), \cdots, \xi^k(\theta)$ is a topological transformation of Ω into itself.*

(b) *The first and second order partial derivatives of $\xi^1(\theta), \cdots, \xi^k(\theta)$ are uniformly continuous and bounded functions of θ.*

(c) *The greatest lower bound of the absolute value of the Jacobian $\partial(\xi^1, \cdots, \xi^k)/\partial(\theta^1, \cdots, \theta^k)$ is positive.*

Let $\xi = (\xi^1, \cdots, \xi^k)$ denote a variable point of the parameter space Ω.

Since according to Assumption VI the transformation

$$\xi^i = \xi^i(\theta) \qquad (i = 1, \cdots, k) \tag{135}$$

is topological, we can solve the equations (135) and we obtain

$$\theta^i = \theta^i(\xi^1, \cdots, \xi^k) \qquad (i = 1, \cdots, k). \tag{136}$$

From conditions (b) and (c) of Assumption VI it follows that the first and second order derivatives of $\theta^i(\xi^1, \cdots, \xi^k)$ are uniformly continuous and bounded functions of ξ and the absolute value of the Jacobian $\partial(\theta^1, \cdots, \theta^k)/\partial(\xi^1, \cdots, \xi^k)$ has a positive lower bound.

Let $f^*(x, \xi)$ be the probability density function we obtain from the probability density function $f(x, \theta)$ of x by substituting the right-hand side of (136) for θ^i. Hence $f^*(x, \xi)$ is the probability density function of x in the transformed parameter space. It is clear that the maximum likelihood estimate of ξ^i is equal to $\dot{\xi}_n^i = \xi^i(\dot{\theta}_n^1, \cdots, \dot{\theta}_n^k)$, where $\dot{\theta}_n$ is the maximum likelihood estimate of θ.

Denote by I*, II*, $\cdots$, V* the assumptions which we obtain from Assumptions I–V respectively by substituting $f^*(x, \xi)$ for $f(x, \theta)$, ξ for θ and $\dot{\xi}_n$ for $\dot{\theta}_n$. We shall show that Assumptions I*–V* can be derived from Assumptions I–VI.

Assumption I* is an immediate consequence of Assumptions I and VI. Since according to Assumption VI the first derivatives of $\xi^i(\theta)$ are continuous and bounded functions of θ, the transformation (135) is uniformly continuous. Hence, for each positive ϵ^* there exists a positive ϵ such that the inequality $|\dot{\theta}_n - \theta| \leqq \epsilon$ implies the inequality $|\dot{\xi}_n - \xi| \leqq \epsilon^*$. From this and Assumption II we obtain Assumption II*.

Denote by $\nu_i(x, \theta_1, \delta)$ the least upper bound, and by $\mu_i(x, \theta_1, \delta)$ the greatest lower bound of $\partial \log f(x, \theta)/\partial\theta^i$ in the interval $\theta_1 - \delta \leqq \theta \leqq \theta_1 + \delta$. Using the Taylor expansion we obtain

$$\frac{\partial \log f(x, \overset{*}{\theta_1})}{\partial\theta^i} = \frac{\partial \log f(x, \theta_1)}{\partial\theta_1} + \sum_j (\theta_1^{*j} - \theta_1^j) \frac{\partial^2 \log f(x, \bar{\theta}_1)}{\partial\theta^i\partial\theta^j}, \tag{137}$$

where $\bar{\theta}_1$ lies in the interval $[\theta_1, \theta_1^*]$. From (137) it follows that

$$\left| \frac{\partial \log f(x, \overset{*}{\theta_1})}{\partial\theta^i} - \frac{\partial \log f(x, \theta_1)}{\partial\theta^i} \right| \leqq 2\delta \sum_j [|\phi_{ij}(x, \theta, \delta)| + |\psi_{ij}(x, \theta, \delta)|] \tag{138}$$

for any θ and δ for which

$$\theta - \delta \leqq \theta_1 \leqq \theta + \delta \quad \text{and} \quad \theta - \delta \leqq \overset{*}{\theta_1} \leqq \theta + \delta.$$

From (138) we obtain for any positive δ

$$|\nu_i(x, \theta, \delta) - \mu_i(x, \theta, \delta)| \leqq 2\delta \sum_j [|\phi_{ij}(x, \theta, \delta)| + |\psi_{ij}(x, \theta, \delta)|]. \tag{139}$$

Let $\{\theta_{1n}\}$ and $\{\theta_{2n}\}$ ($n=1, 2, \cdots$, ad inf.) be two sequences of parameter points such that

$$\lim_{n=\infty} (\theta_{1n} - \theta_{2n}) = 0. \tag{140}$$

According to Assumption III the expectations $E_{\theta_1}[\phi_{ij}(x, \theta_2, \delta)]^2$ and $E_{\theta_1}[\psi_{ij}(x, \theta_2, \delta)]^2$ are bounded functions of θ_1, θ_2 and δ in the domain D_ϵ defined in Assumption III. Hence also the expectations $E_{\theta_1}|\phi_{ij}(x, \theta_2, \delta)|$ and $E_{\theta_1}|\psi_{ij}(x, \theta_2, \delta)|$ are bounded functions of θ_1, θ_2 and δ in the domain D_ϵ. From this and relations (139) and (140) it follows that for any sequence $\{\delta_n\}$ of positive numbers for which $\lim_{n=\infty} \delta_n = 0$ we have

$$\lim_{n=\infty} \{E_{\theta_{1n}}\nu_i(x, \theta_{2n}, \delta_n) - E_{\theta_{1n}}\mu_i(x, \theta_{2n}, \delta_n)\} = 0. \tag{141}$$

Since

$$\mu_i(x, \theta_{2n}, \delta_n) \leqq \frac{\partial \log f(x, \theta_{2n})}{\partial \theta^i} \leqq \nu_i(x, \theta_{2n}, \delta_n)$$

it follows from (141) that

$$\begin{aligned} &\lim_{n=\infty} \left\{E_{\theta_{1n}}\nu_i(x, \theta_{2n}, \delta_n) - E_{\theta_{1n}} \frac{\partial \log f(x, \theta_{2n})}{\partial \theta^i}\right\} = 0, \\ &\lim_{n=\infty} \left\{E_{\theta_{1n}}\mu_i(x, \theta_{2n}, \delta_n) - E_{\theta_{1n}} \frac{\partial \log f(x, \theta_{2n})}{\partial \theta^i}\right\} = 0. \end{aligned} \tag{142}$$

Using the Taylor expansion we have

$$\frac{\partial \log f(x, \theta_{2n})}{\partial \theta^i} = \frac{\partial \log f(x, \theta_{1n})}{\partial \theta^i} + \sum_j (\overset{j}{\theta}_{2n} - \overset{j}{\theta}_{1n}) \frac{\partial^2 \log f(x, \bar{\theta}_n)}{\partial \theta^i \partial \theta^j}, \tag{143}$$

where $\bar{\theta}_n$ lies in the interval $[\theta_{1n}, \theta_{2n}]$. Since the expectations $E_{\theta_1}|\phi_{ij}(x, \theta_2, \delta)|$ and $E_{\theta_2}|\psi_{ij}(x, \theta_2, \delta)|$ are bounded functions of θ_1, θ_2, and δ in the domain D_ϵ, we obtain

$$\lim_{n=\infty} E_{\theta_{1n}} \left\{\sum_j (\overset{j}{\theta}_{2n} - \overset{j}{\theta}_{1n}) \frac{\partial^2 \log f(x, \bar{\theta}_n)}{\partial \theta^i \partial \theta^j}\right\} = 0.$$

Hence it follows from (143) and Assumption IV that

$$\lim_{n=\infty} E_{\theta_{1n}} \frac{\partial \log f(x, \theta_{2n})}{\partial \theta^i} = \lim_{n=\infty} E_{\theta_{1n}} \frac{\partial \log f(x, \theta_{1n})}{\partial \theta^i} = 0. \tag{144}$$

We obtain from (142) and (144)

$$\lim E_{\theta_{1n}}\nu_i(x, \theta_{2n}, \delta_n) = \lim E_{\theta_{1n}}\mu_i(x, \theta_{2n}, \delta_n) = 0. \tag{145}$$

Denote by $\phi_{ij}^*(x, \xi_1, \delta^*)$ the least upper bound, and by $\psi_{ij}^*(x, \xi, \delta^*)$ the greatest lower bound of $\partial^2 \log f^*(x, \xi)/\partial\xi^i\partial\xi^j$ in the interval $\xi_1 - \delta^* \leqq \xi \leqq \xi_1 + \delta^*$. We have

$$\tag{146} \begin{aligned} \frac{\partial^2 \log f^*(x, \xi)}{\partial\xi^i\partial\xi^j} &= \sum_l \sum_m \frac{\partial^2 \log f(x, \theta)}{\partial\theta^l\partial\theta^m} \frac{\partial\theta^l}{\partial\xi^i} \frac{\partial\theta^m}{\partial\xi^j} \\ &\quad + \sum_l \frac{\partial \log f(x, \theta)}{\partial\theta^l} \frac{\partial^2\theta^l}{\partial\xi^i\partial\xi^j}. \end{aligned}$$

Since

$$E_\theta\left(\frac{\partial \log f(x, \theta)}{\partial\theta}\right) = 0,$$

we obtain from (146)

$$\tag{147} E_\xi \frac{\partial^2 \log f^*(x, \xi)}{\partial\xi^i\partial\xi^j} = \sum_l \sum_m E_\theta \left[\frac{\partial^2 \log f(x, \theta)}{\partial\theta^l\partial\theta^m} \frac{\partial\theta^l}{\partial\xi^i} \frac{\partial\theta^m}{\partial\xi^j}\right].$$

Hence the determinant

$$\tag{148} \left| E_\xi \frac{\partial^2 \log f^*(x, \xi)}{\partial\xi^i\partial\xi^j} \right| = \left| E_\theta \frac{\partial^2 \log f(x, \theta)}{\partial\theta^i\partial\theta^j} \right| \left(\frac{\partial(\theta^1, \cdots, \theta^k)}{\partial(\xi^1, \cdots, \xi^k)}\right)^2.$$

Since the determinant $|-E_\theta\partial^2 \log f(x, \theta)/\partial\theta^i\partial\theta^j|$ has a positive lower bound. it follows from (148) and Assumption VI that

$$\tag{149} |-E_\xi\partial^2 \log f^*(x, \xi)/\partial\xi^i\partial\xi^j|$$

has a positive lower bound.

For any positive δ^* let $\delta(\delta^*)$ be the smallest positive number such that for any two points ξ_1 and ξ_2 for which $|\xi_1 - \xi_2| \leqq \delta^*$ we have $|\theta_1 - \theta_2| \leqq \delta(\delta^*)$ where θ_1 and θ_2 are the image points of ξ_1 and ξ_2 by transformation (136). From (146) we obtain

$$\tag{150} \begin{aligned} \overset{*}{\phi}_{ij}(x, \xi, \overset{*}{\delta}) &\leqq \sum_l \sum_m \bar{\phi}_{lm}[x, \theta, \delta(\overset{*}{\delta})] \frac{\partial\theta^l}{\partial\xi^i} \frac{\partial\theta^m}{\partial\xi^j} \\ &\quad + \sum_l \bar{\nu}_l[x, \theta, \delta(\overset{*}{\delta})] \frac{\partial^2\theta^l}{\partial\xi^i\partial\xi^j}, \\ \overset{*}{\psi}_{ij}(x, \xi, \overset{*}{\delta}) &\geqq \sum_l \sum_m \bar{\psi}_{lm}[x, \theta, \delta(\overset{*}{\delta})] \frac{\partial\theta^l}{\partial\xi^i} \frac{\partial\theta^m}{\partial\xi^j} \\ &\quad + \sum_l \bar{\mu}_l[x, \theta, \delta(\overset{*}{\delta})] \frac{\partial^2\theta^l}{\partial\xi^i\partial\xi^j}, \end{aligned}$$

where θ is the image point of ξ by transformation (136), and the derivatives $\partial\theta^l/\partial\xi^i$, $\partial^2\theta^l/\partial\xi^i\partial\xi^j$ are taken at some points in the interval $[\xi - \delta^*, \xi + \delta^*]$, and the functions $\bar{\phi}_{lm}(x, \theta, \delta)$, $\bar{\psi}_{lm}(x, \theta, \delta)$, $\bar{\mu}_l(x, \theta, \delta)$ and $\bar{\nu}_l(x, \theta, \delta)$ satisfy the inequalities

$$\psi_{lm}(x, \theta, \delta) \leqq \bar{\phi}_{lm}(x, \theta, \delta) \leqq \phi_{lm}(x, \theta, \delta);$$

$$\psi_{lm}(x, \theta, \delta) \leqq \bar{\psi}_{lm}(x, \theta, \delta) \leqq \phi_{lm}(x, \theta, \delta);$$

$$\mu_l(x, \theta, \delta) \leqq \bar{\nu}_l(x, \theta, \delta) \leqq \nu_l(x, \theta, \delta);$$

$$\mu_l(x, \theta, \delta) \leqq \bar{\mu}_l(x, \theta, \delta) \leqq \nu_l(x, \theta, \delta).$$

Let $\{\xi_{1n}\}$ and $\{\xi_{2n}\}$ be two sequences of parameter points such that $\lim \xi_{1n} = \lim \xi_{2n} = \xi$. Let θ_{1n} be the image point of ξ_{1n}, θ_{2n} the image point of ξ_{2n}, and θ the image point of ξ (by transformation (136)). Let furthermore $\{\delta_n^*\}$ be a sequence of positive numbers such that $\lim \delta_n^* = 0$. Then we obviously have $\lim \delta(\delta_n^*) = 0$ and therefore using (145) and Assumption III we obtain

$$\lim_{n=\infty} E_{\theta_{1n}} \phi_{lm}[x, \theta_{2n}, \delta(\overset{*}{\delta_n})] = E_\theta \partial^2 \log f(x, \theta)/\partial\theta^l \partial\theta^m,$$

$$\lim_{n=\infty} E_{\theta_{1n}} \psi_{lm}[x, \theta_{2n}, \delta(\overset{*}{\delta_n})] = E_\theta \partial^2 \log f(x, \theta)/\partial\theta^l \partial\theta^m,$$

$$(151) \qquad \lim_{n=\infty} E_{\theta_{1n}} \nu_l \; [x, \theta_{2n}, \delta(\overset{*}{\delta_n})] = 0,$$

$$\lim_{n=\infty} E_{\theta_{1n}} \mu_l \; [x, \theta_{2n}, \delta(\overset{*}{\delta_n})] = 0$$

uniformly in θ. From (150) and (151) and the uniform continuity of the derivatives $\partial\theta^l/\partial\xi^i$ and $\partial^2\theta^l/\partial\xi^i\partial\xi^j$ we obtain

$$(152) \qquad \lim_{n=\infty} E_{\xi_{1n}} \overset{*}{\phi}_{ij}(x, \xi_{2n}, \overset{*}{\delta_n}) = \lim_{n=\infty} E_{\xi_{1n}} \overset{*}{\psi}_{ij}(x, \xi_{2n}, \overset{*}{\delta_n}) = E_\xi \frac{\partial^2 \log f^*(x, \xi)}{\partial\xi^i \partial\xi^j}$$

uniformly in ξ.

Because of (150) we have both

$$| \overset{*}{\phi}_{ij}(x, \xi, \overset{*}{\delta}) | \quad \text{and} \quad | \overset{*}{\psi}_{ij}(x, \xi, \overset{*}{\delta}) |$$

$$\leqq \sum_l \sum_m \Big\{ [| \psi_{lm}[x, \theta, \delta(\delta^*)] |$$

$$(153) \qquad + | \phi_{lm}[x, \theta, \delta(\delta^*)] |] \Big[\underset{\xi}{\text{l.u.b.}} \Big| \frac{\partial\theta^l}{\partial\xi^i} \frac{\partial\theta^m}{\partial\xi^j} \Big| \Big] \Big\}$$

$$+ \sum_l \Big\{ [| \nu_l[x, \theta, \delta(\delta^*)] | + | \mu_l[x, \theta, \delta(\delta^*)] |] \Big[\underset{\xi}{\text{l.u.b.}} \Big| \frac{\partial^2\theta^l}{\partial\xi^i \partial\xi^j} \Big| \Big] \Big\}$$

where the least upper bound with respect to ξ is to be taken over the interval $[\xi - \delta^*, \xi + \delta^*]$.

We shall show that $E_{\theta_1}[\nu_i(x, \theta_2, \delta)]^2$ and $E_{\theta_1}[\mu_i(x, \theta_2, \delta)]^2$ are bounded functions of θ_1, θ_2 and δ in the domain D_ϵ for sufficiently small ϵ. Our statement is proved if we show that $E_{\theta_1}[\partial \log f(x, \theta_1)/\partial\theta^i]^2$, $E_{\theta_1}[\nu_i(x, \theta_2, \delta) - \partial \log f(x, \theta_1)/\partial\theta^i]^2$ and $E_{\theta_1}[\mu_i(x, \theta_2, \delta) - \partial \log f(x, \theta_1)/\partial\theta^i]^2$ are bounded in D_ϵ. The first of these expressions is bounded because of Assumption V. From (138) it follows that

$$\left| \nu_i(x, \theta_2, \delta) - \frac{\partial \log f(x, \theta_1)}{\partial\theta^i} \right| \leqq 2\delta' \sum_j \left[\left| \phi_{ij}(x, \theta_2, \delta') \right| + \left| \psi_{ij}(x, \theta_2, \delta') \right| \right]$$

and

$$\left| \mu_i(x, \theta_2, \delta) - \frac{\partial \log f(x, \theta_1)}{\partial\theta^i} \right| \leqq 2\delta' \sum_j \left[\left| \phi_{ij}(x, \theta_2, \delta') \right| + \left| \psi_{ij}(x, \theta_2, \delta') \right| \right]$$

where $\delta' = \delta + \sum_j |\theta_1^j - \theta_2^j|$. From the above inequalities and the fact that $E_{\theta_1}[\phi_{ij}(x, \theta_2, \delta)]^2$ and $E_{\theta_1}[\psi_{ij}(x, \theta_2, \delta)]^2$ are bounded in D_ϵ it follows that for sufficiently small ϵ the expressions $E_{\theta_1}[\nu_i(x, \theta_2, \delta) - \partial \log f(x, \theta_1)/\partial\theta^i]^2$ and $E_{\theta_1}[\mu_i(x, \theta_2, \delta) - \partial \log f(x, \theta_1)/\partial\theta^i]^2$ are bounded in D_ϵ. Hence our statement is proved.

Since the derivatives $\partial\theta^l/\partial\xi^i$ and $\partial^2\theta^l/\partial\xi^i\partial\xi^j$ are bounded functions of ξ, and since $E_{\theta_1}[\phi_{ij}(x, \theta_2, \delta)]^2$ and $E_{\theta_1}[\psi_{ij}(x, \theta_2, \delta)]^2$ are bounded functions of θ_1, θ_2 and δ in the domain D_ϵ, it follows from (153) that there exists a positive ϵ^* such that $E_{\xi_1}[\phi_{ij}^*(x, \xi_2, \delta^*)]^2$ and $E_{\xi_1}[\psi_{ij}^*(x, \xi_2, \delta^*)]^2$ are bounded functions of ξ_1, ξ_2 and δ^* in the domain defined by $|\xi_1 - \xi_2| \leqq \epsilon^*$ and $|\delta^*| \leqq \epsilon^*$. Assumption III* follows from the latter statement and the relations (152) and (149).

Assumption IV* is an immediate consequence of Assumption IV.

We have

$$\frac{\partial \log f^*(x, \xi)}{\partial\xi^i} = \sum_j \frac{\partial \log f(x, \theta)}{\partial\theta^j} \frac{\partial\theta^j}{\partial\xi^i}. \tag{154}$$

For any points x and θ denote the maximum of the k expressions $|\partial \log f(x, \theta)/\partial\theta^1|, \cdots, |\partial \log f(x, \theta)/\partial\theta^k|$ by $\rho(x, \theta)$. From Assumption V it follows easily that

$$E_\theta[\rho(x, \theta)]^{2+\eta} \tag{155}$$

is a bounded function of θ. Since the derivatives $\partial\theta^j/\partial\xi^i$ are bounded functions of ξ, Assumption V* follows from (154) and (155).

Denote by $c_{ij}^*(\theta)$ the function of θ we obtain from $-E_\xi\partial^2 \log f^*(x, \xi)/\partial\xi^i\partial\xi^j$ by substituting $\xi(\theta)$ for ξ. Then we obtain from (147)

$$\overset{*}{c}_{ij}(\theta) = \sum_l \sum_m c_{lm}(\theta) \frac{\partial\theta^l}{\partial\xi^i} \frac{\partial\theta^m}{\partial\xi^j}. \tag{156}$$

Denote by A the matrix $\|\partial\theta^i/\partial\xi^j\|$ $(i, j = 1, \cdots, k)$ and let $\bar{A}$ be the transposed of A. Then equation (156) can be written as

$$\|\overset{*}{c}_{ij}(\theta)\| = \bar{A}\|c_{ij}(\theta)\|A. \tag{157}$$

Let $\|\overset{*}{\sigma}_{ij}(\theta)\|$ be the inverse of the matrix $\|\overset{*}{c}_{ij}(\theta)\|$. Futhermore let B be the matrix $\|\partial\xi^i/\partial\theta^j\|$ $(i, j=1, \cdots, k)$ and denote by $\bar{B}$ the transposed of B. Since $B=A^{-1}$, we obtain from (157)

$$\|\overset{*}{\sigma}_{ij}(\theta)\| = B\|\sigma_{ij}(\theta)\|\bar{B}, \tag{158}$$

where $\|\sigma_{ij}(\theta)\| = \|c_{ij}(\theta)\|^{-1}$. Equation (158) can be written as

$$\overset{*}{\sigma}_{ij}(\theta) = \sum_l \sum_m \frac{\partial\xi^i}{\partial\theta^l}\frac{\partial\xi^j}{\partial\theta^m}\sigma_{lm}(\theta). \tag{159}$$

Let

$$\|\overset{*}{\bar{c}}_{pq}(\theta)\| = \|\overset{*}{\sigma}_{pq}(\theta)\|^{-1} \qquad (p, q = 1, \cdots, r). \tag{160}$$

Denote by W_n^* the critical region defined by the inequality

$$n\sum_{q=1}^{r}\sum_{p=1}^{r}\xi^p(\dot{\theta}_n)\xi^q(\dot{\theta}_n)\overset{*}{\bar{c}}_{pq}(\dot{\theta}_n) \geqq d_n, \tag{161}$$

where the constant d_n is chosen so that

$$\underset{\boldsymbol{\theta}}{\text{l.u.b.}}\ P(W^* \mid \boldsymbol{\theta}) = \alpha.$$

The point $\boldsymbol{\theta}$ is restricted to points of the set ω defined by equations (134).

For each positive c and for each point $\boldsymbol{\theta}$ of ω we define the surface $S_c(\boldsymbol{\theta})$ by the equations

$$\begin{aligned} &\sum_{q=1}^{r}\sum_{p=1}^{r}\xi^p(\theta)\xi^q(\theta)\overset{*}{\bar{c}}_{pq}(\boldsymbol{\theta}) = c, \\ &\sum_{j=1}^{k}\gamma_{tj}(\boldsymbol{\theta})\xi^j(\theta) = \sum_{j=1}^{k}\gamma_{tj}(\boldsymbol{\theta})\xi^j(\boldsymbol{\theta}) \qquad (t = r+1, \cdots, k), \end{aligned} \tag{162}$$

where the coefficients $\gamma_{tj}(\boldsymbol{\theta})$ satisfy the following condition: There exists a matrix $\|\beta_{pq}(\boldsymbol{\theta})\|$ $(p, q=1, \cdots, r)$ such that if we form the matrix $A(\boldsymbol{\theta})$ given in (120) then

$$A(\boldsymbol{\theta})\|\overset{*}{\sigma}_{ij}(\boldsymbol{\theta})\|\bar{A}(\boldsymbol{\theta}) = I, \tag{163}$$

where $\bar{A}(\boldsymbol{\theta})$ denotes the transposed of $A(\boldsymbol{\theta})$ and I denotes the unit matrix.

Consider the transformation of the parameter space given by

$$\begin{aligned} \theta'^p &= \beta_{p1}(\boldsymbol{\theta})\xi^1(\theta) + \cdots + \beta_{pr}(\boldsymbol{\theta})\xi^r(\theta) \qquad (p = 1, \cdots, r), \\ \theta'^t &= \gamma_{t1}(\boldsymbol{\theta})\xi^1(\theta) + \cdots + \gamma_{tk}(\boldsymbol{\theta})\xi^k(\theta) \qquad (t = r+1, \cdots, k), \end{aligned} \tag{164}$$

where $\boldsymbol{\theta}$ denotes the point of ω for which θ lies on $S_c(\boldsymbol{\theta})$ for some value of c. The matrix $\|\beta_{pq}(\boldsymbol{\theta})\|$ is chosen so that (163) is fulfilled. The transformation (164) transforms $S_c(\boldsymbol{\theta})$ into the sphere $S_c'(\boldsymbol{\theta})$ given by

$$\sum_{p=1}^{r} (\theta'^p)^2 = c, \qquad \theta'^t = \sum_{j=1}^{k} \gamma_{tj}(\boldsymbol{\theta})\xi^j(\boldsymbol{\theta}).$$

We define a weight function $\eta(\theta)$ as follows:

$$\eta(\theta) = \lim_{\rho=0} A[\omega'(\theta, \rho)]/A[\omega(\theta, \rho)], \tag{165}$$

where the symbols on the right-hand side of (165) are defined as in (109).

Since Assumptions I*–V* are fulfilled if Assumption VI holds, we obtain from Theorems IV, V and VI the following theorem.

THEOREM VII. *Let W_n^* be the region defined in* (161) *and let $S_c(\boldsymbol{\theta})$ be the surface defined in* (162). *Furthermore let $\eta(\theta)$ be the weight function defined in* (165). *If Assumption* VI *holds, then for testing the hypothesis* $\xi^1(\theta) = \cdots = \xi^r(\theta) = 0$ *the sequence* $\{W_n^*\}$

(a) *has asymptotically best average power with respect to the surfaces $S_c(\boldsymbol{\theta})$ and the weight function $\eta(\theta)$;*

(b) *has asymptotically best constant power on the surfaces $S_c(\boldsymbol{\theta})$;*

(c) *is an asymptotically most stringent test.*

13. **Optimum properties of the likelihood ratio test.** For testing a composite hypothesis H_ω, Neyman and Pearson introduced a statistic(5), called likelihood ratio, defined as follows: The density function in the sample space is given by $\prod_{\alpha=1}^{n} f(x_\alpha, \theta)$. Denote by $P(x_1, \cdots, x_n)$ the maximum of this function with respect to $\theta^1, \cdots, \theta^k$, and let $P_\omega(x_1, \cdots, x_n)$ be the conditional maximum with respect to $\theta^1, \cdots, \theta^k$, subject to the condition that θ must be a point of ω. Then the likelihood ratio for testing the hypothesis H_ω is given by $\lambda_n(\omega, E_n) = P_\omega(x_1, \cdots, x_n)/P(x_1, \cdots, x_n)$. It is obvious that the value of $\lambda_n(\omega, E_n)$ always lies between 0 and 1. Neyman and Pearson recommend the use of the left tail as critical region, that is the hypothesis H_ω is rejected if the value of $\lambda_n(\omega, E_n)$ is less than a certain constant $\bar{\lambda}_n(\omega)$. Denote the region $\lambda_n(\omega, E_n) < \bar{\lambda}_n(\omega)$ by $L_n(\omega)$. In all that follows we choose the constant $\bar{\lambda}_n(\omega)$ so that

$$\underset{\boldsymbol{\theta}}{\text{l.u.b.}}\ P[L_n(\omega) \mid \boldsymbol{\theta}] = \alpha.$$

We shall prove that there exists a finite value B such that $-2 \log \bar{\lambda}_n(\omega) < B$ for all n and for all ω. Consider the Taylor expansion

(5) See in this connection J. Neyman and E. S. Pearson, *On the use and interpretation of certain test criteria for purposes of statistical inference*, Biometrika vol. 20A (1928).

$$\sum_\alpha \log f(x_\alpha, \theta) = \sum_\alpha \log f(x_\alpha, \dot{\theta}_n) + \frac{1}{2} \sum \sum n(\dot{\theta}_n^i - \theta^i)(\dot{\theta}_n^j - \theta^j) \sum_\alpha \frac{\partial^2 \log f(x_\alpha, \bar{\theta}_n)}{\partial\theta^i\partial\theta^j} \frac{1}{n}$$

where $\bar{\theta}_n$ lies in the interval $[\dot{\theta}_n, \theta]$. Since

$$\sum_\alpha \log f(x_\alpha, \dot{\theta}_n) - \sum_\alpha \log f(x_\alpha, \theta) \geqq -\log \lambda_n(\omega, E_n),$$

we have

$$-2 \log \lambda_n(\omega, E_n) \leqq -\sum \sum n(\dot{\theta}_n^i - \theta^i)(\dot{\theta}_n^j - \theta^j) \frac{1}{n} \sum_\alpha \frac{\partial^2 \log f(x_\alpha, \bar{\theta}_n)}{\partial\theta^i\partial\theta^j}.$$

Since $n^{-1}\sum_\alpha \partial^2 \log f(x_\alpha, \bar{\theta}_n)/\partial\theta^i\partial\theta^j$ converges stochastically to $-c_{ij}(\theta)$ under the assumption that θ is the true parameter point, it follows easily from Proposition I that for any $\epsilon > 0$ there exists a positive value $A(\epsilon)$ such that for any ω

$$\limsup_{n=\infty} \left\{ \underset{\theta}{\text{l.u.b.}}\, P\left[-\sum \sum (\dot{\theta}_n^i - \theta^i)(\dot{\theta}_n^j - \theta^j) \sum_\alpha \frac{\partial^2 \log f(x_\alpha, \bar{\theta}_n)}{\partial\theta^i\partial\theta^j} \geqq A(\epsilon) \mid \theta \right] \right\} \leqq \epsilon$$

and $\lim_{\epsilon=0} A(\epsilon) = +\infty$. Hence

$$\limsup_{n=\infty} \left\{ \underset{\theta,\omega}{\text{l.u.b.}}\, P[-2 \log \lambda_n(\omega, E_n) \geqq A(\epsilon) \mid \theta] \right\} \leqq \epsilon. \tag{166}$$

This proves the existence of a finite number B with the required property.

For any subsets Γ and Γ' of the parameter space we denote by $\delta(\Gamma, \Gamma')$ the greatest lower bound of the distance between θ and θ' where θ is restricted to points of Γ and θ' is restricted to points of Γ'. We shall call $\delta(\Gamma, \Gamma')$ the distance of the sets Γ and Γ'.

Let $\{\theta_n\}$ be a sequence of parameter points such that $\lim \delta(\theta_n, \omega) = 0$ and $\lim_{n=\infty} n^{1/2}\delta(\theta_n, \omega) = +\infty$. We shall prove that there exists a positive ν_0 such that for any constant A

$$\lim_{n=\infty} P\left[\sum_\alpha \log f(x_\alpha, \dot{\theta}_n) - \sum_\alpha \log f(x_\alpha, \theta^*) > A \mid \theta_n \right] = 1, \tag{167}$$

where θ^* denotes a point of ω for which $|\theta^* - \theta_n| \leqq \nu_0$ and $\sum_\alpha \log f(x_\alpha, \theta^*) \geqq \sum_\alpha \log f(x_\alpha, \theta)$ for all θ in the domain $|\theta - \theta_n| \leqq \nu_0$. Consider the Taylor expansion

$$\sum_\alpha \log f(x_\alpha, \theta^*) - \sum_\alpha \log f(x_\alpha, \dot{\theta}_n) = \frac{n}{2} \sum \sum (\dot{\theta}_n^i - \theta^{*i})(\dot{\theta}_n^j - \theta^{*j}) \frac{1}{n} \sum_\alpha \frac{\partial^2 \log f(x_\alpha, \bar{\theta}_n)}{\partial\theta^i\partial\theta^j}, \tag{168}$$

where $\bar{\theta}_n$ lies in the interval $[\dot{\theta}_n, \boldsymbol{\theta}^*]$. Because of conditions (a) and (b) of Assumption III for any $\epsilon > 0$ the positive number ν_0 can be chosen so that

$$(169) \qquad \lim_{n=\infty} P\left[\left| c_{ij}(\boldsymbol{\theta}^*) + \frac{1}{n} \sum_{\alpha} \frac{\partial^2 \log f(x_\alpha, \bar{\theta}_n)}{\partial\theta^i \partial\theta^j} \right| < \epsilon \,\Big|\, \theta_n \right] = 1.$$

Since $\|c_{ij}(\boldsymbol{\theta})\|$ is positive definite and the determinant $|c_{ij}(\boldsymbol{\theta})|$ has a positive lower bound, we obtain from Proposition I, (168) and (169) that for some positive, ν_0, (167) holds. Hence our statement is proved.

We say that the likelihood ratio test is uniformly consistent if for any positive ν

$$\lim_{n=\infty} P[L_n(\omega) \mid \theta] = 1$$

uniformly in ω and θ over the domain $\delta(\theta, \omega) \geqq \nu$. We postulate the following assumption.

ASSUMPTION VII. *The likelihood ratio test is uniformly consistent.*

This assumption together with the uniform consistency of the maximum likelihood estimate $\dot{\theta}_n$ will be proved in a forthcoming paper on the basis of some weak assumptions on the density function $f(x, \theta)$.

Let $\omega_\nu(\theta)$ be the intersection of ω with the set of all points θ' for which $\sum_{i=1}^{k} |\theta^i - \theta'^i| \geqq \nu$. From Assumption VII it follows that for any positive ν

$$(170) \qquad \lim_{n=\infty} P\{-2 \log \lambda_n[\omega_\nu(\theta), E_n] \geqq -2 \log \bar{\lambda}_n[\omega_\nu(\theta)] \mid \theta\} = 1$$

uniformly in θ.

Let $\{\theta_n\}$ be a sequence of parameter points such that $\lim_{n=\infty} \delta(\theta_n, \omega) = 0$ and $\lim_{n=\infty} n^{1/2}\delta(\theta_n, \omega) = +\infty$. Denote by ω_n the set of all points $\boldsymbol{\theta}$ for which $|\boldsymbol{\theta} - \theta_n| \leqq \nu_0$. Since $-2 \log \bar{\lambda}_n[\omega_\nu(\theta)]$ has a finite upper bound it follows from (167) that for a sufficiently small ν_0

$$(171) \qquad \lim_{n=\infty} P\{-2 \log \lambda_n[\omega_n, E_n] \geqq -2 \log \bar{\lambda}_n[\omega_{\nu_0}(\theta_n)] \mid \theta_n\} = 1.$$

Obviously

$$-2 \log \lambda_n(\omega, E_n) = \text{minimum } \{-2 \log \lambda_n[\omega_{\nu_0}(\theta_n), E_n], -2 \log \lambda_n(\omega_n, E_n)\}.$$

From (170) and (171) we obtain

$$(172) \qquad \lim_{n=\infty} P\{-2 \log \lambda_n(\omega, E_n) \geqq -2 \log \bar{\lambda}_n[\omega_{\nu_0}(\theta_n)] \mid \theta_n\} = 1.$$

Since $-2 \log \lambda_n[\omega_{\nu_0}(\theta_n), E_n] \geqq -2 \log \lambda_n(\omega, E_n)$, we have $-2 \log \bar{\lambda}_n[\omega_{\nu_0}(\theta_n)] \geqq -2 \log \bar{\lambda}_n(\omega)$. Hence from (172) we obtain

$$(173) \qquad \lim_{n=\infty} P[L_n(\omega) \mid \theta_n] = 1.$$

From Assumption VII it follows that (173) holds for any sequence $\{\theta_n\}$ for which $\lim n^{1/2}\delta(\theta_n, \omega) = +\infty$.

Now let us consider the case where ω is the set given by ${}_1\theta = {}_1\theta_0$. Let $\{\theta_n\}$ be a sequence of parameter points for which the sequence $n^{1/2}\delta(\theta_n, \omega)$ is bounded. Denote by $\dot{\theta}_n(\omega)$ the parameter point in ω for which

$$\prod_{\alpha=1}^{n} f[x_\alpha, \dot{\theta}_n(\omega)] = P_\omega(x_1, \cdots, x_n).$$

Let $\boldsymbol{\theta}_n$ be the point for which $\delta(\theta_n, \omega) = \delta(\theta_n, \boldsymbol{\theta}_n)$. Denote by $T_n(\nu)$ the region given by the inequalities

$$(174)\qquad \begin{aligned} |\dot{\theta}_n - \boldsymbol{\theta}_n| &< \nu, \\ |\dot{\theta}_n(\omega) - \boldsymbol{\theta}_n| &< \nu. \end{aligned}$$

We shall prove that

$$(175)\qquad \lim P[T_n(\nu) \mid \theta_n] = 1 \qquad \text{for any } \nu > 0.$$

Consider the Taylor expansion

$$(176)\qquad \begin{aligned} \sum_\alpha \log f(x_\alpha, \boldsymbol{\theta}_n) &= \sum_\alpha \log f(x_\alpha, \dot{\theta}_n) \\ &+ \frac{n}{2} \sum \sum (\dot{\theta}_n^i - \boldsymbol{\theta}_n^i)(\dot{\theta}_n^j - \boldsymbol{\theta}_n^j) \frac{1}{n} \sum_\alpha \frac{\partial^2 \log f(x_\alpha, \bar{\theta}_n)}{\partial\theta^i \partial\theta^j}. \end{aligned}$$

Since $[n^{-1}\sum_\alpha \partial^2 \log f(x_\alpha, \bar{\theta}_n)/\partial\theta^i\partial\theta^j + c_{ij}(\boldsymbol{\theta}_n)]$ converges stochastically to zero, and since $n^{1/2}|\theta_n^i - \boldsymbol{\theta}_n^i|$ is bounded, we easily obtain from Proposition I that for any $\epsilon > 0$ there exists a constant B_ϵ such that

$$\limsup_{n=\infty} P[\sum \log f(x_\alpha, \dot{\theta}_n) - \sum \log f(x_\alpha, \boldsymbol{\theta}_n) \geqq B_\epsilon \mid \theta_n] = \epsilon.$$

Since $-2 \log \lambda_n(\omega, E_n) \leqq 2\sum \log f(x_\alpha, \dot{\theta}_n) - 2\sum \log f(x_\alpha, \boldsymbol{\theta}_n)$ we have

$$(177)\qquad \limsup_{n=\infty} P[-2 \log \lambda_n(\omega, E_n) \geqq 2B_\epsilon \mid \theta_n] \leqq \epsilon.$$

Denote by ω_n the subset of ω in which $|\boldsymbol{\theta} - \boldsymbol{\theta}_n| \geqq \nu$. From Assumption VII it follows that

$$(178)\qquad \lim_{n=\infty} P[-2 \log \lambda_n(\omega_n, E_n) \geqq -2 \log \bar{\lambda}_n(\omega_n) \mid \theta_n] = 1.$$

Since $-2 \log \bar{\lambda}_n(\omega) \leqq -2 \log \bar{\lambda}_n(\omega_n)$ we have

$$(179)\qquad \lim_{n=\infty} P[-2 \log \lambda_n(\omega_n, E_n) \geqq -2 \log \bar{\lambda}_n(\omega) \mid \theta_n] = 1.$$

For any given constant B there exists a positive $\alpha < 1$ such that $-2 \log \bar{\lambda}_n(\omega)$

$\geqq B$ if l.u.b.$_\theta$ $P[L_n(\omega)|\theta]=\alpha$. Hence from (179) we obtain

$$\lim_{n=\infty} P[-2\log\lambda_n(\omega_n, E_n) \geqq B \mid \theta_n] = 1 \tag{180}$$

for any constant B. From (180) it follows that

$$\lim_{n=\infty} \{P[-2\log\lambda_n(\omega, E_n) \geqq B \mid \theta_n] - P[\,|\dot\theta_n(\omega) - \theta_n| \geqq \nu \mid \theta_n]\} \geqq 0 \tag{181}$$

for arbitrary values of B. From (181) and (177) we obtain

$$\lim_{n=\infty} P[\,|\dot\theta_n(\omega) - \theta_n| \geqq \nu \mid \theta_n] = 0.$$

Hence (175) is proved.

Consider the Taylor expansion

$$\begin{aligned}\sum_\alpha \log f(x_\alpha, \theta_n) &= \sum_\alpha \log f[x_\alpha, \dot\theta_n(\omega)]\\ &+ \frac{n}{2}\sum_{r+1}^{k}\sum_{r+1}^{k}[\dot\theta_n^i(\omega) - \theta_n^i][\dot\theta_n^j(\omega) - \theta_n^j]\frac{1}{n}\sum_\alpha \frac{\partial^2 \log f(x_\alpha, \ddot\theta_n)}{\partial\theta^i\partial\theta^j}\end{aligned} \tag{182}$$

where $\ddot\theta_n$ lies in the interval $[\theta_n, \dot\theta_n(\omega)]$. In the following arguments we shall use the following lemma: *Let* $\|\lambda_{ij}\|$ $(i, j=1, \cdots, k)$ *be a definite matrix and for each integer s let* λ_{ij}^s *be a real number such that* $\lim_{s=\infty}\lambda_{ij}^s=\lambda_{ij}$. *Then*

$$\lim_{s=\infty}\left(\sum\sum \nu_i\nu_j\lambda_{ij}\Big/\sum\sum \nu_i\nu_j\lambda_{ij}^s\right) = 1$$

uniformly in $\nu_1, \cdots, \nu_k$. From (175), the Taylor expansions (176) and (182), and the above lemma it follows that for any $\epsilon>0$

$$\begin{aligned}\lim_{n=\infty} P[(1+\epsilon)q_n - (1-\epsilon)\bar q_n &\geqq -2\log\lambda_n(\omega, E_n)\\ &\geqq (1-\epsilon)q_n - (1+\epsilon)\bar q_n \mid \theta_n] = 1,\end{aligned} \tag{183}$$

where

$$q_n = \sum_1^k\sum_1^k n(\dot\theta_n^i - \theta_n^i)(\dot\theta_n^j - \theta_n^j)c_{ij}(\theta_n) \tag{184}$$

and

$$\bar q_n = \sum_{r+1}^k\sum_{r+1}^k n[\dot\theta_n^i(\omega) - \theta_n^i][\dot\theta_n^j(\omega) - \theta_n^j]c_{ij}(\theta_n). \tag{185}$$

Since $-2\log\lambda_n(\omega, E_n)\geqq 0$, we obtain from (183)

$$\lim_{n=\infty} P[(1+\epsilon)q_n - (1-\epsilon)\bar q_n \geqq 0 \mid \theta_n] = 1. \tag{186}$$

Since the sequence $\{n^{1/2}\delta(\theta_n, \boldsymbol{\theta}_n)\}$ is bounded, the expression q_n is bounded in the probability sense, that is for each positive ρ there exists a positive value A_ρ such that

$$\lim_{n=\infty} P(q_n > A_\rho \mid \theta_n) \leqq \rho.$$

From (186) it follows that $\bar{q}_n$ is also bounded in the probability sense. Hence because of (183) we have

$$\lim_{n=\infty} P[q_n - \bar{q}_n + \epsilon \geqq -2 \log \lambda_n(\omega, E_n) \geqq q_n - \bar{q}_n - \epsilon \mid \theta_n] = 1 \tag{187}$$

for any $\epsilon > 0$. From the Taylor expansions

$$\begin{aligned} \overset{i}{y}_n(\boldsymbol{\theta}_n) &= \frac{1}{n^{1/2}} \sum_\alpha \frac{\partial \log f(x_\alpha, \boldsymbol{\theta}_n)}{\partial \theta^i} \\ &= -\sum_{j=1}^{k} n^{1/2}(\overset{j}{\hat{\theta}}_n - \overset{j}{\boldsymbol{\theta}}_n) \frac{1}{n} \sum_\alpha \frac{\partial^2 \log f(x_\alpha, \overset{*}{\theta}_n)}{\partial\theta^i \partial\theta^j} \qquad (i = 1, \cdots, k) \end{aligned} \tag{188}$$

and

$$\overset{i}{y}_n(\boldsymbol{\theta}_n) = -\sum_{j=r+1}^{k} n^{1/2}[\overset{j}{\hat{\theta}}_n(\omega) - \overset{j}{\boldsymbol{\theta}}_n] \frac{1}{n} \sum_\alpha \frac{\partial^2 \log f(x_\alpha, \overset{**}{\theta}_n)}{\partial\theta^i \partial\theta^j} \qquad (i = r+1, \cdots, k) \tag{189}$$

we obtain

$$\overset{i}{y}_n(\boldsymbol{\theta}_n) = +\sum_{j=1}^{k} n^{1/2}(\overset{j}{\hat{\theta}}_n - \overset{j}{\boldsymbol{\theta}}_n)[c_{ij}(\boldsymbol{\theta}_n) + \epsilon_{ijn}(E_n)], \tag{190}$$

$$n^{1/2}(\overset{i}{\hat{\theta}}_n - \overset{i}{\boldsymbol{\theta}}_n) = \sum_{j=1}^{k} [\sigma_{ij}(\boldsymbol{\theta}_n) + \eta_{ijn}(E_n)]\overset{j}{y}_n(\boldsymbol{\theta}_n) \qquad (i = 1, \cdots, k), \tag{191}$$

and

$$n^{1/2}[\overset{i}{\hat{\theta}}_n(\omega) - \overset{i}{\boldsymbol{\theta}}_n] = \sum_{j=r+1}^{k} [\bar{\sigma}_{ij}(\boldsymbol{\theta}_n) + \zeta_{ijn}(E_n)]\overset{j}{y}_n(\boldsymbol{\theta}_n) \quad (i = r+1, \cdots, k), \tag{192}$$

where

$$\|\sigma_{ij}(\boldsymbol{\theta}_n)\| = \|c_{ij}(\boldsymbol{\theta}_n)\|^{-1} \qquad (i, j = 1, \cdots, k),$$
$$\|\bar{\sigma}_{ij}(\theta)\| = \|c_{ij}(\theta)\|^{-1} \qquad (i, j = r+1, \cdots, k),$$

and for any positive ν we have

$$\begin{aligned} \lim_{n=\infty} P[\mid \epsilon_{ijn}(E_n) \mid < \nu \mid \theta_n] &= \lim P[\mid \eta_{ijn}(E_n) \mid < \nu \mid \theta_n] \\ &= \lim_{n=\infty} P[\mid \zeta_{ijn}(E_n) \mid < \nu \mid \theta_n] = 1. \end{aligned}$$

Hence

$$\sum_{r+1}^{k}\sum_{r+1}^{k} n[\hat{\theta}_n^{i}(\omega) - \theta_n^{i}][\hat{\theta}_n^{j}(\omega) - \theta_n^{j}]c_{ij}(\theta_n)$$

$$= \sum_{r+1}^{k}\sum_{r+1}^{k}\sum_{r+1}^{k}\sum_{r+1}^{k} [c_{ij}(\theta_n)\bar{\sigma}_{il}(\theta_n)\bar{\sigma}_{jm}(\theta_n) + \rho_{ijlmn}(E_n)]y_n^{l}(\theta_n)y_n^{m}(\theta_n)$$

$$= \sum_{r+1}^{k}\sum_{r+1}^{k} [\bar{\sigma}_{ij}(\theta_n) + \rho'_{ijn}(E_n)]y_n^{i}(\theta_n)y_n^{j}(\theta_n)$$

$$= \sum_{j=r+1}^{k}\ \sum_{i=r+1}^{k}\ \sum_{m=1}^{k}\sum_{l=1}^{k} [\bar{\sigma}_{ij}(\theta_n)c_{il}(\theta_n)c_{jm}(\theta_n) + \rho''_{ijmln}(E_n)]n(\hat{\theta}_n^{l} - \theta_n^{l})(\hat{\theta}_n^{m} - \theta_n^{m}),$$

where

$$\lim_{n=\infty} P[\,|\,\rho''_{ijmln}(E_n)\,| < \nu \mid \theta_n] = 1$$

for any positive ν. If at least one of the integers l and m is greater than r, we have

$$\sum_{j=r+1}^{k}\ \sum_{i=r+1}^{k} \bar{\sigma}_{ij}(\theta_n)c_{il}(\theta_n)c_{jm}(\theta_n) = c_{ml}(\theta_n).$$

Hence

$$\sum_{m=1}^{k}\sum_{l=1}^{k} v_m v_l c_{lm}(\theta_n) - \sum_{j=r+1}^{k}\ \sum_{i=r+1}^{k}\ \sum_{l=1}^{k}\sum_{m=1}^{k} \bar{\sigma}_{ij}(\theta_n)c_{im}(\theta_n)c_{jm}(\theta_n)v_l v_m$$

$$= \sum_{m=1}^{r}\sum_{l=1}^{r} v_l v_m\left[c_{lm}(\theta_n) - \sum_{j=r+1}^{k}\ \sum_{i=r+1}^{k} \bar{\sigma}_{ij}(\theta_n)c_{im}(\theta_n)c_{jl}(\theta_n)\right].$$

The coefficient

$$A_{lm}(\theta_n) = c_{lm}(\theta_n) - \sum_{r+1}^{k}\sum_{r+1}^{k} \bar{\sigma}_{ij}(\theta_n)c_{im}(\theta_n)c_{jl}(\theta_n)$$

can be written as the following ratio:

$$(193)\qquad A_{lm}(\theta_n) = \frac{\begin{vmatrix} c_{lm}(\theta_n) & c_{l\,r+1}(\theta_n) & \cdots & c_{lk}(\theta_n) \\ c_{m\,r+1}(\theta_n) & c_{r+1\,r+1}(\theta_n) & \cdots & c_{r+1\,k}(\theta_n) \\ \cdot & \cdot & \cdots & \cdot \\ c_{mk}(\theta_n) & c_{k\,r+1}(\theta_n) & \cdots & c_{kk}(\theta_n) \end{vmatrix}}{\begin{vmatrix} c_{r+1\,r+1}(\theta_n) & \cdots & c_{r+1\,k}(\theta_n) \\ \cdot & \cdots & \cdot \\ c_{k\,r+1}(\theta_n) & \cdots & c_{kk}(\theta_n) \end{vmatrix}}.$$

It is known that if A' is the adjoint of any determinant A, and M and M'

are corresponding m-rowed minors of A and A' respectively, then M' is equal to the product of A^{m-1} by the algebraic complement of M.

Let A be the k-rowed determinant $|\sigma_{ij}(\theta_n)|$ $(i, j=1, \cdots, k)$ and let M be the $(k-r+1)$-rowed minors

$$M = \begin{vmatrix} \sigma_{lm}(\theta_n) & \sigma_{l\, r+1}(\theta_n) & \cdots & \sigma_{lk}(\theta_n) \\ \sigma_{r+1\, m}(\theta_n) & \sigma_{r+1\, r+1}(\theta_n) & \cdots & \sigma_{r+1\, k}(\theta_n) \\ \cdot & \cdot & \cdots & \cdot \\ \sigma_{km}(\theta_n) & \sigma_{k\, r+1}(\theta_n) & \cdots & \sigma_{kk}(\theta_n) \end{vmatrix}. \tag{194}$$

Then we have

$$M' = A^{k-r+1} \begin{vmatrix} c_{lm}(\theta_n) & c_{l\, r+1}(\theta_n) & \cdots & c_{lk}(\theta_n) \\ c_{m\, r+1}(\theta_n) & c_{r+1\, r+1}(\theta_n) & \cdots & c_{r+1\, k}(\theta_n) \\ \cdot & \cdot & \cdots & \cdot \\ c_{mk}(\theta_n) & c_{k\, r+1}(\theta_n) & \cdots & c_{kk}(\theta_n) \end{vmatrix} = A^{k-r}\overline{M}, \tag{195}$$

where $\overline{M}$ denotes the algebraic complement of M. Let M_1 be the $(k-r)$-rowed minor $|\sigma_{ij}(\theta)|$ $(i, j=r+1, \cdots, k)$. Then we obtain

$$M_1' = A^{k-r} \begin{vmatrix} c_{r+1\, r+1}(\theta_n) & \cdots & c_{r+1\, k}(\theta_n) \\ \cdot & \cdots & \cdot \\ c_{k\, r+1}(\theta_n) & \cdots & c_{kk}(\theta_n) \end{vmatrix} = A^{k-r-1}\overline{M}_1, \tag{196}$$

where $\overline{M}_1$ denotes the algebraic complement of M_1.

From (193), (195) and (196) we obtain

$$A_{lm}(\theta_n) = \overline{M}/\overline{M}_1 = \bar{c}_{lm}(\theta_n) \qquad (l, m = 1, \cdots, r),$$

where $\|\bar{c}_{lm}(\theta_n)\| = \|\sigma_{lm}(\theta_n)\|^{-1}$ $(l, m=1, \cdots, r)$. Hence

$$q_n - \bar{q}_n = n\sum_{q=1}^{r}\sum_{p=1}^{r}(\hat{\theta}_n^p - \theta_n^p)(\hat{\theta}_n^q - \theta_n^q)[\bar{c}_{pq}(\theta_n) + \eta_{pqn}(E_n)] + \rho_n(E_n),$$

where for any positive ν

$$\lim_{n=\infty} P[|\eta_{pqn}(E_n)| < \nu \mid \theta_n] = \lim_{n=\infty} P[|\rho_n(E_n)| < \nu \mid \theta_n] = 1.$$

From (187) it follows that for any positive ϵ

$$\begin{aligned} \lim_{n=\infty} P\Big\{ -2\log\lambda_n(\omega, E_n) - \epsilon \leqq \sum_1^r\sum_1^r n(\hat{\theta}_n^p - \theta_n^p)(\hat{\theta}_n^q - \theta_n^q) \\ \cdot [\bar{c}_{pq}(\theta_n) + \eta_{pqn}(E_n)] \leqq -2\log\lambda_n(\omega, E_n) + \epsilon \mid \theta_n \Big\} = 1. \end{aligned} \tag{197}$$

Since $\boldsymbol{\theta}_n^p = \theta_0^p$ $(p=1, \cdots, r)$ from (197) we obtain because of the boundedness of the sequence $\{n^{1/2}(\theta_n - \boldsymbol{\theta}_n)\}$

$$\lim_{n=\infty} P\left[-2 \log \lambda_n(\omega, E_n) - \epsilon \leqq \sum \sum n(\dot\theta_n^p - \theta_0^p)(\dot\theta_n^q - \theta_0^q)\bar{c}_{pq}(\boldsymbol{\theta}_n) \leqq -2 \log \lambda_n(\omega, E_n) + \epsilon \mid \theta_n\right] = 1.$$

The above equation remains valid if we substitute $\dot\theta_n$ for $\boldsymbol{\theta}_n$, that is

$$\lim_{n=\infty} P\left[-2 \log \lambda_n(\omega, E_n) - \epsilon \leqq n\sum \sum (\dot\theta_n^p - \theta_0^p)(\dot\theta_n^q - \theta_0^q)\bar{c}_{pq}(\dot\theta_n) \leqq -2 \log \lambda_n(\omega, E_n) + \epsilon \mid \theta_n\right] = 1. \tag{198}$$

Let W_n^* be the critical region defined in (115). Since (173) holds for any sequence $\{\theta_n\}$ for which $\lim n^{1/2}\delta(\theta_n, \omega) = \infty$, we obtain from (198)

$$\lim_{n=\infty} \{P(W_n^* \mid \theta) - P[L_n(\omega) \mid \theta]\} = 0 \tag{199}$$

uniformly in θ.

Now we consider the general case where ω is given by r equations

$$\xi^1(\theta) = \cdots = \xi^r(\theta) = 0$$

such that Assumption VI is satisfied. As we have seen in §12, the whole theory remains valid if we replace the parameters $\theta^1, \cdots, \theta^k$ by the new parameters

$$\xi^1 = \xi^1(\theta), \cdots, \xi^k = \xi^k(\theta),$$

where the functions $\xi^1(\theta), \cdots, \xi^k(\theta)$ satisfy Assumption VI. Hence from (199) it follows that

$$\lim_{n=\infty} \{P(W_n^* \mid \theta) - P[L_n(\omega) \mid \theta]\} = 0 \tag{200}$$

uniformly in θ, where W_n^* denotes the region defined in (161).

From (200) and Theorem VII we obtain the following theorem.

THEOREM VIII. *Let $S_c(\boldsymbol{\theta})$ be the surface defined in* (162) *and $\eta(\theta)$ be the weight function defined in* (165). *If Assumption* VI *holds, then for testing the hypothesis $\xi^1(\theta) = \cdots = \xi^r(\theta) = 0$ the likelihood ratio test*

(a) *has asymptotically best average power with respect to the surfaces $S_c(\boldsymbol{\theta})$ and weight function $\eta(\theta)$;*

(b) *has asymptotically best constant power on the surfaces $S_c(\boldsymbol{\theta})$;*

(c) *is an asymptotically most stringent test.*

14. Large sample distribution of the likelihood ratio. S. S. Wilks(6) has derived the large sample distribution of the likelihood ratio $\lambda_n(\omega, E_n)$ if ω is

(6) S. S. Wilks, *Distribution of the likelihood ratio in large samples*, Ann. Math. Statist. 1938.

a linear subspace of the parameter space and if the hypothesis to be tested is true. Here we derive the large sample distribution of $\lambda_n(\omega, E_n)$ for any set ω satisfying Assumption VI in both cases, when the hypothesis to be tested is true, and when it is not true.

Let $u_1, \cdots, u_r$ be r independently and normally distributed variates with unit variances. Denote the expected value of u_p by μ_p. The distribution of the statistic

$$U^2 = u_1^2 + \cdots + u_r^2$$

is known(7). The only parameter involved in this distribution is $\lambda^2 = \mu_1^2 + \cdots + \mu_r^2$. Let us denote the cumulative distribution of U^2 by $F_r(\lambda^2, t)$, that is,

$$P[(U^2 < t)] = F_r(\lambda^2, t) \qquad (\lambda^2 = \mu_1^2 + \cdots + \mu_r^2). \tag{201}$$

Obviously $F_r(0, t)$ is the χ^2-distribution with r degrees of freedom.

Let $v_1, \cdots, v_r$ be r variates which have a joint normal distribution. Denote by μ_p the mean value of v_p and by σ_{pq} the covariance between v_p and v_q. Consider the statistic

$$V^2 = \sum_{q=1}^{r} \sum_{p=1}^{r} \lambda_{pq} v_p v_q, \tag{202}$$

where $\|\lambda_{pq}\| = \|\sigma_{pq}\|^{-1}$. It is easy to verify that the distribution of V^2 is given by

$$P(V^2 < t) = F_r(\lambda^2, t), \tag{203}$$

where

$$\lambda^2 = \sum \sum \lambda_{pq} \mu_p \mu_q. \tag{204}$$

We will now derive the limit distribution of the expression on the left-hand side of (161), that is of the statistic

$$Q_n = n \sum_{q=1}^{r} \sum_{p=1}^{r} \xi^p(\dot{\theta}_n) \xi^q(\dot{\theta}_n) \overset{*}{c}_{pq}(\dot{\theta}_n). \tag{205}$$

The joint distribution of the variates $n^{1/2}[\xi^1(\dot{\theta}_n) - \xi^1(\theta)], \cdots, n^{1/2}[\xi^r(\dot{\theta}_n) - \xi^r(\theta)]$ converges with $n \to \infty$ uniformly towards the cumulative normal distribution with zero means and covariance matrix $\|\sigma_{pq}^*(\theta)\| = \|\overset{*}{c}_{pq}(\theta)\|^{-1}$. Since $\dot{\theta}_n$ is a uniformly consistent estimate of θ and since $c_{ij}(\theta)$ is a uniformly continuous function of θ, the statistic

$$\overline{Q}_n = n \sum \sum \xi^p(\dot{\theta}_n) \xi^q(\dot{\theta}_n) \overset{*}{c}_{pq}(\theta) \tag{206}$$

---

(7) See for instance P. C. Tang, *The power function of the analysis of variance tests*, Statistical Research Memoirs vol. 2 (1938).

has the same limit distribution as Q_n, that is,

$$\lim_{n=\infty} \{P[\bar{Q}_n < t \mid \theta] - P[Q_n < t \mid \theta]\} = 0 \tag{207}$$

uniformly in θ and t.

It is easy to see that

$$\lim_{n=\infty} \{P[\bar{Q}_n < t \mid \theta] - F_r[\lambda_n^2(\theta), t] = 0 \tag{208}$$

uniformly in θ and t, where

$$\lambda_n^2(\theta) = n \sum \sum \xi^p(\theta)\xi^q(\theta)\bar{c}_{pq}^*(\theta). \tag{209}$$

Hence, because of (207) we have

$$\lim_{n=\infty} \{P[Q_n < t \mid \theta] - F_r[\lambda_n^2(\theta), t]\} = 0 \tag{210}$$

uniformly in θ and t.

Let $\{\theta_n\}$ $(n=1, 2, \cdots$, ad inf.) be a sequence of parameter points for which $n^{1/2}\delta(\theta_n, \omega)$ is bounded. Then we obtain from (198)

$$\lim_{n=\infty} P[-2 \log \lambda_n(\omega, E_n) - \epsilon \leqq Q_n \leqq -2 \log \lambda_n(\omega, E_n) + \epsilon \mid \theta_n] = 1 \tag{211}$$

for any positive ϵ. From (210) and (211) it follows that

$$\lim_{n=\infty} \{P[-2 \log \lambda_n(\omega, E_n) < t \mid \theta_n] - F_r[\lambda_n^2(\theta_n), t]\} = 0 \tag{212}$$

uniformly in t. Since (173) holds for any sequence $\{\theta_n\}$ for which $\lim n^{1/2}\delta(\theta_n, \omega) = +\infty$, we obtain from (212)

$$\lim_{n=\infty} \{P[-2 \log \lambda_n(\omega, E_n) < t \mid \theta] - F_r[\lambda_n^2(\theta), t]\} = 0 \tag{213}$$

uniformly in θ and t. Hence we have proved the following theorem.

THEOREM IX. *Let $F_r(\lambda^2, t)$ be the distribution function defined in* (201) *and let $\lambda_n(\omega, E_n)$ be the likelihood ratio statistic for testing the hypothesis $\xi^1(\theta) = \cdots = \xi^r(\theta) = 0$. Let furthermore $\lambda_n^2(\theta)$ be the expression defined in* (209). *Then, if Assumption* VI *holds, we have*

$$\lim_{n=\infty} \{P[-2 \log \lambda_n(\omega, E_n) < t \mid \theta] - F_r[\lambda_n^2(\theta), t]\} = 0$$

uniformly in t and θ. If the hypothesis to be tested is true, that is if θ is a point of ω, $\lambda_n^2(\theta) = 0$ and therefore the limit distribution of $-2 \log \lambda_n(\omega, E_n)$ is the χ^2-distribution with r degrees of freedom.

15. **Summary**. Let $f(x^1, \cdots, x^m, \theta^1, \cdots, \theta^k)$ be the joint probability density function of the variates $x^1, \cdots, x^m$ involving k unknown parameters $\theta^1, \cdots, \theta^k$. Any set of values $\theta^1, \cdots, \theta^k$ can be represented by a point θ of the k-dimensional Cartesian space with the coordinates $\theta^1, \cdots, \theta^k$. Let ω be the subset of the parameter space defined by the equations

$$\xi^1(\theta) = \xi^2(\theta) = \cdots = \xi^r(\theta) = 0 \qquad (r \leqq k),$$

that is, ω is the set of all points θ for which the above equations are fulfilled. Denote by H_ω the hypothesis that the true parameter point θ is an element of ω. In this paper the question of an appropriate test of the hypothesis H_ω is discussed when the number of observations is large.

The following notations have been introduced. The point $\dot{\theta}_n$ denotes the point with the coordinates $\dot{\theta}_n^1, \cdots, \dot{\theta}_n^k$ where $\dot{\theta}_n^i$ is the maximum likelihood estimate of θ^i based on n independent observations on $x^1, \cdots, x^m$. The expected value of $-\partial^2 \log f(x^1, \cdots, x^m, \theta^1, \cdots, \theta^k)\partial\theta^i\partial\theta^j$ is denoted by $c_{ij}(\theta)$ and $\|\sigma_{ij}(\theta)\| = \|c_{ij}(\theta)\|^{-1}$. Furthermore $\sum_{l=1}^k \sum_{m=1}^k (\partial\xi^p/\partial\theta^l)(\partial\xi^q/\partial\theta^m)\sigma_{lm}(\theta)$ $(p, q=1, \cdots, r)$ is denoted by $\overset{*}{\sigma}_{pq}(\theta)$ and

$$\|\overset{*}{\bar{c}}_{pq}(\theta)\| = \|\overset{*}{\sigma}_{pq}(\theta)\|^{-1} \qquad (p, q = 1, \cdots, r).$$

The region W_n^* denotes the critical region defined by the inequality

$$n \sum_{q=1}^r \sum_{p=1}^r \xi^p(\dot{\theta}_n)\xi^q(\dot{\theta}_n)\overset{*}{\bar{c}}_{pq}(\dot{\theta}_n) \geqq d_n,$$

where n is the number of independent observations on $x^1, \cdots, x^m$, and the constant d_n is chosen so that the least upper bound of the probability that the sample point falls within W_n^*, calculated under the restriction that the true parameter point lies in ω, is equal to a given positive $\alpha < 1$.

Let λ_n be the likelihood ratio statistic for testing H_ω and let L_n be the critical region defined by the inequality

$$\lambda_n \leqq \bar{\lambda}_n,$$

where the constant $\bar{\lambda}_n$ is chosen so that the least upper bound of the probability that the sample point falls within L_n, calculated under the restriction that the true parameter point lies in ω, is equal to α.

Under certain assumptions on $f(x^1, \cdots, x^m, \theta^1, \cdots, \theta^k)$ and the functions $\xi^1(\theta), \cdots, \xi^r(\theta)$ the following results have been obtained:

I. For testing the hypothesis H_ω the critical regions W_n^* and L_n both: (1) have asymptotically best average power over a family of surfaces defined in (162); (2) have asymptotically best constant power along the surfaces defined in (162); (3) are asymptotically most stringent tests. The exact definitions of these notions are given in Definitions VIII, X and XII, respectively.

II. The statistics $-2 \log \lambda_n$ and $n \sum_{q=1}^r \sum_{p=1}^r \xi^p(\dot{\theta}_n)\xi^q(\dot{\theta}_n)\overset{*}{\bar{c}}_{pq}(\dot{\theta}_n)$ have the same limit distribution. The limit distribution of $-2 \log \lambda_n$ is the χ^2-distribution with r degrees of freedom if the hypothesis to be tested is true. If the true parameter point θ_n is not an element of ω, the distribution of $-2 \log \lambda_n$ approaches the distribution of a sum of non-central squares

$$U^2 = u_1^2 + \cdots + u_r^2,$$

where the variates $u_1, \cdots, u_r$ are independently and normally distributed with unit variances and

$$\sum_{p=1}^{r} (Eu_p)^2 = n \sum \sum \xi^p(\theta_n)\xi^q(\theta_n)\bar{c}^*_{pq}(\theta_n).$$

Columbia University,
New York, N. Y.

Made in United States of America

Reprinted from The Annals of Mathematical Statistics
Vol. XIV, No. 4, December, 1943

AN EXACT TEST FOR RANDOMNESS IN THE NON-PARAMETRIC CASE BASED ON SERIAL CORRELATION[1]

By A. Wald and J. Wolfowitz

Columbia University

1. Introduction. A sequence of variates $x_1, \cdots, x_N$ is said to be a random series, or to satisfy the condition of randomness, if $x_1, \cdots, x_N$ are independently distributed with the same distribution; i.e., if the joint cumulative distribution function (c.d.f.) of $x_1, \cdots, x_N$ is given by the product $F(x_1) \cdots F(x_N)$ where $F(x)$ may be any c.d.f.

The problem of testing randomness arises frequently in quality control of manufactured products. Suppose that x in some quality character of a product and that $x_1, x_2, \cdots, x_N$ are the values of x for N consecutive units of the product arranged in some order (usually in the order they were produced). The production process is said to be in a state of statistical control if the sequence $(x_1, \cdots, x_N)$ satisfies the condition of randomness. A number of tests of randomness have been devised for purposes of quality control, all having the following features in common: 1) They are based on runs in the sequence $x_1, \cdots, x_N$. 2) The test procedure is invariant under topologic transformation of the x-axis, i.e., the test procedure leads to the same result if the original variates $x_1, \cdots, x_N$ are replaced by $x'_1, \cdots, x'_N$ where $x'_\alpha = f(x_\alpha)$ and $f(t)$ is any continuous and strictly monotonic function of t. 3) The size of the critical region, i.e., the probability of rejecting the hypothesis of randomness when it is true, does not depend on the common c.d.f. $F(x)$ of the variates $x_1, \cdots, x_N$. Condition (3) is *a fortiori* fulfilled if condition (2) is satisfied and if $F(x)$ is continuous. The fulfillment of condition (3) is very desirable, since in many practical applications the form of the c.d.f. $F(x)$ is unknown.

Tests of randomness are of importance also in the analysis of time series (particularly of economic time series) where they are frequently based on the so-called serial correlation. The serial correlation coefficient with lag h is defined by the expression[2] (see, for instance, Anderson [1])

$$R_h = \frac{\sum_{\alpha=1}^{N} x_\alpha x_{h+\alpha} - \left(\sum_{\alpha=1}^{N} x_\alpha\right)^2 \Big/ N}{\sum_{\alpha=1}^{N} x_\alpha^2 - \left(\sum_{\alpha=1}^{N} x_\alpha\right)^2 \Big/ N} \tag{1}$$

where $x_{h+\alpha}$ is to be replaced by $x_{h+\alpha-N}$ for all values of α for which $h + \alpha > N$. The distribution of R_h has recently been studied by R. L. Anderson [1], T. Koopmans [2], L. C. Young [3], J. v. Neumann [4, 5], B. I. Hart and J. v. Neu-

[1] Presented to the Institute of Mathematical Statistics and the American Mathematical Society at a joint meeting at New Brunswick, New Jersey, on September 13, 1943.

[2] Some authors (see, for instance, [2] p. 27, equation (61)) use a non-circular definition,

mann [6], and J. D. Williams [7], under the assumption that $x_1, \cdots, x_N$ are independently distributed with the same normal distribution. Thus, in addition to the randomness of the series $(x_1, \cdots, x_N)$ it is assumed that the common c.d.f. of the variates $x_1, \cdots, x_N$ is normal. This is a restrictive assumption since frequently the form of the common c.d.f. $F(x)$ of the variates $x_1, \cdots, x_N$ is unknown.

The purpose of this paper is to develop a test procedure based on R_h such that (*a*) if $F(x)$ is continuous the size of the critical region does not depend on the common c.d.f. $F(x)$ of the variates $x_1, \cdots, x_N$, thus making an exact test of significance possible also when nothing is known about $F(x)$ except its continuity; (*b*) if $F(x)$ is not continuous, but all its moments are finite and its variance is positive, the size of the critical region approaches, as $N \to \infty$, the value it would have if $F(x)$ were continuous. Thus in the limit an exact test is possible in this case as well. We will refer to the case where the form of $F(x)$ is unknown as the non-parametric case, in contrast to the case when it is known that $F(x)$ is a member of a finite parameter family of c.d.f.'s.

The test based on the serial correlation seems to be suitable if the alternative to randomness is the existence of a trend[3] or of some regular cyclical movement in the data. In the analysis of time series it is frequently assumed that this is the case and this is perhaps the reason why tests based on serial correlation are widely used in the analysis of time series. In quality control of manufactured products the existence of a trend is often considered as the alternative to randomness, caused perhaps by the steady deterioration of a machine in the production process. Thus, tests of randomness based on serial correlation could also be used in quality control.

2. An exact test procedure based on R_h. Let a_α be the observed value of $x_\alpha(\alpha = 1, \cdots, N)$. Consider the subpopulation where the set $(x_1, \cdots, x_N)$ is restricted to permutations of $a_1, \cdots, a_N$. In this subpopulation the probability that $(x_1, \cdots, x_N)$ is any particular permutation $(a_1', \cdots, a_N')$ of $(a_1, \cdots, a_N)$ is equal to $1/N!$ if the hypothesis to be tested, i.e., that of randomness, is true. (If two of the a_i $(i = 1, 2, \cdots, N)$ are identical we assume that some distinguishing index is attached to each so that they can then be regarded as distinct and so that there still are $N!$ permutations of the elements $a_1, \cdots, a_N$.)

The probability distribution of R_h in this subpopulation can be determined as follows: Consider the set of $N!$ values of R_h which are obtained by substituting for $(x_1, \cdots, x_N)$ all possible permutations of $(a_1, \cdots, a_N)$. (A value which occurs more than once is counted as many times as it occurs.) Each of these values of R_h has the probability $1/N!$. On the basis of this distribution of R_h an exact test of significance can be carried out. Suppose that α is the level of significance, i.e., the size of the critical region. We choose as critical region a subset of M values out of the set of $N!$ values of R_h where $M/N! = \alpha$. The sub-

[3] If the existence of a trend is feared it may be preferable to use the non-circular statistic discussed, for example, in [2].

set of M values which constitute the critical region will depend in each particular problem on the possible alternatives to randomness. For example, if a linear trend is the only possible alternative to randomness, then the critical region will consist of the M largest values[4] of R_h. The value of the lag h will also be chosen on the basis of the alternatives under consideration. For instance, if some cyclical movement in the data is suspected the choice of h will depend on the form of these cycles. The general idea underlying the choice of the subset of M values and of the lag is to make the power of the test with respect to the alternatives which are particularly feared as high as possible.

If R_h has the same value for several permutations of $(a_1, \cdots, a_N)$, it may be impossible to have a critical region consisting of exactly M values of R_h. For example, if $a_1 = a_2 = \cdots = a_N$, then all the $N!$ values of R_h are equal, and the number of values of R_h included in the critical region must be either 0 or $N!$. If $F(x)$ is continuous the probability that two values of R_h be equal is zero. This explains why an exact test is always possible when $F(x)$ is continuous. On the other hand, if $F(x)$ is not continuous, the probability that several values of R_h be equal is positive. However, the theorem we shall prove in Section 4 shows that in the limit an exact test is possible even when $F(x)$ is not continuous, but has finite moments and a positive variance. For if the latter is true, the probability is one that the weaker conditions for the validity of our theorem (given at the end of Section 4) will be fulfilled.

Consider the statistic

$$\bar{R}_h = \sum_{\alpha=1}^{N} x_\alpha x_{h+\alpha} \tag{2}$$

where $x_{h+\alpha}$ is to be replaced by $x_{h+\alpha-N}$ for all values of α for which $h + \alpha > N$. Since in the subpopulation under consideration $\sum_{\alpha=1}^{N} x_\alpha$ and $\sum_{\alpha=1}^{N} x_\alpha^2$ are constants, the statistic $\bar{R}_h$ is a linear function of R_h in this subpopulation. Hence, the test based on $\bar{R}_h$ is equivalent to the test based on R_h. Since $\bar{R}_h$ is simpler than R_h, in what follows we shall restrict ourselves to the statistic $\bar{R}_h$.

We shall now show that, if h is prime to N, the totality T_h of the $N!$ values taken by $\bar{R}_h$ is the same as T_1, the totality of the $N!$ values taken by $\bar{R}_1$.

In the argument which follows it is to be understood that, whenever a positive integer is greater than N, it is to be replaced by that positive integer less than or equal to N which differs from it by an integral multiple of N.

Clearly it will be sufficient to show the existence of a permutation $p_1, p_2, \cdots, p_N$ of the first N integers such that

$$p_i + 1 = p_{i+h} \qquad (i = 1, 2, \cdots, N).$$

Such a permutation is given by

$$j = p_{(j-1)h+1} \qquad (j = 1, 2, \cdots, N).$$

For if $j \neq j'$ then $(j-1)h + 1 \neq (j'-1)h + 1$ because h is prime to N. Hence to every positive integer i there is a unique positive integer j, $(i, j \leq N)$ such

[4] See footnote 3.

that

$$i = (j - 1)h + 1$$

Now

$$p_i + 1 = p_{(j-1)h+1} + 1 = j + 1 = p_{jh+1} = p_{i+h},$$

which is the required result.

In what follows we shall restrict ourselves to the case when h is prime to N. This is not a very restrictive assumption since in practice h will be small as compared with N and by omitting a few observations we can always make N prime to h. Since T_h is the same as T_1 we shall deal with the statistic $\bar{R}_1$ only. To simplify the notation we shall write R instead of $\bar{R}_1$. Thus, the test procedure will be based on the statistic

$$R = \sum_{\alpha=1}^{N-1} x_\alpha x_{\alpha+1} + x_N x_1. \tag{4}$$

If N is very small an exact test of significance can be carried out by actually calculating the $N!$ possible values of R. However, this procedure is practically impossible if N is not small. In Section 3 the exact mean value and variance of R will be calculated, and in section 4 the normality of the limiting distribution of R will be proved. Thus, if N is sufficiently large so that the limiting distribution of R can be used, a test of significance can easily be carried out. Difficulties in carrying out the test arise if N is neither sufficiently small to make the computation of the $N!$ values of R practically possible, nor sufficiently large to permit the use of the limiting distribution. In such cases it may be helpful to determine the third and fourth, and perhaps higher, moments of R, on the basis of which upper and lower limits for the cumulative distribution of R can be derived. (For a description of the Tchebycheff inequalities by which this can be done see, for example, Uspensky, [8], pp. 373–380.) Since the limiting distribution is normal it may be useful to approximate the distribution by a Gram-Charlier series or to employ similar methods.

3. Mean value and variance of R.[5] It is clear that

$$\begin{aligned} E(R) &= NE(x_1 x_2) = \frac{N}{N(N-1)} \sum_{\alpha \neq \beta}\sum a_\alpha a_\beta \\ &= \frac{1}{N-1} [(a_1 + \cdots + a_N)^2 - (a_1^2 + \cdots + a_N^2)]. \end{aligned} \tag{5}$$

To calculate the variance of R we first calculate the second moment of R about the origin. We have

$$\begin{aligned} E(R^2) &= E(x_1x_2 + \cdots + x_{N-1}x_N + x_Nx_1)^2 \\ &= NEx_1^2x_2^2 + 2NEx_1x_2^2x_3 + (N^2 - 3N)Ex_1x_2x_3x_4. \end{aligned} \tag{6}$$

[5] The first four moments of a similar statistic have been obtained by Young [3].

To express the expected values $Ex_1^2x_2^2$, $Ex_1x_2^2x_3$, and $Ex_1x_2x_3x_4$ we shall introduce the following notations for the symmetric functions of $a_1, \cdots, a_N$: For any set of positive integers $i_1, i_2, \cdots, i_k$ the symbol $S_{i_1 i_2 \cdots i_k}$ denotes the symmetric function $\sum_{\alpha_k} \cdots \sum_{\alpha_1} a_{\alpha_1}^{i_1} \cdots a_{\alpha_k}^{i_k}$ where the summation is to be taken over all possible sets of k positive integers $\alpha_1, \cdots, \alpha_k$ subject to the restriction that $\alpha_u \leq N$ and $\alpha_u \neq \alpha_v$ $(u, v = 1, \cdots, k)$.

From (6) we easily obtain

$$E(R^2) = \frac{N}{N(N-1)} S_{22} + \frac{2N}{N(N-1)(N-2)} S_{121} + \frac{N^2 - 3N}{N(N-1)(N-2)(N-3)} S_{1111} \tag{7}$$

$$= \frac{S_{22}}{(N-1)} + \frac{2S_{121}}{(N-1)(N-2)} + \frac{S_{1111}}{(N-1)(N-2)}.$$

It will probably facilitate computation to express each of the symmetric functions in the right member of (7) by a sum of terms, each a product of factors S_r $(r = 1, 2, \cdots)$. One can easily verify the relationships

$$S_{11} = S_1^2 - S_2 \tag{8}$$

$$S_{12} = S_{21} = S_1S_2 - S_3 \tag{9}$$

$$S_{13} = S_{31} = S_1S_3 - S_4 \tag{10}$$

$$S_{22} = S_2^2 - S_4 \tag{11}$$

$$S_{111} = S_{11}S_1 - 2S_{12} = (S_1^2 - S_2)S_1 - 2(S_1S_2 - S_3) \tag{12}$$

$$= S_1^3 - 3S_1S_2 + 2S_3$$

$$S_{112} = S_{121} = S_{211} = S_{11}S_2 - 2S_{13} \tag{13}$$

$$= (S_1^2 - S_2)S_2 - 2(S_1S_3 - S_4)$$

$$= S_1^2S_2 - S_2^2 - 2S_1S_3 + 2S_4$$

$$S_{1111} = S_{111}S_1 - 3S_{112} \tag{14}$$

$$= S_1^4 - 3S_1^2S_2 + 2S_1S_3 - 3S_1^2S_2 + 3S_2^2 + 6S_1S_3 - 6S_4$$

$$= S_1^4 - 6S_1^2S_2 + 8S_1S_3 + 3S_2^2 - 6S_4.$$

It follows from (5) that

$$E(R) = \frac{1}{N-1}(S_1^2 - S_2), \tag{15}$$

and from (7), (11), (13), (14), and (15) that the variance of R is given by

$$\sigma^2(R) = E(R^2) - [E(R)]^2$$

$$= \frac{S_2^2 - S_4}{N-1} + \frac{S_1^4 - 4S_1^2S_2 + 4S_1S_3 + S_2^2 - 2S_4}{(N-1)(N-2)} - \frac{1}{(N-1)^2}(S_1^2 - S_2)^2. \tag{16}$$

The mean value and variance of R can easily be computed from (15) and (16) as soon as the values of S_1, S_2, S_3, and S_4 have been determined.

The formulas (15) and (16) are considerably simplified if $S_1 = 0$. In the special case that $S_1 = 0$ we have

$$E(R) = -\frac{S_2}{N-1} \tag{15'}$$

and

$$\sigma^2(R) = \frac{S_2^2 - S_4}{N-1} + \frac{S_2^2 - 2S_4}{(N-1)(N-2)} - \frac{S_2^2}{(N-1)^2}. \tag{16'}$$

We can always make S_1 equal to zero by replacing a_α by $b_\alpha = a_\alpha - N^{-1} \Sigma a_\alpha$. This substitution is permissible, since it changes the statistic R only by an additive constant and consequently leaves the test procedure unaffected. Thus, in practical applications it may be convenient to replace a_α by b_α and to use formulas (15′) and (16′).

4. Limiting distribution of R. Let $\{a_\alpha\}$ ($\alpha = 1, 2, \cdots$ ad inf.) be a sequence of real numbers with the following properties:

a) There exists a sequence of numbers $A_1, A_2, \cdots, A_r, \cdots$ such that

$$\frac{1}{N}\left|\sum_{\alpha=1}^{N} a_\alpha^r\right| \leq A_r \qquad (r = 1, 2, \cdots \text{ ad inf.}) \tag{17}$$

for all N. (This condition means that the moments about the origin of the sequence $a_1, a_2, \cdots, a_N$ are bounded functions of N.)

b) If

$$\delta(N) = \frac{1}{N}\left[\sum_{\alpha=1}^{N} a_\alpha^2 - \frac{1}{N}\left(\sum_{\alpha=1}^{N} a_\alpha\right)^2\right],$$

then

$$\liminf_N \delta(N) > 0. \tag{18}$$

(This condition means that the dispersion of the N values $a_1, a_2, \cdots, a_N$ is eventually bounded below.)

Let $R(N)$ be the serial correlation coefficient R as defined in (4), where $x_1, \cdots, x_N$ is a random permutation of $a_1, a_2, \cdots, a_N$. We shall prove the following

THEOREM: *As $N \to \infty$, the probability that*

$$\frac{R(N) - E(R(N))}{\sigma(R(N))} < t$$

approaches the limit

$$\frac{1}{\sqrt{2\pi}} \int_{-\infty}^{t} e^{-\frac{1}{2}x^2}\, dx.$$

For any function $f(N)$ and any positive function $\phi(N)$ let

$$f(N) = O(\phi(N))$$

mean that $|f(N)/\phi(N)|$ is bounded from above for all N, and let

$$f(N) = \Omega(\phi(N))$$

mean that

$$f(N) = O(\phi(N))$$

and that $\liminf_{N} |f(N)/\phi(N)| > 0$. Also let

$$f(N) = o(\varphi(N))$$

mean that

$$\lim_{N\to\infty} \frac{f(N)}{\phi(N)} = 0.$$

Let $[\rho]$ denote the largest integer less than or equal to ρ.

To simplify the proof we shall temporarily assume:

c) There exists a positive constant K such that, for every positive integral N,

$$-K \leq S_1 = \sum_{\alpha=1}^{N} a_\alpha \leq K. \tag{19}$$

This restriction will be removed later.

LEMMA 1:

$$\sum_{\alpha_1<\cdots<\alpha_k} \cdots \sum a_{\alpha_1} a_{\alpha_2} \cdots a_{\alpha_k} = O(N^{[\frac{1}{2}k]}).$$

PROOF: $\sum_{\alpha_1<\cdots<\alpha_k} \cdots \sum a_{\alpha_1} \cdots a_{\alpha_k}$ can be written as the sum of a finite number of terms where each term is a product of factors S_r ($r = 1, 2, \cdots$). This representation will be called the normal representation of $\sum \cdots \sum a_{\alpha_1} \cdots a_{\alpha_k}$. Since $S_1 = O(1)$ by (19) and $S_r = O(N)$ by (17) and since the number of factors S_r ($r > 1$) in a single term of the normal representation of $\sum \cdots \sum a_{\alpha_1} \cdots a_{\alpha_k}$ is at most $[\frac{1}{2}k]$, the equation $\sum \cdots \sum a_{\alpha_1} \cdots a_{\alpha_k} = O(N^{[\frac{1}{2}k]})$ must hold.

LEMMA 2: *Let* $y = x_1 \cdots x_k z$, *where* $z = x_{k+1}^{i_1} \cdots x_{k+r}^{i_r}$ *and* $i_j > 1$ ($j = 1, \cdots, r$). *If* $(x_1, \cdots, x_N)$ *is a random permutation of* $a_1, \cdots, a_N$, *and if* $k, r, i_1, \cdots, i_r$ *are fixed values independent of* N, *then* $E(y) = O(N^{[\frac{1}{2}k]-k})$.

PROOF: Let $E(y \mid x_{k+1}, \cdots, x_{k+r})$ be the conditional expected value of y when $x_{k+1}, \cdots, x_{k+r}$ are fixed. It follows easily from Lemma 1 that

$$E(y \mid x_{k+1}, \cdots, x_{k+r}) = O(N^{[\frac{1}{2}k]-k}).$$

Hence also $E(y) = O(N^{[\frac{1}{2}k]-k})$ and Lemma 2 is proved.

Denote $x_\alpha x_{\alpha+1}$ by $y_\alpha(\alpha = 1, \cdots, N-1)$ and $x_N x_1$ by y_N, and consider the expansion of $(y_1 + \cdots + y_N)^r$. Let y be a term of this expansion, i.e., $y = \frac{N!}{i_1! \cdots i_u!} y_{\alpha_1}^{i_1} \cdots y_{\alpha_u}^{i_u}$ $(\alpha_1 < \alpha_2 < \cdots < \alpha_u)$. We will say that two factors y_α and y_β are neighbors if $| \alpha - \beta + 1 |$ or $| \alpha - \beta - 1 |$ is either 0 or N. The set of u factors $y_{\alpha_1}, \cdots, y_{\alpha_u}$ can be subdivided into cycles as follows: The first cycle contains y_{α_1} and all those y_α which can be reached from y_{α_1} by a succession of neighboring y_α. The second cycle contains the first y_α of the remaining sequence and all those which can be reached from the first y_α by a succession of neighboring y_α. The third cycle is similarly constructed from the remaining sequence, etc. After a finite number of cycles have been withdrawn the sequence will be exhausted. If m is the number of such cycles we will say that y has m cycles.

LEMMA 3: *Let y be a term of the expansion $(x_1x_2 + \cdots + x_Nx_1)^r = (y_1 + \cdots + y_N)^r$ (r fixed). Let m be the number of cycles in y and k be the number of linear factors in y if y is written as a function of $x_1, \cdots, x_N$ (i.e., if we replace y_α by $x_\alpha x_{\alpha+1}$). Then the maximum value of $m + [\frac{1}{2}k] - k$ is equal to $[\frac{1}{2}r]$.*

PROOF: First we maximize $m + [\frac{1}{2}k] - k$ with respect to k when m is fixed. If $m \leq [\frac{1}{2}r]$, then the minimum value of k is obviously zero. Let $m = [\frac{1}{2}r] + r'$ $(r' > 0)$. The minimum value of k is reached if each cycle consists of a single factor y_α and if each factor y_α in y is either linear or squared. If r is even, then the minimum value of k is $4r'$ and if r is odd then the minimum value of k is $4r' - 2$. Hence for $m = [\frac{1}{2}r] + r'$ we have

$$\max_k \,(m + [\tfrac{1}{2}k] - k) = [\tfrac{1}{2}r] - r' \qquad \text{if } r \text{ is even}$$

and

$$= [\tfrac{1}{2}r] - r' + 1 \text{ if } r \text{ is odd.}$$

Hence maximizing with respect to m and k we obtain

$$\max \,(m + [\tfrac{1}{2}k] - k) = [\tfrac{1}{2}r],$$

and Lemma 3 is proved.

LEMMA 4: *The expected value of the sum of all those terms in the expansion of $(x_1x_2 + \cdots + x_Nx_1)^r$ for which m is the number of cycles and k the number of linear factors (if y is expressed in terms of $x_1, \cdots, x_N$) is equal to $O(N^{m+[\frac{1}{2}k]-k})$.*

This Lemma follows from Lemma 2 and the fact that the number of terms y with the required properties is $O(N^m)$.

LEMMA 5:

$$E(x_1x_2 + \cdots + x_Nx_1)^r = O(N^{[\frac{1}{2}r]}).$$

This follows from Lemmas 3 and 4.

LEMMA 6: *If r is even then*

$$E(x_1x_2 + \cdots + x_Nx_1)^r = \left(C^N_{\frac{1}{2}r}\left(\frac{r!}{2^{\frac{1}{2}r}}\right) E(x_1^2 x_2^2 \cdots x_r^2)\right) + o(N^{\frac{1}{2}r}).$$

PROOF: It follows easily from our considerations in proving Lemma 3 that $m + [\frac{1}{2}k] - k < \frac{1}{2}r$ for all terms in the expansion of $(x_1x_2 + \cdots + x_Nx_1)^r$ which are not of the type $x_1^2 \cdots x_r^2$. Hence it follows from Lemma 4 that the expected value of the sum of all those terms in the expansion of $[x_1x_2 + \cdots + x_Nx_1]^r$ which are not of the type $x_1^2 \cdots x_r^2$ is equal to $o(N^{\frac{1}{2}r})$. Lemma 6 follows from the fact that $2^{-\frac{1}{2}r}r!$ is the coefficient of the terms of the type $x_1^2 \cdots x_r^2$ in the expansion of $(x_1x_2 + \cdots + x_Nx_1)^r$ and that the number of terms of such type is equal to $C^N_{\frac{1}{2}r}$.

LEMMA 7. $\lim\limits_{N\to\infty} \dfrac{E(x_1x_2 + \cdots + x_Nx_1)^r}{\{E(x_1x_2 + \cdots + x_Nx_1)^2\}^{\frac{1}{2}r}} = 0$ *if r is odd and* $= 2^{-\frac{1}{2}r}r!/(\frac{1}{2}r)!$ *if r is even.*

PROOF: From Lemma 6 it follows that

$$E(x_1x_2 + \cdots + x_Nx_1)^2 = NE(x_1^2x_2^2) + o(N) = \Omega(N). \tag{20}$$

The first half of Lemma 7 follows from Lemma 5 and equation (20). If r is even then it follows from (20) that

$$\begin{aligned}\lim \frac{E(x_1x_2 + \cdots + x_Nx_1)^r}{\{E(x_1x_2 + \cdots + x_Nx_1)^2\}^{\frac{1}{2}r}} &= \lim_{N\to\infty} \frac{2^{-\frac{1}{2}r}C^N_{\frac{1}{2}r}\, r!\, E(x_1^2 \cdots x_r^2)}{N^{\frac{1}{2}r}(Ex_1^2x_2^2)^{\frac{1}{2}r}} \\ &= \lim \frac{r!}{2^{\frac{1}{2}r}(\frac{1}{2}r)!}\, \frac{E(x_1^2 \cdots x_r^2)}{(E(x_1^2x_2^2))^{\frac{1}{2}r}}.\end{aligned} \tag{21}$$

It follows from (17), (19), and the normal representation of symmetric functions that

$$k! \sum_{a_{\alpha_1} < a_{\alpha_2} < \cdots < a_{\alpha_k}} \cdots \sum a_{\alpha_1}^2 \cdots a_{\alpha_k}^2 = S_2^k + O(N^{k-1}).$$

From (17) and (18) we have $S_2 = \Omega(N)$. Since

$$E(x_1^2 \cdots x_r^2) = r!(\sum_{a_{\alpha_1} < a_{\alpha_2} < \cdots < a_{\alpha_r}} \cdots \sum a_{\alpha_1}^2 \cdots a_{\alpha_r}^2)[N(N-1) \cdots (N-r+1)]^{-1},$$

we obtain

$$\lim_{N\to\infty} \frac{E(x_1^2 \cdots x_r^2)}{(E(x_1^2x_2^2))^{\frac{1}{2}r}} = 1. \tag{22}$$

The second half of Lemma 7 follows from (21) and (22).

LEMMA 8:

$$\lim_{N\to\infty} \frac{E(R(N))}{\sigma(R(N))} = 0, \tag{23}$$

$$\lim_{N\to\infty} \frac{E(R^2(N))}{\sigma^2(R(N))} = 1. \tag{24}$$

PROOF: Equation (24) is a trivial consequence of (23). From (15) $E(R) = O(1)$ and from (16) $\sigma(R) = \Omega(N^{\frac{1}{2}})$. The lemma follows easily from these relations.

PROOF OF THE THEOREM: According to Lemma 7 the r-th moment of $R[E(R^2)]^{-\frac{1}{2}}$ approaches the r-th moment of the normal distribution as $N \to \infty$. From this and Lemma 8 the required result follows if condition (c) holds. It remains therefore merely to remove condition (c). Assume now only that $a_1, a_2, \cdots, a_\alpha, \cdots$ satisfy conditions (a) and (b).

$R(N)$ is formed from the population of values $a_1, a_2, \cdots, a_N$. Addition of a constant q to $a_1, \cdots, a_N$ adds the same constant to all the values of $R(N)$ and hence leaves $[R(N) - E(R(N))]/\sigma(R(N))$ unaltered. Let $q^{(N)}$ be $-\sum_{\alpha=1}^{N} a_\alpha/N$ and write $b_\alpha^{(N)} = a_\alpha + q^{(N)}$. Consider the sequences

$$B^{(i)} = b_1^{(i)}, b_2^{(i)}, \cdots, b_i^{(i)} \qquad (i = 1, 2, \cdots, \text{ad inf.}).$$

From (17) it follows that the $| q^{(N)} |$ are bounded for all N. Hence the sequences $B^{(i)}$ satisfy condition (a). They obviously satisfy condition (c). Since $\delta(j)$ is invariant under addition of a constant we have

$$\liminf_j \frac{1}{j}\left(\sum_{\alpha=1}^{j} (b_\alpha^{(j)})^2 - \frac{1}{j}\left(\sum_{\alpha=1}^{j} b_\alpha^{(j)}\right)^2\right) > 0,$$

so that the $B^{(i)}$ satisfy condition (b). Since $[R(N) - E(R(N))]/\sigma(R(N))$ has the same distribution in the sequence $a_1, a_2, \cdots, a_N$ as in the sequence $B^{(N)}$, the theorem follows.

It should be remarked that the theorem remains valid if conditions (a) and (b) are replaced by the weaker condition

$$\mu_r/\mu_2^{\frac{1}{2}r} = O(1) \qquad (r = 3, 4, \cdots, \text{ad inf.})$$

where

$$\mu_r = \frac{1}{N}\sum_{\alpha=1}^{N}\left(a_\alpha - \frac{1}{N}\sum_{\alpha=1}^{N} a_\alpha\right)^r.$$

This follows easily from the fact that $[R(N) - E(R(N))]/\sigma(R(N))$ remains unaltered if we replace the sequence $a_1, \cdots, a_N$ by the sequence $c_1^N, c_2^N, \cdots, c_N^N$ where

$$c_\alpha^N = \left(a_\alpha - \frac{1}{N}\sum_1^N a_\alpha\right) \Big/ \left[\frac{1}{N}\sum\left(a_\alpha - \frac{1}{N}\sum a_\alpha\right)^2\right]^{\frac{1}{2}}.$$

Conditions (a) and (b) are obviously satisfied by the sequence $c_1^N, \cdots, c_N^N$.

5. Transformation of the original observations.

Let $f(t)$ be a continuous and strictly monotonic function of t $(-\infty < t < +\infty)$. Suppose we replace the original observations $a_1, \cdots, a_N$ by $d_1, \cdots, d_N$, where $d_\alpha = f(a_\alpha)$ $(\alpha = 1, \cdots, N)$. We obtain a valid test of significance if we carry out the test procedure as if $d_1, \cdots, d_N$ were the observed values instead of $a_1, \cdots, a_N$. We could also replace the observed values $a_1, \cdots, a_N$ by their ranks. The question arises whether there is any advantage in making the test on the transformed values instead of on the original observations. It may well be that by certain transformations we could considerably increase the power of the test with respect to alternatives under consideration. This problem needs further study.

6. Summary. A test procedure based on serial correlation is given for testing the hypothesis that $x_1, \cdots, x_N$ are independent observations from the same population, i.e., that $x_1, \cdots, x_N$ is a random series. By considering the distribution of the serial correlation coefficient in the subpopulation consisting of all permutations of the actually observed values a test procedure is obtained such that

a) if the common c.d.f. $F(x)$ is continuous, the size of the critical region, i.e., the probability of rejecting the hypothesis of randomness when it is true, does not depend upon $F(x)$,

b) if $F(x)$ is not continuous but all its moments are finite and its variance is positive, the size of the critical region approaches, as $N \to \infty$, the value it would have if $F(x)$ were continuous. Thus in the limit an exact test is possible in this case as well.

It is shown that the test based on the serial correlation with lag h is equivalent to the test based on the statistic[6]

$$\sum_{\alpha=1}^{N} x_\alpha x_{h+\alpha}$$

where $x_{h+\alpha}$ is to be replaced by $x_{h+\alpha-N}$ for all values of α for which $h + \alpha > N$. If h is prime to N, the distribution of $\sum_1^N x_\alpha x_{h+\alpha}$ is exactly the same as the distribution of $R = \sum_1^N x_\alpha x_{1+\alpha}$.

The mean value and variance of R are given by the following expressions:

$$E(R) = (S_1^2 - S_2)/(N - 1)$$

and

$$\sigma^2(R) = \frac{S_2^2 - S_4}{N - 1} + \frac{S_1^4 - 4S_1^2 S_2 + 4S_1 S_3 + S_2^2 - 2S_4}{(N - 1)(N - 2)} - \frac{(S_1^2 - S_2)^2}{(N - 1)^2}$$

where $S_r = x_1^r + \cdots + x_N^r$.

It is shown that under some mild restrictions the limiting distribution of R is normal. The test procedure can therefore be easily carried out when N is sufficiently large to permit the use of the limiting distribution of R.

REFERENCES

[1] R. L. Anderson, *Annals of Math. Stat.*, Vol. 13 (1942), p. 1.
[2] T. Koopmans, *ibid.*, p. 14.
[3] L. C. Young, *Annals of Math. Stat.*, Vol. 12 (1941), p. 293.
[4] J. v. Neumann, *Annals of Math. Stat.*, Vol. 12 (1941), p. 367.
[5] J. v. Neumann, *Annals of Math. Stat.*, Vol. 13 (1942), p. 86.
[6] B. I. Hart and J. v. Neumann, *Annals of Math. Stat.*, Vol. 13 (1942), p. 207.
[7] J. D. Williams, *Annals of Math. Stat.*, Vol. 12 (1943), p. 239.
[8] J. Uspensky, *Introduction to Mathematical Probability*, New York, 1937.

[6] If the non-circular definition of the serial correlation coefficient is used, the term $x_N x_{N+h}$ should be omitted.

Reprinted from THE ANNALS OF MATHEMATICAL STATISTICS
Vol. XV, No. 2, June, 1944

ON A STATISTICAL PROBLEM ARISING IN THE CLASSIFICATION OF AN INDIVIDUAL INTO ONE OF TWO GROUPS[1]

BY ABRAHAM WALD

Columbia University

1. Introduction. In social, economic and industrial problems we are often confronted with the task of classifying an individual into one of two groups on the basis of a number of test scores. For example, in the case of personnel selection the acceptance or rejection of an applicant is frequently based on a number of test scores obtained by the applicant. A similar situation arises in connection with college entrance examinations. Again, on the basis of a number of test scores, the admission or rejection of a student has to be decided. In all such problems it is assumed that there are two populations, say π_1 and π_2, one representing the population of individuals fit, and the other the population of individuals unfit for the purpose under consideration. The problem is that of classifying an individual into one of the populations π_1 and π_2 on the basis of his test scores. Often, some statistical data from past experience are available which can be utilized in making the classification. Suppose that from past experience we have the test scores of N_1 individuals who *are known* to belong to population π_1, and also the test scores of N_2 individuals who *are known* to belong to population π_2. These data will be utilized in classifying a new individual on the basis of his test scores.

In this paper we shall deal with the statistical problem of classifying an individual into one of the populations π_1 and π_2 on the basis of his test scores and on the basis of past experience, given in the form of two samples, one drawn from π_1 and the other from π_2. In the next section we give a precise formulation of the statistical problem and state the assumptions we make about the populations π_1 and π_2.

2. Statement of the problem. We consider two sets of p variates $(x_1, \cdots, x_p)$ and $(y_1, \cdots, y_p)$. It is assumed that each of the sets $(x_1, \cdots, x_p)$ and $(y_1, \cdots, y_p)$ has a p-variate normal distribution and the two sets are independent of each other. It is furthermore assumed that the covariance matrix of the variates $x_1, \cdots, x_p$ is equal to the covariance matrix of the variates $y_1, \cdots, y_p$, i.e. $\sigma_{x_i x_j} = \sigma_{y_i y_j}$ $(i, j = 1, \cdots, p)$. We will denote this common covariance by σ_{ij}. Let us denote the mean value of x_i by μ_i and the mean value of y_i by ν_i. Furthermore we will denote the normal population with mean values $\mu_1, \cdots, \mu_p$ and covariance matrix $|| \sigma_{ij} ||$ by π_1, and the normal population with mean values $\nu_1, \cdots, \nu_p$ and covariance matrix $|| \sigma_{ij} ||$ by π_2.

A sample of size N_1 is drawn from the population π_1 and a sample of size N_2 is

---

[1] The author wishes to thank Dr. Irving Lorge, Columbia University, for calling his attention to this problem.

drawn from the population π_2. Denote by $x_{i\alpha}$ the α-th observation on x_i $(i = 1, \cdots, p; \alpha = 1, \cdots, N_1)$ and $y_{i\beta}$ the β-th observation on y_i $(i = 1, \cdots, p; \beta = 1, \cdots, N_2)$. Let z_i $(i = 1, \cdots, p)$ be a single observation on the i-th variate drawn from a p-variate population π, where it is known a priori that π is either identical with π_1 or with π_2. The set $(z_1, \cdots, z_p)$ is assumed to be distributed independently of $(x_1, \cdots, x_p)$ and $(y_1, \cdots, y_p)$.

We will deal here with the following statistical problem: On the basis of the observations $x_{i\alpha}, y_{i\beta}, z_i$ $(i = 1, \cdots, p; \alpha = 1, \cdots, N_1; \beta = 1, \cdots, N_2)$ we test the hypothesis H_1 that the population π, from which the set $(z_1, \cdots, z_p)$ has been drawn, is equal to π_1. The parameters $\mu_1, \cdots, \mu_p, \nu_1, \cdots, \nu_p$ and $\| \sigma_{ij} \|$ are assumed to be unknown.

3. The statistic to be used for testing the hypothesis H_1. In this problem there exists only a single alternative hypothesis to the O-hypothesis H_1 to be tested, i.e. the hypothesis H_2 that π is equal to π_2. If the parameters $\mu_1, \cdots, \mu_p, \nu_1, \cdots, \nu_p$ and $\| \sigma_{ij} \|$ were known we could easily find (on the basis of a lemma by Neyman and Pearson) the critical region which is most powerful with respect to the alternative H_2. Let us assume for the moment that the parameters $\mu_1, \cdots, \mu_p, \nu_1, \cdots, \nu_p$ and $\| \sigma_{ij} \|$ are known and let us compute the critical region for testing H_1 which is most powerful with respect to the alternative H_2. According to a lemma by Neyman and Pearson[2] this critical region is given by the inequality

$$\frac{p_2(z_1, \cdots, z_p)}{p_1(z_1, \cdots, z_p)} \geq k, \tag{1}$$

where $p_1(z_1, \cdots, z_p)$ denotes the joint probability density function of $z_1, \cdots, z_p$ under the hypothesis H_1, $p_2(z_1, \cdots, z_p)$ denotes the joint probability density function of $(z_1, \cdots, z_p)$ under the hypothesis H_2, and k is a constant determined so that the critical region should have the required size.

Denote the determinant value $| \sigma_{ij} |$ of the matrix $\| \sigma_{ij} \|$ by σ^2. Then

$$p_1(z_1, \cdots, z_p) = \frac{1}{(2\pi)^{p/2}\sigma} e^{-\frac{1}{2}\sum_{j=1}^{p}\sum_{i=1}^{p}\sigma^{ij}(z_i-\mu_i)(z_j-\mu_j)}, \tag{2}$$

and

$$p_2(z_1, \cdots, z_p) = \frac{1}{(2\pi)^{p/2}\sigma} e^{-\frac{1}{2}\sum_{j=1}^{p}\sum_{i=1}^{p}\sigma^{ij}(z_i-\nu_i)(z_j-\nu_j)}, \tag{3}$$

where the matrix $\| \sigma^{ij} \|$ denotes the inverse matrix of the matrix $\| \sigma_{ij} \|$. Taking logarithms of both sizes of the inequality (1), we obtain the inequality

$$-\tfrac{1}{2}\Big\{\sum_j \sum_i \sigma^{ij}[(z_i - \nu_i)(z_j - \nu_j) - (z_i - \mu_i)(z_j - \mu_j)]\Big\} \geq \log k. \tag{4}$$

[2] J. Neyman and E. S. Pearson, "Contributions to the theory of testing statistical hypotheses," *Stat. Res. Mem.*, Vol. 1, London, 1936.

Multiplying both sides of (4) by 2, we have

$$\sum_j \sum_i \sigma^{ij}[(z_i - \mu_i)(z_j - \mu_j) - (z_i - \nu_i)(z_j - \nu_j)] \geq 2 \log k. \tag{5}$$

The critical region (5) is most powerful with respect to the alternative H_2, but it cannot be used for our purposes since the parameters $\mu_1, \cdots, \mu_p, \nu_1, \cdots, \nu_p$ and $\| \sigma_{ij} \|$ are unknown. The optimum estimate of σ_{ij} on the basis of the observations $x_{i\alpha}$ and $y_{i\beta}$ is given by the sample covariance

$$s_{ij} = \frac{\sum_{\alpha=1}^{N_1} (x_{i\alpha} - \bar{x}_i)(x_{j\alpha} - \bar{x}_j) + \sum_{\beta=1}^{N_2} (y_{i\beta} - \bar{y}_i)(y_{j\beta} - \bar{y}_j)}{N_1 + N_2 - 2} \tag{6}$$

where $\bar{x}_i = \frac{\sum_\alpha x_{i\alpha}}{N_1}$ and $\bar{y}_i = \frac{\sum_\beta y_{i\beta}}{N_2}$. The optimum estimates of μ_i and ν_i are given by $\bar{x}_i$ and $\bar{y}_i$ respectively ($i = 1, \cdots, p$). Hence for testing H_1 it seems reasonable to use the statistic R which we obtain from the left hand side of (5) by substituting the optimum estimates for the unknown parameters. Thus R is given by

$$R = \sum_j \sum_i s^{ij}[(z_i - \bar{x}_i)(z_j - \bar{x}_j) - (z_i - \bar{y}_i)(z_j - \bar{y}_j)], \tag{7}$$

where $\| s^{ij} \| = \| s_{ij} \|^{-1}$. The critical region for testing H_1 is given by the inequality

$$R \geq C, \tag{8}$$

where C is a constant determined in such a way that the critical region should have the required size. It is interesting to notice that R is proportional to the difference $T_1^2 - T_2^2$ where T_i ($i = 1, 2$) denotes the generalized Student's ratio[3] for testing the hypothesis that the set $(z_1, \cdots, z_p)$ is drawn from the population π_i. In our case the statistic T_1 cannot be used for testing H_1, since T_1 is appropriate for this purpose if the class of alternative hypotheses contains *all* p-variate normal populations having the same covariance matrix as π_1. In our case the class of alternatives consists merely of a single alternative, namely, the alternative π_2.

For the sake of certain simplifications we shall propose the use of a statistic U which differs slightly from the statistic R. In order to obtain U, we consider the inequality (5). Since $\sigma^{ij} = \sigma^{ji}$ this inequality can be reduced to

$$\sum_j \sum_i \sigma^{ij} z_i(\nu_j - \mu_j) \geq k', \tag{9}$$

where k' denotes a certain constant. The statistic U is obtained from the left hand side of (9) by substituting the optimum estimates for the unknown para-

[3] See. in this connection H. Hotelling, "The generalization of Student's ratio," *Annals of Math. Stat.*, Vol. 2, and R. C. Bose and S. N. Roy, "The exact distribution of the Studentized D^2 statistic," *Sankhya*, Vol. 3.

meters. Thus

$$U = \Sigma\Sigma s^{ij} z_i(\bar{y}_j - \bar{x}_j), \tag{10}$$

and the critical region is given by the inequality

$$U \geq d, \tag{11}$$

where the constant d is chosen so that the critical region should have the required size. The statistic U differs from R merely by a term which does not depend on the quantities $z_1, \cdots, z_p$. If N_1 and N_2 are large the difference $U - R$ is practically constant and therefore the critical regions (8) and (11) are identical. The use of U seems to be as justifiable as that of R and because of certain simplifications we propose the use of the critical region (11).

The statistic U is closely connected with the so called discriminant function[4] introduced by R. A. Fisher for discriminating between the two populations π_1 and π_2. The discriminant function D is given by

$$D = b_1 d_1 + b_2 d_2 + \cdots + b_p d_p \tag{12}$$

where $d_i = \bar{y}_i - \bar{x}_i$ and the coefficient b_i is proportional to $\sum_{j=1}^{p} s^{ij} d_j$. The coefficients $b_1, \cdots, b_p$ are called the coefficients of the discriminant function. We see that U is proportional to the statistic $\sum_{i=1}^{p} b_i z_i$ which is obtained from the right hand side of (12) by substituting z_i for d_i.

4. Solution of the problem when N_1 and N_2 are large. Denote by $F(U, N_1, N_2 \mid \pi_i)$ the cumulative probability distribution of U under the hypothesis that the set $(z_1, \cdots, z_p)$ has been drawn from the population π_i $(i = 1, 2)$. If N_1 and N_2 approach infinity the distribution $F(U, N_1, N_2 \mid \pi_i)$ converges to a normal distribution, since the variates s_{ij}, $\bar{x}_i$ and $\bar{y}_i$ converge stochastically to the constants σ_{ij}, μ_i and ν_i respectively $(i, j = 1, \cdots, p)$. Let us denote $\lim_{N_1 = N_2 = \infty} F(U, N_1, N_2 \mid \pi_i)$ by $\Phi(U \mid \pi_i)$ $(i = 1, 2)$. Furthermore denote by α_i the mean value, and by σ_i the standard deviation of the distribution $\Phi(U \mid \pi_i)$ $(i = 1, 2)$. It is obvious that $\sigma_1 = \sigma_2 = \sigma$ (say). It is easy to verify that the variates

$$\bar{\alpha}_1 = \Sigma\Sigma s^{ij} \bar{x}_i(\bar{y}_j - \bar{x}_j), \tag{13}$$

$$\bar{\alpha}_2 = \Sigma\Sigma s^{ij} \bar{y}_i(\bar{y}_j - \bar{x}_j), \tag{14}$$

$$\begin{aligned} \bar{\sigma}^2 &= \sum_{i=1}^{p} \sum_{j=1}^{p} \sum_{k=1}^{p} \sum_{l=1}^{p} s^{ik} s^{jl} (\bar{y}_k - \bar{x}_k)(\bar{y}_l - \bar{x}_l) s_{ij} \\ &= \sum_{k=1}^{p} \sum_{l=1}^{p} s^{kl} (\bar{y}_k - \bar{x}_k)(\bar{y}_l - \bar{x}_l), \end{aligned} \tag{15}$$

converge stochastically to the constants α_1, α_2 and σ^2 respectively.

---

[4] R. A. Fisher, "The statistical utilization of multiple measurements," *Annals of Eugenics*, 1938.

Hence for large values of N_1 and N_2 we can assume that U is normally distributed with mean value $\bar{\alpha}_i$ and standard deviation $\bar{\sigma}$ if the hypothesis H_i ($i =$ 1, 2) is true. Thus the critical region for testing H_1 is given by the inequality

$$U \geq \bar{\alpha}_1 + \lambda\bar{\sigma}, \tag{16}$$

where the constant λ is chosen in such a way that $\frac{1}{\sqrt{2\pi}} \int_\lambda^\infty e^{-t^2/2}\, dt$ is equal to the required size of the critical region.

Finally, some remarks about the proper choice of the size of the critical region may be of interest. Two kinds of error may be committed. H_1 may be rejected when it is true, and H_1 may be accepted when H_2 is true. Suppose that W_1 and W_2 are two positive numbers expressing the importance of an error of the first kind and an error of the second kind respectively. If the purpose of the statistical investigation is given it will usually be possible to determine the values of W_1 and W_2. We shall deal here with the question of determining the size of the critical region as a function of the weights W_1 and W_2. Denote by P_i the probability that (16) holds under the assumption that H_i is true ($i = 1, 2$). Then P_1 is the size of the critical region (also the probability of an error of the first kind), and $1 - P_2$ is the probability of an error of the second kind. Both probabilities P_1 and P_2 are functions of λ and are given by the following expressions:

$$P_1 = \frac{1}{\sqrt{2\pi}} \int_\lambda^\infty e^{-t^2/2}\, dt, \tag{17}$$

and

$$P_2 = \frac{1}{\sqrt{2\pi}} \int_{((\bar{\alpha}_1 - \bar{\alpha}_2)/\bar{\sigma})+\lambda}^\infty e^{-t^2/2}\, dt. \tag{18}$$

From (13) and (14) we obtain

$$\bar{\alpha}_2 - \bar{\alpha}_1 = \sum_j \sum_i s^{ij}(\bar{y}_i - \bar{x}_i)(\bar{y}_j - \bar{x}_j). \tag{19}$$

Since the right hand side of (19) is positive definite, we have $\bar{\alpha}_2 > \bar{\alpha}_1$. Hence because of (17) and (18) we also have $P_2 > P_1$. By the risk of committing a certain error we understand the probability of that error multiplied by its weight. Hence the risk of committing an error of the first kind is given by W_1P_1, and the risk of committing an error of the second kind is given by $W_2(1 - P_2)$. It seems reasonable to choose the value of λ so that the two risks become equal to each other, i.e. such that

$$W_1P_1 = W_2(1 - P_2). \tag{20}$$

Hence using (17) and (18) we obtain the following equation in λ

$$W_1 \frac{1}{\sqrt{2\pi}} \int_\lambda^\infty e^{-t^2/2}\, dt - W_2 \frac{1}{\sqrt{2\pi}} \int_{-\infty}^{((\bar{\alpha}_1 - \bar{\alpha}_2)/\bar{\sigma})+\lambda} e^{-t^2/2}\, dt = 0. \tag{21}$$

Using a table of the normal distribution, the value of λ which satisfies the equation (21) can easily be found. For $W_1 = W_2$ the solution of (21) is given by

$$\lambda = \frac{\bar{\alpha}_2 - \bar{\alpha}_1}{2\bar{\sigma}},$$

and the critical region is given by the inequality

$$U \geq \bar{\alpha}_1 + \lambda\bar{\sigma} = \bar{\alpha}_1 + \frac{\bar{\alpha}_2 - \bar{\alpha}_1}{2} = \frac{\bar{\alpha}_1 + \bar{\alpha}_2}{2}.$$

5. Some results concerning the exact sampling distribution of the statistic U. If N_1 and N_2 are not large the solution given in section 4 cannot be used and it is necessary to derive the exact sampling distribution of U. Let

$$(\bar{y}_i - \bar{x}_i)\sqrt{\frac{N_1 N_2}{N_1 + N_2}} = z_i' \qquad (i = 1, \cdots, p). \tag{22}$$

Then

$$U = \sqrt{\frac{N_1 + N_2}{N_1 N_2}} \sum_i \sum_j s^{ij} z_i z_j' \tag{23}$$

where the variates $z_1', \cdots, z_p'$ are distributed independently of the set $(z_1, \cdots, z_p)$, the mean value of z_i' is equal to $(\nu_i - \mu_i)\sqrt{\frac{N_1 N_2}{N_1 + N_2}}$ and the covariance between z_i' and z_j' is equal to σ_{ij}. It is known that the set of covariances s_{ij} is distributed independently of the set $(z_1, \cdots, z_p, z_1', \cdots, z_p')$ and therefore the distribution of U remains unchanged if instead of (6) we have

$$s_{ij} = \frac{\sum_{\alpha=1}^{n} t_{i\alpha}^2}{n} \qquad (n = N_1 + N_2 - 2), \tag{24}$$

where the variates $t_{i\alpha}$ are distributed independently of the set $(z_1, \cdots, z_p, z_1', \cdots, z_p')$, have a joint normal distribution with mean values zero, $\sigma_{t_{i\alpha}t_{j\alpha}} = \sigma_{ij}$ and $\sigma_{t_{i\alpha}t_{j\beta}} = 0$ if $\alpha \neq \beta$. It is necessary to derive the distribution of U under both hypotheses H_1 and H_2. In both cases the mean values of $z_1, \cdots, z_p, z_1', \cdots, z_p'$ are not zero. Instead of U we will consider the statistic

$$U' = \sum_{i=1}^{p} \sum_{j=1}^{p} s^{ij} z_i z_j'$$

which differs from U only in the proportionality factor $\sqrt{\frac{N_1 + N_2}{N_1 N_2}}$. The distributions of U' under the hypotheses H_1 and H_2 are contained as special cases in the distribution of the statistic

$$V = \sum_j \sum_i s^{ij} t_{i,n+1} t_{j,n+2}, \tag{25}$$

where s_{ij} is given by (24) and the joint distribution of the variates $t_{i\beta}$ $(i = 1, \cdots, p; \beta = 1, \cdots, n + 2)$ is given by

$$\frac{1}{(2\pi)^{p(n+2)/2}\sigma^{n+2}} e^{-\frac{1}{2}\sum_{j=1}^{p}\sum_{i=1}^{p}\sigma^{ij}\left[\sum_{\alpha=1}^{n} t_{i\alpha}t_{j\alpha}+(t_{i,n+1}-\xi_i)(t_{j,n+1}-\xi_j)+(t_{i,n+2}-\eta_i)(t_{j,n+2}-\eta_j)\right]} \tag{26}$$

$$\times \prod_{\beta=1}^{n+2}\sum_{i=1}^{p} dt_{i\beta}.$$

The quantities $\xi_1, \cdots, \xi_p, \eta_1, \cdots, \eta_p$ are constants and σ^2 denotes the determinant value of the matrix $\| \sigma_{ij} \|$.

We will deal here with the distribution of the statistic V given in (25) under the assumption that the joint distribution of the variates $t_{i\beta}$ $(i = 1, \cdots, p; \beta = 1, \cdots, n + 2)$ is given by (26).

In order to derive the distribution of V we shall have to prove several lemmas.

LEMMA 1. *Let* $\| \lambda_{ij} \|$ $(i, j = 1, \cdots, p)$ *be an arbitrary non-singular matrix, and let*

$$t'_{i\beta} = \sum_{j=1}^{p} \lambda_{ij} t_{j\beta} \qquad (i = 1, \cdots, p; \beta = 1, \cdots, n + 2).$$

Let furthermore s'_{ij} *be given by*

$$s'_{ij} = \frac{\sum_{\alpha=1}^{n} t'_{i\alpha} t'_{j\alpha}}{n}.$$

Then $\sum_j \sum_i s^{ij} t_{i,n+1} t_{j,n+2} = \sum_j \sum_i s'^{ij} t'_{i,n+1} t'_{j,n+2}$, *i.e. the statistic V is invariant under non-singular linear transformations.*

PROOF. We obviously have

$$t'_{i,n+1} t'_{j,n+2} = \sum_{k=1}^{p}\sum_{l=1}^{p} \lambda_{ik}\lambda_{jl} t_{k,n+1} t_{l,n+2}. \tag{27}$$

Furthermore we have

$$s'_{ij} = \sum_{k=1}^{p}\sum_{l=1}^{p} \lambda_{ik}\lambda_{jl} s_{k.}. \tag{28}$$

Hence

$$\| s'_{ij} \| = \| \lambda_{ij} \| \, \| s_{ij} \| \, \| \bar{\lambda}_{ij} \| \tag{29}$$

where $\bar{\lambda}_{ij} = \lambda_{ji}$.

From (29) we obtain

$$\| s'^{ij} \| = \| \bar{\lambda}^{ij} \| \, \| s^{ij} \| \, \| \lambda^{ij} \|, \tag{30}$$

and therefore

$$s'^{ij} = \sum_{k=1}^{p}\sum_{l=1}^{p} \lambda^{ki}\lambda^{lj} s^{kl}. \tag{31}$$

Hence from (27) and (31) we obtain

$$(32) \quad \sum_j \sum_i s'^{ij} t'_{i,n+1} t'_{j,n+2} = \sum_j \sum_i \sum_k \sum_l \sum_u \sum_v \lambda^{ki} \lambda^{lj} s^{kl} \lambda_{iu} \lambda_{jv} t_{u,n+1} t_{v,n+2} .$$

The coefficient of $t_{u,n+1} t_{v,n+2}$ on the right hand side of (32) is given by

$$(33) \quad \sum_j \sum_i \sum_k \sum_l \lambda^{ki} \lambda^{lj} s^{kl} \lambda_{iu} \lambda_{jv} = \sum_k \sum_l \{(\sum_i \lambda^{ki} \lambda_{iu})(\sum_j \lambda^{lj} \lambda_{jv}) s^{kl}\} = s^{uv}.$$

Lemma 1 follows from (32) and (33).

LEMMA 2. *The distribution of* V *remains unchanged if we assume that the covariance matrix* $\| \sigma_{ij} \|$ *is equal to the unit matrix, i.e. the joint distribution of the variates* $t_{i\beta}$ $(i = 1, \cdots, p; \beta = 1, \cdots, n + 2)$ *is given by*

$$(34) \quad \frac{1}{(2\pi)^{p(n+2)/2}} e^{-\frac{1}{2}\left[\sum_{i=1}^{p} \sum_{\alpha=1}^{n} t_{i\alpha}^2 + \sum_i (t_{i,n+1} - \rho_i)^2 + \sum_i (t_{i,n+2} - \zeta_i)^2\right]},$$

where the constants ρ_i *and* ζ_i *are functions of the constants* $\xi_1, \cdots, \xi_p, \eta_1, \cdots, \eta_p$ *and of the* σ_{ij}.

Lemma 2 is an immediate consequence of Lemma 1. Hence we have to derive the distribution of V under the assumption that the variates $t_{i\beta}$ have the joint distribution given in (34).

Let R_i $(i = 1, \cdots, p)$ be the point of the $n + 2$ dimensional Cartesian space with the coordinates $t_{i1}, \cdots, t_{i,n+2}$. Let $P = (u_1, \cdots, u_{n+2})$ and $Q = (v_1, \cdots, v_{n+2})$ be two arbitrary points such that $\sum_{\beta=1}^{n+2} u_\beta v_\beta = 0$ and $\sum u_\beta^2 = \sum v_\beta^2 = 1$. Denote by 0 the origin of the coordinate system and let $\bar{t}_{i,n+1}$ be the projection of the vector $0R_i$ on the vector $0P$. We have

$$(35) \quad \bar{t}_{i,n+1} = \sum_{\beta=1}^{n+2} t_{i\beta} u_\beta \qquad (i = 1, \cdots, p).$$

Similarly, the projection $\bar{t}_{i,n+2}$ of the vector $0R_i$ on $0Q$ is given by

$$(36) \quad \bar{t}_{i,n+2} = \sum_{\beta=1}^{n+2} t_{i\beta} v_\beta .$$

Let $\bar{R}_i$ $(i = 1, \cdots, p)$ be the projection of the point R_i on the n-dimensional hyperplane through 0 and perpendicular to the vectors $0P$ and $0Q$. Denote the coordinates of $\bar{R}_i$ by $r_{i1}, \cdots, r_{i,n+2}$ respectively and let $\bar{s}_{ij}$ be defined by

$$(37) \quad \bar{s}_{ij} = \frac{\sum_{\beta=1}^{n+2} r_{i\beta} r_{j\beta}}{n}.$$

If we rotate the coordinate system so that the $(n + 1)$-axis coincides with $0P$ and the $(n + 2)$-axis coincides with $0Q$, and if $\bar{t}_{i1}, \cdots, \bar{t}_{i,n+2}$ denote the coordinates of R_i $(i = 1, \cdots, p)$ referred to the new system, then we have

$$(38) \quad \bar{s}_{ij} = \frac{1}{n} \sum_{\beta=1}^{n+2} r_{i\beta} r_{j\beta} = \frac{1}{n} \sum_{\alpha=1}^{n} \bar{t}_{i\alpha} \bar{t}_{j\alpha}, \text{ and}$$

$$(39) \quad \sum_{\beta=1}^{n+2} t_{i\beta} t_{j\beta} = \sum_{\beta=1}^{n+2} \bar{t}_{i\beta} \bar{t}_{j\beta} .$$

From (38) and (39) we obtain

$$\text{(40)}\qquad \bar{s}_{ij} = \frac{\sum_{\beta=1}^{n+2} t_{i\beta} t_{j\beta} - \bar{t}_{i,n+1}\bar{t}_{j,n+1} - \bar{t}_{i,n+2}\bar{t}_{j,n+2}}{n}.$$

We will now prove

LEMMA 3. *Let $\bar{V}$ be defined by*

$$\text{(41)}\qquad \bar{V} = \sum_j \sum_i \bar{s}^{ij}\bar{t}_{i,n+1}\bar{t}_{j,n+2},$$

where $\bar{t}_{i,n+1}$, $\bar{t}_{i,n+2}$ and $\bar{s}_{ij}$ are given by the formulas (35), (36) *and* (40) *respectively. Let furthermore the joint probability distribution of the variates $t_{i\beta}$ $(i = 1, \cdots, p;\ \beta = 1, \cdots, n + 2)$ be given by*

$$\text{(42)}\qquad \frac{1}{(2\pi)^{p(n+2)/2}} e^{-\frac{1}{2}\left[\sum_{i=1}^{p}\sum_{\beta=1}^{n+2}(t_{i\beta}-\rho_i u_\beta - \zeta_i v_\beta)^2\right]} \prod_i \prod_\beta dt_{i\beta}.$$

Then the distribution of $\bar{V}$ calculated under the assumption that the quantities $u_1, \cdots, u_{n+2}, v_1, \cdots, v_{n+2}$ are constants and the joint probability distribution of the variates $t_{i\beta}$ is given by (42), *is the same as the distribution of V calculated under the assumption that the joint probability distribution of the variates $t_{i\beta}$ is given by* (34).

PROOF. If we rotate the coordinate system so that the $(n + 1)$-axis coincides with $0P$ and the $(n + 2)$-axis coincides with $0Q$, and if $\bar{t}_{i1}, \cdots, \bar{t}_{i,n+2}$ denote the coordinates of R_i $(i = 1, \cdots, p)$ in the new system, then $\bar{t}_{i,n+1}$ and $\bar{t}_{i,n+2}$ are given by the right hand sides of (35) and (36) respectively. Furthermore

$$\bar{s}_{ij} = \frac{\sum_{\alpha=1}^{n} \bar{t}_{i\alpha}\bar{t}_{j\alpha}}{n}.$$

Hence the distribution of $\bar{V}$ is certainly the same as that of V if the joint probability distribution of the variates $\bar{t}_{i\beta}$ $(i = 1, \cdots, p;\ \beta = 1, \cdots, n + 2)$ is given by the expression which we obtain from (34) by substituting $\bar{t}_{i\beta}$ for $t_{i\beta}$. Thus, in order to prove Lemma 3 we have merely to show that if the variates $\bar{t}_{i\beta}$ have the joint probability distribution (34), the variates $t_{i\beta}$ have the joint probability distribution (42). Since the variates $t_{i1}, \cdots, t_{i,n+2}$ are obtained by an orthogonal transformation of the variates $\bar{t}_{i1}, \cdots, \bar{t}_{i,n+2}$, it follows that the variates $t_{i\beta}$ $(i = 1, \cdots, p;\ \beta = 1, \cdots, n + 2)$ are independently and normally distributed with unit variances. We have

$$\text{(43)}\qquad t_{i\beta} = \sum_{\gamma=1}^{n+2} \lambda_{\beta\gamma}\bar{t}_{i\gamma}$$

where $\lambda_{\beta\gamma}$ is equal to the cosine of the angle between the β-th axis of the original system and γ-th axis of the new system. Since

$$\lambda_{\beta,n+1} = u_\beta \quad \text{and} \quad \lambda_{\beta,n+2} = v_\beta,$$

and since $E(\bar{t}_{i\gamma}) = 0$ for $\gamma = 1, \cdots, n$, $E(\bar{t}_{i,n+1}) = \rho_i$ and $E(\bar{t}_{i,n+2}) = \zeta_i$, it follows from (43) that

$$(44) \qquad E(t_{i\beta}) = \rho_i u_\beta + \zeta_i v_\beta .$$

Hence Lemma 3 is proved.

We will now prove

LEMMA 4. *Let P be a point with the coordinates $u_1, \cdots, u_{n+2}$ and Q a point with the coordinates $v_1, \cdots, v_{n+2}$ such that $\Sigma u_\beta v_\beta = 0$ and $\Sigma u_\beta^2 = \Sigma v_\beta^2 = 1$. Denote by L_p the flat space determined by the vectors $0R_1, \cdots, 0R_p$ ($R_i = (t_{i1}, \cdots, t_{i,n+2})$) and let $\bar{P}$ be the projection of P on L_p and $\bar{Q}$ the projection of Q on L_p. Denote furthermore by θ_1 the angle between the vectors $0P$ and $0\bar{P}$, by θ_1' the angle between $0P$ and $0\bar{Q}$, by θ_2 the angle between $0Q$ and $0\bar{Q}$, by θ_2' the angle between $0Q$ and $0\bar{P}$, and finally by θ_3 the angle between $0\bar{P}$ and $0\bar{Q}$. Then the statistic $\bar{V}$ defined in (41) is equal to*

$$(45) \qquad \bar{V} = - \frac{\begin{vmatrix} 0 & a_1 & a_2 \\ b_1 & a_{11} & a_{12} \\ b_2 & a_{12} & a_{22} \end{vmatrix}}{\begin{vmatrix} a_{11} & a_{12} \\ a_{12} & a_{22} \end{vmatrix}},$$

where

$$(46) \quad a_1 = \cos^2 \theta_1 ; \quad a_2 = \cos \theta_1' \cos \theta_2 ; \quad b_1 = \cos \theta_1 \cos \theta_2' ; \quad b_2 = \cos^2 \theta_2 ;$$

$$(47) \quad a_{11} = \frac{\cos^2 \theta_1 - a_1^2 - b_1^2}{n}, \qquad a_{22} = \frac{\cos^2 \theta_2 - a_2^2 - b_2^2}{n}$$

$$\text{and} \quad a_{12} = \frac{\cos \theta_1 \cos \theta_2 \cos \theta_3 - a_1 a_2 - b_1 b_2}{n}.$$

PROOF. If we rotate the coordinate system in such a way that the $(n + 1)$-axis coincides with $0P$ and the $(n + 2)$-axis coincides with $0Q$, and if $\bar{t}_{i1}, \cdots, \bar{t}_{i,n+2}$ are the coordinates of R_i in the new system, then

$$\bar{s}_{ij} = \frac{\sum_{\alpha=1}^{n} \bar{t}_{i\alpha} \bar{t}_{j\alpha}}{n} .$$

According to Lemma 1 the statistic V is invariant under linear transformations of the variables $t_{i\beta}$. Hence $\bar{V}$ is also invariant under linear transformations of the variables $\bar{t}_{i\beta}$. Thus the value of $\bar{V}$ remains unchanged if the points $R_1, \cdots, R_p$ are replaced by arbitrary points $R_1', \cdots, R_p'$ of L_p subject to the condition that the vectors $0R_1', \cdots, 0R_p'$ be linearly independent. Hence we may assume that the vectors $0R_3, \cdots, 0R_p$ are perpendicular to each other and lie in the intersection of L_p with the n-dimensional flat space which goes through 0 and is perpendicular to $0P$ and $0Q$. Furthermore we may assume that $R_1 = \bar{P}$ and $R_2 = \bar{Q}$. Then $0R_i$ is perpendicular to $0P$, $0Q$, $0R_1$ and $0R_2$ ($i = 3, \cdots, p$).

The statistic $\bar{V}$ can obviously be written in the form:

$$\bar{V} = -\frac{\begin{vmatrix} 0 & \bar{t}_{1,n+1} & \cdots & \bar{t}_{p,n+1} \\ \bar{t}_{1,n+2} & \bar{s}_{11} & \cdots & \bar{s}_{1p} \\ \vdots & \vdots & & \vdots \\ \bar{t}_{p,n+2} & \bar{s}_{p1} & \cdots & \bar{s}_{pp} \end{vmatrix}}{\begin{vmatrix} \bar{s}_{11} & \cdots & \bar{s}_{1p} \\ \vdots & & \vdots \\ \bar{s}_{p1} & \cdots & \bar{s}_{pp} \end{vmatrix}}. \tag{48}$$

Because of our choice of the points $R_1, \cdots, R_p$, we have

$$\bar{t}_{i,n+1} = \bar{t}_{i,n+2} = 0 \qquad (i = 3, \cdots, p) \tag{49}$$

and

$$\sum_{\beta=1}^{n+2} \bar{t}_{i\beta}\,\bar{t}_{j\beta} = 0 \text{ if } i \neq j \qquad (i = 3, \cdots, p; j = 1, \cdots, p). \tag{50}$$

From (49) and (50) it follows that $\bar{s}_{ij} = 0$ for $i \neq j$ except $\bar{s}_{12}$ which is not necessarily zero. Hence $\bar{V}$ reduces to the expression

$$\bar{V} = -\frac{\begin{vmatrix} 0 & \bar{t}_{1,n+1} & \bar{t}_{2,n+1} \\ \bar{t}_{1,n+2} & \bar{s}_{11} & \bar{s}_{12} \\ \bar{t}_{2,n+2} & \bar{s}_{12} & \bar{s}_{22} \end{vmatrix}}{\begin{vmatrix} \bar{s}_{11} & \bar{s}_{12} \\ \bar{s}_{12} & \bar{s}_{22} \end{vmatrix}}. \tag{51}$$

We obviously have $\bar{t}_{1,n+1} = a_1$, $\bar{t}_{2,n+1} = a_2$, $\bar{t}_{1,n+2} = b_1$ and $\bar{t}_{2,n+2} = b_2$.

For any two points A and B denote the length of the vector AB by $\overline{AB}$. Since $n\bar{s}_{11} + (\bar{t}_{1,n+1})^2 + (\bar{t}_{1,n+2})^2 = \overline{0P}^2$, $n\bar{s}_{22} + (\bar{t}_{2,n+1})^2 + (\bar{t}_{2,n+2})^2 = \overline{0Q}^2$ and $n\bar{s}_{12} + \bar{t}_{1,n+1}\bar{t}_{2,n+1} + \bar{t}_{1,n+2}\bar{t}_{2,n+2} = \overline{0P}\cdot\overline{0Q}\cdot\cos\theta_3$, we can easily verify that $\bar{s}_{11} = a_{11}$, $\bar{s}_{12} = a_{12}$ and $\bar{s}_{22} = a_{22}$. Hence Lemma 4 is proved.

The angles θ_1' and θ_2' can be expressed in terms of the angles θ_1, θ_2 and θ_3. In order to show this, let us rotate the coordinate system so that the first p coordinates lie in the flat space L_p defined in Lemma 4. Let $u_1', \cdots, u_{n+2}'$ be the coordinates of P and $v_1', \cdots, v_{n+2}'$ the coordinates of Q referred to the new axes. Then, since $\overline{0P} = \overline{0Q} = 1$, we have

$$\cos\theta_1 = \sqrt{u_1'^2 + \cdots + u_p'^2}; \qquad \cos\theta_1' = \frac{u_1'v_1' + \cdots + u_p'v_p'}{\sqrt{v_1'^2 + \cdots + v_p'^2}};$$

$$\cos\theta_2 = \sqrt{v_1'^2 + \cdots + v_p'^2}; \qquad \cos\theta_2' = \frac{u_1'v_1' + \cdots + u_p'v_p'}{\sqrt{u_1'^2 + \cdots + u_p'^2}};$$

and

$$\cos\theta_3 = \frac{u_1'v_1' + \cdots + u_p'v_p'}{\sqrt{u_1'^2 + \cdots + u_p'^2}\sqrt{v_1'^2 + \cdots + v_p'^2}}.$$

Hence

$$\cos\theta_1' = \cos\theta_1\cos\theta_3 \quad\text{and}\quad \cos\theta_2' = \cos\theta_2\cos\theta_3.$$

Introducing the notations

$$m_1 = \cos^2 \theta_1, \quad m_2 = \cos^2 \theta_2 \quad \text{and} \quad m_3 = \cos \theta_1 \cos \theta_2 \cos \theta_3,$$

we have

$$a_1 = m_1, \qquad a_2 = m_3, \qquad b_1 = m_3, \qquad b_2 = m_2;$$

$$\begin{cases} a_{11} = \dfrac{m_1 - m_1^2 - m_3^2}{n}, & a_{12} = \dfrac{m_3(1 - m_1 - m_2)}{n} \\ \text{and} \qquad a_{22} = \dfrac{m_2 - m_2^2 - m_3^2}{n} \end{cases}$$

Substituting the above values in (45) we obtain

$$\begin{aligned} \bar{V} &= -n \frac{m_3}{m_3^2 - 1 + m_1 + m_2 - m_1 m_2} \\ &= -n \frac{\cos \theta_1 \cos \theta_2 \cos \theta_3}{\cos^2 \theta_1 \cos^2 \theta_2 \cos^2 \theta_3 - \sin^2 \theta_1 \sin^2 \theta_2}. \end{aligned}$$

Hence, Lemma 4 can be written as

LEMMA 4′. *Let P be a point with the coordinates $u_1, \cdots, u_{n+2}$ and Q a point with the coordinates $v_1, \cdots, v_{n+2}$. Denote by L_p the flat space determined by the vectors $0R_1, \cdots, 0R_p$ and let $\bar{P}$ be the projection of P on L_p and $\bar{Q}$ the projection of Q on L_p. Denote furthermore by θ_1 the angle between $0P$ and $0\bar{P}$, by θ_2 the angle between $0Q$ and $0\bar{Q}$ and by θ_3 the angle between $0\bar{P}$ and $0\bar{Q}$. Then the statistic $\bar{V}$ defined in* (41) *is equal to*

$$\bar{V} = -n \frac{\cos \theta_1 \cos \theta_2 \cos \theta_3}{\cos^2 \theta_1 \cos^2 \theta_2 \cos^2 \theta_3 - \sin^2 \theta_1 \sin^2 \theta_2}. \tag{45′}$$

If P is a point of the $(n + 1)$-axis and Q a point of the $(n + 2)$-axis, then $\bar{V}$ is identical with the statistic V given in (25). Hence we obtain the following

Geometric interpretation of the statistic V defined in (25). If θ_1 denotes the angle between the $(n + 1)$-axis and the flat space L_p determined by the vectors $0R_1, \cdots, 0R_p$, θ_2 the angle between the $(n + 2)$-axis and the flat space L_p, and if θ_3 denotes the angle between the projections of the last two coordinate axes on L_p, then the statistic V is equal to the right hand side of (45′).

Denote by S the $2n + 1$-dimensional surface in the $2n + 4$-dimensional space of the variables $u_1, \cdots, u_{n+2}, v_1, \cdots, v_{n+2}$ defined by the following equations

$$\sum_{\beta=1}^{n+2} u_\beta^2 = \sum_{\beta=1}^{n+2} v_\beta^2 = 1; \qquad \sum_{\beta=1}^{n+2} u_\beta v_\beta = 0. \tag{52}$$

denote by C the $2n + 1$-dimensional volume of the surface S, i.e.

$$C = \int_S dS. \tag{53}$$

Now we will assume that $u_1, \cdots, u_{n+2}, v_1, \cdots, v_{n+2}$ are random variables and the joint probability distribution function is defined as follows: the point $(u_1, \cdots, u_{n+2}, v_1, \cdots, v_{n+2})$ is restricted to points of S and the probability density function of S is defined by

$$\frac{dS}{C}. \tag{54}$$

Hence for any subset A of S the probability of A is equal to the $2n + 1$-dimensional volume of A divided by the $2n + 1$-dimensional volume of S. It should be remarked that the probability density function (54) is identical with the probability density function we would obtain if we were to assume that $u_1, \cdots, u_{n+2}, v_1, \cdots, v_{n+2}$ are independently, normally distributed with zero means and unit variances and calculate the conditional density function under the restriction that $(u_1, \cdots, u_{n+2}, v_1, \cdots, v_{n+2})$ is a point of S.

LEMMA 5. *The probability distribution of $\bar{V}$ defined in* (41), *calculated under the assumption that the joint probability density of the variables $u_1, \cdots, u_{n+2}$, $v_1, \cdots, v_{n+2}$ $t_{i\beta}$ $(i = 1, \cdots, p; \beta = 1, \cdots, n + 2)$ is given by the product of* (54) *and* (42), *is the same as the distribution of the statistic V calculated under the assumption that the variables $t_{i\beta}$ have the joint probability density function given in* (34).

Lemma 5 is an immediate consequence of lemma 3.

LEMMA 6. *Let L_p be an arbitrary p-dimensional flat space in the $n + 2$ dimensional Cartesian space, and let M_p be the flat space determined by the first p coordinate axes. Assuming that the joint probability density function of u_β, v_β, $t_{i\beta}$ $(i = 1, \cdots, p; \beta = 1, \cdots, n + 2)$ is given by the product of* (54) *and* (42), *the conditional distribution of $\bar{V}$ calculated under the restriction that the points $R_1, \cdots, R_p$ lie in L_p, is the same as the conditional distribution of $\bar{V}$ calculated under the restriction that the points $R_1, \cdots, R_p$ lie in M_p. The point R_i denotes the point with the coordinates $t_{i1}, \cdots, t_{i,n+2}$.*

PROOF. Let P be the point with the coordinates $u_1, \cdots, u_{n+2}$ and let Q be the point with the coordinates $v_1, \cdots, v_{n+2}$. Let us rotate the coordinate system so that the first p axes lie in the flat space L_p. Denote the coordinates of P in the new system by $u_1', \cdots, u_{n+2}'$, those of Q by $v_1', \cdots, v_{n+2}'$, and those of R_i by $t_{i1}', \cdots, t_{i,n+2}'$ $(i = 1, \cdots, p)$. Let S' be the surface defined by

$$\Sigma u_\beta'^2 = \Sigma v_\beta'^2 = 1 \quad \text{and} \quad \Sigma u_\beta' v_\beta' = 0. \tag{55}$$

It is clear that the surface S' is identical with the surface S defined in (52). It is furthermore clear that if the joint density function of $u_1, \cdots, u_{n+2}, v_1, \cdots$, v_{n+2} is given by $\frac{dS}{C}$, the joint density function of $u_1', \cdots, u_{n+2}', v_1', \cdots, v_{n+2}'$ is the same, i.e. it is given by $\frac{dS'}{C}$. It can readily be seen that for any given set of values $u_1', \cdots, u_{n+2}', v_1', \cdots, v_{n+2}'$ the conditional joint probability density of the variates $t_{i\beta}'$ is given by the function obtained from (42) by substituting

$t'_{i\beta}$ for $t_{i\beta}$, u'_β for u_β and v'_β for v_β, provided that for any given set of values $u_1, \cdots, u_{n+2}, v_1, \cdots, v_{n+2}$ the joint conditional distribution of the variates $t_{i\beta}$ is given by (42). Hence, if the joint distribution of $u_1, \cdots, u_{n+2}$, $v_1, \cdots, v_{n+2}$ and $t_{i\beta}$ $(i = 1, \cdots, p; \beta = 1, \cdots, n+2)$ is given by the product of (54) and (42), the joint probability density function of the variates u'_β, v'_β, $t'_{i\beta}$ $(i = 1, \cdots, p; \beta = 1, \cdots, n+2)$ is obtained from that of u_β, v_β, $t_{i\beta}$ by substituting S' for S and $t'_{i\beta}$ for $t_{i\beta}$.

According to Lemma 4′, $\bar{V}$ can be expressed as a function of the angles θ_1, θ_2 and θ_3 defined in Lemma 4′. Each angle θ_k $(k = 1, 2, 3)$ can be expressed as a function of the variables $t_{i\beta}$, u_β, v_β. It is obvious that the value of θ_k remains unchanged if we substitute $t'_{i\beta}$ for $t_{i\beta}$, u'_β for u_β and v'_β for v_β. Hence also the value of $\bar{V}$ remains unchanged if we substitute $t'_{i\beta}$ for $t_{i\beta}$, u'_β for u_β and v'_β for v_β. Lemma 6 is a consequence of this fact and of the fact that the joint probability density of the variates $t'_{i\beta}$, u'_β and v'_β is identical with that of the variates $t_{i\beta}$, u_β and v_β.

Lemma 7. *Assuming that the joint probability distribution of the variates* $u_\beta, v_\beta, t_{i\beta}$ $(i = 1, \cdots, p; \beta = 1, \cdots, n+2)$ *is given by the product of* (54) *and* (42), *the conditional joint probability distribution of* $u_1, \cdots, u_{n+2}, v_1, \cdots, v_{n+2}$, *calculated under the restriction that the points* $R_i = (t_{i1}, \cdots, t_{i,n+2})$ $(i = 1, \cdots, p)$ *lie in the flat space determined by the first p coordinate axes, is given by*

$$\frac{e^{-\frac{1}{2}\sum_{\gamma=p+1}^{n+2}\sum_{i=1}^{p}(\rho_i u_\gamma+\zeta_i v_\gamma)^2} f(u_1, \cdots, u_{n+2}, v_1, \cdots, v_{n+2})\, dS}{\int_S e^{-\frac{1}{2}\sum_{\gamma=p+1}^{n+2}\sum_{i=1}^{p}(\rho_i u_\gamma+\zeta_i v_\gamma)^2} f(u_1, \cdots, u_{n+2}, v_1, \cdots, v_{n+2})\, dS}, \tag{56}$$

where S denotes the surface defined in (52), *and* $f(u_1, \cdots, u_{n+2}, v_1, \cdots, v_{n+2})$ *denotes the expected value of*

$$\left\|\begin{matrix} r_{11} & \cdots & r_{1p} \\ r_{21} & \cdots & r_{2p} \\ \vdots & & \vdots \\ r_{p1} & \cdots & r_{pp} \end{matrix}\right\|^{\frac{n+2-p}{2}} \qquad \left(r_{ij} = \sum_{\alpha=1}^{p} t_{i\alpha} t_{j\alpha}\right) \tag{57}$$

calculated under the assumption that the joint distribution of the variates $t_{i\beta}$ is given by (42).

Proof. Denote by $\bar{R}_i$ the projection of R_i on the flat space determined by the first p coordinate axes, i.e. $\bar{R}_i = (t_{i1}, \cdots, t_{ip}, 0, \cdots, 0)$. Let l_1 be the length of $\bar{R}_1$, and let l_i be the distance of $\bar{R}_i$ from the flat space determined by the vectors $0\bar{R}_1, \cdots, 0\bar{R}_{i-1}$ $(i = 2, \cdots, p)$. Then, as is known,

$$l_1 l_2 \cdots l_i = \sqrt{\begin{vmatrix} r_{11} & \cdots & r_{1i} \\ r_{21} & \cdots & r_{2i} \\ \cdot & \cdots & \cdot \\ r_{i1} & \cdots & r_{ii} \end{vmatrix}} \qquad (i = 1, \cdots, p), \tag{58}$$

where $r_{kl} = \sum_{\alpha=1}^{p} t_{k\alpha} t_{l\alpha}$.

We introduce the new variables

$$t^*_{i\gamma} = \frac{t_{i\gamma}}{l_i} \qquad (i = 1, \cdots, p;\ \gamma = p + 1, \cdots, n + 2). \tag{59}$$

Then the joint probability density function of the variates u_β, v_β, $t_{i\alpha}$, $t^*_{i\gamma}$ $(i = 1, \cdots, p;\ \beta = 1, \cdots, n + 2,\ \alpha = 1, \cdots, p,\ \gamma = p + 1, \cdots, n + 2)$ is given by

$$\frac{(l_1 \cdots l_p)^{n+2-p}}{C(2\pi)^{p(n+2)/2}} e^{-\frac{1}{2}\left[\sum_{i=1}^{p}\sum_{\alpha=1}^{p}(t_{i\alpha}-\rho_i u_\alpha-\zeta_i v_\alpha)^2+\sum_{i=1}^{p}\sum_{\gamma=p+1}^{n+2}(l_i t^*_{i\gamma}-\rho_i u_\gamma-\zeta_i v_\gamma)^2\right]} \times \left(\prod_i \prod_\alpha dt_{i\alpha}\right)\left(\prod_i \prod_\gamma dt^*_{i\gamma}\right) dS. \tag{60}$$

Substituting zero for $t^*_{i\gamma}$ $(i = 1, \cdots, p,\ \gamma = p + 1, \cdots, n + 2)$ in (60), we obtain an expression which is proportional to the conditional joint probability density of the variates u_β, v_β, $t_{i\alpha}$ $(\beta = 1, \cdots, n + 2;\ i = 1, \cdots, p,\ \alpha = 1, \cdots, p)$, calculated under the restriction that the points R_i $(i = 1, \cdots, p)$ fall in the flat space determined by the first p coordinate axes. Hence this conditional density function is given by

$$A e^{-\frac{1}{2}\sum_{\gamma=p+1}^{n+2}\sum_{i=1}^{p}(\rho_i u_\gamma+\zeta_i v_\gamma)^2}(l_1 l_2 \cdots l_p)^{n+2-p} \times e^{-\frac{1}{2}\left[\sum_{i=1}^{p}\sum_{\alpha=1}^{p}(t_{i\alpha}-\rho_i u_\alpha-\zeta_i v_\alpha)^2\right]} dS \prod_i \prod_\alpha dt_{i\alpha} \tag{61}$$

where A denotes a constant. The conditional distribution of the variates u_β, v_β $(\beta = 1, \cdots, n + 2)$ is obtained from (61) by integrating it with respect to the variables $t_{i\alpha}$ $(i = 1, \cdots, p;\ \alpha = 1, \cdots, p)$. Because of (58), we see that the resulting formula is identical with (56). Hence Lemma 7 is proved.

LEMMA 8. *Let* $m_1 = u_1^2 + \cdots + u_p^2$; $m_2 = v_1^2 + \cdots + v_p^2$, *and* $m_3 = u_1 v_1 + \cdots + u_p v_p$. *If the joint distribution of the variates* $u_1, \cdots, u_{n+2}$, $v_1, \cdots, v_{n+2}$ *is given by* (54), *then the joint distribution of* m_1, m_2, m_3 *is given by*

$$\frac{B}{\sqrt{m_1 m_2(1 - m_1)(1 - m_2)}} F_p(m_1)F_p(m_2)\Phi_p\left(\frac{m_3}{\sqrt{m_1 m_2}}\right)F_{n+2+p}(1 - m_1) \times F_{n+2-p}(1 - m_2)\Phi_{n+2-p}\left(\frac{-m_3}{\sqrt{(1 - m_1)(1 - m_2)}}\right) dm_1\, dm_2\, dm_3 \tag{62}$$

where B *denotes a constant,*

$$F_k(t) = \frac{1}{2^{k/2}\Gamma\left(\frac{k}{2}\right)}(t)^{(k-2)/2}e^{-\frac{1}{2}t} \quad \text{and} \quad \Phi_k(t) = \frac{\Gamma\left(\frac{k}{2}\right)}{\sqrt{\pi}\,\Gamma\left(\frac{k-1}{2}\right)}(1 - t^2)^{(k-3)/2}. \tag{63}$$

PROOF. Let $m_1' = u_{p+1}^2 + \cdots + u_{n+2}^2$, $m_2' = v_{p+1}^2 + \cdots + v_{n+2}^2$,

$m_3' = u_{p+1}v_{p+1} + \cdots + u_{n+2}v_{n+2}$, $\bar{m}_3 = \dfrac{m_3}{\sqrt{m_1 m_2}}$ and $\bar{m}_3' = \dfrac{m_3'}{\sqrt{m_1' m_2'}}$. First we calculate the joint distribution of m_1, m_2, $\bar{m}_3$ m_1', m_2', $\bar{m}_3'$ under the assumption that $u_1, \cdots, u_{n+2}, v_1, \cdots, v_{n+2}$ are normally independently distributed with zero means and unit variances. This joint distribution is given by

$$\begin{aligned} F_p(m_1)F_p(m_2)\Phi_p(\bar{m}_3)F_{n+2-p}(m_1')F_{n+2-p}(m_2') \\ \times \Phi_{n+2-p}(\bar{m}_3')\, dm_1\, dm_2\, d\bar{m}_3\, dm_1'\, dm_2'\, d\bar{m}_3'. \end{aligned} \tag{64}$$

Hence the joint distribution of m_1, m_2, m_3, m_1', m_2', m_3' is given by

$$\begin{aligned} \frac{1}{\sqrt{m_1 m_2 m_1' m_2'}} F_p(m_1)F_p(m_2)\Phi_p\left(\frac{m_3}{\sqrt{m_1 m_2}}\right) F_{n+2-p}(m_1')F_{n+2-p}(m_2') \\ \times \Phi_{n+2-p}\left(\frac{m_3'}{\sqrt{m_1' m_2'}}\right) dm_1\, dm_2\, dm_3\, dm_1'\, dm_2'\, dm_3'. \end{aligned} \tag{65}$$

The required conditional distribution of m_1, m_2, m_3 is equal to the conditional distribution of m_1, m_2, m_3 obtained from the joint distribution (65) under the restrictions $m_1 + m_1' = 1$, $m_2 + m_2' = 1$ and $m_3 + m_3' = 0$. Hence if in (65) we substitute $1 - m_1$ for m_1', $1 - m_2$ for m_2' and $-m_3$ for m_3' we obtain an expression proportional to the conditional distribution of m_1, m_2, m_3. This proves Lemma 8.

LEMMA 9. *For any point $(u_1, \cdots, u_{n+2}, v_1, \cdots, v_{n+2})$ of the surface S defined in (52) the expected value of (57) (calculated under the assumption that (42) is the joint distribution of $t_{i\beta}$) is a function of m_1 m_2, and m_3 only, where m_1, m_2 and m_3 are defined in Lemma 8.*

PROOF. Let $\| \lambda_{\alpha\beta} \|$ $(\alpha, \beta = 1, \cdots, p)$ be an orthogonal matrix such that

$$\lambda_{1\beta} = \frac{u_\beta}{\sqrt{u_1^2 + \cdots + u_p^2}} \qquad (\beta = 1, \cdots, p) \tag{66}$$

and

$$\lambda_{2\beta} = \frac{u_\beta + \lambda v_\beta}{\sqrt{\sum_{\beta=1}^{p} (u_\beta + \lambda v_\beta)^2}} \qquad (\beta = 1, \cdots, p) \tag{67}$$

where

$$\lambda = \frac{-\sum_1^p u_\beta^2}{\sum_1^p u_\beta v_\beta}.$$

Let

$$t_{i\alpha}' = \sum_{\beta=1}^{p} \lambda_{\alpha\beta} t_{i\beta} \qquad (\alpha = 1, \cdots, p). \tag{68}$$

Then the variates $t'_{i\alpha}$ are independently and normally distributed with unit variances. Since for any point of S, $E(t_{i\alpha}) = \rho_i u_\alpha + \zeta_i v_\alpha$, we have because of (66), (67) and (68)

$$E(t_{i\gamma}) = 0 \qquad (i = 1, \cdots, p, \gamma = 3, 4, \cdots, p),$$

$$E(t_{i1}) = \varphi_{i1}(m_1, m_2, m_3),$$

and $$E(t_{i2}) = \varphi_{i2}(m_1, m_2, m_3).$$

Hence the joint distribution of the variates $t'_{i\alpha}$ $(i = 1, \cdots, p; \alpha = 1, \cdots, p)$ depends merely on m_1, m_2 and m_3. Since $r_{ij} = \sum_{\alpha=1}^{p} t_{i\alpha} t_{j\alpha} = \sum_{\alpha=1}^{p} t'_{i\alpha} t'_{j\alpha}$, the expression (57) can be expressed as a function of the variables $t'_{i\alpha}$. Hence the distribution of the expression (57) depends merely on the parameters m_1, m_2, and m_3. This proves Lemma 9.

The main result of this section is the following

THEOREM. *Let V be the statistic given in* (25) *and let the joint distribution of the variates $t_{i\beta}$ $(i = 1, \cdots, p; \beta = 1, \cdots, n + 2)$ be given by* (34). *Then the probability distribution of V is the same as the distribution of*

$$(69) \qquad -n \frac{m_3}{m_3^2 - (1 - m_1)(1 - m_2)}$$

where the joint distribution of m_1, m_2 and m_3 is equal to a constant multiple of the product of the following three factors: the expression (62), *the exponential $e^{\frac{1}{2}(m_1 \Sigma \rho_i^2 + 2m_3 \Sigma \rho_i \zeta_i + m_2 \Sigma \zeta_i^2)}$ and the expected value of*

$$(70) \qquad \left(\begin{vmatrix} r_{11} & \cdots & r_{1p} \\ \cdot & & \cdot \\ \cdot & & \cdot \\ \cdot & & \cdot \\ r_{p1} & \cdots & r_{pp} \end{vmatrix} \right)^{(n+2-p)/2} \qquad \left(r_{ij} = \sum_{\alpha=1}^{p} t_{i\alpha} t_{j\alpha} \right).$$

The expected value of (70) *is calculated under the assumption that the variates $t_{i\alpha}$ are normally and independently distributed with unit variances and $E(t_{i\alpha}) = \rho_i u_\alpha + \zeta_i v_\alpha$ $(i = 1, \cdots, p; \alpha = 1, \cdots, p)$ where $\sum_{\alpha=1}^{p} u_\alpha^2 = m_1$, $\sum_{\alpha=1}^{p} v_\alpha^2 = m_2$ and $\sum_{\alpha=1}^{p} v_\alpha u_\alpha = m_3$. The domain of the variables m_1, m_2 and m_3 is given by the inequalities: $0 \leq m_1 \leq 1; 0 \leq m_2 \leq 1; -\sqrt{m_1 m_2} \leq m_3 \leq \sqrt{m_1 m_2}$.*

PROOF. First we note that the expected value of (70) is a function of m_1, m_2 and m_3 only. Let P be the point with the coordinates $u_1, \cdots, u_{n+2}$, and Q the point with the coordinates $v_1, \cdots, v_{n+2}$. Assume that the points $R_i = (t_{i1}, \cdots, t_{i,n+2})$ $(i = 1, \cdots, p)$ lie in the flat space determined by the first p coordinate axes. Assume furthermore that $u_1 v_1 + \cdots + u_{n+2} v_{n+2} = 0$ and that the lengths of the vectors $0P$ and $0Q$ are equal to 1. Then

$$\cos \theta_1 = \sqrt{u_1^2 + \cdots + u_p^2}; \qquad \cos \theta_2 = \sqrt{v_1^2 + \cdots + v_p^2}$$

and

$$\cos \theta_3 = \frac{u_1 v_1 + \cdots + u_p v_p}{\sqrt{u_1^2 + \cdots + u_p^2}\,\sqrt{v_1^2 + \cdots + v_p^2}},$$

where θ_1 denotes the angle between $0P$ and the flat space L_p determined by the vectors $0R_1, \cdots, 0R_p$; θ_2 denotes the angle between $0Q$ and L_p, and θ_3 denotes the angle between the projections of $0P$ and $0Q$ on L_p. According to Lemma 4′ the statistic $\bar{V}$ defined in (41) is equal to

$$\bar{V} = -n \frac{\cos\theta_1 \cos\theta_2 \cos\theta_3}{\cos^2\theta_1 \cos^2\theta_2 \cos^2\theta_3 - \sin^2\theta_1 \sin^2\theta_2} \tag{71}$$

$$= -n \frac{m_3}{m_3^2 - (1 - m_1)(1 - m_2)}$$

where

$$\text{(72)} \quad m_1 = \cos^2\theta_1 = u_1^2 + \cdots + u_p^2, \qquad m_2 = \cos^2\theta_2 = v_1^2 + \cdots + v_p^2$$

$$\text{and} \quad m_3 = \cos\theta_1 \cos\theta_2 \cos\theta_3 = u_1v_1 + \cdots + u_pv_p .$$

It follows from Lemmas 5 and 6 that the distribution of V is the same as the conditional distribution of $\bar{V}$ calculated under the assumption that the unconditional joint probability density of the variates u_β, v_β and $t_{i\beta}$ is given by the product of (54) and (42) and under the restriction that the points R_i $(i = 1, \cdots, p)$ fall in the flat space determined by the first p coordinate axes. Since $e^{-\frac{1}{2}\sum_{\gamma=p+1}^{n+2}\sum_{i=1}^{p}(\rho_i u_\gamma + \zeta_i v_\gamma)^2}$ is a constant multiple of

$$e^{\frac{1}{2}(m_1\sum\rho_i^2 + 2m_3\sum\rho_i\zeta_i + m_2\sum\zeta_i^2)} \tag{73}$$

from Lemmas 7, 8 and 9 it follows readily that the joint conditional distribution of $m_1 = u_1^2 + \cdots + u_p^2$, $m_2 = v_1^2 + \cdots + v_p^2$ and $m_3 = u_rv_1 + \cdots + u_pv_p$ is equal to a constant multiple of the product of (62), (73) and the expected value of 70. This proves our theorem.

It can be shown that the variates m_1, m_2 and m_3 are of the order $\frac{1}{n}$ in the probability sense. Hence

$$-n \frac{m_3}{m_3^2 - (1 - m_1)(1 - m_2)} = nm_3(1 + \epsilon) \tag{74}$$

where ϵ is of the order $\frac{1}{n}$. Hence we can say: *even for moderately large n the distribution of the statistic V is well approximated by the distribution of nm_3, where the joint distribution of m_1, m_2 and m_3 is equal to a constant multiple of the product of* (62), (73) *and the expected value of* (70).

If $n + 2 - p$ is an even integer, the expected value of (70) is obviously an elementary function of m_1, m_2 and m_3. Hence, if $n + 2 - p$ is even, the joint distribution of m_1, m_2 and m_3 is also an elementary function of m_1, m_2 and m_3.

If the constants ρ_i and ζ_i $(i = 1, \cdots, p)$ in formula (34) are equal to zero, the expected value of (70) is a constant and the joint distribution of m_1, m_2 and m_3 is given by (62).

Reprinted from The Annals of Mathematical Statistics
Vol. XV, No. 3, September, 1944

NOTE ON A LEMMA

By Abraham Wald

Columbia University

In a previous paper on the power function of the analysis of variance test[1], the author stated the following lemma (designated there as Lemma 2):

Lemma 2. *Let $v_1, \cdots, v_k$ be k normally and independently distributed variates with a common variance σ^2. Denote the mean value of v_i by α_i $(i = 1, \cdots, k)$ and let $f(v_1, \cdots, v_k, \sigma)$ be a function the variables $v_1, \cdots, v_k$ and σ which does not involve the mean values $\alpha_1, \cdots, \alpha_k$. Then, if the expected value of $f(v_1, \cdots, v_k, \sigma)$ is equal to zero, $f(v_1, \cdots, v_k, \sigma)$ is identically equal to zero, except perhaps on a set of measure zero.*

In the paper mentioned above it was intended to state this lemma for bounded functions $f(v_1, \cdots, v_k)$ and the lemma was used there only in a case where $f(v_1, \cdots, v_k)$ is bounded. Through an oversight this restriction on $f(v_1, \cdots, v_k)$ was not stated explicitly.[2] The published proof of Lemma 2 is adequate if $f(v_1, \cdots, v_k)$ is assumed to be bounded. From the fact that the moments of a multivariate normal distribution determine uniquely the distribution it is concluded there that if for any set $(r_1, \cdots, r_k)$ of non-negative integers

$$(1) \qquad \int_{-\infty}^{+\infty} \cdots \int_{-\infty}^{+\infty} v_1^{r_1} \cdots v_k^{r_k} f(v_1, \cdots, v_k) e^{-\frac{1}{2}\Sigma(v_i - \alpha_i)^2} \, dv_1 \cdots dv_k = 0$$

identically in the parameters $\alpha_1, \cdots, \alpha_k$ then $f(v_1, \cdots, v_k)$ must be equal to zero except perhaps on a set of measure zero. This conclusion is obvious if $f(v_1, \cdots, v_k)$ is bounded. In fact, from (1) and the boundedness of $f(v_1, \cdots, v_k)$ it follows that there exists a finite value A such that

$$\varphi(v_1, \cdots, v_k) = \frac{1}{(2\pi)^{k/2}} \left[1 - \frac{1}{A} f(v_1, \cdots, v_k) \right] e^{-\frac{1}{2}\Sigma(v_i - \alpha_i)^2}$$

is a probability density function with moments equal to those of the normal distribution

$$\psi(v_1, \cdots, v_k) = \frac{1}{(2\pi)^{k/2}} e^{-\frac{1}{2}\Sigma(v_i - \alpha_i)^2}.$$

Hence $f(v_1, \cdots, v_k)$ must be equal to zero except perhaps on a set of measure zero. However, this conclusion is not so immediate if no restriction is imposed on $f(v_1, \cdots, v_k)$ except that

$$(2) \qquad \int_{-\infty}^{+\infty} \cdots \int_{-\infty}^{+\infty} | f(v_1, \cdots, v_k) | e^{-\frac{1}{2}\Sigma(v_i - \alpha_i)^2} \, dv_1 \cdots dv_k < \infty$$

for all values of the parameters $\alpha_1, \cdots, \alpha_k$. It is the purpose of this note to prove this. In other words, we shall prove the following proposition:

---

[1] A. Wald, "On the power function of the analysis of variance test," *Annals of Math. Stat.*, Vol. 13 (1942), pp. 434.

[2] I wish to thank Prof. J. Neyman for calling my attention to this omission.

PROPOSITION I. *If* (2) *holds for all values of the parameters* $\alpha_1, \cdots, \alpha_k$ *and if for any set* $(r_1, \cdots, r_k)$ *of non-negative integers equation* (1) *holds identically in* $\alpha_1, \cdots, \alpha_k$, *then* $f(v_1, \cdots, v_k)$ *must be equal to zero except perhaps on a set of measure zero.*

On the basis of Proposition I and the arguments given on p. 438 of the paper mentioned before, it can be seen that restriction (2) on the function $f(v_1, \cdots, v_k)$ is sufficient for the validity of Lemma 2.

To prove Proposition I, we shall first show that the following lemma holds.

LEMMA A. *If* $h(v_1, \cdots, v_k)$ *is a probability density function and if*

$$\int_{\infty}^{+\infty} \cdots \int_{\infty}^{+\infty} h(v_1, \cdots, v_k) e^{\delta \sum_i |v_i|} \, dv_1 \cdots dv_k < \infty \tag{3}$$

for some $\delta > 0$, *then the problem of moments is determined for the moments of the distribution* $h(v_1, \cdots, v_k)$.

This lemma was proved by G. H. Hardy for $k = 1$.[3] I shall prove it for $k > 1$. Since

$$\sum_{n=0}^{\infty} \frac{\delta^{2n} (\sum_i |v_i|)^{2n}}{(2n)!} < e^{\delta \Sigma |v_i|} \tag{4}$$

we obtain from (3)

$$\int_{\infty}^{+\infty} \cdots \int_{\infty}^{+\infty} h(v_1, \cdots, v_k) \left[\sum_{n=0}^{\infty} \frac{\delta^{2n} (\sum |v_i|)^{2n}}{(2n)!} \right] dv_1 \cdots dv_k < \infty. \tag{5}$$

Hence

$$\int_{\infty}^{+\infty} \cdots \int_{\infty}^{+\infty} h(v_1, \cdots, v_k) \left[\sum_{n=0}^{\infty} \frac{\delta^{2n} \left(\sum_{i=1}^{k} v_i^{2n} \right)}{(2n)!} \right] dv_1 \cdots dv_k < \infty. \tag{6}$$

Denote the $2n$th moment of v_i by $\mu_{2n}^{(i)}$. Because of (3) the moments $\mu_{2n}^{(i)}$ are finite. Furthermore, denote $\sum_{i=1}^{k} \mu_{2n}^{(i)}$ by λ_{2n}. Then we obtain from (6)

$$\sum_{n=0}^{\infty} \frac{\delta^{2n} \lambda_{2n}}{(2n)!} < \infty. \tag{7}$$

From (7) it follows that

$$\limsup_{n=\infty} \frac{\delta^{2n} \lambda_{2n}}{(2n)!} < 1. \tag{8}$$

Hence

$$\limsup_{n=\infty} \left(\frac{\delta^{2n} \lambda_{2n}}{(2n)!} \right)^{1/2n} \leq 1. \tag{9}$$

---

[3] See for instance, SHOHAT and TAMARKIN, "The problem of moments," *Math. Surveys No.* 1, *Amer. Math. Soc.*, New York, 1943, p. 20.

Since according to Stirling's formula

$$\lim_{n=\infty} (2n)!/(2n)^{2n}e^{-2n}\sqrt{4\pi n} = 1$$

we obtain from (9)

$$\limsup_{n=\infty} \frac{\delta\lambda_{2n}^{1/2n}}{2ne^{-1}} \leq 1. \tag{10}$$

Taking reciprocals we obtain

$$\liminf_{n=\infty} \frac{2n\lambda_{2n}^{-1/2n}}{e\delta} \geq 1 \tag{11}$$

or

$$\liminf_{n=\infty} n\lambda_{2n}^{-1/2n} \geq \frac{e\delta}{2} > 0. \tag{12}$$

But (12) implies the existence of a positive value ρ so that

$$\lambda_{2n}^{-1/2n} \geq \frac{\rho}{n} \qquad (n = 1, 2, \cdots, \text{ad inf.}) \tag{13}$$

From (13) it follows that

$$\sum_{n=1}^{\infty} \lambda_{2n}^{-1/2n} = \infty. \tag{14}$$

But (14) is Carleman's sufficient condition for the determinateness of the problem of moments. Hence Lemma A is proved.

On the basis of Lemma A we can prove Proposition I as follows: From (2) we obtain

$$\int_{-\infty}^{+\infty} \cdots \int_{-\infty}^{+\infty} |f(v_1, \cdots, v_k)| e^{-\frac{1}{2}\Sigma v_i^2 + \Sigma \alpha_i v_i} dv_1 \cdots dv_k < \infty \tag{15}$$

for all values $\alpha_1, \cdots, \alpha_k$. Let $f_1(v) = f(v)$ for all points $v = (v_1, \cdots, v_k)$ for which $f(v) \geq 0$, and $f_1(v) = 0$ for all points v for which $f(v) < 0$. Similarly, let $f_2(v) = -f(v)$ for all points v for which $f(v) \leq 0$, and $f_2(v) = 0$ for all points v for which $f(v) > 0$. Then $f_1(v)$ and $f_2(v)$ are non-negative functions and

$$f(v) = f_1(v) - f_2(v). \tag{16}$$

From (15) it follows that

$$\int_{-\infty}^{+\infty} \cdots \int_{-\infty}^{+\infty} f_1(v) e^{-\frac{1}{2}\Sigma v_i^2 + \Sigma \alpha_i v_i} dv_1 \cdots dv_k < \infty \tag{17}$$

and

$$\int_{-\infty}^{+\infty} \cdots \int_{-\infty}^{+\infty} f_2(v) e^{-\frac{1}{2}\Sigma v_i^2 + \Sigma \alpha_i v_i} dv_1 \cdots dv_k < \infty. \tag{18}$$

Let

(19) $$f_j^*(v) = f_j(v)e^{-\frac{1}{2}\Sigma v_i^2} \qquad (j = 1, 2).$$

Now we shall show that for any positive values $\beta_1, \cdots, \beta_k$

(20) $$\int_{-\infty}^{+\infty} \cdots \int_{-\infty}^{+\infty} f_j^*(v_1, \cdots, v_k)e^{\beta_1|v_1|+\cdots+\beta_k|v_k|}\, dv_1 \cdots dv_k < \infty.$$

In fact, consider the 2^k sets $(a_1, \cdots, a_k)$ where $a_i = \pm 1$ $(i = 1, \cdots, k)$. Denote by $R_{a_1 \cdots a_k}$ the subset of the k-dimensional Cartesian space which consists of all points $v = (v_1, \cdots, v_k)$ for which v_i is either zero or signum $v_i =$ signum a_i $(i = 1, \cdots, k)$. Putting $\alpha_i = a_i\beta_i$, it follows from (17) and (18) that

(21) $$\int_{R_{a_1 \cdots a_k}} f_j^*(v_1, \cdots, v_k)e^{\beta_1|v_1|+\cdots+\beta_k|v_k|}\, dv_1 \cdots dv_k < \infty.$$

Since (21) holds for any of the 2^k sets $R_{a_1 \cdots a_k}$, equation (20) is proved.

From (1) it follows that

$$\int_{-\infty}^{+\infty} \cdots \int_{-\infty}^{+\infty} v_1^{r_1} \cdots v_k^{r_k}[f_1^*(v_1, \cdots, v_k) - f_2^*(v_1, \cdots, v_k)]\, dv_1 \cdots dv_k = 0,$$

for all non-negative integers $r_1, \cdots, r_k$. Hence, because of (21) and Lemma A we see that

(22) $$f_1^*(v_1, \cdots, v_k) = f_2^*(v_1, \cdots, v_k),$$

except perhaps on a set of measure zero. From (22) it follows that

$$f(v_1, \cdots, v_k) = f_1(v_1, \cdots, v_k) - f_2(v_1, \cdots, v_k) = 0,$$

except perhaps on a set of measure zero. Hence Proposition I is proved.

On a Statistical Generalization of Metric Spaces[1]*

ABRAHAM WALD

In a very interesting paper Karl Menger has recently introduced a statistical generalization of semi-metric and metric spaces.[(2)] According to Menger a set R of elements (points) is called a statistical semi-metric space if with each pair of points p and q of the space R a real function $F(x;\, p,\, q)$ is associated satisfying the following conditions:

(1) $F(x;p,q)=0$ for $x\leqslant 0$ and $\lim_{x=\infty} F(x;p,q)=1$.

(2) $F(x;p,q)$ is a non-decreasing function of x and continuous to the left.

(3) $F(x;p,q)=F(x;q,p)$ for any pair of points p and q.

(4) $F(x;p,p)=1$ for any $x>0$.

The function $F(x;p,q)$ can be interpreted as the probability distribution function of the distance of p and q; i.e., for any value x, $F(x;p,q)$ denotes the probability that the distance of p and q is less than x. In all that follows a distribution function will mean a function of a real variable x which satisfies conditions (1) and (2).

As a statistical generalization of the triangular inequality in metric spaces the following inequality has been proposed by Menger: For any three points p, q and r we have

(5) $$T[(Fx;p,q),\, F(y;q,r)]\leqslant F(x+y;\, p,r)$$

where $T(a,b)$ is a function of two variables satisfying certain conditions. A statistical semi-metric space is called a statistical metric space if inequality (5) is satisfied for all triples p, q and r.

Menger's generalization of the triangular inequality has the drawback that it involves an unspecified function $T(a,b)$ and one can hardly find sufficient justification for a particular choice of this function. Furthermore the notion of "between" introduced by Menger on the basis of inequality (5) has the properties of the between relationship in metric spaces only under restrictive conditions on the distribution functions $F(x;p,q)$. Here we propose another statistical generalization of the triangular inequality which is free from the above mentioned difficulties.

---

(1) An abstract of this paper was published in the Proceedings of the Nat. Academy of Science, 29, (1943) pp. 196-197.

(2) K. Menger, Proc. Nat. Acad. Sci., 28, (1942) p. 535.

* Reprinted from *Reports of a Mathematical Colloquium, Notre Dame*, Ser. 2, Vol. 5–6, 1944, pp. 76–79.

By the symbolic sum $F(x)+'G(x)$ of two distribution functions $F(x)$ and $G(x)$ we mean the distribution function $H(x)$ given by the Stieltjes integral

$$H(x)=\int\int d[F(u)G(v)]$$

where the integration is to be taken over the domain of the (u,v) plane given by the inequality $u+v<x$. Thus if X and Y are independently distributed random variables with the distribution functions $F(x)$ and $G(x)$, respectively, then $F(x)+'G(x)$ is the distribution function of $X+Y$.

In all that follows for any two distribution functions F and G the symbol $F\leqslant G$ will mean that the inequality $F(x)\leqslant G(x)$ holds for all values of x. The symbol $F<G$ will mean that $F\leqslant G$ and F is not identically equal to G. The symbols $F\geqslant G$ and $F>G$ are synonymous with the symbols $G\leqslant F$ and $G<F$, respectively.

As a statistical generalization of the triangular inequality we propose the following inequality: For any three points p, q, r we have

$$(5')\qquad F(x;p,q)+'F(x;q,r)\leqslant F(x;p,r).$$

We will say that two points p and q are different, in symbol $p\neq q$, if there exists a positive value x such that $F(x;p,q)<1$. We will say that a point q lies between the points p and r if

$$(6)\qquad p\neq q\neq r\neq p, \text{ and}$$

$$(7)\qquad F(x;p,q)+'F(x;q,r)=F(x;p,r).$$

Let the symbol pqr denote the relationship that q lies between p and r. We will show that the "between" relationship defined here has the same properties as the between relationship in metric spaces;(3) i.e. we will prove the following propositions:

I. From pqr it follows that rqp.

II. From pqr it follows that qrp and rpq cannot hold.

III. From pqr and prs it follows that pqs and qrs.

First we prove the following

Lemma: If G *and* G* *are two distribution functions for which* $G>G^*$ *then* $F+'G>F+'G^*$ *for any distribution function* F.

Proof: Since any distribution function is continuous to the left, it follows from $G>G^*$ that there exists a positive δ and a finite interval $[a,b]$ such that

$$(8)\qquad G(x)\geqslant G^*(x)+\delta$$

for all values x in the interval $[a,b]$ $(a<b)$.

(3) K. Menger, Mathematische Annalen, 100, (1928) p. 77.

Denote the distribution function $F+'G$ by $H(x)$ and the distribution function $F+'G^*$ by $H^*(x)$. From the definition of the symbolic sum $+'$ it follows easily that

$$H(x)=\int_{-\infty}^{+\infty} G(x-v)\, dF(v) \tag{9}$$

and

$$H^*(x)=\int_{-\infty}^{+\infty} G^*(x-v)\, dF(v). \tag{10}$$

Obviously there exists a real value c such that $c-a$ and $c-b$ are continuity points of $F(v)$ and

$$\int_{c-b}^{c-a} dF(v)>0. \tag{11}$$

From (9) and (10) we obtain

$$H(c)=\int_{-\infty}^{c-b} G(c-v)dF(v)+\int_{c-b}^{c-a} G(c-\ )dF(v)+\int_{c-a}^{+\infty} G(c-v)dF(v) \tag{12}$$

and

$$H^*(c)=\int_{-\infty}^{c-b} G^*(c-v)dF(v)+\int_{c-b}^{c-a} G^*(c-v)dF(v)+\int_{c-a}^{+\infty} G^*(c-v)dF(v). \tag{13}$$

Because of (8) and (11) the middle term of the right hand side of (12) is greater than the middle term of the right hand side of (13). Thus, since $G>G^*$, we obviously have $H(c)>H^*(c)$ and $H(x)\geqslant H^*(x)$ for all values of x. This proves our Lemma.

Now we are able to prove Propositions I–III. Proposition I is an immediate consequence of condition (3) and the fact that the operation $+'$ is commutative.

Let F_0 be the distribution function defined as follows:

$$F_0(x)=0 \text{ for } x\leqslant 0 \text{ and } F_0(x)=1 \text{ for } x>0.$$

In order to prove Proposition II we merely have to show that if F, G and H are three distribution functions such that $F\neq F_0$, $G\neq F_0$, $H\neq F_0$, and $F+'G=H$, then $F+'H\neq G$. Let us assume that $F+'H=G$ and we will derive a contradiction. Substituting $F+'G$ for H we obtain $F+'(F+'G)=G$. Since the operation $+'$ is commutative and associative we have

$G+'(F+'F)=G$. Since $G+'F_0=G$, the above equation can be written as $G+'(F+'F)=G+'F_0$. Since $F+'F<F_0$, the last equation contradicts our Lemma. Thus Proposition II is proved.

For any two points u and v of the space R let the symbol uv denote the distribution function $F(x;u,v)$ associated with the points u and v. Let p,q,r

and s be four mutually different points of R. In order to prove Proposition III we have merely to show that the equations

(14) $$pq+'qr=pr \text{ and } pr+'rs=ps$$

imply the equations

(15) $$pq+'qs=ps, \text{ and}$$

(16) $$qr+'rs=qs.$$

From (14) we obtain

(17) $$pq+'(qr+'rs)=ps.$$

Because of the triangular inequality we have

(18) $$qs \geqslant qr+'rs.$$

If in (18) the sign $>$ were valid, it would follow from (17) and our Lemma that $pq+'qs>ps$ which contradicts the triangular inequality. Thus, the equality sign must hold in (18) and therefore (16) is proved. Equation (15) is an immediate consequence of (16) and (17). Hence Proposition III is proved.

Made in United States of America

Reprinted from The Annals of Mathematical Statistics
Vol. XV, No. 4, December, 1944

STATISTICAL TESTS BASED ON PERMUTATIONS OF THE OBSERVATIONS

A. Wald and J. Wolfowitz

Columbia University

1. Introduction. One of the problems of statistical inference is to devise exact tests of significance when the form of the underlying probability distribution is unknown. The idea of a general method of dealing with this problem originated with R. A. Fisher [13, 14]. The essential feature of this method is that a certain set of permutations of the observations is considered, having the property that each permutation is equally likely under the hypothesis to be tested. Thus, an exact test on the level of significance α can be constructed by choosing a proportion α of the permutations as critical region. In an interesting paper H. Scheffé [2] has shown that for a general class of problems this is the only possible method of constructing exact tests of significance.

Tests based on permutations of the observations have been proposed and studied by R. A. Fisher, E. J. G. Pitman, B. L. Welch, the present authors, and others. Pitman and Welch derived the first few moments of the statistics used in their test procedures. However, it is desirable to derive at least the limiting distributions of these statistics and make it practicable to carry out tests of significance when the sample is large. Such a large sample distribution was derived for a statistic considered elsewhere [1] by the present authors.

In this paper a general theorem on the limiting distribution of linear forms in the universe of permutations of the observations is derived. As an application of this theorem, the limiting distributions of the rank correlation coefficient and that of several statistics considered by Pitman and Welch, are obtained. In the last section the limiting distribution of Hotelling's generalized T in the universe of permutations of the observations is derived.

2. A theorem on linear forms. Let $H_N = (h_1, h_2 \dots, h_N)$ $(N = 1, 2, \dots,$ ad inf.) be sequences of real numbers and let

$$\mu_r(H_N) = N^{-1} \sum_{\alpha=1}^{N} \left((h_\alpha - N^{-1} \sum_{\beta=1}^{N} h_\beta \right)^r$$

for all integral values of r. We define the following symbols in the customary manner: For any function $f(N)$ and any positive function $\varphi(N)$ let $f(N) = O(\varphi(N))$ mean that $| f(N)/\varphi(N) |$ is bounded from above for all N and let

$$f(N) = \Omega(\varphi(N))$$

mean that

$$f(N) = O(\varphi(N))$$

and that

$$\lim_N \inf | f(N)/\varphi(N) | > 0.$$

Also let

$$f(N) = o(\varphi(N))$$

mean that

$$\lim_{N \to \infty} f(N)/\varphi(N) = 0.$$

Let $[\rho]$ denote the largest integer $\leq \rho$.

We shall say that the sequences $H_N (N = 1, 2, \cdots, \text{ad inf.})$ satisfy the condition W if, for all integral $r > 2$,

$$\frac{\mu_r(H_N)}{[\mu_2(H_N)]^{r/2}} = O(1). \tag{2.1}$$

For any value of N let

$$X = (x_1, x_2, \cdots, x_N)$$

be a chance variable whose domain of definition is made up of the $N!$ permutations of the elements of the sequence $A_N = (a_1, a_2, \cdots, a_N)$. (If two of the $a_i (i = 1, 2, \cdots, N)$ are identical we assume that some distinguishing index is attached to each so that they can then be regarded as distinct and so that there still are $N!$ permutations of the elements $a_1, \cdots, a_N$). Let each permutation of A_N have the same probability $(N!)^{-1}$. Let $E(Y)$ and $\sigma^2(Y)$ denote, respectively, the expectation and variance of any chance variable Y.

We now prove the following:

THEOREM. *Let the sequences* $A_N = (a_1, a_2, \cdots, a_N)$ *and* $D_N = (d_1, d_2, \cdots, d_N)$ $(N = 1, 2, \cdots, \text{ad inf.})$ *satisfy the condition* W. *Let the chance variable* L_N *be defined as*

$$L_N = \sum_{i=1}^{N} d_i x_i .$$

Then as $N \to \infty$, *the probability of the inequality*

$$L_N - E(L_N) < t \, \sigma \, (L_N)$$

for any real t, approaches

$$\frac{1}{\sqrt{2\pi}} \int_{-\infty}^{t} e^{-\frac{1}{2}x^2} \, dx.$$

For convenience the proof will be divided into several lemmas.

Since

$$L_N^* = \frac{L_N - E(L_N)}{\sigma(L_N)}$$

remains invariant if a constant is added to all the elements of D_N or of A_N, or if the elements of either of the latter are multiplied by any constant other than zero, we may, in the formation of L_N^*, replace A_N and D_N by the sequences A_N'

and D'_N, respectively, whose ith elements a'_i and $d'_i (i = 1, 2, \cdots, N)$ are, respectively

$$a'_i = [\mu_2(A_n)]^{-\frac{1}{2}} \left(a_i - N^{-1} \sum_{j=1}^{N} a_j \right) \tag{2.2}$$

and

$$d'_i = [\mu_2(D_N)]^{-\frac{1}{2}} \left(d_i - N^{-1} \sum_{j=1}^{N} d_j \right). \tag{2.3}$$

The sequences A'_N and D'_N satisfy the condition W. Furthermore,

$$\mu_1(A'_N) \equiv \mu_1(D'_N) \equiv 0 \tag{2.4}$$

and

$$\mu_2(A'_N) \equiv \mu_2(D'_N) \equiv 1. \tag{2.5}$$

LEMMA 1.

$$\sum_{\alpha_1 < \alpha_2 < \cdots < \alpha_k \leq N} \cdots \sum a'_{\alpha_1} a'_{\alpha_2} \cdots a'_{\alpha_k} = O(N^{[k/2]}) \tag{2.6}$$

$$\sum_{\alpha_1 < \alpha_2 < \cdots < \alpha_k \leq N} \cdots \sum d'_{\alpha_1} d'_{\alpha_2} \cdots d'_{\alpha_k} = O(N^{[k/2]}). \tag{2.7}$$

From (2.4), (2.5), and the fact that the A'_N and D'_N satisfy condition W, it follows that the A'_N and D'_N satisfy conditions a), b), and c) of the theorem on page 383 of [1]. Our lemma 1 is the same as lemma 1 of [1].

LEMMA 2. *Let*

$$V = (v_1, v_2, \cdots, v_N)$$

be the same permutation of the elements of A'_N *that* X *is of the elements of* A_N. *Let* $y = v_1 \cdots v_k z$ *where* $z = v_{(k+1)}^{i_2} \cdots v_{(k+r)}^{i_r}$, $i_j > 1$ $(j = 1, 2, \cdots, r)$, *and* $k, r, i_1, \cdots, i_r$ *are fixed values independent of* N.
Then

$$E(y) = O(N^{[k/2]-k}). \tag{2.8}$$

This is Lemma 2 of [1].

In a similar manner we obtain that

$$\begin{aligned} \sum_{\alpha_1, \alpha_2, \cdots, \alpha_{(k+r)}} \cdots \sum d'_{\alpha_1} \cdots d'_{\alpha_k} d'^{i_1}_{\alpha_{(k+1)}} \cdots d'^{i_r}_{\alpha_{(k+r)}} \\ = O(N^{[k/2]-k}) \cdot O(N^{k+r}) = O(N^{[k/2]+r}). \end{aligned} \tag{2.9}$$

The summation in the above formula is to be taken over all possible sets of $k + r$ distinct positive integers $\leq N$.

LEMMA 3. *Let* $\alpha_1, \cdots, \alpha_{(k+r)}$ *be* $(k + r)$ *distinct positive integers* $\leq N$. *Then*

$$E(v_1 v_2 \cdots v_k v_{(k+1)}^{i_1} \cdots v_{(k+r)}^{i_r}) = E(v_{\alpha_1} v_{\alpha_2} \cdots v_{\alpha_k} v_{\alpha_{(k+1)}}^{i_1} \cdots v_{\alpha_{(k+r)}}^{i_r}). \tag{2.10}$$

This follows from the fact that all permutations of A'_N have the same probability.

LEMMA 4. *Let*

$$L'_N = \sum_{i=1}^{N} d'_i v_i .$$

Then

$$E(L'^p_N) = O(N)^{[p/2]}). \tag{2.11}$$

PROOF: Expand L'^p_N and take the expected value of the individual terms. The contribution to $E(L'^p_N)$ of all the terms which are multiples of the type appearing in the right member of (2.10) with fixed k, r, $i_1, \cdots, i_r (k + i_1 + \cdots + i_r = p)$, is, by Lemmas 2 and 3

$$O(N^{[k/2]-k}) \cdot \sum_{\substack{\alpha_1, \cdots, \alpha_{(k+r)} \\ \text{all different}}} \cdots \sum d'_{\alpha_1} \cdots d'_{\alpha_k} d'^{i_1}_{\alpha(k+1)} \cdots d'^{i_r}_{\alpha(k+r)} = O(N^{[k/2]-k}) O(N^{[k/2]+r})$$

$$= O(N^{2[k/2]-k+r}).$$

Since $i_j > 1 (j = 1, \cdots, r)$, it follows from the fact that $k + i_1 + \cdots + i_r = p$ that $2r \le p - k$ and that $2r = p - k$ only if $i_1 = \cdots = i_r = 2$. Now

$$2\left[\frac{k}{2}\right] - k + r \le r \le \frac{p-k}{2} \le \frac{p}{2}. \tag{2.12}$$

Hence the maximum value of $2\left[\frac{k}{2}\right] - k + r$ is reached when $r = \left[\frac{p}{2}\right]$ and $k = 0$. This proves the lemma.

From the last remarks of the preceding paragraph we obtain

LEMMA 5.

$$E(L'^{2j}_N) - \frac{(2j)!}{j!2^j} \Big(\sum_{\substack{\alpha_1, \cdots, \alpha_j \\ \text{all different}}} \cdots \sum d'^2_{\alpha_1} \cdots d'^2_{\alpha_j}\Big) E(v_1^2 \cdots v_j^2) = o(N^j). \tag{2.13}$$

We now prove

LEMMA 6.

$$E(L'_N) = 0 \tag{2.14}$$

$$E(L'^2_N) = NE(v_1^2) + o(N) = N + o(N). \tag{2.15}$$

Equation (2.14) follows from (2.2). Consider the expectations of the various terms in the expansion of L'^2_N. The sum of all the terms of the type

$$d'_i d'_j E(v_i v_j)$$

is

$$\Big(\sum_{i \ne j} d'_i d'_j\Big) E(v_1 v_2) = O(N) O(N^{-1}) = O(1),$$

by Lemmas 1 and 2. The sum of all the terms of the type

$$d'^2_i E(v_i^2)$$

is

$$\left(\sum_{i=1}^{N} d_i'^2\right) E(v_1^2) = NE(v_1^2) = N,$$

by (2.2) and (2.3). This proves the lemma.

LEMMA 7.

$$E(v_1^2 \cdots v_j^2) = 1 + o\,(1) \tag{2.16}$$

$$\sum_{\substack{\alpha_1 \cdots \alpha_j \\ \text{all different}}} \cdots \sum d_{\alpha_1}'^2 \cdots d_{\alpha_j}'^2 = N^j + o(N^j). \tag{2.17}$$

From (2.2) and (2.3), and Lemma 3, it follows that it will be sufficient to prove (2.17), because (2.16) follows in the same manner. Consider the relation

$$N^j = \left(\sum_{i=1}^{N} d_i'^2\right)^j = \sum_{\substack{\alpha_1, \cdots, \alpha_j \\ \text{all different}}} \cdots \sum d_{\alpha_1}'^2 \cdots d_{\alpha_j}'^2 + \text{other terms}.$$

By (2.9) the sum of these other terms must be not larger than $O(N^{j-1})$. From this follows the lemma.

PROOF of the theorem: Since

$$L_N^* = \frac{L_N'}{\sigma(L_N')} = \frac{L_N - E(L_N)}{\sigma(L_N)},$$

it will be sufficient to show that the moments of L_N^* approach those of the normal distribution as $N \to \infty$. From (2.14), (2.15), and (2.11) we see that, when p is odd, the pth moment of L_N^* is $O(N^{-\frac{1}{2}})$ and hence approaches zero as $N \to \infty$. When p is even and $= 2s$ (say), it follows from Lemma 5 that

$$E(L_N'^{2s}) - \frac{(2s)!}{s!2^s} (\sum_{\substack{\alpha_1, \cdots, \alpha_s \\ \text{all different}}} \cdots \sum d_{\alpha_1}'^2 \cdots d_{\alpha_s}'^2) E(v_1^2 \cdots v_s^2) = o(N^s).$$

Hence from (2.16) and (2.17)

$$E(L_N'^{2s}) = \frac{(2s)!}{s!2^s} N^s + o(N^s). \tag{2.18}$$

From (2.18) and (2.15) we obtain that

$$\lim_{N \to \infty} E(L_N^{*2s}) = \frac{(2s)!}{s!2^s}.$$

This completes the proof of the theorem.

It will be noticed that nothing in the foregoing proof requires that, when $N < N'$, the sequences A_N and D_N be subsequences of $A_{N'}$ and $D_{N'}$. Indeed, the sequences were written as they were simply for typographic brevity. We have therefore

COROLLARY 1. *The theorem is valid for sequences*

$$A_N = (a_{N1}, \cdots, a_{NN})$$
$$D_N = (d_{N1}, \cdots, d_{NN})$$
$$(N = 1, 2, \cdots \text{ ad inf.})$$

provided they fulfill condition W.

COROLLARY 2. *If the elements* $a_j(i = 1, 2, \cdots$ ad inf.) *are all independent observations on the same chance variable, all of whose moments are finite and whose variance is positive, the sequences* $A_N(N = 1, 2, \cdots,$ ad inf.) *will fulfill condition W with probability one.*

3. The rank correlation coefficient. For this well known statistic (see [3])

$$A_N \equiv D_N \equiv (1, 2, 3, \cdots, N).$$

The sequences A_N and D_N satisfy the condition W. For

$$\sum_{i=1}^{N} i^r = O(N^{r+1})$$

and hence, for $r \geq 3$

$$\mu_r(A_N) = \mu_r(D_N) = O(N^r).$$

Also

$$\mu_2(A_N) = \mu_2(D_N) = \Omega(N^2).$$

Hence the distribution of the rank correlation coefficient is asymptotically normal in the case of statistical independence. This result was first proved by Hotelling and Pabst [3].

4. Pitman's test for dependence between two variates. The distribution of the correlation coefficient in the population of permutations of the observations was used by Pitman [4] in a test for dependence between two variates which "involves no assumptions" about the distributions of these variates. In our notation, let

$$(a_i, d_i)(i = 1, 2, \cdots, N)$$

be N observations on the pair of variates A and D whose dependence it is desired to test. Then the value of the correlation coefficient is

$$N^{-1} \sum_{i=1}^{N} d_i' a_i' .$$

At the level β the observations are considered to be significant if the probability that $N^{-1} \mid L_N' \mid$ be equal to or greater than the absolute value of the actually observed correlation coefficient is $\leq \beta$.

In his paper ([4], page 227] Pitman points out that if the ratios of certain sample cumulants are "not too large," then, as $N \to \infty$, the first four moments of $N^{-\frac{1}{2}} L'_N$ will approach 0, 1, 0, and 3, respectively (the first moment is always zero). Our theorem and the relation (2.15) make clear that under proper circumstances all the moments will approach those of the normal distribution.

5. Pitman's procedure for testing the hypothesis that two samples are from the same population. For testing the hypothesis that two samples came from the same population Pitman [5] proposed the following procedure:
Let one sample be

$$a_1, a_2, \cdots, a_m$$

and the other

$$a_{m+1}, a_{m+2}, \cdots, a_{m+n}.$$

Write $m + n = N$, and construct the sequences A_N and A'_N as before defined. Let

$$d_i = 1 \qquad (i = 1, \cdots, m)$$

$$d_i = 0 \qquad (i = m + 1, \cdots, N)$$

and construct the sequences D_N and D'_N. Then the value of the statistic considered by Pitman is, except for a constant factor,

$$N^{-\frac{1}{2}} \left(\sum_{i=1}^{N} d'_i a'_i \right). \tag{5.1}$$

At the level β the observations are considered significant if the probability that $N^{-\frac{1}{2}} \mid L'_N \mid$ be equal to or greater than the observed absolute value of the expression (5.1) is $\leq \beta$.

Let $N \to \infty$, while $\frac{m}{n}$ is constant. Then the sequences D_N are seen to satisfy condition W. If then the sequences A_N satisfy condition W we may, for large N, employ the result of our theorem and expeditiously determine the critical value of Pitman's statistic.

6. Analysis of variance in randomized blocks. Welch [7] and Pitman [6] consider the following problem: Each of n different "varieties of a plant" is planted in one of the n cells which constitute a "block." It is desired to test, on the basis of results from m blocks, the null hypothesis that there is no difference among the varieties. In order to eliminate a possible bias caused by variations in fertility among the cells of a block, the varieties are assigned at random to the cells of a block. If the cells of the jth block are designated by $(j1), (j2), \cdots, (jn)$, a permutation of the integers $1, 2, \cdots, n$ is allocated to the jth block by a chance process, each permutation having the same probability $(n!)^{-1}$.

Let x_{ijk} be the yield of the ith variety in the kth cell of the jth block to which it was assigned by the randomization process. It is assumed that

$$x_{ijk} = y_{jk} + \delta_i + \epsilon_{jk},$$

where y_{jk} is the "effect" of the kth cell in the jth block, δ_i is the "effect" of the ith variety, and ϵ_{jk} are chance variables about whose distribution we assume nothing. The null hypothesis states that

$$\delta_1 = \delta_2 = \cdots = \delta_n = 0.$$

Let a_{jk} be the yield in the kth cell of the jth block and x_{ij} the yield of the ith variety in the jth block. If the null hypothesis is true then, because of the randomization within each block described above, the conditional probability that, given the set $\{a_{jk}\}(k = 1, 2, \cdots, n)$, the sequence $x_{1j}, x_{2j} \cdots, x_{nj}$, be any given permutation of the elements of $\{a_{jk}\}$ is $(n!)^{-1}$. Permuting in all the blocks simultaneously we have that, under the null hypothesis, given the set of mn values $\{a_{jk}\}$ ($j = 1, 2, \cdots m$; $k = 1, 2, \cdots, n$), the conditional probability of any of the permutations is the same, $(n!)^{-m}$. This permits an exact test of the null hypothesis.

The classical analysis of variance statistic that would be employed in the conventional two-way classification with independent normally distributed observations is

$$F = \frac{(m-1)m \sum (x_{i\cdot} - x)^2}{\sum \sum (x_{ij} - x_{i\cdot} - x_{\cdot j} + x)^2}$$

where

$$x_{i\cdot} = m^{-1} \sum_j x_{ij} \qquad x = (mn)^{-1} \sum \sum x_{ij}.$$

$$x_{\cdot j} = n^{-1} \sum_i x_{ij}$$

The statistic W used by Welch and Pitman is

$$W = F(m - 1 + F)^{-1}.$$

Since W is a monotonic function of F and the critical regions are the upper tails, the two tests are equivalent. The distribution of F or W is to be determined in the same manner as that of the other statistics discussed in this paper, i.e., over the equally probable permutations of the values actually observed. The critical region is, as usual, the upper tail.

Since x_{ij} takes any of the values $a_{j1}, \cdots, a_{jn}$ with probability $1/n$, we have

$$E(x_{ij}) = n^{-1} \sum_k a_{jk} = a_j \quad \text{(say)}. \tag{6.1}$$

$$\sigma^2(x_{ij}) = n^{-1} \sum_k (a_{jk} - a_j)^2 = b_j \quad \text{(say)}. \tag{6.2}$$

$$\begin{aligned} \sigma(x_{i_1 j} x_{i_2 j}) &= [n(n-1)]^{-1} \sum_{k_1 \neq k_2} a_{jk_1} a_{jk_2} - a_j^2 \\ &= [n(n-1)]^{-1}[(\sum_k a_{jk})^2 - \sum_k a_{jk}^2] - a_j^2 \\ &= [n^2 a_j^2 - \sum_k a_{jk}^2][n(n-1)]^{-1} - a_j^2 \\ &= (n-1)^{-1}[a_j^2 - n^{-1} \sum_k a_{jk}^2] = -(n-1)^{-1} b_j. \end{aligned} \tag{6.3}$$

Hence

$$(6.4)\qquad E(x_{i\cdot}) = m^{-1} \sum a_j .$$

$$(6.5)\qquad \sigma^2(x_{i\cdot}) = m^{-2} \sum b_j = b \quad \text{(say)}.$$

$$(6.6)\qquad \sigma(x_{i_1\cdot}\, x_{i_2\cdot}) = -[m^2(n-1)]^{-1} \sum b_j = c \quad \text{(say)}. \qquad i_1 \neq i_2$$

Let

$$x_{ij}^* = \sum_v \lambda_{iv} x_{vj} \qquad (i, v = 1, \cdots, n)$$

where $\| \lambda_{iv} \|$ is an orthogonal matrix and

$$\lambda_{n1} = \lambda_{n2} = \cdots = \lambda_{nn} = n^{-\frac{1}{2}}.$$

Then it follows that

$$(6.7)\qquad \begin{aligned} E(x_{i\cdot}^*) &= 0 \\ \sigma^2(x_{i\cdot}^*) &= b - c \qquad (i = 1, 2, \cdots, n-1) \\ \sigma(x_{i_1\cdot}^*\, x_{i_2\cdot}^*) &= 0 \qquad (i_1 \neq i_2 ; i_1, i_2 = 1, \cdots, n-1). \end{aligned}$$

Furthermore, we have

$$(6.8)\qquad \sum_{i=1}^{n-1} x_{i\cdot}^{*2} = \sum_{i=1}^{n} (x_{i\cdot} - x)^2.$$

Applying the well known identity

$$\Sigma\Sigma(x_{ij} - x_{i\cdot} - x_{\cdot j} + x)^2 = \Sigma\Sigma(x_{ij} - x_{\cdot j})^2 - m\Sigma(x_{i\cdot} - x)^2$$

to the definitions of F and W we obtain

$$(6.9)\qquad W = \frac{m \sum_i (x_{i\cdot} - x)^2}{\sum_i \sum_j (x_{ij} - x_{\cdot j})^2}.$$

The denominator of the right member of (6.9) is invariant under permutations *within* each block and equals

$$\sum_j \sum_k (a_{jk} - a_j)^2 = (n-1)m^2(b-c).$$

Hence

$$(6.10)\qquad \begin{aligned} W &= [m(n-1)(b-c)]^{-1} \sum_{i=1}^{n} (x_{i\cdot} - x)^2 \\ &= [m(n-1)(b-c)]^{-1} \sum_{i=1}^{n-1} x_{i\cdot}^{*2}. \end{aligned}$$

If the joint distribution of the $x_{i\cdot}^*$ $(i = 1, 2, \cdots, n-1)$ over the set of admissible permutations approaches a normal distribution with non-singular correlation matrix as m, the number of blocks, becomes large, it follows from (6.7) and (6.10) that the distribution of $m\,(n - 1)\,W$ approaches the x^2 distribution with $n - 1$ degrees of freedom. Hence it remains to indicate conditions on the set $\{a_{jk}\}$ which would make the distribution of the $x_{i\cdot}^*$ approach normality. Each $x_{i\cdot}^*$ is the mean of independent variables, so these conditions need not be very restrictive.

According to Cramér [8], Theorem 21a, page 113, if the variances and covariances fulfill certain requirements (the limiting correlation matrix should also be non-singular) and if a generalized Lindeberg condition holds, normality in the limit will follow. Somewhat more restrictive conditions which are simpler to state and which will be satisfied in most statistical applications are that $o < c' < b_j < c''$ for all j, where c' and c'' are positive constants. Since the variance of x_{ij}^* is $(n - 1)^{-1}nb_j$, it can be seen that the above inequalities imply the fulfillment of the conditions of the Laplace-Liapounoff theorem (see, for example, Uspensky [9], page 318). By [6.7] the correlation matrix is non-singular.

7. Hotelling's generalized T for permutation of the observations. In this section we shall restrict ourselves to bivariate populations, the extension to more than two variables being straightforward. Let $(u_{11}, u_{21}), \cdots, (u_{1m}, u_{2m})$ be m pairs of observations on the chance variables U_1, U_2, and $(u_{1(m+1)}, u_{2(m+1)}), \cdots, (u_{1N}, u_{2N})$, be n pairs of observations on the chance variables U_1', U_2', where $m + n = N$. If each of the pairs U_1, U_2, and U_1', U_2' is jointly normally distributed with the same convariance matrix, the Hotelling generalized T for testing the null hypothesis that

$$E(U_1) = E(U_1') \tag{7.1}$$

and

$$E(U_2) = E(U_2'), \tag{7.2}$$

is given (Hotelling [10]) by

$$T^2 = N^{-1}(mn) \sum_{j=1}^{2} \sum_{i=1}^{2} q_{ij}(\bar{u}_i - \bar{u}_i')(\bar{u}_j - \bar{u}_j')$$

where

$$m\bar{u}_i = \sum_{l=1}^{m} u_{il} \quad n\bar{u}_i' = \sum_{l=m+1}^{N} u_{il}$$

and the matrix $\| q_{ij} \|$ is the inverse of the matrix $\| b_{ij} \|$ with b_{ij} given by

$$(N - 2)b_{ij} = \sum_{l=1}^{m} (u_{il} - \bar{u}_i)(u_{jl} - \bar{u}_j) + \sum_{l=m+1}^{N} (u_{il} - \bar{u}_i')(u_{jl} - \bar{u}_j').$$

In Hotelling's procedure the b_{ij} are sample estimates of the population covariances whose distribution is independent of that of the sample means. A constant multiple of the statistic T^2 has the analysis of variance distribution under the null hypothesis. If the population covariances were known and used in place of the b_{ij}, T^2 would have the χ^2 distribution with two degrees of freedom.

Let us now apply the generalized T over the permutations of the actually observed values, as was done with other statistics in previous sections. If we do this literally we will find that the b_{ij} are no longer independent of the sample means. To avoid this complication we shall use a slightly different statistic T' which, as will be shown later, is a monotonic function of T, so that the test based on T' is identical with that based on T. The statistic T' is defined as follows: Let

$$\bar{U}_i = N^{-1} \sum_{k=1}^{N} u_{ik}$$

$$c'_{ij} = N[(N-1)mn]^{-1} \sum_{k=1}^{N} (u_{ik} - \bar{U}_i)(u_{jk} - \bar{U}_j) \qquad (i, j, = 1, 2)$$

and

$$\| q'_{ij} \| = \| c'_{ij} \|^{-1}.$$

Then

$$T'^2 = \sum_{i=1}^{2} \sum_{j=1}^{2} q'_{ij}(\bar{u}_i - \bar{u}'_i)(\bar{u}_j - \bar{u}'_j). \tag{7.3}$$

The expression T'^2 is much simpler than T^2, since the coefficients q'_{ij} are constants in the population of permutations of the observations. We shall show that T'^2 is a monotonic function of T^2. Let

$$Q_{ij} = \sum_{k=1}^{m} (u_{ik} - \bar{u}_i)(u_{jk} - \bar{u}_j) + \sum_{k=m+1}^{N} (u_{ik} - \bar{u}'_i)(u_{jk} - \bar{u}'_j)$$

$$Q'_{ij} = \sum_{k=1}^{N} (u_{ik} - \bar{U}_i)(u_{jk} - \bar{U}_j)$$

$$\| Q^{ij} \| = \| Q_{ij} \|^{-1}$$

$$\| Q'^{ij} \| = \| Q'_{ij} \|^{-1}.$$

Then the expressions

$$T_1^2 = \sum_{i=1}^{2} \sum_{j=1}^{2} Q^{ij}(\bar{u}_i - \bar{u}'_i)(\bar{u}_j - \bar{u}'_j) \tag{7.4}$$

and

$$T_2^2 = \sum_{i=1}^{2} \sum_{j=1}^{2} Q'^{ij}(\bar{u}_i - \bar{u}'_i)(\bar{u}_j - \bar{u}'_j), \tag{7.5}$$

are constant multiples of T^2 and T'^2, respectively. Hence it is sufficient to show that T_2^2 is a monotonic function of T_1^2. We have

$$Q'_{ij} = Q_{ij} + m(\bar{u}_i - \bar{U}_i)(\bar{u}_j - \bar{U}_j) + n(\bar{u}'_i - \bar{U}_i)(\bar{u}'_j - \bar{U}_j). \tag{7.6}$$

Furthermore, we have

$$\bar{u}_i - \bar{U}_i = \bar{u}_i - \frac{m\bar{u}_i + n\bar{u}'_i}{m+n} = \frac{n(\bar{u}_i - \bar{u}'_i)}{m+n}. \tag{7.7}$$

Similarly

$$\bar{u}_i' - \bar{U}_i = \bar{u}_i' - \frac{m\bar{u}_i + n\bar{u}_i'}{m+n} = \frac{m(\bar{u}_i' - \bar{u}_i)}{m+n}. \tag{7.8}$$

From (7.6), (7.7) and (7.8) it follows that

$$\begin{aligned} Q_{ij}' &= Q_{ij} + \frac{mn^2}{(m+n)^2}(\bar{u}_i - \bar{u}_i')(\bar{u}_j - \bar{u}_j') + \frac{nm^2}{(m+n)^2}(\bar{u}_i - \bar{u}_i')(\bar{u}_j - \bar{u}_j') \\ &= Q_{ij} + \frac{mn}{m+n}(\bar{u}_i - \bar{u}_i')(\bar{u}_j - \bar{u}_j'). \end{aligned} \tag{7.9}$$

Denote $\dfrac{mn}{m+n}$ by λ and $\bar{u}_i - \bar{u}_i'$ by h_i. Then we have

$$Q_{ij}' = Q_{ij} + \lambda h_i h_j . \tag{7.10}$$

Denote the cofactor of Q_{ij} in $\| Q_{ij} \|$ by R_{ij} and the cofactor of Q_{ij}' in $\| Q_{ij}' \|$ by R_{ij}'. Then

$$\frac{|Q_{ij}|}{|Q_{ij}'|} = \frac{|Q_{ij}|}{|Q_{ij} + \lambda h_i h_j|} = \frac{|Q_{ij}|}{|Q_{ij}| + \lambda\Sigma\Sigma R_{ij} h_i h_j} = \frac{1}{1 + \lambda T_1^2}. \tag{7.11}$$

Furthermore, we have

$$\frac{|Q_{ij}|}{|Q_{ij}'|} = \frac{|Q_{ij}' - \lambda h_i h_j|}{|Q_{ij}'|} = \frac{|Q_{ij}'| - \lambda\Sigma\Sigma R_{ij}' h_i h_j}{|Q_{ij}'|} = 1 - \lambda T_2^2 . \tag{7.12}$$

From (7.11) and (7.12) it follows that T_2^2 is a monotonic function of T_1^2. Hence also T'^2 is a monotonic function of T^2 and, therefore, we do not change our test procedure by using T'^2 instead of T^2.

Let the sequence of pairs

$$(x_{11} . x_{21}), \cdots , (x_{1N} , x_{2N})$$

be a permutation of the actually observed pairs

$$(u_{11} , u_{21}), \cdots , (u_{1N} , u_{2N})$$

where to each permutation is ascribed the same probability $(N!)^{-1}$. Then one obtains for $i = 1, 2$,

$$E(\bar{x}_i - \bar{x}_i') = 0 \tag{7.13}$$

$$\sigma^2(\bar{x}_i - \bar{x}_i') = N[(N-1)mn]^{-1} \sum_{j=1}^{N} (u_{ij} - \bar{U}_i)^2 = c_{ii}' \tag{7.14}$$

$$E(\bar{x}_1 - \bar{x}_1')(\bar{x}_2 - \bar{x}_2') = N[(N-1)mn]^{-1} \sum_{j=1}^{N} (u_{1j} - \bar{U}_1)(u_{2j} - \bar{U}_2) = c_{12}' . \tag{7.15}$$

Hence $\| c_{ij}' \|$ is the covariance matrix of the variates

$$(\bar{x}_1 - \bar{x}_1') \qquad \text{and} \qquad (\bar{x}_2 - \bar{x}_2').$$

Now we shall show that the limiting distribution of T'^2, as $N \to \infty$, is the χ^2 distribution with 2 degrees of freedom, provided that the observation u_{ik} ($i = 1, 2$; $k = 1, \cdots, N$) satisfy some slight restrictions. Since $\| q'_{ij} \|$ is the inverse of the covariance matrix $\| c'_{ij} \|$ our statement about the limiting distribution of T'^2 is obviously proved if we can show that $\bar{x}_1 - \bar{x}'$ and $\bar{x}_2 - \bar{x}'_2$ have a joint normal distribution in the limit.

Let $N \to \infty$ while m/n remains constant. Let the sequences A_N and D_N of Section II be defined as follows:
There are two sequences A_N, denoted respectively by A_{1N} and A_{2N}, such that

$$a_{ij} = u_{ij} \qquad (i = 1, 2; j = 1, \cdots, N).$$

Also

$$d_j = \frac{1}{m} \qquad (j = 1, \cdots, m)$$

$$d_j = -\frac{1}{n} \qquad (j = m + 1, \cdots, N).$$

Then the sequences D_N satisfy the condition W. If also the sequences A_{iN} satisfy the condition W, the distribution of $\bar{x}_i - \bar{x}'_i$ approaches the normal distribution as N increases, by the theorem of Section 2. If the joint distribution of $\bar{x}_1 - \bar{x}'_i$ and $\bar{x}_2 - \bar{x}'_2$ approaches a normal distribution with non-singular correlation matrix, the distribution of T'^2 approaches that of χ^2 with two degrees of freedom.

The correlation matrix of $(\bar{x}_1 - \bar{x}'_1)$ and $(\bar{x}_2 - \bar{x}'_2)$ will be of rank two in the limit if the correlation coefficient between $(\bar{x}_1 - \bar{x}'_1)$ and $(\bar{x}_2 - \bar{x}'_2)$ approaches a limit ρ, where $| \rho | < 1$. By (7.14) and (7.15) this is equivalent to saying that the absolute value of the angle between the vectors A'_{1N} and A'_{2N} is eventually greater than a positive lower bound. We shall show that, if the correlation coefficient approaches, as $N \to \infty$, a limit ρ whose absolute value is less than one, and if A_{1N} and A_{2N} satisfy the condition W, then $(\bar{x}_1 - \bar{x}'_1)$ and $(\bar{x}_2 - \bar{x}'_2)$ are *jointly* normally distributed in the limit.

Let δ_1 and δ_2 be any two real numbers not both zero. Then the sequence

$$A_N^* = (a_1^*, \cdots, a_N^*)$$

where

$$a_j^* = \delta_1 a_{1j} + \delta_2 a_{2j}$$

will be shown to satisfy the condition W. If either δ_1 or δ_2 is zero this is trivial; assume therefore that neither is zero. Without loss of generality we may assume that $\sum_{j=1}^{N} a_{ij} = 0$, for if this were not so we could replace the original a_{ij} by $a'_{ij} = a_{ij} - N^{-1} \sum_j a_{ij}$ as was done in Section 2. Let ρ' be such that $1 > \rho' > | \rho |$.

For N sufficiently large we have

$$\begin{aligned}\mu_2(A_N^*) &\geq N^{-1}(\delta_1^2 \sum_j a_{1j}^2 - 2 \mid \delta_1 \delta_2 \sum a_{1j} a_{2j} \mid + \delta_2^2 \sum_j a_{2j}^2) \\ &\geq N^{-1}(\delta_1^2 \sum_j a_{1j}^2 - 2\rho' \mid \delta_1 \delta_2 \mid \sqrt{(\sum_j a_{1j}^2)(\sum_j a_{2j}^2)} + \delta_2^2 \sum_j a_{2j}^2) \\ &= N^{-1}[(\mid \delta_1 \mid \sqrt{\sum_j a_{1j}^2} - \mid \delta_2 \mid \sqrt{\sum_j a_{2j}^2})^2 \\ &\qquad + 2(1 - \rho') \mid \delta_1 \delta_2 \mid \sqrt{(\sum_j a_{1j}^2)(\sum_j a_{2j}^2)}]\end{aligned}$$

and

$$\mu_2(A_N^*) \leq 2(\delta_1^2 \mu_2(A_{1N}) + \delta_2^2 \mu_2(A_{2N})).$$

Hence

$$\mu_2(A_N^*) = \Omega[\max \{\mu_2(A_{1N}), \mu_2(A_{2N})\}]. \tag{7.16}$$

Also $\mu_r(A_N^*)$ is a sum of constant multiples of terms of the type

$$N^{-1} \sum_j a_{1j}^i a_{2j}^{r-i}.$$

By Schwarz' inequality

$$N^{-1} \sum_j a_{1j}^i a_{2j}^{r-i} \leq N^{-1}(\sum_j a_{1j}^{2i})^{\frac{1}{2}}(\sum_j a_{2j}^{2(r-i)})^{\frac{1}{2}} = (\mu_{2i}(A_{1N})\mu_{2(r-i)}(A_{2N}))^{\frac{1}{2}}. \tag{7.17}$$

The required result follows from (7.16) and (7.17).

Since the sequences A_N^* satisfy the condition W, the limiting distribution of

$$\delta_1(\bar{x}_1 - \bar{x}_1') + \delta_2(\bar{x}_2 - \bar{x}_2'),$$

for any pair δ_1, δ_2 not both zero, is normal. From this and a theorem of Cramér and Wold ([11] Theorem 1; see also [8], Theorem 31) it follows that if the joint distribution of $(\bar{x}_1 - \bar{x}_1')$ and $(\bar{x}_2 - \bar{x}_2')$ approaches a limit, this limit must be the normal distribution. From a theorem of Radon ([12]; see also Cramér [8], page 101) it follows that if the joint distribution of $(\bar{x}_1 - \bar{x}_1')$ and $(\bar{x}_2 - \bar{x}_2')$ does not approach a limit as $N \to \infty$ it is possible to find two subsequences of the sequence $(1, 2, \cdots, N, \cdots$ ad inf.) for each of which the joint distribution approaches a different limit. This contradicts the previous result. Hence the limit exists and is the normal distribution. This proves our statement that the limiting distribution of T'^2 is the χ^2 distribution with two degrees of freedom.

The statistic T'^2 seems to be appropriate for testing the null hypothesis that two bivariate distributions Π_1 and Π_2 are identical if the alternatives are restricted to the case where Π_2 differs from Π_1 only in the mean values, i.e., the distribution Π_2 can be obtained from Π_1 by a translation. This is no restriction as compared with Hotelling's T-test since also the T-test is based on the assumption that the two normal populations differ at most in their mean values, i.e., the covariance matrices in the two populations are assumed to be equal.

REFERENCES

[1] A. Wald and J. Wolfowitz, *Annals of Math. Stat.*, Vol. 14 (1943), p. 378.
[2] Henry Scheffé, *Annals of Math. Stat.*, Vol. 14 (1943), p. 305.
[3] H. Hotelling and M. R. Pabst, *Annals of Math. Stat.*, Vol. 7 (1936), p. 29.
[4] E. J. G. Pitman, *Supp. Jour. Roy. Stat. Soc.*, Vol. 4 (1937), p. 225.
[5] E. J. G. Pitman, *Supp. Jour. Roy. Stat. Soc.*, Vol. 4 (1937), p. 119.
[6] E. J. G. Pitman, *Biometrika*, Vol. 29 (1938), p. 322.
[7] B. L. Welch, *Biometrika*, Vol. 29 (1937), p. 21.
[8] Harald Cramér, *Random Variables and Probability Distributions*, Cambridge, 1937.
[9] J. V. Uspensky, *Introduction to Mathematical Probability*, New York, 1937.
[10] H. Hotelling, *Annals Math. Stat.*, Vol. 2 (1931), p. 359.
[11] H. Cramér and H. Wold, *Journal London Math. Soc.*, Vol. 11 (1936), pp. 290–4.
[12] J. Radon, *Sitzungsberichte Akademie Wien*, Vol. 122 (1913), pp. 1295–1438.
[13] R. A. Fisher, *The Design of Experiments*, Edinburgh, 1937. (Especially Section 21.)
[14] R. A. Fisher, *Statistical Methods for Research Workers*, Edinburgh, 1936. (Especially Section 21.02.)

Made in United States of America

Reprinted from The Annals of Mathematical Statistics
Vol. XVI, No. 1, March, 1945

SAMPLING INSPECTION PLANS FOR CONTINUOUS PRODUCTION WHICH INSURE A PRESCRIBED LIMIT ON THE OUTGOING QUALITY

A. Wald and J. Wolfowitz
Columbia University

1. Introduction. This paper discusses several plans for sampling inspection of manufactured articles which are produced by a continuous production process, the plans being designed to insure that the long-run proportion of defectives shall not exceed a prescribed limit. The plans are applicable to articles which can be classified as "defective" or "non-defective" and which are submitted for inspection either continuously or in lots. In Section 2 the notions of "average outgoing quality limit" and "local stability" are discussed. The valuable concept of average outgoing quality limit for lot inspection is due to Dodge and Romig [4], and that for inspection of continuous production to Dodge [1]. Section 3 contains a description of a simple inspection plan (SPA) applicable to to continuous production and a proof that the plan will insure a prescribed average outgoing quality limit. Section 4 contains a proof that this inspection plan also has the important property that it requires minimum inspection when the production process is in statistical control. In Section 5 is contained the description of a general class of plans which possess both these important properties.

The problem of adapting SPA to the case when the articles are submitted for inspection in lots instead of continuously, is treated in Section 6. Some methods of achieving local stability are discussed in Section 7 and a specific plan is developed there. Finally Section 8 discusses the relationship between the present work and that of the earlier and very interesting paper of H. F. Dodge [1], mentioned above.

If a quick first reading is desired the reader may omit the second half of Section 3 (which contains a proof of the fact that SPA guarantees the prescribed average outgoing quality limit) and the entire Section 4 except for its title (the proof of the statement made in the title of Section 4 occupies the whole section).

2. Fundamental notions. In this paper we shall deal only with a product whose units can be classified as "defective" or "non-defective." We shall assume that the units of the product are submitted for inspection continuously, except in Section 6, where we assume that they are submitted in lots. Throughout the paper we shall assume that the inspection process is non-destructive, that it invariably classifies correctly the units examined, and that defective units, when found, are replaced by non-defectives. By the "quality" of a sequence of units is meant the proportion of defectives in the sequence as produced. By the "outgoing quality" (OQ) of a sequence is meant the proportion of defectives after whatever inspection scheme which is in use has been applied. If this scheme involves random sampling, then in general the OQ is a chance variable.

(It depends on the variations of random sampling.) If the OQ converges to a constant p_a with probability one as the number of units produced increases indefinitely, p_a is called the "average outgoing quality" (AOQ). The AOQ when it exists is therefore the average quality, in the long run, of the production process after inspection. It is a function of both the production process and the inspection scheme. These definitions are due to Dodge [1].

The "average outgoing quality limit" (AOQL) is a number which is to depend only on the inspection scheme and not at all on the production process. Roughly speaking, it is a number, characteristic of an inspection scheme, such that no matter what the variations or eccentricities of the production process, the AOQ never exceeds it. For the purposes of this paper we shall need the following precise definition: Let c_i be zero or one according as the ith unit of the product, before application of the inspection scheme, is a non-defective or a defective, respectively. Let d_i have a similar definition *after* application of the inspection scheme. (We note that if the ith item was inspected, then $d_i = 0$; if the ith item was not inspected, then $c_i = d_i$.) The sequence $c = c_1, c_2, \cdots, c_N, \cdots,$ ad inf. characterizes the production process[1]. The elements of $d = d_1, d_2, \cdots,$ ad inf. are in general chance variables. The number L is called the AOQL if it is the smallest[2] number with the property that the probability is zero that

$$\limsup_N \frac{\sum_{i=1}^{N} d_i}{N} > L,$$

no matter what the sequence c.

It should be noted that this definition of AOQL places no restrictions whatever on the production process, since *all* sequences c are admitted. It is too much to expect a production process to remain always in control; indeed, doubt as to whether statistical control always exists may cause a manufacturer to institute an inspection scheme. The inspection schemes which we shall give below will yield a specified AOQL no matter what the variations in production are. If these schemes are employed, then, even if Maxwell's demon of gas theory fame were to transfer his activities to the production process, he would be unsuccessful in an effort to cause the AOQL to be exceeded. A dishonest manufacturer might sometimes essay to do this. If we imposed restrictions on the sequence c and

---

[1] This use of an infinite sequence to describe the production process deserves a few words. What we consider in this paper are schemes applicable when the number of units produced is large and operate mathematically as if the production sequence were of infinite length. Naturally the latter is never the case in actuality. However, the larger the number of units produced the more nearly will the reality conform to the results derived from the mathematical model. While the present definition uses explicitly the notion of an infinite sequence, such a commonplace statement as "the probability is 1/2 that a coin will fall heads up" uses this notion implicitly. It is also implicit in the intuitive meaning we ascribe to such a word as "average," which is in every day use.

[2] It is not difficult to see that such a number always exists, for it is the lower bound of a set which is non-empty (it contains the point one), bounded from below (zero is a lower bound), and closed.

determined the AOQL on that basis, we would run the danger that the relative frequency of defects in the sequence of outgoing units might exceed the AOQL if it happened that the actual sequence c did not satisfy the restrictions imposed.

After we discuss below various possible sampling inspection plans which insure that the AOQL does not exceed a predetermined value L, it will be seen that for any given $L > 0$ there are many sampling inspection schemes which do this. To choose a particular sampling plan from among them the following considerations may be advanced: If two inspection plans S and S' both insure the inequality AOQL $\leq L$ and if for any sequence c the average number of inspections required by S is not greater than that required by S' and if for some sequences c the average number of inspections required by S is actually smaller than that required by S', then S may be considered, in general, a better inspection plan than S'. However, the amount of inspection required by a sampling plan is not always the *only* criterion for the selection of a proper sampling scheme. There may be also other features of a sampling plan which make it more or less desirable. We shall mention here one such feature, called "local stability," which will play a role in our discussions later. Consider the sequence d obtained from the sequence c by applying a sampling inspection scheme. Even if the AOQL does not exceed L, it may still happen that there will be many large segments of the sequence d within which the relative frequency of ones is considerably higher than L. For instance, it may happen that in the segment $(d_1, \cdots, d_m)$ the relative frequency of ones is equal to $\frac{3}{2}L$, in the segment $(d_{m+1}, \cdots, d_{2m})$ the relative frequency is equal to $\frac{1}{2}L$, in the segment $(d_{2m+1}, \cdots, d_{3m})$ the relative frequency is again equal to $\frac{3}{2}L$, and this is followed again by a segment of m elements where the relative frequency of ones is equal to $\frac{1}{2}L$, and so forth. If m is large, such a sequence d is not very desirable, since each second segment will contain too many defects. A sequence d is said to be not locally stable if there exists a large fixed integer m such that the relative frequency of ones in $(d_{k+1}, \cdots, d_{k+m})$ is considerably greater than L for *many* integral values k. On the other hand, the sequence d is said to be locally stable if for any large m the relative frequency of ones in $(d_{k+1}, \cdots, d_{k+m})$ is not substantially above L for nearly all integral values k. This is clearly not a precise definition of "local stability," but merely an intuitive indication of what we want to understand by the term, since we did not define what we mean by "large m," "many values of k," "considerably above L," etc. A precise definition of local stability will not be needed in this paper, since it is not our intention to develop a complete theory for the choice of the sampling plan. The idea of local stability will be used in this paper merely for making it plausible that some schemes we shall consider behave reasonably in this respect. A similar idea, called "protection against spotty quality," is discussed by Dodge [1]. A possible precise definition of local stability could be given in terms of the frequency with which $F(N) = \frac{1}{(k+1)} \sum_{i=N}^{N+k} d_i$ (k being fixed) lies within given limits.

3. A sampling inspection plan which insures a given AOQL no matter what the variations in the production process. The only feature of the sampling (inspection) plan (SP) studied in this section and hereafter referred to as SPA which we shall consider here is that it insures the achievement of a specified AOQL. Considerations leading to a choice among several schemes are postponed to later sections.

For convenience, let f be the reciprocal of a positive integer. SPA calls for alternating partial inspection and complete inspection. Partial inspection is performed by inspecting one element chosen at random from each of successive groups of $\frac{1}{f}$ elements. Complete inspection means the inspection of every element in the order of production. SPA is completely defined when a rule is given for ending one kind of inspection and beginning the other.

It is clear that all SP need not be of the above class. Thus, for example, a scheme might consist of partial inspection with various f's employed in various sequences. We make no attempt in this paper to examine all possible schemes. For simplicity in practical operation, alternation of complete inspection and partial inspection with fixed f would seem reasonable. The Dodge scheme [1] is of this type.

We shall also not discuss the question of a choice of the constant f, but will assume that a particular value has been chosen for various reasons and is a datum of our problem. Reasons which might influence a manufacturer in his choice of f could be contract specifications which impose a minimum on the amount of inspection, or psychological grounds to the same effect. The manufacturer may desire a certain minimum amount of inspection in order to detect malfunctioning of his production process. Also f controls local stability to some extent. The consequences of a choice of f as they appear in the theory below may also play a role.

Returning to SPA, we begin with partial inspection. Let L be the specified AOQL. Denote by k_N the number of groups of $\frac{1}{f}$ units in which defectives were found as the result of *partial* inspection from the beginning of production through the Nth unit. SPA is as follows:

(a) Begin with partial inspection.

(b) Begin full inspection whenever

$$e_N = \frac{k_N\left(\frac{1}{f} - 1\right)}{N} > L.$$

(c) Resume partial inspection when

$$e_N \leq L.$$

(d) Repeat the procedure. (It will be recalled that defective units, when found, are always to be replaced with non-defectives.)

It is to be observed that in this plan the number of partial inspections increases without limit. For, while complete inspection is going on, the value of k_N remains constant, so that after a long enough period of complete inspection the denominator N of the expression which defines e_N will have increased sufficiently for e_N to be not greater than L. On the other hand, complete inspection may never occur. This will be the case if, for example, no defectives or very few defectives are produced.

We shall now show that the AOQL of the above SP is L. We first note that, at N, e_N can increase only by $\frac{\left(\frac{1}{f} - 1\right)}{N}$. Hence, for sufficiently large N, $e_N < L + \epsilon$, where $\epsilon > 0$ may be arbitrarily small.

Suppose now that the production process is subject to any variations whatsoever, i.e., the sequence

$$c = c_1, c_2, \cdots, c_N, \cdots, \text{ad inf.}$$

is any arbitrary sequence whatever (by their definition the c_i are all zero or one). Our result is therefore proved if we show that, with probability one,

$$\lim_{N \to \infty} \left(e_N - \frac{1}{N} \sum_{i=1}^{N} d_i \right) = 0 \tag{3.1}$$

for this arbitrary c, and that for at least one c

$$\lim_{N \to \infty} e_N = L. \tag{3.2}$$

Let $S(N)$ be the number of groups of $\frac{1}{f}$ units which have been partially inspected through the Nth unit. Define x_i as zero if in the ith partially inspected group a non-defective was found and as one if a defective was found. We have

$$k_N = \sum_{i=1}^{S(N)} x_i .$$

Since the number of times partial inspection takes place increases indefinitely, $S(N) \to \infty$ as $N \to \infty$. Also $S(N) \leq fN < N$. Let α_j be the serial number of the last unit in the jth partially inspected group. Then for all j the expected value $E(x_j)$ of x_j is given by

$$E(x_j) = f\left(\sum_{i=(\alpha_j - (1/f)+1)}^{\alpha_j} c_i \right).$$

We have, for all j

$$\sum_{i=(\alpha_j-(1/f)+1)}^{\alpha_j} (c_i - d_i) = x_j \tag{3.3}$$

so that

$$E\left(\left[\frac{1}{f} - 1\right] x_j - \sum_{\alpha_j-(1/f)+1}^{\alpha_j} d_i \right) = 0.$$

Also from (3.3) it follows, since x_j is the value of a binomial chance variable from a population of fixed number $\left(\frac{1}{f}\right)$, that there exists a positive constant β such that

$$\sigma^2\left(\left[\frac{1}{f}-1\right]x_j - \sum_{\alpha j-(1/f)+1}^{\alpha j} d_i\right) < \beta \tag{3.4}$$

where $\sigma^2(x)$ is the variance of a chance variable x. Now a theorem of Kolmogoroff (Kolmogoroff [2], Fréchet [3], p. 254) states:
A sequence of chance variables with zero means and variances σ_1^2, σ_2^2, $\cdots$ converges with probability one towards zero in the sense of Cèsaro if

$$\sum_{i=1}^{\infty} \frac{\sigma_i^2}{i^2} \tag{3.5}$$

converges. The inequality (3.4) permits us to apply this theorem to the sequence of chance variables of which the jth ($j = 1, 2, \cdots$ ad inf.) is

$$\left(\left[\frac{1}{f}-1\right]x_j - \sum_{\alpha j-(1/f)+1}^{\alpha j} d_i\right),$$

since the series $\sum_{i=1}^{\infty} \frac{1}{i^2}$ is well known to be convergent. We therefore obtain that, with probability one,

$$\lim_{S(N)\to\infty} \frac{\left(\left[\frac{1}{f}-1\right]\sum_{j=1}^{S(N)} x_j - \sum_{j=1}^{N} d_j\right)}{S(N)} = \lim_{N\to\infty} \frac{N}{S(N)}\left(e_N - \frac{1}{N}\sum_{i=1}^{N} d_i\right) = 0,$$

since the units which are fully inspected contribute nothing to Σd_i. Since $S(N) < N$, the desired result (3.1) is a fortiori true.

If c is such that all the c_i are one, it is readily seen that (3.2) holds. If many (this adjective can be precisely defined) defectives are produced, this will also be the case. This completes the proof of the fact that the AOQL of SPA is L no matter how capriciously the production process may vary.

4. When the production process is in statistical control, SPA requires minimum inspection. The production process is said to be in statistical control if there is a positive constant $p \leq 1$ such that, for every i, the probability that $c_i = 1$ is p and is independent of the values taken by the other c's. We shall see that if the process is in statistical control and if SPA is applied to it, the specified AOQL is guaranteed with a minimum amount of inspection.

The number of units inspected through the Nth unit produced is

$$I(N) = N - \left(\frac{1}{f} - 1\right)S(N). \tag{4.1}$$

If the process is in statistical control we have, with probability one,

$$\lim_{N\to\infty} \frac{\sum_{i=1}^{N} c_i}{N} = p \tag{4.2}$$

by the strong law of large numbers. Shortly we shall prove the existence of a constant L^* such that, with probability one,

$$\lim_{N\to\infty} \frac{\sum_{i=1}^{N} d_i}{N} = L^*. \tag{4.3}$$

Assume for the moment that this is so. Since it is only by inspection that defectives are removed, and the units selected for inspection are in statistical control like the original sequence, it follows that, with probability one,

$$\lim_{N\to\infty} \frac{I(N)}{N} = \frac{1}{p}(p - L^*) = 1 - \frac{L^*}{p} \tag{4.4}$$

because, with probability one,

$$\lim_{N\to\infty} \frac{\sum_{i=1}^{N} (c_i - d_i)}{N} = p - L^*.$$

Inspection is therefore at a minimum when L^* is at a maximum compatible with the specified AOQL. By (4.3) the latter means that

$$L^* \leq L. \tag{4.5}$$

SPA has been shown to guarantee this requirement. The optimum situation from the point of view of the amount of inspection would therefore be to have $L^* = L$, but this cannot always be achieved. The absolute minimum amount of inspection clearly is f, i.e., partial inspection exclusively. Consequently from (4.4)

$$1 - \frac{L^*}{p} \geq f$$

so that

$$L^* \leq p(1 - f). \tag{4.6}$$

Combining (4.5) and (4.6) we see that we have to consider three cases:

Case a. If

$$p > \frac{L}{1 - f} \tag{4.7}$$

we have to show that

$$L = L^*. \tag{4.8}$$

Case b. If

$$p < \frac{L}{1 - f} \tag{4.9}$$

we have to show, by (4.4), that

$$1 - \frac{L^*}{p} = f,$$

that is,

(4.10) $$L^* = p(1 - f).$$

Case c. If

(4.11) $$p = \frac{L}{1 - f}$$

we have to show that

(4.12) $$L = L^* = p(1 - f).$$

PROOF of (4.8): We have already remarked in Section 3 that in SPA partial inspection always recurs, but complete inspection need never occur. We shall show in a moment that (4.7) implies that no matter how large an integer γ is chosen, the probability of temporarily stopping partial inspection for some $N > \gamma$ is one. Assume that this is so. Choose an arbitrarily small positive ϵ, and let $\gamma > \dfrac{\left(\frac{1}{f} - 1\right)}{\epsilon}$. For a sequence where complete and partial inspection alternate infinitely many times let

$$A = \alpha_1, \alpha_2, \cdots, \text{ ad inf.}$$

be the sequence of integers at which partial inspection ends, and let

$$B = \beta_1, \beta_2, \cdots, \text{ ad inf.}$$

be the sequence of integers at which complete inspection ends. Then, for all j,

$$\alpha_{j+1} > \beta_j > \alpha_j .$$

From the description of SPA it follows that, for all $N > \gamma$ which belong to either A or B,

(4.13) $$| e_N - L | < \epsilon.$$

In Section 3 we proved

(3.1) $$\lim_{N \to \infty} \left(e_N - \frac{1}{N} \sum_{i=1}^{N} d_i \right) = 0$$

with probability one. Since ϵ is arbitrarily small it follows that, with probability one,

(4.14) $$\lim_{\substack{N \to \infty \\ (N \text{ in } A \text{ or } B)}} \frac{\sum_{i=1}^{N} d_i}{N} = L.$$

To complete the proof of (4.8) we have still to show that L^* exists and that the probability is one that complete inspection will occur infinitely many times. First we prove that L^* exists.

As N increases during an interval of complete inspection, $D(N) = \sum_{i=1}^{N} d_i$ remains constant. Hence $\frac{D(N)}{N}$ decreases monotonically. Since for the ends of such intervals (4.14) holds, it follows that (4.14) holds as $N \to \infty$ and is a member of A, B, or an interval (α_j, β_j) for all j.

Let $N \to \infty$ while always being in the interior of an interval $(\beta_j, \alpha_{j+1}]$, $j = 1, 2, \cdots$, ad inf., which contains α_{j+1} but not β_j. Let N^* be the total number of units in these intervals through the Nth unit produced. Let N_1 and N_2 be such that

$$\beta_j = N_1 < N_2 < \alpha_{j+1}.$$

Then

$$N_2^* - N_1^* = N_2 - N_1.$$

Since the production process is in statistical control, we have, by the strong law of large numbers,

$$\lim_{N \to \infty} \frac{D(N)}{N^*} = p(1 - f) = p' \tag{4.15}$$

with probability one. Let δ^* be the general designation for numbers $<\epsilon$ in absolute value, so that all δ^* are not the same. With probability one for almost all N, we have by (4.15)

$$\frac{D(N_1)}{N_1^*} = p' + \delta^*$$

$$\frac{D(N_2)}{N_2^*} = p' + \delta^*.$$

Write

$$\frac{[D(N_2) - D(N_1)]}{(N_2 - N_1)} = K.$$

Now

$$\frac{D(N_2)}{N_2^*} = \frac{D(N_1) + [D(N_2) - D(N_1)]}{N_1^* + (N_2^* - N_1^*)} = \frac{D(N_1) + [D(N_2) - D(N_1)]}{N_1^* + (N_2 - N_1)}$$

$$= \frac{(p' + \delta^*)N_1^* + K(N_2 - N_1)}{N_1^* + (N_2 - N_1)} = p' + \delta^*.$$

Hence

$$K(N_2 - N_1) = 2\delta^* N_1^* + (p' + \delta^*)(N_2 - N_1). \tag{4.16}$$

Now suppose (4.3) does not hold. From the definition of AOQL it follows that for some $\eta > \epsilon$ there exist sequences (whose totality has a positive probability) so that, for infinitely many N_2 we have

$$\frac{D(N_2)}{N_2} = \frac{D(N_1) + [D(N_2) - D(N_1)]}{N_1 + (N_2 - N_1)} < L - 4\eta. \tag{4.17}$$

For large enough N_1, from (4.14),

$$\frac{D(N_1)}{N_1} = L + \delta^*$$

with probability one and hence, using (4.16) in (4.17)

$$\begin{aligned} N_1(L + \delta^*) + 2\delta^* N_1^* + (p' + \delta^*)(N_2 - N_1) \\ < LN_1 + L(N_2 - N_1) - 4\eta N_2 \end{aligned} \tag{4.18}$$

from which, using the fact that $p' \geq L$ (from (4.7)), we get

$$N_1\delta^* + 2N_1^*\delta^* + \delta^*(N_2 - N_1) < -4\eta N_2 . \tag{4.19}$$

((4.18) and (4.19) hold for the sequences for which (4.17) holds, except perhaps on a set of sequences whose probability is zero.) Since $N_1^* \leq N_1$ and $| \delta^* | < \eta$, we have, on the other hand,

$$\begin{aligned} N_1\delta^* + 2N_1^*\delta^* + \delta^*(N_2 - N_1) &\geq -3\eta N_1 - \eta(N_2 - N_1) \\ &> -4\eta N_1 - 4\eta(N_2 - N_1) = -4\eta N_2 \end{aligned} \tag{4.20}$$

which contradicts (4.19) and proves the desired result ((4.3) and (4.8)), except that it remains to prove that, no matter how large γ, the probability of temporarily stopping partial inspection at some $N > \gamma$ is one. Let $\gamma_0 \geq \gamma$ be some integer at which partial inspection is going on. From (4.2) and (4.7) it would follow, if partial inspection never ceased on a set of sequences with positive probability, that, on this set, with conditional probability one, for N sufficiently large and ϵ sufficiently small,

$$\frac{k_N - k_{\gamma_0}}{f(N - \gamma_0)} > \frac{L}{1 - f} + \epsilon,$$

$$\frac{N}{N - \gamma_0} \frac{k_N(1 - f)}{fN} > L + (1 - f)\epsilon,$$

$$e_N > L \frac{N - \gamma_0}{N} + \frac{(N - \gamma_0)(1 - f)\epsilon}{N},$$

$$e_N > L + \frac{(1 - f)\epsilon}{2}.$$

This contradiction proves that complete inspection is eventually resumed and completes the proof of minimum inspection in Case a.

PROOF of (4.10): We shall prove that (4.9) implies that, with probability one, complete inspection will cease, never to be resumed. For, from (4.15) and (4.9) it follows that for N sufficiently large and ϵ sufficiently small,

$$\frac{D(N)}{N^*} = p' + \delta^* < L - 2\epsilon. \tag{4.21}$$

Hence, a fortiori,

$$\frac{D(N)}{N} < L - 2\epsilon. \tag{4.22}$$

((4.21) and (4.22) hold with probability one.)
(3.1) states that, with probability one,

$$\lim_{N\to\infty}\left(e_N - \frac{D(N)}{N}\right) = 0.$$

Hence for all N sufficiently large, with probability one,

$$e_N < L - \epsilon,$$

i.e., with probability one complete inspection is never resumed.

When (4.9) holds, therefore, with probability one and with a finite number of exceptions SPA will require only partial inspection.

PROOF of (4.12): If $p = \frac{L}{1-f}$ and complete inspection finally never resumes, then (4.12) follows easily. If $p = \frac{L}{1-f}$ and partial and complete inspection alternate infinitely many times, then the proof is similar to that of (4.8) and is therefore omitted. In either case the desired result follows.

5. A class of SP all of which insure both a given AOQL and minimum inspection. Let the definition of SPA be modified in the following particulars:

(b) Begin full inspection whenever

$$e_N = \frac{k_N\left(\frac{1}{f} - 1\right)}{N} > L + \phi(N).$$

(c) Resume partial inspection when

$$e_N \leq L - \psi(N).$$

Let $\phi(N)$ and $\psi(N)$ be such that

$$-\psi(N) \leq \phi(N)$$

$$\lim_{N\to\infty}\phi(N) = \lim_{N\to\infty}\psi(N) = 0.$$

(SPA corresponds to the case $\phi(N) \equiv \psi(N) \equiv 0$.) Then all the SP of this class have the property that the AOQL is L and that inspection is at a minimum in the sense of Section 4. The proofs are essentially the same as those for SPA and hence will be omitted.

6. The inspection plans of Section 5 can also be applied to lot inspection. We shall carry on the discussion of this section in terms of SPA, but the results apply to all the members of the class of plans described in Section 5. We shall show that SPA can also be applied when the product is submitted for inspection in lots. Although we assumed previously that the units of the product are arranged in order of production, the results obtained for SPA remain valid for any arbitrary arrangement of the units. If the product is submitted in lots we may arrange the units as follows: Let $l_1, l_2, \cdots$, etc. be the successive lots in the order of their submission for inspection. Within each lot we consider the units arranged in the order in which they are chosen for inspection. In this way we have arranged all units in an ordered sequence and the inspection can be applied as described before. Thus, we start with partial inspection, i.e., we take out groups of $\frac{1}{f}$ elements in l_1 and inspect one unit (selected at random) from each of these groups. When $e_N > L$, we start complete inspection and revert to partial inspection as soon as $e_N \leq L$. When the units in l_1 are used up in the process of inspection, we continue, using the units of l_2, etc.

If it is found inconvenient to take out a group of $\frac{1}{f}$ units and then to select one unit for inspection, we could modify the sampling inspection plan as follows: Instead of taking out a group of $\frac{1}{f}$ units and then selecting at random one unit from it, we select at random *one* unit from the uninspected part of the lot and look upon this unit as the unit selected at random from a hypothetical group of $\frac{1}{f}$ units. Thus we can proceed exactly as before, except that we have to keep in mind that with each unit inspected under "partial inspection" we have used up another set of $\frac{1}{f} - 1$ units. Thus, as soon as $\left(\frac{1}{f} - 1\right)$ times the number of units inspected under "partial inspection" becomes equal to or greater than the number of units in the uninspected part of the lot, the inspection of that lot is already terminated, and we have to start using the units of the next lot. The inconvenience caused by the necessity of keeping track of the number of units inspected under "partial inspection" and of the number of units in the uninspected part of the lot can be eliminated by further modifying the inspection plan as follows: Instead of beginning complete inspection as soon as $e_N > L$, we continue "partial inspection" until $E_N = e_N - L$ is so large that complete inspection of all the units of the lot not yet used up has to be made in order to bring e_N down to L at the end of the lot. This leads to the following sampling procedure, to be known as SPB: Let N_0 be the number of units in the lot, let N_L be the serial number of the last unit in the preceding lot, and let $E(N_L) =$

$N_L E_{N_L} = N_L(e_{N_L} - L)$ be the "excess" carried over from the preceding lot. For simplicity assume that the following are all integers:

$$LN_0 = M$$

$$\frac{fM}{1-f} = M^*$$

$$fN_0 = N^*$$

and

$$\frac{fE(N_L)}{1-f} = E^*.$$

The inspection procedure is then as follows: Inspect successive units drawn at random until either

(a) $M^* - E^*$ defectives have been found in the first $N' < N^*$ units inspected. In this case inspect further an additional $N_0 - \frac{N'}{f}$ units and this terminates the inspection of the lot. The excess to be carried over to the next lot is then zero.

Or

(b) N^* units have been inspected and the number of defectives found is $H \leq M^* - E^*$. In this case the inspection of the lot is terminated and the present negative excess

$$E(N_L + N_0) = [H - (M^* - E^*)] \frac{(1-f)}{f}$$

is carried over to the next lot. (The serial number of the last element in the present lot is $N_L + N_0$ and

$$e_{(N_L+N_0)} = \frac{N_L e_{N_L} + H \dfrac{(1-f)}{f}}{N_L + N_0}.$$

Hence the present excess is

$$\begin{aligned}(N_L + N_0)[e_{(N_L+N_0)} - L] &= N_L e_{N_L} + H \frac{(1-f)}{f} - LN_L - LN_0 \\ &= N_L(e_{N_L} - L) + H \frac{(1-f)}{f} - M \\ &= \frac{(1-f)}{f} [H - M^* + E^*],\end{aligned}$$

as given above.)

We note an important property of SPB: The excess carried over from a preceding lot is never positive.

7. Possible modifications of the SP to achieve local stability. Although the sampling plans discussed in previous sections are optimum in the sense that they guarantee the desired AOQL with a minimum of inspection when the production process is in statistical control, they do not always behave very favorably as far as local stability is concerned. To make this point clear, consider the following example: Suppose that during a very long initial time period the production process functions very well and the relative frequency of defectives produced is well below L. Thus, applying SPA, say, $e_N - L$ will be considerably less than zero at the end of this period. Now suppose that then the production process suddenly deteriorates and the number of defectives produced during the next period of time is considerably higher than L. In spite of that, complete inspection will not begin for quite some time because e_N became so small during the initial period. Thus there will be a long segment in the sequence of outgoing units within which the relative frequency of defectives will be larger than the prescribed AOQL. Of course, this segment will be counterbalanced by other segments where the relative frequency of defectives will be below the AOQL, so that the AOQL will not be violated. Nevertheless, the occurrence of long segments with too many defectives, i.e., a lack of local stability, is not desirable.

It should be noted that, even though SPA was not designed to achieve considerable local stability, drastic lack of local stability cannot occur when the production process is in statistical control and SPA is employed. In the example given above where the outgoing quality was not locally stable, it was assumed that there were variations in the production process. The existence of statistical control acts as an important stabilizing factor on the quality.

In this section we want to discuss several possible modifications of SPA which will insure a greater degree of local stability. One such modification is the following: We choose a positive constant A and we define the excess E_N^* for each value N as follows: $E^*(N)$ is equal to the excess $E(N)$ as originally defined $(= N[e_N - L])$ as long as for all $N' \leq N$, $E(N') \geq -A$. The difference $E^*(N+1) - E^*(N) = E(N+1) - E(N)$ for all N for which $E(N+1) - E(N) \geq 0$. If $E(N+1) - E(N) < 0$, then $E^*(N+1) = \max[E^*(N) + \{E(N+1) - E(N)\}, -A]$. In other words, with this modification of the sampling inspection plan we set a lower bound $-A$ for the excess. When the excess is positive we begin complete inspection, and revert to partial inspection when the excess becomes non-positive. The effect of this is that, if the proportion of defectives produced becomes large, complete inspection will not be delayed very long, although the proportion of defectives produced in the preceding period may have been considerably below L. It is clear that this modification of SPA does not increase the AOQL. However, the amount of inspection will be somewhat increased, especially when the quality of the product is less than or only slightly greater than L. If the constant A is large, the increase in the amount of inspection is only slight, but also the degree of local stability achieved is not very high. On the other hand, if A is small, the increase in the amount of inspection may be considerable, but a high degree of local stability is achieved. Thus, the choice of A should be made so that a proper balance between local stability and amount of inspection is achieved.

Modifying SPA by setting a lower limit for the excess has the disadvantage that the mathematical treatment of this case is involved. We shall, therefore, consider another modification of the inspection plan which will have largely the same effect, but whose mathematical treatment appears to be much simpler. A fixed positive integer N_0 is chosen and the inspection scheme is designed so that $E_{N_0} \leq 0$ is assured. If E_{N_0} is negative, we replace it by zero. In other words, no excess is carried over from the first segment of N_0 units to the next segment of N_0 units. Thus, the second segment of N_0 units is treated exactly the same way as if it were the first segment, and this is repeated for each consecutive segment of N_0 units. This modification of SPA (the resulting plan is to be known as SPC) has essentially the same effect as setting a lower bound for the excess. Again it is clear that by this modification the AOQL is not increased, but the amount of inspection may be increased. The latter is particularly true when N_0 is small, which corresponds to very high local stability requirements. More efficient plans than SPC can probably be devised for this situation.

Undoubtedly, there are many other possible modifications of the inspection plan by which a greater degree of local stability can be achieved at the price of somewhat increased inspection. It is not the purpose of this paper to enumerate all these possibilities or to develop a theory as to which of them may be considered an optimum procedure. We shall restrict ourselves to a discussion of the mathematical consequences of SPC. First we define it precisely. If it is to be applied to inspection of lots of size N_0 then SPC is simply SPB with $E(N_L)$ and E^* always zero. When applied to continuous production it will operate as follows: Assume for convenience that $M = LN_0$, $N^* = fN_0$, and $\frac{fM}{1-f} = M^*$ are all integers.

(a) Begin each segment of N_0 units with partial inspection, i.e., inspect one unit chosen at random from each successive group of $\frac{1}{f}$ units. Continue partial inspection until one of the following events occurs: either

(b) M^* defectives are found. In this case begin complete inspection with the first unit which follows the group in which the last of the M^* defectives was found and continue until the end of the segment of N_0 units.

or

(b′) N^* groups of $\frac{1}{f}$ units are partially inspected.

(c) Repeat with the next segment of N_0 units.

Comparison with SPB shows that, in SPC, if (b) occurs earlier or at the same time as (b′), then $E_{N_0} = 0$, while if (b′) occurs before (b) we have $E_{N_0} < 0$. In contradistinction to SPB, in SPC there is no carrying over of the excess.

Let us determine the AOQ for SPC when the production process is in a state of statistical control. Denote by p the probability that a unit produced will be defective. Let the chance variable H denote the number of defectives found during partial inspection. The probability that $H = i < M^*$ is

$$\binom{N^*}{i} p^i (1 - p)^{N^* - i}.$$

$H \leq M^*$ always. We have, when $H = i$,

$$E(N_0) = \frac{(1 - f)i}{f} - LN_0 ,$$

and hence

$$N_0 e_{N_0} = \frac{(1 - f)i}{f}.$$

The AOQ is therefore $\frac{(1 - f)}{fN_0}$ multiplied by the expected value of H and is therefore

$$\frac{(1 - f)}{fN_0} \left[M^* - \sum_{i=0}^{M^*-1} (M^* - i) \binom{N^*}{i} p^i (1 - p)^{N^* - i} \right]$$

(7.1)

$$= L \left[1 - \frac{1}{M^*} \sum_{i=0}^{M^*-1} (M^* - i) \binom{N^*}{i} p^i (1 - p)^{N^* - i} \right].$$

The reduction from the original quality p to the AOQ was achieved by inspecting a fraction of units which is $\frac{1}{p}$ times the reduction in the frequency of defectives. Hence, with probability one, the fraction of units inspected when the production process is in statistical control is

(7.2) $$I = 1 - \frac{L}{p} + \frac{(1 - f)}{pN^*} \sum_{i=0}^{M^*-1} (M^* - i) \binom{N^*}{i} p^i (1 - p)^{N^* - i}.$$

When $p \geq \frac{L}{1 - f}$ we see from Section 4 that the third term of the right member of (7.2) represents the price paid in fraction of inspection above the minimum in return for the local stability achieved. When $p < \frac{L}{1 - f}$ the additional inspection is of course $I - f$.

As N_0 becomes larger, SPC becomes more and more like SPA, and consequently the amount of inspection tends to the minimum. As N_0 becomes smaller, the degree of local stability achieved becomes higher and must be paid for by an increasing amount of inspection. An illustrative example will be given in the next section. It has already been pointed out that the mere existence of statistical control implies a considerable amount of local stability even when SPA is applied.

The only practical difficulty which may arise in evaluating the formulas in (7.1) and (7.2) might come from attempting to evaluate

$$T' = \sum_{i=0}^{M^*-1} (M^* - i) \binom{N^*}{i} p^i (1 - p)^{N^*-i}.$$

For those values of the parameters which are likely to occur in application, a good approximation to T' (exactly how good we shall not investigate here) is given by

$$T = \sum_{i=0}^{M^*-1} (M^* - i) \frac{e^{-N^*p}(N^*p)^i}{i!}.$$

A table of T for integral values of M^* from 2 to 16 and for integral values of N^*p from 1 to 25 is given below. The computations were performed under the direction of Mr. Mortimer Spiegelman of the Metropolitan Life Insurance Company, to whom the authors are deeply obliged.

Table of $T = \sum_{i=0}^{M^*-1} (M^* - i) \frac{e^{-N^*p}(N^*p)^i}{i!}$

$M^* - 1$	N^*p											
	1	2	3	4	5	6	7	8	9	10	11	12
1	1.10	.54	.25	.11	.05	.02	.01	.00	.00	.00	.00	.00
2	2.02	1.22	.67	.35	.17	.08	.04	.02	.01	.00	.00	.00
3	3.00	2.08	1.32	.78	.44	.23	.12	.06	.03	.01	.01	.00
4	4.00	3.02	2.13	1.41	.88	.52	.29	.16	.08	.04	.02	.01
5	5.00	4.01	3.05	2.20	1.49	.96	.59	.35	.20	.11	.06	.03
6	6.00	5.00	4.02	3.08	2.26	1.57	1.04	.66	.41	.24	.14	.08
7	7.00	6.00	5.01	4.03	3.12	2.31	1.64	1.12	.73	.46	.28	.17
8	8.00	7.00	6.00	5.01	4.05	3.16	2.37	1.71	1.19	.79	.51	.32
9	9.00	8.00	7.00	6.00	5.02	4.08	3.20	2.43	1.77	1.25	.85	.56
10	10.00	9.00	8.00	7.00	6.01	5.03	4.10	3.24	2.48	1.83	1.31	.91
11	11.00	10.00	9.00	8.00	7.00	6.01	5.05	4.13	3.28	2.53	1.89	1.37
12	12.00	11.00	10.00	9.00	8.00	7.01	6.02	5.07	4.16	3.32	2.58	1.95
13	13.00	12.00	11.00	10.00	9.00	8.00	7.01	6.03	5.08	4.19	3.36	2.63
14	14.00	13.00	12.00	11.00	10.00	9.00	8.00	7.01	6.04	5.10	4.22	3.40
15	15.00	14.00	13.00	12.00	11.00	10.00	9.00	8.01	7.02	6.05	5.12	4.25

8. The SP of H. F. Dodge. H. F. Dodge [1] has proposed a very interesting SP for continuous production. The plan is defined by two constants i and f and may be described as follows: Begin with complete inspection of the units consecutively as produced and continue such inspection until i units in succession are found non-defective. Thereafter inspect a fraction f of the units. Continue partial inspection until a defect is found. Then start complete inspection again and continue until i units in succession are found non-defective. Repeat the procedure.

Dodge [1] derived formulas for determining the AOQL corresponding to any pair i and f, under the assumption that the production process is in a state of statistical control. Dodge's formulas for the AOQL are not necessarily valid if we do not make this restriction on the production process, i.e., if we admit that the probability p that a unit will be defective may vary in any arbitrary way during the production process. This, of course, is not a criticism of the derivation of the formulas; it cannot be considered surprising that a formula is not valid under assumptions different from those under which it was derived. However, it is relevant to point out the fact that the Dodge SP does not guarantee the AOQL under all circumstances, so that care must be taken to ensure that certain requirements are met. Exactly what these requirements are is not known; statistical control is a sufficient condition, but is probably not necessary and could be weakened. It seems likely to the authors that, if p varies only slowly (with N) with infrequent "jumps," the Dodge SP will produce results which will exceed the AOQL by little, if at all. But if the "jumps" are numerous and appropriately spaced it is possible to exceed the AOQL by substantial amounts, as the example below will show. The Dodge plan was intended to serve as an aid to the detection and correction of malfunctioning of the production process and this use would tend to prevent the occurrence of such a phenomenon. Parenthetically, it should be remarked that the information obtained in the course of inspection according to either the plans discussed in this paper or any reasonable scheme should, if possible, be sent at once to the producing divisions for their guidance.

Table of $T = \sum_{i=0}^{M^*-1} (M^* - i) \dfrac{e^{-N^*p}(N^*p)^i}{i!}$

(Continued)

$M^* - 1$	N^*p												
	13	14	15	16	17	18	19	20	21	22	23	24	25
1	.00	.00	.00	.00	.00	.00	.00	.00	.00	.00	.00	.00	.00
2	.00	.00	.00	.00	.00	.00	.00	.00	.00	.00	.00	.00	.00
3	.00	.00	.00	.00	.00	.00	.00	.00	.00	.00	.00	.00	.00
4	.01	.00	.00	.00	.00	.00	.00	.00	.00	.00	.00	.00	.00
5	.02	.01	.00	.00	.00	.00	.00	.00	.00	.00	.00	.00	.00
6	.04	.02	.01	.01	.00	.00	.00	.00	.00	.00	.00	.00	.00
7	.10	.05	.03	.02	.01	.00	.00	.00	.00	.00	.00	.00	.00
8	.20	.12	.07	.04	.02	.01	.01	.00	.00	.00	.00	.00	.00
9	.36	.23	.14	.08	.05	.03	.01	.01	.00	.00	.00	.00	.00
10	.61	.40	.26	.16	.10	.06	.03	.02	.01	.01	.00	.00	.00
11	.97	.66	.44	.29	.18	.11	.07	.04	.02	.01	.01	.00	.00
12	1.43	1.02	.71	.48	.32	.20	.13	.08	.05	.03	.02	.01	.01
13	2.00	1.48	1.07	.75	.52	.35	.23	.15	.09	.06	.03	.02	.01
14	2.68	2.05	1.54	1.12	.80	.55	.38	.25	.16	.10	.07	.04	.02
15	3.44	2.72	2.10	1.59	1.17	.84	.59	.41	.27	.18	.12	.07	.05

An example to show that the AOQL can be exceeded can be constructed as follows: Let $i = 54$ and $f = 0.1$. Then according to the graphs of [1], page 272, the AOQL should be 0.02. Define a sequence of 60 successive units free of defectives as a segment of type 1, and a sequence of 60 successive units where the production process is in statistical control with $p = 0.1$, as a segment of type 2. Suppose that the sequence of units produced consists of segments of types 1 and 2 always alternating. Then it follows that the first item inspected in a segment of type 2 is always inspected on a partial inspection basis. We now assume that, unless the occurrence of a defective has previously terminated partial inspection, the 1st, 11th, 21st, 31st, 41st, and 51st items in a segment of type 2 will be chosen for partial inspection, and if the 1st item is found defective, the entire segment of type 2 will be cleared of defectives. (Both of these assumptions favor the Dodge SP.) Then the situation is as described in the following table:

	(1) *Probability of first terminating partial inspection at each item*	(2) *Expected number of defectives remaining in segment of type 2 after partial inspection has been terminated*	(3) (1) x (2)
1st	.1	0	0
11th	$(.9)(.1) = .09$	.9	.081
21st	$(.9)^2(.1) = .081$	1.8	.1458
31st	$(.9)^3(.1) = .0729$	2.7	.19683
41st	$(.9)^4(.1) = .06561$	3.6	.236196
51st	$(.9)^5(.1) = .059049$	4.5	.2657205

Probability that an entire segment of type 2 will be partially inspected	*Expected number of defectives left in a segment of type 2 which has been inspected only partially*	*Product*
$(.9)^6 = .531441$	5.4	2.8697814

Sum = 3.7953279

The AOQ is therefore $\frac{3.7953279}{120} = .0316+$, while $L = .02$.

It is therefore difficult to compare the Dodge plan with any of the plans described in this paper with respect to their effect on a production process not in statistical control. If the production process is in statistical control, then, as we have already seen, SPA requires minimum inspection (and, incidentally, because of the existence of statistical control, produces a fair degree of local stability). If, when statistical control exists, one requires both maintenance of a given AOQL and a higher degree of local stability than is produced by SPA, the relevant comparison is between the Dodge plan and SPC. Both will probably give good results as regards local stability, but it is not possible at present to make these intuitive notions precise, as we have not given an exact definition of local stability. The following example (in which statistical control is assumed) may not be unrepresentative of what the situation is with regard to the amount of inspection required.

Fraction of product inspected under the Dodge plan and under SPC when
$L = .045$ $f = .1$

p	Fraction of product inspected under the Dodge plan	Fraction of product inspected under SPC when		
		$N_0 = 400$	$N_0 = 1000$	$N_0 = 2000$
.01	.12	.12	.10	.10
.02	.15	.17	.11	.10
.03	.19	.22	.14	.11
.04	.23	.28	.19	.15
.05	.28	.34	.26	.21
.06	.33	.40	.33	.29
.07	.39	.45	.39	.37
.08	.45	.50	.46	.44
.09	.52	.54	.51	.50
.10	.58	.57	.55	.55

The decrease in inspection required by SPC as N_0 increases is evident in this table. When $N_0 = 2000$ SPC requires less inspection than the Dodge plan, when $N_0 = 400$ it requires more inspection than the Dodge plan. How the various degrees of local stability achieved compare remains an open question. The case when $N_0 = 400$ probably lies in the region where SPC is inefficient (as regards amount of inspection) and corresponds to a high degree of local stability.

We note that both plans call for increased inspection as the quality worsens (p increases). If the manufacturer is required to pay for the inspection this serves as an added incentive to improve quality of output.

REFERENCES

[1] H. F. Dodge, *Annals of Math. Stat.*, 14 (1943), p. 264.
[2] A. Kolmogoroff, *Comptes Rendus Acad. Sciences*, 191 (1930), p. 910.
[3] Maurice Fréchet, *Généralités sur les Probabilités. Variables aléatoires.* Paris, 1937.
[4] H. F. Dodge and H. G. Romig, *Bell System Tech. Journal*, 20 (1941), p. 1.

Made in United States of America

Reprinted from THE ANNALS OF MATHEMATICAL STATISTICS
Vol. XVII, No. 2, June, 1946

TOLERANCE LIMITS FOR A NORMAL DISTRIBUTION[1]

BY A. WALD AND J. WOLFOWITZ

Columbia University and *University of North Carolina*

Summary. The problem of constructing tolerance limits for a normal universe is considered. The tolerance limits are required to be such that the probability is equal to a preassigned value β that the tolerance limits include at least a given proportion γ of the population. A good approximation to such tolerance limits can be obtained as follows: Let $\bar{x}$ denote the sample mean and s^2 the sample estimate of the variance. Then the approximate tolerance limits are given by

$$\bar{x} - \sqrt{\frac{n}{\chi^2_{n,\beta}}}\, rs \quad \textit{and} \quad \bar{x} + \sqrt{\frac{n}{\chi^2_{n,\beta}}}\, rs$$

where n is one less than the number N of observations, $\chi^2_{n,\beta}$ denotes the number for which the probability that χ^2 with n degrees of freedom will exceed this number is β, and r is the root of the equation

$$\frac{1}{\sqrt{2\pi}} \int_{1/\sqrt{N}-r}^{1/\sqrt{N}+r} e^{-t^2/2}\, dt = \gamma.$$

The number $\chi^2_{n,\beta}$ can be obtained from a table of the χ^2 distribution and r can be determined with the help of a table of the normal distribution.

1. Introduction. The problem of setting tolerance limits for a distribution on the basis of an observed sample was discussed by S. S. Wilks [1], [2] and by one of the present authors [3], [4]. For a univariate distribution the problem may be formulated briefly as follows: Let x be the chance variable under consideration and let $x_1, \cdots, x_N$ be a sample of N independent observations on x. Two functions, L_1 and L_2, of the sample are to be constructed such that the probability that the limits L_1 and L_2 will include at least a given proportion γ of the population is equal to a preassigned value β. The limits L_1 and L_2 are called tolerance limits.

The following two cases have been treated in the literature: (1) Nothing is known about the distribution of x, except perhaps that it is continuous, or that it admits a continuous probability density function. (2) The functional form of the distribution of x is known and only the values of a finite number of parameters involved in the distribution of x are unknown. We shall refer to (1) as the non-

[1] This paper reports work done by the authors in the Statistical Research Group, Division of War Research, Columbia University, under contract OEMsr-618 with the Applied Mathematics Panel, National Defense Research Committee. The work was first reported in an unpublished memorandum, "Tolerance Limits for a Normal Distribution" (SRG number 392, 3 January 1945) written by the authors, of whom one was a staff member and the other a consultant of the Group. The problem was suggested by W. Allen Wallis on the grounds that the limits previously proposed (see [4], section 5) are unsatisfactory for most practical purposes.

parametric case and to (2) as the parametric case. An exact solution of the problem for univariate distributions in the non-parametric case has been given by S. S. Wilks [1]. His results have been extended to multivariate distributions by one of the present authors [3]. An asymptotic solution of the problem in the parametric case, which may be used for large samples, was given in [4].[2]

In the present paper we shall deal with the problem of setting tolerance limits for a normal distribution with unknown mean and variance. Approximation formulas are obtained which differ from the exact values by a magnitude of the order $1/N^2$. They give much closer approximations to the exact values than those which can be obtained by applying the general asymptotic results in [4] to the normal distribution. In addition, the approximation formulas in the present paper have the advantage of considerable simplicity and can easily be computed with the help of tables of the normal and χ^2 distributions. To estimate the closeness of the approximation of the formulas given in this paper, a method of computing upper and lower limits for the exact values has been derived. Computations show that the approximation is good even for small values of N. A few numerical examples are given in section 7.

2. Precise formulation of the problem and notation. Let $x_1, \cdots, x_N$ be N independent observations from a normal population with mean μ and variance σ^2, both unknown. We shall denote by $\bar{x}$ the arithmetic mean of the observations and by s^2 the sample estimate of the population variance σ^2, i.e.,

$$\bar{x} = \frac{\sum_{i=1}^{N} x_i}{N} \tag{2.1}$$

and

$$s^2 = \frac{\sum (x_i - \bar{x})^2}{n}, \quad \text{where } n = N - 1 . \tag{2.2}$$

For any positive λ we shall denote by $A(\bar{x}, s, \lambda)$, or more briefly by A, the proportion of the normal universe included between the limits $\bar{x} - \lambda s$ and $\bar{x} + \lambda s$, i.e.,

$$A = A(\bar{x}, s, \lambda) = \frac{1}{\sqrt{2\pi}\,\sigma}\int_{\bar{x}-\lambda s}^{\bar{x}+\lambda s} e^{-(1/2\sigma^2)(t-\mu)^2}\,dt . \tag{2.3}$$

A is a chance variable, since the limits of integration are chance variables. In this paper we shall deal with the problem of determining the value of λ so that the probability that A exceeds a preassigned value γ is equal to a preassigned value β. The desired tolerance limits will then be given by $\bar{x} - \lambda s$ and $\bar{x} + \lambda s$, respectively. In practice, the values β and γ will usually be chosen near unity, frequently $\geq .95$.

[2] Although the results obtained in the non-parametric case could be applied to the parametric case as well, it would not be satisfactory to do so, since for the parametric case methods having greater efficiency can be devised by taking into account the available information regarding the functional form of the distribution.

It can be verified that the distribution of A does not depend on the unknown parameters μ and σ. Thus we can assume without loss of generality that $\mu = 0$ and $\sigma = 1$.

For any given positive value λ we shall denote by $P(\gamma,\lambda)$ the probability that $A > \gamma$. For a given value $\bar{x}$ we shall denote by $P(\gamma,\lambda \mid \bar{x})$ the conditional probability that $A > \gamma$ under the condition that the sample mean has a given value $\bar{x}$. It is clear that $P(\gamma,\lambda)$ is equal to the expected value of $P(\gamma,\lambda \mid \bar{x})$, i.e., ·

$$P(\gamma,\lambda) = \frac{\sqrt{N}}{\sqrt{2\pi}} \int_{-\infty}^{+\infty} P(\gamma, \lambda \mid \bar{x})\, e^{-\frac{1}{2}N\bar{x}^2}\, d\bar{x}\,. \tag{2.4}$$

3. Method of computing $P(\gamma,\lambda \mid \bar{x})$ for any given values γ,λ and $\bar{x}$. Since $A = A(\bar{x},s,\lambda)$ is a strictly increasing function of s, the equation in s

$$A(\bar{x},s,\lambda) = \gamma \tag{3.1}$$

has exactly one root in s. Denote this root by

$$s = r(\bar{x},\gamma,\lambda). \tag{3.2}$$

Thus, $r(\bar{x},\gamma,\lambda)$ is that value for which

$$\frac{1}{\sqrt{2\pi}} \int_{\bar{x}-\lambda r(\bar{x},\gamma,\lambda)}^{\bar{x}+\lambda r(\bar{x},\gamma,\lambda)} e^{-t^2/2}\, dt = \gamma. \tag{3.3}$$

It is clear that $\lambda r(\bar{x},\gamma,\lambda)$ does not depend on λ. We shall write

$$\lambda r(\bar{x},\gamma,\lambda) = r(\bar{x},\gamma). \tag{3.4}$$

Obviously $r(\bar{x},\gamma)$ is that value for which

$$\frac{1}{\sqrt{2\pi}} \int_{\bar{x}-r(\bar{x},\gamma)}^{\bar{x}+r(\bar{x},\gamma)} e^{-t^2/2}\, dt = \gamma. \tag{3.5}$$

For given values of $\bar{x}$ and γ the value $r(\bar{x},\gamma)$ can be obtained from a table of the normal distribution.

Since $A(\bar{x},s,\lambda)$ is a strictly increasing function of s, the inequality $A(\bar{x},s,\lambda) > \gamma$ is equivalent to the inequality $s > r(\bar{x},\gamma,\lambda) = r(\bar{x},\gamma)/\lambda$. Hence, since $\bar{x}$ and s are independently distributed, we have

$$P(\gamma,\lambda \mid \bar{x}) = P(s > r(\bar{x},\gamma)/\lambda) \tag{3.6}$$

where $P(s > c)$ denotes the probability that $s > c$ for any constant c. In genral, for any relation R we shall denote by $P(R)$ the probability that R holds.

Since ns^2 has the χ^2 distribution with $n = N - 1$ degrees of freedom, we have

$$P\left(s > \frac{r(\bar{x}, \gamma)}{\lambda}\right) = P\left(\chi_n^2 > \frac{nr^2(\bar{x}, \gamma)}{\lambda^2}\right) \tag{3.7}$$

where χ_n^2 stands for a random variable which has the χ^2 distribution with n degrees of freedom. The probability on the right-hand side of (3.7) can be obtained from a table of the χ^2 distribution.

Hence, we see that the computation of $P(\gamma,\lambda \mid \bar{x})$ for given values γ,λ and $\bar{x}$ can be carried out in two simple steps. First we determine the value of $r(\bar{x},\gamma)$ from a table of the normal distribution and then read the value of

$$P\left(\chi_n^2 > \frac{nr^2(\bar{x},\gamma)}{\lambda^2}\right)$$

from a table of the χ^2 distribution.

4. Proof that the difference $P\left(\gamma,\lambda \,\middle|\, \frac{1}{\sqrt{N}}\right) - P(\gamma,\lambda)$ is of the order $1/N^2$. It is clear that $P(\gamma,\lambda \mid \bar{x})$ is an even function of $\bar{x}$. Hence, in the expansion of $P(\gamma,\lambda \mid \bar{x})$ in a power series in $\bar{x}$, only even powers will occur. Terminating the Taylor expansion (in section 8 we prove its validity) at the fourth term, we have

$$(4.1)\quad P(\gamma,\lambda \mid \bar{x}) = P(\gamma,\lambda \mid 0) + \frac{\bar{x}^2}{2}\,\frac{\partial^2 P(\gamma,\lambda \mid \bar{x})}{\partial \bar{x}^2}\bigg|_{\bar{x}=0} + \frac{\bar{x}^4}{4!}\,\frac{\partial^4 P(\gamma,\lambda \mid \bar{x})}{\partial \bar{x}^4}\bigg|_{\bar{x}=\xi}$$

where $0 \leq \xi \leq \bar{x}$.

The expected value of $P(\gamma,\lambda \mid \bar{x})$ (considering $\bar{x}$ as a random variable) is equal to $P(\gamma,\lambda)$. Since the expected value of $\bar{x}^2$ is $1/N$ and the expected value of

$$\frac{\bar{x}^4}{4!}\,\frac{\partial^4 P}{\partial \bar{x}^4}\bigg|_{\bar{x}=\xi}$$

is of the order $1/N^2$ (this is proved in section 9), we obtain from (4.1)

$$(4.2)\qquad P(\gamma,\lambda) = P(\gamma,\lambda \mid 0) + \frac{1}{2N}\,\frac{\partial^2 P}{\partial \bar{x}^2}\bigg|_{\bar{x}=0} + 0\left(\frac{1}{N^2}\right).$$

On the other hand, substituting $1/\sqrt{N}$ for $\bar{x}$ in (4.1) we obtain

$$(4.3)\qquad P\left(\gamma,\lambda \,\middle|\, \frac{1}{\sqrt{N}}\right) = P(\gamma,\lambda \mid 0) + \frac{1}{2N}\,\frac{\partial^2 P}{\partial \bar{x}^2}\bigg|_{\bar{x}=0} + \frac{1}{4!N^2}\,\frac{\partial^4 P}{\partial \bar{x}^4}\bigg|_{\bar{x}=\xi'}$$

where $0 \leq \xi' \leq 1/\sqrt{N}$. Hence, since the second term of the right member of (4.3) is of the order $1/N^2$,

$$(4.4)\qquad P\left(\gamma,\lambda \,\middle|\, \frac{1}{\sqrt{N}}\right) = P(\gamma,\lambda \mid 0) + \frac{1}{2N}\,\frac{\partial^2 P}{\partial \bar{x}^2}\bigg|_{\bar{x}=0} + 0\left(\frac{1}{N^2}\right).$$

From (4.2) and (4.4) it follows that

$$(4.5)\qquad P(\gamma,\lambda) - P\left(\gamma,\lambda \,\middle|\, \frac{1}{\sqrt{N}}\right) = 0\left(\frac{1}{N^2}\right).$$

Thus, this difference approaches zero rapidly as $N \to \infty$.

5. Computation of the value λ for which $P\left(\gamma,\lambda \,\middle|\, \frac{1}{\sqrt{N}}\right)$ takes a preassigned value β. Denote by $\chi^2_{n,\beta}$ that value for which $P(\chi_n^2 > \chi^2_{n,\beta}) = \beta$. This value can be obtained from a table of the χ^2 distribution. From (3.6) and (3.7) it follows that the required value λ^* of λ is given by the root of the equation

$$\frac{n}{\lambda^2} r^2\left(\frac{1}{\sqrt{N}}, \gamma\right) = \chi^2_{n,\beta}. \tag{5.1}$$

Thus, the desired value of λ^* is given by

$$\lambda^* = \sqrt{\frac{n}{\chi^2_{n,\beta}}}\, r\left(\frac{1}{\sqrt{N}}, \gamma\right). \tag{5.2}$$

The value $r\left(\frac{1}{\sqrt{N}}, \gamma\right)$ is defined by (3.5) and can be obtained from a table of the normal distribution.[3]

6. Lower and upper limits for $P(\gamma,\lambda)$. As mentioned in section 2, $P(\gamma,\lambda)$ is equal to the expected value of $P(\gamma,\lambda \mid \bar{x})$. Thus,

$$P(\gamma,\lambda) = \frac{\sqrt{N}}{\sqrt{2\pi}} \int_{-\infty}^{+\infty} P(\gamma,\lambda \mid \bar{x}) e^{-\frac{1}{2}N\bar{x}^2}\, d\bar{x}. \tag{6.1}$$

To obtain upper and lower limits for $P(\gamma,\lambda)$, we shall construct upper and lower limits for the integral on the right-hand side of (6.1). It can easily be seen that $P(\gamma,\lambda \mid \bar{x})$ is a strictly decreasing function of $\bar{x}^2$. Hence, to obtain lower and upper limits for the integral in the right member of (6.1) we can proceed as follows: Choose a positive constant d and a positive integer k. Denote by a_i the probability that $id \leq \bar{x} \leq (i+1)d$, $(i = 0, 1, \cdots, k-1)$, and let a_k be the probability that $\bar{x} > kd$. Then $2\sum_{i=0}^{k} a_i P(\gamma,\lambda \mid id)$ is an upper bound, and $2\sum_{i=1}^{k} a_{i-1} P(\gamma,\lambda \mid id)$ is a lower bound of the integral in question. Thus

$$P(\gamma,\lambda) \geq 2\sum_{i=1}^{k} a_{i-1} P(\gamma,\lambda \mid id) \tag{6.2}$$

and

$$P(\gamma,\lambda) \leq 2\sum_{i=0}^{k} a_i P(\gamma,\lambda \mid id). \tag{6.3}$$

The two limits can be brought arbitrarily close to each other by choosing d sufficiently small and k sufficiently large. A method of computing $P(\gamma,\lambda \mid \bar{x})$ for any given value $\bar{x}$ has been described in section 3 and the quantities a_i can be obtained from a table of the normal distribution. The amount of computational work, however, increases rapidly with increasing k.

[3] The Statistical Research Group computed, under the supervision of Albert H. Bowker, a table of tolerance limit factors λ (see formula 5.2) for β = .75, .90, .95, .99; γ = .75, .90, .95, .99, .999; N = 2 (1) 102 (2) 180 (5) 300 (10) 400 (25) 750 (50) 1000. Mr. Bowker also developed an asymptotic formula for λ (published elsewhere in this issue of the *Annals*) which, when $\beta \leq .99$, $\gamma \leq .999$, and N $\geq$ 160, agrees with (5.2) to within 1 unit in the third significant figure. The Applied Mathematics Panel plans to publish the table and a brief explanation of tolerance limits in the volume entitled *Techniques of Statistical Analysis* described in the footnote on page 217.

7. Approximate determination of the tolerance limits. The exact tolerance limits are given by $\bar{x} - \lambda s$ and $\bar{x} + \lambda s$ where λ is the root of the equation in λ

$$P(\gamma,\lambda) = \beta. \tag{7.1}$$

This equation has exactly one root in λ, since $P(\gamma,\lambda)$ is a strictly increasing function of λ. Denote this root by $\lambda = \lambda(\beta,\gamma)$. Thus, the exact tolerance limits are given by $\bar{x} - \lambda(\beta,\gamma)s$ and $\bar{x} + \lambda(\beta,\gamma)s$.

We have seen in section 4 that $P\left(\gamma,\lambda \mid \frac{1}{\sqrt{N}}\right)$ closely approximates $P(\gamma,\lambda)$, the difference being of the order $1/N^2$. Thus, a close approximation to $\lambda(\beta,\gamma)$ can be obtained by solving the equation in λ,

$$P\left(\gamma, \lambda \mid \frac{1}{\sqrt{N}}\right) = \beta. \tag{7.2}$$

This equation has again exactly one root in λ, since $P\left(\gamma,\lambda \mid \frac{1}{\sqrt{N}}\right)$ is a strictly increasing function of λ. Denote the root of equation (7.2) by $\lambda = \lambda^*(\beta,\gamma)$. Thus approximate tolerance limits are given by $\bar{x}-\lambda^*(\beta,\gamma)s$ and $\bar{x}+\lambda^*(\beta,\gamma)s$. In section 5 it has been shown that

$$\lambda^*(\beta,\gamma) = \sqrt{\frac{n}{\chi^2_{n,\beta}}}\, r \tag{7.3}$$

where $n = N-1$, $\chi^2_{n,\beta}$ is that number for which the probability that χ^2 with n degrees of freedom exceeds this number is β, and r is the root of the equation

$$\frac{1}{\sqrt{2\pi}} \int_{1/\sqrt{N}-r}^{1/\sqrt{N}+r} e^{-t^2/2}\, dt = \gamma\,. \tag{7.4}$$

The number $\chi^2_{n,\beta}$ can be obtained from a table of the χ^2 distribution and r can be determined from a table of the normal distribution.

Since $\lambda^*(\beta,\gamma)$ is only an approximation to $\lambda(\beta,\gamma)$, $P[\gamma,\lambda^*(\beta,\gamma)]$ will differ slightly from β. To judge the goodness of the approximation of $\lambda^*(\beta,\gamma)$ to the exact value $\lambda(\beta,\gamma)$, it is desirable to derive upper and lower limits for the difference $P[\gamma,\lambda^*(\beta,\gamma)] - \beta$. Such limits can be obtained by computing upper and lower limits for $P[\gamma,\lambda^*(\beta,\gamma)]$ using the method described in section 6.

We cite here a few numerical examples to show the goodness of the approximation.

N	γ	β	$\lambda^*(\beta,\gamma)$	Upper limit of $P[\gamma,\lambda^*(\beta,\gamma)]$	Lower limit of $P[\gamma,\lambda^*(\beta,\gamma)]$
2	.95	.95	37.674	.95202	.95077
9	.95	.99	4.550	.98989	.98908
25	.95	.95	2.631	.95161	.94393
25	.95	.99	2.972	.99024	.98813

8. Validity of the Taylor expansion of $P(\gamma,\lambda \mid \bar{x})$. We shall show that $P(\gamma,\lambda \mid \bar{x})$ has derivatives of all orders at every point $\bar{x}$, γ and λ being fixed. This is sufficient to validate the Taylor expansion used in section 4.

For typographical convenience write

$$r(\bar{x},\gamma) = R.$$

We have

$$\frac{1}{\sqrt{2\pi}} \int_{\bar{x}-R}^{\bar{x}+R} e^{-\frac{1}{2}t^2}\, dt = \gamma. \tag{8.1}$$

Differentiating (8.1) with respect to $\bar{x}$ we obtain

$$\left(1 + \frac{dR}{d\bar{x}}\right) e^{-\frac{1}{2}(\bar{x}+R)^2} = \left(1 - \frac{dR}{d\bar{x}}\right) e^{-\frac{1}{2}(\bar{x}-R)^2} \tag{8.2}$$

whence

$$\frac{dR}{d\bar{x}} = \tanh \bar{x}R\,. \tag{8.3}$$

Now the analytic function $\tanh z$ of the complex variable z has only purely imaginary singularities. Hence R possesses derivatives of all orders for all real values of $\bar{x}$.

Now

$$P(\gamma,\lambda \mid \bar{x}) = P\left(s > \frac{R}{\lambda}\right) = 1 - k \int_0^R t^{n-1} e^{-nt^2/(2\lambda^2)}\, dt$$

where k is a constant. Hence from (8.3)

$$\frac{\partial P}{\partial \bar{x}} = -kR^{n-1} e^{-nR^2/(2\lambda^2)} \tanh \bar{x}R\,. \tag{8.4}$$

The right member of (8.4) is a product of functions which are analytic in the entire (complex) R plane by a function which possesses derivatives of all orders for every real $\bar{x}$. Since R possesses a derivative (with respect to $\bar{x}$) for all real $\bar{x}$, it follows that P possesses derivatives of all orders for every real $\bar{x}$.

9. Proof that

$$E\left[\frac{\bar{x}^4}{4!} \frac{\partial^4 P}{\partial \bar{x}^4}\bigg|_{\bar{x}=\xi}\right] = 0\left(\frac{1}{N^2}\right).$$

Since R is a minimum at $\bar{x} = 0$ it follows that $P(\gamma,\lambda \mid \bar{x})$ has a maximum there. Hence, from (4.1), the quantity

$$\bar{x}^2 \left(\frac{1}{2}\right) \frac{\partial^2 P}{\partial \bar{x}^2}\bigg|_{\bar{x}=0} + \frac{\bar{x}^4}{4!} \frac{\partial^4 P}{\partial \bar{x}^4}\bigg|_{\bar{x}=\xi}$$

is never positive. Therefore

$$\left.\frac{\partial^4 P}{\partial \bar{x}^4}\right|_{\bar{x}=\xi} \leq -\left.\frac{12}{x^2}\frac{\partial^2 P}{\partial \bar{x}^2}\right|_{\bar{x}=0}.$$

Consequently $\left.\frac{\partial^4 P}{\partial \bar{x}^4}\right|_{\bar{x}=\xi}$ is bounded above for $|\bar{x}| \geq \delta$, where $\delta > 0$ is arbitrarily small. Since P possesses everywhere derivatives of all orders, the fourth derivative is continuous and hence bounded above for $|\bar{x}| \leq \delta$. From this we obtain that $\left.\frac{\partial^4 P}{\partial \bar{x}^4}\right|_{\bar{x}=\xi}$ is bounded above for every real $\bar{x}$.

Since $P(\gamma,\lambda \mid \bar{x})$ is always positive we have, from (4.1), that

$$\left.\frac{\partial^4 P}{\partial \bar{x}^4}\right|_{\bar{x}=\xi} \geq -\frac{12\left(2P + \bar{x}^2 \left.\frac{\partial^2 P}{\partial \bar{x}^2}\right|_{\bar{x}=0}\right)}{\bar{x}^4}.$$

For $|\bar{x}|$ greater than a sufficiently large number C, the left member of the above inequality is thus bounded below. For $|\bar{x}| \leq C$ we have that $\left.\frac{\partial^4 P}{\partial \bar{x}^4}\right|_{\bar{x}=\xi}$ is bounded below because $\frac{\partial^4 P}{\partial \bar{x}^4}$ is continuous. Hence $\left.\frac{\partial^4 P}{\partial \bar{x}^4}\right|_{\bar{x}=\xi}$ is bounded below for every real $\bar{x}$.

Since $\left.\frac{\partial^4 P}{\partial \bar{x}^4}\right|_{\bar{x}=\xi}$ is bounded above and below for every real $\bar{x}$, the desired result follows.

REFERENCES

[1] S. S. Wilks, "Determination of sample sizes for setting tolerance limits," *Annals of Math. Stat.*, Vol. 12 (1941), pp. 91–96.

[2] S. S. Wilks, "Statistical prediction with special reference to the problem of tolerance limits," *Annals of Math. Stat.*, Vol. 13 (1942), pp. 400–409.

[3] A. Wald, "An extension of Wilks' method for setting tolerance limits," *Annals of Math. Stat.*, Vol. 14 (1943), pp. 45–55.

[4] A. Wald, "Setting of tolerance limits when the sample is large," *Annals of Math. Stat.*, Vol. 13 (1942), pp. 389–399.

Made in United States of America

Reprinted from The Annals of Mathematical Statistics
Vol. XVII, No. 4, December, 1946

SOME IMPROVEMENTS IN SETTING LIMITS FOR THE EXPECTED NUMBER OF OBSERVATIONS REQUIRED BY A SEQUENTIAL PROBABILITY RATIO TEST

By Abraham Wald

Columbia University

Summary. Upper and lower limits for the expected number n of observations required by a sequential probability ratio test have been derived in a previous publication [1]. The limits given there, however, are far apart and of little practical value when the expected value of a single term z in the cumulative sum computed at each stage of the sequential test is near zero. In this paper upper and lower limits for the expected value of n are derived which will, in general, be close to each other when the expected value of z is in the neighborhood of zero. These limits are expressed in terms of limits for the expected values of certain functions of the cumulative sum Z_n at the termination of the sequential test.

In section 7 a general method is given for determining limits for the expected value of any function of Z_n.

1. Introduction. Let x be a random variable and let $f(x, \theta)$ be the elementary probability law of x involving an unknown parameter θ. Let H_0 denote the hypothesis that $\theta = \theta_0$, and H_1 the hypothesis that $\theta = \theta_1$, where θ_0 and θ_1 are given specified values. The sequential probability ratio test for testing H_0 against H_1, as defined in [1], is given as follows: Put

$$z_i = \log \frac{f(x_i, \theta_1)}{f(x_i, \theta_0)} \tag{1.1}$$

where x_i denotes the i-th observation on x. Two constants, a and b are chosen where $a > 0$ and $b < 0$. At each stage of the experiment, at the m-th trial for each positive integral value m, the cumulative sum

$$Z_m = z_1 + \cdots + z_m \tag{1.2}$$

is computed. Experimentation is continued as long as $b < Z_m < a$. The first time that Z_m does not lie between b and a, experimentation is terminated. The hypothesis H_1 is accepted if $Z_m \geqq a$, and H_0 is accepted if $Z_m \leqq b$.

Let n denote the smallest value of m for which Z_m does not lie between b and a. Then n is the number of observations required by the sequential test. The expected value of n is a function of the true parameter value θ and is denoted by $E_\theta(n)$.

Upper and lower limits for $E_\theta(n)$ have been derived in section 4 of [1]. These limits, however, are of little practical value when the expected value of

$$z = \log \frac{f(x, \theta_1)}{f(x, \theta_0)} \tag{1.3}$$

is in the neighborhood of zero, for they converge to $+\infty$ and $-\infty$, respectively, as the expected value of z approaches zero. It can be shown that the expected value of z is negative when $\theta = \theta_0$, and positive when $\theta = \theta_1$.[1] Thus, if the expected value of z is a continuous function of θ, there will be a value θ' between θ_0 and θ_1 such that the expected value of z is zero when $\theta = \theta'$. Hence, the limits for $E_\theta(n)$, as given in [1], are of no practical value when θ is near θ'.

The purpose of this paper is to derive upper and lower limits for $E_\theta(n)$ which will be, in general, close to each other when θ is in the neighborhood of θ'. Thus, it will generally be possible to obtain close limits for $E_\theta(n)$ over the whole range of θ, if the limits given here are used for values in a certain small interval containing θ', and the limits given in [1] are used when θ is outside this interval.

2. Notation. We shall use the following notations throughout the paper. For any random variable u, the symbol $E_\theta(u)$ will denote the expected value of u when θ is the true value of the parameter. The conditional expected value of u, under the restriction that some relationship R is fulfilled will be denoted by $E_\theta(u \mid R)$. The symbol $P(R \mid \theta)$ will denote the probability that the relationship R holds when θ is true.

The cumulative distribution function of z will be denoted by $F(z, \theta)$ when θ is the true value of the parameter. The moment generating function of z, when θ is true, will be denoted by $\varphi(t, \theta)$, i.e.

$$\varphi(t, \theta) = \int_{-\infty}^{\infty} e^{tz} \, dF(z, \theta). \tag{2.1}$$

3. Assumptions concerning the family of distribution functions $F(z, \theta)$. In this section we shall formulate two assumptions concerning $F(z, \theta)$ which will then be used to prove various lemmas and theorems. Since we are interested in values of θ near θ', we shall restrict the domain of θ to a finite closed interval I containing θ' in its interior. It will be understood throughout the paper that any statements concerning θ refer to the domain I, even if this is not explicitly stated.

ASSUMPTION 1. *The moment generating function $\varphi(t, \theta)$ exists for any point t in the complex plane and any value θ, and is a continuous function of θ.*

ASSUMPTION 2. *There eists a positive δ such that $P(e^z > 1 + \delta \mid \theta)$ and $P(e^z < 1 - \delta \mid \theta)$ have positive lower bounds with respect to θ.*

4. *Proof that $\varphi(t, \theta)$ is continuous in t and θ jointly and that all moments of z are continuous functions of θ.*[2] In this section we shall prove the following theorem:

[1] This follows easily from Lemma 1 in [1], p. 156.

[2] The original proof of the author was somewhat lengthy. The present proof was suggested by T. E. Harris.

THEOREM 4.1. *It follows from Assumption 1 that $\varphi(t, \theta)$ is continuous in t and θ jointly and all moments of z are continuous functions of θ.*

PROOF: First we show that $\varphi(t, \theta)$ is a bounded function of t and θ in the domain $|t| \leqq t_0$, for any finite positive value t_0. Clearly,

$$0 \leqq |\varphi(t, \theta)| \leqq 2[\varphi(t_0, \theta) + \varphi(-t_0, \theta)] \tag{4.1}$$

for all values t for which $|t| \leqq t_0$. The boundedness of $\varphi(t_0, \theta)$ and $\varphi(-t_0, \theta)$ follows from Assumption 1. Hence $\varphi(t, \theta)$ is a bounded function of θ and t over any bounded t-domain.

Let $\{t_m, \theta_m\}$ ($m = 1, 2, \cdots$, ad inf.) be a sequence of pairs converging to the pair (t', θ'). We have

$$\varphi(t_m, \theta_m) - \varphi(t', \theta') = [\varphi(t_m, \theta_m) - \varphi(t', \theta_m)] + [\varphi(t', \theta_m) - \varphi(t', \theta')]. \tag{4.2}$$

The second expression in brackets converges to zero by continuity in θ. Thus the first part of Theorem 4.1 is proved if we show that

$$\lim_{m \to \infty} [\varphi(t_m, \theta_m) - \varphi(t', \theta_m)] = 0. \tag{4.3}$$

It follows from Assumption 1 that for any given θ, $\varphi(t, \theta)$ is an analytic function with no singularities in any finite t-domain. Hence we can expand $\varphi(t_m, \theta_m)$ in a Taylor series around $t = t'$, i.e.

$$\varphi(t_m, \theta_m) - \varphi(t', \theta_m) = \sum_{k=1}^{\infty} \frac{1}{k!} \left(\frac{\partial^k \varphi(t, \theta_m)}{\partial t^k} \bigg|_{t=t'} \right) (t_m - t')^k. \tag{4.4}$$

Let r be a given positive value. Because of the boundedness of $\varphi(t, \theta)$ in any finite t-domain, there exists a constant M such that $|\varphi(t, \theta)| < M$ for all θ and for all t in the domain $|t - t'| \leqq r$. From the Cauchy integral formula for an analytic function it follows that

$$\frac{1}{k!} \left| \frac{\partial^k \varphi(t, \theta_m)}{\partial t^k} \bigg|_{t=t'} \right| \leqq \frac{M}{r^k}. \tag{4.5}$$

From (4.4) and (4.5) we obtain

$$|\varphi(t_m, \theta_m) - \varphi(t', \theta_m)| \leqq M \sum_{k=1}^{\infty} \frac{|t_m - t'|^k}{r^k}. \tag{4.6}$$

Equation (4.3) is an immediate consequence of (4.6). This proves the first half of Theorem 4.1.

Let C be a circle in the complex t-plane with finite radius and center at the origin. According to the Cauchy integral formula we have

$$\frac{1}{2\pi i} \int_C \frac{\varphi(t, \theta)}{t^{k+1}}\, dt = \frac{1}{k!} \frac{\partial^k \varphi(t, \theta)}{\partial t^k} \bigg|_{t=0} = \frac{1}{k!} E_\theta(z^k). \tag{4.7}$$

Since $\varphi(t, \theta)$ is continuous in t and θ jointly, the integral on the left hand side of (4.7) is a continuous function of θ. This proves the second half of Theorem 4.1.

5. Some lemmas. In this section we shall prove several lemmas which will then be used to derive the results contained in sections 6 and 8.

LEMMA 5.1. *It follows from assumptions* 1 *and* 2 *that for any given* θ *the equation in* t

$$\varphi(t, \theta) = 1 \tag{5.1}$$

has exactly two real roots, one of which is zero. The other real root is different from zero if $E_\theta(z) \neq 0$. *If* $E_\theta(z) = 0$, *both roots are equal to zero, i.e., zero is a double root of* (5.1).

This lemma is essentially the same as Lemma 2 in [2] and the proof is therefore omitted.[3]

Let $h(\theta)$ denote the non-zero root of (5.1), if $E_\theta(z) \neq 0$. If $E_\theta(z) = 0$, we put $h(\theta) = 0$.

In what follows the variable t will be restricted to real values, unless the contrary is explicitly stated.

LEMMA 5.2. *It follows from assumptions* 1 *and* 2 *that* $h(\theta)$ *is a continuous function of* θ.

PROOF: It follows from assumption 2 that

$$\lim_{t \to \pm\infty} \varphi(t, \theta) = +\infty \tag{5.2}$$

uniformly in θ. Hence, since by definition

$$\varphi[h(\theta), \theta] = 1$$

identically in θ, $h(\theta)$ must be a bounded function of θ.

Let $\{\theta_m\}$ be a sequence of parameter values which converges to θ^*. From Theorem 4.1 it follows that

$$\lim_{m \to \infty} [\varphi(t, \theta_m) - \varphi(t, \theta^*)] = 0 \tag{5.3}$$

uniformly in t over any finite interval. Since $h(\theta)$ is bounded, we obtain from (5.3)

$$\lim_{m \to \infty} \{\varphi[h(\theta_m), \theta_m] - \varphi[h(\theta_m), \theta^*]\} = 0. \tag{5.4}$$

Since $\varphi[h(\theta_m), \theta_m] = 1$, it follows from (5.4) that

$$\lim_{m \to \infty} \varphi[h(\theta_m), \theta^*] = 1.$$

It follows from assumption 1 that for any limit point h of the bounded sequence $\{h(\theta_m)\}$ ($m = 1, 2, \cdots$, ad inf.) we have

---

[3] Condition IV of Lemma 2 in [2] is not postulated here, since the validity of this condition is implied by assumption 1. Condition IV could have been omitted also in [2], since it follows from condition III.

$$\varphi(h, \theta^*) = 1 \tag{5.5}$$

If $h(\theta^*) = 0$, then equation $\varphi(t, \theta^*) = 1$ has the only root $t = 0$. Consequently, all limit points of $\{h(\theta_m)\}$ must be equal to zero, that is

$$\lim_{m\to\infty} h(\theta_m) = 0 \quad \text{if} \quad h(\theta^*) = 0. \tag{5.6}$$

Now let us assume that $h(\theta^*) \neq 0$. Since the second derivative of $\varphi(t, \theta)$ with respect to t is positive, it can be seen that $\varphi(t, \theta) < 1$ for values t in the open interval $(0, h(\theta))$, and $\varphi(t, \theta) > 1$ for any t outside the closed interval $[0, h(\theta)]$. Hence, $\varphi(t, \theta) < 1$ implies that $|h(\theta)| > |t|$ and $h(\theta)$ and t have the same sign. Now let t_0, be a value in the open interval $(0, h(\theta^*))$. Then we have

$$\varphi(t_0, \theta^*) < 1 \tag{5.7}$$

It follows from assumption 1 that

$$\varphi(t_0, \theta_m) < 1 \tag{5.8}$$

for sufficiently large m. Hence $h(\theta_m)$ and t_0 have the same sign and

$$|h(\theta_m)| > |t_0| \tag{5.9}$$

Inequality (5.9) implies that zero cannot be a limit point of the sequence $\{h(\theta_m)\}$. Since $\varphi(t, \theta^*) = 1$ has only the roots $t = 0$ and $t = h(\theta^*)$, it follows from (5.5) that the sequence $\{h(\theta_m)\}$ cannot have a limit point different from $h(\theta^*)$. Thus,

$$\lim_{m\to\infty} h(\theta_m) = h(\theta^*) \tag{5.10}$$

and Lemma 5.2 is proved.

LEMMA 5.3. *It follows from assumption 1 that for any given t, $E_\theta(e^{|zt|})$ is a bounded function of θ.*

PROOF: We have

$$E_\theta(e^{|tz|}) \leqq E_\theta(e^{tz} + e^{-tz}) = \varphi(t, \theta) + \varphi(-t, \theta) \tag{5.11}$$

It follows from assumption 1 that $\varphi(t, \theta)$ and $\varphi(-t, \theta)$ are bounded functions of θ. Hence Lemma 5.3 is proved.

LEMMA 5.4. *Let θ' be a value of θ such that $E_{\theta'}(z) = 0$, but $E_\theta(z) \neq 0$ for all $\theta \neq \theta'$ in an open interval containing θ'. It follows from assumptions 1 and 2 that*

$$\lim_{\theta\to\theta'} \left(-\frac{2E_\theta(z)}{h(\theta)}\right) = E_{\theta'}(z^2). \tag{5.12}$$

PROOF: We have

$$e^{h(\theta)z} = 1 + h(\theta)z + \frac{[h(\theta)]^2}{2} z^2 + \frac{[h(\theta)]^3}{6} z^3 e^{uh(\theta)z} \tag{5.13}$$

where $0 \leqq u \leqq 1$. Hence

(5.14) $$E_\theta(e^{h(\theta)z}) = 1 + h(\theta)E_\theta(z) + \frac{[h(\theta)]^2}{2} E_\theta(z^2) + \frac{[h(\theta)]^3}{6} E_\theta(z^3 e^{uh(\theta)z}).$$

Since $E_\theta(e^{h(\theta)z}) = 1$, we obtain from (5.14)

(5.15) $$h(\theta)E_\theta(z) + \frac{[h(\theta)]^2}{2} E_\theta(z^2) + \frac{[h(\theta)]^3}{6} E_\theta(z^3 e^{uh(\theta)z}) = 0.$$

We shall consider only values θ for which $h(\theta) \neq 0$. For such values of θ, also $E_\theta(z) \neq 0$. Dividing (5.15) by $h(\theta)E_\theta(z)$, we obtain

(5.16) $$1 + \frac{h(\theta)}{2E_\theta(z)}\left[E_\theta(z^2) + \frac{h(\theta)}{3} E_\theta(z^3 e^{uh(\theta)z})\right] = 0.$$

Let t_0 be an upper bound of $|h(\theta)|$ with respect to θ. Then for a suitably chosen constant C we have

(5.17) $$|z^3 e^{uh(\theta)z}| < Ce^{|t_0 z|}.$$

From this and Lemma 5.3 it follows that $E_\theta(z^3 e^{uh(\theta)z})$ is a bounded function of θ.

Because of the continuity of $h(\theta)$ we have

(5.18) $$\lim_{\theta \to \theta'} h(\theta) = 0.$$

Lemma 5.4 follows from (5.16), (5.18), the boundedness of $E_\theta(z^3 e^{uh(\theta)z})$ and the fact that $E_\theta(z^2)$ is a continuous function of θ and $E_{\theta'}(z^2) > 0$.

LEMMA 5.5. *From assumptions 1 and 2 it follows that for any given t, $E_\theta(e^{|tZ_n|})$ exists and is a bounded function of θ.*

PROOF: It is sufficient to show that $E_\theta(e^{tZ_n})$ is a bounded function of θ for any t, since

(5.19) $$e^{|tZ_n|} \leqq e^{tZ_n} + e^{-tZ_n}$$

Clearly, e^{tZ_n} lies between $e^{bt+z_n t}$ and $e^{at+z_n t}$. Hence Lemma 5.5 is proved if we show that $E_\theta(e^{z_n t})$ is a bounded function of θ.

It follows from Assumption 2 that there exists a positive integer k and a positive constant g such that

(5.20) $$P(|z_1 + \cdots + z_k| \geqq a - b \mid \theta) \geqq g$$

for all θ. For any positive integer m and for any real values $\lambda_1 < \lambda_2$ we have

(5.21) $$\frac{P[(m-1)k < n \leqq mk \mid \theta]}{P[(m-1)k < n \mid \theta]} \geqq g \qquad (m = 1, 2, \cdots, \text{ad inf.})$$

and

(5.22) $$\frac{P[(m-1)k < n \leqq mk \ \&\ \lambda_1 \leqq z_n < \lambda_2 \mid \theta]}{P[(m-1)k < n \mid \theta]} \leqq 1 - [1 - P(\lambda_1 \leqq z < \lambda_2 \mid \theta)]^k.$$

Hence

$$(5.23)\quad \frac{P[(m-1)k < n \leqq mk \;\&\; \lambda_1 \leqq z_n < \lambda_2 \mid \theta]}{P[(m-1)k < n \leqq mk \mid \theta]} \leqq \frac{1 - [1 - P(\lambda_1 \leqq z < \lambda_2 \mid \theta]^k}{g}.$$

Multiplying (5.23) by $P[(m-1)k < n \leqq mk \mid \theta]$ and summing with respect to m we obtain

$$(5.24)\quad P(\lambda_1 \leqq z_n < \lambda_2 \mid \theta) \leqq \frac{1 - [1 - P(\lambda_1 \leqq z < \lambda_2 \mid \theta)]^k}{g}.$$

From (5.24) it follows readily that

$$(5.25)\quad \frac{P(\lambda_1 \leqq z_n < \lambda_2 \mid \theta)}{P(\lambda_1 \leqq z < \lambda_2 \mid \theta)}$$

is a bounded function of λ_1, λ_2 and θ. Let A be an upper bound of the ratio (5.25). Then

$$(5.26)\quad E_\theta(e^{tz_n}) \leqq AE_\theta(e^{tz}) = A\varphi(t, \theta).$$

Because of Assumption 1, $\varphi(t, \theta)$ is a bounded function of θ. Hence also $E_\theta(e^{tz_n})$ is bounded and Lemma 5.5 is proved.

6. The limiting value of $E_\theta(n)$ when θ approaches a value θ' for which $E_{\theta'}(z) = 0$. In this section we shall prove the following theorem:

THEOREM 6.1. *Let θ' be a value of θ such that $E_{\theta'}(z) = 0$, but $E_\theta(z) \neq 0$ for all $\theta \neq \theta'$ in an open interval containing θ'. If assumptions* 1 *and* 2 *hold, we have*

$$(6.1)\quad \lim_{\theta \to \theta'} \left[E_\theta(n) - \frac{E_\theta(Zn^2)}{E_{\theta'}(z^2)} \right] = 0.$$

PROOF: Consider the Taylor expansion

$$(6.2)\quad e^{h(\theta)Z_n} = 1 + h(\theta)Z_n + \frac{[h(\theta)]^2}{2} Z_n^2 + \frac{[h(\theta)]^3}{6} Z_n^3 e^{\lambda h(\theta) Z_n}$$

where $0 \leqq \lambda \leqq 1$. It was shown in [2] (p. 286) that

$$(6.3)\quad E_\theta e^{h(\theta) Z_n} = 1.$$

Hence, taking expected values on both sides of (6.2), we obtain

$$(6.4)\quad h(\theta)E_\theta(Z_n) + \frac{[h(\theta)]^2}{2} E_\theta(Z_n^2) + \frac{[h(\theta)]^3}{6} E_\theta(Z_n^3 e^{\lambda h(\theta) Z_n}) = 0.$$

We consider only values of θ for which $E_\theta(z) \neq 0$. For such values, also $h(\theta) \neq 0$. Thus, we can divide both sides of (6.4) by $h(\theta)E_\theta(z)$. We then obtain

$$\frac{E_\theta(Z_n)}{E_\theta(z)} + \frac{h(\theta)}{2E_\theta(z)}\left[E_\theta Z_n^2 + \frac{h(\theta)}{3} E_\theta(Z_n^3 e^{\lambda h(\theta) Z_n})\right] = 0. \tag{6.5}$$

It was shown in [1] (p. 142) that

$$E_\theta(n) = \frac{E_\theta(Z_n)}{E_\theta(z)}. \tag{6.6}$$

Hence

$$E_\theta(n) + \frac{h(\theta)}{2E_\theta(z)}\left[E_\theta(Z_n^2) + \frac{h(\theta)}{3} E_\theta(Z_n^3 e^{\lambda h(\theta) Z_n})\right] = 0. \tag{6.7}$$

Let t_0 be an upper bound of $| h(\theta) |$. Then for a properly chosen constant C we have

$$| Z_n^3 e^{\lambda h(\theta) Z_n} | \leqq Ce^{| t_0 Z_n |} \tag{6.8}$$

From this and Lemma 5.5 it follows that $E_\theta(Z_n^3 e^{\lambda h(\theta) Z_n})$ is a bounded function of θ. Since $\lim_{\theta \to \theta'} h(\theta) = 0$ and $E_\theta(Z_n^2)$ has a positive lower bound, Theorem 6.1 follows from 6.7, Lemma 5.4 and Theorem 4.1.

If $\lim_{\theta \to \theta'} E_\theta Z_n^2 = E_{\theta'} Z_n^2$, Theorem 6.1 gives[4]

$$E_{\theta'}(n) = \frac{E_{\theta'}(Z_n^2)}{E_{\theta'}(z^2)}. \tag{6.9}$$

Limits for $E_{\theta'}(n)$ can be obtained by computing limits for $E_\theta(Z_n^2)$. In the next section we shall give a general method for obtaining limits for $E_\theta[\psi(Z_n)]$, where $\psi(Z_n)$ is any function of Z_n.

7. Determination of lower and upper limits for the expected value of any function of Z_n. Let $\psi(Z_n)$ be a function of Z_n. Limits for $E_\theta[\psi(Z_n)]$ may be determined as follows: First we determine limits for $E_\theta[\psi(Z_n) \mid Z_n \geqq a]$. Let r be a positive variable. Clearly, for any given value r we have

$$E\text{-}_\theta(\psi Z_n) \mid Z_{n-1} = a - r \text{ and } Z_n \geqq a] = E_\theta[\psi(a - r + z) \mid z \geqq r] \tag{7.1}$$

From (7.1) we obtain the limits

$$\begin{aligned} &\underset{0<r<a-b}{\text{g.l.b.}}\ E_\theta[\psi(a - r + z) \mid z \geqq r] \leqq E_\theta[\psi(Z_n) \mid Z_n \geqq a] \\ &\qquad \leqq \underset{0<r<a-b}{\text{l.u.b.}}\ E_\theta[\psi(a - r + z) \mid z \geqq r]. \end{aligned} \tag{7.2}$$

Limits for $E_\theta[\psi(Z_n) \mid Z_n \leqq b]$ can be obtained in a similar way. Again, let r be a positive variable. For any value of r we have

$$E_\theta[\psi(Z_n) \mid Z_n \leqq b \text{ and } Z_{n-1} = b + r] = E_\theta[\psi(b + r + z) \mid z \leqq -r \tag{7.3}$$

Hence we obtain the limits

[4] The validity of (6.9) was shown by the author [3] using an entirely different method.

$$(7.4)\quad \underset{0<r<a-b}{\text{g.l.b.}} E_\theta[\psi(b + r + z) \mid z \leqq -r] \leqq E_\theta[\psi(Z_n) \mid Z_n \leqq b] \leqq \underset{0<r<a-b}{\text{l.u.b.}} E_\theta[\psi(b + r + z) \mid z \leqq -r].$$

Since

$$(7.5)\quad E_\theta[\psi(Z_n)] = P(Z_n \geqq a)E_\theta[\psi(Z_n) \mid Z_n \geqq a] + P(Z_n \leqq b)E_\theta[\psi(Z_n) \mid Z_n \leqq b],$$

a lower (upper) limit for $E_\theta[\psi(Z_n)]$ can be obtained, by replacing the conditional expected values on the right hand side of (7.5) by their lower (upper) limits given in (7.2) and (7.4).

8. Limits for $E_\theta(n)$ when $h(\theta)$ is near but unequal to zero. Let θ' be a value of θ for which $h(\theta') = 0$. In this section we shall derive limits for $E_\theta(n)$ which will generally be close to each other for values θ in a small neighborhood of θ'.

From equation (6.7) we obtain

$$(8.1)\quad E_\theta(n) = -\frac{h(\theta)}{2E_\theta(z)}\left[E_\theta Z_n^2 + \frac{h(\theta)}{3} E_\theta(Z_n^3 e^{\lambda h(\theta) Z_n})\right]$$

where $0 \leqq \lambda \leqq 1$. Thus, limits for $E_\theta(n)$ can be obtained by deriving limits for $E_\theta Z_n^2$ and $E_\theta(Z_n^3 e^{\lambda h(\theta) Z_n})$. Limits for $E_\theta Z_n^2$ can be obtained by using the method described in section 7.

If θ is near θ', any crude limits for $E_\theta(Z_n^3 e^{\lambda h(\theta) Z_n})$ will serve the purpose, since, as has been shown in section 6, $E_\theta(Z_n^3 e^{\lambda h(\theta) Z_n})$ is bounded and $\lim_{\theta \to \theta'} h(\theta) = 0$.

Limits for $E_\theta(Z_n^3 e^{\lambda h(\theta) Z_n})$ can be obtained as follows: For simplicity, let us assume that $h(\theta) > 0$. Then

$$(8.2)\quad Z_n^3 \leqq Z_n^3 e^{\lambda h(\theta) Z_n} \leqq Z_n^3 e^{h(\theta) Z_n} \qquad (h(\theta) > 0)$$

Thus, to determine limits for $E_\theta(Z_n^3 e^{\lambda h(\theta) Z_n})$, it is sufficient to determine a lower limit for $E_\theta(Z_n^3)$ and an upper limit for $E_\theta(Z_n^3 e^{h(\theta) Z_n})$. The latter limits may be derived by using the method given in section 7.

If $h(\theta) < 0$, we have

$$(8.3)\quad Z_n^3 \geqq Z_n^3 e^{\lambda h(\theta) Z_n} \geqq Z_n^3 e^{h(\theta) Z_n}$$

and a similar procedure will yield the desired limits for $E_\theta(Z_n^3 e^{\lambda h(\theta) Z_n})$.

It should be emphasized that the limits of $E_\theta(n)$, as given in this section, can be expected to be close only if $h(\theta)$ is near zero. For values of θ for which $h(\theta)$ is not near zero, the limits of $E_\theta(n)$ given in [1] can be used.

REFERENCES

[1] A. WALD, "Sequential tests of statistical hypotheses," *Annals of Math. Stat.*, Vol. 16, (1945), pp. 117–186.

[2] A. WALD, "On cumulative sums of random variables," *Annals of Math. Stat.*, Vol. 15, (1944), pp. 283–296.

[3] A. WALD, "Differentiation under the expected sign in the fundamental identity of sequential analysis," *Annals of Math. Stat.*, Vol. 17 (1946), pp. 493–497.

Made in United States of America

Reprinted from THE ANNALS OF MATHEMATICAL STATISTICS
Vol. XVII, No. 4, December, 1946

DIFFERENTIATION UNDER THE EXPECTATION SIGN IN THE FUNDAMENTAL IDENTITY OF SEQUENTIAL ANALYSIS

BY ABRAHAM WALD

Columbia University

1. Introduction. Let $\{z_\alpha\}$ ($\alpha = 1, 2, \cdots$, ad inf.) be a sequence of random variables which are independently distributed with identical distributions. Let a be a positive, and b a negative constant. For each positive integral value m, let Z_m denote the sum $z_1 + \cdots + z_m$. Denote by n the smallest integral value for which Z_n does not lie in the open interval (b, a). For any random variable u, let the symbol $E(u)$ denote the expected value of u. The following identity, which plays a fundamental role in sequential analysis, has been proved in [1].

$$E[e^{Z_n t}\varphi(t)^{-n}] = 1, \tag{1.1}$$

where

$$\varphi(t) = E(e^{zt}) \tag{1.2}$$

and the distribution of z is equal to the common distribution of $z_1, z_2, \cdots$, etc. Identity (1.1) holds for all points t in the complex plane for which $\varphi(t)$ exists and $|\varphi(t)| \geq 1$.

The purpose of this paper is to formulate conditions under which we may differentiate (1.1) with respect to t under the expectation sign. This is of interest, since various results in sequential analysis can easily be established by differentiating (1.1) under the expectation sign. For example, the formula for $E(n)$ can immediately be obtained by differentiating (1.1) at $t = 0$. The derivative of $e^{Z_n t}\varphi(t)^{-n}$ at $t = 0$ is given by

$$(1.3) \qquad Z_n - \frac{\varphi'(0)}{\varphi(0)} n = Z_n - E(z)n$$

where $\varphi'(t)$ denotes the derivative of $\varphi(t)$. Hence, if we may differentiate (1.1) under the expectation sign, we obtain the basic formula

$$(1.4) \qquad E(Z_n) = E(z)E(n).$$

If $E(z) \neq 0$, the above equation has been used [2] to derive lower and upper limits for $E(n)$. If, however, $E(z) = 0$, formula (1.4) is of little value. It will be shown in section 3 that

$$(1.5) \qquad E(n) = \frac{E(Z_n^2)}{E(Z^2)} \quad \text{when} \quad E(z) = 0.$$

This result is obtained, as will be seen in section 3, by differentiating identity (1.1) twice at $t = 0$.

2. A sufficient condition for the differentiability of (1.1) under the expectation sign. In what follows, the parameter t in (1.1) will be restricted to real values, even if this is not stated explicitly. For any random variable u and any relation R, the symbol $E(u \mid R)$ will denote the conditional expected value of u under the restriction that R holds. In this section we shall establish the following theorem.

THEOREM 2.1. *If $\varphi(t)$ exists for all real values t, identity* (1.1) *may be differentiated under the expectation sign any number of times with respect to t at any value t in the domain $\varphi(t) \geq 1$.*

PROOF: First we shall derive an upper bound for $E(e^{tZ_n} \mid n = m)$ for any given integral value m. Consider the case when $t > 0$. Then

$$(2.1) \qquad E(e^{tZ_n} \mid n = m) \leq E(e^{tZ_n} \mid Z_n \geq a, n = m) \qquad (t > 0).$$

Clearly,

$$(2.2) \qquad E(e^{tZ_n} \mid Z_n \geq a, n = m, e^{tZ_{n-1}} = \rho e^{at}) = e^{at}\rho E\left(e^{zt} \mid e^{zt} \geq \frac{1}{\rho}\right).$$

Let $l(t)$ denote the least upper bound of the expression

$$(2.3) \qquad \rho E\left(e^{zt} \mid e^{zt} \geq \frac{1}{\rho}\right)$$

with respect to ρ over the interval $(e^{-(a-b)|t|}, 1)$. The existence of $\varphi(t)$ implies that $l(t)$ is finite. It follows from (2.1) and (2.2) that

$$E(e^{tZ_n} \mid n = m) \leq e^{at} l(t) \qquad (t > 0) \tag{2.4}$$

and, therefore, also

$$E(e^{tZ_n}) \leq e^{at} l(t) \qquad (t > 0). \tag{2.5}$$

If $t < 0$, one can show in a similar way that

$$E(e^{tZ_n} \mid n = m) \leq e^{bt} l(t) \qquad (t < 0) \tag{2.6}$$

and

$$E(e^{tZ_n}) \leq e^{bt} l(t) \qquad (t < 0). \tag{2.7}$$

To prove Theorem 2.1, it is sufficient to show that the following two propositions hold.[1]

PROPOSITION 2.1. *All derivatives of $e^{Z_n t}\varphi(t)^{-n}$ with respect to t exist in the domain $\varphi(t) \geq 1$.*

PROPOSITION 2.2. *For any positive integral value r and for any finite interval I in which $\varphi(t) \geq 1$, it is possible to find a function $D(Z_n, n)$ such that*

$$D(Z_n, n) \geq \left| \frac{d^r}{dt^r} [e^{Z_n t} \varphi(t)^{-n}] \right| \tag{2.8}$$

for all values t in I and

$$E[D(Z_n, n)] < \infty. \tag{2.9}$$

Proposition 2.1 is clearly true, if all derivatives of $\varphi(t)$ exist. The existence of these derivatives follows from the existence of $\varphi(t)$ for all values t.

Since $\frac{d^r}{dt^r} e^{Z_n t}\varphi(t)^{-n}$ is equal to the sum of a finite number of terms of the type $Z_n^{r_1} n^{r_2} e^{Z_n t} \varphi(t)^{-n}$, Proposition 2.2 is proved if we can show that for any given integral values r_1 and r_2 there exists a function $D_{r_1 r_2}(Z_n, n)$ such that

$$D_{r_1 r_2}(Z_n, n) \geq | Z_n^{r_1} n^{r_2} e^{Z_n t} \varphi(t)^{-n} | \tag{2.10}$$

for all t in I and

$$E[D_{r_1 r_2}(Z_n, n)] < \infty. \tag{2.11}$$

Clearly, since $\varphi(t) \geq 1$ in I,

$$| Z_n^{r_1 r_2} e^{Z_n t} \varphi(t)^{-n} | \leq | Z_n^{r_1} | n^{r_2} e^{|Z_n| t_0} \tag{2.12}$$

where t_0 is an upper bound of $|t|$ in I. Let t_1 be a value $> t_0$. Then for a properly chosen constant C we have

$$| Z_n^{r_1} | e^{|Z_n| t_0} < C e^{|Z_n| t_1}. \tag{2.13}$$

[1] See, for example, E. J. McShane, *Integration*, Princeton University Press (1944), p. 216, 217 and 276.

Hence, it follows from (2.12) and (2.13) that

$$|Z_n^{r_1} n^{r_2} e^{Z_n t} \varphi(t)^{-n}| \leq C n^{r_2} e^{|Z_n| t_1} \leq C n^{r_2}(e^{Z_n t_1} + e^{-Z_n t_1}) \tag{2.14}$$

for all t in I.

We put

$$D_{r_1 r_2}(Z_n, n) = C n^{r_2}(e^{Z_n t_1} + e^{-Z_n t_1}). \tag{2.15}$$

We have

$$E[D_{r_1 r_2}(Z_n, n)] = C \sum_{m=1}^{\infty} p_m m^{r_2}[E(e^{Z_n t_1} \mid n = m) + E(e^{-Z_n t_1} \mid n = m)] \tag{2.16}$$

where p_m denotes the probability that $n = m$.

Hence, because of (2.4) and (2.6), we obtain

$$\begin{aligned} E[D_{r_1 r_2}(Z_n, n)] &\leq C(e^{a t_1} l(t_1) + e^{-b t_1} l(-t_1))[\Sigma p_m m^{r_2}] = \\ &= C[e^{a t_1} l(t_1) + e^{-b t_1} l(-t_1)] E(n^{r_2}). \end{aligned} \tag{2.17}$$

Since all moments of n are finite,[2] Proposition 2.2 is proved. This completes the proof of Theorem 2.1.

3. The expected value of n when $E(z) = 0$. It will be shown in this section that

$$E(n) = \frac{E(Z_n^2)}{E(z^2)} \quad \text{when} \quad E(z) = 0, \tag{3.1}$$

if identity (1.1) can be differentiated twice under the expectation sign at $t = 0$. The second derivative of $e^{tZ_n}\varphi(t)^{-n}$ with respect to t is given by

$$\left\{\left[Z_n - n \frac{\varphi'(t)}{\varphi(t)}\right]^2 - n \frac{\varphi''(t)\varphi(t) - [\varphi'(t)]^2}{[\varphi(t)]^2}\right\} e^{Z_n t} \varphi(t)^{-n} \tag{3.2}$$

where $\varphi'(t)$ denotes the first, and $\varphi''(t)$ the second derivative of $\varphi(t)$.

Since $\varphi(0) = 1$, $\varphi'(0) = E(z) = 0$ and $\varphi''(0) = E(z^2)$, putting $t = 0$, expression (3.2) becomes

$$Z_n^2 - n\varphi''(0) = Z_n^2 - nE(z^2) \tag{3.3}$$

Hence, if (1.1) may be differentiated twice under the expectation sign at $t = 0$, we obtain

$$E[Z_n^2 - nE(z^2)] = 0 \tag{3.4}$$

from which (3.1) follows.

An approximate value of $E(n)$ can be obtained from (3.1) by neglecting the excess of Z_n over the boundaries. Then Z_n can take only the values a and b. Hence

$$E(Z_n^2) \sim a^2 P(Z_n \geq a) + b^2 P(Z_n \leq b) \tag{3.5}$$

where the sign $\sim$ denotes approximate equality.

---

[2] See the paper by C. Stein, "A note on cumulative suns," in this issue of the *Annals of Mathematical Statistics*.

It was shown in [1] (equation 28) that neglecting the excess of Z_n over the boundaries, the approximation formula

$$P(Z_n \geq a) \sim \frac{1 - e^{bh}}{e^{ah} - e^{bh}} \tag{3.6}$$

holds, where h is the non-zero root of the equation $\varphi(t) = 1$. This formula was derived there under the assumption that $E(z) \neq 0$. If $E(z)$ approaches zero, $h \to 0$ and the right hand member of (3.6) converges to $\frac{-b}{a - b}$.

Putting $P(Z_n \geq a) = \frac{-b}{a - b}$ and $P(Z_n \leq b) = 1 - \frac{-b}{a - b} = \frac{a}{a - b}$, we obtain from (3.5)

$$E(Z_n^2) \sim a^2 \left(\frac{-b}{a - b}\right) + b^2 \frac{a}{a - b} = -ab. \tag{3.7}$$

Hence[3]

$$E(n) \sim \frac{-ab}{E(z^2)}. \tag{3.8}$$

Limits for $E(n)$ can be obtained by deriving limits for $E(Z_n^2)$. Let r be a non-negative real variable. One can verify that

$$a^2 \leqq E(Z_n^2 \mid Z_n \geq a) \leqq \underset{0<r<a-b}{\text{l.u.b.}} E[(a - r + z)^2 \mid z \geq r] \tag{3.9}$$

and

$$b^2 \leqq E(Z_n^2 \mid Z_n \leqq b) \leq \underset{0<r<a-b}{\text{l.u.b.}} E[(b + r + z)^2 \mid z + r \leq 0]. \tag{3.10}$$

We have

$$E(Z_n^2) = P(Z_n \geq a)E(Z_n^2 \mid Z_n \geq a) + P(Z_n \leq b)E(Z_n^2 \mid Z_n \leq b). \tag{3.11}$$

Limits for $E(Z_n^2)$ can be obtained by replacing the conditional expected values in the right hand member of (3.11) by their limits given in (3.9) and (3.10).

REFERENCES

[1] A. Wald, "On cumulative sums of random variables," *Annals of Math. Stat.*, Vol. 15 (1944), pp. 285, 287.

[2] A. Wald, "Sequential tests of statistical hypotheses," *Annals of Math. Stat.*, Vol. 16 (1946), p. 143.

---

[3] This approximation formula was obtained also by W. A. Wallis independently of the author. It is included in the publication of the Statistical Research Group of Columbia Univ., *Techniques of Statistical Analysis*, Chapter 17, Section 7.2, McGraw Hill, New York (1946).

Reprinted from the
BULLETIN OF THE AMERICAN MATHEMATICAL SOCIETY
Vol. 53, No. 2, pp. 142-153
February, 1947

LIMIT DISTRIBUTION OF THE MAXIMUM AND MINIMUM OF SUCCESSIVE CUMULATIVE SUMS OF RANDOM VARIABLES

ABRAHAM WALD

1. **Introduction.** For any positive integral value N, let $X_{N1}, X_{N2}, \cdots, X_{NN}$ be independent and identically distributed random variables each having standard deviation 1. Let μ_N denote the mean value of X_{Ni}, and let

$$S_{Nk} = X_{N1} + X_{N2} + \cdots + X_{Nk}. \tag{1.1}$$

Two cases will be considered: (1) the sequence $\{N^{1/2}\mu_N\}$ converges to a finite value as $N \to \infty$; (2) $\lim_{N=\infty} N^{1/2}\mu_N = \infty$. In case (1) we shall obtain for any positive constants a and b the limit values of

$$P_N(a) = \text{prob}\,\{\max(S_{N1}, \cdots, S_{NN}) < aN^{1/2}\} \tag{1.2}$$

and

$$\begin{aligned} P_N^*(a, b) = \text{prob}\,\{-bN^{1/2} &< \min(S_{N1}, \cdots, S_{NN}) \\ &\leqq \max(S_{N1}, \cdots, S_{NN}) < aN^{1/2}\} \end{aligned} \tag{1.3}$$

as $N \to \infty$. In case (2), we shall obtain for any real value c the limit of

$$Q_N(c) = \text{prob}\,\{\max(S_{N1}, \cdots, S_{NN}) < N\mu_N + cN^{1/2}\} \tag{1.4}$$

as $N \to \infty$.

In the particular case when $\mu_N = 0$ and $a = b$, the limit values of (1.2) and (1.3) were recently obtained by Erdös and Kac [1].[1] The case when $\mu_N \neq 0$, especially when $\mu_N N^{1/2}$ converges to a finite value, is of particular importance in the theory of sequential tests of statistical hypotheses. It will be seen in §3 that the limit distribution of the number of observations required by a sequential probability ratio test can immediately be obtained from the limit values of (1.2) and (1.3), and vice versa.

2. **Proof that the limit values of** (1.2) **and** (1.3) **do not depend on the distribution of the** X**'s.** It will be assumed in this section that $\mu_N N^{1/2}$ converges to a finite value as $N \to \infty$. The independence of the limit values of (1.2) and (1.3) of the distribution of the X's was proved by Erdös and Kac [1] in the special case when $\mu_N = 0$ and $a = b$. To deal with the more general case considered here, we shall

Received by the editors, July 29, 1946.

[1] Numbers in brackets refer to the references cited at the end of the paper.

follow essentially their method of proof. Let k be a positive integer, and let

$$N_j = \left[j \frac{N}{k} \right] \qquad (j = 1, 2, \cdots, k), \tag{2.1}$$

$$P_{N,k}(a) = \text{prob}\,\{\max\,(S_{NN_1}, \cdots, S_{NN_k}) < aN^{1/2}\}, \tag{2.2}$$

and

$$\begin{aligned} E_{Nr} = \text{prob}\,\{S_{Nr} \geqq aN^{1/2}, S_{N1} \\ < aN^{1/2}, \cdots, S_{N,r-1} < aN^{1/2}\} \, (r = 1, \cdots, N). \end{aligned} \tag{2.3}$$

Let, furthermore, ϵ be a positive number. Following Erdös and Kac, for $N_i < r \leqq N_{i+1}$ we write

$$\begin{aligned} E_{Nr} = \text{prob}\,\{S_{Nr} \geqq aN^{1/2}, S_{N1} < aN^{1/2}, \cdots, S_{N,r-1} \\ < aN^{1/2}, \left| S_{NN_{i+1}} - S_{Nr} \right| \geqq \epsilon N^{1/2}\} \\ + \text{prob}\,\{S_{Nr} \geqq aN^{1/2}, S_{N1} < aN^{1/2}, \cdots, S_{N,r-1} \\ < aN^{1/2}, \left| S_{NN_{i+1}} - S_{Nr} \right| < \epsilon N^{1/2}\}. \end{aligned} \tag{2.4}$$

Clearly, the first of these probabilities is equal to E_{Nr} prob $\{ |S_{NN_{i+1}} - S_{Nr}| \geqq \epsilon N^{1/2}\}$. Since

$$E(S_{NN_{i+1}} - S_{Nr})^2 \leqq (N_{i+1} - N_i) + (N_{i+1} - N_i)^2 \mu_N^2, \tag{2.5}$$

by Tchebychef's inequality we have

$$\begin{aligned} \text{prob}\,\{ \left| S_{NN_{i+1}} - S_{Nr} \right| \geqq \epsilon N^{1/2}\} \\ \leqq \frac{(N_{i+1} - N_i) + (N_{i+1} - N_i)^2 \mu_N^2}{\epsilon^2 N}. \end{aligned} \tag{2.6}$$

Since

$$\begin{aligned} \frac{N_{i+1} - N_i}{N} &= \frac{1}{N}\left\{\left[(i + 1)\frac{N}{k}\right]\right\} - \frac{1}{N}\left\{\left[i\frac{N}{k}\right]\right\} \\ &\leqq \frac{1}{k} + \frac{1}{N} \leqq \frac{2}{k}, \end{aligned}$$

we obtain from (2.6)

$$\text{prob}\,\{ \left| S_{NN_{i+1}} - S_{Nr} \right| \geqq \epsilon N^{1/2}\} \leqq \frac{2}{\epsilon^2 k}\left(1 + \frac{2A}{k}\right) \tag{2.7}$$

where A is an upper bound of the sequence $\{N\mu_N^2\}$. From (2.4), (2.7) and the equation

$$\sum_{r=1}^{N} E_{Nr} = 1 - P_N(a) \leqq 1, \tag{2.8}$$

we easily obtain

$$\begin{aligned} 1 - P_N(a) &= \sum_{r=1}^{N} E_{Nr} \leqq \frac{2}{\epsilon^2 k}\left(1 + \frac{2A}{k}\right) \\ &+ \sum_i \sum_{N_i < r \leqq N_{i+1}} \text{prob}\,\{S_{Nr} \geqq aN^{1/2}, \\ &S_{N1} < aN^{1/2}, \cdots, S_{N,r-1} < aN^{1/2}, \\ &\left| S_{NN_{i+1}} - S_{Nr} \right| < \epsilon N^{1/2}\}. \end{aligned} \tag{2.9}$$

Clearly, the double sum is less than the probability that at least one of the sums $S_{NN_1}, \cdots, S_{NN_k}$ exceeds $(a-\epsilon)N^{1/2}$. Hence

$$1 - P_N(a) \leqq \frac{2}{\epsilon^2 k}\left(1 + \frac{2A}{k}\right) + 1 - P_{N,k}(a - \epsilon). \tag{2.10}$$

This inequality can be written as

$$P_{N,k}(a - \epsilon) - \frac{2}{\epsilon^2 k}\left(1 + \frac{2A}{k}\right) \leqq P_N(a). \tag{2.11}$$

Let $G_{k1}, \cdots, G_{kk}$ be normally and independently distributed random variables with mean $\mu/k^{1/2}$ and variance 1, where $\mu = \lim_{N\to\infty} N^{1/2}\mu_N$. Let, furthermore,

$$R_{ki} = G_{k1} + \cdots + G_{ki} \qquad (i = 1, \cdots, k).$$

It follows from the central limit theorem that

$$\lim_{N=\infty} P_{N,k}(a) = \text{prob}\,\{\max(R_{k1}, \cdots, R_{kk}) < ak^{1/2}\}. \tag{2.12}$$

From (2.11), (2.12) and the relation $P_N(a) < P_{N,k}(a)$ we obtain[2]

$$\begin{aligned} \text{prob}\,&\{\max(R_{k1}, \cdots, R_{kk}) < (a - \epsilon)k^{1/2}\} - \frac{2}{\epsilon^2 k}\left(1 + \frac{2A}{k}\right) \\ &\leqq \liminf_{N=\infty} P_N(a) \leqq \limsup_{N=\infty} P_N(a) \\ &\leqq \text{prob}\,\{\max(R_{k1}, \cdots, R_{kk}) < ak^{1/2}\}. \end{aligned} \tag{2.13}$$

If the distribution of the X's is such that $|X_{Ni}|$ is constant, $\lim_{N\to\infty} P_N(a)$ exists and is a continuous function of a, as will be seen

[2] This inequality corresponds to inequality (1) in [1].

in §3. Let $P(a)$ denote this limit. Using the arguments given by Erdös and Kac [1, pp. 295–296], one can show that inequality (2.13) implies that for any arbitrary distribution of the X's we have

$$\lim_{N=\infty} P_N(a) = P(a). \tag{2.14}$$

It will be seen in §3 that also $\lim_{N=\infty} P_N^*(a, b) = P^*(a, b)$ exists and is a continuous function of a and b when the X's are distributed such that $|X_{ni}|$ is constant. The proof that $\lim_{N=\infty} P_N^*(a, b) = P^*(a, b)$ for any arbitrary distribution of the X's can be carried out in exactly the same manner as that of (2.14).

3. **Determination of the limit values of (1.2) and (1.3) when the distribution of the X's is such that $|X_{Ni}|$ is constant.**[3] In this section it will be assumed that $|X_{Ni}|$ is constant and that $N^{1/2}\mu_N$ converges to a finite value as $N \to \infty$. The limits of (1.2) and (1.3) as $N \to \infty$ can easily be obtained from some results in the theory of sequential tests of statistical hypotheses (see [2] and [3]).

The sequential probability ratio test for testing a statistical hypothesis H_0 against an alternative hypothesis H_1 is defined as follows: Let H_i be the hypothesis that the elementary probability law of the random variable X under consideration is equal to $f_i(x)$ $(i=0, 1)$. Let

$$z = \log \frac{f_1(x)}{f_0(x)}$$

and

$$z_k = \log \frac{f_1(x_k)}{f_0(x_k)}$$

where x_k denotes the kth observation on x. Thus, $z_1, z_2, \cdots$, ad inf.

---

[3] This problem is intimately connected with a discrete model for the Brownian motion of a particle moving in a field of constant force (that is, gravity). Let a particle starting from the origin move along the x-axis in such a way that in each step it can move Δx to the right or Δx to the left with respective probabilities p and $q=1-p$. Let the duration of each step be Δt. This random walk becomes a model of the Brownian motion in a field of constant force in the limit when $\Delta x \to 0$, $\Delta t \to 0$, $2p-1 \to 0$ in such a way that $(\Delta x)^2/2\Delta t = D$, $(2p-1)/2\Delta x = c/4D$ where c and D are physical constants. The problem of finding the probability that the particle should remain in an interval around the origin during a time interval $(0, t)$ is equivalent to the problem of finding the limit value of (1.3) when $|X_{Ni}|$ is constant.

The limit values of (1.2) and (1.3) are obtained here without difficulty from some previous results of the author [2]. In a subsequent publication [4], M. Kac treated the special case of a free particle $(c=0)$ using an interesting and entirely different method of attack. His method could be extended to treat also the case when $c \neq 0$

are independent and identically distributed random variables. The test procedure is carried out as follows: two positive constants a and b are chosen. At each stage of the experiment, at the ith trial for each integral value i, the cumulative sum

$$Z_i = z_1 + z_2 + \cdots + z_i \qquad (i = 1, 2, \cdots, \text{ad inf.})$$

is computed. Additional observations are taken as long as $-b < Z_i < a$. The first time that this inequality does not hold, the test procedure is terminated. Let n denote the smallest integral value of i for which Z_i does not lie in the open interval $(-b, a)$. H_1 is accepted if $Z_n \geqq a$, and H_0 is accepted if $Z_n \leqq -b$.

It has been shown in [2] that for all points t in the complex plane for which the absolute value of $\phi(t) = E(e^{zt})$ is not less than 1, the following identity holds:

$$(3.1) \qquad E[e^{Z_n t}\phi(t)^{-n}] = 1.$$

Assume now that z can take only two values, g and $-g$ $(g>0)$. Let p denote the probability that $z = g$. Then the expected value of z is equal to

$$(3.2) \qquad \mu = g(2p - 1)$$

and the variance of z is given by

$$(3.3) \qquad \sigma^2 = g^2[1 - (2p - 1)^2].$$

Let

$$(3.4) \qquad m = \frac{\mu^2}{2\sigma^2} n,$$

$$(3.5) \qquad t_1(\tau) = \frac{1}{g} \log \frac{e^{-(\mu^2/2\sigma^2)\tau} - (e^{-(\mu^2/\sigma^2)\tau} - 4p(1 - p))^{1/2}}{2p}$$

and

$$(3.6) \qquad t_2(\tau) = \frac{1}{g} \log \frac{e^{-(\mu^2/2\sigma^2)\tau} + (e^{-(\mu^2/\sigma^2)\tau} - 4p(1 - p))^{1/2}}{2p}$$

where τ is a purely imaginary variable. Since the absolute value of

$$\phi[t_i(\tau)] = e^{-(\mu^2/2\sigma^2)\tau}$$

is equal to 1, we may substitute $t_i(\tau)$ for t in (3.1). We then obtain

$$(3.7) \qquad E(e^{Z_n t_i(\tau)} e^{m\tau}) = 1 \qquad (i = 1, 2).$$

For any random variable u and any relation R let $E(u \mid R)$ denote the conditional expected value of u when R holds. For any value r, let $[r]$ denote the smallest integer not less than r. Since Z_n can take only the values $g[a/g]$ and $-g[b/g]$, equation (3.7) can be written as follows

$$\begin{aligned} &\text{prob}\left\{Z_n = -g\left[\frac{b}{g}\right]\right\} e^{-g[b/g]t_i(\tau)} E\left(e^{m\tau} \,\middle|\, Z_n = -g\left[\frac{b}{g}\right]\right) \\ (3.8) \quad & \\ &+ \text{prob}\left\{Z_n = g\left[\frac{a}{g}\right]\right\} e^{g[a/g]t_i(\tau)} E\left(e^{m\tau} \,\middle|\, Z_n = g\left[\frac{a}{g}\right]\right) = 1 \ (i = 1, 2). \end{aligned}$$

Solving the two linear equations (3.8) in the unknowns $\psi_1(\tau) = \text{prob}\,\{Z_n = -g[b/g]\}E(e^{m\tau} | Z_n = -g[b/g])$ and $\psi_2(\tau) = \text{prob}\{Z_n = g[a/g]\}E(e^{m\tau} | Z_n = g[a/g])$, we obtain

$$\begin{aligned} E(e^{m\tau}) &= \psi_1(\tau) + \psi_2(\tau) \\ (3.9) \qquad &= \frac{e^{g[a/g]t_1(\tau)} + e^{-g[b/g]t_2(\tau)} - e^{g[a/g]t_2(\tau)} - e^{-g[b/g]t_1(\tau)}}{e^{g[a/g]t_1(\tau) - g[b/g]t_2(\tau)} - e^{g[a/g]t_2(\tau) - g[b/g]t_1(\tau)}}. \end{aligned}$$

We shall be interested in the limiting case when μ and σ take a sequence of values such that

$$(3.10) \qquad \lim \mu = 0, \qquad \lim \sigma = 0 \quad \text{and} \quad \lim \frac{\mu}{\sigma^2} = d,$$

where d is a finite value not equal to 0. It follows from (3.2) and (3.3) that (3.10) is equivalent with

$$(3.11) \qquad \lim (2p - 1) = 0, \qquad \lim g = 0 \quad \text{and} \quad \lim \frac{2p - 1}{g} = d.$$

It can easily be verified that

$$\frac{1}{g} \log \frac{e^{-(\mu^2/2\sigma^2)\tau}(1 \pm (1 - 4p(1 - p)e^{(\mu^2/\sigma^2)\tau})^{1/2})}{2p}$$

converges to

$$(3.12) \qquad -d(1 \pm (1 - \tau)^{1/2})$$

uniformly over any finite τ-interval as μ, σ and μ/σ^2 approach the limit values given in (3.10). Hence the characteristic function of m given in (3.9) converges to

$$(3.13)\quad \psi(\tau)=\frac{e^{-ad(1-(1-\tau)^{1/2})}+e^{bd(1+(1-\tau)^{1/2})}-e^{-ad(1+(1-\tau)^{1/2})}-e^{bd(1-(1-\tau)^{1/2})}}{e^{-ad(1-(1-\tau)^{1/2})+bd(1+(1-\tau)^{1/2})}-e^{-ad(1+(1-\tau)^{1/2})+bd(1-(1-\tau)^{1/2})}}$$

uniformly over any finite τ-interval as μ, σ and μ/σ^2 approach the limit values given in (3.10).

The characteristic function $\psi(\tau)$ has been inverted in [2] yielding the limit distribution of m. Denote this limit distribution by $F(u)$, that is, $F(u)=\text{prob}\{m<u\}$ where m is a random variable whose characteristic function is equal to $\psi(\tau)$. The value of $F(u)$ depends on the constants ad and bd, since these constants are involved in the characteristic function $\psi(\tau)$. To put this dependence in evidence, we shall also use the symbol $F(u|ad, bd)$.

We shall now express the limit value of $P_N^*(a, b)$ in terms of $F(u|ad, bd)$. It is assumed that X_{Ni} can take only two values, g_N and $-g_N$ ($g_N>0$). The values g_N and $\text{prob}\{X_{Ni}=g_N\}$ are chosen so that the standard deviation of X_{Ni} is equal to 1 and the mean value of X_{Ni} has the prescribed value μ_N. We consider the case when $\lim_{N=\infty}\mu_N N^{1/2}=d\neq 0$. Let the distribution of z be equal to that of $X_{Ni}/N^{1/2}$. The mean and standard deviation of z are then equal to

$$(3.14)\qquad \mu = \mu_N/N^{1/2} \quad\text{and}\quad \sigma = 1/N^{1/2}$$

respectively. Hence, the limit values of μ, σ and μ/σ^2 as $N\to\infty$ are equal to those given in (3.10). Clearly,

$$(3.15)\qquad \begin{aligned} P_N^*(a, b) &= \text{prob}\ \{n > N\} \\ &= \text{prob}\ \left\{m > \frac{\mu^2}{2\sigma^2}N\right\} = \text{prob}\ \left\{m > \frac{\mu_N^2 N}{2}\right\}. \end{aligned}$$

Since $\lim_{N=\infty}\mu_N N^{1/2}=d$ and since the limit distribution $F(u)$ of m is a continuous function of u, we obtain from (3.15)

$$(3.16)\qquad \lim_{N=\infty} P_N^*(a, b) = 1 - F\left(\frac{d^2}{2}\middle|\, ad, bd\right) \qquad (d \neq 0).$$

The above formula is valid for $d\neq 0$ .We shall now determine the limit of $P_N^*(a, b)$ when $d=0$, that is, when $\lim_{N=\infty}\mu_N N^{1/2}=0$. Since the value of $P_N^*(a, b)$ depends on the value of $d_N=\mu_N N^{1/2}$, we shall put this in evidence by writing $P_N^*(a, b|d_N)$.

Clearly, for any $d\neq 0$, we have

$$(3.17)\qquad \begin{aligned} P_N^*(a + 2|d|, b + 2|d||0) &> P_N^*(a, b|d) \\ &> P_N^*(a - 2|d|, b - 2|d||0). \end{aligned}$$

It follows from (3.17) that

$$(3.18)\quad \begin{aligned}\liminf_{N=\infty} P_N^*(a+2|d|, b+2|d| \mid 0) &\geqq 1 - F\left(\frac{d^2}{2} \,\middle|\, ad, bd\right) \\ &\geqq \limsup_{N=\infty} P_N^*(a-2|d|, b-2|d| \mid 0).\end{aligned}$$

If $\mu_N=0$, it follows from the first limit theorem of Erdös and Kac [1] that $\lim_{N=\infty} \text{prob}\{\max(S_{N1}, \cdots, S_{NN}) < aN^{1/2}\}$ and $\lim_{N=\infty} \text{prob}\{\min(S_{N1}, \cdots, S_{NN}) > -bN^{1/2}\}$ are continuous functions of a and b. This implies that $\limsup_{N=\infty} P_N^*(a, b|0)$ and $\liminf_{N=\infty} P_N^*(a, b|0)$ are continuous functions of a and b, and that

$$(3.19)\quad \begin{aligned}\lim_{d=0} \limsup_{N=\infty} [&P_N^*(a+2|d|, b+2|d| \mid 0) \\ &- P_N^*(a-2|d|, b-2|d| \mid 0)] = 0.\end{aligned}$$

It follows from (3.18) that

$$\begin{aligned}&\limsup_{N=\infty} P_N^*(a+2|d|, b+2|d| \mid 0) \\ &\quad - \liminf_{N=\infty} P_N^*(a+2|d|, b+2|d| \mid 0) \\ &\qquad \leqq \limsup_{N=\infty} [P_N^*(a+2|d|, b+2|d| \mid 0) \\ &\qquad\quad - P_N^*(a-2|d|, b-2|d| \mid 0)].\end{aligned}$$

Hence, because of (3.19), we have

$$\begin{aligned}\lim_{d=0} [&\limsup_{N=\infty} P_N^*(a+2|d|, b+2|d| \mid 0) \\ &- \liminf_{N=\infty} P_N^*(a+2|d|, b+2|d| \mid 0)] = 0.\end{aligned}$$

The existence and continuity of $P^*(a, b|0) = \lim_{N=\infty} P_N^*(a, b|0)$ follows from the above equation and the continuity of $\limsup_{N=\infty} P_N^*(a, b|0)$ and $\liminf_{N=\infty} P_N^*(a, b|0)$. Hence, because of (3.18), we have

$$(3.20)\quad P^*(a, b \mid 0) = \lim_{d=0}\left[1 - F\left(\frac{d^2}{2} \,\middle|\, ad, bd\right)\right].$$

Let $\{d_N\}$ be a sequence of values such that $\lim_{N=\infty} d_N = 0$. Substituting d_N for d in (3.17) and letting $N \to \infty$, we obtain

$$(3.21)\quad \lim_{N=\infty} P_N^*(a, b \mid d_N) = P^*(a, b \mid 0) = \lim_{d=0}\left[1 - F\left(\frac{d^2}{2} \,\middle|\, ad, bd\right)\right].$$

We shall now show that

$$\lim_{N=\infty} P_N(a) = \lim_{b=\infty} \lim_{N=\infty} P_N^*(a, b). \tag{3.22}$$

To prove (3.22), it is sufficient to show that

$$\lim_{b=\infty} \liminf_{N=\infty} \text{prob} \{\min (S_{N1}, \cdots, S_{NN}) > -bN^{1/2}\} = 1. \tag{3.23}$$

Let r be an upper bound of $|N^{1/2}\mu_N|$. Clearly,

$$\begin{aligned} &\text{prob} \{\min (\bar{S}_{N1}, \cdots, \bar{S}_{NN}) > (-b + r)N^{1/2}\} \\ &\qquad \leqq \text{prob} \{\min (S_{N1}, \cdots, S_{NN}) > -bN^{1/2}\} \end{aligned} \tag{3.24}$$

where $\bar{S}_{Ni} = S_{Ni} - i\mu_N$. Since according to the first limit theorem of Erdös and Kac [1] we have

$$\begin{aligned} &\lim_{b=\infty} \lim_{N=\infty} \text{prob} \{\min (\bar{S}_{N1}, \cdots, \bar{S}_{NN}) > (-b + r)N^{1/2}\} \\ &\qquad = \lim_{b=\infty} \left(\frac{2}{\pi}\right)^{1/2} \int_0^{b-r} e^{-t^2/2} dt = 1, \end{aligned}$$

(3.23) follows from (3.24). Hence (3.22) is proved.

If $\lim (\mu/\sigma^2) = d > 0$ and $b \to \infty$, the characteristic function $\psi(\tau)$ of m given in (3.13) converges to

$$\psi^*(\tau) = e^{ad(1-(1-\tau)^{1/2})}. \tag{3.25}$$

This characteristic function has been inverted in [2] and the corresponding distribution function of m is given by

$$H(m)dm = \frac{ad}{2\Gamma(1/2)m^{3/2}} e^{-(a^2d^2/4m)-m+ad} dm \qquad (0 \leqq m < \infty).$$

Hence

$$\lim_{N=\infty} P_N(a) = \text{prob} \left\{m > \frac{d^2}{2}\right\} = \int_{d^2/2}^{\infty} H(m)dm. \tag{3.26}$$

We can summarize the results of this section in the following theorem.

THEOREM. *If* $\lim_{N=\infty} \mu_N N^{1/2} = d \neq 0$, *the limit value of the probability* (1.3) *is given by*

$$P^*(a, b \mid d) = \text{prob} \left\{m > \frac{d^2}{2}\right\}$$

where m is a random variable whose characteristic function is given in (3.13). *If $d=0$, the limit value of* (1.3) *is equal to* $\lim_{d=0} P^*(a, b \mid d)$. *For any finite value d, the limit value of* (1.2) *is equal to* $\lim_{b=\infty} P^*(a, b \mid d)$. *If $d>0$, the limit value of* (1.2) *is given explicitly in* (3.26).

4. **Derivation of the limit value of** (1.4) **when** $\lim \mu_N N^{1/2} = \infty$. In this section we shall determine the limit value of the probability $Q_N(c)$ defined in (1.4) assuming that

$$(4.1) \qquad \lim \mu_N N^{1/2} = \infty.$$

We can assume without loss of generality that $\mu_N > 0$ for all N. Let r be a positive number and let $\lambda(N)$ be a positive integral-valued function of N such that

$$(4.2) \qquad \lambda(N) < N, \qquad \lim_{N=\infty} \frac{N - \lambda(N)}{N} = 0,$$

$$(4.3) \qquad \lim_{N=\infty} \frac{N - \lambda(N)}{N^{1/2}} \mu_N = r'$$

where $\infty \geqq r' > r$. It follows from (4.1) that such a function $\lambda(N)$ exists. Because of (4.2) and (4.3) we have for sufficiently large N

$$(4.4) \qquad \begin{aligned} &\text{prob}\left\{\max(S_{N1}, \cdots, S_{N\lambda(N)}) < N\mu_N + cN^{1/2}\right\} \\ &> \text{prob}\left\{\max(S_{N1}, \cdots, S_{N\lambda(N)}) < \lambda(N)\mu_N + (c + r)N^{1/2}\right\} \\ &> \text{prob}\left\{\max(\bar{S}_{N1}, \cdots, \bar{S}_{N\lambda(N)}) < (c + r)N^{1/2}\right\} \end{aligned}$$

where

$$(4.5) \qquad \bar{S}_{Ni} = S_{Ni} - i\mu_N.$$

Let $\epsilon > 0$. Since for $\mu_N = 0$ we have $\lim_{c=\infty} \lim_{N=\infty} P_N(c) = 1$, there exists a fixed value r_0 (independent of N) such that

$$(4.6) \qquad \lim_{N=\infty} \text{prob}\left\{\max(\bar{S}_{N1}, \cdots, \bar{S}_{N\lambda(N)}) < (c + r_0)N^{1/2}\right\} = 1 - \frac{\epsilon}{2}.$$

Putting $r = r_0$, we obtain from (4.4) and (4.6)

$$(4.7) \qquad \text{prob}\left\{\max(S_{N1}, \cdots, S_{N\lambda(N)}) < N\mu_N + cN^{1/2}\right\} \geqq 1 - \epsilon$$

for sufficiently large N. Hence

$$(4.8) \qquad \begin{aligned} &\text{prob}\left\{\max(S_{N1}, \cdots, S_{NN}) < N\mu_N + cN^{1/2}\right\} \\ &\leqq \text{prob}\left\{\max(S_{N,\lambda(N)+1}, \cdots, S_N) < N\mu_N + cN^{1/2}\right\} \\ &\leqq \text{prob}\left\{\max(S_{N1}, \cdots, S_{NN}) < N\mu_N + cN^{1/2}\right\} + \epsilon \end{aligned}$$

for sufficiently large N. From (4.2) and the first limit theorem of Erdös and Kac [1, p. 292] it follows easily that

$$\lim_{N=\infty} \text{prob} \left\{\max \left(-\bar{S}_{NN}+\bar{S}_{N,N-1}, \cdots, -\bar{S}_{NN} + \bar{S}_{N,\lambda(N)+1}\right) < \delta N^{1/2}\right\} = 1 \tag{4.9}$$

for any positive δ. Since the inequality

$$\max\left(-\bar{S}_{NN}+\bar{S}_{N,N-1}, \cdots, -\bar{S}_{NN}+\bar{S}_{N,\lambda(N)+1}\right) < \delta N^{1/2}$$

implies the validity of $\max (S_{N)\lambda(N,+1}, \cdots, S_{NN}) \leqq S_{NN}+\delta N^{1/2}$, we obtain from (4.9)

$$\lim_{N=\infty} \text{prob} \left\{S_{NN} \leqq \max (S_{N,\lambda(N)+1}, \cdots, S_{NN}) \leqq S_{NN}+\delta N^{1/2}\right\} = 1. \tag{4.10}$$

Since

$$\lim_{N=\infty} \text{prob} \left\{S_{NN} < N\mu_N + cN^{1/2}\right\} = \frac{1}{(2\pi)^{1/2}}\int_{-\infty}^{c} e^{-t^2/2}dt, \tag{4.11}$$

we obtain from (4.10)

$$\begin{aligned} \frac{1}{(2\pi)^{1/2}}\int_{-\infty}^{c} e^{-t^2/2}dt & \\ &\geqq \limsup_{N=\infty} \text{prob} \left\{\max (S_{N,\lambda(N)+1}, \cdots, S_{NN}) < N\mu_N + cN^{1/2}\right\} \\ &\geqq \liminf_{N=\infty} \text{prob} \left\{\max (S_{N,\lambda(N)+1}, \cdots, S_{NN}) < N\mu_N + cN^{1/2}\right\} \\ &\geqq \frac{1}{(2\pi)^{1/2}}\int_{-\infty}^{c-\delta} e^{-t^2/2}dt. \end{aligned} \tag{4.12}$$

Since δ can be chosen arbitrarily small, it follows from (4.12) that

$$\lim_{N=\infty} \text{prob} \left\{\max (S_{N,\lambda(N)+1}, \cdots, S_{NN}) < N\mu_N + cN^{1/2}\right\} = \frac{1}{(2\pi)^{1/2}}\int_{-\infty}^{c} e^{-t^2/2}dt. \tag{4.13}$$

Finally, it follows from (4.8) and (4.13), since ϵ can be chosen arbitrarily small, that

$$\lim_{N=\infty} \text{prob} \left\{\max (S_{N1}, \cdots, S_{NN}) < N\mu_N + cN^{1/2}\right\} = \frac{1}{(2\pi)^{1/2}}\int_{-\infty}^{c} e^{-t^2/2}dt. \tag{4.14}$$

References

1. P. Erdös and M. Kac, *On certain limit theorems of the theory of probability.* Bull. Amer. Math. Soc. vol. 52 (1946) pp. 292–302.

2. A. Wald, *On cumulative sums of random variables*, Ann. Math. Statist. vol. 15 (1944).

3. ———, *Sequential tests of statistical hypotheses*, Ann. Math. Statist. vol. 16 (1945).

4. M. Kac, *Random walk in the presence of absorbing barriers*, Ann. Math. Statist. vol. 16 (1945).

Columbia University

Reprinted from THE ANNALS OF MATHEMATICAL STATISTICS
Vol. XVIII, No. 3, September, 1947

SEQUENTIAL CONFIDENCE INTERVALS FOR THE MEAN OF A NORMAL DISTRIBUTION WITH KNOWN VARIANCE

BY CHARLES STEIN AND ABRAHAM WALD

Columbia University

1. Summary. We consider sequential procedures for obtaining confidence intervals of prescribed length and confidence coefficient for the mean of a normal distribution with known variance. A procedure achieving these aims is called optimum if it minimizes the least upper bound (with respect to the mean) of the expected number of observations. The result proved is that the usual non-sequential procedure is optimum.

2. Introduction. The problem of sequential confidence sets in general has been considered briefly by one of the authors [1]. Let $\{X_i\}$, $(i = 1, 2, \cdots)$, be a sequence of random variables whose distribution is specified except for the value of a parameter θ whose range is a space Ω. Sequential confidence sets are determined by a rule as to when to stop sampling, together with a function of the sample whose value is one of a specified class of subsets of Ω. The class of subsets is chosen in advance depending on the purpose of the estimation. For example, it may be the class of all intervals of prescribed length or the class of all sets whose diameter does not exceed a given value. It is required that the probability that this (random) set covers θ should be greater than or equal to a specified confidence coefficient α for all θ. A procedure for finding sequential confidence intervals is considered optimum if it minimizes some specified function of the expected numbers of observations. Here this function is taken to be the least upper bound. In contrast with the result of this paper, a case where sequential confidence intervals may have an advantage over non-sequential procedures has been given by one of the authors [2]. The X_i are independently normally distributed with unknown mean and unknown variance, and the problem is to find confidence intervals of fixed length for the unknown mean. As was first shown by Dantzig [3] this cannot be accomplished by a non-sequential procedure. Another case where this is true is the problem of finding confidence intervals of the form (p_0, kp_0) where k is a specified number greater than 1, for the probability in a binomial distribution.

Let $\{X_i\}$, $(i = 1, 2, \cdots)$, be independently normally distributed with unknown mean ξ and known variance σ_1^2. It is desired to specify a sequential procedure for obtaining confidence intervals of fixed length l for the mean ξ. This is provided by a rule according to which at each stage of the experiment, after obtaining the first m observations $X_1, \cdots, X_m$ for each integral value m, one makes one of the following decisions:

a) Take an $(m + 1)$st observation.

b) Terminate the procedure and state that the mean lies in the interval

$(Y - \frac{1}{2}l, Y + \frac{1}{2}l)$, where $Y = \mathfrak{C}_m(X_1, \cdots, X_m)$, $\mathfrak{C}_m$ being a measurable real-valued function. The serial number m of the observation on which the procedure terminates is, of course, a random variable and will be denoted by n.

For any relation R the symbol $P(R \mid \xi)$ will denote the probability that R holds when ξ is the true mean of X_i. The confidence coefficient of a sequential procedure S is defined by

$$\alpha(S) = \underset{\xi}{g.l.b.}\ P(Y - \tfrac{1}{2}l < \xi < Y + \tfrac{1}{2}l \mid \xi). \tag{1}$$

Denote by $n_0(S)$ the maximum expected number of observations, i.e.

$$n_0(S) = \underset{\xi}{\text{l.u.b.}}\ E(n \mid \xi, S) \tag{2}$$

where $E(n \mid \xi, S)$ denotes the expected value of n when ξ is the true mean and the procedure S is used.

A procedure S will be considered optimum if, for all S' such that $\alpha(S') = \alpha(S)$,

$$n_0(S) \leq n_0(S'). \tag{3}$$

It will be shown that an optimum procedure $S(\nu, c)$ can be obtained as follows:

a) For all $m < \nu$, a fixed positive integer, take another observation.

b) For $m = \nu$, terminate the procedure if

$$\sum_1^\nu X_i^2 - \frac{1}{\nu}\left(\sum_1^\nu X_i\right)^2 > c\sigma_1^2 \tag{4}$$

and let $Y = \frac{1}{\nu}\sum_1^\nu X_i$. (The inequality (4) is used merely as a device for fixing the probability of taking ν observations, this random event to be independent of whether $(Y - \frac{1}{2}l, Y + \frac{1}{2}l)$ covers ξ, given ν.)

c) Otherwise take a $(\nu + 1)$st observation, terminating the process, and let

$$Y = \frac{1}{\nu + 1}\sum_1^{\nu+1} X_i,$$

When $c = 0$, this is the usual non-sequential procedure.

Clearly,

$$\alpha[S(\nu, c)] = P\{\chi_{\nu-1}^2 > c\}H\left(\frac{\sqrt{\nu}\,l}{2\sigma_1}\right) + [1 - P\{\chi_{\nu-1}^2 > c\}]H\left(\frac{\sqrt{\nu + 1}\,l}{2\sigma_1}\right), \tag{5}$$

where

$$H(u) = \frac{1}{\sqrt{2\pi}}\int_{-u}^{u} e^{-\frac{1}{2}x^2}\,dx = \sqrt{\frac{2}{\pi}}\int_0^u e^{-\frac{1}{2}x^2}\,dx. \tag{6}$$

Also

$$n_0[S(\nu, c)] = \nu + 1 - P\{\chi_{\nu-1}^2 > c\}, \tag{7}$$

By a proper choice of ν and c we can achieve any desired confidence coefficient

$\alpha \geq H\left(\frac{l}{\sqrt{2}\sigma_1}\right)$. There is no essential loss of generality in considering only the case $\sigma_1 = 1$, and this will be done in the remainder of this paper.

3. A lower bound for $n_0(S)$ and an upper bound for $\alpha(S)$. Consider any sequential procedure S for obtaining confidence intervals of length l. Put

$$\alpha(\xi, S) = P\{Y - \tfrac{1}{2}l < \xi < Y + \tfrac{1}{2}l \mid \xi\}. \tag{8}$$

That is, $\alpha(\xi, S)$ is the probability that the confidence interval will cover the true mean ξ when the procedure S is used. According to (1),

$$\alpha(S) = \underset{\xi}{\text{g.l.b.}}\ \alpha(\xi, S). \tag{9}$$

In order to obtain a lower bound for $n_0(S)$ and an upper bound for $\alpha(S)$, we suppose that the procedure S is applied when ξ is not a fixed number but a random variable normally distributed with mean 0 and variance σ^2. Then the probability that the confidence interval covers ξ is

$$\bar{\alpha}(\sigma, S) = \frac{1}{\sqrt{2\pi}\sigma}\int_{-\infty}^{+\infty} e^{-\xi^2/2\sigma^2}\alpha(\xi, S)\,d\xi \geqq \alpha(S) \tag{10}$$

and the expected number of observations is

$$\bar{E}(n \mid \sigma, S) = \frac{1}{\sqrt{2\pi}\sigma}\int_{-\infty}^{+\infty} e^{-\xi^2/2\sigma^2}E(n \mid \xi, S)\,d\xi \leqq n_0(S). \tag{11}$$

Let $p_m(\xi, S)$, ($m = 1, 2, \cdots$, ad. inf.), denote the probability that $n = m$ when ξ is the true mean and procedure S is used. Put

$$\bar{p}_m(\sigma, S) = \frac{1}{\sqrt{2\pi}\sigma}\int_{-\infty}^{+\infty} e^{-\xi^2/2\sigma^2}p_m(\xi, S)\,d\xi. \tag{12}$$

Since

$$\bar{E}(n \mid \sigma, S) = \sum_{m=1}^{\infty} m\bar{p}_m(\sigma, S) \tag{13}$$

we obtain from (11)

$$\sum_{m=1}^{\infty} m\bar{p}_m(\sigma, S) \leqq n_0(S). \tag{14}$$

We shall now derive an upper bound for $\bar{\alpha}(\sigma, S)$. Since $X_i = \xi + \epsilon_i$ where the ϵ_i are independently normally distributed with mean 0 and variance 1, the joint distribution of ξ and X_i, ($i = 1, \cdots, m$), is a multivariate normal distribution with

$$E\xi = EX_i = 0 \tag{15}$$

and covariance matrix

$$(16) \qquad E\begin{pmatrix} \xi \\ X_1 \\ \vdots \\ X_m \end{pmatrix}(\xi, X_1, \cdots, X_m) = \begin{pmatrix} \sigma^2 & \sigma^2 & \cdots & \cdots & \sigma^2 \\ \sigma^2 & \sigma^2+1 & \sigma^2 & \cdots & \sigma^2 \\ \vdots & \sigma^2 & \sigma^2+1 & \cdots & \sigma^2 \\ \vdots & & & & \\ \vdots & \vdots & & & \vdots \\ \sigma^2 & \sigma^2 & \cdots & \cdots & \sigma^2+1 \end{pmatrix}.$$

Thus the conditional distribution of ξ given $X_1, \cdots, X_m$ is normal with mean

$$E(\xi \mid X_1, \cdots, X_m) = (\sigma^2, \cdots, \sigma^2)\begin{pmatrix} \sigma^2+1 & \sigma^2 & \cdots & \sigma^2 \\ \sigma^2 & \sigma^2+1 & \cdots & \sigma^2 \\ \vdots & \vdots & & \vdots \\ \sigma^2 & \sigma^2 & \cdots & \sigma^2+1 \end{pmatrix}^{-1}\begin{pmatrix} X_1 \\ \vdots \\ X_m \end{pmatrix}$$

$$(17) \qquad = \sigma^2(1, 1, \cdots, 1)\begin{pmatrix} \frac{(m-1)\sigma^2+1}{m\sigma^2+1} & -\frac{\sigma^2}{m\sigma^2+1} & \cdots & -\frac{\sigma^2}{m\sigma^2+1} \\ -\frac{\sigma^2}{m\sigma^2+1} & \frac{(m-1)\sigma^2+1}{m\sigma^2+1} & \cdots & -\frac{\sigma^2}{m\sigma^2+1} \\ \vdots & \vdots & & \vdots \\ -\frac{\sigma^2}{m\sigma^2+1} & -\frac{\sigma^2}{m\sigma^2+1} & \cdots & \frac{(m-1)\sigma^2+1}{m\sigma^2+1} \end{pmatrix}$$

$$\times \begin{pmatrix} X_1 \\ \vdots \\ X_m \end{pmatrix} = \frac{\sigma^2}{m\sigma^2+1}\sum_1^m X_i$$

and variance

$$(18) \qquad \sigma^2 - \frac{\sigma^4}{(m\sigma^2+1)^2}E\left(\sum_{i=1}^m X_i\right)^2 = \frac{\sigma^2}{m\sigma^2+1}.$$

If $X_1, \cdots, X_m$ is a sequence for which the process is terminated on the mth trial, the conditional probability that the interval of length l will cover ξ is clearly maximized by taking

$$(19) \qquad Y = E(\xi \mid X_1, \cdots, X_m) = \frac{\sigma^2}{m\sigma^2+1}\sum_1^m X_i$$

and, by (18) this probability has the value $H(c_m)$ where H is defined by (6) and

$$(20) \qquad c_m = \sqrt{m + \frac{1}{\sigma^2}}\,\frac{l}{2}.$$

Hence,

$$\bar{\alpha}(\sigma, S) \leq \sum_{m=1}^{\infty} \bar{p}_m(\sigma, S)H(c_m). \tag{21}$$

From this and (10) we obtain

$$\alpha(S) \leq \sum_{1}^{\infty} \bar{p}_m(\sigma, S)H(c_m). \tag{22}$$

This upper limit of $\alpha(S)$ and the lower limit of $n_0(S)$ given in (14) will be used later to prove that $S(\nu, c)$ is an optimum procedure.

4. Maximum value of $\sum_{1}^{\infty} \bar{p}_m(\sigma, S)H(c_m)$ subject to the condition that $\sum_{1}^{\infty} m\bar{p}_m(\sigma, S)$ does not exceed a given bound. We shall show that the maximum of $\sum_{1}^{\infty} \bar{p}_m(\sigma, S)H(c_m)$ subject to

$$E(n \mid \sigma, S) = \sum_{1}^{\infty} m\bar{p}_m(\sigma, S) \leqq \nu + a,$$

where ν is a positive integer and $0 \leq a < 1$, is obtained by choosing $\bar{p}_m(\sigma, S) = p_m^*$ defined by

$$\begin{aligned} p_m^* &= 0 \text{ for } m < \nu \text{ or } m > \nu + 1 \\ p_\nu^* &= 1 - \text{a} \\ p_{\nu+1}^* &= a. \end{aligned} \tag{23}$$

For, suppose to the contrary that there exists a sequence $\{p_m\}$ such that the following conditions hold:

$$\begin{gathered} p_m \geq 0, \qquad \sum_{1}^{\infty} p_m = 1 \\ \sum_{1}^{\infty} mp_m \leq \nu + a = \sum_{1}^{\infty} mp_m^* \\ \sum_{1}^{\infty} p_m H(c_m) > \sum_{1}^{\infty} p_m^* H(c_m). \end{gathered} \tag{24}$$

We have

$$H(u) = \sqrt{\frac{2}{\pi}} \int_0^u e^{-\frac{1}{2}x^2}\, dx = \frac{1}{\sqrt{2\pi}} \int_0^{u^2} y^{-\frac{1}{2}} e^{-\frac{1}{2}y}\, dy. \tag{25}$$

Put

$$C = H(c_{\nu+1}) - H(c_\nu) = \frac{1}{\sqrt{2\pi}} \int_{c_\nu^2}^{c_{\nu+1}^2} y^{-\frac{1}{2}} e^{-\frac{1}{2}y}\, dy. \tag{26}$$

With the aid of $p_\nu = 1 - \sum_{m \neq \nu} p_m$, we obtain from the last two inequalities in (24)

$$(27) \qquad 0 < \sum_1^\infty (p_m - p_m^*)H(c_m) - C \sum_1^\infty (p_m - p_m^*)m = \sum_{m \neq \nu} (p_m - p_m^*)K_m$$

where

$$(28) \qquad K_m = H(c_m) - H(c_\nu) - (m - \nu)[H(c_{\nu+1}) - H(c_\nu)].$$

Clearly $K_{\nu+1} = 0$. Also, for $m < \nu$, since the integrand is a strictly decreasing function of y,

$$(29) \qquad \begin{aligned} K_m &= (\nu - m) \int_{c_\nu^2}^{c_\nu^2+1} y^{-\frac{1}{2}} e^{-\frac{1}{2}y}\, dy - \int_{c_m^2}^{c_\nu^2} y^{-\frac{1}{2}} e^{-\frac{1}{2}y}\, dy \\ &< (\nu - m) \frac{l^2}{4} y^{-\frac{1}{2}} e^{-\frac{1}{2}y} \Big|_{y=c_\nu^2} - (\nu - m) \frac{l^2}{4} y^{-\frac{1}{2}} e^{-\frac{1}{2}y} \Big|_{y=c_\nu^2} = 0. \end{aligned}$$

Similarly for $m > \nu + 1$, $K_m < 0$. But $p_m^* = 0$ for $m \neq \nu, \nu + 1$ so that

$$(30) \qquad \sum_{m \neq \nu, 1\, \nu+1} (p_m - p_m^*)K_m \leq 0$$

which contradicts (27) since $K_{\nu+1} = 0$.

Thus, we have shown that the inequality

$$(31) \qquad \bar{E}(n \mid \sigma, S) \leq \nu + a$$

implies the inequality

$$(32) \qquad \sum_1^\infty \bar{p}_m(\sigma, S)H(c_m) \leq (1 - a)H(c_\nu) + aH(c_{\nu+1}).$$

5. Proof that $S(\nu, c)$ is an optimum procedure. Since, according to (14) and (22)

$$(33) \qquad n_0(S) \geq \bar{E}(n \mid \sigma, S) \quad \text{and} \quad \alpha(S) \leq \sum_1^\infty \bar{p}_m(\sigma, S)H(c_m),$$

it follows from the result expressed in (31) and (32) that, for any procedure S satisfying the inequality

$$(34) \qquad n_0(S) \leq \nu + a,$$

we must have

$$(35) \qquad \alpha(S) \leq (1 - a)H(c_\nu) + aH(c_{\nu+1})$$

identically in σ. Since $H(u)$ is continuous, it follows that

$$(36) \qquad \alpha(S) \leq (1 - a)H\left(\sqrt{\nu}\, \frac{l}{2}\right) + aH\left(\sqrt{\nu + 1}\, \frac{l}{2}\right)$$

for any procedure S satisfying (34)..

The right hand side of (36) is $\alpha[S(\nu, c)]$ where c is chosen so that

$$1 - a = P\{\chi^2_{\nu-1} > c\}. \tag{37}$$

We use an indirect proof to show that $S(\nu, c)$ is an optimum procedure. Suppose to the contrary that there is a procedure S' such that

$$\alpha(S') = \alpha[S(\nu, c)] \tag{38}$$

but

$$n_0(S') < n_0[S(\nu, c)]. \tag{39}$$

By (5) and (7), $\alpha[S(\nu, c)]$ is a continuous strictly increasing function of

$$\nu + 1 - P\{\chi^2_{\nu-1} > c\}$$

and this latter is $n_0[S(\nu, c)]$. If we choose ν', c' so that

$$\begin{aligned} n_0(S') &< \nu' + 1 - P\{\chi^2_{\nu'-1} > c'\} \\ &< \nu + 1 - P\{\chi^2_{\nu-1} > c\}, \end{aligned} \tag{40}$$

it follows that

$$\alpha[S(\nu', c')] < \alpha[S(\nu, c)] = \alpha(S'). \tag{41}$$

But (41) and the first part of (40) contradict the result expressed in (34) and (36).

REFERENCES

[1] A. WALD, *Sequential Analysis*, John Wiley and Sons, 1947, section 11.2.
[2] CHARLES STEIN, "A two-sample test for a linear hypothesis whose power is independent of the variance", *Annals of Math. Stat.*, Vol. 16 (1945), pp. 243–258.
[3] G. B. DANTZIG, "On the non-existence of tests of 'Student's' hypothesis having power functions independent of σ". *Annals of Math. Stat.*, Vol. 11 (1940), p. 186.

Made in United States of America

Reprinted from The Annals of Mathematical Statistics
Vol. XVIII, No. 4, December, 1947

A NOTE ON REGRESSION ANALYSIS

By Abraham Wald

Columbia University

1. Introduction. In regression analysis a set of variables $y, x_1, \cdots, x_p$ is considered where y is called the dependent variable and $x_1, \cdots, x_p$ are the independent variables. Let y_α denote the αth observation on y and $x_{i\alpha}$ the αth observation on x_i, $(i = 1, \cdots, p; \alpha = 1, \cdots, N)$. The observations $x_{i\alpha}$ are treated as given constants, while the observations $y_1, \cdots, y_N$ are regarded as chance variables. The following two assumptions are usually made concerning the joint distribution of the variates $y_1, \cdots, y_N$:

(a) The variates $y_1, \cdots, y_N$ are normally and independently distributed with a common unknown variance σ^2.

(b) The expected value of y_α is equal to $\beta_1 x_{1\alpha} + \cdots + \beta_p x_{p\alpha}$ where $\beta_1, \cdots, \beta_p$ are unknown constants.

In some problems it seems reasonable to assume that the regression coefficients $\beta_1, \cdots, \beta_p$ are not constants, but chance variables. This leads to a different probability model for regression analysis and the object of this note is to discuss certain aspects of this model. In what follows in this note we shall make the following assumptions concerning the joint distribution of the chance variables $y_1, \cdots, y_N; \beta_1, \cdots, \beta_p$.

Assumption 1. For given values of $\beta_1, \cdots, \beta_p$ the joint conditional probability density function of $y_1, \cdots, y_N$ is given by

$$(1.1) \qquad \frac{1}{(2\pi)^{N/2}\sigma^N} \exp\left[-\frac{1}{2\sigma^2}\sum_{\alpha=1}^{N}(y_\alpha - \beta_1 x_{1\alpha} - \cdots - \beta_p x_{p\alpha})^2\right]$$

Assumption 2. The regression coefficients $\beta_1, \cdots, \beta_p$ are independently distributed.

Assumption 3. The regression coefficients $\beta_1, \cdots, \beta_r$, $(r \leq p)$, are normally distributed with zero means and a common variance σ'^2.

The purpose of this note is to derive confidence limits for the ratio $\frac{\sigma'^2}{\sigma^2}$. Such confidence limits have been derived by the author [1] for analysis of variance problems assuming that there are only main effects but no interactions. The regression problem treated in the present note is much more general and includes all the analysis of variance problems with or without interactions as special cases.

It should be remarked that Assumptions 2 and 3 do not exclude the case where $\beta_{r+1}, \cdots, \beta_p$ are constants.

2. Derivation of confidence limits for the ratio $\frac{\sigma'^2}{\sigma^2}$. Let $b_1, \cdots, b_p$ be the sample estimates of $\beta_1, \cdots, \beta_p$ obtained by the method of least squares. We shall denote the difference $b_i - \beta_i$ by ϵ_i, $(i = 1, \cdots, p)$. It is known that for given values of $\beta_1, \cdots, \beta_p$ the conditional joint distribution of $\epsilon_1, \cdots, \epsilon_p$ is normal with zero means and variance-covariance matrix $\| c_{ij} \| \sigma^2$ where

$$\| c_{ij} \| = \| a_{ij} \|^{-1} \tag{2.1}$$

and

$$a_{ij} = \sum_{\alpha=1}^{N} x_{i\alpha} x_{j\alpha}, \qquad (i, j = 1, \cdots, p). \tag{2.2}$$

Since the conditional distribution of $\epsilon_1, \cdots, \epsilon_p$ does not depend on the values of $\beta_1, \cdots, \beta_p$, the unconditional distribution of $\epsilon_1, \cdots, \epsilon_p$ is the same as the conditional one, and the set of variates $(\beta_1, \cdots, \beta_p)$ is independently distributed of the set $(\epsilon_1, \cdots, \epsilon_p)$. From this and Assumptions 2 and 3 it follows that $b_1, \cdots, b_r$ have a joint normal distribution and that

$$Eb_i = 0, \qquad (i = 1, \cdots, r) \tag{2.3}$$

and

$$Eb_i b_j = \left(c_{ij} + \delta_{ij} \frac{\sigma'^2}{\sigma^2}\right) \sigma^2, \qquad (i, j = 1, \cdots, r) \tag{2.4}$$

where $\delta_{ij} = 0$ for $i \neq j$ and $= 1$ for $i = j$.

We shall denote $\frac{\sigma'^2}{\sigma^2}$ by λ and the elements of the inverse of $\| c_{ij} + \delta_{ij}\lambda \|$ by $d_{ij}(\lambda)$, i.e.,

$$\| d_{ij}(\lambda) \| = \| c_{ij} + \delta_{ij}\lambda \|^{-1}, \qquad (i, j = 1, \cdots, r). \tag{2.5}$$

Then the quadratic form

$$Q(\lambda) = \frac{1}{\sigma^2} \sum_{j=1}^{r} \sum_{i=1}^{r} d_{ij}(\lambda) b_i b_j \tag{2.6}$$

has the χ^2 distribution with r degrees of freedom.

It is known that for any given values of $\beta_1, \cdots, \beta_p, b_1, \cdots, b_p$ the quadratic form

$$Q_a = \frac{1}{\sigma^2} \sum_{\alpha=1}^{N} (y_\alpha - b_1 x_{1\alpha} - \cdots - b_p x_{p\alpha})^2 \tag{2.7}$$

has the χ^2 distribution with $N - p$ degrees of freedom provided that the rank of the matrix $\| x_{i\alpha} \|$ is p. Hence Q_a and $Q(\lambda)$ are independently distributed and the ratio

$$F = \frac{N - p}{r} \frac{Q(\lambda)}{Q_a} \tag{2.8}$$

has the F-distribution with r and $N - p$ degrees of freedom.

Let F_1 and F_2 be two values chosen so that

$$\text{Prob. } \{F_1 \leqq F \leqq F_2\} = c \tag{2.9}$$

where c is a given positive constant less than 1. Then the set of all values λ for which the inequality

$$F_1 \leq \frac{N-p}{r} \frac{Q(\lambda)}{Q_a} \leq F_2 \tag{2.10}$$

holds forms a confidence set for λ with the confidence coefficient c.

We shall now show that $Q(\lambda)$ is a monotonic function of λ and, therefore, the confidence set determined by (2.10) is an interval. Let $\| g_{ij} \|$, $(i, j = 1, \cdots, r)$, be an orthogonal matrix and let

$$b_i^* = \sum_{j=1}^{r} g_{ij} b_j . \tag{2.11}$$

It then follows from (2.3) and (2.4) that

$$E(b_i^*) = 0, \qquad (i = 1, \cdots, r) \tag{2.12}$$

and

$$E(b_i^* b_j^*) = (c_{ij}^* + \delta_{ij}\lambda)\sigma^2, \qquad (i, j = 1, \cdots, r) \tag{2.13}$$

where

$$c_{ij}^* = \sum_{l=1}^{r} \sum_{k=1}^{r} g_{ik} g_{jl} c_{kl} . \tag{2.14}$$

Let

$$\| d_{ij}^*(\lambda) \| = \| c_{ij}^* + \delta_{ij}\lambda \|^{-1}, \qquad (i, j = 1, \cdots, r) \tag{2.15}$$

and put

$$Q^*(\lambda) = \frac{1}{\sigma^2} \Sigma\Sigma\, d_{ij}^*(\lambda) b_i^* b_j^* .$$

It is easy to verify that $Q^*(\lambda)$ is identically equal to $Q(\lambda)$. Hence, to prove the monotonicity of $Q(\lambda)$, it is sufficient to show that $Q^*(\lambda)$ is a monotonic function of λ. Since no restrictions as to the choice of the orthogonal matrix $\| g_{ij} \|$ are made, we shall choose it so that the matrix $\| c_{ij}^* \|$ becomes diagonal, i.e., $c_{ij}^* = 0$ for $i \neq j$, $(i, j = 1, \cdots, r)$. Then

$$d_{ij}^*(\lambda) = 0 \qquad \text{for } i \neq j \tag{2.16}$$

and

$$d_{ii}^*(\lambda) = \frac{1}{c_{ii}^* + \lambda} . \tag{2.17}$$

Hence

$$Q(\lambda) = Q^*(\lambda) = \frac{1}{\sigma^2} \sum_{i=1}^{r} \frac{b_i^{*2}}{c_{ii}^* + \lambda} \tag{2.18}$$

is a monotonically decreasing function of λ. The confidence set determined by (2.10) is, therefore, an interval.

The upper end point of the confidence interval is the root in λ of the equation

$$\frac{N - p}{r}\frac{Q(\lambda)}{Q_a} = F_1 \tag{2.19}$$

and the lower end point is the root in λ of the equation

$$\frac{N - p}{r}\frac{Q(\lambda)}{Q_a} = F_2 . \tag{2.20}$$

If equation (2.20) has no root, the lower end point of the confidence interval is put equal to zero.

REFERENCE

[1] A. WALD, "On the analysis of variance in case of multiple classifications with unequal class frequencies", *Annals. of Math. Stat.*, Vol. 12 (1941).

Made in United States of America

Reprinted from THE ANNALS OF MATHEMATICAL STATISTICS
Vol. XIX, No. 1, March, 1948

ASYMPTOTIC PROPERTIES OF THE MAXIMUM LIKELIHOOD ESTIMATE OF AN UNKNOWN PARAMETER OF A DISCRETE STOCHASTIC PROCESS

BY ABRAHAM WALD

Columbia University

Summary. Asymptotic properties of maximum likelihood estimates have been studied so far mainly in the case of independent observations. In this paper the case of stochastically dependent observations is considered. It is shown that under certain restrictions on the joint probability distribution of the observations the maximum likelihood equation has at least one root which is a consistent estimate of the parameter θ to be estimated. Furthermore, any root of the maximum likelihood equation which is a consistent estimate of θ is shown to be asymptotically efficient. Since the maximum likelihood estimate is always a root of the maximum likelihood equation, consistency of the maximum likelihood estimate implies its asymptotic efficiency.

1. Introduction. Let $\{X_i\}$, $(i = 1, 2, \cdots$, ad. inf.), be a sequence of chance variables. It is assumed that for any positive integral value n the first n chance variables $X_1, \cdots, X_n$ admit a joint probability density function $p_n(x_1, \cdots, x_n, \theta)$ involving an unknown parameter θ. The consistency relations

$$\int_{-\infty}^{+\infty} p_{n+1}(x_1, \cdots, x_{n+1}, \theta)\, dx_{n+1} = p_n(x_1, \cdots, x_n, \theta) \tag{1.1}$$

are assumed to hold.

In what follows, for any chance variable u the symbol $E(u \mid \theta)$ will denote the expected value of u when θ is the true parameter value.

Let $t_n(x_1, \cdots, x_n)$ be an unbiassed estimate of θ. Cramér [1] and Rao [2] have shown that under some weak regularity conditions on the distribution function $p_n(x_1, \cdots, x_n, \theta)$, the variance of t_n cannot fall short of the value

$$\frac{1}{c_n(\theta)} = \frac{1}{E\left[\left(\frac{\partial \log p_n}{\partial \theta}\right)^2 \middle| \theta\right]} \tag{1.2}$$

Thus, for any unbiassed estimate t_n the variate $\sqrt{c_n(\theta)}(t_n - \theta)$ has mean value zero and variance $\geqq 1$. An estimate t_n is called efficient if $\sqrt{c_n(\theta)}(t_n - \theta)$ has mean value zero and variance 1.

A sequence $\{t_n\}$, $(n = 1, 2, \cdots$, ad. inf.), of estimates is said to be asymptotically efficient if the mean of $\sqrt{c_n(\theta)}\,(t_n - \theta)$ is zero and the variance of $\sqrt{c_n(\theta)}\,(t_n - \theta)$ is 1 in the limit as $n \to \infty$. In the literature usually the additional requirement is made that the limiting distribution of $\sqrt{c_n(\theta)}\,(t_n - \theta)$ be normal.

To make a distinction between the two cases when the condition concerning the limiting distribution of $\sqrt{c_n(\theta)}\ (t_n - \theta)$ is fulfilled or not, we shall say that $\{t_n\}$ is asymptotically efficient in the wide sense if it satisfies the conditions concerning the mean and the variance of $\sqrt{c_n(\theta)}\ (t_n - \theta)$. If, in addition, the limiting distribution of $\sqrt{c_n(\theta)}\ (t_n - \theta)$ is normal, we shall say that $\{t_n\}$ is asymptotically efficient in the strict sense. Clearly, if $\{t_n\}$ is asymptotically efficient in the strict sense, it is also asymptotically efficient in the wide sense.

A word of clarification is needed as to the meaning of the conditions concerning the mean and variance of $\sqrt{c_n(\theta)}\ (t_n - \theta)$. One interpretation would be that the requirement is that

$$\lim_{n=\infty} E[\sqrt{c_n(\theta)}\ (t_n - \theta) \mid \theta] = 0 \tag{1.3}$$

and

$$\lim_{n=\infty} E[c_n(\theta)\ (t_n - \theta)^2 \mid \theta] = 1. \tag{1.4}$$

Another interpretation would be that the requirement is that the limiting distribution of $\sqrt{c_n(\theta)}\ (t_n - \theta)$, provided that the limit distribution exists as $n \to \infty$, should have zero mean and unit variance. These two interpretations are certainly not equivalent. It seems to the author that the mean and variance of the limiting distribution is more relevant than the limits of the mean and the variance. We shall, therefore, adopt the following definition of asymptotic efficiency:

Definition: A sequence $\{t_n\}$ of estimates is said to be asymptotically efficient in the wide sense if a sequence $\{u_n\}$, $(n = 1, 2, \cdots, \text{ad. inf.})$, of chance variables exists such that

$$\lim_{n=\infty} E(u_n \mid \theta) = 0, \qquad \lim_{n=\infty} E(u_n^2 \mid \theta) = 1 \tag{1.5}$$

and

$$\sqrt{c_n(\theta)}(t_n - \theta) - u_n \tag{1.6}$$

converges stochastically to zero as $n \to \infty$. If, in addition, the limiting distribution of $\sqrt{c_n(\theta)}\ (t_n - \theta)$ exists and is normal, $\{t_n\}$ is said to be asymptotically efficient in the strict sense.

The reason that a sequence $\{u_n\}$ of chance variables is considered in the above definition, instead of the limiting distribution of $\sqrt{c_n(\theta)}\ (t_n - \theta)$, is that the existence of a limiting distribution of $\sqrt{c_n(\theta)}\ (t_n - \theta)$ is not postulated. If a limiting distribution of $\sqrt{c_n(\theta)}\ (t_n - \theta)$ exists and if this limiting distribution has zero mean and unit variance, a sequence $\{u_n\}$ of chance variables satisfying the conditions (1.5) and (1.6) always exists. This can be seen as follows: Let T_n denote the chance variable $\sqrt{c_n(\theta)}\ (t_n - \theta)$ and let $F_n(t) = \text{prob.}\ \{T_n < t\}$. If a limiting distribution of T_n exists and if this limiting distribution has zero mean and unit variance, then

$$(1.7)\qquad \lim_{a=\infty}\left[\lim_{n=\infty}\int_{-a}^{a} t\,dF_n(t)\right] = 0 \quad\text{and}\quad \lim_{a=\infty}\left[\lim_{n=\infty}\int_{-a}^{a} t^2\,dF_n(t)\right] = 1.$$

From (1.7) it follows that there exists a sequence $\{a_n\}$, ($n = 1, 2, \cdots$, ad. inf.), of positive values such that the following conditions are fulfilled:

$$(1.8)\quad \lim_{n=\infty}\int_{-a_n}^{a_n} t\,dF_n(t) = 0;\ \lim_{n=\infty}\int_{-a_n}^{a_n} t^2\,dF_n(t) = 1;\ \lim_{n=\infty}\text{Prob}\,\{|\,T_n\,| > a_n\} = 0.$$

Let u_n be a chance variable which is equal to T_n whenever $|\,T_n\,| \leqq a_n$, and equal to zero otherwise. Clearly, the sequence $\{u_n\}$ will satisfy conditions (1.5) and (1.6).

In the following section we shall formulate some assumptions concerning the probability density function $p_n(x_1, \cdots, x_n, \theta)$. It will then be shown in section 3 that there exists a root of the maximum likelihood equation

$$(1.9)\qquad \frac{\partial \log p_n}{\partial\theta} = 0$$

which is asymptotically efficient at least in the wide sense.

2. Assumptions concerning the probability density $p_n(x_1, \cdots, x_n, \theta)$. We shall assume that there exists a finite non-degenerate interval A on the θ-axis such that the following conditions hold:

Condition 1. The derivatives $\frac{\partial^i p_n}{\partial\theta^i}$, ($i = 1, 2, 3$), exist for all θ in A and for all samples $(x_1, \cdots, x_n)$ except perhaps for a set of measure zero. We have furthermore,

$$(2.1)\qquad \int_{-\infty}^{+\infty}\cdots\int_{-\infty}^{+\infty} \underset{\theta\,\epsilon\,A}{\text{l.u.b.}}\left|\frac{\partial^i p_n}{\partial\theta^i}\right| dx_1\cdots dx_n < \infty, \qquad (i = 1, 2).$$

Condition 2. For any θ in A we have $\lim_{n=\infty} c_n(\theta) = \infty$.

Condition 3. For any θ in A the standard deviation of $\frac{\partial^2 \log p_n}{\partial\theta^2}$ divided by the expected value of $\frac{\partial^2 \log p_n}{\partial\theta^2}$ (both computed under the assumption that θ is true) converges to zero as $n \rightarrow \infty$.

Condition 4. There exists a positive δ such that for any θ in A the expression

$$(2.2)\qquad \frac{1}{c_n(\theta)}\,E\left[\underset{\theta'}{\text{l.u.b.}}\left|\frac{\partial^3 \log p_n(x_1, \cdots, x_n, \theta')}{\partial\theta'^3}\right|\,\middle|\,\theta\right]$$

is a bounded function of n where θ' is restricted to the interval $|\,\theta' - \theta\,| \leqq \delta$.

In what follows in this section, as well as in section 3, the domain of θ will be restricted to interior points of the interval A unless a statement to the contrary is explicitly made.

Clearly

$$E\left(\frac{\partial \log p_n}{\partial \theta} \middle| \theta\right) = \int_{-\infty}^{+\infty} \cdots \int_{-\infty}^{+\infty} \frac{\partial p_n}{\partial \theta}\, dx_1 \cdots dx_n . \tag{2.3}$$

It follows from Condition 1 that

$$\int_{-\infty}^{+\infty} \cdots \int_{-\infty}^{+\infty} \frac{\partial p_n}{\partial \theta}\, dx_1 \cdots dx_n = \frac{\partial}{\partial \theta} \int_{-\infty}^{+\infty} p_n\, dx_1 \cdots dx_n = 0. \tag{2.4}$$

Hence,

$$E\left(\frac{\partial \log p_n}{\partial \theta} \middle| \theta\right) = 0. \tag{2.5}$$

We have

$$\frac{\partial^2 \log p_n}{\partial \theta^2} = \frac{1}{p_n} \frac{\partial^2 p_n}{\partial \theta^2} - \left(\frac{\partial \log p_n}{\partial \theta}\right)^2 . \tag{2.6}$$

Hence

$$E\left(\frac{\partial^2 \log p_n}{\partial \theta^2} \middle| \theta\right) = E\left(\frac{1}{p_n} \frac{\partial^2 p_n}{\partial \theta^2} \middle| \theta\right) - c_n(\theta). \tag{2.7}$$

But

$$E\left(\frac{1}{p_n} \frac{\partial^2 p_n}{\partial \theta^2} \middle| \theta\right) = 0, \tag{2.8}$$

because of Condition 1. From (2.7) and (2.8) we obtain

$$E\left(\frac{\partial^2 \log p_n}{\partial \theta^2} \middle| \theta\right) = -c_n(\theta). \tag{2.9}$$

Conditions 3 and 4 will generally be fulfilled when the stochastic dependence of x_j on x_i decreases sufficiently fast with increasing value of $|i - j|$. For, in such cases, the following order relations will generally hold: The standard deviation of $\frac{\partial^2 \log p_n}{\partial \theta^2}$ will, in general, be of the order $\sqrt{n}$, the expected value of

$$\underset{|\theta'-\theta| \leq \delta}{\text{l.u.b.}} \left| \frac{\partial^3 \log p_n}{\partial \theta'^3} \right|$$

will usually be of the order n, and $\frac{c_n(\theta)}{n}$ will generally have a positive lower bound and a finite upper bound.

3. Proof that the maximum likelihood equation has a root which is an asymptotically efficient estimate of θ (at least in the wide sense). Let θ_0 denote the true parameter value and let θ be any other value. We put

$$\frac{\partial \log p_n}{\partial \theta} = \Phi_n, \qquad \frac{\partial^2 \log p_n}{\partial \theta^2} = \Phi_n' \qquad \text{and} \qquad \frac{\partial^3 \log p_n}{\partial \theta^3} = \Phi_n''. \tag{3.1}$$

Expanding $\Phi_n(x_1, \cdots, x_n, \theta)$ in a Taylor expansion around $\theta = \theta_0$ we obtain

$$\begin{aligned} \Phi_n(x_1, \cdots, x_n, \theta) &= \Phi_n(x_1, \cdots, x_n, \theta_0) + (\theta - \theta_0)\Phi_n'(x_1, \cdots, x_n, \theta_0) \\ &\quad + \tfrac{1}{2}(\theta - \theta_0)^2\Phi_n''(x_1, \cdots, x_n, \theta_n^*) \end{aligned} \tag{3.2}$$

where θ_n^* is some value between θ_0 and θ. Dividing both sides of (3.2) by $c_n(\theta_0)$ we obtain

$$\begin{aligned} \frac{\Phi_n(x_1, \cdots, x_n, \theta)}{c_n(\theta_0)} &= \frac{\Phi_n(x_1, \cdots, x_n, \theta_0)}{c_n(\theta_0)} \\ &\quad + (\theta - \theta_0)\frac{\Phi_n'(x_1, \cdots, x_n, \theta_0)}{c_n(\theta_0)} + \tfrac{1}{2}(\theta - \theta_0)^2 \frac{\Phi''(x_1, \cdots, x_n, \theta_n^*)}{c_n(\theta_0)}. \end{aligned} \tag{3.3}$$

From Condition 3 and equation (2.9) it follows that

$$\operatorname*{plim}_{n=\infty} \frac{\Phi_n'(x_1, \cdots, x_n, \theta_0)}{c_n(\theta_0)} = -1 \tag{3.4}$$

where the operator plim stands for convergence in probability (stochastic convergence).

According to equation (2.5) the expected value of $\Phi_n(x_1, \cdots, x_n, \theta_0)$ is zero. Since the variance of $\Phi_n(x_1, \cdots, x_n, \theta_0)$ is equal to $c_n(\theta_0)$, and since $\lim_{n=\infty} c_n(\theta) = \infty$, we have

$$\operatorname*{plim}_{n=\infty} \frac{\Phi_n(x_1, \cdots, x_n, \theta_0)}{c_n(\theta_0)} = 0. \tag{3.5}$$

It follows from Condition 4 that for any θ with $|\theta - \theta_0| \leqq \delta$ we have

$$\frac{1}{c_n(\theta_0)} E(|\Phi_n''(x_1, \cdots, x_n, \theta_n^*|) = 0(1). \tag{3.6}$$

According to Markoff's inequality the probability that a positive random variable will exceed λ-times its expected value is not greater than $\frac{1}{\lambda}$. Hence, it follows from (3.6) that for any $\epsilon > 0$ we can find a positive value k_ϵ such that

$$\limsup_{n=\infty} \text{Prob}\left\{\frac{1}{c_n(\theta_0)} |\Phi_n''(x_1, \cdots, x_n, \theta_n^*)| \geqq k_\epsilon\right\} \leqq \epsilon. \tag{3.7}$$

Let ρ be any given positive number. The probability that the maximum likelihood equation

$$\Phi_n(x_1, \cdots, x_n, \theta) = 0 \tag{3.8}$$

will have a root in the interval $(\theta_0 - \rho, \theta_0 + \rho)$ converges to one as $n \to \infty$. This follows easily from (3.3), (3.4), (3.5) and (3.7). Thus, we have shown that the maximum likelihood equation has a root $\bar{\theta}_n$ which is a consistent estimate, i.e. it satisfies the relation

$$\text{plim}\ (\bar{\theta}_n - \theta_0) = 0. \tag{3.9}$$

We shall now show that if $\bar{\theta}_n$ is a root of the maximum likelihood equation (3.8) and if $\bar{\theta}_n$ is a consistent estimate, then $\bar{\theta}_n$ is also asymptotically efficient, at least in the wide sense. For this purpose we substitute $\bar{\theta}_n$ for θ in (3.3) and multiply both sides of the equation by $\sqrt{c_n(\theta_0)}$. We then obtain

$$0 = \frac{\Phi_n(x_1, \cdots, x_n, \theta_0)}{\sqrt{c_n(\theta_0)}} + \sqrt{c_n(\theta_0)}\,(\bar{\theta}_n - \theta_0)\,\frac{\Phi_n'(x_1, \cdots, x_n, \theta_0)}{c_n(\theta_0)} + \sqrt{c_n(\theta_0)}\,(\bar{\theta}_n - \theta_0)^2\, v_n \tag{3.10}$$

where

$$v_n = \frac{1}{2}\,\frac{\Phi_n''(x_1, \cdots, x_n, \theta_n^*)}{c_n(\theta_0)}. \tag{3.11}$$

Let

$$y_n = \frac{\Phi_n(x_1, \cdots, x_n, \theta_0)}{\sqrt{c_n(\theta_0)}} \quad \text{and} \quad z_n = \sqrt{c_n(\theta_0)}\,(\bar{\theta}_n - \theta_0). \tag{3.12}$$

Then (3.10) given

$$-y_n = z_n\,\frac{\Phi_n'(x_1, \cdots, x_n, \theta_0)}{c_n(\theta_0)} + z_n(\bar{\theta}_n - \theta_0)\, v_n. \tag{3.13}$$

It follows from (3.7) and (3.9) that

$$\underset{n=\infty}{\text{plim}}\ (\bar{\theta}_n - \theta_0)\, v_n = 0. \tag{3.14}$$

From (3.4), (3.13) and (3.14) we obtain

$$- y_n = z_n(-1 + \xi_n) \tag{3.15}$$

where

$$\underset{n=\infty}{\text{plim}}\ \xi_n = 0. \tag{3.16}$$

Since $Ey_n = 0$ and $Ey_n^2 = 1$, it follows from (3.15) and (3.16) that

$$\underset{n=\infty}{\text{plim}}\ (z_n - y_n) = 0. \tag{3.17}$$

The asymptotic efficiency (in the wide sense) of $\bar{\theta}_n$ is an immediate consequence of (3.17). Our main result may be summarized in the following theorem:

THEOREM. *If the true value of the parameter θ is an interior point of an inter-*

val A satisfying the conditions 1 — 4, then the maximum likelihood equation (1.9) *has a root*[1] *which is a consistent estimate of* θ. *Furthermore, any root of* (1.9) *which is a consistent estimate of* θ *is also asymptotically efficient at least in the wide sense.*

Since the maximum likelihood estimate is a root of (1.9), it follows from the above theorem that whenever the maximum likelihood estimate is consistent, it is also asymptotically efficient at least in the wide sense.

REFERENCES

[1] H. Cramér, *Mathematical Methods of Statistics*, Princeton Univ. Press, 1946.

[2] C. R. Rao, "Information and the accuracy attainable in the estimation of statistical parameters", *Bull. Calcutta Math. Soc.*, Vol. 37 (1945).

[1] The probability that (1.9) has at least one root converges to unity as $n \to \infty$.

Reprinted from the
Bulletin of the American Mathematical Society
Vol. 54, No. 4, pp. 422-430
April, 1948

ON THE DISTRIBUTION OF THE MAXIMUM OF SUCCESSIVE CUMULATIVE SUMS OF INDEPENDENTLY BUT NOT IDENTICALLY DISTRIBUTED CHANCE VARIABLES

ABRAHAM WALD

1. **Introduction.** Let $X_1, X_2, \cdots$, and so on be a sequence of chance variables and let S_i denote the sum of the first i X's, that is,

$$S_i = X_1 + \cdots + X_i \quad (i = 1, 2, \cdots, \text{ad inf}). \tag{1.1}$$

Let M_N denote the maximum of the first N cumulative sums $S_1, \cdots, S_N$, that is,

$$M_N = \max (S_1, \cdots, S_N). \tag{1.2}$$

The distribution of M_N, in particular the limiting distribution of a suitably normalized form of M_N, has been studied by Erdös and Kac [1][1] and by the author [2] in the special case when the X's are independently distributed with identical distributions.

In this note we shall be concerned with the distribution of M_N when the X's are independent but not necessarily identically distributed. In particular, the mean and variance of X_i may be any functions of i.

In §2 lower and upper limits for M_N are obtained which yield particularly simple limits for the distribution of M_N when the X's are symmetrically distributed around zero.

In §3 the special case is considered when X_i can take only the values 1 and -1 but the probability p_i that $X_i=1$ may be any function of i. The exact probability distribution of M_N for this case is derived and expressed as the first row of a product of N matrices.

The limiting distribution of $M_N/N^{1/2}$ is treated in §4. Since the interesting limiting case arises when the mean of X_i $(i \leqq N)$ is not only a function of i but also a function of N, we have to introduce a double sequence of chance variables. That is, for any N we consider a sequence of N chance variables $X_{N1}, \cdots, X_{NN}$. Let μ_{Ni} denote the mean and σ_{Ni} the standard deviation of X_{Ni}. Let, furthermore, S_{Ni} denote the sum $X_{N1} + \cdots + X_{Ni}$ and M_N the maximum of $S_{N1}, \cdots, S_{NN}$. With the help of a method used by Erdös and Kac [1], the following theorem is established in §4:

Presented to the Society, September 4, 1947; received by the editors June 27, 1947.

[1] Numbers in brackets refer to the references cited at the end of the paper.

THEOREM 1.1 *Let* $\{X_{Ni}\}$ *and* $\{X^*_{Ni}\}$ $(i=1, \cdots, N; N=1, 2, \cdots,$ *ad inf.*) *be two sequences of chance variables such that the following conditions are fulfilled*:

(a) *The X's are independently distributed.*

(b) *The sequence* $\{\sigma_{Ni}\}$ $(i=1, \cdots, N;\ N=1,\ 2, \cdots,$ *ad inf.*) *has a positive lower bound and a finite upper bound.*

(c) $\mu_{Ni}N^{1/2}$ *is a bounded function of i and N.*

(d) *The third absolute moment of* X_{Ni} *is a bounded function of i and N.*

(e) *The conditions* (a)–(d) *remain valid if we replace* X_{Ni} *by* X^*_{Ni}.

(f) *The equation*

$$
\lim_{N=\infty}\left[\frac{\mu^*_{N1}+\cdots+\mu^*_{Nj_i}}{\sigma^{*2}_{N1}+\cdots+\sigma^{*2}_{Nj_i}}\right.
\tag{1.3}
$$

$$
\left.-\frac{\mu_{N1}+\cdots+\mu_{Ni}}{\sigma^2_{N1}+\cdots+\sigma^2_{Ni}}\left(\frac{\sigma^2_{N1}+\cdots+\sigma^2_{NN}}{\sigma^{*2}_{N1}+\cdots+\sigma^{*2}_{NN}}\right)^{1/2}\right]=0
$$

holds for all i and N where μ^*_{Ni} *is the mean and* σ^*_{Ni} *is the standard deviation* X^*_{Ni} *and* j_i *is the smallest positive integer for which*

$$
\frac{\sigma^{*2}_{N1}+\cdots+\sigma^{*2}_{Nj_i}}{\sigma^{*2}_{N1}+\cdots+\sigma^{*2}_{NN}} \geq \frac{\sigma^2_{N1}+\cdots+\sigma^2_{Ni}}{\sigma^2_{N1}+\cdots+\sigma^2_{NN}}.
$$

Let

$$
\overline{M}_N = M^*_N\left(\frac{\sigma^2_{N1}+\cdots+\sigma^2_{NN}}{\sigma^{*2}_{N1}+\cdots+\sigma^{*2}_{NN}}\right)^{1/2}
\tag{1.4}
$$

where M^*_N *is the same function of the* X^**'s as* M_N *is of the X's. Then for any positive* ϵ *we have*

$$
\liminf_{N=\infty}\left[\text{prob}\left\{M_N < cN^{1/2}\right\} - \text{prob}\left\{\overline{M}_N < (c-\epsilon)N^{1/2}\right\}\right] \geqq 0
\tag{1.5}
$$

and

$$
\liminf_{N=\infty}\left[\text{prob}\left\{\overline{M}_N < (c+\epsilon)N^{1/2}\right\} - \text{prob}\left\{M_N < cN^{1/2}\right\}\right] \geqq 0.
\tag{1.6}
$$

The following corollary is a simple consequence of Theorem 1.1:

COROLLARY 1.1. *Let* N' *be any positive integral valued and strictly increasing function of N for which prob* $\{\overline{M}_{N'} < cN'^{1/2}\}$ *converges to a limit function* $P(c)$ *at all continuity points c of* $P(c)$ *as* $N\to\infty$. *Then also*

(1.7) $$\lim_{N=\infty} \text{prob}\left\{M_{N'} < cN'^{1/2}\right\} = P(c)$$

at all continuity points c of P(c).

The validity of Corollary 1.1 can be derived from that of Theorem 1.1 as follows: Let $c=c_0$ be a continuity point of $P(c)$ and substitute N' for N in (1.5) and (1.6). For any positive ρ all limit points of $\text{prob}\left\{\overline{M}_{N'}<(c_0-\epsilon)N'^{1/2}\right\}$ and $\text{prob}\left\{\overline{M}_{N'}<(c_0+\epsilon)N'^{1/2}\right\}$ will lie in the interval $[P(c_0)-\rho, P(c_0)+\rho]$ for sufficiently small ϵ. Hence, equations (1.5) and (1.6) imply that

(1.8) $$\begin{aligned} P(c_0) - \rho &\leqq \liminf_{N=\infty} \text{prob}\left\{M_{N'} < c_0N'^{1/2}\right\} \\ &\leqq \limsup_{N=\infty} \text{prob}\left\{M_{N'} < c_0N'^{1/2}\right\} \leqq P(c_0) + \rho. \end{aligned}$$

Since (1.8) is true for any positive number ρ, Corollary 1.1 is proved.

The result in Corollary 1.1 can be expressed also by saying that for any subsequence $\{N'\}$ of $\{N\}$ for which $\overline{M}_{N'}/N'^{1/2}$ has a limiting distribution as $N\rightarrow\infty$, also $M_{N'}/N'^{1/2}$ has a limiting distribution which is equal to that of $\overline{M}_{N'}/N'^{1/2}$.

It can easily be verified that the conditions (e) and (f) can always be satisfied for chance variables X^*_{Ni} which take only the values 1 and -1 with properly chosen probabilities. Thus, the results of §3 may be used to compute

$$\text{prob}\left\{M^*_N < N^{1/2}c\left(\frac{\overset{*2}{\sigma}_{N1} + \cdots + \overset{*2}{\sigma}_{NN}}{\sigma^2_{N1} + \cdots + \sigma^2_{NN}}\right)^{1/2}\right\}.$$

2. **Derivation of upper and lower bounds for M_N.** Let $X_1, \cdots, X_N$ be a set of N variables and let

(2.1) $$\tilde{X}_i = X_{N-i+1} \qquad (i = 1, 2, \cdots, N).$$

Let, furthermore,

(2.2) $$\tilde{M}_i = \max(\tilde{X}_i, \tilde{X}_i + \tilde{X}_{i-1}, \cdots, \tilde{X}_i + \cdots + \tilde{X}_1), \qquad (i = 1, \cdots, N).$$

Clearly

(2.3) $$\tilde{M}_N = M_N = \max(X_1, X_1 + X_2, \cdots, X_1 + \cdots + X_N).$$

If $X_1, \cdots, X_N$ are independent chance variables, the chance variables $\tilde{M}_1, \tilde{M}_2, \cdots, \tilde{M}_N$ form a simple Markoff chain, that is, the conditional distribution of $\tilde{M}_{i+1}$, given $\tilde{M}_1, \cdots, \tilde{M}_i$, depends only on $\tilde{M}_i$. This is an immediate consequence of the relations:

$$\tilde{M}_{i+1} = \tilde{M}_i + \tilde{X}_{i+1} \qquad \text{if } \tilde{M}_i > 0 \tag{2.4}$$

and

$$\tilde{M}_{i+1} = \tilde{X}_{i+1} \qquad \text{if } \tilde{M}_i \leqq 0. \tag{2.5}$$

We shall now prove the following theorem:

THEOREM 2.1. *The inequality*

$$\tilde{M}_i \leqq |\epsilon_1 \tilde{X}_1 + \cdots + \epsilon_i \tilde{X}_i| \qquad (i = 1, \cdots, N) \tag{2.6}$$

holds where $\epsilon_1 = 1$, $\epsilon_i = 1$ *if* $\epsilon_1 \tilde{X}_1 + \cdots + \epsilon_{i-1}\tilde{X}_{i-1} > 0$ *and* $\epsilon_i = -1$, *if* $\epsilon_1 \tilde{X}_1 + \cdots + \epsilon_{i-1}\tilde{X}_{i-1} \leqq 0$.

PROOF. Clearly, (2.6) holds for $i=1$. We shall prove (2.6) for $i+1$ assuming that it holds for i. For this purpose it is sufficient to show, because of (2.4) and (2.5), that

$$|\epsilon_1 \tilde{X}_1 + \cdots + \epsilon_{i+1}\tilde{X}_{i+1}| - |\epsilon_1 \tilde{X}_1 + \cdots + \epsilon_i \tilde{X}_i| \geqq \tilde{X}_{i+1}. \tag{2.7}$$

Denote $|\epsilon_1, \tilde{X}_1 + \cdots + \epsilon_i \tilde{X}_i|$ by c_i. If $c_i > 0$, then $\epsilon_{i+1} = 1$ and inequality (2.7) goes over into

$$|c_i + \tilde{X}_{i+1}| - c_i \geqq \tilde{X}_{i+1}, \tag{2.8}$$

which is obviously true. If $c_i \leqq 0$, $\epsilon_{i+1} = -1$ and inequality (2.7) is equivalent with

$$||c_i| + \tilde{X}_{i+1}| - |c_i| \geqq \tilde{X}_{i+1}, \tag{2.9}$$

which is obviously true. Hence, Theorem 2.1 is proved.

We shall now prove a theorem giving a lower bound for $\tilde{M}_i$.

THEOREM 2.2. *The inequality*

$$\tilde{K}_i = |\epsilon_1 \tilde{X}_1 + \cdots + \epsilon_i \tilde{X}_i| - 2 \max_{j \leqq i} |\tilde{X}_j| \leqq \tilde{M}_i \qquad (i = 1, \cdots, N) \tag{2.10}$$

holds where the ϵ*'s are defined as in Theorem* 2.1.

PROOF. Theorem 2.2 is obviously true for $i=1$. We shall assume that it is valid for i and we shall prove it for $i+1$. It follows from (2.4) and (2.5) that

$$\tilde{M}_{i+1} - \tilde{M}_i \geqq \tilde{X}_{i+1}, \tag{2.11}$$

$$\tilde{M}_{i+1} \geqq \tilde{X}_{i+1}. \tag{2.12}$$

Hence, to prove (2.10) for $i+1$ assuming that it is true for i, it is sufficient to show that at least one of the following two inequalities holds:

$$\tilde{K}_{i+1} - \tilde{K}_i \leqq \tilde{X}_{i+1}, \tag{2.13}$$

$$\tilde{K}_{i+1} \leqq \tilde{X}_{i+1}. \tag{2.14}$$

Consider first the case when $|\tilde{X}_{i+1}| \leqq |\epsilon_1\tilde{X}_1 + \cdots + \epsilon_i\tilde{X}_i|$. In this case (2.13) always holds, as can easily be verified. If $|\tilde{X}_{i+1}| > |\epsilon_1\tilde{X}_1 + \cdots + \epsilon_i\tilde{X}_i|$ and $\tilde{X}_{i+1} \geqq 0$, then (2.13) holds again. If $|\tilde{X}_{i+1}| > |\epsilon_1\tilde{X}_1 + \cdots + \epsilon_i\tilde{X}_i|$ and $\tilde{X}_{i+1} < 0$, then $|\epsilon_1\tilde{X}_1 + \cdots + \epsilon_i\tilde{X}_i + \epsilon_{i+1}\tilde{X}_{i+1}| \leqq |\tilde{X}_{i+1}|$ and, therefore, $\tilde{K}_{i+1} \leqq |\tilde{X}_{i+1}| - 2 \max_{j \leqq i+1} |\tilde{X}_j| \leqq -|\tilde{X}_{i+1}| = \tilde{X}_{i+1}$. Thus, in this case the inequality (2.14) holds. This completes the proof of Theorem 2.2.

Since $\tilde{M}_N = M_N$, Theorems 2.1 and 2.2 yield the following limits for M_N

$$\begin{aligned} |\epsilon_1\tilde{X}_1 + \cdots + \epsilon_N\tilde{X}_N| - 2 \max_{i \leqq N} |\tilde{X}_i| \\ \leqq M_N \leqq |\epsilon_1\tilde{X}_1 + \cdots + \epsilon_N\tilde{X}_N|. \end{aligned} \tag{2.15}$$

Suppose now that $X_1, \cdots, X_N$ are chance variables such that the conditional distribution of X_i $(i=1, \cdots, N)$ for any given values of $X_{i+1}, \cdots, X_N$ is symmetric around the origin. Then the probability distribution of $|\epsilon_1\tilde{X}_1 + \cdots + \epsilon_N\tilde{X}_N|$ is the same as that of $|X_1 + \cdots + X_N|$, and the distribution of $|\epsilon_1\tilde{X}_1 + \cdots + \epsilon_N\tilde{X}_N| - 2 \max_{i \leqq N} |\tilde{X}_i|$ equals that of $|X_1 + \cdots + X_N - 2\max_{i \leqq N}|X_i|$. It then follows from (2.15) that the following theorem holds:

THEOREM 2.3. *If the conditional distribution of X_i $(i=1, 2, \cdots, N)$, for any given value of $X_{i+1}, \cdots, X_N$ is symmetric around the origin, the inequality*

$$\begin{aligned} \text{prob}\,\{|X_1 + \cdots + X_N| < c\} &\leqq \text{prob}\,\{M_N < c\} \\ &\leqq \text{prob}\,\{|X_1 + \cdots + X_N| - 2 \max_{i \leqq N} |X_i| < c\} \end{aligned} \tag{2.16}$$

holds for any value c.

Inequality (2.15) has also some interesting implications for the asymptotic distribution theory of M_N. In most cases we shall be concerned with the limiting distribution of $M_N/N^{1/2}$ as $N \to \infty$ (this is the case discussed in §4). If $(1/N^{1/2}) \max_{i \leqq N} |X_i|$ converges stochastically to zero, as will usually be the case, inequality (2.15) implies that the limiting distribution of $M_N/N^{1/2}$ is the same as that of $(1/N^{1/2})|\epsilon_1\tilde{X}_1 + \cdots + \epsilon_N\tilde{X}_N|$.

3. The distribution of M_N when X_i can take only the values 1 and -1. Let $X_1, \cdots, X_N$ be independently distributed chance variables such that X_i can take only the values 1 and -1. Let p_i denote the probability that $X_i = 1$. The probability that $X_i = -1$ is then equal to $1 - p_i = q_i$.

Let $\tilde{X}_i$ and $\tilde{M}_i$ $(i=1, \cdots, N)$ be defined by (2.1) and (2.2), respectively. One can easily verify that $\tilde{M}_i$ can take only the values $-1, 0, 1, 2, \cdots, i$. Let c_{ij} denote the probability that $\tilde{M}_i = j$ for $j = 1, \cdots, i$, and let c_{i0} be the probability that $\tilde{M}_i \leqq 0$. It follows from the definition of the $\tilde{M}$'s that the following recursion formulas hold:

$$c_{i+1,0} = q_{i+1}c_{i0} + q_{i+1}c_{i1}, \tag{3.1}$$

$$c_{i+1,j} = p_{i+1}c_{i,j-1} + q_{i+1}c_{i,j+1} \qquad (j = 1, 2, \cdots, i+1). \tag{3.2}$$

Since $\tilde{M}_N = M_N$, we have

$$\text{prob}\,\{M_N = j\} = c_{Nj} \qquad \text{for } j = 1, \cdots, N, \tag{3.3}$$

$$\text{prob}\,\{M_N \leqq 0\} = c_{N0}. \tag{3.4}$$

We shall now construct N square matrices $A_1, \cdots, A_N$, each having $N+1$ rows and $N+1$ columns, such that the first row of the product matrix $A_1A_2 \cdots A_N$ is equal to $(c_{N0}, c_{N1}, \cdots, c_{NN})$. Let $\overset{k}{a}_{ij}$ denote the element in the ith row and jth column of the matrix A_k $(i, j = 1, \cdots, N+1;\ k = 1, \cdots, N)$. We put

$$\begin{aligned} \overset{k}{a}_{11} &= q_k; \qquad \overset{k}{a}_{i,i+1} = p_k \qquad (i = 1, 2, \cdots, N); \\ \overset{k}{a}_{i,i-1} &= q_k \qquad\qquad (i = 2, 3, \cdots, N+1) \end{aligned} \tag{3.5}$$

and all other elements $\overset{k}{a}_{ij}$ equal to zero. It then follows easily from the recursion formulas (3.1) and (3.2) that the first row of the product $A_1A_2 \cdots A_N$ is equal to $(c_{N0}, c_{N1}, \cdots, c_{NN})$. Thus, the first row of the product $A_1A_2 \cdots A_N$ yields the exact probability distribution of M_N.

Starting with the initial values $c_{10} = q_1$, $c_{11} = p_1$, $c_{1j} = 0$ for $j > 1$, the final values $c_{N0}, c_{N1}, \cdots, c_{NN}$ can be best computed by repeated application of the recursion formulas (3.1) and (3.2).

4. Proof of Theorem 1.1. Let $\{X_{Ni}\}$ and $\{X^*_{Ni}\}$ be two double sequences of chance variables for which conditions (a)–(f) of Theorem 1.1 are fulfilled. Let k be a positive integer and $N_1, \cdots, N_k$ a set of positive integers such that $N_1 < N_2 < \cdots < N_k = N$. Let, furthermore,

$$(4.1)\qquad P_{N,k}(c) = \text{prob}\,\{\max\,(S_{NN_1}, S_{NN_2}, \cdots, S_{NN_k}) < cN^{1/2}\}.$$

Because of conditions (b) and (c) of Theorem 1.1, there exist two finite values A and B such that $A \geqq N\mu_{Ni}^2$ and $B \geqq \sigma_{Ni}^2$ for all N and i. Let $\phi(k)$ be an upper bound of the values

$$(4.2)\qquad \frac{N_1}{N},\ \frac{N_2 - N_1}{N},\ \cdots,\ \frac{N_k - N_{k-1}}{N}.$$

For any positive ϵ the following inequality holds:

$$(4.3)\qquad P_{N,k}(c - \epsilon) - \frac{\phi(k)}{\epsilon^2}\,[B + A\phi(k)] \leqq P_N(c) \leqq P_{N,k}(c),$$

where $P_N(c) = \text{prob}\,\{M_N < cN^{1/2}\}$. Using a method given by Erdös and Kac [1], the author [2] has proved the above inequality when $\mu_{Ni} = \mu_N$, $\sigma_{Ni} = 1$ and $N_j = [jN/k]$. To adapt the proof given in [2] to the more general case treated here, it is sufficient to replace the right-hand member of (2.6) in [2] by

$$(4.4)\qquad \frac{(N_{i+1} - N_i)B + (N_{i+1} - N_i)^2\mu_N^2}{\epsilon^2 N},$$

where $\mu_N^2 = \max\,(\mu_{N1}^2, \cdots, \mu_{NN}^2)$.

For the purpose of proving Theorem 1.1, we shall choose N_j to be the smallest positive integer for which

$$(4.5)\qquad \sigma_{N1}^2 + \cdots + \sigma_{NN_j}^2 \geqq \frac{j(\sigma_{N1}^2 + \cdots + \sigma_{NN}^2)}{k}.$$

Since σ_{Ni}^2 has a positive lower bound and a finite upper bound, there exists a positive constant h, independent of k, such that h/k is an upper bound of the values (4.2). It then follows from (4.3) that

$$(4.6)\qquad P_{N,k}(c - \epsilon) - \frac{1}{\epsilon^2 k}\,(a + b/k) \leqq P_N(c) \leqq P_{N,k}(c)$$

when a and b are positive constants independent of N, k, c and ϵ.

Clearly, if Theorem 1.1 is true for the special case when $\sigma_{N1}^2 + \cdots + \sigma_{NN}^2 = \sigma_{N1}^{*2} + \cdots + \sigma_{NN}^{*2}$, it must be true also in the general case. Hence, it is sufficient to prove Theorem 1.1 when $\sigma_{N1}^2 + \cdots + \sigma_{NN}^2 = \sigma_{N1}^{*2} + \cdots + \sigma_{NN}^{*2}$. In what follows we shall therefore restrict ourselves to this special case.

Let N_j^*, $P_{N,k}^*(c)$, and $P_N^*(c)$ have the same meaning with reference to the X^*'s as N, $P_{N,k}(c)$, and $P_N(c)$ with reference to the X's. Then we have

$$P^*_{N,k}(c-\epsilon) - \frac{1}{\epsilon^2 k}(a^* + b^*/k) \leqq P^*_N(c) \leqq P^*_{N,k}(c), \tag{4.7}$$

where a^* and b^* are positive constants independent of N, k, c and ϵ.

Let $G^N_{k1}, G^N_{k2}, \cdots, G^N_{kk}$ be independently and normally distributed chance variables and let the mean and standard deviation of G^N_{ki} be equal to the mean and standard deviation of $(k/N)^{1/2}(S_{NN_i} - S_{NN_{i-1}})$, respectively. Let, furthermore,

$$\begin{aligned} &Q_{N,k}(c) \\ &= \text{prob}\,\{\max(G^N_{k1}, G^N_{k1}+G^N_{k2}, \cdots, G^N_{k1}+\cdots+G^N_{kk}) < ck^{1/2}\}. \end{aligned} \tag{4.8}$$

Clearly, the mean and standard deviation of G^N_{ki} are bounded functions of N, k and i. Furthermore, the standard deviation of G^N_{ki} has a positive lower bound. It then follows from condition (d) and the central limit theorem that

$$\lim_{N=\infty} [Q_{N,k}(c) - P_{N,k}(c)] = 0. \tag{4.9}$$

Let G^{*N}_{ki} and $Q^*_{N,k}(c)$ have the same meaning with reference to the X^*'s as G^N_{ki} and $Q_{N,k}(c)$ with reference to the X's. We then have

$$\lim_{N=\infty} [Q^*_{N,k}(c) - P^*_{N,k}(c)] = 0. \tag{4.10}$$

It follows from condition (f) of Theorem 1.1 that

$$\lim_{N=\infty} E(G^N_{ki} - G^{*N}_{ki}) = 0, \tag{4.11}$$

$$\lim_{N=\infty} E[(G^N_{ki})^2 - (G^{*N}_{ki})^2] = 0. \tag{4.12}$$

Hence

$$\lim_{N=\infty} [Q_{N,k}(c) - Q^*_{N,k}(c)] = 0. \tag{4.13}$$

From (4.9) and (4.10) and (4.13) we obtain

$$\lim_{N=\infty} [P_{N,k}(c) - P^*_{N,k}(c)] = 0. \tag{4.14}$$

Equations (4.6) and (4.14) give

$$\liminf_{N=\infty} \left[P_N(c) - P^*_{N,k}(c-\epsilon) + \frac{1}{\epsilon^2 k}\left(a + \frac{b}{k}\right)\right] \geqq 0 \tag{4.15}$$

and

$$\liminf_{N=\infty} [P^*_{N,k}(c) - P_N(c)] \geqq 0. \tag{4.16}$$

Since

$$P^*_{N,k}(c - \epsilon) \geqq P^*_N(c - \epsilon) \tag{4.17}$$

and since, because of (4.7),

$$P^*_{N,k}(c) - \frac{1}{\epsilon^2 k}(a^* + b^*/k) \leqq P^*_N(c + \epsilon), \tag{4.18}$$

we obtain from (4.15) and (4.16)

$$\liminf_{N=\infty} \left[P_N(c) - P^*_N(c - \epsilon) + \frac{1}{\epsilon^2 k}\left(a + \frac{b}{k}\right)\right] \geqq 0 \tag{4.19}$$

and

$$\liminf_{N=\infty} \left[P^*_N(c + \epsilon) + \frac{1}{\epsilon^2 k}\left(a^* + \frac{b^*}{k}\right) - P_N(c)\right] \geqq 0. \tag{4.20}$$

Hence, since k can be chosen arbitrarily large, we obtain

$$\liminf_{N=\infty} [P_N(c) - P^*_N(c - \epsilon)] \geqq 0 \tag{4.21}$$

and

$$\liminf_{N=\infty} [P^*_N(c + \epsilon) - P_N(c)] \geqq 0. \tag{4.22}$$

This concludes the proof of Theorem 1.1. It may be of interest to note that (4.21) and (4.22) imply that for any subsequence $\{N'\}$ of the sequence $\{N\}$ we have

$$\begin{aligned} \liminf_{N=\infty} P^*_{N'}(c - \epsilon) &\leqq \liminf_{N=\infty} P_{N'}(c) \leqq \limsup_{N=\infty} P_{N'}(c) \\ &\leqq \limsup_{N=\infty} P^*_{N'}(c + \epsilon). \end{aligned} \tag{4.23}$$

References

1. P. Erdös and M. Kac, *On certain limit theorems of the theory of probability*, Bull. Amer. Math. Soc. vol. 52 (1946) pp. 292–302.

2. A. Wald, *Limit distribution of the maximum and minimum of successive cumulative sums of random variables*, Bull. Amer. Math. Soc. vol. 53 (1947) pp. 142–153.

Columbia University

Made in United States of America

Reprinted from The Annals of Mathematical Statistics
Vol. XIX, No. 2, June, 1948

ESTIMATION OF A PARAMETER WHEN THE NUMBER OF UNKNOWN PARAMETERS INCREASES INDEFINITELY WITH THE NUMBER OF OBSERVATIONS

By Abraham Wald

Columbia University

Summary. Necessary and sufficient conditions are given for the existence of a uniformly consistent estimate of an unknown parameter θ when the successive observations are not necessarily independent and the number of unknown parameters involved in the joint distribution of the observations increases indefinitely with the number of observations. In analogy with R. A. Fisher's information function, the amount of information contained in the first n observations regarding θ is defined. A sufficient condition for the non-existence of a uniformly consistent estimate of θ is given in section 3 in terms of the information function. Section 4 gives a simplified expression for the amount of information when the successive observations are independent.

2. Introduction. J. Neyman has recently treated the following estimation problem[1]: Let $X_1, X_2, \cdots$, etc. be a sequence of independent chance variables the distribution of each of which depends on some unknown parameters. Two kinds of parameters are distinguished, structural and incidental parameters. A parameter θ is called structural if there exists an infinite subsequence of the sequence $\{X_i\}$ such that the distribution of each of the chance variables in the subsequence depends on θ. Any parameter which is not structural is called incidental. Neyman has considered the case when there are a finite number of structural parameters, say $\theta_1, \cdots, \theta_s$ and an infinite sequence $\{\xi_i\}$, $(i = 1, 2, \cdots, \text{ad inf.})$, of incidental parameters. He has studied the problem of consistent and efficient estimation of the structural parameters and has obtained several interesting results. He has shown, among others, that the maximum likelihood estimate of a structural parameter θ need not be consistent, even when consistent estimates of θ exist. Neyman has also given a method for obtaining consistent estimates of the structural parameters. This method, however, is applicable only under certain restrictive conditions.

In this paper we shall consider a more general case than that treated by Neyman, but we shall concentrate on one aspect of the problem, namely that of the existence of consistent estimates.

Let $\{X_i\}$, $(i = 1, 2, \cdots, \text{ad inf.})$, be a sequence of chance variables, not necessarily independent of each other. It is assumed that for each n the chance variables $X_1, \cdots, X_n$ admit a joint probability density function $p_n(x_1, \cdots, x_n \mid \theta, \xi_1, \cdots, \xi_n)$ where $\theta, \xi_1, \xi_2, \cdots$, etc. are unknown parameters.[2]

[1] Address given by J. Neyman at the meeting of the Institute of Mathematical Statistics in Atlantic City, January, 1947.

[2] While θ is assumed to be a real variable, we admit ξ_i to be a finite dimensional vector, i.e., $\xi_i = (\xi_{i1}, \ldots, \xi_{ik_i})$ where k_i may be any finite positive integer.

We shall require that the consistency relations among the density functions $p_1, p_2, \cdots$,etc. be fulfilled, i.e.,

$$\int_{-\infty}^{+\infty} p_{n+1}\, dx_{n+1} = p_n, \qquad (n = 1, 2, \cdots, \text{ad inf.}). \tag{1.1}$$

It should be remarked that it is not postulated that p_n actually depends on all the parameters that appear as arguments in p_n. It is merely assumed that p_n does not depend on any parameter that does not appear as an argument in p_n, i.e., p_n does not depend on ξ_i for any $i > n$. It follows, however, from (1.1) that if p_n depends on a parameter ξ, then also p_m depends on ξ for any $m > n$.

Neyman's definition of structural and incidental parameters can be extended to the case of dependent observations considered here by saying that the distribution of X_i does not depend on a parameter ξ if and only if the conditional distribution of X_i for any given values of $X_1, \cdots, X_{i-1}$ does not depend on ξ. It is not postulated that each of the parameters $\xi_1, \xi_2, \cdots$, etc. is incidental; some of them may be structural. We shall not make an explicit distinction between structural and incidental parameters, since for the purposes of the present paper this does not seem to be necessary.

In this paper we shall deal with the problem of formulating conditions under which a uniformly consistent estimate of θ exists. A statistic $t_n(x_1, \cdots, x_n)$ is said to be a uniformly consistent estimate of θ if for any positive δ

$$\lim_{n=\infty} \text{prob. } \{|t_n - \theta| < \delta\} = 1 \tag{1.2}$$

uniformly in θ and the ξ's.

In section 2 a necessary and sufficient condition is given for the existence of a uniformly consistent estimate of θ. In section 3 the amount of information supplied by the first n observations concerning θ is defined. It is then shown that if the amount of information is a bounded function of n over a non-degenerate θ-interval, no uniformly consistent estimate of θ exists. Section 4 gives a simplified formula for the amount of information in the case when the X's are independently distributed.

2. A necessary and sufficient condition for the existence of a uniformly consistent estimate of θ. In deriving a necessary and sufficient condition for the existence of a uniformly consistent estimate of θ, use will be made of some results contained in a publication of the author [1] dealing with statistical decision functions which minimize the maximum risk. In [1] it is assumed that the domain of each of the unknown parameters is a closed and bounded set and that p_n is continuous jointly in all of its arguments. Thus, in order to be able to use the results obtained in [1], we shall have to make the same assumptions here. In what follows we shall, therefore, assume that each of the parameters $\theta, \xi_1, \xi_2, \cdots$, etc. is restricted to a finite closed interval and that p_n is a continuous function of $x_1, \cdots, x_n, \theta, \xi_1, \cdots, \xi_n$.

Let $[a, b]$ $(a < b)$ be the θ-interval to which the values of θ are restricted. Clearly, if $t_n(x_1, \cdots, x_n)$, $(n = 1, 2, \cdots, \text{ad inf.})$, is a uniformly consistent estimate of θ, then also t_n^* is a uniformly consistent estimate of θ when $t_n^* = t_n$ when $a \leqq t_n \leqq b$, $t_n^* = a$ when $t_n < a$ and $t_n^* = b$ when $t_n > b$. Thus, without loss of generality, we can restrict ourselves to estimates t_n which can take values only in the interval $[a, b]$. Uniform consistency of t_n is then equivalent with the condition

$$\lim_{n=\infty} E[(t_n - \theta)^2 \mid \theta, \xi_1, \cdots, \xi_n] = 0 \tag{2.1}$$

uniformly in θ and the ξ's. For any chance variable u the symbol $E(u \mid \theta, \xi_1, \xi_2, \cdots)$ denotes the expected value of u when $\theta, \xi_1, \xi_2, \cdots$ are the true parameter values.

In [1] a non-negative function $W(t_n, \theta)$, called weight function, is introduced which expresses the loss suffered when t_n is the value of the estimate and θ is the true value of the parameter. The risk is defined in [1] as the expected value of the loss, i.e., the risk is given by

$$r_n(\theta, \xi_1, \cdots, \xi_n) = E[W(t_n, \theta) \mid \theta, \xi_1, \cdots, \xi_n]. \tag{2.2}$$

If we put $W(t_n, \theta) = (t_n - \theta)^2$, we have

$$r_n(\theta, \xi_1, \cdots, \xi_n) = E[(t_n - \theta)^2 \mid \theta, \xi_1, \cdots, \xi_n]. \tag{2.3}$$

It can easily be verified that Assumptions 1–4 in section 3 of [1] are fulfilled for the weight function $W(t_n, \theta) = (t_n - \theta)^2$.[3] Thus, all results obtained in [1] can be applied to the risk function given in (2.3). According to Theorem 4.1 in [1] the risk function given in (2.3) is a continuous function of $\theta, \xi_1, \cdots, \xi_n$ for any arbitrary estimate t_n. We shall denote the maximum of (2.3) with respect to $\theta, \xi_1, \cdots, \xi_n$ by $r_n[t_n]$. Thus $r_n[t_n]$ is a functional which associates a non-negative value with any estimate function t_n.

It follows from (2.1) that t_n is a uniformly consistent estimate of θ if and only if

$$\lim_{n=\infty} r_n[t_n] = 0. \tag{2.4}$$

For any θ and for any n let $F_n(\xi_1, \cdots, \xi_n \mid \theta)$ be a cumulative distribution function of $\xi_1, \cdots, \xi_n$. Let, furthermore,

$$\begin{aligned} &q_n(x_1, \cdots, x_n \mid \theta, F_n) \\ &= \int_{-\infty}^{+\infty} \cdots \int_{-\infty}^{+\infty} p_n(x_1, \cdots, x_n \mid \theta, \xi_1 \cdots, \xi_n)\, dF_n(\xi_1, \cdots, \xi_n \mid \theta). \end{aligned} \tag{2.5}$$

We do not require that $F_1, F_2, \cdots$, etc. satisfy the consistency relations, i.e., $\lim_{\xi_{n+1}=\infty} F_{n+1}(\xi_1, \cdots, \xi_{n+1} \mid \theta)$ is not necessarily equal to $F_n(\xi_1, \cdots, \xi_n \mid \theta)$.

[3] In verifying Assumption 4, we may assume that p_n is always > 0, since for any given values $\theta, \xi_1, \ldots, \xi_n$ we may restrict the domain of $(x_1, \ldots, x_r)$ to the subset of the sample space where $p_n > 0$.

Hence, also the distributions q_n do not necessarily satisfy the consistency relations. Clearly

$$(2.6)\qquad r_n[t_n] \geqq \int_{-\infty}^{+\infty} \cdots \int_{-\infty}^{+\infty} (t_n - \theta)^2 q_n(x_1, \cdots, x_n \mid \theta, F_n)\, dx_1, \cdots, dx_n$$

for any θ and any F_n. Hence, (2.4) and (2.6) imply that if t_n is a uniformly consistent estimate of θ, then t_n remains a uniformly consistent estimate of θ also when q_n is the distribution of $X_1, \cdots, X_n$ for any arbitrary choice of F_n.

For each n let $C_n(\theta, \xi_1, \cdots, \xi_n)$ be a joint cumulative distribution function of $\theta, \xi_1, \cdots, \xi_n$. If this is regarded as an a priori distribution of $\theta, \xi_1, \cdots, \xi_n$, and if our aim is to choose t_n so that

$$(2.7)\qquad \begin{aligned} &E(t_n - \theta)^2 \\ &= \int_{-\infty}^{+\infty} \cdots \int_{-\infty}^{+\infty} (t_n - \theta)^2 p_n(x_1, \cdots, x_n \mid \theta, \xi_1, \cdots, \xi_n)\, dC_n\, dx_1 \cdots dx_n \end{aligned}$$

is a minimum, then the best choice of t_n is to put it equal to the a posteriori mean value of θ. Let $t_n^*(x_1, \cdots, x_n; C_n)$ denote the a posteriori mean value of θ when C_n is the a priori distribution, i.e.,

$$(2.8)\qquad t_n^*(x_1, \cdots, x_n; C_n) = \frac{\int \theta p_n(x_1, \cdots, x_n \mid \theta, \xi_1, \cdots, \xi_n)\, dC_n}{\int p_n(x_1, \cdots, x_n \mid \theta, \xi_1 \cdots, \xi_n)\, dC_n}$$

where the integration is to be taken over the whole domain of the parameters $\theta, \xi_1, \cdots, \xi_n$. Let, furthermore, $\bar{r}_n[C_n]$ denote the value of (2.7) when $t_n = t_n^*(x_1, \cdots, x_n; C_n)$. According to Theorem 4.4 in [1] there exists a particular distribution C_n^0, called a least favorable distribution, such that

$$(2.9)\qquad \bar{r}_n[C_n] \leqq \bar{r}_n[C_n^0]$$

for all C_n. Let

$$(2.10)\qquad t_n^0(x_1, \cdots, x_n) = t_n^*(x_1, \cdots x, {}_n; C_n^0).$$

It follows from Theorems (4.5) and (5.1) in [1] that for any estimate t_n we have

$$(2.11)\qquad r_n[t_n] \geqq r_n[t_n^0] = \bar{r}_n(C_n^0).$$

Hence, a necessary and sufficient condition for the existence of a uniformly consistent estimate of θ is that

$$(2.12)\qquad \lim_{n=\infty} \bar{r}_n[C_n^0] = 0.$$

Let $F_n(\xi_1, \cdots, \xi_n \mid \theta)$ denote the conditional cumulative distribution of $\xi_1, \cdots, \xi_n$ for given θ that results from the joint distribution $C_n(\theta, \xi_1, \cdots \xi_n)$ and let $F_n^0(\xi_1, \cdots, \xi_n \mid \theta)$ correspond to $C_n^0(\theta, \xi_1, \cdots, \xi_n)$. Clearly, any uniformly consistent estimate of θ with respect to $p_n(x_1, \cdots, x_n \mid \theta, \xi_1, \cdots, \xi_n)$

is a uniformly consistent estimate also with respect to $q_n(x_1, \cdots, x_n \mid \theta, F_n)$ for any F_n. On the other hand, if $q_n(x_1, \cdots, x_n \mid \theta, F_n^0)$ admits a uniformly consistent estimate of θ, equation (2.12) must hold and, therefore, $p_n(x_1, \cdots, x_n \mid \theta, \xi_1, \cdots, \xi_n)$ admits a uniformly consistent estimate of θ. Hence we arrive at the following theorem:

THEOREM 2.1. *A necessary and sufficient condition that*

$$p_n(x_1, \cdots, x_n \mid \theta, \xi_1, \cdots, \xi_n)$$

admit a uniformly consistent estimate of θ is that $q_n(x_1, \cdots, x_n \mid \theta, F_n)$ admit a uniformly consistent estimate of θ for any arbitrary choice of F_n.

3. Amount of information contained in the first n observations concerning the parameter θ. We shall make the following assumptions:

Assumption 1. The first two derivatives of $p_n(x_1, \cdots, x_n \mid \theta, \xi_1, \cdots, \xi_n)$ with respect to θ exist.

Assumption 2. We have

$$\int_{-\infty}^{+\infty} \cdots \int_{-\infty}^{+\infty} \operatorname{Max}_{\theta} \left| \frac{\partial p_n}{\partial \theta} \right| dx_1 \cdots dx_n < \infty \tag{3.1}$$

and

$$\int_{-\infty}^{\infty} \cdots \int_{-\infty}^{\infty} \operatorname{Max}_{\theta} \left| \frac{\partial^2 p_n}{\partial \theta^2} \right| dx_1 \cdots dx_n < \infty \tag{3.2}$$

for any n.

Assumption 3. The integral

$$\int_{-\infty}^{\infty} \cdots \int_{-\infty}^{\infty} \frac{\partial^2 \log q_n(x_1, \cdots, x_n \mid \theta, F_n)}{\partial \theta^2} q_n(x_1, \cdots, x_n \mid \theta, F_n)\, dx_1 \cdots dx_n$$

exists for any θ, F_n and n where q_n is defined by (2.5).

Since

$$\frac{\partial^2 \log q_n}{\partial \theta^2} = \frac{1}{q_n} \frac{\partial^2 q_n}{\partial \theta^2} - \left(\frac{\partial \log q_n}{\partial \theta} \right)^2$$

and since, because of Assumptions 1 and 2,

$$\int_{-\infty}^{\infty} \cdots \int_{-\infty}^{\infty} \frac{\partial^2 q_n}{\partial \theta^2} dx_1 \cdots dx_n = 0,$$

we have

$$\int_{-\infty}^{\infty} \cdots \int_{-\infty}^{\infty} \frac{\partial^2 \log q_n}{\partial \theta^2} q_n\, dx_1 \cdots dx_n = - \int_{-\infty}^{\infty} \cdots \int_{-\infty}^{+\infty} \left(\frac{\partial \log q_n}{\partial \theta} \right)^2 q_n\, dx_1 \cdots dx_n.$$

Let

$$(3.4)\quad c_n(\theta) = \underset{F_n}{\text{g.l.b.}} \left\{ -\int_{-\infty}^{\infty} \cdots \int_{-\infty}^{\infty} \left(\frac{\partial^2 \log q_n}{\partial \theta^2} \right) q_n \, dx_1 \cdots dx_n \right\}.$$

Clearly $c_n(\theta) \geqq 0$. We shall now show that

$$(3.5)\qquad c_{n+1}(\theta) \geqq c_n(\theta) \qquad \text{for } n = 1, 2, \cdots, \text{ad inf.}$$

In fact, we can write

$$(3.6)\qquad \begin{aligned} \frac{-\partial^2 \log q_{n+1}(x_1, \cdots, x_{n+1} \mid \theta, F_{n+1})}{\partial \theta^2} &= -\frac{\partial^2 \log q_n(x_1, \cdots, x_n \mid \theta, F_n^*)}{\partial \theta^2} \\ &\quad - \frac{\partial^2 \log f_{n+1}(x_{n+1} \mid x_1, \cdots, x_n, \theta, F_{n+1})}{\partial \theta^2} \end{aligned}$$

where $F_n^* = \lim_{\xi_{n+1}=\infty} F_{n+1}(\xi_1, \cdots, \xi_{n+1} \mid \theta)$ and $f_{n+1}(x_{n+1} \mid x_1, \cdots, x_n, \theta, F_{n+1})$ is the conditional probability density function of X_{n+1} given the values of $x_1, \cdots, x_n$ and assuming that the joint density function of $X_1, \cdots, X_{n+1}$ is given by $q_{n+1}(x_1, \cdots, x_{n+1} \mid \theta, F_{n+1})$. Since $c_n(\theta) \leqq$ expected value of

$$-\frac{\partial^2 \log q_n(x_1, \cdots, x_n \mid \theta, F_n^*)}{\partial \theta^2}$$

and since the expected value of $-\dfrac{\partial^2 \log f_{n+1}}{\partial \theta^2}$ is $\geqq 0$, inequality (3.5) must hold.

In analogy with R. A. Fisher's information function, we shall call $c_n(\theta)$ the amount of information contained in the first n observations regarding θ. We shall now prove the following theorem:

THEOREM 3.1. *If* $\lim_{n=\infty} c_n(\theta) \leqq c < \infty$ *over a finite non-degenerate* θ*-interval* I, *then there is no uniformly consistent estimate of* θ.

PROOF. If for any n, $c_n(\theta) \leqq c < \infty$ over the interval I, for each n there exists a distribution $F_n(\xi_1, \cdots, \xi_n \mid \theta)$ such that

$$(3.7)\qquad \begin{aligned} 0 \leqq -\int_{-\infty}^{\infty} \cdots \int_{-\infty}^{\infty} &\frac{\partial^2 \log q_n(x_1, \cdots, x_n \mid \theta, F_n)}{\partial \theta^2} \\ &\cdot q_n(x_1, \cdots, x_n \mid \theta, F_n) \, dx_1 \cdots dx_n \leqq c + 1 \end{aligned}$$

for all n and for all θ in I. Let t_n be any estimate and let

$$(3.8)\qquad \begin{aligned} b_n(\theta) = E(t_n - \theta) &= \int_{-\infty}^{\infty} \cdots \int_{-\infty}^{\infty} (t_n - \theta) q_n(x_1, \cdots, x_n \mid \theta, F_n) \, dx_1 \cdots dx_n \\ &= \int_{-\infty}^{\infty} \cdots \int_{-\infty}^{\infty} t_n q_n(x_1, \cdots, x_n \mid \theta, F_n) \, dx_1 \cdots dx_n - \theta. \end{aligned}$$

Since t_n is bounded, it follows from Assumptions 1 and 2 that $\dfrac{db_n(\theta)}{d\theta}$ exists and is a continuous function of θ. According to a theorem by Cramér [2] we have

$$(3.9) \qquad E(t_n - \theta)^2 = \int_{-\infty}^{\infty} \cdots \int_{-\infty}^{\infty} (t - \theta)^2 q_n \, dx_1 \cdots dx_n \geqq \frac{\left(1 + \frac{db_n}{\partial\theta}\right)^2}{c + 1}$$

for all θ in I. Thus, in order that $\lim_{n=\infty} E(t_n - \theta)^2 = 0$ uniformly in θ, we must have

$$(3.10) \qquad \lim_{n=\infty} \frac{db_n(\theta)}{d\theta} = -1$$

uniformly in θ over I. Let I be the interval ranging from g to h $(g < h)$. From (3.10) it follows that

$$(3.11) \qquad \lim_{n=\infty} [b_n(h) - b_n(g)] = g - h.$$

Hence

$$\liminf_{n=\infty} \max_{\theta \text{ in } I} [b_n(\theta)]^2 \geqq \frac{(g - h)^2}{4}.$$

Since $E(t_n - \theta)^2 \geqq [b_n(\theta)]^2$, $E(t_n - \theta)^2$ cannot converge to zero uniformly in θ and Theorem 3.1 is proved.

4. Formula for $c_n(\theta)$ when $p_n(x_1, \cdots, x_n \mid \theta, \xi_1, \cdots, \xi_n)$ is equal to $\varphi_1(x_1 \mid \theta, \xi_1) \varphi_2(x_2 \mid \theta, \xi_2) \cdots \varphi_n(x_n \mid \theta, \xi_n)$. Let $g_i(x_i \mid x_1, \cdots, x_{i-1}, \theta, F_n)$ be the conditional probability density of X_i given $x_1, \cdots, x_{i-1}$ when the joint density function of $x_1, \cdots, x_n$ is given by $q_n(x_1, \cdots, x_n \mid \theta, F_n)$, $(i \leqq n)$. Clearly,

$$(4.1) \qquad -E\left(\frac{\partial^2 \log q_n}{\partial\theta^2}\right) = -\sum_{i=1}^{n} E\left(\frac{\partial^2 \log g_i}{\partial\theta^2}\right).$$

Now

$$(4.2) \qquad g_i(x_i \mid x_1, \cdots, x_{i-1}, \theta, F_n) = \int_{-\infty}^{\infty} \varphi_i(x_i \mid \theta, \xi_i) \, dH_i(\xi_i \mid x_1, \cdots, x_{i-1}, \theta, F_n)$$

where $H_i(\xi_i \mid x_1, \cdots, x_{i-1}, \theta, F_n)$ denotes the conditional cumulative distribution of ξ_i given $x_1, \cdots, x_{i-1}$, assuming that $F_n(\xi_1, \cdots, \xi_n \mid \theta)$ is the joint cumulative distribution of $\xi_1, \cdots, \xi_n$ and $p_n(x_1, \cdots, x_n \mid \theta, \xi_1, \cdots, \xi_n)$ is the joint density of $X_1, \cdots, X_n$ for any given values of $\theta, \xi_1, \cdots, \xi_n$.

It follows from (4.2) that

$$-\int_{-\infty}^{+\infty} \frac{\partial^2 \log g_i}{\partial\theta^2} g_i \, dx_i \geqq c_{ni}(\theta)$$

$$= \underset{C_i(\xi_i)}{\text{g.l.b.}} \left\{ -\int_{-\infty}^{+\infty} \left[\frac{\partial^2 \log \int_{-\infty}^{+\infty} \varphi_i(x_i \mid \theta, \xi_i) \, dC_i(\xi_i)}{\partial\theta^2} \int_{-\infty}^{\infty} \varphi_i \, dC_i \right] dx_i \right\}$$

where $C_i(\xi_i)$ may be any cumulative distribution of ξ_i. Hence

$$\text{g.l.b.}_{F_n}\left[-E\left(\frac{\partial^2 \log g_i}{\partial\theta^2}\right)\right] = c_{ni}(\theta) \tag{4.3}$$

and, therefore,

$$c_n(\theta) = \sum_{i=1}^{n} c_{ni}(\theta). \tag{4.4}$$

The quantity $c_{ni}(\theta)$ is simply the amount of information contained in the ith observation alone. Thus, formula (4.4) says that if $X_1, \cdots, X_n$ are independent, the total information contained in the first n observations is equal to the sum of the amounts of information contained in each of these observations singly.

REFERENCES

[1] A. Wald, "Statistical decision functions which minimize the maximum risk", *Annals of Math.*, Vol. 46 (1945).

[2] H. Cramér, *Mathematical Methods of Statistics*, Princeton Univ. Press, 1946.

Made in United States of America

Reprinted from THE ANNALS OF MATHEMATICAL STATISTICS
Vol. XIX, No. 3, September, 1948

OPTIMUM CHARACTER OF THE SEQUENTIAL PROBABILITY RATIO TEST

A. WALD AND J. WOLFOWITZ

Columbia University

1. Summary. Let S_0 be any sequential probability ratio test for deciding between two simple alternatives H_0 and H_1, and S_1 another test for the same purpose. We define $(i, j = 0, 1)$:

$\alpha_i(S_j)$ = probability, under S_j, of rejecting H_i when it is true;

$E_i^j(n)$ = expected number of observations to reach a decision under test S_j when the hypothesis H_i is true. (It is assumed that $E_i^1(n)$ exists.)

In this paper it is proved that, if

$$\alpha_i(S_1) \leq \alpha_i(S_0) \qquad (i = 0,1),$$

it follows that

$$E_i^0(n) \leq E_i^1(n) \qquad (i = 0, 1).$$

This means that of all tests with the same power the sequential probability ratio test requires on the average fewest observations. This result had been conjectured earlier ([1], [2]).

2. Introduction. Let $p_i(x)$, $i = 0, 1$, denote two different probability density functions or (discrete) probability functions. (Throughout this paper the index i will always take the values 0, 1). Let X be a chance variable whose distribution can only be either $p_0(x)$ or $p_1(x)$, but is otherwise unknown. It is required to decide between the hypotheses H_0, H_1, where H_i states that $p_i(x)$ is the distribution of X, on the basis of n independent observations $x_1, \cdots, x_n$ on X, where n is a chance variable defined (finite) on almost every infinite sequence

$$\omega = x_1, x_2, \cdots$$

i.e., n is finite with probability one according to both $p_0(x)$ and $p_1(x)$. The definition of $n(\omega)$ together with the rule for deciding on H_0 or H_1 constitute a sequential test.

A sequential probability ratio test is defined with the aid of two positive numbers, $A^* > 1$, $B^* < 1$, as follows: Write for brevity

$$p_{ij} = \prod_{k=1}^{j} p_i(x_k).$$

Then $n = j$ if

$$\frac{p_{1j}}{p_{0j}} \geq A^* \qquad \text{or} \qquad \leq B^*$$

and

$$B^* < \frac{p_{1k}}{p_{0k}} < A^*, \qquad k < j.$$

If

$$\frac{p_{1n}}{p_{0n}} \geq A^*, \qquad \text{the hypothesis } H_1 \text{ is accepted,}$$

if

$$\frac{p_{1n}}{p_{0n}} \leq B^* \text{ the hypothesis } H_0 \text{ is accepted.}$$

In this paper we limit consideration to sequential tests for which $E_i(n)$ exists, where $E_i(n)$ is the expected value of n when H_i is true (i.e., when $p_i(x)$ is the distribution of X). It has been proved in [3] that all sequential probability ratio tests belong to this class. The purpose of the paper is to prove the result stated in the first section. Throughout the proof we shall find it convenient to assume that there is an a priori probability g_i that H_i is true ($g_0 + g_1 = 1$; we shall write $g = (g_0, g_1)$). We are aware of the fact that many statisticians believe that in most problems of practical importance either no a priori probability distribution exists, or that even where it exists the statistical decision must be made in ignorance of it; in fact we share this view. Our introduction of the a priori probability distribution is a purely technical device for achieving the proof which has no bearing on statistical methodology, and the reader will verify that this is so. We shall always assume below that $g_0 \neq 0, 1$.

Let W_0, W_1, c be given positive numbers. We define

$$R = g_0(W_0\alpha_0 + cE_0(n)) + g_1(W_1\alpha_1 + cE_1(n)),$$

and call R the average risk associated with a test S and a given g (obviously R is a function of both). We shall say that H_i is accepted when the decision is made that $p_i(x)$ is the distribution of X. We shall say that H_0 is rejected when H_1 is accepted, and vice versa. The reader may find it helpful to regard W_i as a weight which measures the loss caused by rejecting H_i when it is true, c as the cost of a single observation, and R as the average loss associated with a given g and a test S. For mathematical purposes these are simply quantities which we manipulate in the course of the proof.

3. Role of the probability ratio. Let g, $W = (W_0, W_1)$, and c be fixed. Let S be a given sequential test, with $R(S)$ the associated risk and $n(\omega, S)$ the associated "sample size" function. Let $\psi(x_1, \cdots, x_n)$ be the "decision" function; this is a function which takes only the values 0 and 1, and such that, when $x_1, \cdots, x_n$ is the sample point, the hypothesis with index $\psi(x_1, \cdots, x_n)$ is *rejected*. Define the following decision function $\varphi(x_1, \cdots, x_n)$: $\varphi = 0$ when

$$\lambda = \frac{W_1 g_1 p_{1n}}{W_0 g_0 p_{0n}}$$

is greater than 1, and $\varphi = 1$ when $\lambda < 1$. When $\lambda = 1$, φ may be 0 or 1 at pleasure.

It must be remembered that an actual decision function is a single-valued function of $(x_1, \cdots, x_n)$. We note, however, that

a) the relevant properties of a test are not affected by changing the test on a set T of points ω whose probability is zero according to both H_0 and H_1, i.e., changing the definition on T of n and/or of the decision function, leaves α_0, α_1, $E_0(n)$ and $E_1(n)$ unaltered. In particular, the average risk R remains unchanged.

b) the set of points for which $p_{0n} = p_{1n} = 0$ and λ is indeterminate, has probability zero according to both H_0 and H_1.

In view of the above we decide arbitrarily, in all sequential tests which we shall henceforth consider, to define $n = j$, and $\psi = 0$, whenever $p_{0j} = p_{1j} = 0$, and $n \neq 1, \cdots, (j - 1)$. By this arbitrary action $R(S)$ will not be changed.

Let now

$$L_{in} = \frac{W_i g_i p_{in}}{g_0 p_{0n} + g_1 p_{1n}};$$

$$L_n = cn + \min(L_{0n}, L_{1n}).$$

We have

$$EL_{\psi n} = \Sigma g_i W_i \alpha_i$$

where the operator E denotes the expected value with respect to the joint distribution of H_i and $(x_1, \cdots, x_n)$, i.e., E is the operator $g_0E_0 + g_1E_1$. If now the event $\{\psi(S) \neq \varphi \text{ and } \lambda \neq 1\}$ has positive probability according to either H_0 or H_1, we would have, for $n = n(\omega, S)$,

$$EL_{\varphi n} < EL_{\psi n}.$$

Hence, if the decision function ψ connected with the test S were replaced by the decision function φ, R would be decreased. Since our object throughout this proof will be to make R as small as possible, we shall confine ourselves henceforth, except when the contrary is explicitly stated, to tests for which φ is the decision function. This will be assumed even if not explicitly stated.

The function φ has not yet been uniquely defined when $\lambda = 1$. A definition convenient for later purposes will be given in the next section. R is the same for all definitions.

We thus have that φ is a function only of λ, or, what comes to the same thing when W is fixed, of $r_n = \dfrac{p_{1n}}{p_{0n}}$. Define

$$r_j = \frac{p_{1j}}{p_{0j}}, \qquad j = 1, 2, \cdots.$$

We shall now prove

LEMMA 1. *Let g, W, and c be fixed. There exists a sequential test S^* for which the average risk is a minimum. Its sample size function $n(\omega, S^*)$ can be defined by means of a properly chosen subset K of the non-negative half-line as follows: For any ω consider the associated sequence*

$$r_1, r_2, \cdots$$

and let j be the smallest integer for which $r_j \in K$. Then $n = j$. The function n may be undefined on a set of points ω whose probability according to H_0 and H_1 is zero.

Let $a = (a_1, \cdots, a_d)$ be any point in some finite d-dimensional Euclidean space, provided only that $p_{0d}(a)$ and $p_{1d}(a)$ are not both zero. Let $b = \frac{p_{1d}(a)}{p_{0d}(a)}$ and let $l(a) = cd + \min(L_{0d}, L_{1d})$. Let D be any sequential test whatever for which $n(\omega, D) > d$ for any ω whose first d coördinates are the same as those of a, and for which $E(n \mid a, D) < \infty$, where $E(n \mid a, D)$ is the conditional expected value of n according to the test D under the condition that the first d coördinates of ω are the same as those of a. For brevity let G represent the set of points ω which fulfill this last condition, i.e., that the first d coördinates of ω are the same as those of a. Finally, let $E(L_n \mid a, D)$ be the conditional expected value of L_n according to D under the condition that ω is in the set G. We know that $\min(L_{0d}, L_{1d})$ depends only on $r_d(a) = b$.

Write

$$\nu(a) = \sup_D \, [l(a) - E(L_n \mid a, D)].$$

Let $a_0 = (a_{01}, \cdots, a_{0k})$ be any point such that

$$\frac{p_{1d}(a)}{p_{0d}(a)} = \frac{p_{1k}(a_0)}{p_{0k}(a_0)}.$$

Let D_0 be any sequential test whatever for which $n(\omega, D_0) > k$ for any ω whose first k coordinates are the same as those of a_0, and for which $E(n \mid a_0, D_0) < \infty$ Let

$$\nu(a_0) = \sup_{D_0} \, [l(a_0) - E(L_n \mid a_0, D_0)].$$

We shall prove that $\nu(a) = \nu(a_0)$. Thus we shall be justified in writing

$$\gamma(b) = \nu(a) = \nu(a_0).$$

Suppose, therefore that $\nu(a) > \nu(a_0)$. Let D_1 be a test of the type D such that

$$l(a) - E(L_n \mid a, D_1) > \frac{\nu(a) + \nu(a_0)}{2}.$$

We now partially define another sequential test D_{10} of the type D_0 as follows: Let

$$\bar{a} = a_1, \cdots, a_d, y_1, \cdots, y_t,$$

be any sequence such that $n(\bar{a}, D_1) = d + t$. Then for the sequence

$$\bar{a}_0 = a_{01}, \cdots, a_{0k}, y_1, \cdots, y_t,$$

let $n(\bar{a}_0, D_{10}) = k + t$. The decision function ψ_0 associated with D_{10} will be partially defined as follows:

$$\psi_0(\bar{a}_0) = \varphi(\bar{a}).$$

(The reader will observe that it may happen that $\psi_0(\bar{a}_0) \neq \varphi(\bar{a}_0)$). Since $r_d(a) = r_k(a_0)$ it follows that

$$l(a) - E(L_n \mid a, D_1) = l(a_0) - E(L_n \mid a_0, D_{10}) > \frac{\nu(a) + \nu(a_0)}{2} > \nu(a_0),$$

in violation of the definition of $\nu(a_0)$. A similar contradiction is obtained if $\nu(a) < \nu(a_0)$. Hence $\nu(a) = \nu(a_0)$ as was stated above.

We define K to consist of all numbers b which are such that there exist points a with $r_d(a) = b$, and for which $\gamma(b) \leq 0$. We shall now prove that the test S^* defined in the statement of the lemma is such that $R(S^*)$ is a minimum. Recall that the average risk is the expected value of L_n. Let S be any other test. Let $a^* = (\overset{*}{a_1}, \cdots, \overset{*}{a_{d^*}})$ be any sequence such that either $n(a^*, S^*) = d^*$, or $n(a^*, S) = d^*$, but $n(a^*, S^*) \neq n(a^*, S)$. We exclude the trivial case that the probability of the occurrence of such a sequence, under both H_0 and H_1, is zero. Let $r_{d^*}(a^*) = b^*$. The sequence a^* may be one of three types:

1) $\gamma(b^*) < 0$. Hence $b^* \in K$, $n(a^*, S) > d^*$. It is more advantageous, from the point of view of diminishing the average risk, to terminate the sequential process at once, since $E(L_n \mid a^*, S) > l(a^*)$.

2) $\gamma(b^*) = 0$. Hence $b^* \in K$, $n(a^*, S) > d^*$. If $l(a^*) - E(L_n \mid a^*, S) = 0$, i.e., the supremum is actually attained by S, then, as far as the average risk is concerned, it makes no difference whether the sequential process is terminated with a^* or continued according to S. If, however, $l(a^*) - E(L_n \mid a^*, S) < 0$, it is clearly disadvantageous to proceed according to S. It is impossible that $l(a^*) - E(L_n \mid a^*, S) > 0$, since $\gamma(b^*) = 0$.

3) $\gamma(b^*) > 0$. Hence $b^* \notin K$, $n(a^*, S) = d^*$. Clearly it is more advantageous from the point of view of diminishing the average risk not to terminate the sequential process, but to continue with at least one more observation. After one more observation we are either in case 1 or 2, where it is advantageous to terminate the sequential process, or again in case 3, where it is advantageous to take yet another observation.

We conclude that $R(S^*)$ is a minimum, as was to be proved.

4. A fundamental lemma. Consider the complement of K with respect to the non-negative half-line, and from it delete all points b' for which there exists no point a in some d-dimensional Euclidean space such that $r_d(a) = b'$. The point 1 is never to be considered as of the type of b', i.e., 1 is never to be deleted. Designate the resulting set by $\bar{K}$.

Our proof of the theorem to which this paper is devoted hinges on the following lemma:

LEMMA 2. *Let W, g, c be fixed, and $\bar{K}$ be as defined above. There exist two positive numbers A and B, with $B \leq \frac{W_0 g_0}{W_1 g_1} \leq A$, such that*

a) if $b \in K$, then either $b \geq A$ or $b \leq B$

b) if $b \in \bar{K}$, $B \leq b \leq A$.

Two remarks may be made before proceeding with the proof:

1) We may now complete the definition of φ for tests of the type of S^*. The reader will recall that φ was not uniquely defined when $\lambda = 1$, i.e., when $r_n = \frac{W_0 g_0}{W_1 g_1}$. Lemma 2 shows that it is necessary to define $\varphi(\lambda)$ only when $\lambda = \frac{W_0 g_0}{W_1 g_1} \in K$ and λ is therefore either A or B. We will define $\varphi\left(\frac{W_0 g_0}{W_1 g_1}\right)$ as 0 or 1, according as $\frac{W_0 g_0}{W_1 g_1}$ is A or B, and $A \neq B$. This is simply a convenient definition which will give uniqueness. When $A = B = \frac{W_0 g_0}{W_1 g_1} \in K$, the situation is completely trivial, and we may take $\varphi = 0$ arbitrarily.

2) If $1 \in K$ the above lemma shows that the average risk is minimized (for fixed W, g, c, of course) by taking no observations at all. We have $\varphi = 0$ or 1 according as $1 \geq A$ or $1 \leq B$.

PROOF OF THE LEMMA: Let $h > \frac{W_0 g_0}{W_1 g_1}$ be a point in $\bar{K}$. We will prove that any point h' such that $\frac{W_0 g_0}{W_1 g_1} \leq h' < h$, and such that there exists a point a' in some d'-dimensional Euclidean space for which $r_{d'}(a') = h'$, is also in $\bar{K}$. In a similar way it can be shown that, if $h_0 < \frac{W_0 g_0}{W_1 g_1}$ is any point in $\bar{K}$, any point h_0' such that $h_0 < h_0' \leq \frac{W_0 g_0}{W_1 g_1}$, and such that there exists a point a_0' in some d''-dimensional Euclidean space for which $r_{d''}(a_0') = h_0'$, is also in $\bar{K}$. This will prove the lemma.

Let therefore h and h' be as above. Let S^* be the sequential test based on K, with the decision function φ. Let a be a point in d-space such that $r_d(a) = h$. Since $h \in \bar{K}$ we have $\gamma(h) > 0$.

We now wish to define partially another sequential test $\bar{S}$, with a decision function which may be different from φ, as follows: Let a' be defined as above. Write

$$a = (a_1, \cdots, a_d)$$

$$a' = (a_1', \cdots, a_{d'}').$$

Let

$$\bar{a} = a_1, \cdots, a_d, y_1, \cdots, y_t$$

be any sequence such that $n(\bar{a}, S^*) = d + t$. Then for the sequence

$$\bar{a}' = a_1', \cdots, a_{d'}', y_1, \cdots, y_t$$

let $n(\bar{a}', \bar{S}) = d' + t$. The decision function ψ associated with $\bar{S}$ will be partially defined as follows:

$$\psi(\bar{a}') = \varphi(\bar{a}).$$

Clearly

$$E_i(n \mid a, S^*) - d = E_i(n \mid a', \bar{S}) - d' \qquad (i = 0, 1) \tag{4.1}$$

and

$$E_i(\varphi \mid a, S^*) = E_i(\psi \mid a', \bar{S}) \qquad (i = 0, 1). \tag{4.2}$$

Furthermore, we have

$$\begin{aligned} l(a) - E(L_n \mid a, S^*) \\ = \frac{g_0}{g_0 + g_1 h} \{W_0 + cd - cE_0(n \mid a, S^*) - W_0[1 - E_0(\varphi \mid a, S^*)]\} \\ + \frac{g_1 h}{g_0 + g_1 h} \{cd - cE_1(n \mid a, S^*) - W_1 E_1(\varphi \mid a, S^*)\}. \end{aligned} \tag{4.3}$$

Since $\gamma(h) > 0$, and since

$$cd - cE_1(n \mid a, S^*) - W_1 E_1(\varphi \mid a, S^*) < 0, \tag{4.4}$$

we must have

$$W_0 + cd - cE_0(n \mid a, S^*) - W_0[1 - E_0(\varphi \mid a, S^*)] > 0. \tag{4.5}$$

From $h' < h$ it follows that

$$\frac{g_0}{g_0 + g_1 h'} > \frac{g_0}{g_0 + g_1 h}, \qquad \text{and} \qquad \frac{g_1 h'}{g_0 + g_1 h'} < \frac{g_1 h}{g_0 + g_1 h}. \tag{4.6}$$

Relations (4.1), (4.2), (4.4), (4.5) and (4.6) imply that the value of the right hand member of (4.3) is increased by replacing φ, h, a, S^* and d by ψ, h', a', $\bar{S}$, and d', respectively. This proves our lemma.

If there are values which r_j cannot assume the pair B, A might not be unique. For convenience we shall define A and B uniquely in the manner described below. We will always adhere to this definition thereafter.

We shall first define $\gamma(h)$ for all positive h in a manner consistent with the previous definition, which defined $\gamma(h)$ only for those values of h which could be assumed by r_j. Let h be any positive number and $D(h)$ be any sequential test with the following properties:

(4.7) there exists a set $Q(h)$ of positive numbers such that $n = j$ if and only if the j-th member of the sequence

$$hr_1, hr_2, hr_3, \cdots$$

is the first element of the sequence to be in $Q(h)$

$$E_i(n \mid D(h)) < \infty \qquad (i = 0, 1). \tag{4.8}$$

We define, for $h \geq \frac{W_0 g_0}{W_1 g_1}$,

$$\gamma(h \mid D(h)) = \frac{g_0}{g_0 + g_1 h} \{W_0 E_0(\varphi \mid D(h)) - cE_0(n \mid D(h))\} + \frac{g_1 h}{g_0 + g_1 h} \{-W_1 E_1(\varphi \mid D(h)) - cE_1(n \mid D(h))\}, \tag{4.9}$$

$$\gamma(h) = \sup_{D(h)} \gamma(h \mid D(h)) \tag{4.10}$$

with a corresponding definition for $h \leq \frac{W_0 g_0}{W_1 g_1}$. Thus $\gamma(h)$ is defined for all positive h. This definition coincides with the previous definition whenever the latter is applicable. It is true that the supremum operation in (4.10) is limited to tests which depend only on the probability ratio, as (4.7) implies, but the argument of Lemma 1 shows that this limitation does not diminish the supremum. (It might appear that, for $h = \frac{W_0 g_0}{W_1 g_1}$, $\gamma(h)$ is not uniquely defined. We shall shortly see that this is not the case.)

The quantity $\gamma(h)$ depends, of course, on g_0 and g_1. To put this in evidence, we shall also write $\gamma(h, g_0, g_1)$. One can easily verify that

$$\gamma(h, g_0, g_1) = \gamma\left(1, \frac{g_0}{g_0 + g_1 h}, \frac{g_1 h}{g_0 + g_1 h}\right).$$

More generally, for any positive values h and h', we have $\gamma(h, g_0, g_1) = \gamma(h', \bar{g}_0, \bar{g}_1)$, where $\bar{g}_0$ and $\bar{g}_1$ are suitable functions of g_0, g_1, h, and h'. Thus, if h is not an admissible value of the probability ratio and h' is any admissible value, we can interpret the value of $\gamma(h, g_0, g_1)$ as the value of γ corresponding to h' and some properly chosen a priori probabilities $\bar{g}_0$ and $\bar{g}_1$.

We now define A as the greatest lower bound of all points $h \geq \frac{W_0 g_0}{W_1 g_1}$ for which $\gamma(h) \leq 0$. We define B as the least upper bound of all points $h \leq \frac{W_0 g_0}{W_1 g_1}$ for which $\gamma(h) \leq 0$. If $\gamma(h) \leq 0$ for all h the above definition implies $A = B = \frac{W_0 g_0}{W_1 g_1}$.

The argument of Lemma 2 shows that $\gamma(h)$ is monotonically increasing in the interval $\left(B, \frac{W_0 g_0}{W_1 g_1}\right)$, and that $\gamma(h)$ is monotonically decreasing in the interval $\left(\frac{W_0 g_0}{W_1 g_1}, A\right)$.

We shall now define a sequential test $S^*(h)$ for every positive h. The decision function of $S^*(h)$ will be φ, and $n = j$ if and only if the j-th member of the sequence

$$\gamma(hr_1), \quad \gamma(hr_2), \quad \gamma(hr_3), \cdots$$

is the first element to be ≤ 0. We see that

$$\gamma(h) = \gamma(h \mid S^*(h)) \tag{4.11}$$

for *all* h. Incidentally, this proves that $\gamma(h)$ was uniquely defined at $h = \frac{W_0 g_0}{W_1 g_1}$.

We shall now prove

LEMMA 3. *The function $\gamma(h)$ has the following properties:*

a) It is continuous for all h.

b) $\gamma(A) = \gamma(B) = 0$

c) $\gamma(h) < 0$ for $h > A$ or $< B$.

Only a) and c) require proof, since b) is a trivial consequence of a) and the definition of A and B.

Let h be any point except $\frac{W_0 g_0}{W_1 g_1}$, and let z be any point in a neighborhood of h. Within a neighborhood of h both $E_0(n \mid S^*(z))$ and $E_1(n \mid S^*(z))$ are bounded. Let Δ be an arbitrarily given, positive number. Let h' and h'' be any two points in a sufficiently small neighborhood of h, to be described shortly. We proceed as in the argument of Lemma 2, with the present h' corresponding to h of Lemma 2, the present h'' corresponding to h' of Lemma 2, and with $S^*(h')$ corresponding to S^* of Lemma 2. Since $\frac{g_0}{g_0 + g_1 z}$ and $\frac{g_1 z}{g_0 + g_1 z}$ are continuous functions of z, and since $E_0(n \mid S^*(z))$ and $E_1(n \mid S^*(z))$ are bounded functions of z, we conclude that, when the neighborhood of h is sufficiently small,

$$\gamma(h'') \geq \gamma(h') - \Delta.$$

Reversing the roles of h' and h'' we obtain that in this neighborhood

$$\gamma(h') \geq \gamma(h'') - \Delta,$$

and conclude that

$$|\gamma(h') - \gamma(h'')| \leq \Delta.$$

Since Δ was arbitrary, this implies the continuity of $\gamma(h)$ everywhere, except perhaps at $h = \frac{W_0 g_0}{W_1 g_1}$.

To deal with the point $h = \frac{W_0 g_0}{W_1 g_1}$, proceed as follows: Using the above argument and the definition (4.9), (4.10), we prove that $\gamma(h)$ is continuous on the right at $h = \frac{W_0 g_0}{W_1 g_1}$. Using, at the point $h = \frac{W_0 g_0}{W_1 g_1}$, the definition of $\gamma(h \mid D(h))$ for $h \leq \frac{W_0 g_0}{W_1 g_1}$ i.e.,

$$
(4.12) \quad \begin{aligned} \gamma(h \mid D(h)) = & \frac{g_0}{g_0 + g_1 h} \{-W_0 E_0(1 - \varphi \mid D(h)) - cE_0(n \mid D(h))\} \\ & + \frac{g_1 h}{g_0 + g_1 h} \{W_1 E_1(1 - \varphi \mid D(h)) - cE_1(n \mid D(h))\}, \end{aligned}
$$

(4.10) and (4.11), we prove that $\gamma(h)$ is continuous on the left at $h = \frac{W_0 g_0}{W_1 g_1}$. This proves a).

To prove c), we proceed as follows: Suppose for $h_0 > A$ we had $\gamma(h_0) = 0$. Since

$$
\{ -W_1 E_1(\varphi \mid S^*(h_0)) - cE_1(n \mid S^*(h_0))\} < 0,
$$

we would have that

$$
\{W_0 E_0(\varphi \mid S^*(h_0)) - cE_0(n \mid S^*(h_0))\} > 0.
$$

An argument like that of Lemma 2 would then show that $\gamma(h) > 0$ for $\frac{W_0 g_0}{W_1 g_1} < h < h_0$. This, however, is impossible, because it is a violation of the definition of A.

In a similar way we prove that if $h < B$, $\gamma(h) < 0$. This proves c) and with it the lemma.

5. The behavior of A and B. LEMMA 4. *Let g and c be fixed. Then A and B are continuous functions of W_0 and W_1.*

PROOF: It will be sufficient to prove that A is continuous, the proof for B being similar. Suppose $A > B$. Let h_1 and h_2 be such that

a) $B < h_1 < A < h_2$;

b) $h_2 - h_1 < \Delta$ for an arbitrary positive Δ.

We write $\gamma(h)$ temporarily as $\gamma(h, W_0, W_1)$ in order to exhibit the dependence on W_0 and W_1. Then

$$
\gamma(h_1, W_0, W_1) > 0;
$$
$$
\gamma(h_2, W_0, W_1) < 0.
$$

It follows from (4.9) that $\gamma(h \mid D(h))$ is continuous in W_0, W_1, uniformly in $D(h)$. Hence $\gamma(h, W_0, W_1) = \sup_{D(h)} \gamma(h \mid D(h))$ is also continuous in W_0, W_1. Hence, for ΔW_0 and ΔW_1 sufficiently small,

$$
\gamma(h_1, W_0 + \Delta W_0, W_1 + \Delta W_1) > 0;
$$
$$
\gamma(h_2, W_0 + \Delta W_0, W_1 + \Delta W_1) < 0.
$$

Therefore

$$h_1 \leq A(W_0 + \Delta W_0, W_1 + \Delta W_1) \leq h_2,$$

which proves continuity, since Δ was arbitrary.

If $\frac{W_0 g_0}{W_1 g_1} = A = B$, we take $h_1 < \frac{W_0 g_0}{W_1 g_1} < h_2$, $h_2 - h_1 < \Delta$, and by a similar argument show that

$$\gamma(h_1, W_0 + \Delta W_0, W_1 + \Delta W_1) < 0;$$

$$\gamma(h_2, W_0 + \Delta W_0, W_1 + \Delta W_1) < 0.$$

Thus

$$h_1 \leq B(W_0 + \Delta W_0, W_1 + \Delta W_1) \leq A(W_0 + \Delta W_0, W_1 + \Delta W_1) \leq h_2.$$

This proves the lemma.

LEMMA 5. *Let g, c, and W_1 be fixed. A is strictly monotonic in W_0. As W_0 approaches 0, A approaches 0; as W_0 approaches $+\infty$, A also approaches $+\infty$.*

PROOF: Since $A \geq \frac{W_0 g_0}{W_1 g_1}$, $A \to +\infty$ as $W_0 \to +\infty$. If $W_0 < c$ no reduction in average risk could compensate for taking even a single observation, no matter what the value of h. Hence $\gamma(h) \leq 0$ for all h when $W_0 < c$, so that $A = B$. Since $B \leq \frac{W_0 g_0}{W_1 g_1}$, $B \to 0$ as $W_0 \to 0$. Hence $A \to 0$ as $W_0 \to 0$. It is evident from (4.9) that $\gamma(h \mid D(h))$ is non-decreasing with increasing W_0 (everything else fixed). Hence also

$$\gamma(h) = \sup_{D(h)} \gamma(h \mid D(h)),$$

is non-decreasing with increasing W_0, for fixed $h > \frac{W_0 g_0}{W_1 g_1}$ and fixed W_1. For a positive Δ sufficiently small and for any h such that $A \leq h < A + \Delta$, we have that

$$E_0 (\varphi \mid S^*(h)) > 0.$$

Hence, for such h, $\gamma(h, W_0, W_1)$ is strictly monotonically increasing with increasing W_0. Therefore A is (strictly) monotonically increasing with increasing W_0.

We now define the function $W_0(W_1, \delta)$ of the two positive arguments W_1, δ so that

$$A(W_0(W_1, \delta), W_1) = \delta.$$

By Lemma 5 such a function exists and is single-valued.

6. Properties of the function $W_0(W_1, \delta)$. LEMMA 6. *$W_0(W_1, \delta)$ is continuous in W_1.*

PROOF: Let

$$\lim_{N \to \infty} W_{1N} = W_1,$$

and suppose that the sequence $\{W_0(W_{1N}, \delta)\}$ did not converge. Suppose W_0' and W_0'' were two distinct limit points of this sequence. From the continuity of A (Lemma 4) it would follow that

$$A(W_0', W_1) = A(W_0'', W_1)$$

This, however, violates Lemma 5. The only remaining possibility to be considered is that

$$\lim_{N \to \infty} W_0(W_{1N}, \delta) = \infty.$$

If that were the case, then, since $A \geq \frac{W_0 g_0}{W_1 g_1}$, it would follow that $A \to \infty$, in violation of the fact that $A \equiv \delta$.

LEMMA 7. *We have, for fixed* δ,

$$\lim_{W_1 \to 0} W_0(W_1) = 0;$$

$$\lim_{W_1 \to \infty} W_0(W_1) = \infty.$$

PROOF: If, for small W_1, $W_0(W_1)$ were bounded below by a positive number, then, since $A \geq \frac{g_0 W_0(W_1, \delta)}{W_1 g_1}$, we could make A arbitrarily large by taking W_1 sufficiently small, in violation of the fact that $A \equiv \delta$. To prove the second half of the lemma, assume that $W_0(W_1)$ is bounded above as $W_1 \to \infty$. Then $B\left(\leq \frac{W_0 g_0}{W_1 g_1}\right)$ will approach zero as $W_1 \to \infty$. Let h be fixed so that $B < h < \delta$. Consider the totality of points ω for which there exists an integer $n^*(\omega)$ such that:

$$hr_{n^*} \leq B;$$

$$B < hr_j < \delta, \qquad j < n^*.$$

The conditional expected value of n^* in this totality, when H_0 is true, may be made arbitrarily large by making B sufficiently small. Hence, when W_1 is sufficiently large, for fixed but arbitrary $h < \delta$, the optimum procedure from the point of minimizing the average risk is to reject H_0 at once without taking any more observations. This, however, contradicts the fact that $h < \delta$, and proves the lemma.

LEMMA 8. *We have, for fixed* $\delta > 1$,

$$\lim_{W_1 \to 0} B(W_0(W_1, \delta), W_1) = \delta;$$

$$\lim_{W_1 \to \infty} B(W_0(W_1, \delta), W_1) = 0.$$

PROOF: By Lemma 7,

$$\lim_{W_1 \to 0} W_0(W_1) = 0.$$

When, for fixed c, both W_0 and W_1 are small enough, then, no matter what the value of h, $\gamma(h) < 0$. Hence $A = B$, which proves the first half of the lemma.

Let now $\{W_{1N}\}$ be a sequence such that $\lim W_{1N} = \infty$. Let $\delta > 1$. For the sake of brevity we write $B(W_{1N})$ instead of

$$B(W_0(W_{1N}\delta),\ W_{1N}).$$

Suppose that, for sufficiently large N, $B(W_{1N})$ is bounded below by a positive number. Hence, for sufficiently large N, the probability of rejecting H_1 when it it is true is bounded below by a positive number. Moreover, since $B \leq \frac{W_0 g_0}{W_1 g_1} \leq A$, it follows that, for N sufficiently large, $\frac{W_{0N} g_0}{W_{1N} g_1}$ is bounded above and below by positive constants. Thus, for large N the average risk of the test defined by $B(W_{1N})$, δ, is greater than $u g_1 W_{1N}$, where u is a positive constant which does not depend on N. Moreover, from the definition of $B(W_{1N})$, this risk is a minimum.

Let ϵ be a positive number such that $\epsilon\left(\frac{W_{0N} g_0}{W_{1N} g_1} + 1\right) < \frac{u}{2}$ for all N sufficiently large. Let V_1, V_2, with $0 < V_1 < 1 < V_2$, be two constants such that, for the sequential probability ratio test determined by them, both α_0 and α_1 are $< \epsilon$. Of course $E_0 n$ and $E_1 n$ are finite and determined by the test. For this test the average risk is less than

$$\begin{aligned} &\epsilon(g_0 W_{0N} + g_1 W_{1N}) + c g_0 E_0 n + c g_1 E_1 n \\ &< \frac{u}{2} g_1 W_{1N} + c g_0 E_0 n + c g_1 E_1 n \\ &< \frac{3u}{4} g_1 W_{1N}, \end{aligned}$$

for W_{1N} large enough. This however contradicts the fact that the minimum risk is $> u g_1 W_{1N}$, and proves the lemma.

7. Proof of the theorem. Let a given sequential probability ratio test S_0 be defined by B^*, A^*; $B^* < 1 < A^*$. Let $\alpha_i(S_0)$ be the probability, according to S_0, of rejecting H_i when it is true. Let c be fixed.

By Lemma 4, B is a continuous function of W_0 and W_1. Let $\delta = A^*$ in Lemma 8. Then there exists a pair $\overline{W}_0$, $\overline{W}_1$, with $\overline{W}_0 = W_0(\overline{W}_1, A^*)$, such that

$$A(\overline{W}_0, \overline{W}_1) = A^*;$$

$$B(\overline{W}_0, \overline{W}_1) = B^*.$$

Hence the average risk

$$\sum_i g_i [\overline{W}_i \alpha_i(S_0) + c E_i^0(n)],$$

corresponding to the sequential test S_0 is a minimum.

Now let S_1 be any other test for deciding between H_0 and H_1 and such that

$$\alpha_i(S_1) \leq \alpha_i(S_0), \quad \text{and} \quad E_i^1(n) \text{ exists } (i = 1, 2).$$

Then

$$\sum_i g_i [\overline{W}_i \alpha_i(S_0) + cE_i^0(n)] \leq \sum_i g_i [\overline{W}_i \alpha_i(S_1) + cE_i^1(n)].$$

Since $\alpha_i(S_1) \leq \alpha_i(S_0)$, we have

$$\sum_i g_i E_i^0(n) \leq \sum_i g_i E_i^1(n).$$

Now g_0, g_1 were arbitrarily chosen (subject, of course, to the obvious restrictions). Hence it must be that

$$E_i^0(n) \leq E_i^1(n).$$

This, however, is the desired result.

REFERENCES

1] A. Wald, "Sequential tests of statistical hypotheses", *Annals of Math. Stat.*, Vol. 16 (1945), pp. 117–186.

[2] A. Wald, *Sequential Analysis*, John Wiley and Sons, Inc., New York, 1947.

[3] Charles Stein, "A note on cumulative sums", *Annals of Math. Stat.*, Vol. 17 (1946), pp. 498–499.

Reprinted from THE ANNALS OF MATHEMATICAL STATISTICS
Vol. XX, No. 1, March, 1949

ON DISTINCT HYPOTHESES

BY AGNES BERGER AND ABRAHAM WALD
Columbia University

1. Introduction. The following problem was suggested to one of the authors by Professor Neyman:

Let $X = (X_1, X_4, \cdots, X_n)$ be a chance vector and let h denote any simple hypothesis specifying its distribution. Let H_i be the composite hypothesis that some element h of a set of simple hypotheses $\{h\}_i$, $(i = 0, 1)$, is true, and assume that H_0 and H_1 are known to be exhaustive. Let h_i denote an element of $\{h\}_i$ $(i = 0, 1)$.

For any region W of the sample space S, let $P(W \mid h)$ be the probability that the sample point falls in W when h is true.

We shall call H_0 and H_1 *distinct*, if a region W exists for which

$$P(W \mid h_0) \neq P(W \mid h_1), \qquad \begin{array}{l} \text{for all } h_0 \in \{h\}_0 \\ \text{and all } h_1 \in \{h\}_1 . \end{array}$$

The problem is to establish necessary and sufficient conditions for two composite hypotheses H_0 and H_1 to be distinct.

For any critical region W for testing H_0 against H_1, let $\gamma(W \mid h)$ be the probability of a wrong decision when h is true, i.e.

$$\gamma(W \mid h) = \begin{cases} P(W \mid h) & \text{for} \quad h \in H_0 \\ 1 - P(W \mid h) & \text{for} \quad h \in H_1 . \end{cases}$$

Suppose now that H_0 and H_1 are not distinct. Then to any W a pair h_0', h_1' exist such that

$$P(W \mid h_0') = P(W \mid h_1'),$$

thus

$$\gamma(W \mid h_0') = 1 - \gamma(W \mid h_1'),$$

and therefore

$$\text{(1.1)} \qquad \text{l.u.b. } \gamma(W \mid h) \geq \tfrac{1}{2} \text{ for any } W.$$

This property of non-distinct hypotheses leads us to investigate the conditions under which 2 hypotheses allow a test where the maximum probability of a wrong decision is $< \frac{1}{2}$.

The result, in turn, will enable us to state, for an important class of hypotheses a necessary and sufficient condition for 2 composite hypotheses to be distinct.

2. A lemma. We shall now prove the following lemma:

LEMMA 2.1. *Assume that X has a density function $p(x)$ and let $H_i = h_i$ be the simple hypothesis that $p(x) = p_i(x)$, $(i = 0, 1)$. Assume that the set R of x's* *satisfying* $p_0(x) \neq p_1(x)$ *has a positive measure. Then there exists a region* W *such that* $\gamma(W \mid p_i) < \frac{1}{2}$, $i = 0, 1$.

PROOF: Let R_0 be defined by $p_0 = p_1$, R_1 by $p_0 < p_1$, R_2 by $p_0 > p_1$. Since $\int_S p_i(x)\, dx = 1$ and $p_i(x) \geq 0$, $(i = 0, 1)$, R_1 and R_2 are of positive measure. Let

$$\phi(x) = \begin{cases} p_1 \text{ in } R_1 \\ p_0 \text{ in } R_2 \\ p_1 = p_0 \text{ in } R_0 . \end{cases}$$

Then $\int_S \phi(x)\, dx > 1$ and either

$$\text{a)} \qquad \int_{R_1+R_0} p_1\, dx > \tfrac{1}{2} \quad \text{or} \quad \text{b)} \quad \int_{R_2} p_0\, dx > \tfrac{1}{2}$$

or both. Assume first a).

Let $R_3 \subset R_1 + R_0$ and such that $\int_{R_3} p_1\, dx = \frac{1}{2}$, but $\int_{R_3} p_0\, dx < \frac{1}{2}$. This can be done by including into R_3 a part of R_1 of non-zero measure. Let $R_4 \subset R_1 + R_0 - R_3$ and such that $0 < \int_{R_4} p_1\, dx < \frac{1}{2} - \int_{R_3} p_0\, dx$. Then

$\int_{R_4} p_0\, dx \leq \int_{R4} p_1 dx < \frac{1}{2} - \int_{R_3} p_0\, dx$, thus $\int_{R_3+R_4} p_0\, dx < \frac{1}{2}$ but $\int_{R_3+R_4} p_1\, dx > \frac{1}{2}$.

Assume now b).

Let $R_5 \subset R_2$ and such that $\int_{R_5} p_0\, dx = \frac{1}{2}$. Then $\int_{R_5} p_1\, dx < \frac{1}{2}$. Let $R_6 \subset R_2 - R_5$ and such that $0 < \int_{R_6} p_0\, dx < \frac{1}{2} - \int_{R_5} p_1\, dx$. Then $\int_{R_5+R_6} p_0\, dx > \frac{1}{2}$ and $\int_{R_5+R_6} p_1\, dx < \frac{1}{2}$.

Thus in case a) $W = R_3 + R_4$, and in case b) $W = S - R_5 - R_6$ is a critical region for which $\gamma(W \mid p_i) < \frac{1}{2}$ $(i = 0, 1)$. This proves the lemma.

3. The main theorem. Assume now X to have a density function $p(x, \mid \theta)$ where $\theta = (\theta_1, \theta_4, \cdots, \theta_k)$ is an unknown parameter point. Let ω_0 and ω_1 be two disjoint, bounded and closed subsets of the k-dimensional θ − space. Let $\Omega = \omega_0 + \omega_1$ and suppose that θ is known to belong to Ω, which therefore will be called the parameter space. Let H_i be the hypothesis that the true parameter point is an element of ω_i, $(i = 0, 1)$.

We shall consider the problem of testing H_0 against H_1. Clearly, $P(W \mid h)$ can now be written as $P(W \mid \theta)$ and $\gamma(W \mid h)$ as $\gamma(W \mid \theta)$.

We shall make the following assumptions concerning $p(x \mid \theta)$:

Assumption 1. $p(x \mid \theta)$ is continuous in θ. This is of course always fulfilled if Ω consists only of a finite number of points.

Assumption 2. For any bounded domain M of the sample space we have

$$\int_M [\operatorname*{Max}_\theta p(x \mid \theta)]\, dx < \infty.$$

It follows from Assumptions 1. and 2. that

$$(3.1) \qquad \lim_{r=\infty} \int_{S-S_r} p(x \mid \theta)\, dx = 0$$

uniformly in θ where S_r is the sphere in the sample space with center at the origin and radius r.

In what follows, whenever we shall speak of cumulative distribution function $g(\theta)$ in the k-dimensional parameter space, we shall always mean a cumulative distribution function satisfying the condition

$$\int_\Omega dg\,(\theta) = 1.$$

For any c.d.f. $g(\theta)$ let W_g denote a critical region which contains any sample point x satisfying the inequality

$$\int_{\omega_1} p(x \mid \theta)\, dg(\theta) > \int_{\omega_0} p(x \mid \theta)\, dg(\theta),$$

and does not contain a sample point x for which

$$\int_{\omega_1} p(x \mid \theta)\, dg(\theta) < \int_{\omega_0} p(x \mid \theta)\, dg(\theta).$$

It can easily be verified that W_g minimizes the average risk

$$(3.2) \quad \int_\Omega \gamma(W \mid \theta)\, dg(\theta), \text{ i.e.,} \qquad \int_\Omega W_g \mid \theta)\, dg(\theta) = \operatorname*{Min}_W \int_\Omega \gamma(W \mid \theta)\, dg(\theta).$$

Let Ω_i $(i = 0, 1)$ be the class of all density functions $p(x) = \int_\Omega p(x \mid \theta)\, dg_i(\theta)$ where $g_i(\theta)$ is subject to the condition

$$\int_{\omega_i} dh_i(\theta) = 1.$$

Two density functions $p(x)$ and $q(x)$ are said to be equal if $p(x) \neq q(x)$ holds only in a set of measure zero.

It follows from (3.1) and Assumptions 1. and 2. that $\gamma(W \mid \theta)$ is a continuous function of θ. Let $\gamma(W)$ denote the maximum of $\gamma(W \mid \theta)$ with respect to θ. We shall prove the following theorem:

THEOREM 3.1. *A necessary and sufficient condition for the existence of a region W such that $\gamma(W) < \frac{1}{2}$ is that the classes Ω_c and Ω_1 be disjoint.*

PROOF. Suppose that Ω_0 and Ω_1 are not disjoint. Then there exist two distribution functions $g_0(\theta)$ and $g_1(\theta)$ such that

$$\int_{\omega_0} dg_0(\theta) = \int_{\omega_1} dg_1(\theta) = 1$$

and

$$\int_{\omega_0} p(x \mid \theta)\, dg_0(\theta) = \int_{\omega_1} p(x \mid \theta)\, dg_1(\theta)$$

(except perhaps for points x in a set of measure 0).

Let $g(\theta) = \frac{1}{2}\, g_0(\theta) + \frac{1}{2}\, g_1(\theta)$. Clearly, $\gamma(W) \geq \int_{\Omega} \gamma(W \mid \theta)\, dg(\theta) = \frac{1}{2}$ for any W. This proves the necessity of our condition.

We shall now assume that Ω_0 and Ω_1 are disjoint. First we shall show that the results of [1] can be applied. On pages 297–8 of [1] there are seven conditions listed for the sequential case. For the non-sequential case (the one considered here) the conditions 6 and 7 drop out and the first five conditions can be reduced to the following conditions:

Condition 1: The weight function $W(\theta, d)$ is bounded.

Condition 2: For any θ, the chance vector X admits a density function $p(x \mid \theta)$.

Condition 3: For any sequence $\{\theta_i\}$ $(i = 1, 2, \cdots$, ad inf.) there exists a subsequence $\{\theta_j\}$ $(j = 1, 2, \cdots)$ and a parameter point θ_0 such that

$$\lim_{i=\infty} p(x \mid \theta_{i_j}) = p(x \mid \theta_0)$$

Condition 4: If $\{\theta_i\}$ $(i = 1, 2, \cdots)$ is a sequence of points and θ_0 a point such that

$$\lim_{i=\infty} p(x \mid \theta_i) = p(x \mid \theta_0)$$

then,

$$\lim_{i=\infty} W(\theta_i\,, d) = W(\theta_0\,, d)$$

uniformly in d.

Condition 5: The same as our Assumption 2.

In our problem d(the decision of the statistician) can take only two values: acceptance or rejection of H_0. Condition 1 is evidently fulfilled, since $W(\theta, d) = 0$ if a correct decision is made, and $= 1$ if a wrong decision is made. Clearly, Conditions 2–5 are also fulfilled in our problem.

A distribution $g(\theta)$ is said to be least favorable, if it maximizes the minimum average risk, i.e., if it maximizes $\int_{\Omega} \gamma(W \mid \theta)\, dg(\theta)$ with respect to g.

It follows from Theorems 4.1 and 4.4 of [1] that there exists a least favorable distribution.

Let $g^*(\theta)$ be a least favorable distribution. Then, as has been shown in [1] there exists a W_{g^*} such that

$$\text{(3.3)} \qquad \operatorname*{Max}_{\theta} \gamma(W_{g^*} \mid \theta) = \int_{\Omega} \gamma(W_{g^*} \mid \theta)\, dg^*(\theta).$$

Thus, our theorem is proved if we can show that

$$\text{(3.4)} \qquad \int_{\Omega} \gamma(W_{g^*} \mid \theta)\, dg^*(\theta) < \tfrac{1}{2}.$$

Let H_0^* be the hypothesis that the true density is given by

$$p_0(x) = \frac{\int_{\omega_0} p(x \mid \theta)\, dg^*(\theta)}{\int_{\omega_0} dg^*(\theta)},$$

and H_1^* the hypothesis that the true density is given by

$$p_1(x) = \frac{\int_{\omega_1} p(x \mid \theta)\, dg^*(\theta)}{\int_{\omega_1} dg^*(\theta)}.$$

Since Ω_0 and Ω_1 are disjoint, $p_0(x)$ and $p_1(x)$ are different density functions. Hence, according to Lemma 2.1, there exists a critical region W^* for testing H_0^* such that $\alpha^* < \frac{1}{2}$ and $\beta^* < \frac{1}{2}$, where α^* is the probability of type I error, and β^* is the probability of type II error. Clearly,

$$\text{(3.5)} \qquad \tfrac{1}{2} > \alpha^* \int_{\omega_0} dg^*(\theta) + \beta^* \int_{\omega_1} dg^*(\theta) = \int_{\Omega} \gamma(W^* \mid \theta)\, dg^*(\theta).$$

Hence, our theorem is proved.

It follows from (1.1) that if H_0 and H_1 are not distinct, Ω_0 and Ω_1 are not disjoint.

On the other hand, suppose that Ω_0 and Ω_1 are not disjoint and let

$$\int_{\omega_0} p(x \mid \theta)\, dg_0(\theta) = \int_{\omega_1} p(x \mid \theta)\, dg_1(\theta).$$

Then for every W

$$\text{(3.6)} \qquad \int_{\omega_0} P(W \mid \theta)\, dg_0(\theta) = \int_{\omega_1} P(W \mid \theta)\, dg_1(\theta).$$

Assume now that ω_i is a connected set ($i = 0, 1$). Then, because of the continuity of $P(W \mid \theta)$ there exist 2 functions $\theta_0(W)$, $\theta_1(W)$, $\theta_i(W)$ belonging to $\omega_i (i = 0, 1)$ such that

$$P(W \mid \theta_0(W)) = \int_{\omega_0} P(W \mid \theta)\, dg_0(\theta)$$

and

$$P(W \mid \theta_1(W)) = \int_{\omega_1} P(W \mid \theta)\, dg_1(\theta)$$

for every W. Hence, because of (3.6),

$$P(W \mid \theta_0(W)) = P(W \mid \theta_1(W))$$

for every W. Thus, we arrive at the following theorem:

THEOREM 3.2. *If ω_i is a connected set ($i = 0, 1$), then, under the assumptions of Theorem 3.1, a necessary and sufficient condition for H_0 and H_1 to be distinct is that the sets Ω_0 and Ω_1 be disjoint.*

REFERENCE

[1] A. WALD, "Foundations of a general theory of sequential decision functions," *Econometrica*, Vol. 15 (1947), pp. 279–313.

Reprinted from THE ANNALS OF MATHEMATICAL STATISTICS
Vol. XX, No. 4, December, 1949

NOTES

This section is devoted to brief research and expository articles and other short items.

NOTE ON THE CONSISTENCY OF THE MAXIMUM LIKELIHOOD ESTIMATE[1]

By ABRAHAM WALD

Columbia University

1. Introduction. The problem of consistency of the maximum likelihood estimate has been treated in the literature by several authors (see, for example, Doob [1][2] and Cramér [2][3]). The purpose of this note is to give another proof of the consistency of the maximum likelihood estimate which may be of interest because of its relative simplicity and because of the easy verifiability of the underlying assumptions. The present proof has some common features with that given by Doob, insofar that both proofs make no differentiability assumptions (thus, not even the existence of the likelihood equation is postulated) and both are based on the strong law of large numbers and an inequality involving the log of a random variable. The assumptions in the present note are stronger in some respects than those made by Doob, but also the results obtained here are stronger. For the sake of simplicity, the author did not attempt to give the most general results or to weaken the underlying assumptions as much as possible. Remarks on possible generalizations are made in Section 4.

Let $X_1, X_2, \cdots$, etc. be independently and identically distributed chance variables. The most frequently considered case in the literature is that where the common distribution is known, except for the values of a finite number of

[1] The author wishes to thank J. L. Doob for several comments and suggestions he made in connection with this note.

[2] According to a communication from Doob, his Theorem 4 is incorrect, but is correct if the class of almost everywhere continuous functions in that theorem is replaced by a suitable class C of functions. The class C can be any one of a variety of classes; for example, the class of bounded almost everywhere continuous functions, or the larger class of almost everywhere continuous functions each of which is less than or equal in modulus to any one of a prescribed sequence of functions with finite expectations. His Theorem 5 on the consistency of the maximum likelihood is then dependent on the class C used in Theorem 4.

[3] The proof given by Cramér [2], pp. 500–504, establishes the consistency of some root of the likelihood equation but not necessarily that of the maximum likelihood estimate when the likelihood equation has several roots. Recently, Huzurbazar [3] showed that under certain regularity conditions the likelihood equation has at most one consistent solution and that the likelihood function has a relative maximum for such a solution. Since there may be several solutions for which the likelihood function has relative maxima, Cramér's and Huzurbazar's results taken together still do not imply that a solution of the likelihood equation which makes the likelihood function an absolute maximum is necessarily consistent.

parameters, $\theta^1, \theta^2, \cdots, \theta^k$. In this note we shall treat the parametric case. For any parameter point $\theta = (\theta^1, \cdots, \theta^k)$, let $F(x, \theta)$ denote the corresponding cumulative distribution function of X_i; i.e., $F(x, \theta) = \text{prob. } \{X_i < x\}$. The totality Ω of all possible parameter points is called the parameter space. Thus, the parameter space Ω is a subset of the k-dimensional Cartesian space.

It is assumed in this note that for any θ, the cumulative distribution function $F(x, \theta)$ admits an elementary probability law $f(x, \theta)$. If $F(x, \theta)$ is absolutely continuous, $f(x, \theta)$ denotes the density at x. If $F(x, \theta)$ is discrete, $f(x, \theta)$ is equal to the probability that $X_i = x$.

Throughout this note the following assumptions will be made.

Assumption 1. *$F(x, \theta)$ is either discrete for all θ or is absolutely continuous for all θ.*

Before formulating the next assumption, we shall introduce the following notations: for any θ and for any positive value ρ let $f(x, \theta, \rho)$ be the supremum of $f(x, \theta')$ with respect to θ' when $|\theta - \theta'| \leqq \rho$. For any positive r, let $\varphi(x, r)$ be the supremum of $f(x, \theta)$ with respect to θ when $|\theta| > r$. Furthermore, let $f^*(x, \theta, \rho) = f(x, \theta, \rho)$ when $f(x, \theta, \rho) > 1$, and $=1$ otherwise. Similarly, let $\varphi^*(x, r) = \varphi(x, r)$ when $\varphi(x, r) > 1$, and $=1$ otherwise.

Assumption 2. *For sufficiently small ρ and for sufficiently larger r the expected values $\int_{-\infty}^{\infty} \log f^*(x, \theta, \rho)\, dF(x, \theta_0)$ and $\int_{-\infty}^{\infty} \log \varphi^*(x, r)\, dF(x, \theta_0)$ are finite where θ_0 denotes the true parameter point.*[4]

Assumption 3. *If $\lim_{i=\infty} \theta_i = \theta$, then $\lim_{i=\infty} f(x, \theta_i) = f(x, \theta)$ for all x except perhaps on a set which may depend on the limit point θ (but not on the sequence θ_i) and whose probability measure is zero according to the probability distribution corresponding to the true parameter point θ_0.*

Assumption 4. *If θ_1 is a parameter point different from the true parameter point θ_0, then $F(x, \theta_1) \not\equiv F(x, \theta_0)$ for at least one value of x.*

Assumption 5. *If $\lim_{i=\infty} |\theta_i| = \infty$, then $\lim_{i=\infty} f(x, \theta_i) = 0$ for any x except perhaps on a fixed set (independent of the sequence θ_i) whose probability is zero according to the true parameter point θ_0.*

Assumption 6. *For the true parameter point θ_0 we have*

$$\int_{-\infty}^{\infty} |\log f(x, \theta_0)|\, dF(x, \theta_0) < \infty.$$

Assumption 7. *The parameter space Ω is a closed subset of the k-dimensional Cartesian space.*

Assumption 8. *$f(x, \theta, \rho)$ is a measurable function of x for any θ and ρ.*

It is of interest to note that if we forbid the dependence of the exceptional set on θ in Assumption 3, Assumption 8 is a consequence of Assumption 3, as can easily be verified.

---

[4] The measurability of the functions $f^*(x, \theta, \rho)$ and $\varphi^*(x, r)$ for any θ, ρ and r follows easily from Assumption 8.

In the discrete case, Assumption 8 is unnecessary. In fact, we may replace $f(x, \theta, \rho)$ everywhere by $\bar{f}(x, \theta, \rho)$ where $\bar{f}(x, \theta, \rho) = f(x, \theta, \rho)$ when $f(x, \theta_0) > 0$, and $\bar{f}(x, \theta, \rho) = 1$ when $f(x, \theta_0) = 0$. Here θ_0 denotes the true parameter point. Since $f(x, \theta_0) > 0$ only for countably many values of $x, \bar{f}(x, \theta, \rho)$ is obviously a measurable function of x.

In the absolutely continuous case, $F(x, \theta)$ does not determine $f(x, \theta)$ uniquely. If Assumptions 3, 5 and 8 hold for one choice of $f(x, \theta)$, they do not necessarily hold for another choice of $f(x, \theta)$. This is in a way undesirable, but assumptions of such nature are unavoidable if we want to insure the consistency of the maximum likelihood estimate. It is, however, possible to formulate assumptions which remain valid for all possible choices of $f(x, \theta)$ and which insure the consistency of the maximum likelihood estimate for a particular choice of $f(x, \theta)$. In this connection the following remark due to Doob is of interest. Let Assumptions 3′ and 5′ be the same as 3 and 5, respectively, except that the exceptional set is permitted to depend on the sequence θ_i. If 3′ and 5′ hold for one choice of $f(x, \theta)$, they also hold for any other choice. Doob has shown that Assumptions 3′ and 5′ insure the existence of a choice of $f(x, \theta)$ for which Assumptions 3, 5 and 8 hold. Thus, one may say that Assumptions 3′ and 5′ are the essential ones and the stronger assumptions 3, 5 and 8 are needed merely to exclude a "bad" choice of $f(x, \theta)$.

2. Some lemmas. In this section we shall prove some lemmas which will be used in the next section to obtain the main theorems. Let θ_0 be the true parameter point. By the expected value Eu of any chance variable u we shall mean the expected value determined under the assumption that θ_0 is the true parameter point. For any chance variable u, u' will denote the chance variable which is equal to u when $u > 0$ and equal to zero otherwise. Similarly, for any chance variable u, the symbol u'' will be used to denote the chance variable which is equal to u when $u < 0$ and equal to zero otherwise. We shall say that the expected value of u exists if $Eu' < \infty$. If the expected value of u' is finite but that of u'' is not, we shall say that the expected value of u is equal to $-\infty$.

LEMMA 1. *For any $\theta \neq \theta_0$ we have*

$$E \log f(X, \theta) < E \log f(X, \theta_0) \tag{1}$$

where X is a chance variable with the distribution $F(x, \theta_0)$.

PROOF. It follows from Assumption 2 that the expected values in (1) exist. Because of Assumption 6, we have

$$E \mid \log f(X, \theta_0) \mid < \infty. \tag{2}$$

If $E \log f(X, \theta) = -\infty$, Lemma 1 obviously holds. Thus, we shall merely consider the case when $E \log f(X, \theta) > -\infty$. Then

$$E \mid \log f(X, \theta) \mid < \infty. \tag{3}$$

Let $u = \log f(X, \theta) - \log f(X, \theta_0)$.[5] Clearly, $E \mid u \mid < \infty$. It i known that for any chance variable u which is not equal to a constant (with probability one) and for which $E|u| < \infty$, we have[6]

$$Eu < \log Ee^u. \tag{4}$$

Since in our case

$$Ee^u \leqq 1, \tag{5}$$

and since u differs from zero on a set of positive probability (due to Assumption 4), we obtain from (4)

$$Eu < 0. \tag{6}$$

Thus, Lemma 1 is proved.

We shall now prove the following lemma.

LEMMA 2. $\lim_{\rho=0} E \log f(X, \theta, \rho) = E \log f(X, \theta)$.

PROOF. Let $f^*(x, \theta, \rho) = f(x, \theta, \rho)$ when $f(x, \theta, \rho) \geqq 1$, and $=1$ otherwise. Similarly, let $f^*(x, \theta) = f(x, \theta)$ when $f(x, \theta) \geqq 1$, and $=1$ otherwise. It follows from Assumption 3 that

$$\lim_{\rho=0} \log f^*(x, \theta, \rho) = \log f^*(x, \theta) \tag{7}$$

except perhaps on a set whose probability measure is zero. Since $\log f^*(x, \theta, \rho)$ is an increasing function of ρ, it follows from (7) and Assumption 2 that

$$\lim_{\rho=0} E \log f^*(X, \theta, \rho) = E \log f^*(X, \theta). \tag{8}$$

Let $f^{**}(x, \theta, \rho) = f(x, \theta, \rho)$ when $f(x, \theta, \rho) \leqq 1$, and $=1$ otherwise. Similarly, let $f^{**}(x, \theta) = f(x, \theta)$ when $f(x, \theta) \leqq 1$, and $=1$ otherwise. Clearly,

$$|\log f^{**}(x, \theta, \rho)| \leqq |\log f^{**}(x, \theta)| \tag{9}$$

and

$$\lim_{\rho=0} \log f^{**}(x, \theta, \rho) = \log f^{**}(x, \theta) \tag{10}$$

for all x except perhaps on a set whose probability measure is zero. The relation

$$\lim_{\rho=0} E \log f^{**}(X, \theta, \rho) = E \log f^{**}(X, \theta) \tag{11}$$

follows from (9) and (10) in both cases, when $E \log f^{**}(X, \theta)$ is finite and when $E \log f^{**}(X, \theta) = -\infty$. Lemma 2 is an immediate consequence of (8) and (11).

LEMMA 3. *The equation*

$$\lim_{r=\infty} E \log \varphi(X, r) = -\infty. \tag{12}$$

holds.

---

[5] It is of no consequence what value is assigned to u when $f(x, \theta)$ or $f(x, \theta_0)$ is zero, since the probability of such an event, because of (3), is zero.

[6] This is a generalization of the inequality between geometric and arithmetic means. See, for example, HARDY, LITTLEWOOD, POLYA, *Inequalities*, Cambridge 1934, p. 137, Theorem 184.

PROOF. It follows from Assumption 5 that

$$\lim_{r=\infty} \log \varphi(x, r) = -\infty, \tag{13}$$

for any x (except perhaps on a set of probability 0). Since according to Assumption 2,

$$E \log \varphi^*(X, r) < \infty, \tag{14}$$

and since $\log \varphi(x, r) - \log \varphi^*(x, r)$ and $\log \varphi^*(x, r)$ are decreasing functions of r, Lemma 3 follows easily from (13).

3. The main theorems. We shall now prove the following theorems.

THEOREM 1. *Let ω be any closed subset of the parameter space Ω which does not contain the true parameter point θ_0. Then*

$$\text{prob.} \left\{ \lim_{n=\infty} \frac{\operatorname{Sup}_{\theta \epsilon \omega} f(X_1, \theta) f(X_2, \theta) \cdots f(X_n, \theta)}{f(X_1, \theta_0) f(X_2, \theta_0) \cdots f(X_n, \theta_0)} = 0 \right\} = 1. \tag{15}$$

PROOF. Let r_0 be a positive number chosen such that

$$E \log \varphi(X, r_0) < E \log f(X, \theta_0). \tag{16}$$

The existence of such a positive number follows from Lemma 3. Let ω_1 be the subset of ω consisting of all points θ of ω for which $| \theta | \leqq r_0$. With each point θ in ω_1 we associate a positive value ρ_θ such that

$$E \log f(X, \theta, \rho_\theta) < E \log f(X, \theta_0). \tag{17}$$

The existence of such a ρ_θ follows from Lemmas 1 and 2. Since the set ω_1 is compact, there exists a finite number of points $\theta_1, \cdots, \theta_h$ in ω_1 such that $S(\theta_1, \rho_{\theta_1}) + \cdots + S(\theta_h, \rho_{\theta_h})$ contains ω_1 as a subset. Here $S(\theta, \rho)$ denotes the sphere with center θ and radius ρ. Clearly,

$$0 \leqq \operatorname{Sup}_{\theta \epsilon \omega} f(x_1, \theta) \cdots f(x_n, \theta) \leqq \sum_{i=1}^{h} f(x_1, \theta_i, \rho_{\theta_i}) \cdots f(x_n, \theta_i, \rho_{\theta_i}) + \varphi(x_1, r_0) \cdots \varphi(x_n, r_0).$$

Hence, Theorem 1 is proved if we can show that

$$\text{prob} \left\{ \lim_{n=\infty} \frac{f(X_1, \theta_i, \rho_{\theta_i}) \cdots f(X_n, \theta_i, \rho_{\theta_i})}{f(X_1, \theta_0) \cdots f(X_n, \theta_0)} = 0 \right\} = 1 \quad (i = 1, \cdots, h) \tag{18}$$

and

$$\text{prob} \left\{ \lim_{n=\infty} \frac{\varphi(X_1, r_0) \cdots \varphi(X_n, r_0)}{f(X_1, \theta_0) \cdots f(X_n, \theta_0)} = 0 \right\} = 1. \tag{19}$$

The above equations can be written as

$$\text{(20)} \qquad \text{prob}\left\{\lim_{n=\infty} \sum_{\alpha=1}^{n} [\log f(X_\alpha, \theta_i, \rho_{\theta_i}) - \log f(X_\alpha, \theta_0)] = -\infty\right\} = 1$$

$$(i = 1, \cdots, h)$$

and

$$\text{(21)} \qquad \text{prob}\left\{\lim_{n=\infty} \sum_{\alpha=1}^{n} [\log \varphi(X_\alpha, r_0(- \log f(X_\alpha, \theta_0)] = -\infty\right\} = 1.$$

These equations follow immediately from (16), (17) and the strong law of large numbers. This completes the proof of Theorem 1.

THEOREM 2. *Let* $\bar{\theta}_n(x_1, \cdots, x_n)$ *be a function of the observations* $x_1, \cdots, x_n$ *such that*

$$\text{(22)} \qquad \frac{f(x_1, \bar{\theta}_n) \cdots f(x_n, \bar{\theta}_n)}{f(x_1, \theta_0) \cdots f(x_n, \theta_0)} \geqq c > 0 \text{ for all } n \text{ and for all } x_1, \cdots, x_n.$$

Then

$$\text{(23)} \qquad \text{prob}\,\{\lim_{n=\infty} \bar{\theta}_n = \theta_0\} = 1.$$

PROOF. It is sufficient to prove that for any $\epsilon > 0$ the probability is one that all limit points $\bar{\theta}$ of the sequence $\{\bar{\theta}_n\}$ satisfy the inequality $|\bar{\theta} - \theta_0| \leqq \epsilon$. The event that there exists a limit point $\bar{\theta}$ of the sequence $\{\bar{\theta}_n\}$ such that $|\bar{\theta} - \theta_0| > \epsilon$ implies that $\operatorname{Sup}_{|\theta - \theta_0| \geqq \epsilon} f(x_1, \theta) \cdots f(x_n, \theta) \geqq f(x_1, \bar{\theta}_n) \cdots f(x_n, \bar{\theta}_n)$ for infinitely many n. But then

$$\text{(24)} \qquad \frac{\operatorname{Sup}_{|\theta-\theta_0| \geqq \epsilon} f(x_1, \theta) \cdots f(x_n, \theta)}{f(x_1, \theta_0) \cdots f(x_n, \theta_0)} \geqq c > 0$$

for infinitely many n. Since, according to Theorem 1, this is an event with probability zero, we have shown that the probability is one that all limit points $\bar{\theta}$ of $\{\bar{\theta}_n\}$ satisfy the inequality $|\bar{\theta} - \theta_0| \leqq \epsilon$. This completes the proof of Theorem 2.

Since a maximum likelihood estimate $\hat{\theta}_n(x_1, \cdots, x_n)$, if it exists, obviously satisfies (22) with $c = 1$, Theorem 2 establishes the consistency of $\hat{\theta}_n(x_1, \cdots, x_n)$ as an estimate of θ.

4. Remarks on possible generalizations. The method given in this note can be extended to establish the consistency of the maximum likelihood estimates for certain types of dependent chance variables for which the strong law of large numbers remains valid.

The assumption that the parameter space Ω is a subset of a finite dimensional Cartesian space is unnecessarily restrictive. Let Ω be any abstract space. All of our results can easily be shown to remain valid if Assumptions 3, 5 and 7 are replaced by the following one:

ASSUMPTION 9. *It is possible to introduce a distance $\delta(\theta_1, \theta_2)$ in the space Ω such that the following four conditions hold:*

(i) *The distance $\delta(\theta_1, \theta_2)$ makes Ω to a metric space*

(ii) $\lim_{i=\infty} f(x, \theta_i) = f(x, \theta)$ *if* $\lim_{i=\infty} \theta_i = \theta$ *for any x except perhaps on a set which may depend on θ (but not on the sequence θ_i) and whose probability measure is zero according to the probability distribution corresponding to the true parameter point θ_0.*

(iii) *If θ_0 is a fixed point in Ω and* $\lim_{i=\infty} \delta(\theta_i, \theta_0) = \infty$, *then* $\lim_{i=\infty} f(x, \theta_i) = 0$ *for any x.*

(iv) *Any closed and bounded subset of Ω is compact.*

REFERENCES

[1] J. L. DOOB, "Probability and statistics," *Trans. Amer. Math. Soc.*, Vol. 36 (1934).

[2] H. CRAMÉR, *Mathematical Methods of Statistics*, Princeton University Press, Princeton, 1946.

[3] V. S. HUZURBAZAR, "The likelihood equation, consistency and the maxima of the likelihood function," *Annals of Eugenics*, Vol. 14 (1948).

Reprinted from THE ANNALS OF MATHEMATICAL STATISTICS
Vol. XX, No. 4, December, 1949

A SEQUENTIAL DECISION PROCEDURE FOR CHOOSING ONE OF THREE HYPOTHESES CONCERNING THE UNKNOWN MEAN OF A NORMAL DISTRIBUTION

BY MILTON SOBEL AND ABRAHAM WALD[1]

Columbia University

1. Introduction. In this paper a multi-decision problem is investigated from a sequential viewpoint and compared with the best non-sequential procedure available. Multi-decision problems occur often in practice but methods to deal with such problems are not yet sufficiently developed.

The problem under consideration here is a 3-decision problem: Given a chance variable which is normally distributed with known variance σ^2, but unknown mean θ, and given two real numbers $a_1 < a_2$, the problem is to choose one of the three mutually exclusive and exhaustive hypotheses

$$H_1 : \theta < a_1 \qquad H_2 : a_1 \leqq \theta \leqq a_2 \qquad H_3 : \theta > a_2 .$$

In order to select a proper sequential decision procedure, the parameter space is subdivided into 5 mutually exclusive and exhaustive zones in the following manner. Around a_1 there exists an interval (θ_1, θ_2) in which we have no strong preference between H_1 and H_2 but prefer (strongly) to reject H_3. Around a_2 there exists an interval (θ_3, θ_4) in which we have no strong preference between H_2 or H_3 but prefer (strongly) to reject H_1. For $\theta \leqq \theta_1$ we prefer to accept H_1. For $\theta_2 \leqq \theta \leqq \theta_3$ we prefer to accept H_2. For $\theta \geqq \theta_4$ we prefer to accept H_3.

The intervals (θ_1, θ_2) and (θ_3, θ_4) will be called indifference zones. The determination of these indifference zones is not a statistical problem but should be made on practical considerations concerning the consequences of a wrong decision.

In accordance with the above we define a wrong decision in the following way. For $\theta \leqq \theta_1$, acceptance of H_2 or H_3 is wrong. For $\theta_1 < \theta < \theta_2$ acceptance of H_3 is wrong. For $\theta_2 \leqq \theta \leqq \theta_3$, acceptance of H_1 or H_3 is wrong. For $\theta_3 < \theta < \theta_4$, acceptance of H_1 is wrong. For $\theta \geqq \theta_4$, acceptance of H_1 or H_2 is wrong.

The requirements on our decision procedure necessary to limit the probability of a wrong decision are investigated. Two cases are considered.

Case 1: Prob. of a wrong decision $\leqq \gamma$ for all θ.

$$\text{Case 2: } \begin{cases} \text{Prob. of a wrong decision} \leqq \gamma_1 \text{ for } \theta \leqq \theta_1, \\ \text{Prob. of a wrong decision} \leqq \gamma_2 \text{ for } \theta_1 < \theta < \theta_4, \\ \text{Prob. of a wrong decision} \leqq \gamma_3 \text{ for } \theta \geqq \theta_4 . \end{cases}$$

The decision procedure discussed in the present paper is not an optimum procedure since, as will be seen later, the final decision at the termination of

[1] Work done under the sponsorship of the Office of Naval Research.

experimentation is not in every case a function of only "the sample mean of *all* the observations", although the sample mean is a sufficient statistic for θ. Although the procedure considered is not optimal it is suggested for the following reasons:

1. The decision procedure can be carried out simply. In fact tables can be constructed before experimentation starts that render the procedure completely mechanical.

2. The derivation of the operating characteristic (OC) function, neglecting the excess of the cumulative sum over the boundary, is accomplished with little difficulty. In general, for other multi-decision problems it is unknown how to obtain the OC function.

3. It is believed that the loss of efficiency is not serious; i.e., the suggested sequential procedure is not far from being optimum. In this connection a non-sequential procedure is compared with this sequential procedure. The results show that, for the same maximum probability of making a wrong decision, the sequential procedure requires on the average substantially fewer observations to reach a final decision. In fact, for Case 1 noted above, if $.008 < \gamma < .1$, and if certain symmetrical features are assumed, then the fixed number of observations required by the non-sequential method is greater than the maximum of the average sample number (ASN) function taken over all values of θ.

It was found necessary in the course of the investigation to put an upper bound on the quantity $\frac{\theta_4 - \theta_3}{a_2 - a_1}$ in order that the methods used to obtain upper and lower bounds for the ASN function should give close results. This restriction, however, is likely to be satisfied in practical applications.

All formulas for ASN and OC functions which will be used in this paper will be approximation formulas neglecting the excess of the cumulative sum over the boundaries. Nevertheless, equality signs will be used in these formulas, except when additional approximations are involved.

2. Description of the Decision Procedure.[2] We shall assume that the indifference zones described above have the following properties

$$\text{(i)}\quad \theta_1 < a_1 < \theta_2 \leqq \theta_3 < a_2 < \theta_4$$

$$\text{(ii)}\quad \theta_1 + \theta_2 = 2a_1\ ; \quad \theta_3 + \theta_4 = 2a_2$$

$$\text{(iii)}\quad \theta_2 - \theta_1 = \theta_4 - \theta_3 = \Delta \text{ (say)}.$$

[2] A similar decision procedure was used by P. Armitage [2] as an alternative to the sequential t test (with 2-sided alternatives). The form used there is more restricted as he considers only the case $\theta_2 = \theta_3$. Essential inequalities on the OC function are pointed out but no attempt is made to determine the complete OC and ASN functions. A closely related but somewhat different procedure for dealing with a trichotomy was suggested by Milton Friedman while he was a member of the Statistical Research Group of Columbia University. As far as the authors are aware, no results were obtained concerning the OC and ASN functions of Friedman's procedure.

Let R_1 denote the Sequential Probability Ratio Test for testing the hypothesis that $\theta = \theta_1$ against the hypothesis that $\theta = \theta_2$. We assume for the present that either the proper constants A, B in the probability ratio test are given or that they are approximated from given α, β by the relations

$$A \sim \frac{1-\beta}{\alpha} \qquad B \sim \frac{\beta}{1-\alpha}.$$

Here α and β are upper bounds on the probabilities of first and second types of errors, respectively.

Let R_2 represent the S.P.R.T. for testing the hypothesis that $\theta = \theta_3$ against the alternative that $\theta = \theta_4$. For this test we assume that (α, β, A, B) are replaced by $(\hat{\alpha}, \hat{\beta}, \hat{A}, \hat{B})$ and as above that either $\hat{A}$ and $\hat{B}$ are given or that they are approximated from given $\hat{\alpha}$, $\hat{\beta}$.

The decision procedure is carried out as follows:

Both R_1 and R_2 are computed at each stage of the inspection until

Either: One ratio leads to a decision to stop before the other. Then the former is no longer computed and the latter is continued until it leads to a decision to stop.

Or: Both R_1 and R_2 lead to a decision to stop at the same stage. In this event both computations are discontinued.

The following table gives the rule R for the decisions to be made corresponding to all possible outcomes of R_1 and R_2.

	R_1		R_2		R
If	accepts θ_1	and	accepts θ_3	then	accepts H_1
If	accepts θ_2	and	accepts θ_3	then	accepts H_2
If	accepts θ_2	and	accepts θ_4	then	accepts H_3

We shall show that acceptance of both θ_1 and θ_4 is impossible when $(\hat{A}, \hat{B}) = (A, B)$. For this purpose we need the acceptance number and rejection number formulas. (See page 119 of [1]).

Acceptance Number *Rejection Number*

$$R_1: \frac{\sigma^2}{\Delta} \log B + a_1 n < \sum_{\alpha=1}^{n} x_\alpha < \frac{\sigma^2}{\Delta} \log A + a_1 n$$

$$R_2: \frac{\sigma^2}{\Delta} \log B + a_2 n < \sum_{\alpha=1}^{n} x_\alpha < \frac{\sigma^2}{\Delta} \log A + a_2 n.$$

We shall assume *for convenience* that "between observations" R_1 is tested before R_2 and let the term "initial decision" refer to the first decision made.

Assume θ_1 and θ_4 are both accepted. Then if θ_1 is accepted initially at the mth stage

$$\sum_{\alpha=1}^{m} x_\alpha \leqq \frac{\sigma^2}{\Delta} \log B + a_1 m.$$

Since

$$\frac{\sigma^2}{\Delta} \log B + a_1 m < \frac{\sigma^2}{\Delta} \log B + a_2 m$$

it follows that θ_4 is rejected at the same stage, contradicting the hypothesis. Similarly if θ_4 is accepted initially at the mth stage, then

$$\sum_{\alpha=1}^{m} x_\alpha \geqq \frac{\sigma^2}{\Delta} \log A + a_2 m.$$

Since

$$\frac{\sigma^2}{\Delta} \log A + a_2 m > \frac{\sigma^2}{\Delta} \log A + a_1 m$$

it follows that θ_1 is rejected at the same or at an earlier stage, contradicting the assumption that the acceptance of θ_4 is an initial decision. Hence θ_1 and θ_4 cannot both be accepted.

A geometrical representation of the rule R is given in Figure 1.

R can now be described as follows: Continue taking observations until an acceptance region (shaded area) is reached or both dashed lines are crossed. In the former case, stop and accept as shown above. In the latter case stop and accept H_2.

The proof above that θ_1 and θ_4 cannot both be accepted consists of noting that a point below the acceptance line for θ_1 is already below the rejection line for θ_4 and that a point above the acceptance line for θ_4 is already above the rejection line for θ_1.

If $(\hat{A}, \hat{B}) \neq (A, B)$, a necessary and sufficient condition for the impossibility of accepting θ_1 and θ_4 is that at $n = 1$ the following inequalities should hold.

Rejection Number (of θ_1) for $R_1 \leqq$ Rejection Number (of θ_3) for R_2

and

Acceptance Number (of θ_1) for $R_1 \leqq$ Acceptance Number (of θ_3) for R_2.

In symbols

$$\frac{\sigma^2}{\Delta} \log A + a_1 \leqq \frac{\sigma^2}{\Delta} \log \hat{A} + a_2$$

and

$$\frac{\sigma^2}{\Delta} \log B + a_1 \leqq \frac{\sigma^2}{\Delta} \log \hat{B} + a_2.$$

These can be written as

$$\frac{A}{\hat{A}} \leqq e^{d\Delta/\sigma^2} \quad \text{and} \quad \frac{B}{\hat{B}} \leqq e^{d\Delta/\sigma}$$

respectively, where $d = a_2 - a_1$.

Since $\frac{d\Delta}{\sigma^2} > 0$, the above inequalities are certainly fulfilled when

$$\frac{B}{\bar{B}} \leqq 1 \quad \text{and} \quad \frac{A}{\bar{A}} \leqq 1. \tag{2.1}$$

In what follow in this paper, we shall restrict ourselves to cases where acceptance of both θ_1 and θ_4 is impossible, even if this is not stated explicitly.

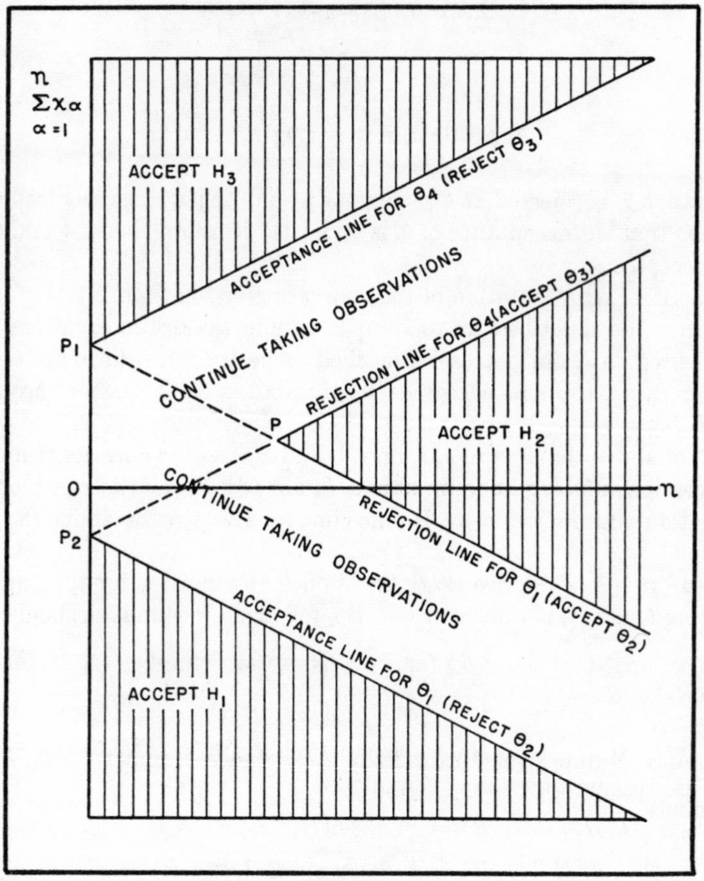

FIGURE 1

3. Derivation of OC Functions. Let $L(H_i \mid \theta, R)$ denote the probability of accepting H_i when θ is the true mean and R is the sequential rule used. Let H_{θ_i} denote the hypothesis that $\theta = \theta_i$. Since, as shown above, H_1 is accepted if and only if θ_1 is accepted, we have

$$L(H_1 \mid \theta, R) = L(H\theta_1 \mid \theta, R_1). \tag{3.1}$$

Similarly,

$$L(H_3 \mid \theta, R) = L(H_{\theta_4} \mid \theta, R_2). \tag{3.2}$$

From the fact that R_1 and R_2 each terminate at some finite stage with probability one, it follows that R will terminate at some finite stage with probability one. Hence

$$L(H_2 \mid \theta, R) = 1 - L(H_1 \mid \theta, R) - L(H_3 \mid \theta, R). \tag{3.3}$$

From pp. 50–52 of [1], the following equations are obtained.

$$L(H_1 \mid \theta, R) = L(H_{\theta_1} \mid \theta, R_1) = \frac{A^{h_1} - 1}{A^{h_1} - B^{h_1}} \tag{3.4}$$

where

$$h_1 = h_1(\theta) = \frac{\theta_2 + \theta_1 - 2\theta}{\theta_2 - \theta_1} = \frac{a_1 - \theta}{\frac{\Delta}{2}}$$

and

$$L(H_{\theta_3} \mid \theta, R_2) = \frac{\hat{A}^{h_2} - 1}{\hat{A}^{h_2} - \hat{B}^{h_2}} \tag{3.5}$$

where

$$h_2 = h_2(\theta) = \frac{\theta_4 + \theta_3 - 2\theta}{\theta_4 - \theta_3} = \frac{a_2 - \theta}{\frac{\Delta}{2}}.$$

These equations involve an approximation, as explained in [1].

Hence

$$L(H_3 \mid \theta, R) = L(H_{\theta_4} \mid \theta, R_2) = 1 - L(H_{\theta_3} \mid \theta, R_2) = \frac{1 - \hat{B}^{h_2}}{\hat{A}^{h_2} - \hat{B}^{h_2}} \tag{3.6}$$

and

$$L(H_2 \mid \theta, R) = 1 - \frac{A^{h_1} - 1}{A^{h_1} - B^{h_1}} - \frac{1 - \hat{B}^{h_2}}{\hat{A}^{h_2} - \hat{B}^{h_2}} = \frac{1 - B^{h_1}}{A^{h_1} - B^{h_1}} - \frac{1 - \hat{B}^{h_2}}{\hat{A}^{h_2} - \hat{B}^{h_2}}. \tag{3.7}$$

Since $L(H_1 \mid \theta, R) = L(H_{\theta_1} \mid \theta, R_1)$, it follows that $L(H_1 \mid \theta, R)$ is a monotonically decreasing function of θ and that

$$L(H_1 \mid -\infty, R) = 1; \qquad L(H_1 \mid \infty, R) = 0$$

$$L(H_1 \mid \theta_1, R) = 1 - \alpha; \qquad L(H_1 \mid \theta_2, R) = \beta$$

$$L(H_1 \mid a_1, R) = \frac{\log A}{\log A + |\log B|}.$$

Similarly, since $L(H_3 \mid \theta, R) = 1 - L(H_{\theta_3} \mid \theta, R_2)$, it follows that $L(H_3 \mid \theta, R)$ is a monotonically increasing function of θ and that

$$L(H_3 \mid -\infty, R) = 0; \qquad L(H_3 \mid \infty, R) = 1$$

$$L(H_3 \mid \theta_3, R) = \hat{\alpha}; \qquad L(H_3 \mid \theta_4, R) = 1 - \hat{\beta}$$

$$L(H_3 \mid a_2, R) = \frac{|\log \hat{B}|}{\log \hat{A} + |\log \hat{B}|}.$$

Since $L(H_2 \mid \theta, R) = 1 - L(H_1 \mid \theta, R) - L(H_3 \mid \theta, R)$ it follows easily from the above results that

$$L(H_2 \mid -\infty, R) = 0; \qquad L(H_2 \mid \infty, R) = 0$$

$$L(H_2 \mid \theta, R) < \alpha \quad \text{for} \quad \theta < \theta_1; \qquad L(H_2 \mid \theta, R) < \hat{\beta} \quad \text{for} \quad \theta > \theta_4$$

$$\frac{|\log B|}{\log A + |\log B|} - \hat{\alpha} < L(H_2 \mid a_1, R) < \frac{|\log B|}{\log A + |\log B|}$$

$$\frac{\log \hat{A}}{\log \hat{A} + |\log \hat{B}|} - \beta < L(H_2 \mid a_2, R) < \frac{\log \hat{A}}{\log \hat{A} + |\log \hat{B}|}$$

$$1 - \beta - \hat{\alpha} < L(H_2 \mid \theta, R) < 1 \quad \text{for} \quad \theta_2 \leqq \theta \leqq \theta_3.$$

4. Probability of Correct Decision. Denote the probability of a correct decision by $L(\theta/R)$. It is defined as follows:

Interval	Correct Decisions	$L(\theta|R)$
$\theta \leqq \theta_1$	acceptance of H_1	$L(H_1 \mid \theta, R)$
$\theta_1 < \theta < \theta_2$	acceptance of H_1 or H_2	$L(H_1 \mid \theta, R) + L(H_2 \mid \theta, R)$
$\theta_2 \leqq \theta \leqq \theta_3$	acceptance of H_2	$L(H_2 \mid \theta, R)$
$\theta_3 < \theta < \theta_4$	acceptance of H_2 or H_3	$L(H_2 \mid \theta, R) + L(H_3 \mid \theta, R)$
$\theta_4 \leqq \theta$	acceptance of H_3	$L(H_3 \mid \theta, R)$

It should be noted that at points of discontinuity, $L(\theta, \mid R)$ is defined as the smaller of the two limiting values.

We shall now discuss some monotonicity properties of the function $L(\theta \mid R)$. From the fact that $L(H_{\theta_1} \mid \theta, R_1)$ and $L(H_{\theta_3} \mid \theta, R_2)$ are continuous with continuous first and second derivatives and are monotonically decreasing for all θ with a single point of inflection in the intervals $\theta_1 < \theta < \theta_2$ and $\theta_3 < \theta < \theta_4$ respectively, it follows that

(i) $L(\theta \mid R)$ is monotonically decreasing with negative curvature for $-\infty < \theta \leqq \theta_1$.

(ii) $L(\theta \mid R)$ is monotonically increasing with negative curvature for $\theta_4 \leqq \theta < \infty$.

Making use of (3.3) we have further

(iii) $L(\theta \mid R)$ is monotonically decreasing with negative curvature for $\theta_1 < \theta < \theta_2$.

(iv) $L(\theta \mid R)$ is monotonically increasing with negative curvature for $\theta_3 < \theta < \theta_4$.

(v) For $\theta_2 \leqq \theta \leqq \theta_3$, $\dfrac{d}{d\theta} L(\theta \mid R) = -\left[\dfrac{d}{d\theta} L(H_1 \mid \theta, R) + \dfrac{d}{d\theta} L(H_3 \mid \theta, R)\right]$ is decreasing, since $\dfrac{d}{d\theta} L(H_1 \mid \theta, R)$ and $\dfrac{d}{d\theta} L(H_3 \mid \theta, R)$ are increasing. In other words $L(\theta \mid R)$ has negative curvature for $\theta_2 \leqq \theta \leqq \theta_3$.

In the special case when $A = \hat{A} = \frac{1}{B} = \frac{1}{\hat{B}}$ and the origin is taken at $\frac{a_1 + a_2}{2}$ for the sake of convenience, it is easy to see that $L(\theta \mid R)$ is symmetric with respect to the origin and, because of (v), has a local maximum at $\theta = 0$.

5. Choice of the constants A, B, $\hat{A}$, $\hat{B}$ to insure prescribed Lower Bounds for $L(\theta \mid R)$. We shall deal here with the question of choosing A, B, $\hat{A}$ and $\hat{B}$ such that $L(\theta \mid R) \geqq 1 - \gamma_1$ when $\theta \leqq \theta_1$, $L(\theta \mid R) \geqq 1 - \gamma_2$ when $\theta_1 < \theta < \theta_4$, and $L(\theta \mid R) \geqq 1 - \gamma_3$ when $\theta \geqq \theta_4$. From the monotonic properties of the correct decision function it is only necessary to insure that

$$(5.1) \quad L(\theta_1 \mid R) = 1 - \gamma_1, L(\theta_2 \mid R) = L(\theta_3 \mid R) = 1 - \gamma_2 \text{ and } L(\theta_4 \mid R) = 1 - \gamma_3.$$

The following relations will be needed:

$$h_1(\theta_1) = h_2(\theta_3) = 1 = -h_1(\theta_2) = -h_2(\theta_4)$$

$$h_2(\theta_2) = \frac{\theta_3 + \theta_4 - 2\theta_2}{\Delta} = \frac{d - \frac{\Delta}{2}}{\frac{\Delta}{2}} = r \quad \text{(say)}$$

$$h_1(\theta_3) = \frac{\theta_1 + \theta_2 - 2\theta_3}{\Delta} = \frac{-d + \frac{\Delta}{2}}{\frac{\Delta}{2}} = -r$$

where $d = \theta_4 - \theta_2 = \theta_3 - \theta_1 = a_2 - a_1$.

The following four equations are obtained from (5.1):

$$(5.2) \quad 1 - L(H_1 \mid \theta_1, R) = L(H_{\theta_2} \mid \theta_1, R_1) = \frac{1 - B}{A - B} = \gamma_1$$

$$(5.3) \quad \begin{aligned} 1 - L(H_2 \mid \theta_2, R) &= L(H_1 \mid \theta_2, R) + L(H_3 \mid \theta_2, R) \\ &= \frac{B(A - 1)}{A - B} + \left[\frac{1 - \hat{B}^r}{\hat{A}^r - \hat{B}^r}\right] = \gamma_2 \end{aligned}$$

$$(5.4) \quad \begin{aligned} 1 - L(H_2 \mid \theta_3, R) &= L(H_3 \mid \theta_3, R) + L(H_1 \mid \theta_3, R) \\ &= \frac{1 - \hat{B}}{\hat{A} - \hat{B}} + \left[\frac{B^r(A^r - 1)}{A^r - B^r}\right] = \gamma_2 \end{aligned}$$

$$(5.5) \quad 1 - L(H_3 \mid \theta_4, R) = L(H_{\theta_3} \mid \theta_4, R_2) = \frac{\hat{B}(\hat{A} - 1)}{\hat{A} - \hat{B}} = \gamma_3.$$

The "bracketed terms" represent quantities less than $\hat{\alpha}$ and β respectively and if r is sufficiently large they can be neglected. This will be made more precise but first let us note the results of neglecting the bracketed terms.

From (5.2) and (5.3) we obtain

$$(5.6) \qquad B(1 - \gamma_1) = \gamma_2, \quad \text{whence} \quad B = \frac{\gamma_2}{1 - \gamma_1}.$$

From (5.2) and (5.6)

$$(5.7) \qquad A = \frac{1 - B(1 - \gamma_1)}{\gamma_1} \quad \text{whence} \quad A = \frac{1 - \gamma_2}{\gamma_1}.$$

Since the last two equations are obtained from the first two by the permutation $A \to \hat{A}$, $B \to \hat{B}$, $\gamma_1 \to \gamma_2$, $\gamma_2 \to \gamma_3$, we have

$$\hat{B} = \frac{\gamma_3}{1 - \gamma_2}$$

$$\hat{A} = \frac{1 - \gamma_3}{\gamma_2}.$$

If $\gamma_1 = \gamma_2 = \gamma_3 = \gamma$ (say) then $A = \hat{A} = \dfrac{1}{B} = \dfrac{1}{\hat{B}} = \dfrac{1 - \gamma}{\gamma}$.

We shall consider the bracketed quantities negligible if the result of neglecting them produces a change of less than 20% in $[1 - L(\theta \mid R)]$ at $\theta = \theta_2, \theta_3$ respectively, i.e., if

$$(5.8) \qquad \frac{1 - \hat{B}^r}{\hat{A}^r - {}^r\hat{B}} = \frac{1 - \left(\dfrac{\gamma_3}{1 - \gamma_2}\right)^r}{\left(\dfrac{1 - \gamma_3}{\gamma_2}\right)^r - \left(\dfrac{\gamma_3}{1 - \gamma_2}\right)^r} \leqq \frac{\gamma_2}{5}$$

and

$$(5.9) \qquad \frac{B^r(A^r - 1)}{A^r - B^r} = \frac{\left(\dfrac{\gamma_2}{1 - \gamma_1}\right)^r \left[\left(\dfrac{1 - \gamma_2}{\gamma_1}\right)^r - 1\right]}{\left(\dfrac{1 - \gamma_2}{\gamma_1}\right)^r - \left(\dfrac{\gamma_2}{1 - \gamma_1}\right)^r} \leqq \frac{\gamma_2}{5}.$$

Inequality, (5.9) can be written as

$$\frac{\gamma_2^r[(1 - \gamma_2)^r - \gamma_1^r]}{(1 - \gamma_2)^r(1 - \gamma_1)^r - \gamma_1^r \gamma_2^r} \leqq \frac{\gamma_2}{5}$$

or

$$(1 - \gamma_2)^r \left[\gamma_2^r - \frac{\gamma_2}{5}(1 - \gamma_1)^r\right] \leqq (\gamma_1 \gamma_2)^r \left(1 - \frac{\gamma_2}{5}\right).$$

This will certainly hold if

$$\gamma_2^r \leqq \frac{\gamma_2}{5}(1 - \gamma_1)^r$$

or if

$$\left(\frac{\gamma_2}{1 - \gamma_1}\right)^r \leqq \frac{\gamma_2}{5}.$$

Assume that γ_1, γ_2 and γ_3 are each less than $\frac{1}{2}$. Then the last inequality can be written as

$$(5.10) \qquad r \geqq \frac{\log\left(\frac{5}{\gamma_2}\right)}{\log\left(\frac{1-\gamma_1}{\gamma_2}\right)}.$$

Starting with (5.8) the same relation is obtained except that γ_1 is replaced by γ_3, namely

$$(5.11) \qquad r \geqq \frac{\log \frac{5}{\gamma_2}}{\log \frac{1-\gamma_3}{\gamma_2}}.$$

Let

$$k = \frac{\log \frac{5}{\gamma_2}}{\log \frac{1-\bar{\gamma}}{\gamma_2}}$$

where $\bar{\gamma}$ is the larger of γ_1 and γ_3. Then k is the larger of the right hand members of (5.10) and (5.11). Then for (5.8) and (5.9) to hold it is sufficient that

$$r \geqq k.$$

If $\gamma_2 = .05$ and $0 < \gamma_1, \gamma_3 < .1$ then k is approximately $\frac{2}{1.3} = 1.54$. If $\gamma_2 = .01$ and $0 < \gamma_1, \gamma_3 < .1$ then k is approximately $\frac{2.7}{2} = 1.35$.

We shall now investigate under what conditions the approximate solution obtained above for A, B, $\hat{A}$, $\hat{B}$ are such that acceptance of both θ_1 and θ_4 is impossible.

It follows from (2.1) that the following pair of inequalities are sufficient for the impossibility of accepting both θ_1 and θ_4:

$$(5.12) \qquad \frac{A}{\hat{A}} = \frac{\gamma_2}{\gamma_1}\frac{1-\gamma_2}{1-\gamma_3} \leqq 1; \qquad \frac{B}{\hat{B}} = \frac{\gamma_2}{\gamma_3}\frac{1-\gamma_2}{1-\gamma_1} \leqq 1.$$

If $\gamma_1 \neq \gamma_3$ let the smaller and larger of the pair (γ_1, γ_3) be denoted by $\underline{\gamma}$ and $\bar{\gamma}$ respectively. Since $1 - \underline{\gamma} > 1 - \bar{\gamma}$, then

$$\frac{\gamma_2(1-\gamma_2)}{\bar{\gamma}(1-\underline{\gamma})} < \frac{\gamma_2(1-\gamma_2)}{\underline{\gamma}(1-\bar{\gamma})}$$

and we need only consider one of the two inequalities in (5.12). The condition $\gamma_2 < \underline{\gamma}$ will in general satisfy (5.12). More precisely if all the γ's are restricted to the interval (0, .1) then

$$\frac{9}{10} \leqq \frac{1-\gamma_2}{1-\underline{\gamma}} < \frac{1-\gamma_2}{1-\bar{\gamma}} \leqq \frac{10}{9}$$

and it is sufficient for the validity of (5.12) that $\gamma_2 \leqq (.9)\,\underline{\gamma}$.

If $\gamma_1 = \gamma_3 = \gamma$ (say) then the two inequalities reduce to one

$$\gamma_2^2 - \gamma_2 + \gamma - \gamma^2 \geqq 0$$

which can be written as

$$(\gamma_2 - \gamma)(\gamma_2 - 1 + \gamma) \geqq 0.$$

Since the inequality $\gamma_2 \geqq 1 - \gamma$ is impossible when all γ's are $< \frac{1}{2}$, we see that $\gamma_2 \leq \gamma$ is sufficient for the validity of (5.12) when $\gamma_1 = \gamma_3 = \gamma < \frac{1}{2}$.

There remains the problem of finding an approximate solution for equations (5.2) to (5.5) when $r < k$. Since

$$r = \frac{d - \frac{\Delta}{2}}{\frac{\Delta}{2}} = \frac{\theta_3 - \theta_2 + \frac{\Delta}{2}}{\frac{\Delta}{2}} \geqq 1$$

we merely have to consider the interval $1 \leqq r < k$.

The following approximations are used

$$(5.13) \qquad \begin{aligned} &\frac{1 - B}{A - B} \sim \frac{1}{A}; \quad &&\frac{B(A - 1)}{A - B} \sim B; \quad &&\frac{1 - \hat{B}^r}{\hat{A}^r - \hat{B}^r} \sim \frac{1}{\hat{A}^r} \\ &\frac{1 - \hat{B}}{\hat{A} - \hat{B}} \sim \frac{1}{\hat{A}}; \quad &&\frac{B^r(A^r - 1)}{A^r - B^r} \sim B^r; \quad &&\frac{\hat{B}(\hat{A} - 1)}{\hat{A} - \hat{B}} \sim \hat{B}, \end{aligned}$$

which upon substitution yield

$$(5.14) \qquad A = \frac{1}{\gamma_1}$$

$$(5.15) \qquad \hat{B} = \gamma_3$$

$$(5.16) \qquad B + \frac{1}{\hat{A}^r} = \gamma_2$$

$$(5.17) \qquad \frac{1}{\hat{A}} + B^r = \gamma_2 .$$

Subtraction of (5.17) from (5.16) shows that $B = \frac{1}{\hat{A}}$ is a solution. Substituting this result back in (5.16) leads to the equation

$$(5.18) \qquad B + B^r = \gamma_2 .$$

It can easily be verified that between zero and unity this equation has exactly one root. Since $1 \leqq r < \infty$, the root of the above equation lies between $\frac{\gamma_2}{2}$ and γ_2.

Taking γ_2 as a first approximation for B and substituting $\gamma_2 + \epsilon$ for B in (5.18), we obtain

$$\epsilon + (\gamma_2 + \epsilon)^r = 0.$$

Expanding $(\gamma_2 + \epsilon)^r$ in a power series in ϵ and neglecting second and higher order terms, the above equation gives

$$\epsilon \sim \frac{\gamma_2^r}{1 + r\gamma_2^{r-1}}.$$

Thus,

$$B = \frac{1}{A} \sim \gamma_2 - \frac{\gamma_2^r}{1 + r\gamma_2^{r-1}} = \frac{\gamma_2[1 + (r-1)\gamma_2^{r-1}]}{1 + r\gamma_2^{r-1}}. \tag{5.19}$$

It is necessary to investigate under what conditions the above approximate solution satisfies (5.2) to (5.5) to within a 20% error in $[1 - L(\theta/R)]$, i.e., such that

$$-\frac{\gamma_1}{5} < \frac{\gamma_1(1-B)}{1-\gamma_1 B} - \gamma_1 < \frac{\gamma_1}{5} \tag{5.20}$$

$$-\frac{\gamma_3}{5} < \frac{\gamma_3(1-B)}{1-\gamma_3 B} - \gamma_3 < \frac{\gamma_3}{5} \tag{5.21}$$

$$-\frac{\gamma_2}{5} < \frac{B(1-\gamma_1)}{1-\gamma_1 B} + \frac{B^r(1-\gamma_3^r)}{1-(\gamma_3 B)^r} - \gamma_2 < \frac{\gamma_2}{5} \tag{5.22}$$

$$-\frac{\gamma_2}{5} < \frac{B(1-\gamma_3)}{1-\gamma_3 B} + \frac{B^r(1-\gamma_1^r)}{1-(\gamma_1 B)^r} - \gamma_2 < \frac{\gamma_2}{5} \tag{5.23}$$

where for B the value in (5.19) is understood.

It can be shown that if γ_1, γ_2, γ_3, are each between zero and .1 then the inequalities (5.20) to (5.23) hold. Furthermore it can be shown that if, in addition $\gamma_2 \leqq \min(\gamma_1, \gamma_3)$ then also the inequalities (2.1) hold. The latter inequalities are sufficient to ensure the impossibility of accepting both θ_1 and θ_4.

6. Bounds for the ASN Function. First we shall derive lower bounds for the ASN function. Let $E(n/\theta, R)$ denote the expected value of n when θ is the true mean and R is the sequential rule employed. For $\theta < \theta_2$ the probability of coming to a decision first with R_2 is large and therefore

$$E(n/\theta, R) \sim E(n/\theta, R_1)\ \theta < \theta_2.$$

From the definition of R it follows that

$$E(n/\theta, R) > E(n/\theta, R_1) \qquad \text{for all } \theta.$$

Hence $E(n/\theta, R_1)$ serves as a close lower bound when $\theta < \theta_2$.

Similarly

$$E(n/\theta, R) \sim E(n/\theta, R_2) \qquad \text{for } \theta > \theta_3$$

$$E(n/\theta, R) > E(n/\theta, R_2) \qquad \text{for all } \theta.$$

Hence $E(n/\theta, R_2)$ serves as a close lower bound for $\theta > \theta_3$.

Combining the above we have

$$E(n/\theta, R) > \text{Max}\,[E(n/\theta, R_1)\ ,\ E(n/\theta, R_.)] \tag{6.1}$$

where, neglecting the excess over the boundary,

$$E(n/\theta, R_1) = \frac{L(H_{\theta_1}/\theta, R_1) \log B + L(H_{\theta_2}/\theta, R_1) \log A}{\frac{\Delta}{\sigma^2}(\theta - a_1)} \tag{6.2}$$

$$E(n/\theta, R_2) = \frac{L(H_{\theta_3}/\theta, R_2) \log \hat{B} + L(H_{\theta_4}/\theta, R_2) \log \hat{A}}{\frac{\Delta}{\sigma_2}(\theta - a_2)} \tag{6.3}$$

Formula (6.1) gives a valid lower bound over the whole range of θ, but this lower bound will not be very close in the interval (θ_2, θ_3), particularly in the neighbourhood of the mid-point $\frac{\theta_2 + \theta_3}{2}$. The authors were not able to find any simple method for obtaining a closer lower bound in this interval. The upper bound given later in this section will, however, be fairly close also in the interval (θ_2, θ_3) and can be used as an approximation to the exact value.

We shall now derive upper bounds for the ASN function. Let R_1^* be the following rule: "Continue to take observations until R_1 accepts θ_1." Since this implies the rejection of θ_4 at the same or at a previous stage, it follows that R must terminate not later than R_1^*. Hence

$$E(n/\theta, R_1^*) \geqq E(n/\theta, R). \tag{6.4}$$

As a matter of fact one can easily verify that $E(n/\theta, R_1^*) > E(n/\theta, R)$. Thus $E(n/\theta, R_1^*)$ is an upper bound for $E(n/\theta, R)$. This upper bound will be close when the probability of accepting θ_1 is high, i.e., for $\theta \leqq \theta_1$.

By the general formula

$$E(n) = \frac{E\left(\sum_{i=1}^{n} z_i\right)}{E(z)}$$

(see p. 53 [1]) we obtain, upon neglecting the excess over the boundary,

$$E(n/\theta, R_1^*) = \frac{\log B}{\frac{\Delta}{\sigma^2}(\theta - a_1)}. \tag{6.5}$$

This coincides with (6.2) when $L(H_{\theta_2}/\theta, R_1) = 0$.

Similarly, if R_2^* denotes the rule of continuing until R_2 accepts θ_4, then

$$E(n/\theta, R_2^*) > E(n/\theta, R) \tag{6.6}$$

$$E(n/\theta, R_2^*) = \frac{\log \hat{A}}{\frac{\Delta}{\sigma^2}(\theta - a_2)} \tag{6.7}$$

and this will be a close upper bound for $\theta \geqq \theta_4$.

If $A = \hat{A} = \frac{1}{B} = \frac{1}{\hat{B}}$ and if $a_1 + a_2 = 0$ the above results reduce to

$$(6.8)\qquad E(n/\theta, R) \gtrsim E(n/\theta, R_1^*) = \frac{-h}{\lambda + \theta} \qquad \text{for} \qquad \theta \leqq \theta_1$$

$$(6.9)\qquad E(n/\theta, R) \gtrsim E(n/\theta, R_2^*) = \frac{h}{\theta - \lambda} \qquad \text{for} \qquad \theta \geqq \theta_4$$

where the symbol $\gtrsim$ stands for a close inequality, and where

$$h = \frac{\sigma^2}{\Delta} \log A \qquad \text{and} \qquad \lambda = a_2 = -a_1 .$$

To establish an upper bound for the ASN function in the interval $\theta_2 < \theta < \theta_3$ we shall restrict ourself to the case where $A = \hat{A} = \frac{1}{B} = \frac{1}{\hat{B}}$. These relations are fulfilled by the approximate values of A, B, $\hat{A}$, $\hat{B}$ suggested in section 5 when $\gamma_1 = \gamma_2 = \gamma_3$ and $r \geqq k$. We shall choose the origin to be at $\frac{a_1 + a_2}{2}$, i.e., we put $\frac{a_1 + a_2}{2} = 0$. Then the vertex P of the triangle (P_1, P_2, P) in diagram 1 lies on the abscissa axis and $OP_1 = OP_2 = h$. The abscissa of the vertex P is $\frac{h}{\lambda} = N$ (say) where $\lambda = a_2 = -a_1$. Let $y = \sum_{i=1}^{N} X_i$ represent the sum of the first N observations. Let R_{23} denote the rule: "Continue until both θ_2 and θ_3 are accepted". This is tantamount to neglecting the two outer lines in diagram 1, i.e., the acceptance lines for θ_1 and θ_4. Then clearly,

$$(6.10)\qquad E(n/\theta, R_{23}) > E(n/\theta, R).$$

When θ lies between θ_2 and θ_3 this inequality will be close, since the probability of crossing either of the two outer lines is then small.

However $E(n/\theta, R_{23})$ was found difficult to compute and it was necessary to consider instead the rule R'_{23} : "Take N observations. If $y = \sum_{i=1}^{N} X_i < 0$ then continue until θ_2 is accepted. If $y > 0$ then continue until θ_3 is accepted".[3] Clearly,

$$(6.11)\qquad E(n/\theta, R'_{23}) > E(n/\theta, R_{23}).$$

This inequality, however, will be close only if the probability of concluding the test before N observations, given that $\theta_2 < \theta < \theta_3$, is small.

Some investigations by the authors seem to indicate that the inequality (6.11) will be close when $\Delta < \lambda$. This inequality is likely to be fulfilled in practical problems.

We shall now proceed to determine the value of $E(n/\theta, R'_{23})$. Neglecting the excess over the boundary, we have

$$(6.12)\qquad E\left(n/\theta, R'_{23}, \sum_{i=1}^{N} x_i = y\right) = \frac{h}{\lambda} + \frac{y}{\lambda - \theta} \qquad \text{for} \qquad y > 0$$

[3] The event $y = 0$ has probability zero and it is indifferent what rule is adopted for that case.

and

$$(6.13) \qquad E\left(n/\theta, R'_{23}, \sum_{i=1}^{N} x_i = y\right) = \frac{h}{\lambda} - \frac{y}{\lambda + \theta} \qquad \text{for} \qquad y < 0$$

where, for any condition C, $E(n/\theta, R, C)$ denotes the conditional expected value of n given that the true mean is θ, that R is the sequential rule used and that the condition C is fulfilled.

Multiplying with the density of y and then integrating with respect to y, we obtain after simplification

$$(6.14) \quad E(n/\theta, R'_{23}) = \frac{1}{\lambda^2 - \theta^2}\left[h\lambda + 2h\theta\phi\left(\frac{\theta}{\sigma}\sqrt{\frac{h}{\lambda}}\right) + 2\sigma\sqrt{\frac{h\lambda}{2\pi}}\, e^{-(h\theta^2/2\lambda\sigma^2)}\right]$$

where $\phi(x) = \int_0^x \frac{e^{-(y^2/2)}}{\sqrt{2\pi}}\, dy$, and $\theta_2 < \theta < \theta_3$.

In particular, for $\theta = 0$ we get

$$(6.15) \qquad E(n/\theta = 0, R'_{23}) = \frac{h}{\lambda} + \frac{\sigma}{\lambda^2}\sqrt{\frac{2h\lambda}{\pi}}.$$

To establish a close upper bound for $\theta_3 < \theta < \theta_4$ we must bring the line of acceptance of θ_4 into account. The line of acceptance of θ_1 can be disregarded since the probability of accepting θ_1 is very small.

We therefore define the rule R_{34} as follows:

"Continue with R_1 until θ_2 is accepted and with R_2 until either θ_3 or θ_4 is accepted."

Since the ASN function for R_{34} is difficult to compute we define a modified rule R'_{34} as follows:

"Proceed to take $N\left(= \frac{h}{\lambda}\right)$ observations without regard to any rule. If $y = \sum_{i=1}^{N} X_i < 0$ then continue only with R_1 until θ_2 is accepted. If $0 < y < 2h$ then continue only with R_2 until either θ_3 or θ_4 is accepted. If $y \geqq 2h$ then stop taking observations and accept H_3."

It is clear that the following inequalities hold

$$(6.16) \qquad E(n/\theta, R'_{34}) > E(n/\theta, R_{34}) > E(n/\theta, R).$$

The proximity of $E(n/\theta, R_{34})$ and $E(n/\theta, R)$, as stated above, is based on the fact that the probability of accepting θ_1, when $\theta_3 < \theta < \theta_4$, is small.

The proximity of $E(n/\theta, R_{34})$ and $E(n/\theta, R'_{34})$ is assured if the probability of terminating with R_{34} (and with R) before N observations is small. It can be shown that the latter condition is fulfilled when $\Delta < \lambda$. In terms of the quantity r defined in Section 5 this can be written as $r > 3$.

To determine the value of $E(n/\theta, R'_{34})$ the following two preliminary results will be needed:

If $0 < y < 2h$,

$$E\left(n/\theta, R'_{34}, \sum_{i=1}^{N} x_i = y\right) = \frac{h}{\lambda} + \frac{2h - y - 2h\left[\dfrac{1 - e^{-(2/\sigma^2)(\lambda-\theta)(2h-y)}}{1 - e^{-(4h/\sigma^2)(\lambda-\theta)}}\right]}{\theta - \lambda} \tag{6.17}$$

$$= C \text{ (say)}.$$

If $y < 0$,

$$E\left(n/\theta, R'_{34}, \sum_{i=1}^{N} x_i = y\right) = \frac{h}{\lambda} - \frac{y}{\lambda + \theta} = D \text{ (say)}. \tag{6.18}$$

Both are easily obtained from formula (7.25) on p. 123 of [1].
Multiplying with the density of y and integrating with respect to y, we obtain after simplification

$$\begin{aligned} E(n/\theta, R'_{34}) = \frac{h}{\lambda} &+ \left[\phi\left(\frac{2\lambda - \theta}{\sigma}\sqrt{\frac{h}{\lambda}}\right) + \phi\left(\frac{\theta}{\sigma}\sqrt{\frac{h}{\lambda}}\right)\right] \\ &\cdot \frac{h}{(\lambda - \theta)}\left(\frac{\theta}{\lambda} - \frac{2e^{-(2h(\lambda-\theta)/\sigma^2)}}{1 + e^{-2h(\lambda-\theta)/\sigma^2}}\right) \\ &+ \frac{\sigma}{\lambda(\lambda - \theta)}\sqrt{\frac{h\lambda}{2\pi}}\left[e^{-(h\theta^2/2\lambda\sigma^2)} - e^{-h(2\lambda-\theta)^2/2\lambda\sigma^2}\right] \\ &- \frac{h\theta}{2\lambda(\lambda + \theta)}\left[1 - 2\phi\left(\frac{\theta}{\sigma}\sqrt{\frac{h}{\lambda}}\right)\right] + \frac{\sigma}{\lambda(\lambda + \theta)}\sqrt{\frac{h\lambda}{2\pi}}\, e^{-(h\theta^2/2\lambda\sigma^2)}. \end{aligned} \tag{6.19}$$

Formula (6.19) is an improvement on (6.14) as it will give for any θ a smaller upper bound, but in the neighborhood of the origin the difference is insignificant.
For $\theta = \lambda$ we obtain from (6.19) using L'Hopital's rule

$$\begin{aligned} E(n/\lambda, R'_{34}) = \frac{h^2}{\sigma^2} &- \frac{h}{4\lambda\sigma^2}(4h\lambda - 3\sigma^2) \\ &\cdot \left[1 - 2\phi\left(\frac{\sqrt{h\lambda}}{\sigma}\right)\right] + \left(\frac{h\lambda + \sigma^2}{2\lambda^2\sigma}\right)\sqrt{\frac{h\lambda}{2\pi}}\, e^{-(h\lambda/2\sigma^2)}. \end{aligned} \tag{6.20}$$

If $\dfrac{\sqrt{h\lambda}}{\sigma} > 2.5$, the above formula can be approximated by

$$E(n/\lambda, R'_{34}) \sim \frac{h^2}{\sigma^2} + \frac{2\sigma}{\lambda^2}\sqrt{\frac{h\lambda}{2\pi}}\, e^{-(h\lambda/2\sigma^2)}. \tag{6.21}$$

Since the right hand member above lies between $\dfrac{h^2}{\sigma^2}$ and $(1.002)\,\dfrac{h^2}{\sigma^2}$ when $\dfrac{\sqrt{h\lambda}}{\sigma} > 2.5$ then for practical purposes

$$E(n/\lambda, R'_{34}) \sim \frac{h^2}{\sigma^2} \qquad \text{when } \left(\frac{\sqrt{h\lambda}}{\sigma} > 2.5\right). \tag{6.22}$$

An upper bound for $E(n/\theta, R)$ for $\theta_1 < \theta < \theta_2$ can be obtained by defining R_{12} and R'_{12} in an analagous way to R_{34} and R'_{34}. Because of reasons of symmetry, $E(n/\theta, R'_{12})$ can be obtained from (6.19) by replacing θ by $-\theta$.

The method used for obtaining upper bounds for $E(n/\theta, R)$ can easily be extended to the more general case when the equalities $A = \hat{A} = \dfrac{1}{B} = \dfrac{1}{\hat{B}}$ do not necessarily hold. However, the resulting formulas are more cumbersome and we shall merely give without proof the upper bound corresponding to (6.14). This upper bound becomes

$$E(n/\theta, R'_{23}) = N + \left(\frac{N\theta - h_3}{\lambda - \theta}\right)\left[\frac{1}{2} - \phi(a)\right] + \left(\frac{h_3 - N\theta}{\lambda + \theta}\right)\left[\frac{1}{2} - \phi(b)\right] + \sigma\sqrt{\frac{N}{2\pi}}\left[\frac{e^{-a^2/2}}{\lambda - \theta} + \frac{e^{-b^2/2}}{\lambda + \theta}\right]$$

where

$$h_{11} = \frac{\sigma^2}{\Delta}\log A \qquad h_{10} = \frac{\sigma^2}{\Delta}\log B$$

$$h_{21} = \frac{\sigma^2}{\Delta}\log \hat{A} \qquad h_{20} = \frac{\sigma^2}{\Delta}\log \hat{B}$$

$$a_2 = -a_1 = \lambda$$

$$N = \frac{h_{11} - h_{20}}{2\lambda}; \qquad a = \frac{h_3 - N\theta}{\sigma\sqrt{N}}; \qquad b = \frac{h_3 + N\theta}{\sigma\sqrt{N}}; \qquad h_3 = \frac{h_{11} + h_{20}}{2}.$$

7. An Example. We shall consider the following example

$\sigma^2 = 1$, $\theta_1 = -\frac{5}{16}$, $\theta_2 = -\frac{3}{16}$, $\theta_3 = \frac{3}{16}$, $\theta_4 = \frac{5}{16}$, $\gamma_1 = \gamma_2 = \gamma_3 = \gamma = .029$

then

$$A = \hat{A} = \frac{1}{B} = \frac{1}{\hat{B}} = \frac{1-\gamma}{\gamma} = 33.5 \qquad r = 7 \gg 3 > k \sim 1.47$$

and

$$h = \frac{\sigma^2}{\Delta}\log A = 28, \ \lambda = \frac{\theta_3 + \theta_4}{2} = \tfrac{1}{4}, \ \Delta = \theta_2 - \theta_1 = \theta_4 - \theta_3 = \tfrac{1}{8}.$$

Using formulas (6.1) and (6.7) the following upper and lower bounds were obtained

θ	$\frac{5}{16}$	$\frac{6}{16}$	$\frac{7}{16}$	$\frac{8}{16}$	$\frac{9}{16}$	$\frac{10}{16}$	$\frac{12}{16}$	$\frac{14}{16}$	$\frac{16}{16}$	$\frac{18}{16}$	$\frac{20}{16}$
Upper bound..........	448	224	149	112	89.6	74.7	56	44.8	37.3	32	28
Lower bound..........	421	224	149	112	89.6	74.7	56	44.8	37.3	32	28

Formulas (6.14) and (6.1) yield

θ	0	$\frac{1}{16}$	$\frac{2}{16}$	$\frac{3}{16}$
Upper Bound.........	146	163	229	450
Lower Bound..........	112	149	224	421

In the neighborhood of the origin the true value is very nearly the upper bound. From formulas (6.19), (6.22) and (6.1) we obtain

θ	$\frac{3}{16}$	$\frac{4}{16}$	$\frac{5}{16}$
Upper Bound..........	422	784.5	423
Lower Bound..........	421	784	421

As shown above for the end points of the indifference zone, (6.19) gives better results than (6.14) or (6.7). This is as it should be since (6.19) takes into account possibilities omitted in (6.14) and (6.7). The greater accuracy of (6.19) is offset by a slight increase in computation.

In the graph of the Bounds of the ASN function shown in Figure 2, a single curve is shown wherever the upper and lower bound are sufficiently close to each other.

Since (6.14) contains an even function of θ and since elsewhere the corresponding bounds are mirror images with respect to $\theta = 0$, the bounds for negative θ are exactly the same as those for the corresponding positive θ.

Consider the following non-sequential rule applied to our problem. With a fixed number N_0 of observations compute the mean $\bar{x}$ and accept H_1 if $\bar{x}$ falls in the interval $(-\infty, a_1)$, accept H_2 if $\bar{x}$ falls in $[a_1, a_2]$ and accept H_3 if $\bar{x}$ falls in (a_2, ∞). This is certainly a reasonable procedure. One can also verify that no other non-sequential rule exists that is uniformly better (for all possible values of θ) than the one under consideration.

The two decision procedures become comparable if we introduce the indifference zones and define a wrong decision in the non-sequential case exactly as was done for our sequential procedure (see Section 1).

For the non-sequential case (just as in the sequential case) the probability of a wrong decision will be discontinuous at θ_1, θ_2, θ_3 and θ_4. At each of these points there will be a left-sided and right-sided limit, different from each other. As in the sequential case we shall take the probability of a wrong decision at a discontinuity point to be equal to the larger of the left and right hand limits. One can easily verify that the maximum probability of a wrong decision occurs at $\theta = \theta_3$ (which is equal to the value at $\theta = \theta_2$).

We then determine N_0 by setting the maximum probability of a wrong decision equal to γ, i.e.

$$\phi\left(\frac{d - \Delta/2}{\sigma} \sqrt{N_0}\right) + \phi\left(\frac{\Delta}{2\sigma} \sqrt{N_0}\right) = 1 - \gamma. \tag{7.1}$$

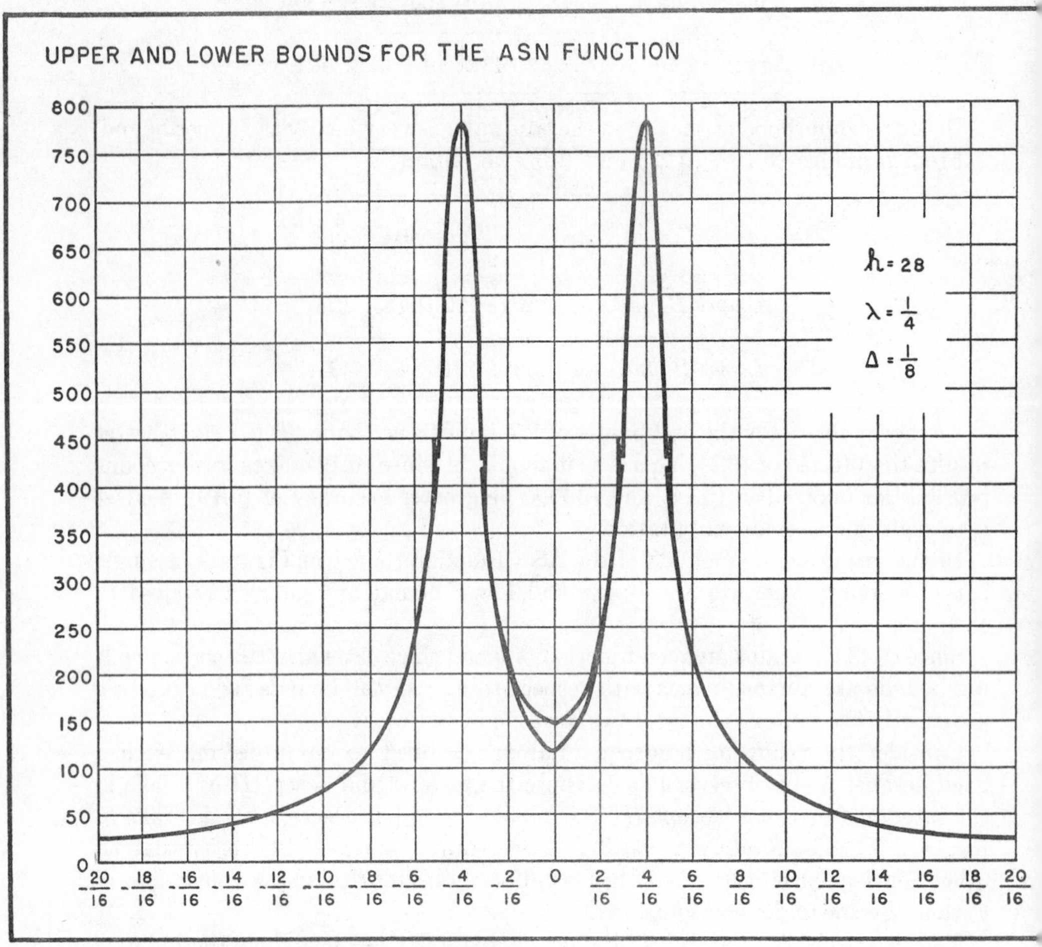

FIGURE 2

For the particular problem considered above, this gives $N_0 = 915.4$. Hence 916 observations are required in order to ensure that this non-sequential procedure will have the maximum probability $\gamma = .029$ of a wrong decision. This is to be compared with the maximum over all θ of the ASN function in the sequential procedure, which was 784.5.

Returning to (7.1) we shall derive lower and upper bounds for the root of that equation. Since

$$\infty > \frac{d - \Delta/2}{\sigma} \sqrt{N_0} \geqq \frac{\Delta}{2\sigma} \sqrt{N_0}$$

it is clear that the root of the equation

$$\phi\left(\frac{\Delta}{2\sigma}\sqrt{N_0}\right) + \phi\left(\frac{\Delta}{2\sigma}\sqrt{N_0}\right) = 1 - \gamma$$

is an upper bound for the root of (7.1) and that the root of the equation

$$\phi(\infty) + \phi\left(\frac{\Delta}{2\sigma}\sqrt{N_0}\right) = 1 - \gamma$$

or

$$\phi\left(\frac{\Delta}{2\sigma}\sqrt{N_0}\right) = \frac{1}{2} - \gamma$$

is a lower bound for the root of (7.1). We shall compare the value of $x = \frac{\Delta}{2\sigma}\sqrt{N_0}$ with the value of $y = \frac{\Delta}{2\sigma}\sqrt{\operatorname*{Max}_\theta \text{ASN}}$. Since

$$\operatorname*{Max}_\theta \text{(ASN function)} \sim \frac{h^2}{\sigma^2} = \frac{\sigma^2}{\Delta^2}\left(\log \frac{1-\gamma}{\gamma}\right)^2 \text{ (for sufficiently small } \frac{\Delta}{d}\text{)}.$$

then

$$y = \frac{\Delta}{2\sigma}\sqrt{\operatorname*{Max}_\theta \text{ASN}} \sim \frac{1}{2}\log\frac{1-\gamma}{\gamma} \text{ (for sufficiently small } \frac{\Delta}{d}\text{)}.$$

The following table gives upper and lower bounds for x and the corresponding value of y for the type of example under consideration, i.e., when $A = \hat{A} = \frac{1}{B} = \frac{1}{\hat{B}}$ and $r \geqq k$.

γ	.001	.002	.005	.008	.01	.05	.1
$\underline{x}$ and $\bar{x}$	3.08–3.31	2.87–3.10	2.57–2.81	2.41–2.65	2.33–2.58	1.64–1.96	1.28–1.65
y	3.45	3.11	2.65	2.41	2.30	1.47	1.10

As the table shows[4] for $.1 > \gamma > .008$

$$x > \underline{x} > y$$

[4] Actually, the inequality in question is shown only for the values of γ given in the table. However it can be verified that the inequality remains valid for all values of γ between .1 and .008.

and hence

$$N_0 > \underset{\theta}{\text{Max}}\ \text{ASN} \qquad \text{(for sufficiently small } \frac{\Delta}{d}\text{)}.$$

The statement and the table above are not meant to delimit the region in which the sequential rule is superior to the non-sequential procedure.

REFERENCES

[1] ABRAHAM WALD, *Sequential Analysis*, John Wiley and Sons, 1947.

[2] P. ARMITAGE, "Some sequential tests of student's hypothesis," *Supplement to the Journal of the Royal Statistical Society*, *Vol.* 9 (1947) No. 2, p. 250.

III. NOTE ON THE IDENTIFICATION OF ECONOMIC RELATIONS *

BY A. WALD

T. C. Koopmans and H. Rubin have discussed the problem of identification of economic relations in [II – 2], and have obtained a number of very interesting results. In this note the problem treated by Koopmans and Rubin is somewhat generalized and a different approach to its solution is briefly discussed.

1. *Definitions and Formulation of the Problem*

Let $x_1, \ldots, x_K$ be a set of K variables[1] and let $A = [\alpha_{gk}]$ $(g = 1, \ldots, G;\ k = 1, \ldots, K)$ be a given matrix of rank G. Denote the linear form $\sum_{k=1}^{K} \alpha_{gk} x_k$ by l_g $(g = 1, \ldots, G)$ and let $\Sigma = [\sigma_{gh}]$ $(g, h = 1, \ldots, G)$ be a given symmetric and positive definite matrix. Furthermore, let

$$\varphi_r(\alpha_{11}, \alpha_{12}, \ldots, \alpha_{GK}, \sigma_{11}, \sigma_{12}, \ldots, \sigma_{GG}) = 0 \qquad (r = 1, \ldots, R) \tag{1.1}$$

be a given system of equations, called a priori restrictions, that are satisfied by the quantities α_{gk} and σ_{gh}. For any nonsingular matrix $\Upsilon = [\upsilon_{gh}]$ $(g, h = 1, \ldots, G)$ we shall denote the matrix ΥA by $A(\Upsilon)$ and the elements of $A(\Upsilon)$ by $\alpha_{gk}(\Upsilon)$ $(g = 1, \ldots, G;$ $k = 1, \ldots, K)$. Thus, $\alpha_{gk}(\Upsilon) = \sum_{h=1}^{G} \upsilon_{gh}\alpha_{hk}$. Furthermore, we shall

---

[1] The integer K corresponds to what was denoted K_x in [II].

* Reprinted with permission from *Cowles Commission Monograph* 10 (1950, John Wiley & Sons, Inc.).

denote the matrix $\Upsilon \Sigma \Upsilon'$ (Υ' is the transpose of Υ) by $\Sigma(\Upsilon)$ and the elements of $\Sigma(\Upsilon)$ by $\sigma_{gh}(\Upsilon)$ $(g, h = 1, \ldots, G)$. Finally, the linear form $\sum_{h=1}^{G} \upsilon_{gh} l_h$ will be denoted by $l_g(\Upsilon)$.

DEFINITION 1.1. *A nonsingular matrix* $\Upsilon = [\upsilon_{gh}]$ $(g, h = 1, \ldots, G)$ *is said to be an admissible transformation if and only if the equations*

$$(1.2)\qquad \varphi_r\{\alpha_{11}(\Upsilon), \alpha_{12}(\Upsilon), \ldots, \alpha_{GK}(\Upsilon), \sigma_{11}(\Upsilon), \sigma_{12}(\Upsilon), \ldots, \sigma_{GG}(\Upsilon)\} = 0 \qquad (r = 1, \ldots, R)$$

are fulfilled.

DEFINITION 1.2. *An element* α_{gk} *of the matrix* A *is said to be identifiable*[1] *if* $\alpha_{gk}(\Upsilon)$ *takes only a finite number of different values over the domain of all admissible transformations* Υ. *Similarly, an element* σ_{gh} *of* Σ *is said to be identifiable if* $\sigma_{gh}(\Upsilon)$ *takes only a finite number of different values over the domain of all admissible transformations* Υ.

DEFINITION 1.3. *The linear form* l_g *is said to be identifiable if the coefficients* $\alpha_{g1}, \ldots, \alpha_{gK}$ *are identifiable.*

The matrix A has GK elements and the matrix Σ has $(G^2 + G)/2$ elements. Thus, the total number of elements in the two matrices A and Σ is equal to $GK + (G^2 + G)/2 = P$ (say). Consider the elements of A and Σ arranged in an ordered sequence and denote them by $\theta_1, \ldots, \theta_P$, respectively. The set $\theta = (\theta_1, \ldots, \theta_P)$ can be represented by a point in the P-dimensional space, called parameter space. For any nonsingular transformation Υ we shall denote the point $(\theta_1(\Upsilon), \ldots, \theta_P(\Upsilon))$ by $\theta(\Upsilon)$.

DEFINITION 1.4. *A coordinate* θ_p *of a point* θ *will be said to be locally identifiable if there exists an open set* ω *containing* θ *such that for any admissible transformation* Υ *either* $\theta_p(\Upsilon) = \theta_p$ *or* $\theta(\Upsilon)$ *lies outside* ω.

---

[1]This concept corresponds to what was called multiple identifiability in [II-2.4.4].

The problem considered in this note is to formulate conditions under which a coordinate θ_p of a point θ of the parameter space is identifiable or is locally identifiable.

2. *Two Lemmas*

In this section we shall prove two lemmas which will then be used for deriving necessary and sufficient conditions for the identification of θ_p.

Consider the quadratic form

$$X = \sum_{h=1}^{G} \sum_{g=1}^{G} \sigma^{gh} l_g l_h, \tag{2.1}$$

where $[\sigma^{gh}]$ is the inverse of $[\sigma_{gh}]$. Let ξ_{kl} $(k, l = 1, \ldots, K)$ denote the coefficient of $x_k x_l$ in X. For any nonsingular transformation Υ we shall put

$$X(\Upsilon) = \sum_{h=1}^{G} \sum_{g=1}^{G} \sigma^{gh}(\Upsilon) l_g(\Upsilon) l_h(\Upsilon), \tag{2.2}$$

where $[\sigma^{gh}(\Upsilon)]$ denotes the inverse of $[\sigma_{gh}(\Upsilon)]$. We shall denote by $\xi_{kl}(\Upsilon)$ the coefficient of $x_k x_l$ in $X(\Upsilon)$.

LEMMA 2.1. *For any nonsingular transformation Υ we have* $\xi_{kl}(\Upsilon) = \xi_{kl}$ $(k, l = 1, \ldots, K)$.

Proof: Denote by l the row vector $[l_1 \ldots l_G]$. Using matrix notation we can write

$$X = l\, \Sigma^{-1} l' \tag{2.3}$$

and

$$X(\Upsilon) = l(\Upsilon)\, \Sigma^{-1}(\Upsilon)\, l'(\Upsilon), \tag{2.4}$$

where l' is the transpose of l and Σ^{-1} is the inverse of Σ. We have

$$l'(\Upsilon) = \Upsilon\, l', \tag{2.5}$$

(2.6) $$l(\Upsilon) = l\,\Upsilon'\,,$$

and

(2.7) $$\Sigma^{-1}(\Upsilon) = \Upsilon'^{-1}\,\Sigma^{-1}\,\Upsilon^{-1}\,.$$

Hence, from (2.4) – (2.7) we obtain

(2.8) $$X(\Upsilon) = l\,\Upsilon'\,\Upsilon'^{-1}\,\Sigma^{-1}\,\Upsilon^{-1}\,\Upsilon\,l = l\,\Sigma^{-1}\,l' = X,$$

and Lemma 2.1 is proved.

Let $\theta^* = (\theta_1^*, \ldots, \theta_p^*)$ be a parameter point different from θ and denote by l_g^*, X^*, and ξ_{kl}^* the expressions we obtain from l_g, X, and ξ_{kl}, respectively, by substituting θ^* for θ. Now we shall prove the following lemma.

LEMMA 2.2. *If θ^* is a point such that $\xi_{kl}^* = \xi_{kl}$ $(k,\ l = 1, \ldots, K)$, then there exists a nonsingular transformation Υ such that*

(2.9) $$\xi_{kl}^* = \xi_{kl}(\Upsilon) \qquad (k,\ l = 1, \ldots, K).$$

Proof: From $\xi_{kl}^* = \xi_{kl}$ it follows that $X^* = X$ identically in $x_1, \ldots, x_K$. Thus we have

(2.10) $$X^* = \sum\sum \sigma^{*gh}\, l_g^*\, l_h^* = \sum\sum \sigma^{gh}\, l_g\, l_h = X.$$

Since $l_1, \ldots, l_G$ are independent linear forms and since $[\sigma_{gh}]$ is nonsingular, the rank of the quadratic form X is equal to G. Hence, the rank of X^* is also equal to G and, therefore, $l_1^*, \ldots, l_G^*$ are independent linear forms. From this and (2.10) it follows that each linear form l_g^* is a linear combination of the forms $l_1, \ldots, l_G$. Hence there exists exactly one nonsingular transformation Υ such that

(2.11) $$l_g(\Upsilon) = l_g^* \qquad (g = 1, \ldots, G).$$

From Lemma 2.1 it follows that

(2.12) $$\sum\sum \sigma^{gh}(\Upsilon)\, l_g(\Upsilon)\, l_h(\Upsilon) = \sum\sum \sigma^{gh}\, l_g\, l_h\,.$$

From (2.10), (2.11), and (2.12) we obtain

$$\sum\sum \sigma^{gh}(\Upsilon)\, l_g^*\, l_h^* = \sum\sum \sigma^{*gh}\, l_g^*\, l_h^* . \tag{2.13}$$

Hence

$$\sigma^{gh}(\Upsilon) = \sigma^{*gh} \qquad (g,\ h = 1,\ \ldots,\ G), \tag{2.14}$$

and, therefore,

$$\sigma_{gh}(\Upsilon) = \sigma^*_{gh} \qquad (g,\ h = 1,\ \ldots,\ G). \tag{2.15}$$ 1

Since (2.11) implies that $\alpha_{gk}(\Upsilon) = \alpha^*_{gh}$, Lemma 2.2 is proved.

The coefficients ξ_{kl} $(k,\ l = 1,\ \ldots,\ K)$ depend, of course, on the parameter point θ. To make this evident, we shall occasionally replace ξ_{kl} by $\xi_{kl}(\theta)$, and $\xi_{kl}(\Upsilon)$ by $\xi_{kl}\{\theta(\Upsilon)\}$. Since $\xi_{kl}\{\theta(\Upsilon)\} = \xi_{kl}(\theta)$, we shall say that the functions $\xi_{kl}(\theta)$ are invariant under nonsingular transformations Υ.

Let $F(\theta)$ be a function of θ. We shall say that $F(\theta)$ is invariant under nonsingular transformations if for any nonsingular transformation Υ we have $F\{\theta(\Upsilon)\} = F(\theta)$. Clearly, if $F(\theta)$ is a function of $\xi_{11}(\theta)$, $\xi_{12}(\theta)$, ..., $\xi_{KK}(\theta)$, then $F(\theta)$ is invariant under nonsingular transformations. We shall show that the converse is also true. Let $F(\theta)$ be a function such that $F\{\theta(\Upsilon)\} = F(\theta)$ for all nonsingular transformations Υ. Suppose that $F(\theta)$ is not a function of $\xi_{11}(\theta)$, $\xi_{12}(\theta)$, ..., $\xi_{KK}(\theta)$. Then there exist two points θ'' and θ''' such that

$$\xi_{kl}(\theta'') = \xi_{kl}(\theta''') \qquad (k,\ l = 1,\ \ldots,\ K) \tag{2.16}$$

and

$$F(\theta'') \neq F(\theta''') . \tag{2.17}$$

From Lemma 2.2 and (2.16) it follows that there exists a nonsingular transformation Υ such that

$$\theta''(\Upsilon) = \theta''' . \tag{2.18}$$

But then

$$F(\theta'') \neq F\{\theta''(\Upsilon)\} , \tag{2.19}$$

which contradicts our assumption about $F(\theta)$. Hence $F(\theta)$ must be a function of $\xi_{11}(\theta)$, $\xi_{12}(\theta)$, ..., $\xi_{KK}(\theta)$. Thus, the functions $\xi_{11}(\theta)$, $\xi_{12}(\theta)$, ..., $\xi_{KK}(\theta)$ form a fundamental set of invariants.

3. Necessary and Sufficient Conditions for the Identification of a Coordinate θ_p of a Parameter Point θ

Let θ be a parameter point satisfying the a priori conditions (1.1). And further, let $\theta^* = (\theta_1^*, \ldots, \theta_p^*)$ be an unknown parameter point and consider the equations in θ^* :

$$(3.1) \qquad \xi_{kl}(\theta^*) = \xi_{kl}(\theta) \qquad (k, l = 1, \ldots, K)$$

and

$$(3.2) \qquad \varphi_r(\theta^*) = 0 \qquad (r = 1, \ldots, R) \text{ (a priori conditions).}$$

The following two theorems are immediate consequences of Lemmas 2.1 and 2.2.

THEOREM 3.1. *A necessary and sufficient condition that θ_p be identifiable is that the equations (3.1) and (3.2) in the unknowns θ_1^*, ..., θ_P^* should admit of only a finite number of solutions for θ_p^*.*

THEOREM 3.2. *A necessary and sufficient condition that θ_p be locally identifiable is that there exists a finite neighborhood ω of θ such that for any solution θ^* in ω of the equations (3.1) and (3.2) we have $\theta_p^* = \theta_p$.*

In what follows in this section we shall assume that the R equations (3.2) have unique solutions in R unknowns, i.e., in R coordinates of θ^*. We may assume without loss of generality that these R coordinates are the last ones, i.e., θ_{P-R+1}^*, ..., θ_P^*. Thus, equations (3.2) can be written as

$$(3.3) \qquad \theta_p^* = \psi_p(\theta_1^*, \ldots, \theta_{P-R}^*) \qquad (p = P - R + 1, \ldots, P).$$

We shall assume that the functions ψ_p admit of continuous first-order partial derivatives. For any parameter point $\theta = (\theta_1, \ldots, \theta_P)$ in the P-dimensional parameter space we shall denote by $\bar{\theta}$ the parameter point in the $(P - R)$-dimensional space obtained from θ by omitting the last R coordinates, i.e., $\bar{\theta} = (\theta_1, \ldots, \theta_{P-R})$.

Denote by $\bar{\xi}_{kl}(\bar{\theta})$ the function we obtain from $\xi_{kl}(\theta)$ by substituting $\psi_p(\bar{\theta})$ for θ_p ($p = P - R + 1, \ldots, P$). Then the system of equations (3.1) and (3.2) is equivalent to the system

(3.4) $$\bar{\xi}_{kl}(\bar{\theta}^*) = \bar{\xi}_{kl}(\bar{\theta}) \qquad (k, l = 1, \ldots, K)$$

and

(3.5) $$\theta_p^* = \psi_p(\bar{\theta}^*) \qquad (p = P - R + 1, \ldots, P).$$

Denote the $(P - R)$-dimensional parameter space by $\bar{\Omega}$. For any point $\bar{\theta}$ of $\bar{\Omega}$ we shall denote by $\Delta(\bar{\theta})$ the Jacobian of the functions $\bar{\xi}_{11}(\bar{\theta})$, $\bar{\xi}_{12}(\bar{\theta})$, ..., $\bar{\xi}_{KK}(\bar{\theta})$ taken at the point $\bar{\theta}$. A point $\bar{\theta}$ of $\bar{\Omega}$ will be called regular if the following condition is satisfied: Any minor of the Jacobian of the $K^2 + P - R$ functions $\bar{\xi}_{11}(\bar{\theta})$, $\bar{\xi}_{12}(\bar{\theta})$, ..., $\bar{\xi}_{KK}(\bar{\theta})$, $d_p(\bar{\theta}) = \theta_p$ ($p = 1, \ldots, P - R$) is either unequal to zero at $\bar{\theta}$ or is *identically* zero in some finite neighborhood of $\bar{\theta}$.

THEOREM 3.3. *Let $\bar{\theta}^0$ be a regular point and denote by $\delta_p(\bar{\theta})$ the Jacobian of the $K^2 + 1$ functions $\bar{\xi}_{11}(\bar{\theta})$, $\bar{\xi}_{12}(\bar{\theta})$, ..., $\bar{\xi}_{KK}(\bar{\theta})$, $d_p(\bar{\theta}) = \theta_p$ for any value of p satisfying $p \leq P - R$. A necessary and sufficient condition that θ_p be locally identifiable for any point $\bar{\theta}$ in a finite neighborhood $\bar{\theta}^0$ is that the rank of $\Delta(\bar{\theta}^0)$ be equal to the rank of $\delta_p(\bar{\theta}^0)$.*

Proof: Since $\bar{\theta}^0$ is a regular point, a necessary and sufficient condition that $\bar{\theta}_p$ be a single valued function of $\bar{\xi}_{11}(\bar{\theta})$, $\bar{\xi}_{12}(\bar{\theta})$, ..., $\bar{\xi}_{KK}(\bar{\theta})$ in a finite neighborhood of $\bar{\theta}^0$ is that the rank of $\Delta(\bar{\theta}^0)$ be equal to that of $\delta_p(\bar{\theta}^0)$. Theorem 3.3 follows from this and the fact that the functions $\bar{\xi}_{11}(\bar{\theta})$, $\bar{\xi}_{12}(\bar{\theta})$, ..., $\bar{\xi}_{KK}(\bar{\theta})$ form a fundamental set of invariants.

VIII. REMARKS ON THE ESTIMATION OF UNKNOWN PARAMETERS IN INCOMPLETE SYSTEMS OF EQUATIONS*

BY A. WALD

1. *Formulation of the Problem*

Let $x_1, \ldots, x_N$ be N variables and denote by x_{nt} the value of x_n at the time point t where t can take any integral value. It is assumed that these variables satisfy a system of H $(H < N)$ stochastic equations

$$f_h(x_{1t}, x_{1,\,t-1}, \ldots, x_{1,\,t-\tau^{\square}}; \ldots;$$

$$x_{Nt}, x_{N,\,t-1}, \ldots, x_{N,\,t-\tau^{\square}}; \alpha_1, \ldots, \alpha_P) = u_{ht}, \qquad (h = 1, \ldots, H) \tag{1}$$

where $\alpha_1, \ldots, \alpha_P$ are unknown parameters, f_h is a given function of the variables $x_{n,\,t-\tau}$ $(n = 1, \ldots, N;\ \tau = 0, 1, \ldots, \tau^{\square})$ and of the parameters α_p $(p = 1, \ldots, P)$, and $u_{1t}, \ldots, u_{Ht}$ are random variables. It is assumed that the distribution of the random vector $u_t = (u_{1t}, \ldots, u_{Ht})$ is independent of t and that the vectors u_1, u_2, ..., are independently distributed. It is also assumed that the unknown distribution function of u_t is known to be an element of a given finite-parameter family Ω of cumulative distribution functions. Denote by $\theta_1, \ldots, \theta_Q$ the unknown parameters involved in the distribution function of u_t.

* Reprinted with permission from *Cowles Commission Monograph* 10 (1950, John Wiley & Sons, Inc.).

The problem considered here is to estimate all or some of the unknown parameters $\alpha_1, \ldots, \alpha_P; \theta_1, \ldots, \theta_Q$ on the basis of the observed values x_{nt} $(n = 1, \ldots, N;\ t = 1, \ldots, T)$.

Since $H < N$, the distribution of the random vectors $u_t = (u_{1t}, \ldots, u_{Ht})$, $t = 1, \ldots, T$, does not determine the distribution of the variables x_{nt}, $n = 1, \ldots, N$, but only imposes a restriction on the latter distribution. For this reason, the title of this note refers to incomplete systems of equations. It will appear from what follows that the estimation problems in incomplete systems are essentially different from those in complete systems discussed in other contributions to this volume.

2. *Existence of Consistent Estimates*

A basic problem is, of course, the question of the existence of consistent estimates of the unknown parameters. We shall say that a parameter point $\alpha = (\alpha_1, \ldots, \alpha_P)$ satisfies condition C if the following conditions are simultaneously fulfilled: (1) the distribution of the vector

$$u_t(\alpha) = \Big[f_1(x_{1t}, \ldots, x_{1,t-\tau^\square}; \ldots; x_{Nt}, \ldots, x_{N,t-\tau^\square}; \alpha_1, \ldots, \alpha_P), \\ \ldots, f_H(x_{1t}, \ldots, x_{1,t-\tau^\square}; \ldots; x_{Nt}, \ldots, x_{N,t-\tau^\square})\Big]$$

is independent of t; (2) the vectors $u_1(\alpha)$, $u_2(\alpha)$, ..., are independently distributed; (3) the distribution of $u_t(\alpha)$ is an element of Ω. It is clear that if there exist two parameter points α' and α'' such that both satisfy condition C, no consistent estimate of the parameter point α exists. On the other hand, if there exists one and only one parameter point α that satisfies condition C, then, under some further mild restrictions that we do not propose to discuss here, a consistent estimate of α will exist.

Whether or not there exist several parameter points α satisfying condition C depends on the joint probability distribution of the observable variables x_{nt} $(n = 1, \ldots, N;\ t = 1, \ldots, T)$. Thus, the existence of consistent estimates depends on the joint probability distribution of the observable variables. Since this distribution is usually unknown a priori, we cannot be sure that a consistent estimate exists. This difficulty, however, is not as serious as it would appear at first sight. In fact, instead of point estimates, we are usually more interested in constructing a confidence region for the unknown parameters corresponding to a given confidence coefficient. We shall see in the next section that a confidence region can be obtained irrespective of the existence of consistent estimates. The only effect of the nonexistence of consistent estimates on the confidence region is that the diameter of the confidence region (maximum distance between two points of the confidence region) will not approach zero as the number of observations approaches infinity.

3. *Construction of Confidence Regions for the Parameter Point* α

To construct a confidence region for the parameter point α we may proceed as follows: Suppose that the prescribed confidence coefficient is δ. For any given parameter point $\alpha^0 = (\alpha_1^0, \ldots, \alpha_p^0)$ we construct a critical region $W(\alpha^0)$ of size $1 - \delta$ for testing the hypothesis that $\alpha = \alpha^0$. The critical region $W(\alpha^0)$ is a subset in the HT-dimensional space of the variables u_{ht} ($h = 1, \ldots, H$; $t = 1, \ldots, T$) and we reject the hypothesis that $\alpha = \alpha^0$ if and only if the observed point $u_{11}(\alpha^0)$, $u_{12}(\alpha^0)$, ..., $u_{Ht}(\alpha^0)$ falls in $W(\alpha^0)$. Of course, $W(\alpha^0)$ is to be constructed in such a way that the probability measure of $W(\alpha^0)$ is equal to $1 - \delta$ under any distribution of u_t that is an element of Ω. Usually such a region $W(\alpha^0)$ can be constructed without any difficulty. Now consider all possible parameter points α^0 and the corresponding family of critical regions $W(\alpha^0)$. The set of all parameter points α^0 that are not rejected on the basis of this test procedure will form a confidence set with confidence coefficient δ.

Thus, we see that there is no serious difficulty in obtaining a confidence region for α. There will be, in general, infinitely many possible confidence regions, since $W(\alpha^0)$ can usually be chosen in infinitely many different ways. The problem of proper choice of $W(\alpha^0)$ is not yet solved. The theory of confidence intervals as developed by J. Neyman [1937] does not apply, since Neyman deals with the parametric case, while in our problem the class of all possible distribution functions of the observable variables x_{nt} ($n = 1, \ldots, N$; $t = 1, \ldots, T$) compatible with the relations (1) cannot be described by a finite number of parameters owing to the fact that $H < N$.

4. *Some Examples*

In this section we shall give a few examples to illustrate the procedure for the construction of confidence regions. It should be emphasized, however, that the confidence regions described later in this section are by no means "best." As a matter of fact they may be very inefficient under certain conditions and it is not our intention to recommend them for practical use.

Consider the following problem: Let x_1, x_2, and x_3 be three observable variables and denote by x_{nt} ($n = 1, 2, 3;\ t = 1, 2, \ldots$) the value of x_n at the time point t. Suppose that x_{1t}, x_{2t}, and x_{3t} satisfy the relation

$$x_{1t} + \alpha_2 x_{2t} + \alpha_3 x_{3t} + \alpha_0 = u_t \ , \tag{2}$$

where α_2, α_3, α_0 are unknown constants and u_1, u_2, ... are independently and normally distributed random variables with zero means and a common variance σ^2. We shall construct a confidence region for the parameters α_2, α_3, and α_0 on the basis of the observed values x_{nt} ($n = 1, 2, 3;\ t = 1, \ldots, T$). For simplicity we shall assume that $T = 3V$, where V is a positive integer. Denote by $\bar{x}_n^m$ the arithmetic mean of the observations on x_n in the mth group of V observations, i.e.,

$$\bar{x}_n^m = \frac{1}{V} \sum_{t=(m-1)V+1}^{mV} x_{nt} \ , \qquad m = 1, 2, 3 \, . \tag{3}$$

Furthermore, let

$$\begin{aligned} s_m^2 = \frac{1}{V-1} \sum_{t=(m-1)V+1}^{mV} \Big[&(x_{1t} - \bar{x}_1^m) + \alpha_2 (x_{2t} - \bar{x}_2^m) \\ &+ \alpha_3 (x_{3t} - \bar{x}_3^m) \Big]^2 , \qquad m = 1, 2, 3 \, . \end{aligned} \tag{4}$$

It is clear that $(V-1)s_1^2/\sigma^2$, $(V-1)s_2^2/\sigma^2$, and $(V-1)s_3^2/\sigma^2$ are independently distributed, each having the χ^2-distribution with $V-1$ degrees of freedom. Thus, each of the expressions

$$(5) \qquad t_m = \frac{(\bar{x}_1^m + \alpha_2 \bar{x}_2^m + \alpha_3 \bar{x}_3^m + \alpha_0) V^{1/2}}{s_m}, \qquad m = 1, 2, 3,$$

has the t-distribution with $V - 1$ degrees of freedom, and t_1, t_2, and t_3 are independently distributed.

A confidence region for $(\alpha_2, \alpha_3, \alpha_0)$ with confidence coefficient δ can be constructed as follows: Let λ_δ be the value for which the probability that $-\lambda_\delta \leq t \leq \lambda_\delta$ is equal to $\delta^{1/3}$. Here t denotes a random variable that has the t-distribution with $V - 1$ degrees of freedom. The set of all parameter points $(\alpha_2, \alpha_3, \alpha_0)$ that satisfy simultaneously the following three inequalities,

$$(6) \qquad | t_m | \leq \lambda_\delta , \qquad m = 1, 2, 3,$$

forms a confidence region with confidence coefficient δ.

The confidence region given by (6) is certainly far from being the best possible one. There is a loss of efficiency in taking three different estimates s_1^2, s_2^2, s_3^2 for σ^2. A better procedure would be to combine these three estimates into a single one given by $s^2 = \frac{1}{3}(s_1^2 + s_2^2 + s_3^2)$. This, however, would make the variables t_1, t_2, and t_3 dependent and would complicate the derivation of a confidence region.

If the determinant

$$(7) \qquad \begin{vmatrix} \bar{x}_2^1 & \bar{x}_3^1 & 1 \\ \bar{x}_2^2 & \bar{x}_3^2 & 1 \\ \bar{x}_2^3 & \bar{x}_3^3 & 1 \end{vmatrix}$$

is bounded away from zero as $T \to \infty$, and if s_1^2, s_2^2, and s_3^2 are bounded functions of T, the diameter of the confidence region given by (6) will approach zero as $T \to \infty$. If the distribution of the vector $v_t = (x_{1t}, x_{2t}, x_{3t})$ is independent of t, if v_1, v_2, ... are independently distributed, and if the first two moments of x_{nt} exist, then the determinant (7) will converge stochastically to zero and the probability is one that the diameter of the confidence set given by (6) will not approach zero as $T \to \infty$.

Another possible procedure for obtaining a confidence region is the following: Consider the expression

$$(8) \qquad F = \frac{V \sum_{m=1}^{3} (\bar{x}_1^m + \alpha_2 \bar{x}_2^m + \alpha_3 \bar{x}_3^m + \alpha_0)^2}{(V-1)(s_1^2 + s_2^2 + s_3^2)} \, \frac{T-3}{3} \, .$$

This has the F-distribution with 3 and $T-3$ degrees of freedom. Let F_δ be a positive value for which the probability that $F \leq F_\delta$ is δ. Then the set of all parameter points α_2, x_3, α_0 that satisfy the inequality

$$(9) \qquad F \leq F_\delta$$

will form a confidence set with confidence coefficent δ.

A confidence region for α_2, α_3, α_0 can also be obtained as follows: Consider the serial correlation

$$(10) \qquad r = \frac{\sum_{t=1}^{T-1} \left[(x_{1t} + \alpha_2 x_{2t} + \alpha_3 x_{3t} + \alpha_0) \times (x_{1,t+1} + \alpha_2 x_{2,t+1} + \alpha_3 x_{3,t+1} + \alpha_0) \right]}{\sum_{t=1}^{T} (x_{1t} + \alpha_2 x_{2t} + \alpha_3 x_{3t} + \alpha_0)^2}$$

and let r_δ be a value such that the probability that $|r| \leq r_\delta$ is equal to δ. Then the set of all parameter points $(\alpha_2, \alpha_3, \alpha_0)$ for which $|r| \leq r_\delta$ will form a confidence set with confidence coefficient δ.

ANNALS OF MATHEMATICS
Vol. 52, No. 3, November, 1950

NOTE ON ZERO SUM TWO PERSON GAMES

BY ABRAHAM WALD

(Received January 19, 1950)

The normalized form of a zero sum two person game has been defined by v. Neumann [1] as follows: There are two players and there is a bounded real valued function $K(x, y)$ given where x may be any element of a given space X and y may be any element of a given space Y. Player 1 chooses an element x of X and player 2 chooses an element y of Y, each choice being made in complete ignorance of the other. After these choices have been made, player 1 receives the amount $K(x, y)$ and player 2 the amount $-K(x, y)$.

In the present note we shall assume that X consists of all points x of the m-dimensional Cartesian space with coordinates between 0 and 1, and Y consists of all points y of the n-dimensional Cartesian space with coordinates between 0 and 1. Here m and n are given positive integers. The only restriction we impose on $K(x, y)$ is that it is a bounded and Borel measurable function of x and y.

It is said that player 1 uses a pure strategy if he chooses a particular element x of X, and a mixed strategy if he merely chooses a probability distribution in X. In the latter case the actual selection of an element x of X is made with the help of a chance mechanism constructed so that the probability distribution of the selected element x is identical with the probability distribution chosen by player 1. Pure and mixed strategies for player 2 are defined similarly. A probability distribution in the space X can be represented by a cumulative distribution function $F(x)$ where $F(x)$ is equal to the probability that the coordinates of the selected element of X will be less than the corresponding coordinates of x. Of course, we have $\int_X dF(x) = 1$. Similarly, a probability distribution in Y can be represented by a cumulative distribution function $G(y)$. If player 1 uses the mixed strategy represented by the c.d.f.F, and player 2 the mixed strategy represented by the c.d.f.G, then the expected value of the outcome $K(x, y)$ is given by

$$K^*(F, G) = \int_Y \int_X K(x, y)\, dF(x)\, dG(y). \tag{1}$$

The game is said to be strictly determined if

$$\operatorname*{Max}_F \operatorname*{Min}_G K^*(F, G) = \operatorname*{Min}_G \operatorname*{Max}_F K^*(F, G). \tag{2}$$

It is known that there are outcome functions $K(x, y)$ for which the game is not strictly determined (see, for example, [2]). Conditions on $K(x, y)$ which are sufficient to insure the validity of (2) have been formulated by J. Ville [3] and the author [4].

In this note we shall deal with a somewhat different question. Instead of imposing conditions on $K(x, y)$, we shall try to formulate restrictions on the choice of F and G which insure the validity of (2) for any bounded and Borel measurable outcome function $K(x, y)$. For any positive value c, let Ω'_c denote the class of all c.d.f.F which admit a density function $f(x)$ satisfying the inequality $f(x) \leqq c$ for all x. Similarly, let Ω''_c denote the class of all c.d.f.$G(x)$ which admit a density function $g(x)$ satisfying the inequality $g(x) \leqq c$ for all x. We shall prove the following theorem.

THEOREM. *Let c' and c'' be two positive numbers such that $\Omega'_{c'}$ and $\Omega''_{c''}$ are not empty. If player 1 is restricted to mixed strategies F which are elements of $\Omega'_{c'}$, and player 2 is restricted to mixed strategies G which are elements of $\Omega''_{c''}$, then the game is strictly determined for any bounded and Borel measurable outcome function $K(x, y)$.*

PROOF. A sequence $\{F_i\}$ ($i = 1, 2, \cdots$, ad inf.) of c.d.f. is said to converge to a c.d.f. F, in symbol $\lim_{i=\infty} F_i = F$, if $\lim_{i=\infty} F_i(x) = F(x)$ for all continuity points of $F(x)$. Since X is a bounded and closed subset of the m-dimensional Cartesian space, the class of all c.d.f. F is a compact in the sense of the above convergence definition. If $\lim F_i = F$ and F_i is an element of $\Omega'_{c'}$ for each i, then also F is an element of $\Omega'_{c'}$. Thus, $\Omega'_{c'}$ is a closed subset of the class of all c.d.f. F, and therefore $\Omega'_{c'}$ is compact. Convergence in the space of the c.d.f. G is defined similarly and the same argument shows that $\Omega''_{c''}$ is compact.

Let $\Omega_{c',c''}$ be the Cartesian product of the spaces $\Omega'_{c'}$ and $\Omega''_{c''}$, i.e., $\Omega_{c',c''}$ is the set of all pairs (F, G) for which F is an element of $\Omega'_{c'}$ and G is an element of $\Omega''_{c''}$. We shall now show that $K^*(F, G)$ is continuous in F and G jointly over the domain $\Omega_{c',c''}$. Let $\{F_i\}$ ($i = 1, 2, \cdots$, ad inf.) be a sequence of elements of $\Omega'_{c'}$ such that F_i converges to an element F_0 of $\Omega'_{c'}$ as $i \to \infty$. Similarly, let $\{G_i\}$ ($i = 1, 2, \cdots$, ad inf.) be a sequence of elements of $\Omega''_{c''}$ such G_i converges to an element G_0 of $\Omega''_{c''}$ as $i \to \infty$. Since F_0 and G_0 are absolutely continuous, we have $\lim_{i=\infty} F_i(x)$ and $\lim_{i=\infty} G_i(y) = G_0(y)$ for all x and y. Hence $\lim_{i=\infty} F_i(x) G_i(y) = F_0(x)G_0(y)$ for all pairs (x, y). Clearly, $F_i(x)G_i(y)$ ($i = 0, 1, 2, \cdots$, ad inf.) is a cumulative distribution function in the space Z of all pairs (x, y). Let γ be any Borel measurable subset of Z. Since $F_0(x)G_0(y)$ is absolutely continuous, we have

$$\lim_{i=\infty} \int_\gamma d[F_i(x)G_i(y)] = \int_\gamma d[F_0(x)\, G_0(y)]. \tag{3}$$

The relation $\lim_{i=\infty} K^*(F_i, G_i) = K^*(F_0, G_0)$ follows easily from (3). Thus, the continuity of $K^*(F, G)$ over the domain $\Omega_{c',c''}$ is proved.

We shall now introduce an intrinsic metric in each of the spaces $\Omega'_{c'}$ and $\Omega''_{c''}$. The distance between two elements F_1 and F_2 of $\Omega'_{c'}$ is defined by

$$\delta(F_1, F_2) = \operatorname*{Max}_G \mid K^*(F_1, G) - K^*(F_2, G) \mid, \tag{4}$$

and the distance between two elements G_1 and G_2 of $\Omega''_{c''}$ by

$$\delta(G_1, G_2) = \operatorname*{Max}_F \mid K^*(F, G_1) - K^*(F, G_2) \mid. \tag{5}$$

Since $\Omega'_{c'}$ and $\Omega''_{c''}$ are compact, continuity of $K^*(F, G)$ implies uniform continuity, that is, if $\lim F_i = F_0$ and $\lim G_i = G_0$, then $\lim K^*(F_i, G) = K^*(F_0, G)$ uniformly in G, and $\lim K^*(F, G_i) = K^*(F, G_0)$ uniformly in F. Hence, $\Omega'_{c'}$ is compact in the sense of the metric (4), and $\Omega''_{c''}$ is compact in the sense of the metric (5).

Let us now consider a game where the elements F and G are the pure strategies of the players 1 and 2, respectively, and the outcome function is given by $K^*(F, G)$. A mixed strategy of player 1 is then given by a probability measure ξ defined over a Borel field B_1 of subsets of $\Omega'_{c'}$, and a mixed strategy η of player 2 is given by a probability measure η defined over a Borel field B_2 of subsets of $\Omega''_{c''}$. We choose B_1 as the smallest Borel field containing all open (in the sense of the metric (4)) subsets of $\Omega'_{c'}$, and B_2 as the smallest Borel field containing all open subsets of $\Omega''_{c''}$. If player 1 uses the mixed strategy ξ and player 2 the mixed strategy η, then the expected outcome is equal to

$$K^{**}(\xi, \eta) = \int_{\Omega''_{c''}} \int_{\Omega'_{c'}} K^*(F, G) \, d\xi \, d\eta. \tag{6}$$

Since the spaces $\Omega'_{c'}$ and $\Omega''_{c''}$ are compact in the sense of the metrics (4) and (5), respectively, it follows from the results given in an earlier publication (see Theorem 3.7 in [4]) that the game is strictly determined, i.e.,

$$\operatorname*{Max}_{\xi} \operatorname*{Min}_{\eta} K^{**}(\xi, \eta) = \operatorname*{Min}_{\eta} \operatorname*{Max}_{\xi} K^{**}(\xi, \eta). \tag{7}$$

Let $\bar{\xi}$ denote a probability measure ξ subject to the restriction that there exists a finite subset ω' of $\Omega'_{c'}$ such that $\bar{\xi}(\omega') = 1$. Similarly, let $\bar{\eta}$ denote a probability measure η subject to the restriction that there exists a finite subset ω'' of $\Omega''_{c''}$ for which $\bar{\eta}(\omega'') = 1$. Because of the compactness of the spaces $\Omega'_{c'}$ and $\Omega''_{c''}$, we have

$$\operatorname*{Sup}_{\bar{\xi}} \operatorname*{Inf}_{\bar{\eta}} K^{**}(\bar{\xi}, \bar{\eta}) = \operatorname*{Max}_{\xi} \operatorname*{Min}_{\eta} K^{**}(\xi, \eta) \quad \text{and} \tag{8}$$

$$\operatorname*{Inf}_{\bar{\xi}} \operatorname*{Sup}_{\bar{\eta}} K^{**}(\bar{\xi}, \bar{\eta}) = \operatorname*{Min}_{\eta} \operatorname*{Max}_{\xi} K^{**}(\xi, \eta),$$

where Sup stands for Supremum with respect to $\bar{\xi}$ and Inf for Infimum with respect to $\bar{\eta}$.

For any $\bar{\xi}$ let $F_{\bar{\xi}}$ denote the distribution function defined by

$$F_{\bar{\xi}}(x) = \sum_{i=1}^{k} \alpha_i F_i(x)$$

where α_i is the probability assigned to F_i by $\bar{\xi}$ and $\sum_1^k \alpha_i = 1$. For any $\bar{\eta}$, the distribution function $G_{\bar{\eta}}$ is similarly defined. Clearly,

$$K^{**}(\bar{\xi}, \bar{\eta}) = K^*(F_{\bar{\xi}}, G_{\bar{\eta}}). \tag{9}$$

From (7), (8), and (9) it follows that

$$\text{(10)} \qquad \underset{\bar{\xi}}{\text{Sup}}\ \underset{\bar{\eta}}{\text{Inf}}\ K^*(F_{\bar{\xi}}, G_{\bar{\eta}}) = \underset{\bar{\eta}}{\text{Inf}}\ \underset{\bar{\xi}}{\text{Sup}}\ K^*(F_{\bar{\xi}}, G_{\bar{\eta}}).$$

Since the set of all $F_{\bar{\xi}}$ corresponding to all possible $\bar{\xi}$ coincides with Ω'_c, and the set of all $G_{\bar{\eta}}$ corresponding to all possible $\bar{\eta}$ coincides with $\Omega''_{c''}$, the following equations hold

$$\text{(11)} \qquad \underset{\bar{\xi}}{\text{Sup}}\ \underset{\bar{\eta}}{\text{Inf}}\ K^*(F_{\bar{\xi}}, G_{\bar{\eta}}) = \underset{F}{\text{Max}}\ \underset{G}{\text{Min}}\ K^*(F, G)$$

and

$$\text{(12)} \qquad \underset{\bar{\eta}}{\text{Inf}}\ \underset{\bar{\xi}}{\text{Sup}}\ K^*(F_{\bar{\xi}}, G_{\bar{\eta}}) = \underset{G}{\text{Min}}\ \underset{F}{\text{Max}}\ K^*(F, G).$$

Strict determinateness of the game (relation (2)) follows from (10), (11), and (12). This completes the proof of our theorem.

Columbia University

References

[1] J. v. Neumann and Oskar Morgenstern, Theory of Games and Economic Behavior, Princeton University Press, 1944.

[2] A. Wald, *Generalization of a Theorem by v. Neumann Concerning Zero Sum Two Person Games*, Annals of Mathematics, Vol. 46 (1945).

[3] J. Ville, Note *Sur la Theorie Generale des Jeux on intervient l'Habileté des Jouers* in the book "Applications aux Jeux de Hasard," by Emile Borel and Jean Ville, Tome IV, Fascicule II in the "Traite du Calcul des Probabilites et de ses Applications" par Emile Borel (1938).

[4] A. Wald, *Foundations of a General Theory of Sequential Decision Functions*, Econometrica, Vol. 15 (1947).

RELATIONS AMONG CERTAIN RANGES OF VECTOR MEASURES

A. DVORETZKY, A. WALD, AND J. WOLFOWITZ

1. Introduction and definitions. The purpose of the present paper is to prove certain measure theoretical results concerning ranges of measures. One of our results (the closure and convexity result implied by Theorem 4) may be regarded as a generalization of a theorem of Liapounoff [5]. The results obtained here have applications to statistics and the theory of games.

Throughout this paper $\{x\} = X$ denotes an arbitrary space, and $\{S\} \equiv \mathfrak{S}$ denotes a Borel field of subsets of X; that is, $\mathfrak{S}$ is a nonempty family of subsets of X which is closed with respect to the operations of complementation (with respect to X) and countable union. The phrase, *S is measurable,* will be used as synonymous with $S \in \mathfrak{S}$.

A real-valued countably additive set function defined for all measurable sets will be called a *measure*. Thus we admit measures assuming negative or infinite values. A measure cannot, however, assume the value $+\infty$ for one measurable set and $-\infty$ for another such set, since in such a case additivity cannot be defined satisfactorily. A measure is called *finite* if it assumes finite values for all measurable sets. It is called *nonnegative* if it assumes nonnegative values for all such sets.

We say that $f(x)$ is a *measurable function* if it is real-valued, defined for all $x \in X$, and if, moreover, the set f_c of all $x \in X$ for which $f(x) < c$ is measurable for every real number c. A *step function* is a measurable function which assumes only a *finite* number of values.

If n is a positive integer and $\eta_j(x)\,(j = 1, \cdots, n)$ are nonnegative measurable functions satisfying

$$\eta_1(x) + \cdots + \eta_n(x) = 1 \qquad \text{for every } x \in X, \tag{1}$$

then $\eta(x) = [\eta_1(x), \cdots, \eta_n(x)]$ will be called a *probability n*-vector. The functions $\eta_j(x)$ are called the *components* of this vector. If all the components of

Received July 20, 1950. This research was sponsored (in part) by the Office of Naval Research; the main results of the paper were announced without proof in an earlier publication [1]. The first author is on leave of absence from the Hebrew University, Jerusalem, Israel.

Pacific J. Math. 1 (1951), 59-74.

$\eta(x)$ are step functions, then $\eta(x)$ is called a (probability) *step n-vector.* We shall occasionally denote such vectors by $\eta^0(x)$. If, in particular, all components of $\eta(x)$ assume only the two values zero and one, that is, if for every $x \in X$ one $\eta_j(x)$ is equal to one and all others vanish, then $\eta(x)$ is called a *pure n-vector.* Such vectors will be denoted by $\eta^*(x)$. If the jth component ($j = 1, \cdots, n$) of $\eta^*(x)$ is considered as the characteristic function of a set S_j, then the sets $S_1, \cdots, S_n$ are measurable and disjoint and their union is X. Conversely, if $S_1, \cdots, S_n$ is a decomposition of X into n disjoint measurable sets, and $\eta_j^*(x)$ is the characteristic function of S_j, then $\eta^*(x) = [\eta_1^*(x), \cdots, \eta_n^*(x)]$ is a pure n-vector. We therefore call $\eta^*(x)$ also a *decomposition n-vector* or, more specifically, a decomposition n-vector corresponding to the decomposition $X = S_1 \cup \cdots \cup S_n$.

Let $\mu_k(S)$ $(k = 1, \cdots, p)$ be a finite set of measures, and let $\eta(x)$ be a probability n-vector. We denote by $v(\eta) = v(\eta; \mu_1, \cdots, \mu_p)$ the np dimensional vector (or point in np space),

$$\left[\int_X \eta_1(x)\, d\mu_1(x), \cdots, \quad \int_X \eta_1(x)\, d\mu_p(x),\right.$$

$$\left.\int_X \eta_2(x)\, d\mu_1(x), \cdots, \quad \int_X \eta_p(x)\, d\mu_p(x)\right].$$

The set of all points $v(\eta) = v(\eta; \mu_1, \cdots, \mu_p)$ corresponding to all probability n-vectors $\eta(x)$ is called the *n-range* of $\mu_1, \cdots, \mu_p$ and is further denoted by $V_n(\mu_1, \cdots, \mu_p)$ or, more concisely, by V_n. In the same way we define the *step n-range* of $\mu_1, \cdots, \mu_p$ as the set of all points $v(\eta^0) = v(\eta^0; \mu_1, \cdots, \mu_p)$ corresponding to all step n-vectors $\eta^0(x)$, and denote it by $V_n^0(\mu_1, \cdots, \mu_p)$ or V_n^0. Similarly V_n^* or $V_n^*(\mu_1, \cdots, \mu_p)$ denotes the set of all points

$$v(\eta^*) = v(\eta^*; \mu_1, \cdots, \mu_p)$$

corresponding to all pure n-vectors $\eta^*(x)$ and is called the *decomposition n-range* of $\mu_1, \cdots, \mu_p$. When no confusion is possible we replace n-range in the above terms by range.

It is shown in Section 2 that if $\mu_1, \cdots, \mu_p$ are finite measures then the range $V_n(\mu_1, \cdots, \mu_p)$ is compact and convex and coincides with the step-range $V_n^0(\mu_1, \cdots, \mu_p)$. Actually a stronger result is proved; this states that the points $v(\eta; \mu_1, \cdots, \mu_p)$ for which the components of $\eta(x)$ assume at most 2^{np-p+1}

different values already fill V_n. Applying a theorem of Liapounoff we deduce, in Section 3, the result that if the measures are atomless then the decomposition range $V_n^*(\mu_1, \cdots, \mu_p)$ is identical with $V_n(\mu_1, \cdots, \mu_p)$. This result is extended in Section 4 to arbitrary (not necessarily finite) atomless measures. Applications of these results to statistics and the theory of games are briefly indicated in Section 5.

2. Identity of the step range and the range for finite measures. First, we prove the following result.

THEOREM 1. *If $\mu_1, \cdots, \mu_p$ are finite measures, then for every n, the range $V_n(\mu_1, \cdots, \mu_p)$ is a compact and convex set in Euclidean np dimensional space.**

Proof. Let $A = v(\eta)$ and $A' = v(\eta')$ be any two points of V_n. Then every point of the segment joining them is represented vectorially by $cA + (1-c)A'$, with $0 < c < 1$. But such a point is clearly $v[c\eta + (1-c)\eta']$ and, since $c\eta + (1-c)\eta'$ is a probability n-vector, the point also belongs to the range. Thus V_n is convex.

The proof of compactness is more difficult. We start by establishing a lemma on sequences of measures.

LEMMA 1. *Let $\{B\} = \mathfrak{B}$ be a Borel field of subsets of X generated by countably many sets. Let μ^t $(t = 1, 2, \cdots)$ and μ be measures over $\mathfrak{B}$ satisfying, for all $B \in \mathfrak{B}$,*

$$0 \le \mu^t(B) \le \mu(B) < \infty \qquad (t = 1, 2, \cdots). \tag{2}$$

Then there exists a measure ν over $\mathfrak{B}$ satisfying

$$0 \le \nu(B) \le \mu(B) \quad \text{for all } B \in \mathfrak{B}, \tag{3}$$

and a sequence of integers t_q $(q = 1, 2, \cdots)$ satisfying

$$0 < t_1 < t_2 < \cdots < t_q < t_{q+1} < \cdots \tag{4}$$

such that

$$\lim_{q=\infty} \mu^{t_q}(B) = \nu(B) \qquad \text{for every } B \in \mathfrak{B}. \tag{5}$$

---

*For the special case when X is a finite-dimensional Euclidean space and all the μ_k are absolutely continuous, this theorem follows from Theorems 3.1 and 3.2a of [6].

The proof of the lemma proceeds as follows. Let $B_0 = X, B_1, \cdots, B_n \cdots$ be a countable basis of $\mathfrak{B}$. Then, according to the well-known diagonal procedure of Cantor, there exists a sequence (4) for which

$$\beta_r = \lim_{q=\infty} \mu^{t_q}(B_r)$$

exists for $r = 0, 1, 2, \cdots$. To prove the existence of the limit in (5) for every B, and the fact that this limit is a measure, it suffices to show that: (a) if $\mu^{t_q}(B)$ tends to a limit as $q \longrightarrow \infty$, then so does $\mu^{t_q}(\overline{B})$ where $\overline{B}$ is the complement of B with respect to X; and (b) if $B^s (s = 1, 2, \cdots)$ are disjoint sets of $\mathfrak{B}$ for which $\lim_{q=\infty} \mu^{t_q}(B^s)$ exists for $s = 1, 2, \cdots$, then we have also

$$(6) \qquad \lim_{q=\infty} \mu^{t_q}\left(\bigcup_{s=1}^{\infty} B^s\right) = \sum_{s=1}^{\infty} \lim_{q=\infty} \mu^{t_q}(B^s) .$$

Now (a) follows immediately when we write $\mu^t(\overline{B}) = \mu^t(X) - \mu^t(B)$ and observe that $\mu^{t_q}(X)$ has the limit β_0. To prove (6) it is sufficient to observe that the functions μ^t are countably additive, that by (2) we have $\mu^{t_q}(B^s) \leq \mu(B^s)$, and that $\sum_{s=1}^{\infty} \mu(B^s)$ is a convergent series of nonnegative terms. (This is the standard bounded convergence argument.) Since (2) and (5) obviously imply (3), the proof of Lemma 1 is completed.

Let now $\eta^t(x) (t = 1, 2, \cdots)$ be any sequence of probability n-vectors. The compactness of V_n will be proved if we show that there exist a probability n-vector and a sequence (4) satisfying

$$(7) \quad \lim_{q=\infty} \int_X \eta_j^{t_q}(x)\, d\mu_k(x) = \int_X \eta_j(x)\, d\mu_k(x) \qquad (j = 1, \cdots, n;\ k = 1, \cdots, p) .$$

Denote by $B_{j,\rho}^t (t = 1, 2, \cdots; j = 1, \cdots, n;\ \rho$ rational with $0 \leq \rho \leq 1)$ the set of all x for which $\eta_j^t(x) \leq \rho$, and let $\{B\} = \mathfrak{B} \subset \mathfrak{S}$ be the smallest Borel field containing these sets. Write $|\mu_k|$ for the absolute measure* associated with μ_k. Put $\mu(B) = |\mu_1|(B) + \cdots + |\mu_p|(B)$ for every $B \in \mathfrak{B}$. Then $\mathfrak{B}$, μ, and

*That is, $|\mu_k|(S) = \sup\ [|\mu_k(S')| + |\mu_k(S'')|]$ for all decompositions of S into two disjoint measurable sets S' and S''.

μ^t, defined by

$$\mu^t(B) = \int_B \eta_1^t(x)\ d\mu(x) \qquad (t = 1, 2, \cdots, B \in \mathfrak{B}),$$

satisfy the conditions of Lemma 1. Hence there exists a nonnegative measure ν_1 over $\mathfrak{B}$ and a sequence (4), for which

$$\text{(8)} \qquad \lim_{q=\infty} \int_B \eta_1^{t_q}(x)\ d\mu(x) = \nu_1(B)$$

for every $B \in \mathfrak{B}$.

Again applying Lemma 1, we can extract from the sequence t_q a further subsequence for which (8) holds with the subscript 1 replaced by 2. Repeating this $n - 1$ times, and again denoting, for simplicity of writing, the final subsequence by t_q, we see that there exist nonnegative measures $\nu_1, \cdots, \nu_n$ over $\mathfrak{B}$ and a sequence (4) satisfying

$$\text{(9)} \qquad \lim_{q=\infty} \int_B \eta_j^{t_q}(x)\ d\mu(x) = \nu_j(B) \qquad (j = 1, \cdots, n)$$

for every $B \in \mathfrak{B}$. Clearly, we have

$$\text{(10)} \qquad \nu_1(B) + \cdots + \nu_n(B) = \mu(B) \qquad (B \in \mathfrak{B})$$

By the Radon-Nikodym theorem there exist $\mathfrak{B}$-measurable functions $f_j(x)$ $(j = 1, \cdots, n)$ such that

$$\text{(11)} \qquad \nu_j(B) = \int_B f_j(x)\ d\mu(x) \qquad (j = 1, \cdots, n)$$

for every $B \in \mathfrak{B}$. Since the ν_j are nonnegative measures, we may assume that the f_j are nonnegative functions; and, because of (10), we may further assume that $f_1(x) + \cdots + f_n(x) = 1$ for every x. The f_j are $\mathfrak{B}$-measurable and are, *a fortiori*, $\mathfrak{S}$-measurable; hence $[f_1(x), \cdots, f_n(x)]$ is a probability n-vector. We denote this vector by $\eta(x)$ and proceed to show that (7) holds with this η and the above constructed sequence (4) satisfying (9).

Let $g_k(x) (k = 1, \cdots, p)$ denote a $\mathfrak{B}$-measurable Radon-Nikodym derivative

$d\mu_k(x)/d\mu(x)$. Then, replacing f_j in (11) by η_j, we have

$$\int_X \eta_j(x)\, d\mu_k(x) = \int_X \eta_j(x)\, g_k(x)\, d\mu(x)$$

$$= \int_X g_k(x)\, d\int \eta_j(x)\, d\mu(x) = \int_X g_k(x)\, d\nu_j(x).$$

Similarly, the left side of (7) may be rewritten as

$$\lim_{q=\infty} \int_X g_k(x)\, d\int \eta_j^{t_q}(x)\, d\mu(x),$$

and thus (7) follows from (9). This completes the proof of Theorem 1.

For any compact convex set C in a Euclidean space, we designate as *extreme points of C*, all those points of C which are not *interior* points of any segment lying in C. Our next result is the following.

THEOREM 2. *If the measures $\mu_1, \cdots, \mu_p$ are finite, and $v(\eta)$ is an extreme point of V_n, then the set of x for which $0 < \eta_j(x) < 1$ for at least one $j (j = 1, \cdots, n)$ is a null-set* for each of the measures $\mu_1, \cdots, \mu_p$. In particular, all extreme points of V_n belong to the decomposition range V_n^*.*

Proof. Let Y denote the set of x defined in the theorem. If Y is not a null-set for μ_{k_0} with $1 \le k_0 \le p$, then there exist integers j_0, j_1 with $1 \le j_0 < j_1 \le n$, a number $\delta > 0$, and a measurable set $Z \subset Y$, such that

(12) $\quad \delta < \eta_j(x) < 1 - \delta \quad$ for $x \in Z$ and $\quad j = j_0, j_1$,

and

$$\mu_{k_0}(Z) \neq 0.$$

Let $\zeta = \zeta(x) = [\zeta_1(x), \cdots, \zeta_n(x)]$ be the vector defined as follows:

$$\zeta_{j_0}(x) = -\zeta_{j_1}(x) = \begin{cases} \delta \text{ if } x \in Z \\ 0 \text{ if } x \notin Z \end{cases}$$

*A measurable S is a *null-set* for the measure μ if $\mu(S') = 0$ for every measurable $S' \subset S$.

and all other components vanish identically.

Because of (12), $\eta(x) + \theta\zeta(x)$ is a probability n-vector whenever $-1 \leq \theta \leq 1$. Since

$$\int_X [\eta_{j_0}(x) + \zeta_{j_0}(x)]\, d\mu_{k_0}(x) - \int_X [\eta_{j_0}(x) - \zeta_{j_0}(x)]\, d\mu_{k_0}(x)$$

$$= 2\delta\mu_{k_0}(Z) \neq 0,$$

the points $v(\eta + \zeta)$ and $v(\eta - \zeta)$ are different. But clearly as θ increases from -1 to $+1$ the point $v(\eta + \theta\zeta)$ moves from $v(\eta - \zeta)$ to $v(\eta + \zeta)$ along the segment connecting them. Moreover, $v(\eta)$ is the middle point of this segment and thus it cannot be an extreme point of V_n.

If $v(\eta)$ is an extreme point, then Y is a null-set for all μ_k. Therefore, if we put $\eta^*(x) = \eta(x)$ for $x \notin Y$ and, say, $\eta_1^*(x) = 1$ for $x \in Y$, the decomposition vector $\eta^*(x)$ thus defined satisfies $v(\eta^*) = v(\eta)$. This proves the last assertion of Theorem 2.

THEOREM 3. *If the measures $\mu_1, \cdots, \mu_p$ are finite, then the step-range V_n^0 coincides with the range V_n. More precisely, every point of V_n may be represented as $v(\eta^0)$, where η^0 is a step n-vector whose components assume not more than 2^{np-p+1} different values.*

Proof. According to Theorem 1, V_n is a compact convex set in Euclidean np-dimensional space. However, because of the p equations

$$\sum_{j=1}^{n} \int_X \eta_j(x)\, d\mu_k(x) = \mu_k(x) \qquad (k = 1, \cdots, p)$$

V_n lies in an $N = np - p$ dimensional linear subspace. Hence, according to well-known facts on convex bodies, every point P of V_n may be represented vectorially by

$$P = c_1 P_1 + \cdots + c_N P_N + c_{N+1} P_{N+1},$$

where $P_1, \cdots, P_{N+1}$ are extreme points of V_n and $c_1, \cdots, c_{N+1}$ are nonnegative constants whose sum is 1. According to Theorem 2, we have $P_r = v(\eta^{*r})$ with η^{*r} a decomposition n-vector $(r = 1, 2, \cdots, N+1)$.

Hence, putting $\eta^0 = \sum_{r=1}^{N+1} c_r \eta^{*r}$, we have $P = v(\eta^0)$. Clearly, for every x, every component of $\eta^0(x)$ equals $\sum_{r\epsilon K} c_r$, where K is a subset of $\{1,2,\cdots,N+1\}$. There being 2^{N+1} such subsets, Theorem 3 is proved.

3. Identity of the range and the decomposition range for finite atomless measures. A measurable set S is called an *atom* of the measure μ if $\mu(S) \neq 0$ and if, moreover, for every measurable $S' \subset S$ we have either $\mu(S') = 0$ or $\mu(S') = \mu(S)$. If the measure $\mu(S)$ has no atoms it is called *atomless*.

For atomless measures we can improve on Theorem 3 by establishing the following result.

THEOREM 4. *If $\mu_1, \cdots, \mu_p$ are finite atomless measures then, for every n, the range V_n and the decomposition range V_n^* are identical.*

According to Theorem 1, the common range is convex and compact.

Proof. In view of Theorem 3 it suffices to prove that, in the present case, $V_n^* = V_n^0$.

For this purpose we shall use the following fact: If $\mu_1, \cdots, \mu_p$ are finite and atomless, then, given $0 \leq c \leq 1$, there exists a measurable set S for which

$$\mu_k(S) = c\mu_k(X) \qquad (k = 1, \cdots, p). \tag{13}$$

The existence of such a set S follows immediately from a result of Liapounoff [5] (see also [3]) according to which, under the above stated conditions, the set of points $\mu_1(S), \cdots, \mu_p(S)$ in Euclidean p-space corresponding to all measurable S is convex. Indeed, the empty set Λ and X are certainly measurable and $(1 - c)\ \mu_k(\Lambda) + c\mu_k(X) = c\mu_k(X)$ for all k.

To complete the proof of Theorem 4, we use the following lemma.

LEMMA 2. *If $\mu_1, \cdots, \mu_p$ are finite and atomless and $c_1, \cdots, c_n$ are non-negative numbers satisfying $c_1 + \cdots + c_n = 1$, then there exists a decomposition of X into n disjoint measurable sets $S_1, \cdots, S_n$ having the property that*

$$\mu_k(S_j) = c_j\mu_k(X) \qquad (j = 1, \cdots, n;\ k = 1, \cdots, p). \tag{14}$$

Indeed, according to (13) there exists a measurable S_1 satisfying (14) for $j = 1$. Similarly, there exists a measurable $S_2 \subset X - S_1$ satisfying

$$\mu_k(S_2) = \frac{c_2}{c_2 + \cdots + c_n}\ \mu_k(X - S_1) = c_2\mu_k(X),$$

where we interpret

$$\frac{c_2}{c_2 + \cdots + c_n}$$

as zero if $c_2 = \cdots = c_n = 0$. That is, S_2 satisfies (14) for $j = 2$. In the same manner $S_j \subset X - \bigcup_{i=1}^{j-1} S_i$ satisfying (14) may be obtained for $j = 1, \cdots, n-1$. But then

$$\mu_k\left(X - \bigcup_{j=1}^{n-1} S_j\right) = 1 - (c_1 + \cdots + c_{n-1})\,\mu_k(X) = c_n \mu_k(X) ;$$

thus

$$S_n = X - \bigcup_{j=1}^{n-1} S_j$$

satisfies (14) for $j = n$ as required. Hence, Lemma 2 holds.

The proof of Theorem 4 can now easily be completed. Let $\eta^0(x)$ be any step n-vector. Then X can be decomposed into a finite number of disjoint measurable subsets Y_t over each of which all the components of $\eta^0(x)$ are constant. According to Lemma 2, Y_t may be decomposed into n disjoint measurable sets $S_{1,t}, \cdots, S_{n,t}$ such that

$$(15) \qquad \mu_k(S_{j,t}) = \int_{Y_t} \eta_j^0(x)\, d\mu_k(x) \qquad (j = 1, \cdots, n;\ k = 1, \cdots, p) .$$

Putting $S_j = \bigcup_t S_{j,t} (j = 1, \cdots, n)$ we have, from (15),

$$\int_X \eta_j^0(x)\, d\mu_k(x) = \mu_k(S_j) \qquad (j = 1, \cdots, n;\ k = 1, \cdots, p) .$$

Thus the point $v(\eta^0; \mu_1, \cdots, \mu_p)$ coincides with $v(\eta^*; \mu_1, \cdots, \mu_p)$, where $\eta_j^*(x) = 1$ if $x \in S_j$ and zero otherwise. In other words, $V_n^0 \subset V_n^*$. Since the converse inclusion is obvious, Theorem 4 is proved.

Remarks. (a) Liapounoff [5] proved that if the conditions of Theorem 4 are satisfied then the set of all points $[\mu_1(S), \cdots, \mu_p(S)]$ in Euclidean p-space corresponding to all measurable S is convex and compact. This result is clearly implied by the convexity and compactness of V_n^*; thus the convexity and compactness of V_n^* may be considered as a generalization of Liapounoff's theorem. If we put $\overline{S} = X - S$, then Liapounoff's result is easily seen to be equivalent to the statement that the set of all points $[\mu_1(S), \cdots, \mu_p(S), \mu_1(\overline{S}), \cdots, \mu_p(\overline{S})]$ in Euclidean $2p$ dimensional space is convex and compact. But this amounts precisely to the assertion that V_n^* is convex and compact for $n = 2$. That this assertion remains valid also for $n > 2$ is precisely the generalization of Liapounoff's result contained in Theorem 4.

We used in our proof the convexity part of Liapounoff's result. This is, however, the easier part (cf. Halmos [3]), and thus our method furnishes also a new proof of Liapounoff's theorem.

(b) The values 0 and 1 are among those which the components of η^0 in Theorem 3 are allowed to assume. Hence, on combining the results of Theorems 3 and 4 we see that, if all but p' of the measures $\mu_1, \cdots, \mu_p$ are atomless, we may replace p by p' in the exponent of 2 in Theorem 3. This estimate is again independent of the number of atoms.

(c) If the measures $\mu_1, \cdots, \mu_p$ in Theorem 4 are not assumed to be atomless, then of course V_n^* need not be convex. It is, however, compact as can easily be seen on decomposing into atomless and purely atomic parts and dealing separately with each (see, for example, [3]).

(d) For some applications the following is of importance: If η is a probability n-vector, then there exists a decomposition n-vector η^* with $v(\eta^*) = v(\eta)$ having the further property that, for every $x \in X$ and $j = 1, \cdots, n$, the vanishing of $\eta_j(x)$ implies that of $\eta_j^*(x)$. This assertion follows easily from Theorem 4. Indeed, X may be decomposed into a finite number of measurable sets Y with the following property: If $\eta_j(x) = 0$ for some $x \in Y$, then $\eta_j(x) = 0$ for all $x \in Y$. Let $j_1, \cdots, j_m$ be those j for which $\eta_j(x) > 0$ when $x \in Y$. We may now define $\eta^*(x)$ for $x \in Y$ by applying Theorem 4 (with X replaced by Y and n by m) to the m-vector formed by these components, and putting $\eta_j(x) = 0$ for all other j and $x \in Y$. Combining these definitions for all sets Y, we obtain an η^* with the required property.

4. Extension to arbitrary atomless measures. The assumption of finiteness in Theorem 4 is unnecessary. Indeed, we shall prove the following result.

THEOREM 5. *If the measures $\mu_1, \cdots, \mu_p$ are atomless, then, for every n,* *the range V_n and the decomposition range V_n^* are identical.*

Since the measures are now allowed to assume infinite values, the components of $V(\eta)$ are no longer necessarily finite and one should look upon V_n and V_n^* as imbedded in Euclidean space extended by allowing each coordinate to assume also infinite values.

Before proceeding to the proof we establish the following lemmas.

LEMMA 3. *If μ is a nonnegative atomless measure with $\mu(X) = \infty$, and u is any finite positive number, then there exists a measurable set T with $\mu(T) = u$.*

Proof. Since μ is nonnegative and atomless, there exists a set S with $0 < \mu(S) < \infty$. We first show that $\alpha = \sup \mu$ for all such sets S is infinite. Indeed, assume α finite; then, for every integer m, there exists a measurable S_m with $\mu(S_m) > \alpha - 1/m$. Put $S' = \bigcup_{m=1}^{\infty} S_m$; then $\mu(S') = \alpha$. But $\mu(X - S') = \infty$; hence, there exists a measurable $S'' \subset X - S'$ with $0 < \mu(S'') = b < \infty$. Thus $\alpha < \mu(S' \cup S'')$, contradicting the assumption that α is finite.

Therefore, given u there exists a measurable T' with $u < \mu(T') < \infty$. But then, according to the intermediary values theorem of Sierpinski (see, for example, [2, 52]), or the one dimensional case of Liapounoff's theorem, there exists a measurable $T \subset T'$ with $\mu(T) = u$.

LEMMA 4. *If μ is a nonnegative atomless measure with $\mu(X) = \infty$, and q is any positive integer, then X may be decomposed into q measurable disjoint sets $X_1, \cdots, X_q$ with $\mu(X_1) = \cdots = \mu(X_q) = \infty$.*

Proof. According to Lemma 3, there exist a set T_1 with $\mu(T_1) = 1$, a set $T_2 \subset X - T_1$ with $\mu(T_2) = 1$, a set $T_3 \subset X - (T_1 \cup T_2)$ with $\mu(T_3) = 1$, and so on. Putting $X_i = \bigcup_{n=0}^{\infty} T_{qn+i}$ for $i = 1, \cdots, q - 1$ and $X_q = X - \bigcup_{i=1}^{q-1} X_i$ we obtain the required result.

LEMMA 5. *If $\nu_1, \cdots, \nu_m$ are nonnegative atomless measures with $\nu_1(X) = \cdots \nu_m(X) = \infty$, and q is any positive integer, then X may be decomposed into q measurable sets $X_1, \cdots, X_q$ satisfying $\nu_i(X_1) = \cdots = \nu_i(X_q) = \infty$ for $i = 1, \cdots, m$.*

Proof. For $m = 1$, this lemma reduces to the preceding one. Assume $m > 1$ and the lemma proved for $m - 1$. According to Lemma 4, X is the union of m disjoint measurable sets $Y_1, \cdots, Y_m$ with $\nu_m(Y_1) = \cdots = \nu_m(Y_m) = \infty$. For every $i (i = 1, \cdots, m-1)$ let i' denote the smallest integer for which $\nu_i(Y_{i'}) = \infty$. (Since $\nu_i(X) = \infty$ we have $1 \leq i' \leq m$.) Put $Y' = \bigcup_{i=1}^{m-1} Y_{i'}$ and $Y'' = X - Y'$. Then $\nu_m(Y'') = \infty$ and Y'' is the union of disjoint measurable sets $Y_1'', \cdots, Y_q''$ with $\nu_m(Y_1'') = \cdots = \nu_m(Y_q'') = \infty$. Also $\nu_i(Y'') = \infty$ for $i = 1, \cdots, m-1$ and hence, by the assumption of induction, it can be decomposed into measurable sets $Y_1', \cdots, Y_q'$ with $\nu_i(Y_1') = \cdots = \nu_i(Y_q')$ for $i = 1, \cdots, m-1$. Putting $X_1 = Y_1' \cup Y_1'', \cdots, X_q = Y_q' \cup Y_q''$, we obtain the required decomposition.

LEMMA 6. *Let μ, ν be nonnegative atomless measures with $\mu(X) < \infty$, $\nu(X) = \infty$. Then either X may be decomposed into countably many measurable sets, each having finite ν measure, or there exists a measurable set T with $\mu(T) = 0$, $\nu(T) = \infty$.*

Proof. For every positive integer t consider the measure μ_t defined by

$$\mu_t(S) = \nu(S) - t\mu(S) .$$

According to Hahn (see for example [2, p.18] or [4, p.121]) X may be decomposed into two disjoint measurable sets Y_t and $\overline{Y}_t$ with $\mu_t(S) \leq 0$ for every measurable $S \subset Y_t$ and $\mu_t(S) > 0$ for every measurable $S \subset \overline{Y}$. Clearly,

$$\nu(Y_t) \leq t\mu(Y_t) \leq t\mu(X) < \infty .$$

Put now $Y_t' = Y_1 \cup \cdots \cup Y_t$ and $Z_1 = Y_1'$, $Z_t = Y_t' - Y_{t-1}'$ for $t = 2, 3, \cdots$, and denote by Z_0 the complement of $\bigcup_{t=1}^{\infty} Z_t$. Then $X = \bigcup_{t=0}^{\infty} Z_t$ and $\nu(Z_t) < \infty$ for $t \geq 1$. If $\nu(Z_0) < \infty$ then this is a decomposition of X into countably many sets of finite ν measure. If, on the other hand, $\nu(Z_0) = \infty$ then, by Lemma 3, there exists for every integer u a measurable $T_u \subset Z_0$ with $\nu(T_u) = u$. Moreover, $\mu(T_u) = 0$ since, according to the construction of Z_0, $\mu(S) > 0$ for $S \subset Z_0$ implies $\nu(S) = \infty$. Thus $T = \bigcup_{u=1}^{\infty} T_u$ has the properties required in Lemma 6.

Proof of Theorem 5. Since every measure is the difference between two nonnegative measures, we may assume throughout the proof that the measures $\mu_k (k = 1, \cdots, p)$ are nonnegative.

Let η be any probability n-vector. For every $j (j = 1, \cdots, n)$ we denote by $Y_{j,0}$ the set of x for which $\eta_j(x) = 0$ and by $Y_{j,t} (t = 1, 2, \cdots)$ the set of x for which

$$\frac{1}{t+1} < \eta_j(x) < \frac{1}{t} .$$

We use Y to denote any set of the form

$$\bigcap_{j=1}^{n} Y_{jt_j} \quad \text{with } t_j = 0, 1, 2, \cdots \qquad (j = 1, \cdots, n) .$$

The space X is thus decomposed into countably many sets Y having the following property: There exists a nonempty subset $J = J(Y)$ of $\{1, \cdots, n\}$ and a *positive* $\delta = \delta(Y)$ such that for all $x \in Y$ we have

$$\eta_j(x) > \delta > 0 \quad \text{if } j \in J, \qquad \eta_j(x) = 0 \quad \text{if } j \notin J. \tag{16}$$

Let Y be any such set and consider the subset K' of $\{1, \cdots, p\}$ consisting of all those k for which Y can be decomposed into countably many sets, all having finite μ_k measure. If K' is empty, we call Y final, if not we decompose Y into countably many measurable sets Y' with $\mu_k(Y') < \infty$ for $k \in K'$. Let Y' be any such set and denote by K'' the subset of $\{1, \cdots, p\}$ consisting of all k for which Y' can be decomposed into countably many sets, all having finite μ_k measure. Clearly, $K' \subset K''$. If $K' = K''$ we call Y' final, if not we decompose it into countably many Y'' with $\mu_k(Y'') < \infty$ for $k \in K''$. Again a $K''' \supset K''$ is defined and Y'' is called final if $K'' = K'''$, and so on. After not more than p steps we always end with a final set Z.

We have thus decomposed Y, and hence X, into countably many sets Z having the following property: To every Z there corresponds a decomposition of $\{1, 2, \cdots, p\}$ into two disjoint sets K and K such that $\mu_k(Z) < \infty$ if $k \in K$, while if $k \in \overline{K}$ then Z cannot be decomposed into countably many sets, all having finite μ_k measure. Furthermore, since Z is contained in some Y, (16) holds for all $x \in Z$.

Next, we show how to decompose Z into disjoint measurable sets $Z_1, \cdots, Z_n$ satisfying

$$\mu_k(Z_j) = \int_Z \eta_j(x)\, d\mu_k(x) \qquad (j = 1, \cdots, n;\ k = 1, \cdots, p) . \tag{17}$$

(If $\eta_j(x) = 0$ for all $x \in Z$, the right side of (17) is understood to be 0 even when $\mu_k(Z) = \infty$.)

If K is empty, then the possibility of such a decomposition is assured by Theorem 4.

If K is empty then, by (16), the integral in (17) is infinite if $j \in J$ and is zero otherwise. By Lemma 5 it is possible to decompose Z into sets Z_j $(j \in J)$ with

$$\mu_k(Z_j) = \infty \quad \text{for} \quad k = 1, \cdots, p .$$

Denoting the empty set, for $j \notin J$, by Z_j, we have a decomposition satisfying (17).

Finally, assume both K and $\bar{K}$ nonempty. We define a nonnegative measure μ by $\mu(S) = \Sigma_{k \in K} \mu_k(S)$. Clearly, μ is atomless and $\mu(Z) < \infty$. According to Lemma 6 there exists, for every $k \in \bar{K}$, a measurable $T_k \subset Z$ with $\mu(T_k) = 0$, $\mu_k(T_k) = \infty$. Let T be the union of $T_k (k \in \bar{K})$. Then (see the treatment of the case when K is empty) it is possible to decompose T into disjoint measurable sets $Z_1', \cdots, Z_n'$ so that Z_j' is empty for $j \notin J$, while for all $j \in J$ and $k \in \bar{K}$ we have $\mu_k(Z_j') = \infty$. Since $\mu(T) = 0$ we have, for all j, $\mu_k(Z_j') = 0$ whenever $k \in K$. Let $T' = Z - T$; then it is possible, by Theorem 4, to decompose T' into disjoint measurable sets $T_1', \cdots, T_n'$ such that T_j' is empty for $j \notin J$, while for $j \in J$ and $k \in K$ we have

$$\mu_k(Y_j) = \int_{T'} \eta_j(x)\, d\mu_k(x) = \int_Z \eta_j(x)\, d\mu_k(x) .$$

Putting $Z_j = T_j' \cup Z_j'$ for $j = 1, \cdots, n$, we have a decomposition satisfying (17).

We now define the decomposition n-vector η^* as follows: For $x \in Z$, put $\eta_j^*(x) = 1$ if $x \in Z_j$, and $\eta_j^*(x) = 0$ for all other $x \in Z$. Because of the countable additivity of the measures and the integrals, (17) implies $v(\eta^*) = v(\eta)$ and the proof is completed.

Remarks. (a) The last remark after Theorem 4 applies also here. Indeed, our construction in the proof of Theorem 5 yields a vector having the properties required of η^* in that remark.

(b) In applications usually X can be decomposed into countably many sets of finite μ_k measure $(k = 1, \cdots, p)$. For this special case Theorem 5 is, of course, an immediate consequence of Theorem 4.

4. Application to statistics and the theory of games.* Theorem 4 (together with its extension mentioned in the last remark of the preceding section) has

* A more detailed discussion and other results, including a discussion of the sequential statistical decision problem, are contained in our paper, *Elimination of randomization in certain statistical decision procedures and zero-sum two-person games*, Annals of Mathematical Statistics, 22, No. 1, March, 1951. A brief discussion of these applications was also given in an earlier publication [1].

immediate applications to the following statistical decision problem: Let $y = \{y_1, \cdots, y_t\}$ be a random vector with t components, where t is a given positive integer. For every point $x = (x_1, \cdots, x_t)$ of the t-dimensional Euclidean space X, let $F(x)$ denote the probability that $y_i < x_i$ for $i = 1, \cdots, t$; that is, $F(x)$ is the distribution function of y. The distribution function $F(x)$ is assumed to be unknown. It is known, however, that $F(x)$ is one of the distribution functions $F_1(x), \cdots, F_m(x)$. An observation x is made on y and according to the observed value x the statistician may adopt any one of n decisions $j (j = 1, \cdots, n)$. Let $W_{i,j}(x)$ denote the loss sustained by the statistician when $F_i(x)$ is the true distribution of y, x is the observed value of y, and the jth decision is adopted. $W_{i,j}(x)$ is assumed to be a finite nonnegative and measurable function of x. If the statistician, on observing the value x, adopts the various decisions with probabilities $\eta_j(x)$, where these are nonnegative measurable functions satisfying (1), then the risk, or expected loss, when $F_i(x)$ is the true distribution function, is given by

$$r_i(\eta) = \sum_{j=1}^{n} \int_X W_{i,j}(x)\, \eta_j(x)\, dF_i(x) .$$

The decision function $\eta_j(x)$ is said to be nonrandomized if for every x all but one of the $\eta_j(x)$ vanish. Theorem 4 yields without difficulty the following result: *If the distribution functions $F_i(x)$ $(i = 1, \cdots, m)$ are atomless then, given any decision function $\eta(x)$, there exists a nonrandomized decision function $\eta^*(x)$ such that $r_i(\eta) = r_i(\eta^*)$ $(i = 1, \cdots, m)$.*

Similar application can be made to the theory of games. In fact, the above described statistical decision problem may be interpreted as a zero-sum two-person game as follows: Player 1 has a finite number of pure strategies $i (i = 1, \cdots, m)$, while a pure strategy of Player 2 is a nonrandomized decision function $\eta^*(x)$ (decomposition n-vector). If i is the pure strategy of Player 1 and $\eta^*(x)$ the pure strategy of Player 2, the outcome is defined by

$$R[i, \eta^*(x)] = r_i(\eta^*) .$$

A mixed strategy of Player 1 is represented by a vector $\xi = (\xi_1, \cdots, \xi_m)$ with nonnegative components whose sum is one, while a mixed strategy of Player 2 is given by a probability n-vector $\eta(x)$. The expected value of the outcome corresponding to the mixed strategies ξ and $\eta(x)$ is given by

$$R[\xi,\eta(x)] = \sum_{i=1}^{m} \xi_i r_i(\eta) .$$

The above stated result for the statistical decision problem can be restated in game terminology as follows: *If the distribution functions $F_i(x)$ $(i = 1, \cdots, m)$ are atomless, then given any mixed strategy $\eta(x)$ of Player 2, there exists a pure strategy $\eta^*(x)$ such that $R[\xi, \eta^*(x)] = R[\xi, \eta(x)]$ for all strategies ξ of Player 1.*

References

1. A. Dvoretzky, A. Wald, and J. Wolfowitz, *Elimination of randomization in certain problems of statistics and of the theory of games*, Proc. Nat. Acad. Sci. U.S.A. 36 (1950), 256-259..

2. H. Hahn and A. Rosenthal, *Set functions*, University of New Mexico Press, Albuquerque (1948).

3. P. R. Halmos, *The range of a vector measure*, Bull. Amer. Math. Soc. 54 (1948), 416-421.

4. ——————, *Measure theory*, Van Nostrand, New York (1950).

5. A. Liapounoff, *Sur les fonctions-vecteurs complètement additives*, Bull. Acad. Sci. URSS Sér. Math. [Izvestia Akad. Nauk SSSR], 4 (1940), 465-478.

6. A. Wald, *Statistical decision functions*, Wiley, New York (1950).

National Bureau of Standards, Los Angeles
Columbia University

Reprinted from THE ANNALS OF MATHEMATICAL STATISTICS
Vol. 22, No. 1, March, 1951

ELIMINATION OF RANDOMIZATION IN CERTAIN STATISTICAL DECISION PROCEDURES AND ZERO-SUM TWO-PERSON GAMES[1]

BY A. DVORETZKY,[2] A. WALD,[3] AND J. WOLFOWITZ[3]

Institute for Numerical Analysis and Columbia University

Summary. The general existence of minimax strategies and other important properties proved in the theory of statistical decision functions (e.g., [3]) and the theory of games (e.g., [5]) depends upon the convexity of the space of decision functions and the convexity of the space of strategies. This convexity can be obtained by the use of randomized decision functions and mixed (randomized) strategies. In Section 2 of the present paper the authors state the extension (first announced in [1]) of a measure theoretical result known as Lyapunov's theorem [2]. This result is applied in Section 3 to the statistical decision problem where the number of distributions and decisions is finite. It is proved that when the distributions are continuous (more generally, "atomless," see footnote 7 below) randomization is unnecessary in the sense that every randomized decision function can be replaced by an equivalent nonrandomized decision function. Section 4 extends this result to the case when the decision space is compact. Section 5 extends the results of Section 3 to the sequential case. Sections 6 and 7 show, by counterexamples, that the results of Section 3 cannot be extended to the case of infinitely many distributions without new restrictions.[4] Section 8 gives sufficient conditions for the elimination of randomization under maintenance of ϵ-equivalence. Section 9 concludes with a restatement of the results in the language of the theory of games.

1. Introduction. We shall consider the following statistical decision problem: Let x be the generic point in an n-dimensional Euclidean[5] space R, and let Ω be a given class of cumulative distribution functions $F(x)$ in R. The cumulative distribution function $F(x)$ of the vector chance variable $X = (X_1, \cdots, X_n)$ with range in R is not known. It is known, however, that F is an element of the given class Ω. There is also given a space D whose elements d represent the possible decisions that can be made by the statistician in the problem under consideration. Let $W(F, d, x)$ denote the "loss" when F is the true distribution of

[1] The main results of this paper were announced without proof in an earlier publication [1] of the authors.

[2] On leave of absence from the Hebrew University, Jerusalem, Israel.

[3] Research under a contract with the Office of Naval Research.

[4] The impossibility of such an extension is related to the failure of Lyapunov's theorem when infinitely many measures are considered. (cf. A. LYAPUNOV, "Sur les fonctions-vecteurs complètement additives," *Izvestiya Akad. Nauk SSSR. Ser. Mat.*, Vol. 10 (1946), pp. 277–279.)

[5] The restriction to a Euclidean space is not essential (see [1]).

X, the decision d is made and x is the observed value of X. We shall define the distance between two elements d_1 and d_2 of D by

$$\rho(d_1, d_2) = \operatorname*{Sup}_{F,x} | W(F, d_1, x) - W(F, d_2, x) |. \tag{1.1}$$

Let B be the smallest Borel field of subsets of D which contains all open subsets of D as elements. Let B_0 be the totality of Borel sets of R. We shall assume that $W(F, d, x)$ is bounded[6] and, for every F, a function of d and x which is measurable $(B \times B_0)$. By a decision function $\delta(x)$ we mean a function which associates with each x a probability measure on D defined for all elements of B. We shall occasionally use the symbol δ_x instead of $\delta(x)$ when we want to emphasize that x is kept fixed. A decision function $\delta(x)$ is said to be nonrandomized if for every x the probability measure $\delta(x)$ assigns the probability one to a single point d of D. For any measurable subset D^* of D (D^* an element of B), the symbol $\delta(D^* \mid x)$ will denote the probability measure of D^* according to the set function $\delta(x)$. It will be assumed throughout this paper that for any given D^* the function $\delta(D^* \mid x)$ is a Borel measurable function ot x. The adoption of a decision function $\delta(x)$ by the statistician means that he proceeds according to the following rule: Let x be the observed value of X. The element d of the space D is selected by an independent chance mechanism constructed in such a way that for any measurable subset D^* of D the probability that the selected element d will be included in D^* is equal to $\delta(D^* \mid x)$.

Given the sample point x and given that $\delta(x)$ is the decision function adopted, the expected value of the loss $W(F, d, x)$ is given by

$$W^*(F, \delta, x) = \int_D W(F, d, x)\, d\delta_x. \tag{1.2}$$

The expected value of the loss $W(F, d, x)$ when F is the true distribution of X and $\delta(x)$ is the decision function adopted (but x is not known) is obviously equal to

$$r(F, \delta) = \int_R W^*(F, \delta, x)\, dF(x). \tag{1.3}$$

The above expression is called the risk when F is true and δ is adopted.

We shall say that the decision functions $\delta(x)$ and $\delta^*(x)$ are equivalent if

$$r(F, \delta^*) = r(F, \delta) \qquad \text{for all } F \text{ in } \Omega. \tag{1.4}$$

We shall say that $\delta(x)$ and $\delta^*(x)$ are strongly equivalent if for every measurable subset D^* of D we have

$$\int_R \delta(D^* \mid x)\, dF(x) = \int_R \delta^*(D^* \mid x)\, dF(x) \qquad \text{for all } F \text{ in } \Omega. \tag{1.5}$$

[6] The restriction of boundedness is not essential (see [1]).

If δ and δ^* are strongly equivalent, they are equivalent for any loss function which is a function of F and d only.

For any positive ϵ, we shall say that $\delta(x)$ and $\delta^*(x)$ are ϵ-equivalent if

$$|r(F, \delta) - r(F, \delta^*)| \leqq \epsilon \quad \text{for all } F \text{ in } \Omega, \tag{1.6}$$

and strongly ϵ-equivalent if

$$\left| \int_R \delta(D^* \mid x)\, dF(x) - \int_R \delta^*(D^* \mid x)\, dF(x) \right| \leqq \epsilon \tag{1.7}$$

for all measurable D^* and for all F in Ω.

In Section 2 we state a measure-theoretical result first announced in [1] and proved in [6]. This result is then used in Section 3 to prove that for every decision function there exists an equivalent, as well as a strongly equivalent, nonrandomized decision function δ^*, if Ω and D are finite and if each element $F(x)$ of Ω is atomless.[7] This result is extended in Section 4 to the case where D is compact. Section 5 deals with the sequential case for which similar results are proved. A precise definition of a sequential decision function is given in Section 5.

The finiteness of Ω is essential for the validity of the results given in Sections 2–5. The examples given in Section 6 show that even when Ω is such a simple class as the class of all univariate normal distributions with unit variance, there exist decision functions δ such that no equivalent nonrandomized decision functions exist. In Section 7, an example is given where a decision function δ and a positive ϵ exist such that no nonrandomized decision function δ^* is ϵ-equivalent to δ.

In Section 8, sufficient conditions are given which guarantee that for every δ and for every $\epsilon > 0$ there exists a nonrandomized decision function δ^* which is ϵ-equivalent to δ.

2. A measure-theoretical result. Let $\{y\} = Y$ be any space and let $\{S\} = \mathfrak{S}$ be a Borel field of subsets of Y. Let $\mu_k(S)(k = 1, \cdots, q)$ be a finite number of real-valued, σ-finite and countably additive set functions defined for all $S \in \mathfrak{S}$. The following theorem was stated by the authors [1]:

THEOREM 2.1. *Let $\delta_j(y)$ $(j = 1, 2, \cdots, m)$ be real non-negative $\mathfrak{S}$-measurable functions satisfying*

$$\sum_{j=1}^{m} \delta_j(y) = 1 \tag{2.1}$$

for all $y \in Y$. Then if the set functions $\mu_k(S)$ are atomless there exists a decomposition of Y into m disjoint subsets $S_1, \cdots, S_m$ belonging to $\mathfrak{S}$ having the property

[7] A set function μ defined on a Borel field $\mathfrak{S}$ is called atomless if it has the following property: If for some $S \in \mathfrak{S}$, $\mu(S) \neq 0$, then there exists an $S' \subset S$ such that $S' \in \mathfrak{S}$ and such that $\mu(S') \neq \mu(S)$ and $\mu(S') \neq 0$. A cumulative distribution function is called atomless if its associated set function is atomless.

that

$$\int_Y \delta_j(y)\, d\mu_k(y) = \mu_k(S_j) \qquad (j = 1, \cdots, m;\ k = 1, \cdots, q). \tag{2.2}$$

If $\delta_j^*(y) = 1$ *for all* $y \in S_j$ *and* $= 0$ *for any other* $y (j = 1, \cdots, m)$, *then the above equation can be written as*

$$\int_Y \delta_j(y)\, d\mu_k(y) = \int_Y \delta_j^*(y)\, d\mu_k(y) \qquad (j = 1, \cdots, m;\ k = 1, \cdots, q). \tag{2.3}$$

This theorem is an extension of a result of A. Lyapunov [2] and is basic for deriving most of the results of the present paper.

3. Elimination of randomization when Ω and D are finite and each element $F(x)$ of Ω is atomless. In this section we shall assume that Ω consists of the elements $F_1(x), \cdots, F_p(x)$ and D of the elements $d_1, \cdots, d_m$. Moreover, we assume that $F_i(x)$ is atomless for $i = 1, \cdots, p$. A decision function $\delta(x)$ is now given by a vector function $\delta(x) = [\delta_1(x), \cdots, \delta_m(x)]$ such that

$$\delta_j(x) \geqq 0, \qquad \sum_{j=1}^{m} \delta_j(x) = 1 \tag{3.1}$$

for all $x \in R$. Here $\delta_j(x)$ is the probability that the decision d_j will be made when x is the observed value of X. The risk when F_i is true and the decision function $\delta(x)$ is adopted is now given by

$$r(F_i, \delta) = \sum_{j=1}^{m} \int_R W(F_i, d_j, x)\delta_j(x)\, dF_i(x). \tag{3.2}$$

A nonrandomized decision function $\delta^*(x)$ is a vector function whose components $\delta_j^*(x)$ can take only the values 0 and 1 for all x.

For any measurable subset S of R let

$$\nu_{ij}(S) = \int_S W(F_i, d_j, x)\, dF_i(x) \qquad (i = 1, \cdots p;\ j = 1, \cdots, m). \tag{3.3}$$

Then the measures $\nu_{ij}(S)$ are finite, atomless and countably additive. Using these set functions, equation (3.2) can be written as

$$r(F_i, \delta) = \sum_{j=1}^{m} \int_R \delta_j(x)\, d\nu_{ij}(x). \tag{3.4}$$

Replacing in Theorem 2.1 the space Y by R, the set of measures $\{\mu_1, \cdots, \mu_q\}$ by the set $\{\nu_{ij}\} (i = 1, \cdots, p;\ j = 1, \cdots, m)$, it follows from Theorem 2.1 that there exists a nonrandomized decision function $\delta^*(x)$ such that

$$\int_R \delta_j(x)\, d\nu_{ij}(x) = \int_R \delta_j^*(x)\, d\nu_{ij}(x) \qquad (i = 1, \cdots, p;\ j = 1, \cdots, m). \tag{3.5}$$

This immediately yields the following theorems:

THEOREM 3.1. *If Ω and D are finite and if each element $F(x)$ of Ω is atomless, then for any decision function $\delta(x)$ there exists an equivalent nonrandomized decision function $\delta^*(x)$.*

Putting $W(F, d, x) = 1$ identically in F, d and x, equation (3.5) immediately yields the following theorem:

THEOREM 3.2. *If Ω and D are finite and if each element $F(x)$ of Ω is atomless, then for any decision function $\delta(x)$ there exists a strongly equivalent nonrandomized decision function $\delta^*(x)$.*

4. Elimination of randomization when Ω is finite, D is compact and each element $F(x)$ of Ω is atomless. Again, let $\Omega = \{F_1, \cdots, F_p\}$ where the distributions F are atomless. If the loss $W(F, d, x)$ does not depend on x, the finiteness of Ω implies that D is at least conditionally compact with respect to the metric (1.1) (see Theorem 3.1 in [3]). We postulate that D is compact (but permit the loss to depend on x), and shall prove that if $\delta(x)$ is any decision function, there exists a nonrandomized decision function $\delta^*(x)$ such that $\delta^*(x)$ is equivalent to $\delta(x)$, i.e..

$$r_i(\delta) = r_i(\delta^*) \qquad (i = 1, \cdots, p), \tag{4.1}$$

where $r_i(\delta)$ stands for $r(F_i, \delta)$.

Since D is compact there exists an infinite sequence of decompositions of the space D into a finite number of disjoint nonempty measurable sets, the lth decomposition to be $C(1, 1, \cdots, 1), \cdots, C(k_1, \cdots, k_l)$ with the properties:

(a) Any two sets C which have the same number of indices not all identical, are disjoint.

(b) The sum of all sets with the same number l of indices is D ($l = 1, 2, \cdots$ ad inf.).

(c) If the sequence of indices of one set C constitutes a proper initial part of the sequence of indices of another set C, the first set includes the second.

(d) The diameters of all sets with l indices are bounded above by $h(l)$ and

$$\lim_{l \to \infty} h(l) = 0.$$

Let l be fixed and define

$$\Delta_{m_1, \cdots, m_l}(x) = \delta[C(m_1, \cdots, m_l \mid x]. \tag{4.2}$$

Define, furthermore,

$$\begin{aligned} W_i[x, C(m_1, \cdots, m_l)] &= \frac{1}{\Delta_{m_1 \cdots m_l}(x)} \int_{C(m_1, \cdots, m_l)} W(F_i, d, x)\, d\delta_x \\ &\qquad \text{if } \Delta_{m_1 \cdots m_l}(x) > 0, \\ &= 0 \qquad \text{if } \Delta_{m_1 \cdots m_l}(x) = 0. \end{aligned} \tag{4.3}$$

Clearly,

$$(4.4)\qquad r_i(\delta) = \sum_{m_l=1}^{k_l} \cdots \sum_{m_1=1}^{k_1} \int_R W_i[x, C(m_1, \cdots, m_l)]\Delta_{m_1\cdots m_l}(x)\, dF_i(x).$$

Considering a decision space D_l with elements $d_{m_1\cdots m_l}$ $(m_i = 1, \cdots, k_i;$ $i = 1, \cdots, l)$ and putting the loss $W(F_i, d_{m_1\cdots m_l}, x) = W_i[x, C(m_1, \cdots, m_l)]$, equations (3.3) and (3.5) imply that there exists a finite sequence of measurable functions $\bar{\Delta}_{m_1\cdots m_l}(x)$ $(m_1 = 1, \cdots, k_1; \cdots; m_l = 1, \cdots, k_l)$ such that

$$(4.5)\qquad \bar{\Delta}_{m_1\cdots m_l}(x) = 0 \text{ or } 1 \qquad \text{for all } x,$$

$$(4.6)\qquad \sum_{m_l} \cdots \sum_{m_1} \bar{\Delta}_{m_1\cdots m_l}(x) = 1 \qquad \text{for all } x,$$

$$(4.7)\qquad \bar{\Delta}_{m_1\cdots m_l}(x) = 0 \qquad \text{whenever } \Delta_{m_1\cdots m_l}(x) = 0,$$

and

$$(4.8)\qquad \begin{aligned} &\int_R W_i[x, C(m_1, \cdots, m_l)]\bar{\Delta}_{m_1\cdots m_l}(x)\, dF_i(x) \\ &\qquad = \int_R W_i[x, C(m_1, \cdots, m_l)\Delta_{m_1\cdots m_l}(x)\, dF_i(x). \end{aligned}$$

Let now $\bar{\delta}(x)$ be the decision function for which

$$(4.9)\qquad \bar{\delta}[C(m_1, \cdots, m_l) \mid x] = \bar{\Delta}_{m_1\cdots m_l}(x)$$

and for any measurable subset $D_{m_1\cdots m_l}$ of $C(m_1, \cdots, m_l)$

$$(4.10)\qquad \bar{\delta}[D_{m_1\cdots m_l} \mid x]\bar{\Delta}_{m_1\cdots m_l}(x) = \frac{\delta(D_{m_1\cdots m_l} \mid x)}{\delta[C(m_1, \cdots, m_l) \mid x]},$$

where $\dfrac{\delta(D_{m_1\cdots m_l} \mid x)}{\delta[C(m_1, \cdots, m_l) \mid x]}$ is defined to be $= 0$ when $\delta[C(m_1, \cdots, m_l) \mid x] = 0$.

It then follows from (4.4) and (4.8) that

$$(4.11)\qquad r_i(\delta) = r_i(\bar{\delta}).$$

Applying the above result for $l = 1$, we conclude that there exists a decision function $\delta^1(x)$ with the following properties: The choice among the C's with one index is nonrandom. The decision, once given the C (with one index) chosen, is made according to $\delta(x)$. We have $\delta^1[C(m_1) \mid x] = 0$ whenever $\delta[C(m_1) \mid x] = 0$ and

$$r_i(\delta) = r_i(\delta^1) \qquad (i = 1, \cdots, p).$$

Repeat the above procedure for every C with two indices, using $W_i\{x, C(m_1, m_2)\}$ as weight function and $\delta^1(x)$ as the decision function. We conclude that there exists a decision function $\delta^2(x)$ with the following properties: The choice among the C's with two indices is nonrandom. $\delta^2[C(m_1, m_2) \mid x] = 0$ whenever $\delta^1[C(m_1, m_2) \mid x] = 0$. The decision, once given the C (with two indices) chosen, is made according to $\delta^1(x)$ and, therefore, in accordance with $\delta(x)$. We have

$$\int_R \int_{C(m_1)} W(F_i, d, x)\, d\delta_x^1\, dF_i(x) = \int_R \int_{C(m_1)} W(F_i, d, x)\, d\delta_x^2\, dF_i(x) \quad \begin{matrix}(m_1 = 1, 2, \cdots, k_1) \\ (i = 1, \cdots, p).\end{matrix}$$

Repeat the above procedure for all C's with l indices, $l = 3, 4, \cdots$ ad inf. At the lth stage we obtain a decision function $\delta^l(x)$ with the following properties: The decision among the C's with l indices is nonrandom. $\delta^l[C(m_1, \cdots, m_l) \mid x] = 0$ whenever $\delta^{l-1}[C(m_1, \cdots, m_l) \mid x] = 0$. The decision, once given the chosen C with l indices, is made according to $\delta(x)$. We have

$$\int_R \int_{C(m_1, \cdots, m_{l-1})} W(F_i, d, x)\, d\delta_x^{l-1}\, dF_i(x) = \int_R \int_{C(m_1, \cdots, m_{l-1})} W(F_i, d, x)\, d\delta_x^l\, dF_i(x)$$

$$\begin{pmatrix} i = 1, \cdots, p \\ m_1 = 1, \cdots, k_1 \\ m_{l-1} = 1, \cdots, k_{l-1} \end{pmatrix}.$$

Hold x fixed and let $C(x; l)$ be that C with l indices for which

$$\int_{C(x;l)} d\delta_x^l = 1.$$

Then $C(x; l + 1)$ is a proper subset of $C(x; l)$ for every positive l. The sequence $C(x; l)$, $l = 1, 2, \cdots$, determines, because D is compact, a unique limit point $c(x)$ such that any neighborhood of $c(x)$ contains almost all sets $C(x; l)$. Hence the sequence of probability measures $\delta_x^l (l = 1, 2, \cdots$, ad inf.) converges to a limit probability measure δ_x^* which assigns probability one to any measurable set which contains the point $c(x)$. Since $W(F_i, d, x)$ is continuous in d, we have

$$\lim_{l=\infty} \int_D W(F_i, d, x)\, d\delta_x^l = \int_D W(F_i, d, x)\, d\delta_x^* \tag{4.12}$$

for any x.

Now let x vary over R. It follows from (4.12) and the boundedness of $W(F, d, x)$ that $\lim_{l=\infty} r_i(\delta^l) = r_i(\delta^*)$. Since $r_i(\delta^l) = r_i(\delta)$, also $r_i(\delta^*) = r_i(\delta)$ $(i = 1, \cdots, p)$. Thus the probability measures $\delta^*(x)$ constitute the desired nonrandomized decision function.

It remains to show that for any measurable subset D^* of D, the function $\delta^*(D^* \mid x)$ is a measurable function of x. The measurability of $\delta^*(D^* \mid x)$ can easily be shown for any D^*, if it is shown for all closed sets D^*, since every measurable set can be attained by a denumerable number of Borel operations (denumerably infinite sums and complements) starting with closed sets. Thus we shall assume that D^* is closed. For any positive ρ let D_ρ^* be the sum of all open spheres with center in D^* and radius ρ. It is easy to see that

$$\delta^*(D_{2\rho}^* \mid x) \geqq \liminf_{l=\infty} \delta^l(D_\rho^* \mid x) \geqq \delta^*(D^* \mid x).$$

Since $\lim_{\rho=0} \delta^*(D_{2\rho}^* \mid x) = \delta^*(D^* \mid x)$, it follows from the above relation that

$$\lim_{\rho=0} \liminf_{l} \delta^l(D_\rho^* \mid x) = \delta^*(D^* \mid x).$$

Since $\delta^l(D_\rho^* \mid x)$ is a measurable function of x, the measurability of $\delta^*(D^* \mid x)$ is proved.

5. Elimination of randomization in the sequential case. In this section we shall consider the following sequential decision problem: Let $X = \{X_n\}$ $(n = 1, 2, \cdots, \text{ad inf.})$ be a sequence of chance variables. Let x be the generic point in the space $\bar{R}$ of all infinite sequences of real numbers, i.e., $x = \{x_n\}$ $(n = 1, 2, \cdots, \text{ad inf.})$ where each x_n is a real number. It is known that the distribution function $F(x)$ of X is an element of Ω, where Ω consists of a finite number of distribution functions $F_1(x), \cdots, F_p(x)$, and that the distribution function of X_1 is continuous according to $F_i(x)$, $i = 1, \cdots, p$. The statistician is assumed to have a choice of a finite number of (terminal) decisions $d_1, \cdots, d_m$, i.e., the space D consists of the elements $d_1, d_2, \cdots, d_m$. A decision rule δ is now given by a sequence of nonnegative, measurable functions $\delta_{\nu t}(x_1, \cdots, x_t)$ $(\nu = 0, 1, \cdots, m; t = 1, 2, \cdots, \text{ad inf.})$ satisfying

$$\sum_{\nu=0}^{m} \delta_{\nu t}(x_1, \cdots, x_t) = 1 \tag{5.1}$$

for $-\infty < x_1, \cdots, x_t < \infty$. The decision rule δ is defined in terms of the functions $\delta_{\nu t}$ as follows: After the value x_1 of X_1 has been observed, the statistician decides either to continue experimentation and take another observation, or to stop further experimentation and adopt a terminal decision $d_j (j = 1, \cdots, m)$ with the respective probabilities $\delta_{01}(x_1)$ and $\delta_{j1}(x_1)$ $(j = 1, \cdots, m)$. If it is decided to continue experimentation, a value x_2 of X_2 is observed and it is again decided either to take a further observation or adopt a terminal decision $d_j (j = 1, \cdots, m)$ with the respective probabilities $\delta_{02}(x_1, x_2)$ and $\delta_{j2}(x_1, x_2) (j = 1, \cdots, m)$, etc. The decision rule is called nonrandomized if each $\delta_{\nu t}$ can take only the values 0 and 1.

Let $v_{i\nu t}(x_1, \cdots, x_t)$ represent the sum of the loss and the cost of experimentation when F_i is true, the terminal decision d_ν is made and experimentation is terminated with the tth observation

$$(\nu = 1, 2, \cdots, m; i = 1, \cdots, p; t = 1, 2, \cdots, \text{ad inf.}).$$

The functions $v_{i\nu t}(x_1, \cdots, x_t)$ are assumed to be finite, nonnegative and measurable. We shall consider only decision rules δ for which the probability is one that experimentation will be terminated at some finite stage. The risk (expected loss plus expected cost of experimentation) when F_i is true and the rule δ is adopted is then given by

$$r_i(\delta) = \sum_{t=1}^{\infty}\sum_{\nu=1}^{m}\int_{R_t} v_{i\nu t}(x_1, \cdots, x_t)\delta_{01}(x_1)\delta_{02}(x_1, x_2) \cdots \delta_{0(t-1)}(x_1, \cdots, x_{t-1}) \cdot \delta_{\nu t}(x_1, \cdots, x_t)\, dF_{it}(x_1, \cdots, x_t), \tag{5.2}$$

where R_t is the t-dimensional space of $x_1, \cdots, x_t$ and $F_{it}(x_1, \cdots, x_t)$ is the cumulative distribution function of $X_1, \cdots, X_t$ when F_i is the distribution function of X.

We shall say that the decision rules δ^1 and δ^2 are equivalent if $r_i(\delta^1) = r_i(\delta^2)$ for $i = 1, \cdots, p$. We shall say that δ^1 and δ^2 are strongly equivalent if

$$\int_{R_t} v_{i\nu t}(x_1, \cdots, x_t)\delta^1_{01}(x_1) \cdots \delta^1_{0(t-1)}(x_1, \cdots, x_{t-1})\delta^1_{\nu t}(x_1, \cdots, x_t)\, dF_{it} = \int_{R_t} v_{i\nu t}(x_1, \cdots, x_t)\delta^2_{01}(x_1) \cdots \delta^2_{0(t-1)}(x_1, \cdots, x_{t-1})\delta^2_{\nu t}(x_1, \cdots, x_t)\, dF_{it} \tag{5.3}$$

for $i = 1, 2, \cdots, p$; $\nu = 1, \cdots, m$ and $t = 1, 2, \cdots$, ad inf.

Clearly, if δ^1 and δ^2 are strongly equivalent and if the functions $v_{i\nu t}(x_1, \cdots, x_t)$ reduce to constants $v_{i\nu t}$, then δ^1 and δ^2 are equivalent for all possible choices of the constants $v_{i\nu t}$.

Let

$$\varphi_i(x, \delta) = \sum_{t=1}^{\infty}\sum_{\nu=1}^{m} v_{i\nu t}(x_1, \cdots, x_t)\delta_{01}(x_1) \cdots \delta_{0(t-1)}(x_1, \cdots, x_{t-1})\delta_{\nu t}(x_1, \cdots, x_t). \tag{5.4}$$

We shall prove the following lemma:

LEMMA 5.1. *Let δ be a decision rule for which $\varphi_i(x, \delta) < \infty$ for all x, except perhaps on a set of x's whose probability is zero according to every distribution function $F_i(x) (i = 1, \cdots, p)$. Let τ and T be given positive integers. Then there exists a decision function $\bar{\delta}$ with the following properties:*

$$\bar{\delta}_{\nu\tau}(x_1, \cdots, x_\tau) = 0 \quad \text{or} \quad 1, \qquad \sum_{\nu=0}^{m} \bar{\delta}_{\nu\tau}(x_1, \cdots, x_\tau) = 1, \tag{5.5}$$

for every point in $R_\tau (\nu = 0, 1, \cdots, m)$,

$$\bar{\delta}_{\nu t}(x_1, \cdots, x_t) = \delta_{\nu t}(x_1, \cdots, x_t) \qquad (\nu = 0, 1, \cdots, m; t \neq \tau), \tag{5.6}$$

$$r_i(\delta) = r_i(\bar{\delta}) \qquad (i = 1, \cdots, p), \tag{5.7}$$

$$\int_{R_t} v_{i\nu t}\delta_{01} \cdots \delta_{0(t-1)}\delta_{\nu t}\, dF_{it} = \int_{R_t} v_{i\nu t}\bar{\delta}_{01} \cdots \bar{\delta}_{0(t-1)}\bar{\delta}_{\nu t}\, dF_{it} \qquad (\nu = 1, \cdots, m; t = 1, \cdots, T), \tag{5.8}$$

$$\varphi_i(x, \bar{\delta}) < \infty, \tag{5.9}$$

for all x except perhaps on a set whose probability is zero according to every distribution $F_i(x)(i = 1, \cdots, p)$.

PROOF. We can write $\varphi_i(x, \delta)$ as follows:

$$\varphi_i(x, \delta) = \sum_{t=1}^{\tau-1} \sum_{\nu=1}^{m} v_{i\nu t}(x_1, \cdots, x_t)\delta_{01} \cdots \delta_{0(t-1)}\delta_{\nu t} + \sum_{t=\tau}^{\infty} \sum_{\nu=0}^{m} g_{i\nu\tau t}(x_1, \cdots, x_t)\delta_{\nu\tau}, \tag{5.10}$$

where $g_{i\nu\tau t}(x_1, \cdots, x_t)$ does not depend on $\delta_{0\tau}, \delta_{1\tau}, \cdots, \delta_{m\tau}$. The first double sum reduces to zero when $\tau = 1$. Clearly, if a $\bar{\delta}$ with the desired properties exists, then

$$\varphi_i(x, \bar{\delta}) = \sum_{t=1}^{\tau-1} \sum_{\nu=1}^{m} v_{i\nu t}(x_1, \cdots, x_t)\delta_{01} \cdots \delta_{0(t-1)}\delta_{\nu t} + \sum_{t=\tau}^{\infty} \sum_{\nu=0}^{m} g_{i\nu\tau t}(x_1, \cdots, x_t)\bar{\delta}_{\nu\tau}. \tag{5.11}$$

For any subset S of R, let

$$\mu_{i\nu\tau t}(S) = \int_S g_{i\nu\tau t}(x_1, \cdots, x_t)\, dF_i \qquad (t = \tau, \tau + 1, \cdots, T), \tag{5.12}$$

and

$$\mu_{i\nu\tau}(S) = \int_S \left[\sum_{t=T+1}^{\infty} g_{i\nu\tau t}(x_1, \cdots x_t) \right] dF_i. \tag{5.13}$$

The measures $\mu_{i\nu\tau t}$ are not defined if $\tau > T$. Clearly, the measures

$$\mu_{i\nu\tau t}(\nu = 0, 1, \cdots, m; t = \tau, \tau + 1, \cdots, T)$$

and the measures $\mu_{i\nu\tau}(\nu = 1, \cdots, m)$ are nonnegative, countably additive and σ-finite. Since for any x for which $\varphi_i(x, \delta) < \infty$ and $\delta_{0\tau} > 0$, the sum

$$\sum_{t=T+1}^{\infty} g_{i0\tau t}(x_1, \cdots, x_t) < \infty,$$

it follows from the assumptions of Lemma 5.1 that $\mu_{i0\tau}$ is σ-finite over the space R' consisting of all x for which $\delta_{0\tau} > 0$. Of course, $\mu_{i0\tau}$ is nonnegative and countably additive. Let R'' be the set of all points x for which $\delta_{0\tau} = 0$. We put

$$\bar{\delta}_{0\tau}(x_1, \cdots, x_\tau) = 0 \quad \text{for all} \quad x \text{ in } R''. \tag{5.14}$$

Application of Theorem 2.1 to each of the spaces R' and R'' shows that there exist measurable functions $\bar{\delta}_{\nu\tau}(x_1, \cdots, x_\tau)(\nu = 0, 1, \cdots, m)$ such that in addiion to (5.14) the following conditions hold:

$$\bar{\delta}_{\nu\tau} = 0 \quad \text{or} \quad 1(\nu = 0, 1, \cdots m) \quad \text{and} \quad \sum_{\nu=0}^{m} \bar{\delta}_{\nu\tau} = 1 \quad \text{for all } x, \tag{5.15}$$

$$\int_R \delta_{\nu\tau}\, d\mu_{i\nu\tau t} = \int_R \bar{\delta}_{\nu\tau}\, d\mu_{i\nu\tau t} \tag{5.16}$$

$$(i = 1, \cdots, p; \nu = 0, 1, \cdots m; t = \tau, \tau + 1, \cdots, T),$$

$$\int_R \delta_{\nu\tau}\, d\mu_{i\nu\tau} = \int_R \bar{\delta}_{\nu\tau}\, d\mu_{i\nu\tau} \qquad (i = 1, \cdots, p; \nu = 0, 1, \cdots m). \tag{5.17}$$

Lemma 5.1 is a simple consequence of the equations (5.14)–(5.17).

For any positive integer u, we shall say that a decision rule δ is truncated at the uth stage if $\delta_{0u'} = 0$ for $u' \geqq u$ identically in x.

THEOREM 5.1. *If δ is truncated at the uth stage there exists a nonrandomized decision rule δ^* that is strongly equivalent to δ.*

PROOF. It is sufficient to prove Theorem 5.1 in the case where $\delta_{\nu t} = 0$ for $t > u$ and $\nu \neq 1$ and $\delta_{1t} = 1$ for $t > u$. Clearly, $\varphi_i(x, \delta) < \infty$ for all x. Putting $\tau = 1$ and $T = u$ in Lemma 5.1, this lemma implies the existence of a decision rule δ^1 with the following properties: (a) δ^1 is strongly equivalent to δ; (b) $\delta^1_{\nu 1} = 0$ or 1 ($\nu = 0, 1, \cdots, m$); (c) $\delta^1_{\nu t} = \delta_{\nu t}$ for $\nu = 0, 1, \cdots, m$ and $t > 1$. Applying Lemma 5.1 to δ^1 and putting $\tau = 2$ and $T = u$, we see that there exists a decision rule δ^2 with the following properties: (a) δ^2 is strongly equivalent to δ^1; (b) $\delta^2_{\nu 2} = 0$ or 1 ($\nu = 0, 1, \cdots, m$); (c) $\delta^2_{\nu t} = \delta^1_{\nu t}$ for $\nu = 0, 1, \cdots, m$ and $t \neq 2$. Continuing this procedure, at the uth step we obtain a decision rule δ^u that is nonrandomized and is strongly equivalent to all the preceding ones. This proves our theorem.

We shall say that two decision rules δ^1 and δ^2 are strongly equivalent up to the Tth stage if

$$\begin{aligned} &\int_{R_t} v_{i\nu t}(x_1, \cdots, x_t)\delta^1_{01} \cdots \delta^1_{0(t-1)}\, \delta^1_{\nu t}\, dF_{it} \\ &\quad = \int_{R_t} v_{i\nu t}(x_1, \cdots, x_t)\delta^2_{01} \cdots \delta^2_{0(t-1)}\delta^2_{\nu t}\, dF_{it} \end{aligned} \tag{5.18}$$

$$\text{for} \quad i = 1, \cdots, p; \nu = 1, \cdots, m \quad \text{and} \quad t = 1, \cdots, T.$$

Furthermore, we shall say that a decision rule δ is nonrandomized up to the stage T if $\delta_{\nu t} = 0$ or 1 for $\nu = 0, 1, \cdots, m$ and $t = 1, \cdots, T$.

We now prove the following theorem.

THEOREM 5.2. *If δ is a decision rule for which $\varphi_i(x, \delta) < \infty$, except perhaps on a set of x's of probability zero according to every $F_i(x)(i = 1, \cdots, p)$, then there exists a nonrandomized decision rule δ^* that is equivalent to δ.*

PROOF. Let $\{\epsilon_i\}$ and $\{\eta_i\}$ ($i = 1, 2, \cdots$, ad inf.) be two sequences of positive numbers such that $\lim_{i=\infty} \epsilon_i = 0$ and $\lim_{i=\infty} \eta_i = \infty$. Let T_1 be a positive integer such that

$$r_i(\delta) - \sum_{t=1}^{T_1} \sum_{\nu=1}^{m} \int_{R_t} v_{i\nu t}(x_1, \cdots, x_t)\delta_{01} \cdots \delta_{0(t-1)}\, \delta_{\nu t}\, dF_{it} < \epsilon_1 \quad \text{if} \quad r_i(\delta) < \infty, \tag{5.19}$$

and

$$(5.20)\qquad \sum_{t=1}^{T_1}\sum_{\nu=1}^{m}\int_{R_t} v_{i\nu t}(x_1, \cdots, x_t)\delta_{01}\cdots\delta_{0(t-1)}\delta_{\nu t}\, dF_{it} > \eta_1 \quad \text{if} \quad r_i(\delta) = \infty.$$

Let δ^1 be a decision rule such that $\varphi_i(x, \delta^1) < \infty$ (except perhaps on a set of probability measure zero); δ^1 is equivalent to δ; δ^1 is strongly equivalent to δ up to the T_1^{th} stage; δ^1 is nonrandomized up to the T_1^{th} stage and $\delta^1_{\nu t} = \delta_{\nu t}$ for $t > T_1$. The existence of such a decision rule follows from a repeated application of Lemma 5.1. In general, after $\delta^1, \cdots, \delta^j$ and $T_1, \cdots, T_j$ are given, let δ^{j+1} be a decision rule such that $\varphi_i(x, \delta^{j+1}) < \infty$ (except perhaps on a set of probability measure zero); δ^{j+1} is equivalent to δ^j; δ^{j+1} is strongly equivalent to δ^j up to the T_{j+1}^{th} stage, where T_{j+1} is a positive integer chosen so that $T_{j+1} > T_j$ and (5.19) and (5.20) hold with δ replaced by δ^j, ϵ_1 replaced by ϵ_{j+1} and η_1 replaced by η_{j+1}; δ^{j+1} is nonrandomized up to the stage T_{j+1}; $\delta^{j+1}_{\nu t} = \delta^j_{\nu t}$ for $t \leqq T_j$ and $\delta^{j+1}_{\nu t} = \delta_{\nu t}$ for $t > T_{j+1}$. The existence of such a decision rule δ^{j+1} follows again from a repeated application of Lemma 5.1.

Let δ^* be the decision rule given by the equations

$$(5.21)\qquad \delta^*_{\nu t} = \delta^t_{\nu t} \qquad (\nu = 0, 1, \cdots, m;\, t = 1, 2, \cdots, \text{ ad inf.}).$$

It follows easily from the above stated properties of the decision rules δ^j ($j = 1, 2, \cdots$, ad inf.) that δ^* is nonrandomized and $r_i(\delta^*) = r_i(\delta)$ $(i = 1, \cdots, p)$. This completes the proof of Theorem 5.2.

6. Examples where admissible[8] decision functions do not admit equivalent nonrandomized decision functions. In this section we shall construct examples which show that there exist admissible decision functions $\delta(x)$ which do not admit equivalent nonrandomized decision functions $\delta^*(x)$.

Example 1. Let X be a normally distributed chance variable with unknown mean θ and variance unity. This means that Ω is the totality of all univariate normal distributions with unit variance. Suppose we wish to test the hypothesis H_0 that the true mean θ is rational on the basis of a single observation x on X. Thus, D consists of two elements d_1 and d_2 where d_1 is the decision to accept H_0 and d_2 is the decision to reject H_0. For any decision function $\delta(x)$, let $\delta_1(x)$ denote the value of $\delta(d_1 \mid x)$. Let the loss be zero when a correct decision is made, and the loss be one when a wrong decision is made. Then the risk when θ is the true mean and the decision function $\delta(x)$ is adopted is given by

$$(6.1)\qquad r(\theta, \delta) = \frac{1}{\sqrt{2\pi}}\int_{-\infty}^{\infty} e^{-\frac{1}{2}(x-\theta)^2}\delta_1(x)\, dx \qquad \text{when } \theta \text{ is irrational,}$$

$$(6.2)\qquad r(\theta, \delta) = \frac{1}{\sqrt{2\pi}}\int_{-\infty}^{\infty} e^{-\frac{1}{2}(x-\theta)^2}(1 - \delta_1(x))\, dx \qquad \text{when } \theta \text{ is rational.}$$

---

[8] A decision function with risk function $r(F)$ is called admissible if there exists no other decision function with risk function $r'(F)$ such that $r'(F) \leqq r(F)$ for every $F \in \Omega$, and the inequality sign holds for at least one $F \in \Omega$.

Let $\delta_1^0(x) = \frac{1}{2}$ for all x. Clearly,

$$r(\theta, \delta^0) = \tfrac{1}{2} \tag{6.3}$$

for all θ. We shall now show that $\delta^0(x)$ is an admissible decision function. For suppose that there exists a decision function $\delta'(x)$ such that

$$r(\theta, \delta') \leqq r(\theta, \delta^0) = \tfrac{1}{2} \tag{6.4}$$

for all θ, and

$$r(\theta_1, \delta') < r(\theta_1, \delta^0) = \tfrac{1}{2} \tag{6.5}$$

for some value θ_1. Suppose first that θ_1 is rational. Since the integrals in (6.1) and (6.2) are continuous functions of θ, for an irrational value θ_2 sufficiently near to θ_1 we shall have $r(\theta_2, \delta') > \frac{1}{2}$ which contradicts (6.4). Thus, θ_1 cannot be rational. In a similar way, one can show that θ_1 cannot be irrational. Hence, the assumption that a decision function $\delta'(x)$ satisfying (6.4) and (6.5) exists leads to a contradiction and the admissibility of $\delta^0(x)$ is proved.

Let now $\delta^*(x)$ be any decision function for which

$$r(\theta, \delta^*) = r(\theta, \delta^0) \tag{6.6}$$

for all θ. Now (6.6) implies that

$$\frac{1}{\sqrt{2\pi}} \int_{-\infty}^{\infty} e^{-\frac{1}{2}(x-\theta)^2} (\delta_1(x) - \delta_1^*(x))\, dx = 0 \tag{6.7}$$

identically in θ. Since $\delta_1(x) - \delta_1^*(x)$ is a bounded function of x, it follows from the uniqueness properties of the Laplace transform that (6.7) can hold only if $\delta_1(x) - \delta_1^*(x) = 0$ except perhaps on a set of measure zero. Hence, no nonrandomized decision function $\delta^*(x)$ can satisfy (6.6).

In the above example, the distributions consistent with the hypothesis H_0 which is to be tested (normal distributions with rational means) are not well separated from the alternative distributions (normal distributions with irrational means). One might think that this is perhaps the reason for the existence of an admissible decision function δ^0 such that no nonrandomized decision function δ^* can have as good a risk function as δ^0 has. That this need not be so, is shown by the following:

EXAMPLE 2. Suppose that X is a normally distributed chance variable with mean θ and variance unity. The value of θ is unknown. It is known, however, that the true value of θ is contained in the union of the two intervals $[-2, -1]$ and $[1, 2]$. Suppose that we want to test the hypothesis that θ is contained in the interval $[-2, -1]$ on the basis of a single observation x on X. Suppose, furthermore, that the chance variable X itself is not observable and only the chance variable $Y = f(X)$ can be observed where $f(x) = x$ when $|x| < 1$, and $= |x|$ when $|x| \geqq 1$. Let the loss be zero when a correct decision is made, and one when a wrong decision is made. For any decision function $\delta(y)$, let

$\delta_1(y)$ denote the value of $\delta(d_1 \mid y)$ where d_1 denotes the decision to accept H_0. Let $\delta^0(y)$ be the following decision function:

$$\begin{aligned} \delta_1^0(y) &= 1 \quad \text{when } -1 < y < 0 \\ &= 0 \quad \text{when } \quad 0 \leqq y < 1 \\ &= \tfrac{1}{2} \quad \text{when } \quad y \geqq 1. \end{aligned} \tag{6.8}$$

First we shall show that $\delta^0(y)$ is an admissible decision function. For this purpose, consider the following probability density function $g(\theta)$ in the parameter space: $g(\theta) = \frac{1}{2}$ when $-2 \leqq \theta \leqq -1$ or $1 \leqq \theta \leqq 2$, $= 0$ for all other θ. If we interpret $g(\theta)$ as the a priori probability distribution of θ, the a posteriori probability of the θ-interval $[-2, -1]$ is greater (less) than the a posteriori probability of the θ-interval $[1, 2]$ when $-1 < y < 0$ $(0 < y < 1)$, and the a posteriori probabilities of the two intervals are equal to each other when $y = 0$ or $y \geqq 1$. Hence, $\delta^0(y)$ is a Bayes solution relative to the a priori distribution $g(\theta)$, i.e.,

$$\int_{-2}^{-1} r(\theta, \delta^0)\, d\theta + \int_1^2 r(\theta, \delta^0)\, d\theta \leqq \int_{-2}^{-1} r(\theta, \delta)\, d\theta + \int_1^2 r(\theta, \delta)\, d\theta \tag{6.9}$$

for any decision function δ. Suppose now that δ is a decision function for which $r(\theta, \delta) \leqq r(\theta, \delta^0)$ for all θ. It then follows from (6.9) that $r(\theta, \delta) < r(\theta, \delta^0)$ can hold at most on a set of θ's of measure zero. Since, as can easily be verified, $r(\theta, \delta)$ and $r(\theta, \delta^0)$ are continuous functions of θ, it follows that $r(\theta, \delta) = r(\theta, \delta^0)$ everywhere and the admissibility of δ^0 is proved.

Let now $\delta'(y)$ be any decision function for which $r(\theta, \delta') = r(\theta, \delta^0)$ for all θ, i.e.,

$$\frac{1}{\sqrt{2\pi}} \int_{-\infty}^{\infty} e^{-\frac{1}{2}(x-\theta)^2} [\delta^0(y) - \delta'(y)]\, dx = 0 \qquad \text{for all } \theta. \tag{6.10}$$

Since $\delta_1^0(y) - \delta_1'(y)$ is a bounded function of x, it follows from the uniqueness properties of the Laplace transform that (6.10) can hold only if $\delta_1^0(y) = \delta_1'(y)$ except perhaps on a set of measure zero. Thus, no nonrandomized decision function δ^* exists such that $r(\theta, \delta^*) = r(\theta, \delta^0)$ for all θ.

7. Compactness of Ω in the ordinary sense is not sufficient for the existence of ϵ-equivalent nonrandomized decision functions. Let $\Omega = \{F\}$ be the totality of density functions[9] on the interval $0 \leqq x \leqq 1$ for which $F(x) \leqq c$ for every x, where c is some positive constant greater than 2. The sample space will be the interval $0 \leqq x \leqq 1$. We shall say that the sequence $F_1, F_2, \cdots$ converges to F if

$$\lim_{n\to\infty} \int_{-\infty}^{x} F_n(y)\, dy = \int_{-\infty}^{x} F(y)\, dy$$

[9] Here $F(x)$ denotes a density function. This represents a change in notation from preceding sections.

for every real x. The set Ω is compact in the sense of the above convergence definition.[10] Let A be a fixed interval $a_1 \leqq x \leqq a_2$ where $0 < a_1 < a_2 < 1$. Let $D = \{d_1, d_2\}$ and define W as follows:

$$W(F, d_1) + W(F, d_2) \equiv 1,$$

$$W(F, d_1) = 0 \text{ or } 1$$

according as the probability of A under F is rational or not. For any decision function $\delta(x)$, let $\delta_1(x)$ denote the probability assigned to d_1 by $\delta(x)$, i.e., $\delta_1(x) = \delta(d_1 \mid x)$.

Let $\delta'(x)$ be the decision function for which $\delta_1'(x) \equiv \frac{1}{2}$. We shall prove that $\delta'(x)$ is an admissible decision function. For suppose there exists a decision function $\delta^0(x)$ such that

$$r(F, \delta^0) \leqq r(F, \delta') = \tfrac{1}{2} \tag{7.1}$$

for every F, and for F_0 we have

$$r(F_0, \delta^0) < r(F_0, \delta'). \tag{7.2}$$

Now, if $F_i \to F_0$ and $W(F_i, d_1) = W(F_0, d_1)$ for every i, then $r(F_i, \delta) \to r(F_0, \delta)$ for every decision function $\delta(x)$, and, in particular, for $\delta^0(x)$. If $F_i \to F_0$ and $W(F_i, d_1) + W(F_0, d_1) = 1$ for every i, then $r(F_i, \delta) \to 1 - r(F_0, \delta)$ for every decision function $\delta(x)$ and, in particular, for $\delta^0(x)$. Clearly, we can construct two sequences of functions F such that each sequence converges to F_0, the probability of A according to every member of the first sequence is rational, and the probability of A according to every member of the second sequence is irrational. Because of (7.2) it follows that inequality (7.1) will be violated for almost every member of one of these two sequences. Hence δ' is admissible.

Let us now prove that there cannot exist a nonrandomized decision function $\delta^*(x)$ such that

$$r(F, \delta^*) \leqq r(F, \delta') + \tfrac{1}{4} = \tfrac{3}{4} \tag{7.3}$$

for every $F \in \Omega$. Suppose there were such a decision function $\delta^*(x)$. Let H be the set of x's where $\delta_1^*(x) = 1$, and let $\bar{H}$ be the complement of H with respect to the interval [0, 1]. If H is a set of measure zero or one then obviously (7.3) is violated for some F. Thus, it is sufficient to consider the case when H is a set of positive measure $\alpha < 1$. Suppose for a moment that $\alpha > \frac{1}{2}$. Let G be the density which is zero on $\bar{H}$ and constant on H. From (7.3) it follows that $P\{A \mid G\}$ is rational. There exists a density $G' \in \Omega$ such that $P\{H \mid G'\} > \frac{3}{4}$ and $P\{A \mid G'\}$ is irrational. But then (7.3) is violated for G'. If $\alpha \leqq \frac{1}{2}$, let $\bar{G}$ be the density which is zero on H and constant on $\bar{H}$. From (7.3) it follows that $P\{A \mid \bar{G}\}$ is irrational. There exists a density $\bar{G}' \in \Omega$ such that $P\{\bar{H} \mid \bar{G}'\} >$

[10] The cumulative distribution functions are well-known to be compact in the usual convergence sense. Since the densities are bounded above the limit cumulative distribution function must be absolutely continuous.

$\frac{3}{4}$ and $P\{A \mid \bar{G}'\}$ is rational. But then (7.3) is violated for $\bar{G}'$. Thus (7.3) can never hold for every $F \in \Omega$ and the desired result is proved.

8. Sufficient conditions for the existence of ϵ-equivalent nonrandomized decision functions. In this section we shall consider the nonsequential decision problem (as described in the introduction), and we shall give sufficient conditions for the existence of ϵ-equivalent nonrandomized decision functions. We shall consider the following four metrics in the space Ω:

$$\rho_1(F_1, F_2) = \operatorname*{Sup}_S \left| \int_S dF_1 - \int_S dF_2 \right| \tag{8.1}$$

when S is any measurable subset of R,

$$\rho_2(F_1, F_2) = \operatorname*{Sup}_{d,x} \left| W(F_1, d, x) - W(F_2, d, x) \right|, \tag{8.2}$$

$$\rho_3(F_1, F_2) = \rho_1(F_1, F_2) + \rho_2(F_1, F_2), \tag{8.3}$$

$$\rho_4(F_1, F_2) = \operatorname*{Sup}_\delta \left| r(F_1, \delta) - r(F_2, \delta) \right|. \tag{8.4}$$

First we prove the following lemma:

LEMMA 8.1. *If Ω is conditionally compact in the sense of the metric ρ_3, then it is conditionally compact in the sense of the metric ρ_4.*

PROOF. Let $\{F_i\}(i = 1, 2, \cdots, \text{ad inf.})$ be a Cauchy sequence in the sense of the metric ρ_3, i.e.,

$$\lim_{i,j=\infty} \rho_3(F_i, F_j) = 0. \tag{8.5}$$

It follows from (8.5) and (8.3) that $W(F_i, d, x)$ converges, as $i \to \infty$, to a limit function $W(d, x)$ uniformly in d and x, i.e.,

$$\lim_{i=\infty} W(F_i, d, x) = W(d, x) \tag{8.6}$$

uniformly in d and x. Hence

$$\lim_{i=\infty} \int_D W(F_i, d, x)\, d\delta_x = \int_D W(d, x)\, d\delta_x \tag{8.7}$$

uniformly in x and δ. Because of (8.5), we have

$$\lim_{i,j=\infty} \rho_1(F_i, F_j) = 0. \tag{8.8}$$

Hence there exists a distribution function $F_0(x)$ (not necessarily an element of Ω) such that

$$\lim_{i=\infty} \rho_1(F_i, F_0) = 0. \tag{8.9}$$

It follows from (8.7) and (8.9) that

$$(8.10)\qquad \lim_{i=\infty} \int_R \left[\int_D W(F_i, d, x)\, d\delta_x\right] dF_i(x) = \int_R \left[\int_D W(d, x)\, d\delta_x\right] dF_0(x)$$

uniformly in δ. Hence $\{F_i\}$ is a Cauchy sequence in the sense of the metric ρ_4 and Lemma 8.1 is proved.

Next we prove

LEMMA 8.2. *If D is conditionally compact in the sense of the metric* (1.1) *and if δ is any decision function, then for any $\epsilon > 0$ there exists a finite subset D^1 of D and a decision function δ^1 such that $\delta^1(D^1 \mid x) = 1$ identically in x and δ^1 is ϵ-equivalent to δ.*

PROOF. Since D is conditionally compact, it is possible to decompose D into a finite number of disjoint subsets $D_1, \cdots, D_u$ such that the diameter of D_j is less then $\epsilon (j = 1, \cdots, u)$. Let d_j be an arbitrary but fixed point of $D_j (j = 1, \cdots, u)$ and let $\delta^1(x)$ be the decision function determined by the condition

$$(8.11)\qquad \delta^1(d_j \mid x) = \delta(D_j \mid x) \qquad (j = 1, \cdots, u).$$

Clearly

$$(8.12)\qquad \left| \int_D W(F, d, x)\, d\delta_x - \int_D W(F, d, x)\, d\delta_x^1 \right| \leqq \epsilon$$

for all F and x. Hence,

$$(8.13)\qquad | r(F, \delta^1) - r(F, \delta) | \leqq \epsilon$$

for all F and our lemma is proved.

We are now in a position to prove the main theorem.

THEOREM 8.1. *If the elements $F(x)$ of Ω are atomless, if Ω is conditionally compact in the sense of the metrics ρ_1 and ρ_2, and if D is conditionally compact in the the sense of the metric* (1.1), *then for any $\epsilon > 0$ and for any decision function $\delta(x)$ there exists an ϵ-equivalent nonrandomized decision function $\delta^*(x)$.*

PROOF. Because of Lemma 8.2, it is sufficient to prove our theorem for finite D. Thus, we shall assume that D consists of the elements $d_1, \cdots, d_m$. It is easy to verify that conditional compactness of Ω in the sense of both metrics ρ_1 and ρ_2 implies conditional compactness in the sense of the metric ρ_3, and because of Lemma 8.1, also in the sense of the metric ρ_4. Thus, conditional compactness of Ω in the sense of the metrics ρ_1 and ρ_2 implies the existence of a finite subset $\Omega^* = \{F_1, \cdots, F_k\}$ of Ω such that Ω^* is $\epsilon/2$-dense in Ω in the sense of the metric ρ_4. Let δ^* be a nonrandomized decision function that is equivalent to δ if Ω is replaced by Ω^*. The existence of such a δ^* follows from Theorem 3.1. Since Ω^* is $\epsilon/2$-dense in Ω (in the sense of the metric ρ_4), we have

$$(8.14)\qquad | r(F, \delta^*) - r(F, \delta) | \leqq \epsilon \quad \text{for all} \quad F \text{ in } \Omega$$

and our theorem is proved.

We shall now introduce some notions with the help of which we shall be able to strengthen Theorem 3.1. For any measurable subset S of R, let

$$r(F, \delta \mid S) = \int_S \left[\int_D W(F, d, x)\, d\delta_x \right] dF(x). \tag{8.15}$$

We shall refer to the above expression as the contribution of the set S to the risk. For any S we shall consider the following four metrics in Ω:

$$\rho_{1S}(F_1, F_2) = \operatorname*{Sup}_S \left| \int_{S^*} dF_1 - \int_{S^*} dF_2 \right| \tag{8.16}$$

where S^* is any measurable subset of S,

$$\rho_{2S}(F_1, F_2) = \operatorname*{Sup}_{d, x \in S} \mid W(F_1, d, x) - W(F_2, d, x) \mid, \tag{8.17}$$

$$\rho_{3S}(F_1, F_2) = \rho_{1S}(F_1, F_2) + \rho_{2S}(F_1, F_2), \tag{8.18}$$

$$\rho_{4S}(F_1, F_2) = \operatorname*{Sup}_\delta \mid r(F_1, \delta \mid S) - r(F_2, \delta \mid S) \mid . \tag{8.19}$$

Finally let the metric $\rho_S(d_1, d_2)$ in D be defined by

$$\rho_S(d_1, d_2) = \operatorname*{Sup}_{F, x \in S} \mid W(F, d_1, x) - W(F, d_2, x) \mid . \tag{8.20}$$

We shall now prove the following stronger theorem:

THEOREM 8.2. *Let all elements F of Ω be atomless. If there exists a decomposition of R into a sequence $\{R_i\}$ ($i = 1, 2, \cdots$, ad inf.) of disjoint subsets such that Ω is conditionally compact in the sense of the metrics ρ_{1R_i} and ρ_{2R_i} for each i, and such that D is conditionally compact in the sense of the metric ρ_{R_i} for each i, then for any $\epsilon > 0$ and for any decision function δ there exists an ϵ-equivalent nonrandomized decision function δ^*.*

PROOF. Let $\{R_i\}$ be a decomposition of R for which the conditions of our theorem are fulfilled. Let $\{\epsilon_i\}$ be a sequence of positive numbers such that $\sum_{i=1}^{\infty} \epsilon_i = \epsilon$. Let $\delta^1(x)$ be a decision function such that $\delta_1(x) = \delta(x)$ for any x not in R_1, $\delta^1(x)$ is nonrandomized over R_1 (for any x in R_1, $\delta^1(x)$ assigns the probability one to a single point d in D) and such that

$$\mid r(F, \delta \mid R_1) - r(F, \delta^1 \mid R_1) \mid \leqq \epsilon_1 \quad \text{for all } F. \tag{8.21}$$

The existence of such a decision function δ^1 follows from Theorem 8.1 (replacing R by R_1). After $\delta^1, \cdots, \delta^{i-1}$ have been defined ($i \geqq 1$), let δ^i be a decision function such that δ^i is nonrandomized over R^i, $\delta^i(x) = \delta^{i-1}(x)$ for all x in $\bigcup_{j=1}^{i-1} R_j$, $\delta^i(x) = \delta(x)$ for all x in $R - \bigcup_{j=1}^{i} R_j$ and such that

$$\mid r(F, \delta^i \mid R_i) - r(F, \delta \mid R_i) \mid \leqq \epsilon_i \quad \text{for all} \quad F \text{ in } \Omega. \tag{8.22}$$

The existence of such a decision function δ^i follows again from Theorem 8.1. Clearly $\delta^i(x)$ converges to a limit $\delta^*(x)$, as $i \to \infty$. This limit decision function $\delta^*(x)$ is obviously nonrandomized and satisfies the conditon

$$(8.23) \qquad | r(F, \delta \mid R_i) - r(F, \delta^* \mid R_i) | \leqq \epsilon_i$$

for all i and F. Theorem 8.2 is an immediate consequence of this.

The conditions of Theorem 8.2 will be fulfilled for a wide class of statistical decision problems. For example, this is true for the decision problems which satisfy the following six conditions:

CONDITION 1. *The sample space R is a finite dimensional Euclidean space. All elements $F(x)$ of Ω are absolutely continuous.*

CONDITION 2. *Ω admits a parametric representation, i.e., each element F of Ω is associated with a parametric point θ in a finite dimensional Euclidean space E.*

We shall denote the density function $p(x)$ corresponding to the parameter point θ by $p(x, \theta)$.

CONDITION 3. *The set of parameter points θ which correspond to all elements F of Ω is a closed subset of E.*

We shall call this set of all parameter points θ the parameter space. Since there is a one-to-one correspondence between the elements F of Ω and the points θ of the parameter space, there is no danger of confusion if we denote the parameter space also by Ω.

CONDITION 4. *The density function $p(x, \theta)$ is continuous in $\theta \in \Omega$ for every x.*

CONDITION 5. *The loss $W(\theta, d)$ when θ is true and the decision d is made does not depend on x. D is conditionally compact in the sense of the metric $\rho(d_1, d_2) = \operatorname{Sup}_\theta | W(\theta, d_1) - W(\theta, d_2) |$.*

CONDITION 6. *For any bounded subset M of R, we have* $\lim_{\left\{ \substack{|\theta| = \infty \\ \theta \in \Omega} \right\}} \int_M p(x, \theta)\, dx = 0$.

We shall now show that the conditions of Theorem 8.2 are fulfilled for any decision problem that satisfies Conditions 1–6. Let S_i be the sphere in R with center at the origin and radius i. Let $R_1 = S_1$ and $R_i = S_i - \bigcup_{j=1}^{i-1} R_j (i = 1, 2, \cdots, \text{ad inf.})$. Condition 5 implies that D is conditionally compact in the sense of the metric ρ_{R_i} for all i. It follows from Condition 5 and Theorem 2.1 in [3] that Ω is conditionally compact in the sense of the metric $\rho(\theta_1, \theta_2) = \operatorname{Sup}_d | W(\theta_1, d) - W(\theta_2, d) |$. Hence Ω is conditionally compact in the sense of the metric ρ_{2R_i} for each i. It remains to be shown that Ω is conditionally compact in the sense of the metric ρ_{1R_i} for each i. For this purpose, consider any sequence $\{\theta_j\} (j = 1, 2, \cdots, \text{ad inf.})$ of parameter points. There are 2 cases possible: (a) $\{\theta_j\}$ admits a subsequence that converges in the Euclidean sense to a finite point θ_0; (b) $\lim_{j=\infty} | \theta_j | = \infty$. Let us consider first the case (a) and let $\{\theta_j'\}$ be a subsequence of $\{\theta_j\}$ which converges to a finite point θ_0. It then follows from Condition 4 and a theorem of Robbins [4] that $\{\theta_j'\}$ is a Cauchy subsequence in the sense of the metric ρ_{1R_i} for each i. In case (b), Condition 6 implies that the sequence $\{\theta_j\}$ is a Cauchy sequence in the sense of the metric ρ_{1R_i} for each i. Thus, Ω is conditionally compact in the sense of the metric ρ_{1R_i}. This completes the proof of our assertion that a decision problem that satisfies Conditions 1–6, satisfies also the conditions of Theorem 8.2.

9. Application to the theory of games. Translation of the results of Section 2 into the language of the theory of games is immediate and we shall do this only very briefly. The function $W(F_i, d_j, x)$ $(i = 1, \cdots, p; j = 1, \cdots, m; x \in R)$, of Section 1 is now called the pay-off function of a zero-sum two-person game. The game is played as follows: Player I selects one of the integers $1, \cdots, p$, say i, without communicating his choice to player II. A random observation $x \in R$ on a chance variable whose distribution function is F_i is obtained and communicated to player II. The latter chooses one of the integers $1, \cdots, m$, say j. The game now ends with the receipt by player I and player II of the respective sums $W(F_i, d_j, x)$ and $-W(F_i, d_j, x)$. Randomized (mixed) and nonrandomized (pure) strategies are defined in the same manner as the corresponding decision functions in Section 1. When the distribution functions $F_i(x)(i = 1, \cdots, p)$ are all atomless the obvious analogues of Theorems 3.1 and 3.2 hold.

It should be remarked that the usual definition of randomized (mixed) strategy is not as general as the one given above. In the usual definition player II chooses, by a random mechanism independent of the random mechanism which yields the point x, some one of a (usually finite) number of nonrandomized (pure) strategies, and then plays the game according to the nonrandomized strategy selected. In our definition (used in [3]) the random choice is allowed to depend on x. Clearly our method of randomization includes the usual one as a special case. The relation between the two methods of randomization will be discussed by two of the authors in a forthcoming paper [7].

Suppose that the number of possible decisions is at most denumerable, and that the decision procedure consists in choosing at random and in advance of the observations, one of a finite number of nonrandomized decision functions. The sample space can be divided into an at most denumerable number of sets in each of which only a finite number of decisions is possible (the possible decisions vary from set to set). In each set our results are applicable. Since the number of sets is denumerable the resultant decision function is measurable. We conclude: It follows from our results that if a decision procedure consists of selecting with preassigned probabilities one of a finite number of nonrandomized decision functions with the number of possible decisions at most denumerably infinite, and if the possible distributions are finite in number and atomless, then there exists an equivalent nonrandomized decision function. More general results can be obtained for this case (where one chooses at random and in advance of the observations, one of a finite number of nonrandomized decision functions). By application of the methods of Sections 4 and 8 the requirement that the number of possible decisions be denumerable can be easily removed. The procedures are straightforward and we omit them.

REFERENCES

[1] A. Dvoretzky, A. Wald, and J. Wolfowitz, "Elimination of randomization in certain problems of statistics and of the theory of games," *Proc. Nat. Acad. Sci.*, Vol. 36 (1950), pp. 256–260.

[2] A. Lyapunov, "Sur les fonctions-vecteurs complètement additives," *Izvestiya Akad. Nauk SSSR. Ser. Mat.*, Vol. 4 (1940), pp. 465–478.

[3] A. Wald, *Statistical Decision Functions*, John Wiley & Sons, 1950.

[4] Herbert Robbins, "Convergence of distributions," *Annals of Math. Stat.*, Vol. 19 (1948), pp. 72–75.

[5] J. von Neumann and O. Morgenstern, *Theory of Games and Economic Behavior*, Princeton University Press, 1944.

[6] A. Dvoretzky, A. Wald, and J. Wolfowitz, "Relations among certain ranges of vector measures," *Pacific Journal of Mathematics* (1951).

[7] A. Wald and J. Wolfowitz, "Two methods of randomization in statistics and the theory of games," *Annals of Mathematics*, to be published.

Reprinted from THE ANNALS OF MATHEMATICAL STATISTICS
Vol. 22, No. 1, March, 1951

ON THE FUNDAMENTAL LEMMA OF NEYMAN AND PEARSON[1]

BY GEORGE B. DANTZIG AND ABRAHAM WALD[2]

Department of the Air Force and Columbia University

1. Summary and introduction. The following lemma proved by Neyman and Pearson [1] is basic in the theory of testing statistical hypotheses:

LEMMA. *Let* $f_1(x), \cdots, f_{m+1}(x)$ *be* $m + 1$ *Borel measurable functions defined over a finite dimensional Euclidean space* R *such that* $\int_R |f_i(x)| \, dx < \infty$ $(i = 1, \cdots, m + 1)$. *Let, furthermore,* $c_1, \cdots, c_m$ *be* m *given constants and* $\mathcal{S}$ *the class of all Borel measurable subsets* S *of* R *for which*

$$\int_S f_i(x) \, dx = c_i \qquad (i = 1, \cdots, m). \tag{1.1}$$

Let, finally, $\mathcal{S}_0$ *be the subclass of* $\mathcal{S}$ *consisting of all members* S_0 *of* $\mathcal{S}$ *for which*

$$\int_{S_0} f_{m+1}(x) \, dx \geqq \int_S f_{m+1}(x) \, dx \quad \text{for all } S \text{ in } \mathcal{S}. \tag{1.2}$$

If S *is a member of* $\mathcal{S}$ *and if there exist* m *constants* $k_1, \cdots, k_m$ *such that*

$$f_{m+1}(x) \geqq k_1 f_1(x) + \cdots + k_m f_m(x) \quad \text{when} \quad x \in S, \tag{1.3}$$

$$f_{m+1}(x) \leqq k_1 f_1(x) + \cdots + k_m f_m(x) \quad \text{when} \quad x \notin S, \tag{1.4}$$

then S *is a member of* $\mathcal{S}_0$.

The above lemma gives merely a sufficient condition for a member S of $\mathcal{S}$ to be also a member of $\mathcal{S}_0$. Two important questions were left open by Neyman and Pearson: (1) the question of existence, that is, the question whether $\mathcal{S}_0$ is non-empty whenever $\mathcal{S}$ is non-empty; (2) the question of necessity of their sufficient condition (apart from the obvious weakening that (1.3) and (1.4) may be violated on a set of measure zero).

The purpose of the present note is to answer the above two questions. It will be shown in Section 2 that $\mathcal{S}_0$ is not empty whenever $\mathcal{S}$ is not empty. In Section 3, a necessary and sufficient condition is given for a member of $\mathcal{S}$ to be also a member of $\mathcal{S}_0$. This necessary and sufficient condition coincides with the Neyman-Pearson sufficient condition under a mild restriction.

2. Proof that $\mathcal{S}_0$ is not empty whenever $\mathcal{S}$ is not empty. Each function $f_i(x)$ determines a finite measure μ_i given by the equation

$$\mu_i(S) = \int_S f_i(x) \, dx \qquad (i = 1, 2, \cdots, m + 1). \tag{2.1}$$

[1] The main results of this paper were obtained by the authors independently of each other using entirely different methods.

[2] Research under contract with the Office of Naval Research.

Let μ be the vector measure with the components $\mu_1, \cdots, \mu_{m+1}$; i.e., for any measurable set S the value of $\mu(S)$ is the vector $(\mu_1(S), \cdots, \mu_{m+1}(S))$. Thus, for each S the value of $\mu(S)$ can be represented by a point in the $m + 1$-dimensional Euclidean space E. A point $g = (g_1, \cdots, g_{m+1})$ of E is said to belong to the range of the vector measure μ if and only if there exists a measurable subset S of R such that $\mu(S) = g$.

It was proved by Lyapunov [2] (see also [4]) that the range M of μ is a bounded, closed and convex subset of E. Let L be the line in E which is parallel to the $(m + 1)$-th axis and goes through the point $(c_1, c_2, \cdots, c_m, 0)$. Suppose that $\mathcal{S}$ is not empty. Then the intersection M^* of L with M is not empty. Because of Lyapunov's theorem, M^* is a finite closed interval (which may reduce to a single point). There exists a subset S of R such that $\mu(S)$ is equal to the upper end point of M^*. Clearly, S is a member of $\mathcal{S}_0$.

3. Necessary and sufficient condition that a member of $\mathcal{S}$ be also a member of $\mathcal{S}_0$. Let $\nu(S)$ be the vector measure with the components $\mu_1(S), \cdots \mu_m(S)$. According to the aforementioned theorem of Lyapunov, the range N of ν is a bounded, closed and convex subset of the m-dimensional Euclidean space.

By the dimension of a convex subset Q of a finite dimensional Euclidean space we shall mean the dimension of the smallest dimensional hyperplane that contains Q. A point q of a convex set Q is said to be an interior point of Q if there exists a sphere V with center at q and positive radius such that $V \cap \Pi \subset Q$, where Π is the smallest dimensional hyperplane containing Q. Any point q that is not an interior point of Q will be called a boundary point. We shall now prove the following theorem.

THEOREM 3.1. *If $(c_1, \cdots, c_m)$ is an interior point of N, then a necessary and sufficient condition for a member S of $\mathcal{S}$ to be a member of $\mathcal{S}_0$ is that there exist m constants $k_1, \cdots, k_m$ such that* (1.3) *and* (1.4) *hold for all x except perhaps on a set of measure zero.*

PROOF. The Neyman-Pearson lemma cited in Section 1 states that our condition is sufficient. Thus, we merely have to prove the necessity of our condition. Assume that $(c_1, \cdots, c_m)$ is an interior point of N. Let c^* be the largest value for which $(c_1, \cdots, c_m, c^*) \in M$, and c^{**} the smallest value for which

$$(c_1, \cdots, c_m, c^{**}) \in M.$$

We shall first consider the case when $c^* = c^{**}$. Let $(\bar{c}_1, \cdots, \bar{c}_m)$ be any other interior point of N. We shall show that there exists exactly one real value $\bar{c}$ such that $(\bar{c}_1, \cdots, \bar{c}_m, \bar{c}) \in M$. For suppose that there are two different values $\bar{c}^*$ and $\bar{c}^{**}$ such that both $(\bar{c}_1, \cdots, \bar{c}_m, \bar{c}^*)$ and $(\bar{c}_1, \cdots, \bar{c}_m, \bar{c}^{**})$ are in M. Since $(c_1, \cdots, c_m)$ and $(\bar{c}_1, \cdots, \bar{c}_m)$ are interior points of N, there exists a point $(c_1', \cdots, c_m')$ in N such that $(c_1, \cdots, c_m)$ lies in the interior of the segment determined by $(c_1', \cdots, c_m')$ and $(\bar{c}_1, \cdots, \bar{c}_m)$. There exists a real value c' such that $(c_1', \cdots, c_m', c') \in M$. Consider the convex set T determined by the 3 points: $(\bar{c}_1, \cdots, \bar{c}_m, \bar{c}^*)$, $(\bar{c}_1, \cdots, \bar{c}_m, \bar{c}^{**})$ and $(c_1', \cdots, c_m', c')$. Obviously, $T \subset M$. But T contains points $(c_1, \cdots, c_m, h)$ and $(c_1, \cdots, c_m, h')$ with

$h \neq h'$, contrary to our assumption that $c^* = c^{**}$. Thus, for any interior point $(\bar{c}_1, \cdots, \bar{c}_m)$ of N there exists exactly one real value $\bar{c}$ such that $(\bar{c}_1, \cdots, \bar{c}_m, \bar{c}) \epsilon M$. Since M is closed and convex, this remains true also when $(\bar{c}_1, \cdots, \bar{c}_m)$ is a boundary point of N. Thus, there exists a single valued function $\varphi(g_1, \cdots, g_m)$ such that $g_{m+1} = \varphi(g_1, \cdots, g_m)$ holds for all points $g = (g_1, \cdots, g_m, g_{m+1})$ in M. Since M is convex, φ must be linear; i.e., $\varphi(g_1, \cdots, g_m) = \sum_{i=1}^{m} k_i g_i + k_0$. Since the origin is obviously contained in M, we have $k_0 = 0$. Thus, we have $g_{m+1} = \sum_{i=1}^{m} k_i g_i$ for all points g in M. But then $f_{m+1}(x) = \sum_{i=1}^{m} k_i f_i(x)$ must hold for all x, except perhaps on a set of measure zero. Thus, for any subset S of R, the inequalities (1.3) and (1.4) are fulfilled for all x, except perhaps on a set of measure zero. This completes the proof of our theorem in the case when $c^* = c^{**}$.

We shall now consider the case when $c^{**} < c^*$. Let c be any value between c^{**} and c^*; i.e., $c^{**} < c < c^*$. We shall show that $(c_1, \cdots, c_m, c)$ is an interior point of M. For this purpose, consider a finite set of points $c^i = (c_1^i, \cdots, c_m^i)$ in $N (i = 1, \cdots, n)$ such that $c^1, \cdots, c^n$ are linearly independent, the simplex determined by $c^1, \cdots, c^n$ has the same dimension as N and contains the point $(c_1, \cdots, c_m)$ in its interior. Such points c^i in N obviously exist. There exist real values $h_i (i = 1, \cdots, n)$ such that $(c_1^i, \cdots, c_m^i, h_i) \epsilon M$ $(i = 1, \cdots, n)$. Let T be the smallest convex set containing the points $(c_1^i, \cdots, c_m^i, h_i)$ $(i = 1, \cdots, n)$, $(c_1, \cdots, c_m, c^*)$ and $(c_1, \cdots, c_m, c^{**})$. Clearly, the dimension of T is the same as that of M and $(c_1, \cdots, c_m, c)$ is an interior point of T. Thus, $(c_1, \cdots, c_m, c)$ is an interior point of M. The point $(c_1, \cdots, c_m, c^*)$ is obviously a boundary point of M. Let $g = (g_1, \cdots, g_{m+1})$ be the generic designation of a point in the $m + 1$-dimensional Euclidean space E. Since $(c_1, \cdots, c_m, c^*)$ is a boundary point of M, there exists an m-dimensional hyperplane Π through $(c_1, \cdots, c_m, c^*)$ such that Π contains only boundary points of M and M lies entirely on one side of Π.[3] Let the equation of Π be given by

$$k_{m+1} g_{m+1} - \sum_{i=1}^{m} k_i g_i = k_{m+1} c^* - \sum_{i=1}^{m} k_i c_i. \tag{3.1}$$

Since Π contains only boundary points of M, and since $(c_1, \cdots, c_m, c)$ is not a boundary point when $c^{**} < c < c^*$, the hyperplane Π cannot be parallel to the $(m + 1)$-th coordinate axis; i.e., $k_{m+1} \neq 0$. We can assume without loss of generality that $k_{m+1} = 1$. Since M lies entirely on one side of Π, and since for $(g_1, \cdots, g_m, g_{m+1}) = (c_1, \cdots, c_m, c^{**})$ the left hand member of (3.1) is smaller than the right hand member, we must have

$$g_{m+1} - \sum_{i=1}^{m} k_i g_i \leqq c^* - \sum_{i=1}^{m} k_i c_i \tag{3.2}$$

for all $g \epsilon M$. Let S be a subset of R such that

[3] This follows from well known results on convex bodies. See, for example, [3], p. 6.

$$(3.3) \qquad (\mu_1(S), \cdots, \mu_m(S), \mu_{m+1}(S)) = (c_1, \cdots, c_m, c^*).$$

It can easily be seen that (3.2) and (3.3) can be fulfilled simultaneously only if S satisfies the conditions (1.3) and (1.4) for all x, except perhaps on a set of measure zero. This completes the proof of our theorem.

It remains to investigate the case when $(c_1, \cdots, c_m)$ is a boundary point of N. For this purpose, we shall introduce some definitions and prove some lemmas.

Let $\xi = (\xi_1, \cdots, \xi_m)$ be an m-dimensional vector with real valued components at least one of which is not zero. We shall say that ξ is maximal relative to the point $c = (c_1, \cdots, c_m)$ if

$$(3.4) \qquad \sum_{i=1}^{m} \xi_i g_i \leqq \sum_{i=1}^{m} \xi_i c_i$$

for all points $(g_1, \cdots, g_m)$ in N.

We shall say that a set $\{\xi^i\}(i = 1, 2, \cdots, r; r > 1)$ of vectors is maximal relative to the point $c = (c_1, \cdots, c_m)$ if the set $\{\xi^i\}(i = 1, \cdots, r - 1)$ is maximal relative to c, not all components of ξ^r are zero and

$$(3.5) \qquad \sum_{j=1}^{m} \xi_j^r g_j \leqq \sum_{j=1}^{m} \xi_j^r c_j$$

holds for all points $(g_1, \cdots, g_m)$ of N for which

$$(3.6) \qquad \sum_{j=1}^{m} \xi_j^i g_j = \sum_{j=1}^{m} \xi_j^i c_j \qquad (i = 1, \cdots, r - 1).$$

A set of vectors $\{\xi^i\}(i = 1, \cdots, r)$ is said to be a complete maximal set relative to $c = (c_1, \cdots, c_m)$ if $\{\xi^i\}(i = 1, 2, \cdots, r)$ is maximal relative to c and no vector ξ^{r+1} exists such that ξ^{r+1} is linearly independent of the sequence $(\xi^1, \cdots, \xi^r)$ and $(\xi^1, \cdots, \xi^r, \xi^{r+1})$ is maximal relative to c.

LEMMA 3.1. *If $c = (c_1, \cdots, c_m)$ is a boundary point of N, then there exists a positive integer r and a set $\{\xi^1, \cdots, \xi^r\}$ of vectors that is a complete maximal set relative to c.*

PROOF. Since c is a boundary point of N, there exists an $(m - 1)$-dimensional hyperplane Π through c such that N lies entirely on one side of Π.[3] Let the equation of Π be given by

$$\sum_{i=1}^{m} \xi_i g_i = \sum_{i=1}^{m} \xi_i c_i .$$

Since N lies entirely on one side of Π, either $\sum_{i=1}^{m} \xi_i g_i \geqq \sum_{i=1}^{m} \xi_i c_i$ for all points $(g_1, \cdots, g_m)$ in N, or $\sum_{i=1}^{m} \xi_i g_i \leqq \sum_{i=1}^{m} \xi_i c_i$ for all $(g_1, \cdots, g_m)$ in N. We put $\xi^1 = -\xi$ if $\Sigma \xi_i g_i \geqq \Sigma \xi_i c_i$ for all points $(g_1, \cdots, g_m)$ in N. Otherwise, we put $\xi^1 = \xi$. Clearly, ξ^1 is maximal relative to c. If ξ^1 is not a complete maximal set relative to c, there exists a vector ξ^2 such that ξ^2 is linearly independent of

ξ^1 and (ξ^1, ξ^2) is maximal relative to c. If (ξ^1, ξ^2) is not a complete maximal set, we can find a vector ξ^3 such that ξ^3 is linearly independent of (ξ^1, ξ^2) and (ξ^1, ξ^2, ξ^3) is a maximal set relative to c, and so on. Continuing this procedure, we shall arrive at a set $(\xi^1, \cdots, \xi^r)(r \leqq m)$ that is a complete maximal set relative to c. This completes the proof of Lemma 3.1.

LEMMA 3.2. *If $(\xi^1, \cdots, \xi^r)$ is a maximal set of vectors relative to $c = (c_1, \cdots, c_m)$ and if $\nu(S) = c$, then the following two conditions are fulfilled for all x (except perhaps on a set of measure zero):*

a) *If x is a point in R for which $\sum_{j=1}^m \xi_j^i f_j(x) = 0$ for $i = 1, 2, \cdots, u - 1$ and $\sum_{j=1}^m \xi_j^u f_j(x) > 0$ $(u = 1, 2, \cdots, r)$, then $x \in S$.*

b) *If x is a point of R for which $\sum_{j=1}^m \xi_j^i f_j(x) = 0$ for $i = 1, 2, \cdots, u - 1$ and $\sum_{j=1}^m \xi_j^u f_j(x) < 0$, then $x \notin S$.*

PROOF. Assume that $(\xi^1, \cdots, \xi^r)$ is maximal relative to c. Then, ξ^1 is maximal relative to c. This implies that for all x (except perhaps on a set of measure zero) the following condition holds: $x \in S$ when $\sum_{j=1}^m \xi_j^1 f_j(x) > 0$ and $x \notin S$ when $\sum_{j=1}^m \xi_j^1 f_j(x) < 0$. Thus, conditions (a) and (b) of our lemma must be fulfilled for $u = 1$. We shall now show that if (a) and (b) hold for $u = 1, \cdots, v$ then (a) and (b) must hold also for $u = v + 1$. For this purpose, consider the set R' of all points x for which $\sum_{j=1}^m \xi_j^i f_j(x) = 0$ for $i = 1, \cdots, v$. If R is replaced by R', then ξ^{v+1} is maximal relative to $c' = (c_1', \cdots, c_m')$ where $c_i' = \int_{S'} f_i(x)\,dx$ and $S' = S \cap R'$. Hence, for any x in R' (except perhaps on a set of measure zero) the following condition holds: $x \in S$ when $\sum_{j=1}^m \xi_j^{v+1} f_j(x) > 0$ and $x \notin S$ when $\sum_{j=1}^m \xi_j^{v+1} f_j(x) < 0$. But this implies that (a) and (b) hold for $u = v + 1$. This completes the proof of our lemma.

LEMMA 3.3. *Let $(\xi^1, \cdots, \xi^r)$ be a complete maximal set of vectors relative to $c = (c_1, \cdots, c_m)$, and let T be the set of all points $g = (g_1, \cdots, g_m)$ of N for which $\sum_{j=1}^m \xi_j^i g_j = \sum_{j=1}^m \xi_j^i c_j$ for $i = 1, 2, \cdots, r$. Then T is a bounded, closed and convex set and c is an interior point of T.*

PROOF. Clearly, T is a bounded, closed and convex set. Suppose that c is a boundary point of T. Then there exists a hyperplane Π of dimension $m - 1$ such that Π goes through c, Π contains only boundary points of T and T lies entirely on one side of Π[3]. Let the equation of Π be given by

$$\sum_{j=1}^m \xi_j g_j = \sum_{j=1}^m \xi_j c_j,$$

where ξ is independent of $\xi^1, \cdots, \xi^r$. Since T lies on one side of Π, we have either $\sum_{j=1}^{m} \xi_j g_j \geqq \sum_{j=1}^{m} \xi_j c_j$ for all $g = (g_1, \cdots, g_m)$ in T, or $\sum_{j=1}^{m} \xi_j g_j \leqq \sum_{j=1}^{m} \xi_j c_j$ for all g in T. Let $\xi_j^{r+1} = \xi_j (j = 1, \cdots, m)$ in the latter case, and $\xi_j^{r+1} = -\xi_j$ in the former case. Then $\sum_{j=1}^{m} \xi_j^{r+1} g_j \leqq \sum_{j=1}^{m} \xi_j^{r+1} c_j$ for all g in T. But then $(\xi^1, \cdots, \xi^r, \xi^{r+1})$ is a maximal set relative to c, contrary to our assumption that $(\xi^1, \cdots, \xi^r)$ is a complete maximal set. Thus, c must be an interior point of T and our lemma is proved.

THEOREM 3.2. *If $c = (c_1, \cdots, c_m)$ is a boundary point of N and if $(\xi^1, \cdots, \xi^r)$ is a complete maximal set of vectors relative to c, then a necessary and sufficient condition for a member S of $\mathcal{S}$ to be a member of $\mathcal{S}_0$ is that there exist m constants $k_1, \cdots, k_m$ such that for all x in R' (except perhaps on a set of measure zero) the inequalities* (1.3) *and* (1.4) *hold, where R' is the set of all points x for which*

$$\sum_{j=1}^{m} \xi_j^i f_j(x) = 0 \quad \text{for} \quad i = 1, 2, \cdots, r.$$

PROOF. Suppose that $c = (c_1, \cdots, c_m)$ is a boundary point of N and that $(\xi^1, \cdots, \xi^r)$ is a complete maximal set of vectors relative to c. Let R^* be the set of all points x for which the following two conditions hold: (1) $\sum_{j=1}^{m} \xi_j^i f_j(x) \neq 0$ for at least one value i; (2) $\sum_{j=1}^{m} \xi_j^i f_j(x) > 0$ where i is the smallest integer for which $\sum_{j=1}^{m} \xi_j^i f_j(x) \neq 0$. For any member S of $\mathcal{S}$ let S^* denote the intersection of S with $R - R'$. It follows from Lemma 3.2 that $R^* - R^* \cap S^*$ and $S^* - R^* \cap S^*$ are sets of measure zero. Thus

$$\int_{S^*} f_i(x)\, dx = \int_{R^*} f_i(x)\, dx \qquad (i = 1, \cdots, m + 1) \tag{3.7}$$

for all $S \in \mathcal{S}$. Let

$$f_i^*(x) = f_i(x) \quad \text{for} \quad x \in R' \qquad (i = 1, \cdots, m + 1) \tag{3.8}$$

and

$$f_i^*(x) = 0 \quad \text{for} \quad x \in R - R' \quad (i = 1, 2, \cdots, m + 1). \tag{3.9}$$

Let, furthermore,

$$c_i^* = c_i - \int_{R^*} f_i(x)\, dx \qquad (i = 1, \cdots, m) \tag{3.10}$$

Let μ^*, ν^*, M^*, N^*, $\mathcal{S}^*$ and $\mathcal{S}_0^*$ have the same meaning with reference to the functions $f_1^*(x), \cdots, f_{m+1}^*(x)$ and the point $c^* = (c_1^*, \cdots, c_m^*)$ as μ, ν, M, N, $\mathcal{S}$ and $\mathcal{S}_0$ have with reference to the functions $f_1(x), \cdots, f_{m+1}(x)$ and the point $c = (c_1, \cdots, c_m)$.

It follows from Lemma 3.2 that for any subset S of R for which $\nu(S)$ is a point of the set T defined in Lemma 3.3 we have

$$\int_S f_i(x)\,dx = \int_S f_i^*(x)\,dx + \int_{R^*} f_i(x)\,dx \qquad (i = 1, \cdots, m+1).$$

Since the range of $\nu^*(S)$ is equal to N^* even when S is restricted to subsets S for which $\nu(S) \in T$, the set N^* is obtained from the set T by a translation. The same translation brings the point $c = (c_1, \cdots, c_m)$ into $c^* = (c_1^*, \cdots, c_m^*)$. It then follows from Lemma 3.3 that c^* is an interior point of N^*. Application of Theorem 3.1 gives the following necessary and sufficient condition for a member S of $\mathcal{S}^*$ to be a member of $\mathcal{S}_0^*$: There exist m constants $k_1, \cdots, k_m$ such that for all x (except perhaps on a set of measure zero)

$$(3.11) \qquad f_{m+1}^*(x) \geqq k_1 f_1^*(x) + \cdots + k_m f_m^*(x) \quad \text{when} \quad x \in S$$

and

$$(3.12) \qquad f_{m+1}^*(x) \leqq k_1 f_1^*(x) + \cdots + k_m f_m^*(x) \quad \text{when} \quad x \notin S.$$

It follows from (3.8) and (3.9) that (3.11) and (3.12) are equivalent to

$$(3.13) \qquad f_{m+1}(x) \geqq k_1 f_1(x) + \cdots + k_m f_m(x) \quad \text{when} \quad x \in S \cap R'$$

and

$$(3.14) \qquad f_{m+1}(x) \leqq k_1 f_1(x) + \cdots + k_m f_m(x) \quad \text{when} \quad x \in (R - S) \cap R'.$$

Theorem 3.2 follows from this and the fact that every member S of $\mathcal{S}$ is a member of $\mathcal{S}^*$ and that a member S of $\mathcal{S}$ is a member of $\mathcal{S}_0^*$ if and only if S is a member of $\mathcal{S}_0$.

It may be of interest to note that if the set R' is of measure zero, the members of $\mathcal{S}$ can differ from each other only by sets of measure zero; i.e., $\mathcal{S}$ consists essentially of one element. This is an immediate consequence of Lemma 3.2.

REFERENCES

[1] J. Neyman and E. S. Pearson, "Contributions to the theory of testing statistical hypotheses," *Stat. Res. Memoirs*, Vol. 1 (1936), pp. 1–37.

[2] A. Lyapunov, "Sur les fonctions-vecteurs complètement additives," *Izvestiya Akad. Nauk SSSR. Ser. Mat.*, Vol. 4 (1940), pp. 465–78.

[3] T. Bonnesen and W. Fenchel, *Theorie der Konvexen Körper*, Chelsea Publishing Company, New York, 1948.

[4] P. R. Halmos, "The range of a vector measure," *Bull. Am. Math. Soc.*, Vol. 54 (1948), pp. 416–421.

Annals of Mathematics
Vol. 53, No. 3, May, 1951

TWO METHODS OF RANDOMIZATION IN STATISTICS AND THE THEORY OF GAMES[1]

By A. Wald and J. Wolfowitz

(Received August 11, 1950)

1. Introduction

The problem of statistical decisions has been formulated by one of the authors and described, for example, in [1]. We proceed to describe such a formulation in semi-intuitive terms before we state our problem precisely in Section 2. The purpose of this rough description is to describe the motivation of the problem.

Ω is a given collection (finite or infinite) of distribution functions (or probability measures) F, defined on a Borel field B of a space X. A chance variable K with range in X is distributed according to one of the distributions in Ω, but this distribution is unknown to the statistician (or player II of a two-person game). The distribution of K in statistics is determined by the actual problem; in a game it is determined by player I. The statistician (henceforth "player II" is always to be understood to follow "statistician" in parentheses) is required to make a decision, i.e., choose a point d in a given space D^*. His loss is a function of d and the actual distribution F_0 of K, and knowledge of the latter would enable him to minimize the loss. In order to obtain information on F_0 the statistician proceeds to take, seriatim, independent observations on K. These observations have a cost which is to be added to the loss of the statistician. Usually this cost increases with the number of observations and in meaningful problems the loss and cost functions are formulated so as to make it unprofitable for the statistician to take infinitely many observations. (The loss and cost functions may also be functions of the observations; this will in no way change the problem of the present paper.) Thus the statistician has to strike a balance between the cost of ignorance of F_0 and the cost of the observations which presumably furnish information about F_0. A non-randomized sequential decision function is a rule (a function of the observations) which tells the statistician a) when to stop taking further observations, b) what decision to make when he has stopped taking observations.

What we shall call "special randomization" was introduced by v. Neumann into the theory of games [2]. (It applies equally well to statistical decisions; the purpose in both cases is to make a certain space convex). A sequential decision function randomized in the special sense may, for the purposes of this section, be described as follows: There is given a collection Γ, finite or not, of nonrandomized sequential decision functions, and a probability distribution η on Γ. (If the collection is not denumerable questions of measure will arise; such questions

[1] Presented to the International Congress of Mathematicians at Boston, Massachusetts, on September 1, 1950. Research under a contract with the Office of Naval Research and Development.

are relegated to the later sections of this paper.) The statistician performs a random experiment with distribution function η which yields, as an observation, a non-randomized sequential decision function, say f. He then proceeds to act according to f, i.e., he makes observations on K, stops and makes a decision according to f.

What we shall call "general randomization" was introduced by one of the authors (see, for example, [1]). Let s be an order to the statistician to take another observation. Denote by D the set consisting of D^* and s. Let C be a Borel field on D such that the set consisting of the single point s is an element of C. A sequential decision function randomized in the general sense (hereafter it may be referred to simply as a randomized sequential decision function or r.s. d.f.) may be described roughly as follows: There is given a set of probability measures on C, each measure being a function of a possible sequence of observations. The statistician obtains the first observation.[2] This gives him the probability measure, say μ_1, associated with this particular observation. He performs a chance experiment with probability distribution μ_1 which gives him a point in D. If this point is $d \neq s$, the statistician terminates taking observations and makes the decision d. If the point obtained is s, the statistician takes another observation. The r.s.d.f. supplies him with another probability measure, say μ_2, which is a function of the first two observations. The statistician then performs a chance experiment with distribution μ_2, etc., etc.

Consider now the space D^*. In many problems it is convenient to define a metric on D^*. (For example, an "intrinsic" metric is defined in [1]. This particular metric can always be defined; it depends on the loss function.) In this paper we shall assume that D^* is metric, separable, and complete (i.e., every Cauchy sequence possesses a limit point). The set of Borel sets of D^* will be called C^*. The smallest Borel field which contains all the elements of C^* and the set which consists of the single element s will be the set C.

We shall say that two sequential decision functions are equivalent if the probability of obtaining the observations $(x_1, \cdots, x_n)$ and then choosing an element in the set $c \in C$ is the same for both decision functions, identically in c, $(x_1, \cdots, x_n)$, and the distribution of K. It follows at once from the definitions that to every sequential decision function randomized in the special sense there exists an equivalent sequential decision function randomized in the general sense. The purpose of the present paper is to prove that, when D^* is metric, separable, and complete, and C is as defined in the preceding paragraph, the converse is true, namely, every sequential decision function randomized in the general sense is equivalent to one randomized in the special sense. A precise statement and the proof are given in the next section. A special case of this result was proved in [1].

The distribution function F_0 of K will not appear in the proof to be given below; the result is valid identically in F_0. Hence F_0 need not be a member of

---

[2] Even the decision whether to take the first observation at all can be made random. If this is so, only a slight and obvious change is necessary in the subsequent arguments.

any particular class. In particular, it is unnecessary that the observations on K be independent; indeed, F_0 can be the distribution function of an infinite sequence of chance variables, not necessarily independent, and with ranges in different spaces.

2. Proof of equivalence

We begin with the necessary definitions. The conditions on D^* and the definition of C have been given above.

Let $X_1, X_2, \cdots$ be an infinite sequence of abstract spaces, with $x_1, x_2, \cdots$, respectively, their generic points. Let Y denote the Cartesian product $X_1 \times X_2 \times \cdots$ (cf. [3], page 82), and let y be the generic designation of the sequence $(x_1, x_2, \cdots)$. Let $B_1, B_2, \cdots$ be Borel fields on $X_1, X_2, \cdots$, respectively, and let B be the smallest Borel field on Y which contains as elements every infinite Cartesian product $b_1 \times b_2 \times \cdots$, where $b_i \epsilon B_i$, $i = 1, 2, \cdots$.

By a randomized sequential decision function (r.s.d.f.) δ is meant a set of probability measures $\delta(y, m; c)$ with the following properties: 1) For every $y \epsilon Y$ and every positive integral m, $\delta(y, m; c)$ is a non-negative, completely additive set function defined for every set $c \epsilon C$ and such that $\delta(y, m; D) = 1$, 2) If the first m coordinates of y and y' are the same then

$$\delta(y, m; c) \equiv \delta(y', m; c),$$

3) The measures $\delta(y, m; c)$ are such that, for any $c \epsilon C$ and any m, $\delta(y, m; c)$, regarded as a function of y, is measurable B.

By a non-randomized sequential decision function (n.r.s.d.f.) γ is meant a set of probability measures $\gamma(y, m; c)$ which fulfill the requirements imposed upon a r.s.d.f. and in addition are such that, for every y and m, there exists a single element $d[y, m]$ of D such that $\gamma(y, m; d[y, m]) = 1$. We will find it typographically simpler therefore on many occasions to write γ as a point function with values in D, thus:

$$\gamma(y, m) = d[y, m].$$

Both forms will be employed without danger of confusion.

Let α be any sequential decision function, randomized or not. We define

$$\pi_\alpha(y, m; c) = \alpha(y, m; c) \prod_{i=1}^{m-1} \alpha(y, i; s) \tag{1}$$

for $m > 1$, and

$$\pi_\alpha(y, 1; c) = \alpha(y, 1; c). \tag{2}$$

Let $\delta(y, m; c)$ be any given r.s.d.f. We shall construct 1) a set of n.r.s.d.f.'s $\{\gamma(y, m \mid a)\}$, where a, which serves as an index to distinguish different functions γ from each other, is a point in a space H^* on which is defined a Borel field H, 2) a probability measure μ defined on H, such that

$$\pi_\delta(y, m; c) = \int_{H^*} \pi_{\gamma(y,m|a)}(y, m; c)\, d\mu \tag{3}$$

identically in y, m, and c.

In our construction the space H^* will be defined as follows: Let I_m be the interval $0 \leqq a_m < 1$, $m = 1, 2, \cdots$. H^* will be the infinite Cartesian product $I_1 \times I_2 \times \cdots$. The generic point of H^* will be

$$a = (a_1, a_2, \cdots).$$

Let L_m be the totality of Lebesgue measurable subsets of I_m. Then H will be the smallest Borel field on H^* which contains as an element every infinite Cartesian product $l_1 \times l_2 \times \cdots$ where $l_m \in L_m$, $m = 1, 2, \cdots$. The measure μ of such a Cartesian product will be defined as the product of the Lebesgue measures of its factors. It is well known that the definition of μ on such Cartesian products is sufficient to determine μ for every element of H.

Because of the assumptions on D we can construct a denumerable number of sets

$$\{A_{j_1 j_2 \cdots j_k}\},$$

$j_i = 1, 2, \cdots$ ad inf. for $i = 1, 2, \cdots k$, $k = 1, 2, \cdots$ ad inf. with the following properties:

a) every A is a member of C,
b) $\{s\} = A_1 = A_{11} = A_{111} = \cdots$,
c) A's with the same number of indices are disjoint,
d) if the indices of one set A form an initial sequence of the indices of another set A, the first set contains the second set,
e) the diameter of a set A with k indices ($k = 1, 2, \cdots$ ad inf.) is $\leqq 2^{-k}$
f) the union of all A's with the same number of indices is D.

We now define, for fixed y and m, a function $\psi_k(y, m, a_m)$ of a_m, $0 \leqq a_m < 1$, with range in D. Let $d(A_{j_1 \cdots j_k})$ be a fixed point in the closure $\bar{A}_{j_1 \cdots j_k}$ of $A_{j_1 \cdots j_k}$. Let

$$\begin{aligned} J(y, m; A_{j_1 \cdots j_k}) &= \sum_{i=1}^{j_1 - 1} \delta(y, m; A_i) + \sum_{i=1}^{j_2 - 1} \delta(y, m; A_{j_1 i}) \\ &\quad + \sum_{i=1}^{j_3 - 1} \delta(y, m; A_{j_1 j_2 i}) + \cdots + \sum_{i=1}^{j_k - 1} \delta(y, m; A_{j_1 j_2 \cdots j_{(k-1)} i}). \end{aligned} \tag{4}$$

For a_m such that

$$J(y, m; A_{j_1 \cdots j_k}) \leqq a_m < J(y, m; A_{j_1 \cdots j_k}) + \delta(y, m; A_{j_1 \cdots j_k}) \tag{5}$$

we define

$$\psi_k(y, m, a_m) = d(A_{j_1 \cdots j_k}) \tag{6}$$

(For $a_m < \delta(y, m; s)$ we define $\psi_k(y, m, a_m) = s$.) We define the function $\psi(y, m, a_m)$ by

$$\psi(y, m, a_m) = \lim_{k \to \infty} \psi_k(y, m, a_m).$$

Finally, we define

$$\gamma(y, m \mid a) = \psi(y, m, a_m) \tag{7}$$

for every y, m, and a.

Waiving for the moment all questions of measurability, let us prove that (3) holds. For this purpose fix y, m, and c at y_0, m_0, and c_0, respectively. We have that

$$\prod\nolimits_{\gamma(y,m|a)}(y_0, m_0; c_0)$$

is one or zero according as the following conditions are or are not satisfied:

$$a_m < \delta(y_0, m; s) \qquad m < m_0 \tag{8}$$

$$\psi(y_0, m_0, a_{m_0}) \in c_0 . \tag{9}$$

These are conditions on

$$a_1, \cdots, a_{m_0} \tag{10}$$

only. Thus the right member of (3) is simply the Lebesgue measure of the points (10) which satisfy conditions (8) and (9). Obviously it remains only to prove that the Lebesgue measure of the points a_{m_0} for which (9) holds is

$$\delta(y_0, m_0; c_0).$$

Let G be an open set such that

$$\delta(y_0, m_0; G) = \delta(y_0, m_0; \bar{G}) \tag{11}$$

where $\bar{G}$ is the closure of G. Let β_k be the union of all those A's with k indices whose closures have at least one point in common with $\bar{G}$. Then

$$\lim_{k\to\infty} \delta(y_0, m_0, \beta_k) = \delta(y_0, m_0, \bar{G}). \tag{12}$$

Since for any a_{m_0} for which

$$\psi(y_0, m_0, a_{m_0}) \in \bar{G} \tag{13}$$

holds, inequality (5) must hold for some term $A_{j_1 \cdots j_k}$ which appears in the sum β_k, the Lebesgue outer measure of the set of points a_{m_0} satisfying (13) cannot exceed $\delta(y_0, m_0, \beta_k)$ for every k. Hence, because of (12), this outer measure is not greater than $\delta(y_0, m_0; \bar{G}) = \delta(y_0, m_0; G)$. Now $G' = D - \bar{G}$ is also an open set such that

$$\delta(y_0, m_0; \bar{G}') = \delta(y_0, m_0; G').$$

Hence the Lebesgue outer measure of the points a_{m_0}, $0 \leqq a_{m_0} < 1$, such that

$$\psi(y_0, m_0, a_{m_0}) \in G',$$

is not greater than $\delta(y_0, m_0, G') = 1 - \delta(y_0, m_0, \bar{G})$. But the function $\psi(y_0, m_0, a_{m_0})$ is defined for every a_{m_0}, $0 \leqq a_{m_0} < 1$. Hence the sum of the two outer measures must be one. Consequently the set of points a_{m_0}, $0 \leqq a_{m_0} < 1$, such that (13) holds, is measurable and has Lebesgue measure $\delta(y_0, m_0; \bar{G})$. This proves (3) for the case that c_0 is a set G which satisfies (11).

Now every closed set can be represented as the limit of a descending sequence of sets of type G which satisfy (11). Hence (3) holds whenever c_0 is a closed set, from which it follows that (3) holds for any c_0 in C.

The function $\psi_k(y, m, a_m)$ is, for fixed m and a_m, a measurable function of y, i.e., those points y for which $\psi_k(y, m, a_m)$ is in some fixed member of C constitute a set which is in B. Hence $\psi(y, m, a_m)$ is a measurable function of y. This shows that the $\gamma(y, m \mid a)$ fulfill the required measurability conditions. Our proof is now complete.

COLUMBIA UNIVERSITY

REFERENCES

[1] WALD, A., *Statistical Decision Functions*, Ann. Math. Statist., Vol. 20 (1949), pp. 165–205.

[2] v. NEUMANN, J., *Zur Theorie der Gesellschaftsspiele*, Math. Ann., Vol. 100 (1928), pp. 295–320.

[3] SAKS, STANISLAW, Theory of the Integral, Stechert & Company, New York, 1937.

Reprinted from
Proceedings of the Second Berkeley Symposium on Mathematical Statistics and Probability
University of California Press, 1951

ASYMPTOTIC MINIMAX SOLUTIONS OF SEQUENTIAL POINT ESTIMATION PROBLEMS

A. WALD
COLUMBIA UNIVERSITY

1. Introduction

Exact minimax solutions for sequential point estimation problems are, in general, very difficult to obtain. As far as the author is aware, such solutions are known, at present, only in two special cases: (1) in estimating the mean of a normal distribution with known variance (see Wolfowitz [1]) and (2) in estimating the mean of a rectangular distribution with unit range (see Wald [2]). The solution in the first case coincides with the classical nonsequential one, while the solution in the second case is truly sequential.

In this note, we shall derive an asymptotic minimax solution for a general class of point estimation problems. The point estimation problem considered here may be stated as follows: Let $\{X_i\}$ $(i = 1, 2, \ldots, \text{ad inf.})$ be a sequence of independently and identically distributed chance variables. Let $F(u|\theta)$ be the common distribution function involving an unknown parameter θ, that is, $Pr\{X < u\} = F(u|\theta)$. We shall assume that $F(u|\theta)$ admits a density function $f(u|\theta)$. A sequential point estimation procedure T can be defined in terms of two sequences of functions $\{\varphi(x_1, \ldots, x_m)\}$ and $\{t(x_1, \ldots, x_m)\}$ $(m = 1, 2, \ldots, \text{ad inf.})$ where $\varphi(x_1, \ldots, x_m)$ can take only the values 0 and 1. The estimation procedure is then given as follows: Let x_i denote the observed value of X_i. We continue taking observations as long as $\varphi(x_1, \ldots, x_m) = 0$. At the first time when $\varphi(x_1, \ldots, x_m) = 1$, we stop experimentation and estimate the unknown parameter value by $t(x_1, \ldots, x_m)$. We shall assume that the cost of experimentation is proportional with the number of observations. Let c denote the cost of a single observation and let the loss due to estimating the true parameter value θ by t be given by $(t - \theta)^2$.

Let $\nu(\theta, T)$ denote the expected number of observations when θ is the true parameter value and the estimation procedure T is adopted. Furthermore, let $\rho(\theta, T)$ be the expected value of $(t - \theta)^2$ when θ is true and T is adopted. This expected value is given by

$$(1.1)\quad \rho(\theta, T) = \sum_{m=1}^{\infty} \int_{R_m} [t(x_1, \ldots, x_m) - \theta]^2 f(x_1|\theta) \ldots \times f(x_m|\theta)\, dx_1 \ldots dx_m$$

where R_m is the totality of all sample points $(x_1, \ldots, x_m)$ for which $\varphi_i(x_1, \ldots, x_i) = 0$

This research was done under contract with the Office of Naval Research.

when $i < m$ and $\varphi_m(x_1, \ldots, x_m) = 1$. The risk when θ is true, T is adopted, and c is the cost of a single observation is then given by

$$r(\theta, T, c) = \rho(\theta, T) + c\nu(\theta, T) . \tag{1.2}$$

An estimation procedure T_0 is said to be a minimax solution for a given value of c if

$$\sup_\theta r(\theta, T_0, c) \leqq \sup_\theta r(\theta, T, c) \text{ for all } T . \tag{1.3}$$

The symbol $\sup_\theta$ means supremum with respect to θ.

For every positive c, let T_c^0 be an estimation procedure. We shall say that T_c^0 is an asymptotic minimax solution if

$$\lim_{c=0} \frac{\sup_\theta r(\theta, T_c^0, c)}{\inf_T \sup_\theta r(\theta, T, c)} = 1 . \tag{1.4}$$

The symbol $\inf_T$ means infimum with respect to T. Clearly, if T_c^0 is an asymptotic minimax solution, for practical purposes T_c^0 may be regarded as a minimax solution when c is sufficiently small.

For any relation H, the symbol $Pr\{H | \theta\}$ will denote the probability that H holds when θ is the true value of the parameter. Furthermore, for any chance variable y, the symbol $E(y | \theta)$ will denote the expected value of y when θ is the true parameter value. Let

$$d(\theta) = E\left[\left(\frac{\partial \log f(x | \theta)}{\partial \theta}\right)^2 | \theta\right] \tag{1.5}$$

and let

$$d_0 = \inf_\theta d(\theta) . \tag{1.6}$$

The main result of this paper is that under certain regularity conditions the estimation procedures T_c^0 and T_c^1 are asymptotic minimax solutions where T_c^0 and T_c^1 are defined as follows:

Estimation rule T_c^0*:* Take N_c observations and estimate θ by the maximum likelihood estimate $\hat{\theta}_{N_c}$ based on the first N_c observations where N_c is the smallest integer $\geqq 1/\sqrt{cd_0}$.

Estimation rule T_c^1*:* Stop experimentation for the smallest positive integral value of n for which

$$\frac{1}{n d(\hat{\theta}_n)} - \frac{1}{(n+1) d(\hat{\theta}_n)} \leqq c \tag{1.7}$$

and estimate θ by $\hat{\theta}_n$. Here $\hat{\theta}_n$ denotes the maximum likelihood estimate of θ based on the first n observations.

Although T_c^0 and T_c^1 both are asymptotic minimax solutions, T_c^1 seems to be preferable to T_c^0 for small c, since

$$\lim_{c=0} \frac{r(\theta, T_c^1, c)}{r(\theta, T_c^0, c)} < 1 \text{ for any } \theta \text{ for which} \tag{1.8}$$

$$d(\theta) > d_0 , \tag{1.9}$$

as will be seen later.

2. Regularity assumptions

In what follows, for any chance variable y, the symbol $\sigma^2(y|\theta)$ will denote the variance of y when θ is the true parameter value. The symbol n will be used to denote the number of observations required for the estimation procedure, that is, n is the smallest integer for which $\varphi_n(x_1, \ldots, x_n) = 1$. Wolfowitz [3] has shown that under some weak regularity conditions the following inequality holds for any estimation procedure T:

$$\sigma^2[t(X_1, \ldots, X_n) \mid \theta] \geqq \frac{\left(1+\frac{\partial b(\theta, T)}{\partial\theta}\right)^2}{\nu(\theta, T)\, d(\theta)} \tag{2.1}$$

where

$$b(\theta, T) = E[t(X_1, \ldots, X_n) - \theta|\theta] . \tag{2.2}$$

Since we shall make use of the above inequality, we shall postulate the following assumption:

ASSUMPTION 2.1. *The regularity conditions postulated by Wolfowitz* [3] *to insure the validity of* (2.1) *are fulfilled.*

In addition to the above assumption, we shall make the following assumptions:

ASSUMPTION 2.2. *The domain of θ is an open (finite or infinite interval) interval of the real axis.*

ASSUMPTION 2.3. *$d(\theta)$ is a continuous function of θ and there exists a value θ_0 for which $d(\theta_0) = d_0 = \min_\theta d(\theta)$.*

ASSUMPTION 2.4. *For any positive integer N and for any θ let $Z_N(\theta) = \sqrt{N}(\hat\theta_N - \theta)$. The following limit relation holds:*

$$\lim_{N=\infty} Pr\{Z_N(\theta)\sqrt{d(\theta)} < \lambda \mid \theta\} = \frac{1}{\sqrt{2\pi}}\int_{-\infty}^{\lambda} e^{-u^2/2}du$$

uniformly in λ and θ.

ASSUMPTION 2.5. *$E[Z_N^{2+\delta}(\theta)|\theta]$ is a bounded function of θ and N for some positive δ.*

It is well known that assumption 2.4 holds under rather general conditions (see, for example, [4, p. 430]). The above assumptions can no doubt be weakened, but for the sake of simplicity the author has not attempted to do so here.

3. Proof that T_c^0 is an asymptotic minimax solution

It follows from (1.2) and (2.1) that

$$r(\theta, T, c) \geqq b^2(\theta, T) + \frac{\left(1+\frac{\partial b(\theta, T)}{\partial\theta}\right)^2}{\nu(\theta, T)\, d(\theta)} + c\nu(\theta, T) . \tag{3.1}$$

Taking the minimum with respect to ν, we obtain from (3.1)

$$r(\theta, T, c) \geqq b^2(\theta, T) + \frac{2\sqrt{c}\left|1+\frac{\partial b}{\partial \theta}\right|}{\sqrt{d(\theta)}}. \tag{3.2}$$

Consider a fixed finite and closed interval I of the θ-axis. Let $l(I)$ be the length of I. If $\frac{\partial b}{\partial \theta} \leqq -\epsilon\ (0 < \epsilon)$ for all θ in I, then

$$\sup_{\theta \in I} r(\theta, T, c) \geqq \sup_{\theta \in I} \rho(\theta, T) \geqq \sup_{\theta \in I} b^2(\theta, T) \geqq \frac{\epsilon^2 l^{2}(I)}{4}. \tag{3.3}$$

If $\frac{\partial b}{\partial \theta} > -\epsilon$ for some θ in I, it follows from (3.2) that

$$\sup_{\theta \in I} r(\theta, T, c) \geqq \frac{2\sqrt{c}(1-\epsilon)}{\sqrt{\max_{\theta \in I} d(\theta)}}. \tag{3.4}$$

Let

$$\epsilon(I, c) = \frac{\sqrt{8\sqrt[4]{c}}}{l(I)\sqrt[4]{\max_{\theta \in I} d(\theta)}}. \tag{3.5}$$

Clearly,

$$\frac{\epsilon^2(I, c)\, l^2(I)}{4} = \frac{2\sqrt{c}}{\sqrt{\max_{\theta \in I} d(\theta)}}. \tag{3.6}$$

Since the right hand member of (3.6) is greater than the right hand member of (3.4), it follows from (3.3) and (3.4) that

$$\sup_{\theta \in I} r(\theta, T, c) \geqq \frac{2\sqrt{c}[1-\epsilon(I, c)]}{\sqrt{\max_{\theta \in I} d(\theta)}}. \tag{3.7}$$

Let Ω denote the whole parameter space. It then follows from (3.7) that

$$\sup_{\theta \in \Omega} r(\theta, T, c) \geqq \sup_{I} \frac{2\sqrt{c}[1-\epsilon(I, c)]}{\sqrt{\max_{\theta \in I} d(\theta)}}. \tag{3.8}$$

Let θ_0 be a value of θ for which $d(\theta_0) = \min_{\theta} d(\theta) = d_0$. The existence of such a value is postulated in assumption 2.3. Let I_0 be the closed interval of length l_0 and midpoint θ_0. We then obtain from (3.8)

$$\sup_{\theta \in \Omega} r(\theta, T, c) \geqq \frac{2\sqrt{c}[1-\epsilon(I_0, c)]}{\sqrt{\max_{\theta \in I_0} d(\theta)}}. \tag{3.9}$$

Since by assumption 2.3 the function $d(\theta)$ is continuous in θ, there exists a positive δ, say δ_{l_0} (depending on l_0) such that

$$\max_{\theta \in I_0} d(\theta) \leqq d_0 + \delta_{l_0} \tag{3.10}$$

and

$$\lim_{l_0=0} \delta_{l_0} = 0 . \tag{3.11}$$

We then obtain from (3.9)

$$\sup_{\theta \in \Omega} r(\theta, T, c) \geqq \frac{2\sqrt{c}\,[1-\epsilon(l_0, c)]}{\sqrt{d_0+\delta_{l_0}}} . \tag{3.12}$$

Since the right hand member of (3.12) does not depend on T, we obtain

$$\inf_T \sup_\theta r(\theta, T, c) \geqq \frac{2\sqrt{c}\,[1-\epsilon(l_0, c)]}{\sqrt{d_0+\delta_{l_0}}} . \tag{3.13}$$

For fixed l_0, we have

$$\lim_{c=0} \epsilon(l_0, c) = 0 . \tag{3.14}$$

Hence, it follows from (3.13) that

$$\liminf_{c=0} \frac{\sqrt{d_0+\delta_{l_0}}}{2\sqrt{c}} \inf_T \sup_\theta r(\theta, T, c) \geqq 1 . \tag{3.15}$$

Since δ_{l_0} can be made arbitrarily small by choosing l_0 sufficiently small, and since $d_0 > 0$, we obtain

$$\liminf_{c=0} \frac{\inf_T \sup_\theta r(\theta, T, c)}{2\sqrt{\frac{c}{d_0}}} \geqq 1 . \tag{3.16}$$

We shall now show that

$$\lim_{c=0} \frac{\sup_\theta r(\theta, T_c^0, c)}{2\sqrt{\frac{c}{d_0}}} = 1 . \tag{3.17}$$

Clearly, for the estimation procedure T_c^0 defined in section 1 we have

$$N_c r(\theta, T_c^0, c) = N_c E[(\hat{\theta}_{N_c} - \theta)^2 \mid \theta] + cN_c^2 . \tag{3.18}$$

Let $\{\theta_N\}$ ($N = 1, 2, \ldots$, ad inf.) be any sequence of parameter points. It follows from assumption 2.4 that the distribution of $Z_N(\theta_N)\sqrt{d(\theta_N)}$, as $N \to \infty$, converges to the normal distribution with zero mean and unit variance. Hence, the Helly-Bray theorem [7, p. 31] and assumption 2.5 give

$$\lim_{N=\infty} E[Z_N^2(\theta_N) \mid \theta_N]\, d(\theta_N) = 1 . \tag{3.19}$$

From this it follows that

$$\lim_{c=0} N_c E(\hat{\theta}_{N_c} - \theta)^2 = \frac{1}{d(\theta)} \tag{3.20}$$

uniformly in θ. Hence, because of (3.18), we have

$$\lim_{c=0} N_c r(\theta, T_c^0, c) = \frac{1}{d(\theta)} + \frac{1}{d_0} \tag{3.21}$$

uniformly in θ. Hence,

$$\lim_{c=0} N_c \sup_\theta r(\theta, T_c^0, c) = \sup_\theta \frac{1}{d(\theta)} + \frac{1}{d_0} = \frac{2}{d_0}. \tag{3.22}$$

Since $N_c = 1/\sqrt{cd_0}$, (3.17) is an immediate consequence of (3.22). From (3.16) and (3.17) it follows that

$$\lim_{c=0} \frac{\inf_T \sup_\theta r(\theta, T, c)}{2\sqrt{\frac{c}{d_0}}} = 1. \tag{3.23}$$

Equations (3.17) and (3.23) imply that T_c^0 is an asymptotic minimax solution.

4. Limiting distribution of the maximum likelihood estimate when the number of observations is determined by a sequential rule

In order to study the risk function associated with the estimation procedure T_c^1, it will be necessary to obtain the limiting distribution of $\sqrt{n}\,(\hat{\theta}_n - \theta)$ when n is determined by a sequential rule.

For any positive value c, let $\{\varphi_c(x_1, \ldots, x_m)\}$ ($m = 1, 2, \ldots$, ad inf.) be a sequence of functions which can take only the values 0 and 1. Let n_c be the smallest positive integer for which

$$\varphi_c(x_1, \ldots, x_m) = 0 \quad \text{for } m < n_c$$

and

$$\varphi_c(x_1, \ldots, x_{nc}) = 1.$$

We shall make the following assumptions:

ASSUMPTION 4.1. *There exist a function $N(c, \theta)$ of c and θ, and a positive function $\epsilon(c)$ of c such that*

$$\lim_{c=0} N(c, \theta) = \infty \text{ uniformly in } \theta, \tag{4.1}$$

$$\lim_{c=0} \epsilon(c) = 0, \tag{4.2}$$

and

$$\lim_{c=0} Pr\{N - \epsilon N \leqq n_c < N + \epsilon N \mid \theta\} = 1 \tag{4.3}$$

uniformly in θ.

ASSUMPTION 4.2. *The derivatives* $\dfrac{\partial \log f(x \mid \theta)}{\partial \theta}$ *and* $\dfrac{\partial^2 \log f(x \mid \theta)}{\partial \theta^2}$ *exist.*

ASSUMPTION 4.3. *For some positive δ,* $E\left[\left(\dfrac{\partial \log f(x \mid \theta)}{\partial \theta}\right)^{2+\delta} \mid \theta\right]$ *is a bounded function of θ.*

ASSUMPTION 4.4.

$$E\left[\left(\frac{\partial \log f(x \mid \theta)}{\partial \theta}\right)^2 \mid \theta\right] = d(\theta)$$

has a positive lower bound and is uniformly continuous in θ.

For any positive ρ, let

$$h(x, \theta, \rho) = \sup_{\theta'} \left| \left(\frac{\partial^2 \log f(x \mid \theta)}{\partial \theta^2}\right)_{\theta=\theta'} \right| \tag{4.4}$$

where θ' is restricted to values in the closed interval $|\theta - \rho, \theta + \rho|$. Furthermore, let

$$h_1(x, \theta, \rho) = \sup_{\theta'} \left[\left(\frac{\partial^2 \log f(x \mid \theta)}{\partial \theta^2} \right)_{\theta=\theta'} \right] \tag{4.4a}$$

and

$$h_2(x, \theta, \rho) = \inf_{\theta'} \left[\left(\frac{\partial^2 \log f(x \mid \theta)}{\partial \theta^2} \right)_{\theta=\theta'} \right] \tag{4.4b}$$

when $|\theta' - \theta| \leqq \rho$.

ASSUMPTION 4.5. *$E[h(x, \theta, \rho) \mid \theta]$ is a bounded function of θ for some positive ρ and*

$$\lim_{\rho=0} E[h_i(x, \theta, \rho) \mid \theta] = E\left[\frac{\partial^2 \log f(x \mid \theta)}{\partial \theta^2} \mid \theta \right] = -d(\theta),$$

uniformly in θ.

For any θ, any positive integer m and for any positive δ, let $Q_{\theta,m,\delta}$ denote the event that

$$|\hat{\theta}_k - \theta| \leqq \delta \quad \text{for all } k \geqq m. \tag{4.5}$$

ASSUMPTION 4.6. *For any positive δ, we have*

$$\lim_{m=\infty} Pr\{Q_{\theta,m,\delta} \mid \theta\} = 1 \tag{4.6}$$

uniformly in θ.[1]

We shall prove the following theorem:

THEOREM 4.1. *If assumptions 4.1 to 4.6 hold, then*

$$\lim_{c=0} Pr\{\sqrt{n_c}(\hat{\theta}_{n_c} - \theta)\sqrt{d(\theta)} < \lambda \mid \theta\} = \frac{1}{\sqrt{2\pi}} \int_{-\infty}^{\lambda} e^{-u^2/2} du \tag{4.7}$$

uniformly in λ and θ.

PROOF. Taylor expansion of $\sum_{\alpha=1}^{n_c} \frac{\partial \log f(x_\alpha \mid \theta)}{\partial \theta}$ at $\theta = \hat{\theta}_{n_c}$ gives

$$\sum_{\alpha=1}^{n_c} \frac{\partial \log f(x_\alpha \mid \theta)}{\partial \theta} = \sum_{\alpha=1}^{n_c} \frac{\partial \log f(x_\alpha \mid \hat{\theta}_{n_c})}{\partial \theta} + (\theta - \hat{\theta}_{n_c}) \sum_{\alpha=1}^{n_c} \frac{\partial^2 \log f(x_\alpha \mid \bar{\theta}_c)}{\partial \theta^2} \tag{4.8}$$

where $\bar{\theta}_c$ lies between $\hat{\theta}_{n_c}$ and θ and $\frac{\partial^i \log f(x \mid \theta^*)}{\partial \theta^i}$ denotes the value of $\frac{\partial^i \log f(x \mid \theta)}{\partial \theta^i}$ at $\theta = \theta^*$. Since the first term on the right hand side of (4.8) is zero, we obtain

$$\sum_{\alpha=1}^{n_c} \frac{\partial \log f(x_\alpha \mid \theta)}{\partial \theta} = -(\hat{\theta}_{n_c} - \theta) \sum_{\alpha=1}^{n_c} \frac{\partial^2 \log f(x_\alpha \mid \bar{\theta}_c)}{\partial \theta^2}. \tag{4.9}$$

[1] This assumption states that the maximum likelihood estimate converges strongly to the true parameter point θ and that this convergence is uniform in θ. The strong convergence of the maximum likelihood estimate was proved under very general conditions (see, for example, [5] and [6]). The uniformity of this convergence in θ can also be proved under some slight additional regularity conditions, by making use of a result by Chung [8] concerning the uniformity of the strong law of large numbers.

Hence,

$$\frac{\frac{1}{\sqrt{N}}\left(\sum_{a=1}^{n_c}\frac{\partial \log f(x_a \mid \theta)}{\partial \theta}\right)\sqrt{d(\theta)}}{-\frac{1}{N}\sum_{a=1}^{n_c}\frac{\partial^2 \log f(x_a \mid \bar{\theta}_c)}{\partial \theta^2}} = \sqrt{N}\,(\hat{\theta}_{n_c}-\theta)\,\sqrt{d(\theta)}\,. \tag{4.10}$$

Let $\{\eta_i\}$ ($i = 1, 2, \ldots$, ad inf.) be a sequence of positive numbers such that $\lim_{i=\infty} \eta_i = 0$. It follows from assumption 4.6 that there exists a sequence $\{k_i\}$ ($i = 1, 2, \ldots$, ad inf.) of positive integers such that $\lim_{i=\infty} k_i = \infty$ and

$$\lim_{i=\infty} Pr\{Q_{\theta, k_i, \eta_i} \mid \theta\} = 1 \text{ uniformly in } \theta\,. \tag{4.11}$$

For any positive $k \geqq k_1$, let $\rho_k = \eta_i$ where i is the largest positive integer for which $k \geqq k_i$. Clearly, we have

$$\lim_{k=\infty} \rho_k = 0 \tag{4.12}$$

and

$$\lim_{k=\infty} Pr\{Q_{\theta, k, \rho_k} \mid \theta\} = 1 \text{ uniformly in } \theta\,. \tag{4.13}$$

It follows from (4.13) and assumption 4.1 that

$$\lim_{c=0} Pr\{\,|\,\hat{\theta}_{n_c} - \theta\,| < \rho_{[N-\epsilon N]} \mid \theta\} = 1 \tag{4.13a}$$

uniformly in θ. The symbol $[a]$ denotes the smallest integer $\geqq a$. Since $\bar{\theta}_c$ lies between $\hat{\theta}_{n_c}$ and θ, the above equation gives

$$\lim_{c=0} Pr\{\,|\,\bar{\theta}_c - \theta\,| \leqq \rho_{[N-\epsilon N]} \mid \theta\} \text{ uniformly in } \theta\,. \tag{4.14}$$

Let n_c^* be defined as follows:

$$\begin{aligned} n_c^* &= n_c \text{ when } N - \epsilon N \leqq n_c \leqq N + \epsilon N \\ n_c^* &= [N - \epsilon N] \text{ when } n_c < N - \epsilon N \\ n_c^* &= [N + \epsilon N] \text{ when } n_c > N + \epsilon N\,. \end{aligned} \tag{4.15}$$

For any sequence $\{u_i\}$ of chance variables the symbol

$$\operatorname*{plim}_{i=\infty}(u_i \mid \theta) = \lambda$$

will mean that $\lim_{i=\infty} Pr\{|u_i - \lambda| > \rho \mid \theta\} = 0$ for any $\rho > 0$. It follows immediately from assumption 4.1 that

$$\operatorname*{plim}_{c=0}\left[\frac{1}{\sqrt{N}}\left(\sum_{a=1}^{n_c}\frac{\partial \log f(X_a \mid \theta)}{\partial \theta} - \sum_{a=1}^{n_c^*}\frac{\partial \log f(X_a \mid \theta)}{\partial \theta}\right)\Big|\,\theta\right] = 0 \tag{4.16}$$

and

$$\operatorname*{plim}_{c=0}\left[\frac{1}{N}\left(\sum_{a=1}^{n_c}\frac{\partial^2 \log f(X_a \mid \bar{\theta}_c)}{\partial \theta^2} - \sum_{a=1}^{n_c^*}\frac{\partial^2 \log f(X_a \mid \bar{\theta}_c)}{\partial \theta^2}\right)\Big|\,\theta\right] = 0 \tag{4.17}$$

uniformly in θ. Clearly,

$$(4.18) \qquad \left| \sum_{a=n_c^*+1}^{[N+\epsilon N]} \frac{\partial^2 \log f(x_a \mid \bar{\theta}_c)}{\partial \theta^2} \right| \leqq \sum_{a=[N-\epsilon N]}^{[N+\epsilon N]} h(x_a, \theta, \rho_{[N-\epsilon N]})$$

holds when $|\bar{\theta}_c - \theta| \leqq \rho_{[N-\epsilon N]}$. Because of (4.12) and assumption 4.1, we have

$$(4.19) \qquad \lim_{c=0} (\rho_{[N-\epsilon N]}) = 0 \text{ uniformly in } \theta .$$

It follows from assumption 4.5 that for some positive ρ

$$(4.20) \qquad \lim_{c=0} \frac{1}{N} E\left[\sum_{a=[N-\epsilon N]}^{[N+\epsilon N]} h(X_a, \theta, \rho) \mid \theta \right] = 0 \text{ uniformly in } \theta .$$

Hence, because of (4.18), (4.19), and (4.14), we have

$$(4.21) \qquad \operatorname*{plim}_{c=0} \left[\frac{1}{N} \sum_{a=n_c^*+1}^{[N+\epsilon N]} \frac{\partial^2 \log f(X_a \mid \bar{\theta}_c)}{\partial \theta^2} \,\middle|\, \theta \right] = 0$$

uniformly in θ.

Since

$$E\left\{ \left[\sum_{a=n_c^*+1}^{[N+\epsilon N]} \frac{\partial \log f(X_a \mid \theta)}{\partial \theta} \right]^2 \,\middle|\, \theta \right\} = d(\theta) E\{ ([N+\epsilon N] - n_c^*) \mid \theta \}$$

it follows easily from assumption 4.1 that

$$(4.22) \qquad \operatorname*{plim}_{c=0} \left[\frac{1}{\sqrt{N}} \sum_{a=n_c^*+1}^{[N+\epsilon N]} \frac{\partial \log f(X_a \mid \theta)}{\partial \theta} \,\middle|\, \theta \right] = 0$$

uniformly in θ. We shall now show that

$$(4.23) \qquad \operatorname*{plim}_{c=0} \left[\frac{1}{N} \sum_{a=1}^{[N+\epsilon N]} \frac{\partial^2 \log f(X_a \mid \bar{\theta}_c)}{\partial \theta^2} \,\middle|\, \theta \right] = -d(\theta)$$

uniformly in θ.

Clearly,

$$(4.24) \qquad \frac{1}{N} \sum_{a=1}^{[N+\epsilon N]} h_2(x_a, \theta, \rho_{[N-\epsilon N]}) \leqq \frac{1}{N} \sum_{a=1}^{[N+\epsilon N]} \frac{\partial^2 \log f(x_a \mid \bar{\theta}_c)}{\partial \theta^2}$$
$$\leqq \frac{1}{N} \sum_{a=1}^{[N+\epsilon N]} h_1(x_a, \theta, \rho_{[N-\epsilon N]})$$

whenever $|\bar{\theta}_c - \theta| \leqq \rho_{[N-\epsilon N]}$. Hence, because of (4.14), equation (4.23) is proved if we can show that

$$(4.25) \qquad \operatorname*{plim}_{c=0} \left[\frac{1}{N} \sum_{a=1}^{[N+\epsilon N]} h_i(X_a, \theta, \rho_{[N-\epsilon N]} \mid \theta \right] = -d(\theta)$$

uniformly in θ. But this follows from (4.19) and assumption 4.5. Thus, (4.23) is proved.

We obtain from (4.10), (4.16), (4.17), (4.21), (4.22), and (4.23) that

$$(4.26)\qquad \frac{\frac{1}{\sqrt{N}}\left(\sum_{\alpha=1}^{N+\epsilon N}\frac{\partial \log f(x_\alpha\mid\theta)}{\partial\theta}\right)\sqrt{d(\theta)}+\xi_c}{d(\theta)+\zeta_c}=\sqrt{N}(\hat{\theta}_{n_c}-\theta)\sqrt{d(\theta)}$$

where

$$(4.27)\qquad \operatorname*{plim}_{c=0}(\xi_c\mid\theta)=\operatorname*{plim}_{c=0}(\zeta_c\mid\theta)=0$$

uniformly in θ.

It follows from assumption 4.3 and the central limit theorem that

$$(4.28)\quad \lim_{c=0} Pr\left\{\frac{1}{\sqrt{N}}\sum_{\alpha=1}^{[N+\epsilon N]}\frac{\partial\log f(X_\alpha\mid\theta)}{\partial\theta}\frac{1}{\sqrt{d(\theta)}}<v\,\Big|\,\theta\right\}=\frac{1}{\sqrt{2\pi}}\int_{-\infty}^{v}e^{-u^2/2}du$$

uniformly in v and θ.

Since according to assumption 4.4 $d(\theta)$ has a positive lower bound, theorem 4.1 follows easily from (4.26), (4.27), (4.28) and assumption 4.1.

5. Proof that T_c^1 is an asymptotic minimax solution and that (1.8) holds

Assumptions 4.2 to 4.6 are assumptions concerning $f(x\mid\theta)$ only. If these assumptions hold, it is not difficult to verify that assumption 4.1 is fulfilled for the sequential procedure T_c^1, where $N(c,\theta)=1/\sqrt{cd(\theta)}$. In fact, it follows from the boundedness of $d(\theta)$ that

$$(5.1)\qquad \lim_{c=0} n_c=\infty\ .$$

From this and assumption 4.6 it follows that for any $\delta>0$

$$(5.2)\quad \lim_{c=0} Pr\left\{n_c \text{ is included in}\left[\inf_{|\theta'-\theta|\leqq\delta}\left(\frac{1}{\sqrt{cd(\theta')}}-1\right),\ \sup_{|\theta'-\theta|\leqq\delta}\left(\frac{1}{\sqrt{cd(\theta')}}+1\right)\right]\Big|\,\theta\right\}=1$$

uniformly in θ. Assumption 4.1 is a simple consequence of (5.1), (5.2) and assumption 4.4. Hence, theorem 4.1 yields

$$(5.3)\quad \lim_{c=0} Pr\left\{\sqrt{n_c}(\hat{\theta}_{n_c}-\theta)\sqrt{d(\theta)}<\lambda\mid\theta\right\}=\frac{1}{\sqrt{2\pi}}\int_{-\infty}^{\lambda}e^{-u^2/2}du$$

uniformly in λ and θ. Clearly,

$$(5.4)\qquad N(c,\theta)\,r(\theta,T_c^1,c)=N(c,\theta)\,E[(\theta_{n_c}-\theta)^2\mid\theta]+N(c,\theta)\,cE(n_c\mid\theta)\ .$$

We shall make the additional assumption:

Assumption 5.1. $[N(c,\theta)]^{1+\delta/2}E[(\hat{\theta}_{n_c}-\theta)^{2+\delta}\mid\theta]$ *is a bounded function of* c *and* θ *for some* $\delta>0$.

Since

$$(5.5)\qquad \operatorname{plim}\left(\frac{n_c}{N(c,\theta)}\,\Big|\,\theta\right)=1$$

uniformly in θ, it follows from (5.3) and assumption 5.1 that

$$\lim_{c=0} N(c, \theta) E[(\hat{\theta}_{n_c} - \theta)^2 | \theta] = \frac{1}{d(\theta)} \tag{5.6}$$

uniformly in θ. Furthermore, it can easily be seen that

$$\lim_{c=0} N(c, \theta)\, cE(n_c | \theta) = \frac{1}{d(\theta)} \tag{5.7}$$

uniformly in θ. Hence

$$\lim_{c=0} N(c, \theta)\, r(\theta, T_c^1, c) = \frac{2}{d(\theta)} \tag{5.8}$$

uniformly in θ, or

$$\lim_{c=0} \frac{r(\theta, T_c^1, c)}{2\sqrt{\dfrac{c}{d(\theta)}}} = 1 \tag{5.9}$$

uniformly in θ. This and (3.21) show that the following theorem holds:

THEOREM 5.1. *If assumptions* 4.2 *to* 4.6 *and* 5.1 *hold, and if* (3.21) *and* (3.23) *hold, then* T_c^1 *is an asymptotic minimax solution and* (1.8) *holds.*

Let T_c^2 be the estimation procedure defined as follows: Take first m_c observations where $m_c = [1/\sqrt{cd_1}]$ and d_1 is the least upper bound of $d(\theta)$ with respect to θ. Then take $n_c - m_c$ additional observations where $n_c = [1/\sqrt{cd(\hat{\theta}_{m_c})}]$. Estimate θ by θ_{n_c}.

One can show in a similar way that if assumptions 4.2 to 4.6 hold, and if assumption 5.1 remains valid when T_c^1 is replaced by T_c^2, then

$$\lim_{c=0} \frac{r(\theta, T_c^2, c)}{2\sqrt{\dfrac{c}{d(\theta)}}} = 1 \tag{5.10}$$

uniformly in θ. Thus, because of (5.9), we have

$$\lim_{c=0} \frac{r(\theta, T_c^2, c)}{r(\theta, T_c^1, c)} = 1 \tag{5.11}$$

uniformly in θ.

REFERENCES

[1] J. WOLFOWITZ, "Minimax estimates of the mean of a normal distribution with known variance," *Annals of Math. Stat.*, Vol. 21 (1950), pp. 218–230.

[2] A. WALD, *Statistical Decision Functions*, Wiley, New York, 1950.

[3] J. WOLFOWITZ, "The efficiency of sequential estimates and Wald's equation for sequential processes," *Annals of Math. Stat.*, Vol. 18 (1947), pp. 215–230.

[4] A. WALD, "Tests of statistical hypotheses concerning several parameters when the number of observations is large," *Trans. Amer. Math. Soc.*, Vol. 54 (1943), pp. 426–482.

[5] J. L. DOOB, "Probability and statistics," *Trans. Amer. Math. Soc.*, Vol. 36 (1934), pp. 759–775.

[6] A. WALD, "Note on the consistency of the maximum likelihood estimate," *Annals of Math. Stat.*, Vol. 20 (1949), pp. 595–601.

[7] D. V. WIDDER, *The Laplace Transform*, Princeton University Press, Princeton, 1946.

[8] K. L. CHUNG, "The strong law of large numbers," *Proceedings of the Second Berkeley Symposium on Mathematical Statistics and Probability*, University of California Press, Berkeley, 1951, pp. 341–352.

Reprinted from
Proceedings of the Second Berkeley Symposium on Mathematical Statistics and Probability
University of California Press, 1951

CHARACTERIZATION OF THE MINIMAL COMPLETE CLASS OF DECISION FUNCTIONS WHEN THE NUMBER OF DISTRIBUTIONS AND DECISIONS IS FINITE

A. WALD AND J. WOLFOWITZ
COLUMBIA UNIVERSITY

1. Introduction

The principal object of the present paper is to prove theorem 2 below. This theorem characterizes the minimal complete class in the problem under consideration, and improves on the result of theorem 1. Theorem 1 has been proved by one of us in much greater generality [1]. The proof given below is new and very expeditious. Another reason for giving the proof of theorem 1 here is that it is the first step in our proof of theorem 2. A different proof of theorem 1, based, like ours, on certain properties of convex bodies in finite Euclidean spaces, was communicated earlier to the authors by Dr. A. Dvoretzky. Theorem 3 gives another characterization of the minimal complete class.

Let x be the generic point of a Euclidean[1] space Z, and $f_1(x), \ldots, f_m(x)$ be any m (> 1) distinct cumulative probability distributions on Z. The statistician is presented with an observation on the chance variable X which is distributed in Z according to an unknown one of $f_1, \ldots, f_m$. On the basis of this observation he has to make one of l decisions, say $d_1, \ldots, d_l$. The loss incurred when x is the observed point, f_i is the actual (unknown) distribution, and the decision d_j is made, is $W_{ij}(x)$, where $W_{ij}(x)$ is a measurable function of x such that

$$\int_Z | W_{ij}(x) | df_i < \infty , \qquad i = 1, \ldots, m ; \; j = 1, \ldots, l .$$

A randomized decision function $\eta(x)$, say, hereafter often called "test" for short, is defined as follows: $\eta(x) = [\eta_1(x), \eta_2(x), \ldots, \eta_l(x)]$ where

(a) $\eta(x)$ is defined for all x,

(b) $0 \leqq \eta_j(x), j = 1, \ldots, l,$

(c) $\sum_{j=1}^{l} \eta_j(x) = 1$ identically in x,

(d) $\eta_j(x)$ is measurable, $j = 1, \ldots, l.$

This research was done under a contract with the Office of Naval Research.

[1] The extension to general abstract spaces is trivial and we forego it. This entire paper could be given an abstract formulation without the least mathematical difficulty.

The statistical application of the function $\eta(x)$ is as follows: After the observation x has been obtained the statistician performs a chance experiment to decide which decision to make. The probability of making decision d_j according to this experiment is $\eta_j(x)$ $(j = 1, \ldots, l)$. The risk of the test $\eta(x)$, which will also be called its associated risk or risk point, is the complex $r(\eta) = (r_1, \ldots, r_m)$, where

$$r_i = \int_Z \left(\sum_{j=1}^{l} \eta_j(x) W_{ij}(x) \right) df_i .$$

Let V be the totality of all risk points (in m-dimensional space) corresponding to all possible tests. It follows from results of Dvoretzky, Wald, and Wolfowitz[2] that the set V is closed and convex.

The test T with risk $r = (r_1, \ldots, r_m)$ is called uniformly better than the test T' with risk $r' = (r_1', \ldots, r_m')$ if $r_i \leqq r_i'$ for all i and the inequality sign holds for at least one i. A test T is called admissible if there exists no test uniformly better than T. A test which is not admissible may also be called inadmissible. A class C_0 of tests is called complete if, for any test T' not in C_0, there exists a test T in C_0 which is uniformly better than T'. A complete class is said to be minimal if no proper subclass of it is complete.

2. Proof of the complete class theorem

We first prove:

LEMMA. *The class C of all admissible tests is a minimal complete class.*[3]

If C is complete it is obviously minimal. Suppose C is not complete. Then there exists an inadmissible test T_1 such that no member of C is uniformly better than T_1. Since T_1 is inadmissible there exists an inadmissible test T_2 which is uniformly better than T_1. Consequently there exits a test T_3 which is uniformly better than T_2. Hence T_3 is uniformly better than T_1, and is therefore inadmissible. Proceeding in this manner we obtain a denumerable sequence $T_1, T_2, \ldots$ of tests, each test inadmissible and uniformly better than all its predecessors. Since the set V is closed it follows that there exists a test T_ω which is uniformly better than every member of the sequence $T_1, T_2, \ldots$. Hence T_ω is inadmissible. Repeating this procedure we obtain a nondenumerable well ordered set of inadmissible tests, each uniformly better than all its predecessors. Since each risk point has m components we can therefore obtain a nondenumerable well ordered set of real numbers, each smaller than any of its predecessors. Since this is impossible the lemma is proved.

Let $\xi_0 = (\xi_{01}, \ldots, \xi_{0m})$ be an *a priori* probability distribution on the set consisting of $f_1, \ldots, f_m$. A test T_0 with the property that it minimizes

$$\sum_{i=1}^{m} \xi_{0i} r_i(T)$$

[2] A statement of some of the results is given in the *Proc. Nat. Acad. Sci. U.S.A.*, April, 1950. The fact that V is closed whatever be the f's follows from the complete results which, it is hoped, will be published shortly. The closure of V was also proved by one of the authors [2] under the assumption that the f's admit elementary probability laws.

[3] The fundamental idea of the proof of this lemma is already present in the proof of theorem 2.22 in the book by Wald [2]. Since the proof is so brief it is given here for completeness.

with respect to all tests T is called a Bayes solution with respect to ξ_0, or simply a Bayes solution when it is not necessary to specify ξ_0. Let $\xi_1, \ldots, \xi_h$ be a sequence of (*a priori*) probability distributions (each with m components). A Bayes solution with respect to the sequence $\xi_1, \ldots, \xi_h$ will be defined inductively as follows: When $h = 1$ it is a Bayes solution with respect to ξ_1. For $h > 1$ it is any test T_0 which minimizes

$$\sum_{i=1}^{m} \xi_{hi} r_i (T)$$

with respect to all tests T which are Bayes solutions with respect to the sequence $\xi_1, \ldots, \xi_{h-1}$. Since the set V of risk points is closed it follows that, for any sequence $\xi_1, \ldots, \xi_h$, a Bayes solution exists.

THEOREM 1. *Every admissible test is a Bayes solution with respect to some a priori distribution.* (Hence the class of Bayes solutions is complete.)

PROOF. Let $b = (b_1, \ldots, b_m)$ be a generic point in an m-dimensional Euclidean space. Let the set $B(b)$ be the set of all points $x = x_1, \ldots, x_m$ such that x is different from b and

$$x_i \leqq b_i, \qquad i = 1, \ldots, m.$$

Let the set $B'(b)$ be the set which consists of b and $B(b)$. Suppose T is an admissible test and $r = (r_1, \ldots, r_m)$ is its associated risk point. Since T is admissible r is a boundary[4] point of V and the set $VB(r)$ is empty.

Now V and $B'(r)$ are closed convex sets with only the boundary point r in common, and $B'(r)$ contains interior points. Hence there exists a plane π_1 through r, given by $\mu_1(b) = 0$, where

$$\mu_1(b) = \sum_{i=1}^{m} t_{1i}(b_i - r_i), \tag{1}$$

such that $\mu_1(b) \geqq 0$ in one of V and $B'(r)$, and $\mu_1(b) \leqq 0$ in the other. Reversing the signs of all the t_{1i}'s, if necessary, we can assume that some t_{1i}, say t_{1e}, is positive. Let $K(e)$ be the point each of whose coordinates is r_i except the e-th, which is K. When K is sufficiently small, $t_{1e}(K - r_e) < 0$. Hence for every point of $B(r)$ we have $\mu_1(b) \leqq 0$. From this it follows that every $t_{1i} \geqq 0$. For suppose that t_{1g}, say, were < 0. The point $K(g)$, with K sufficiently small, would be in $B(r)$ and yet $\mu_1[K(g)] > 0$. Thus every $t_{1i} \geqq 0$. We have

$$\mu_1(b) \geqq 0 \tag{2}$$

for every point in V. Hence the point r minimizes $\mu_1(b)$ for every point in V. Therefore T is a Bayes solution with respect to the *a priori* probability distribution ξ_1 whose i-th component ξ_{1i} $(i = 1, \ldots, m)$ is

$$\xi_{1i} = \frac{t_{1i}}{\sum_{i=1}^{m} t_{1i}}.$$

This proves the theorem.

[4] Here the notions of inner point and boundary point are relative to the surrounding m-dimensional space.

3. First characterization of admissible tests

We now prove the main result:

THEOREM 2. *In order that a test T be admissible it is necessary and sufficient that it be a Bayes solution with respect to a sequence of h ($\leqq m$) a priori probability distribution functions* $(\xi_1, \ldots, \xi_h)$, *such that the matrix* $\{\xi_{ij}\}$, $i = 1, \ldots, h$; $j = 1, \ldots, m$, *has the following properties:* (a) *for any j there exists an i such that* $\xi_{ij} > 0$, (b) *the matrix* $\{\xi_{ij}\}$, $i = 1, \ldots, (h-1)$; $j = 1, \ldots, m$, *does not possess property* (a).

PROOF. The sufficiency of the above condition is easy to see. We proceed at once to the proof of necessity.

Let therefore r be the risk point of an admissible test T. By theorem 1 T is a Bayes solution with respect to ξ_1. We shall carry over the notation of theorem 1, except that, for typographical simplicity, we shall put r at the origin. (We may do this without loss of generality.) The origin will be written for short as the point 0. Let V_1 be the intersection of V with the plane π_1 defined by $\mu_1(b) = 0$. V_1 is convex and closed. Suppose it is of dimensionality $m - c_1$, $2 \leqq c_1 \leqq m$. Let the vector α denote the generic point in V_1. Let the vector β be any point in the plane π_1 and not in $B(0)$ such that the convex hull V_1' of V_1 and β is of dimensionality $m - c_1 + 1$. Let V_1'' be the convex hull of V_1 and $(-\beta)$. We now assert that either V_1' or V_1'' has no points in common with $B(0)$. For suppose to the contrary that

$$q_1\alpha_1 + (1 - q_1)\beta$$

and

$$q_2\alpha_2 - (1 - q_2)\beta$$

with

$$\alpha_1 \in V_1, \quad \alpha_2 \in V_1, \quad 0 \leqq q_1 \leqq 1, \quad 0 \leqq q_2 \leqq 1,$$

are both in $B(0)$. Moreover, $q_1 \neq 0$ since $\beta \notin B(0)$, and $q_1 \neq 1$, $q_2 \neq 1$ since 0 is admissible. Hence

$$q_1 (1 - q_2)\, \alpha_1 + (1 - q_1)(1 - q_2)\, \beta$$

and

$$q_2 (1 - q_1)\, \alpha_2 - (1 - q_1)(1 - q_2)\, \beta$$

are each in $B(0)$. Hence

$$q_1 (1 - q_2)\, \alpha_1 + q_2 (1 - q_1)\, \alpha_2$$

is also in $B(0)$, and consequently

$$\alpha_0 = \frac{q_1 (1 - q_2)\, \alpha_1 + q_2 (1 - q_1)\, \alpha_2}{q_1 (1 - q_2) + q_2 (1 - q_1)}$$

is in $B(0)$. Now α_0 lies in the line segment from α_1 to α_2, and hence is in V_1. This contradicts the fact that 0 is admissible and proves our assertion that either V_1' or V_1'' has no points in common with $B(0)$.

We repeat the above procedure $(c_1 - 1)$ times and conclude: There exists a closed convex set V_1^* which contains V_1, lies entirely in π_1, is of dimensionality $m - 1$, and has no points in common with $B(0)$.

Suppose γ_1 of the t_{1i} are positive. If $\gamma_1 = m$ the theorem is proved. Assume therefore that $\gamma_1 < m$. Without loss of generality we assume

$$t_{1i} > 0, \quad i \leqq \gamma_1$$
$$t_{1i} = 0, \quad i > \gamma_1 .$$

Let $B(0; 1)$ be the set of all points b different from 0 such that

$$b_i = 0\,, \qquad i \leqq \gamma_1$$

$$b_i \leqq 0\,, \qquad i > \gamma_1\,.$$

Let $B'(0; 1)$ be the set of points consisting of 0 and $B(0; 1)$. The closed convex sets V_1^* and $B'(0, 1)$ both lie in π_1 and have only the origin 0 in common. 0 is obviously a boundary[5] point of $B'(0; 1)$. It is also a boundary point of V_1^* because $B'(0; 1)$ is of dimensionality $\geqq 1$ $(\gamma_1 < m)$, and $V_1^* B(0)$ is empty. The set V_1^* has inner[5] points. Hence there exists an $(m - 2)$-dimensional linear subspace π_2 of π_1, defined by $\mu_1(b) = 0$ and $\mu_2(b) = 0$, where

$$\mu_2\,(b) = \sum_{i=1}^{m} t_{2i} b_i \tag{3}$$

such that $\mu_2(b) \geqq 0$ in one of V_1^* and $B'(0, 1)$, and $\mu_2(b) \leqq 0$ in the other.

We now consider two cases:

(a) $t_{2i} \neq 0$ for some $i > \gamma_1$. Without loss of generality we assume $t_{2(\gamma_1+1)} \neq 0$. Hence there exists a real number $\lambda_2 \neq 0$ such that

$$t_{1i} + \lambda_2 t_{2i} > 0\,, \qquad i \leqq \gamma_1 + 1\,. \tag{4}$$

The space π_2 can also be defined by $\mu_1(b) = 0$ and $\mu_1(b) + \lambda_2\mu_2(b) = 0$. We will now show that

$$t_{1i} + \lambda_2 t_{2i} \geqq 0\,, \qquad i > \gamma_1 + 1\,. \tag{5}$$

For suppose that, say,

$$t_{1e} + \lambda_2 t_{2e} < 0\,, \qquad e > \gamma_1 + 1\,. \tag{6}$$

By the definition of π_2 the sign of $\mu_1(b) + \lambda_2\mu_2(b)$ does not change in $B(0; 1)$. Using the point $K(\gamma_1 + 1)$ with K negative we see that $\mu_1(b) + \lambda_2\mu_2(b) \leqq 0$ for b in $B(0; 1)$. If (6) held we would have

$$\mu_1\,[K\,(e)] + \lambda_2\mu_2\,[K\,(e)] > 0$$

for K negative, in violation of what we have just proved. Hence (5) must hold.

(b) $t_{2i} = 0$ for $i > \gamma_1$.

Consider the expressions

$$M_1\,(b) = \mu_1\,(b) + \lambda\mu_2\,(b) \tag{7}$$

and

$$M_2\,(b) = \mu_1\,(b) - \lambda\mu_2\,(b)\,. \tag{8}$$

For sufficiently small positive λ all their coefficients are nonnegative. Both expressions do not change sign in V_1^*, because of the definition of π_2. We assert that either $M_1(b) \geqq 0$ in V_1^* or $M_2(b) \geqq 0$ in V_1^*. For $M_1(b)$ and $M_2(b)$ cannot be identically zero on V_1^*, because V_1^* lies in π_1 and is of dimensionality $(m - 1)$. Let b_0 be some point in V_1^* where $M_1(b_0) \neq 0$. Since $M_1(b_0) + M_2(b_0) = 2\mu_1(b_0) = 0$, it follows that $M_2(b_0) \neq 0$, and either $M_1(b_0)$ or $M_2(b_0)$ is positive.

[5] Here the notions of inner point and boundary point are relative to the surrounding space π_1.

But then either $M_1(b)$ or $M_2(b)$ is nonnegative for every b in V_1^*, which is the assertion to be proved. Let $M(b)$ denote that one of $M_1(b)$ and $M_2(b)$ for which $M(b) \geqq 0$ for every point b in V_1^*, and let λ_2 denote that one of λ, $-\lambda$ which is associated with $M(b)$.

In both case (a) and case (b) we have that the test T with associated risk point 0 is a Bayes solution with respect to ξ_1, ξ_2, where

$$\xi_{2i} = \frac{l_{1i} + \lambda_2 l_{2i}}{\sum_{i=1}^{m} (l_{1i} + \lambda_2 l_{2i})} .$$

We redefine $\mu_2(b)$ so that $l_{2i} = \xi_{2i}$. This will help to simplify the notation.

If ξ_1 and ξ_2 do not fulfill the conditions of the theorem for $h = 2$ the above procedure can be repeated. We shall sketch the procedure which yields π_3, π_1 and π_2 having been previously obtained.

Let V_2 be the intersection of π_2 and V_1^*. If V_2 is of dimensionality less than $(m - 2)$ proceed as before to obtain V_2^* which is closed, convex, contains V_2, has no point in common with $B(0)$, and is of dimensionality $(m - 2)$. Let U be the set of integers $i \leqq m$ such that

$$\xi_{1i} = \xi_{2i} = 0$$

and let $\bar{U}$ be the complementary set. The set U is not empty, for else the theorem would be already proved. Let $B(0; 2)$ be the set of all points b different from 0 such that

$$b_i = 0 , \qquad i \in \bar{U}$$

$$b_i \leqq 0 , \qquad i \in U .$$

Let $B'(0; 2)$ be the set of points consisting of 0 and $B(0; 2)$. The closed convex sets $B'(0; 2)$ and V_2^* are separated by an $(m - 3)$-dimensional linear subspace π_3 of π_2 which passes through 0 and may be defined by $\mu_1(b) = 0$, $\mu_2(b) = 0$, and $\mu_3(b) = 0$, where,

$$\mu_3(b) = \sum_{i=1}^{m} l_{3i} b_i .$$

As before, we distinguish two cases. Case (a) occurs when $l_{3i} \neq 0$ for some index $i \in U$. As before, we prove that for suitable $\lambda \neq 0$ the expression

$$\mu_1(b) + \mu_2(b) + \lambda\mu_3(b) \tag{9}$$

has all coefficients nonnegative. Case (b) occurs when $l_{3i} = 0$ for every $i \in U$. For $|\lambda|$ sufficiently small we have then that (9) has all coefficients nonnegative, and either (9) or

$$\mu_1(b) + \mu_2(b) - \lambda\mu_3(b) \tag{10}$$

can be shown as before to be nonnegative in V_2^*. We obtain ξ_3 in this manner.

The above procedure can be repeated as long as the corresponding set U is not empty. However, when the set U is empty, the theorem is proved. The set U will be empty in at most m steps of the procedure.

Suppose for a moment that the f_i all possess density functions f_i^*. A Bayes solution $\eta(x)$ with respect to ξ_1 may be found as follows: $\eta_j(x) = 0$ for all j for which

$$\nu_{1j}(x) = \sum_{i=1}^{m} \xi_{1i} f_i^*(x) W_{ij}(x)$$

is not a minimum with respect to j ($j = 1, \ldots, l$); $\eta_j(x)$ is defined arbitrarily between zero and one, inclusive, for all other j, provided only that every component of the resulting $\eta(x)$ is measurable and the sum is always one. If a Bayes solution with respect to ξ_1, ξ_2 is desired one can proceed as follows: First, define $\eta_j(x) = 0$ for all j for which $\nu_{1j}(x)$ is not a minimum. Among the remaining j define $\eta_j(x) = 0$ for those j for which

$$\nu_{2j}(x) = \sum_{i=1}^{m} \xi_{2i} f_i^* W_{ij}(x)$$

is not a minimum (for these j). Define $\eta_j(x)$ arbitrarily between zero and one, inclusive, for all other j, subject to the requirements of measurability and the fact that the components must add to one. A Bayes solution with respect to $\xi_1, \ldots, \xi_h$ can be obtained similarly.

If the f_i are not absolutely continuous we can proceed as follows: Let τ be the finite measure defined for any Borel set Σ by

$$\tau(\Sigma) = \sum_{i=1}^{m} P\{\Sigma \mid f_i\} .$$

Then every f_i is absolutely continuous with respect to τ and hence, by the Radon-Nikodym theorem, possesses a density function with respect to τ. We can then proceed as before.

4. Second characterization of admissible tests

We return to the problem of characterizing admissible solutions and shall describe another procedure of doing so. Let $\xi_1, \ldots, \xi_u$ be any sequence of *a priori* distributions with the property that for each j ($j = 1, \ldots, m$), there exists exactly one i ($i = 1, \ldots, u$) such that $\xi_{ij} > 0$. We shall call this property the property U. Let $v[i, 1], \ldots, v[i, h(i)]$ be the set of integers j ($j = 1, \ldots, m$), for which $\xi_{ij} > 0$ ($i = 1, \ldots, u$). Let

$$r(1) = [r(1, 1), \ldots, r(1, m)]$$

be the risk point of any Bayes solution with respect to ξ_1. Let $\bar{V}_1$ be the intersection of V with the planes

$$b_{v[1,1]} = r(1, v[1, 1])$$
$$\cdots\cdots\cdots\cdots$$
$$b_{v[1,h(1)]} = r(1, v[1, h(1)]) .$$

Let

$$r(2) = (r(2, 1), \ldots, r(2, m))$$

be any Bayes solution with respect to ξ_2 among the elements of $\bar{V}_1$. (Since $\bar{V}_1$ is closed and bounded at least one such solution exists.) Let $\bar{V}_2$ be the intersection of $\bar{V}_1$ with the planes

$$b_{v[2,1]} = r(2, v[2, 1])$$

$$\cdots \quad \cdots\cdots$$

$$b_{v[2,h(2)]} = r(2, v[2, h(2)]) .$$

Let $r(3)$ be any Bayes solution with respect to ξ_3 among the elements of $\bar{V}_2$, etc., etc. The end product of this procedure is the set $\bar{V}_u$.

THEOREM 3. *The class C of all admissible tests coincides with the class of tests with risk points in* $\bar{V}_u$ *for all sequences* $(\xi_1, \ldots, \xi_u)$ *with the property U.*

PROOF. First we prove that any risk point z in $\bar{V}_u$ is admissible. Suppose that it is not admissible, that z' is uniformly better than z, and that i $(i = 1, 2, \ldots, u)$ is the smallest integer such that for an index j which is a member of

$$v[i, 1], \ldots, v[i, h(i)]$$

the j-th coordinate of z' is less than that of z. We see that z' must lie in

$$\bar{V}_1, \bar{V}_2, \ldots, \bar{V}_{i-1} .$$

Consequently z cannot lie in $\bar{V}_i$, which contradicts the hypothesis that z is in $\bar{V}_u \subset \bar{V}_i$.

Let now z be any point in C. It must be a Bayes solution with respect to, say, ξ_1. If z is the unique Bayes solution with respect to ξ_1, or if ξ_{1j} is positive for every j from 1 to m, there is nothing left to prove. Assume therefore that neither of these is true. We define ξ_1 as the first member of the sequence $\xi_1, \ldots, \xi_u$ which we want to construct. Define $\bar{V}_1$ as before. A reëxamination of the proof of theorem 1 shows that the only property of V that was used in the proof is that V is a convex body (that is, a closed, convex, bounded set). But $\bar{V}_1$ has this property. Hence the argument of theorem 1 can be applied to $\bar{V}_1$ to obtain an *a priori* distribution ξ_2 such that (a) z is a Bayes solution with respect to ξ_2, if one limits one's self to points of $\bar{V}_1$, (b) $\xi_{2j} = 0$ for any j such that $\xi_{1j} > 0$. Repeating the above procedure we obtain the desired result.

5. Concluding remarks

I) The only property of V used in theorems 1, 2, and 3 is that V is a convex body. Suppose that, for some reason, the statistician is limited to choosing one of a given proper subclass of the class of all tests. If the set of risk points of this subclass is a convex body, theorems 1, 2, and 3 will hold for this subclass.

II) The only use that was made in the preceding arguments of the fact that the number l of possible decisions is finite, was in invoking the result of Dvoretzky, Wald, and Wolfowitz that V is a convex body. Suppose now that the number of possible decisions is no longer finite. For each x, $\eta(x)$ is then a probability measure on a Borel field of subsets of the space D of decisions (see [2], for example). The risk point of a test is defined appropriately. If V is a convex body then theorems 1, 2, and 3 remain valid. If V is a convex body for a subclass of the class of all tests then theorems 1, 2, and 3 are valid for this subclass. If the class of available tests is the class of all possible tests it is obvious that V is convex. Whether V is closed will in general depend upon W and the space D.

REFERENCES

[1] A. WALD, "Foundations of a general theory of sequential decision functions," *Econometrica*, Vol. 15 (1947), pp. 279–313.

[2] ———, *Statistical Decision Functions*, Wiley, New York, 1950.

Reprinted from Vol. I, Proceedings of the
International Congress of Mathematicians, 1950
Printed in U.S.A.

BASIC IDEAS OF A GENERAL THEORY OF STATISTICAL DECISION RULES

Abraham Wald

1. Introduction. The theory of statistics has had an extraordinarily rapid growth during the last 30 years. The line of development has been largely set by two great schools, the school of R. A. Fisher and that of Neyman and Pearson. A basic feature of the theories represented by these schools is the development of various criteria for the best possible use of the observations for purposes of statistical test and estimation procedures. In this connection I would like to mention the basic notions of efficiency and sufficiency introduced by Fisher, and that of the power of a test introduced by Neyman and Pearson. It is unnecessary to dwell on the importance of these notions, since they are well known to all statisticians.

Until about 10 years ago, the available statistical theories, except for a few scattered results, were restricted in two important respects: (1) experimentation was assumed to consist of a single stage, i.e., the number of observations to be made was assumed to be fixed in advance of experimentation; (2) the decision problems treated were restricted to special types known in the literature under the names of testing a hypothesis, point and interval estimation. In the last few years a general decision theory has been developed (see, for example, [1][1]) that is free of both of these restrictions. It allows for multi-stage experimentation and it includes the general multi-decision problem.

I would like to outline the principles of this general theory and some of the results that have been obtained.

A statistical decision problem is defined with reference to a sequence $X = \{X_i\}$ $(i = 1, 2, \cdots, \text{ad inf.})$ of random variables. For any sequence $x = \{x_i\}$ $(i = 1, 2, \cdots, \text{ad inf.})$ of real values, let $F(x)$ denote the probability that $X_i < x_i$ holds for all positive integral values i. The function $F(x)$ is called the distribution function of X. A characteristic feature of any statistical decision problem is that F is unknown. It is merely assumed to be known that F is a member of a given class Ω of distribution functions. The class Ω is to be regarded as a datum of the decision problem. Another datum of the decision problem is a space D, called decision space, whose elements d represent the possible decisions that can be made by the statistician in the problem under consideration.

For the sake of simplicity, we shall assume for the purposes of the present discussion that (1) each element F of Ω is absolutely continuous, i.e., it admits a probability density function; (2) the space D consists of a finite number of elements, $d_1, \cdots, d_k$ (say); (3) experimentation is carried out sequentially, i.e., the first stage of the experiment consists of observing the value of X_1. After the value of X_1 has been observed, the statistician may decide either to terminate

[1] Numbers in brackets refer to the references at the end of the paper.

experimentation with some final decision d, or to observe the value of X_2. In the latter case, after X_1 and X_2 have been observed, the statistician may again decide to terminate experimentation with some final decision d, or to observe the value of X_3, and so on.

The above assumptions on the spaces Ω and D and the method of experimentation are replaced by considerably weaker ones in the general theory given in [1]. These assumptions are made here merely for the purpose of simplifying the discussion.

A decision rule δ, that is, a rule for carrying out experimentation and making a final decision d, can be given in terms of a sequence of real-valued and Borel measurable functions $\delta_{im}(x_1, \cdots, x_m)$ $(i = 0, 1, \cdots, k; m = 0, 1, 2, \cdots, \text{ad inf.})$ where $x_1, x_2, \cdots$, etc. are real variables and the functions δ_{im} are subject to the following two conditions:

$$(1.1) \qquad \delta_{im} \geqq 0; \qquad \sum_{i=0}^{k} \delta_{im} = 1 \; (i = 0, 1, \cdots, k; m = 0, 1, 2, \cdots, \text{ad inf.}).$$

The decision rule δ is defined in terms of the function δ_{im} as follows: Let x_i denote the observed value of X_i. At each stage of the experiment (after the mth observation, for each integral value m) we compute the values $\delta_{0m}(x_1, \cdots, x_m)$, $\delta_{1m}(x_1, \cdots, x_m)$, $\cdots$, $\delta_{km}(x_1, \cdots, x_m)$ and then perform an independent random experiment with the possible outcomes $0, 1, 2, \cdots, k$ constructed so that the probability of the outcome i is δ_{im}. If the outcome is a number $i > 0$, we terminate experimentation with the decision d_i. If the outcome is 0, we make an additional observation (we observe the value of X_{m+1}) and repeat the process with the newly computed values $\delta_{0,m+1}$, $\delta_{1,m+1}$, $\cdots$, $\delta_{k,m+1}$, and so on.

The above described decision rule may be called a randomized decision rule, since at each stage of the experiment an independent chance mechanism is used to decide whether experimentation be terminated with some final decision d or whether an additional observation be made. The special case when the functions δ_{im} can take only the values 0 and 1 is of particular interest, since in this case the decision to be made at each stage of the experiment is based entirely on the observed values obtained and one can dispense with the use of an independent chance mechanism. We shall call a decision rule $\delta = \{\delta_{im}\}$ nonrandomized if the functions δ_{im} can take only the values 0 and 1. The question whether it is sufficient to consider only nonrandomized decision rules for the purposes of statistical decision making is of considerable interest. We shall return to this question later.

2. Loss, cost, and risk functions. A basic problem in statistical decision theory is the problem of a proper choice of a decision rule δ. In order to judge the relative merits of the various possible decision rules, it is necessary to state the cost of experimentation and the relative degree of preference we would have for the various possible final decisions d if some element F of Ω were known to us to be the true distribution. The latter may be described by a non-negative function $W(F, d)$, called loss function, which expresses the loss suffered by the statistician when the decision d is made and F happens to be the true distribution of X. In most decision problems each element d_i of D can be interpreted as the decision to accept the hypothesis that the unknown distribution F is an element of a given subclass ω_i of Ω. In such a case we put $W(F, d_i) = 0$ when $F \in \omega_i$ and >0 when $F \notin \omega_i$. The cost of experimentation can be described by a sequence $\{c_m(x_1, \cdots, x_m)\}$ $(m = 1, 2, \cdots, \text{ad inf.})$ of non-negative functions where $c_m(x_1, \cdots, x_m)$ denotes the cost of experimentation if the experiment consists of m observations and x_i is the observed value of X_i $(i = 1, \cdots, m)$. The loss and cost functions are to be regarded as data of the decision problem. The cost function $c_m(x_1, \cdots, x_m)$ is, of course, assumed to be Borel measurable.

Let $p(m, d_i \mid \delta, x_1, \cdots, x_m)$ denote the conditional probability that experimentation will consist of m observations and the decision d_i will be made when δ is the decision rule adopted and x_j is the observed value of X_j $(j = 1, \cdots, m)$. Clearly,

$$p(m, d_i \mid \delta, x_1, \cdots, x_m) = \delta_{00}\delta_{01}(x_1) \cdots \delta_{0,m-1}(x_1, \cdots, x_{m-1})\delta_{im}(x_1, \cdots, x_m). \tag{2.1}$$

For any positive integral value m, let $f(x_1, \cdots, x_m \mid F)$ denote the joint density function of $X_1, \cdots, X_m$ when F is the true distribution of X. The expected loss, i.e., the expected value of $W(F, d)$ depends only on the true distribution F and the decision rule δ adopted. It is given by the expression

$$\begin{aligned} r_1(F, \delta) &= \sum_{i=1}^{k} W(F, d_i)\delta_{0i} \\ &+ \sum_{m=1}^{\infty}\sum_{i=1}^{k}\int_{R_m} W(F, d_i)p(m, d_i \mid \delta, x_1, \cdots, x_m)f(x_1, \cdots, x_m \mid F)\, dx_1, \cdots, dx_m \end{aligned} \tag{2.2}$$

where R_m denotes the space of all m-tuples $(x_1, \cdots, x_m)$.

The expected cost of experimentation depends only on the true distribution F and the decision rule δ adopted. It is given by

$$r_2(F, \delta) = \tag{2.3}$$

$$\sum_{m=1}^{\infty}\sum_{i=1}^{k}\int_{R_m} c_m(x_1, \cdots, x_m)p(m, d_i \mid \delta, x_1, \cdots, x_m)f(x_1, \cdots, x_m \mid F)\, dx_1, \cdots, dx_m.$$

Let

$$r(F, \delta) = r_1(F, \delta) + r_2(F, \delta). \tag{2.4}$$

The quantity $r(F, \delta)$ is called the risk when F is true and the decision rule δ is adopted. For any fixed decision rule δ^0, the risk is a function of F only. We shall call $r(F, \delta^0)$ the risk function associated with the decision rule δ^0.

It is perhaps not unreasonable to judge the merit of any particular decision rule entirely on the basis of the risk function associated with it. We shall say that the decision rule δ^1 is *uniformly better* than the decision rule δ^2 if $r(F, \delta^1) \leqq r(F, \delta^2)$ for all F and $r(F, \delta^1) < r(F, \delta^2)$ for at least one member F of Ω. A decision rule δ will be said to be *admissible* if there exists no uniformly better decision rule. Two decision rules δ^1 and δ^2 will be said to be *equivalent* if they have identical risk functions, i.e., if $r(F, \delta^1) = r(F, \delta^2)$ for all F in Ω. For any $\epsilon > 0$, two decision rules δ^1 and δ^2 will be said to be ϵ-equivalent if $| r(F, \delta^1) - r(F, \delta^2) | \leqq \epsilon$ for all F in Ω.

3. Elimination of randomization when Ω is finite. It was proved recently by Dvoretzky, Wolfowitz, and the author [2] that if Ω is finite, then for every decision rule δ there exists an equivalent nonrandomized decision rule δ^*. Thus, in this case one can dispense with randomization and it is sufficient to consider only nonrandomized decision rules. A similar result for a somewhat more special type of randomized decision rule than the one described here was obtained independently by Blackwell [3]. The proof is based on an extension of a theorem by Liapounoff [4] concerning the range of a vector measure. The continuity of the distribution of X_1 (implied by our assumption of absolute continuity of F) is essential for the above stated result. In case of discontinuous distributions there may exist randomized decision rules with risk functions having some desirable properties that cannot be achieved by any nonrandomized decision rule.

The finiteness of the space Ω is a very restrictive condition which is seldom fulfilled in statistical decision problems. However, it was shown in [2] that for any decision rule δ and for any $\epsilon > 0$ there exists an ϵ-equivalent and nonrandomized decision rule δ^* under very general conditions which are usually fulfilled in decision problems arising in applications.

An interesting result on the possible elimination of randomization, but of a somewhat different nature, was found recently by Hodges and Lehmann [5]. They proved that if the decision problem is a point estimation problem, if D is a Euclidean space, and if the loss $W(F, d)$ is a convex function of d for every F, then for any randomized decision rule δ (with bounded risk function) there exists a nonrandomized decision rule δ^* such that $r(F, \delta^*) \leqq r(F, \delta)$ for all F in Ω. It is remarkable that neither the finiteness of Ω nor the continuity of its elements F are needed for the validity of this result.

4. A convergence definition in the space of decision rules and some continuity theorems. A natural convergence definition in the space of decision rules would seem to be the following one: $\lim_{j=\infty} \delta^j = \delta^0$ if $\lim_{j=\infty} p(m, d_i \mid \delta^j, x_1, \cdots, x_m) = p(m, d_i \mid \delta^0, x_1, \cdots, x_m)$ for all m, all $i > 0$, and for all $x_1, \cdots, x_m$. This convergence definition is, however, too strong for our purposes. Instead, we shall adopt the following weaker one: We shall say that

$$\lim_{j=\infty} \delta^j = \delta^0 \tag{4.1}$$

if

$$\lim_{j=\infty} \delta_{i0}^{j} = \delta_{i0}^{0} \qquad (i = 0, 1, \cdots, k) \tag{4.2}$$

and

$$\lim_{j=\infty} \int_{S_m} p(m, d_i \mid \delta^j, x_1, \cdots, x_m)\, dx_1 \cdots dx_m = \int_{S_m} p(m, d_i \mid \delta^0, x_1, \cdots, x_m)\, dx_1 \cdots dx_m \tag{4.3}$$

$$(i = 0, 1, \cdots, k; m = 1, 2, \cdots, \text{ad inf})$$

holds for every measurable subset S_m of the space of all m-tuples $(x_1, \cdots, x_m)$.

It was shown (see, for example, Theorem 3.1 in [1]) that adopting the above convergence definition the following theorem holds.

THEOREM 4.1. *The space of all decision rules is compact, i.e., every sequence* $\{\delta^j\}$ $(j = 1, 2, \cdots, \text{ad inf.})$ *of decision rules admits a convergent subsequence.*

The above theorem is a simple consequence of known theorems on the "weak" compactness of a set of functions (see, for example, Theorem 17b (p. 33) of [6]).[2]

Before stating certain continuity theorems, we shall formulate two conditions concerning the loss and cost functions.

Condition I. $W(F, d_i)$ *is a bounded function of* F *for* $i = 1, 2, \cdots, k$.

Condition II. *The cost function has the following properties*: (i) $c_m(x_1, \cdots, x_m) \geqq 0$; (ii) $c_{m+1}(x_1, \cdots, x_{m+1}) \geqq c_m(x_1, \cdots, x_m)$; (iii) $c_m(x_1, \cdots, x_m)$ *is a bounded function of* $x_1, \cdots, x_m$ *for every fixed* m; (iv) $\lim_{m=\infty} c_m(x_1, \cdots, x_m) = \infty$ *uniformly in* $x_1, \cdots, x_m$.

The following continuity theorems have been proved in [1]:

THEOREM 4.2. *Let* $\{\delta^j\}$ $(j = 0, 1, 2, \cdots, \text{ad inf.})$ *be a sequence of decision rules such that* $\lim_{j=\infty} \delta^j = \delta^0$ *and such that*

$$\delta_{00}^{j}\delta_{01}^{j}(x_1)\delta_{02}^{j}(x_1, x_2) \cdots \delta_{0N}^{j}(x_1, \cdots, x_N) = 0 \qquad (j = 0, 1, 2, \cdots, \text{ad inf.})$$

identically in $x_1, \cdots, x_N$ *for some positive integer* N. *Then, if Conditions* I *and* II *hold, we have* $\lim_{j=\infty} r(F, \delta^j) = r(F, \delta^0)$ *for all* F.

THEOREM 4.3. *If* $\lim_{j=\infty} \delta^j = \delta^0$ *and if Conditions* I *and* II *hold, then* $\liminf_{j=\infty} r(F, \delta^j) \geqq r(F, \delta^0)$ *for all* F.

5. Bayes and minimax solutions of the decision problem. In this section we shall discuss the notions of Bayes and minimax solutions and some of their properties. These solutions are not only of intrinsic interest, but they play an important role in the construction and characterization of complete classes of

[2] Theorem 3.1 in [1] is actually much stronger and more difficult to prove, since D is not assumed there to be finite.

decision rules discussed in the next section. We shall start out with some definitions.

By an a priori probability distribution ξ in Ω we shall mean a non-negative and countably additive set function ξ defined over a properly chosen Borel field of subsets of Ω for which $\xi(\Omega) = 1$. The Borel field is chosen such that $r(F, \delta)$ is a measurable function of F for every fixed δ.

For any a priori probability distribution ξ, let

$$r^*(\xi, \delta) = \int_\Omega r(F, \delta)\, d\xi. \tag{5.1}$$

A decision rule δ^0 is said to be a Bayes solution relative to the a priori distribution ξ if

$$r^*(\xi, \delta^0) = \operatorname*{Min}_\delta r^*(\xi, \delta). \tag{5.2}$$

A decision rule δ^0 is said to be a Bayes solution in the strict sense if there exists an a priori distribution ξ such that δ^0 is a Bayes solution relative to ξ.

A rule δ^0 is said to be a Bayes solution relative to the sequence $\{\xi_i\}$ ($i = 1, 2, \cdots$, ad inf.) of a priori distributions if

$$\lim_{i=\infty} [r^*(\xi_i, \delta^0) - \operatorname*{Inf}_\delta r^*(\xi_i . \delta)] = 0 \tag{5.3}$$

where the symbol Inf_δ stands for infimum with respect to δ.

We shall say that a decision rule δ^0 is a Bayes solution in the wide sense if there exists a sequence $\{\xi_i\}$ of a priori distributions such that δ^0 is a Bayes solution relative to $\{\xi_i\}$.

A decision rule δ^0 is said to be a *minimax solution* if

$$\operatorname*{Sup}_F r(F, \delta^0) \leqq \operatorname*{Sup}_F r(F, \delta) \text{ for all } \delta, \tag{5.4}$$

where the symbol Sup_F stands for supremum with respect to F.

An a priori distribution ξ_0 is said to be *least favorable* if the following relation is satisfied:

$$\operatorname*{Inf}_\delta r^*(\xi_0, \delta) \geqq \operatorname*{Inf}_\delta r^*(\xi, \delta) \text{ for all } \xi. \tag{5.5}$$

The reason that an a priori distribution ξ_0 satisfying the above relation is called least favorable is this: If an a priori distribution ξ actually exists and is known to the statistician, a satisfactory solution of the decision problem is to use a Bayes solution δ relative to ξ since δ minimizes the average risk (averaged in accordance with the a priori distribution ξ). The minimum average risk that can be achieved will generally be different for different a priori distributions and an a priori distribution ξ may be regarded the less favorable from the point of view of the statistician the greater the minimum average risk associated with ξ. Thus, an a priori distribution satisfying (5.5) will be least favorable from the point of view of the statistician.

We shall state some of the results obtained concerning Bayes and minimax solutions.[3]

THEOREM 5.1. *If Conditions* I *and* II *hold, for any a priori distribution* ξ, *there exists a decision rule* δ *such that* δ *is a Bayes solution relative to* ξ.

THEOREM 5.2. *If Conditions* I *and* II *hold, there exists a minimax solution.*

The above existence theorems can easily be derived from the theorems stated in §4. With the help of these theorems we can even prove the slightly stronger result that admissible Bayes and admissible minimax solutions always exist.

THEOREM 5.3. *If Conditions* I *and* II *hold, then a minimax solution is always a Bayes solution in the wide sense.*

THEOREM 5.4. *If* δ^0 *is a minimax solution and* ξ_0 *is a least favorable a priori distribution, then, if Conditions* I *and* II *hold,* δ^0 *is a Bayes solution relative to* ξ_0 *and the set* ω *of all members* F *of* Ω *for which* $r(F, \delta^0) = \text{Sup}_F\, r(F, \delta^0)$ *has the probability measure* 1 *according to* ξ_0.

The last part of Theorem 5.4 implies that the risk function of a minimax solution has a constant value over a subset ω of Ω whose probability measure is one according to every least favorable a priori distribution ξ. In many decision problems the risk function of a minimax solution is constant over the whole space Ω.

Some additional results can be stated if the validity of the following additional condition is postulated:

Condition III. *The space* Ω *is compact and the loss function* $W(F, d)$ *is continuous in* F *in the sense of the following convergence definition in* Ω: *We shall say that* $\lim_{i=\infty} F_i = F_0$ *if for every positive integer* m *we have*

$$\lim_{i=\infty} \int_{S_m} f(x_1, \cdots, x_m \mid F_i)\, dx_1 \cdots dx_m = \int_{S_m} f(x_1, \cdots, x_m \mid F_0)\, dx_1 \cdots dx_m$$

uniformly in all measurable subsets S_m *of the space of all* m*-tuples* $(x_1, \cdots, x_m)$.

THEOREM 5.5. *If Conditions* I, II, *and* III *hold, a least favorable a priori distribution exists.*

The proof of this theorem is based on the fact that the space of all probability measures ξ on a compact space Ω is compact in the sense of the following convergence definition: $\lim_{i=\infty} \xi_i = \xi_0$ if $\lim \xi_i(\omega) = \xi_0(\omega)$ for any open subset ω of the space Ω whose boundary has probability measure zero according to ξ_0. This result was proved in [1, Theorem 2.15, p. 50]. A closely related result was

[3] For a detailed discussion and proofs, see §3.5 in [1].

obtained by Kryloff and Bogoliouboff [7]. Their convergence definition in the space of the probability measures ξ is somewhat different from the one used here.

THEOREM 5.6. *If Conditions* I, II, *and* III *hold, a minimax solution is always a Bayes solution in the strict sense.*

The above theorem is an immediate consequence of Theorems 5.4 and 5.5.

6. Complete classes of decision rules. A class C of decision rules δ is said to be *complete* if for any rule δ not in C there exists a rule δ^* in C such that δ^* is uniformly better. We shall say that a class C of decision rules is *essentially complete* if for any rule δ not in C there exists a rule δ^* in C such that $r(F, \delta^*) \leqq r(F, \delta)$ for all F in Ω.

Clearly, if C is a complete or at least an essentially complete class of decision rules, we can disregard all decision rules outside C and the problem of choice is reduced to the problem of choosing a particular element of C. Thus, the construction of complete or essentially complete classes of decision rules is of great importance in any statistical decision problem.

The first result concerning complete classes of decision rules is due to Lehmann [8] who constructed such a class in a special case. Soon after the publication of Lehmann's paper, results of great generality were obtained. To state some of these results, let Δ denote the set of all decision rules δ with bounded risk functions. We shall say that a class C of decision rules δ is complete, or essentially complete, relative to Δ if the corresponding condition is fulfilled for every δ in Δ. Among others, the following results have been proved in [1].

THEOREM 6.1. *If Conditions* I *and* II *hold, then the class of all Bayes solutions in the wide sense is complete relative to* Δ.

THEOREM 6.2. *If Conditions* I *and* II *hold, then the closure of the class of all Bayes solutions in the strict sense is essentially complete relative to* Δ.

THEOREM 6.3. *If Conditions* I, II, *and* III *hold, then the class of all Bayes solutions in the strict sense is complete relative to* Δ.

To avoid any possibility of a misunderstanding, it may be pointed out that the notions of Bayes solutions and a priori distributions are used here merely as mathematical tools to express some results concerning complete classes of decision rules, and in no way is the actual existence of an a priori distribution in Ω postulated here.

7. Relation to von Neumann's theory of games. The statistical decision theory, as outlined here, is intimately connected with von Neumann's theory of zero sum two person games [9]. The normalized form of a zero sum two person game is given by von Neumann as follows: There are two players and there is given a bounded and real-valued function $K(u, v)$ of two variables u and v where u may be any point of a space U and v may be any point of a space V. Player 1 chooses a point u in U and player 2 chooses a point v in V, each choice being made in ignorance of the other. Player 1 then gets the amount $K(u, v)$ and player 2 the amount $-K(u, v)$.

Any statistical decision problem may be viewed as a zero sum two person game. Player 1 is the agency, say Nature, who selects an element F of Ω to be the true distribution of X, and player 2 is the statistician who chooses a decision rule δ. The outcome is then given by the risk $r(F, \delta)$ which depends on both the choice F of Nature and the choice δ of the statistician. The theory of zero sum two person games was developed by von Neumann for finite spaces U and V. In statistical decision problems, however, the number of strategies available to Nature (number of elements of Ω) and the number of strategies (number of decision rules) available to the statistician are usually infinite. Many of the results in statistical decision theory were obtained by extending von Neumann's theory to the case of infinite spaces of strategies. In particular, it has been shown in [1] that if Conditions I and II hold, the statistical decision problem, viewed as a zero sum two person game, is strictly determined in the sense of von Neumann's theory, i.e.,

$$\operatorname*{Sup}_{\xi} \operatorname*{Inf}_{\delta} r^*(\xi, \delta) = \operatorname*{Inf}_{\delta} \operatorname*{Sup}_{\xi} r^*(\xi, \delta). \tag{7.1}$$

The above relation plays a fundamental role in the theory of zero sum two person games. In statistical decision theory, the above relation is basic in deriving the results concerning complete classes of decision rules, but otherwise it is of no particular intrinsic interest.

8. Discussion of some special cases. I would like to discuss briefly application of the general theory to a few special cases.

Suppose that Ω consists of two elements F_1 and F_2. According to F_i the random variables $X_1, X_2, \cdots$, ad inf. are independently distributed with the common density function $f_i(t)$ $(i = 1, 2)$. The decision space D consists of two elements d_1 and d_2 where d_i denotes the decision to accept the hypothesis that F_i is the true distribution $(i = 1, 2)$. Let the loss $W(F_i, d_j) = W_{ij} > 0$ when $i \neq j$ and $=0$ when $i = j$. The cost of experimentation is assumed to be proportional to the number of observations, i.e., $c_m(x_1, \cdots, x_m) = cm$ where c denotes the cost of a single observation.

An a priori distribution is given by a set of two non-negative numbers (ξ_1, ξ_2) such that $\xi_1 + \xi_2 = 1$. The quantity ξ_i denotes the a priori probability that F_i is true. It was shown by Wolfowitz and the author [10] that any Bayes solution must be a decision rule of the following type: Let x_j denote the observed value of X_j and let

$$z_j = \log \frac{f_2(x_j)}{f_1(x_j)}. \tag{8.1}$$

We choose two constants a and b ($b < a$) and at each stage of the experiment (after the mth observation for each integral value m) we compute the cumulative sum $Z_m = z_1 + \cdots + z_m$. At the first time when $b < Z_m < a$ does not hold, we terminate experimentation.[4] We make the decision d_1 (accept the hypothesis that F_1 is true) if $Z_m \leqq b$, and the decision d_2 (accept the hypothesis that F_2 is true) if $Z_m \geqq a$. A decision rule of the above type is called a sequential probability ratio test.

Applying the complete class theorem to this case, we arrive at the following result: The class of all sequential probability ratio tests corrseponding to all possible values of the constants a and b is a complete class. This means that if δ is any decision rule that is not a sequential probability ratio test, then there exist two constants a and b such that the sequential probability ratio test corresponding to the constants a and b is uniformly better than δ.[5]

Due to the completeness of the class of all sequential probability ratio tests, the problem of choosing a decision rule is reduced to the problem of choosing the values of the constants a and b. A method for determining the constants a and b such that the resulting sequential probability ratio test is a minimax solution, or a Bayes solution relative to a given a priori distribution, is discussed by Arrow, Blackwell, and Girshick [11].

The properties of the sequential probability ratio tests have been studied rather extensively. The recently developed sequential analysis (see, for example, [12] and [13]) is centered on the sequential probability ratio test. It may be of interest to mention that the stochastic process represented by the sequential probability ratio test is identical with the one-dimensional random walk process that plays an important role in molecular physics.

We shall now consider the case when Ω contains more than two but a finite number of elements. It will be sufficient to discuss the case when Ω consists of 3 elements F_1, F_2, and F_3, since the extension to any finite number >3 is straightforward. As before, the random variables $X_1, X_2, \cdots$, are independently distributed with the common density function $f_i(t)$ when F_i is true ($i = 1, 2, 3$). The decision space D consists of 3 elements d_1, d_2, and d_3 where d_i denotes the decision to accept the hypothesis that F_i is true. Let $W(F_i, d_j) = W_{ij} = 0$ for $i = j$, and >0 when $i \neq j$. The cost of experimentation is again assumed to be proportional to the number of observations, and let c denote the cost of a single observation. Any a priori distribution $\xi = (\xi^1, \xi^2, \xi^3)$ can be represented by a point with the coordinates ξ^1, ξ^2, and ξ^3. The totality of all possible a priori distributions ξ will fill out the triangle T with the vertices V_1, V_2, V_3 where V_i represents the a priori distribution whose ith component ξ^i is equal to 1 (see fig. 1).

[4] If $Z_m = a$ or $= b$, the statistician may use any chance mechanism to decide whether to terminate experimentation or to take an additional observation.

[5] This result follows also from an optimum property of the sequential probability ratio test proved in [10].

In order to construct a complete class of decision rules for this problem, it is necessary to determine the Bayes solution relative to any given a priori distribution $\xi_0 = (\xi_0^1, \xi_0^2, \xi_3^0)$. Let x_i denote the observed value of X_i. After m observations have been made the a posteriori probability distribution $\xi_m = (\xi_m^1, \xi_m^2, \xi_m^3)$ is given by the following expression:

$$\xi_m^i = \frac{\xi_0^i f_i(x_1) f_i(x_2) \cdots f_i(x_m)}{\sum_{j=1}^{3} \xi_0^j f_j(x_1) f_j(x_2) \cdots f_j(x_m)}. \tag{8.2}$$

At each stage of the experiment, the a posteriori probability distribution ξ_m is represented by a point of the triangle T.

It was shown by Wolfowitz and the author [14] that there exist three fixed (independent of the a priori distribution ξ_0), closed and convex subsets S_1, S_2, and S_3 of the triangle T such that the Bayes solution relative to ξ_0 is given by the following decision rule: At each stage of the experiment (after the mth

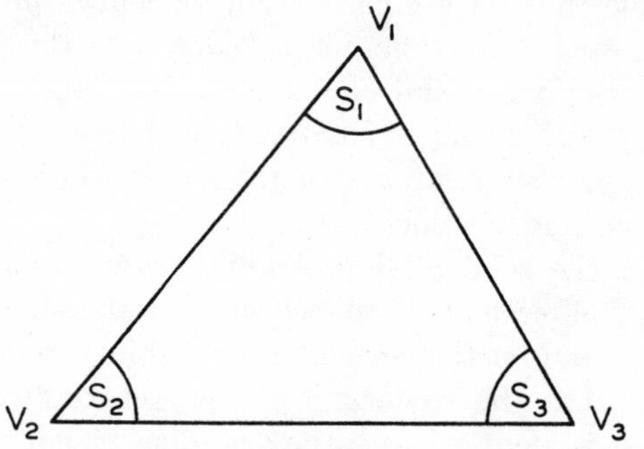

FIGURE 1

observation, for $m = 0, 1, 2, \cdots$) compute the point ξ_m in T. Continue taking additional observations as long as ξ_m does not lie in the union of S_1, S_2, and S_3. If ξ_m lies in the interior of S_i, stop experimentation with the final decision d_i ($i = 1, 2, 3$). If ξ_m lies on the boundary of S_i, an independent chance mechanism may be used to decide whether experimentation be terminated with the final decision d_i or whether an additional observation be made.

The convex sets S_1, S_2, and S_3 depend only on the constants W_{ij} and c. So far no method is available for the explicit computation of the sets S_1, S_2, and S_3 for given values of W_{ij} and c.[6] The development of a method for the explicit determination of the sets S_1, S_2, and S_3 would be of great value, since it would probably indicate a way of dealing with similar difficulties in many other sequential decision problems.

More general results concerning the nature of the Bayes solution for the decision problem described above, admitting also nonlinear cost functions, were obtained by Arrow, Blackwell, and Girshick [11].

[6] The boundary points of the sets S_1, S_2, and S_3 on the periphery of the triangle T and the tangents at these points have been determined in [14].

As a last example, consider the following decision problem: It is known that $X_1, X_2, \cdots$ are independently and identically distributed and the common distribution is known to be rectangular with unit range. Thus, the midpoint θ of the range is the only unknown parameter. The common density function of the chance variables $X_1, X_2, \cdots$ is given by

$$f(t, \theta) = 1 \text{ when } |t - \theta| \leqq 1/2 \qquad (8.3)$$
$$= 0 \text{ otherwise.}$$

For any real value θ^* let d_{θ^*} denote the decision to estimate θ by the value θ^*. The decision space D consists of the elements d_{θ^*} corresponding to all real values θ^*. Let the loss be given by $(\theta - \theta^*)^2$ when θ is the true value of the midpoint of the range and the decision d_{θ^*} is made. The cost of experimentation is assumed to be proportional to the number of observations made. Let c denote the cost of a single observation.

It was shown in [1] that a minimax solution for this problem is given by the following decision rule: Take at least one observation. At each stage of the experiment (after the mth observation, for each positive integral value m) compute the quantity

$$l_m = 1 + \text{Min}(x_1, \cdots, x_m) - \text{Max}(x_1, \cdots, x_m). \qquad (8.4)$$

Continue experimentation as long as $l_m > (24c)^{1/3}$. At the first time when $l_m \leqq (24c)^{1/3}$ stop experimentation and estimate θ by the value

$$\theta^* = \frac{\text{Min}(x_1, \cdots, x_m) + \text{Max}(x_1, \cdots, x_m)}{2}. \qquad (8.5)$$

The risk function associated with this minimax solution is constant over the whole space Ω. The admissibility of the above minimax solution was proved by C. Blyth.

9. Concluding remark. While the general decision theory has been developed to a considerable extent and many results of great generality are available, explicit solutions have been worked out so far only in a relatively small number of special cases. The mathematical difficulties in obtaining explicit solutions, particularly in the sequential case, are still great, but it is hoped that future research will lessen these difficulties and explicit solutions will be worked out in a great variety of problems.

References

1. A. Wald, *Statistical decision functions,* New York, John Wiley & Sons, 1950.
2. A. Dvoretzky, A. Wald, and J. Wolfowitz, *Elimination of randomization in certain statistical decision procedures and zero-sum two person games,* Ann. Math. Statist. vol. 22 (1951).
3. D. Blackwell, *On a theorem of Liapounoff,* Ann. Math. Statist. vol. 22 (1951).
4. A. Liapounoff, *Sur les fonctions-vecteurs complètement additives,* Bull. Acad. Sci. URSS. Ser. Math. vol. 4 (1940).

5. J. L. Hodges and E. L. Lehmann, *Problems in minimax point estimation,* Ann. Math. Statist. vol. 21 (1950).
6. David Vernon Widder, *The Laplace transform,* Princeton, Princeton University Press, 1946.
7. N. Kryloff and N. Bogoliouboff, *La théorie générale de la mesure dans son application à l'étude des systèmes dynamiques de la mécanique non-linéaire,* Ann. of Math. vol. 38 (1937).
8. E. L. Lehmann, *On families of admissible tests,* Ann. Math. Statist. vol. 18 (1947).
9. J. von Neumann and O. Morgenstern, *Theory of games and economic behaviour,* Princeton, Princeton University Press, 1944.
10. A. Wald and J. Wolfowitz, *Optimum character of the sequential probability ratio test,* Ann. Math. Statist. vol. 19 (1948).
11. K. J. Arrow, D. Blackwell, and M. A. Girshick, *Bayes and minimax solutions of sequential decision problems,* Econometrica vol. 17 (1949).
12. A. Wald, *Sequential analysis,* New York, John Wiley & Sons, 1947.
13. Statistical Research Group, Columbia University, *Sequential analysis of statistical data: applications,* New York, Columbia University Press, 1945.
14. A. Wald and J. Wolfowitz, *Bayes solution of sequential decision problems,* Ann. Math. Statist. vol. 21 (1950).

Columbia University,
New York, N. Y., U. S. A.

TESTING THE DIFFERENCE BETWEEN THE MEANS OF TWO NORMAL POPULATIONS WITH UNKNOWN STANDARD DEVIATIONS[1]

BY

ABRAHAM WALD

1. Introduction. Let x_1 and x_2 be two independently and normally distributed random variables. Denote the mean value and standard deviation of x_i $(i = 1, 2)$ by μ_i and σ_i, respectively. All four parameters μ_1, μ_2, σ_1, and σ_2 are assumed to be unknown. In this paper we shall consider the problem of testing the hypothesis that $\mu_1 = \mu_2$ on the basis of a finite number of observations on x_1 and x_2.

The Behrens-Fisher test, which deals with the above problem, is based on the notion of fiducial probability distributions of parameters, introduced by R. A. Fisher. Questions as to the meaning of fiducial probability have arisen in the literature. It is not the intention of the author to discuss this issue. The methods used in the present paper are based exclusively on the classical notion of probability.

In this paper we shall restrict ourselves to the case when the number of observations on x_1 is the same as that on x_2. Let N be the number of observations on x_i and let x_{ij} $(j = 1, \cdots, N)$ denote the jth observation on x_i $(i = 1, 2)$. Furthermore, let

$$\bar{x}_i = \frac{1}{N} \sum_{j=1}^{N} x_{ij} \tag{1.1}$$

and

$$s_i^2 = \frac{\sum_{j=1}^{N} (x_{ij} - \bar{x}_i)^2}{N - 1} \tag{1.2}$$

The four statistics $\bar{x}_1$, $\bar{x}_2$, s_1^2, and s_2^2 form a set of sufficient statistics for the parameters μ_1, μ_2, σ_1^2, and σ_2^2. For testing the hypothesis that $\mu_1 = \mu_2$, we shall consider only critical regions which satisfy the following conditions:

1. The critical region lies in the space of the sufficient statistics $\bar{x}_1$, $\bar{x}_2$, s_1^2, and s_2^2, that is, if the sample $(x_{11}, \cdots, x_{1N}, x_{21}, \cdots, x_{2N})$ belongs to the critical region, then so does any other sample $(x'_{11}, \cdots, x'_{1N}, x'_{21}, \cdots, x'_{2N})$ for which $\bar{x}'_1 = \bar{x}_1$, $\bar{x}'_2 = \bar{x}_2$, $s_1'^2 = s_1^2$ and $s_2'^2 = s_2^2$.

[1] *Editorial note.* This paper was found in Wald's effects after his death and was transmitted to the editors of this volume by Professor J. Wolfowitz.

2. If the sample $(x_{11}, \cdots, x_{1N}, x_{21}, \cdots, x_{2N})$ lies in the critical region, then for any c the sample $(x_{11} + c, \cdots, x_{1N} + c, x_{21} + c, \cdots, x_{2N} + c)$ also lies in the critical region.

3. If the sample $(x_{11}, \cdots, x_{1N}, x_{21}, \cdots, x_{2N})$ lies in the critical region, then for any $k \neq 0$ the sample $(kx_{11}, \cdots, kx_{1N}, kx_{21}, \cdots, kx_{2N})$ also lies in the critical region.

4. If the sample $(x_{11}, \cdots, x_{1N}, x_{21}, \cdots, x_{2N})$ lies in the critical region, then so does any other sample $(x'_{11}, \cdots, x'_{1N}, x'_{21}, \cdots, x'_{2N})$ for which $|\bar{x}'_2 - \bar{x}'_1| > |\bar{x}_2 - \bar{x}_1|$, $s_1'^2 = s_1^2$, and $s_2'^2 = s_2^2$.

It can easily be verified that any closed subset of the sample space satisfying the above four conditions can be represented by an inequality of the form

$$(1.3) \qquad \left| \frac{y}{\sqrt{s_1^2 + s_2^2}} \right| \geqq \phi(l)$$

where

$$(1.4) \qquad y = \sqrt{N}\,(\bar{x}_1 - \bar{x}_2),\ l = \frac{s_2^2}{s_1^2}$$

and $\phi(l)$ is some function of l. Thus the critical region is completely specified by the choice of the function $\phi(l)$. For any given function $\phi(l)$ the size of the critical region (1.3), that is, the probability that the observed sample will satisfy (1.3) when $\mu_1 = \mu_2$, depends only on the single parameter $\lambda = \sigma_2^2/\sigma_1^2$. Let $\alpha(\lambda)$ denote the size of the critical region (1.3). To put the dependence of $\alpha(\lambda)$ on the choice of $\phi(l)$ in evidence, we shall occasionally write $\alpha[\lambda|\phi(l)]$ instead of $\alpha(\lambda)$.

In determining the size of the critical region (1.3), we may put $\sigma_1^2 = 1$ and $\sigma_2^2 = \lambda$. In what follows we shall, therefore, assume that $\sigma_1^2 = 1$ and $\sigma_2^2 = \lambda$. Given the values of s_1^2 and s_2^2, the conditional probability that (1.3) holds when $\mu_1 = \mu_2$ is given by

$$(1.5) \qquad 2G\left(\phi(l)\,\frac{\sqrt{s_1^2 + s_2^2}}{\sqrt{1 + \lambda}}\right)$$

where

$$(1.6) \qquad G(t) = \frac{1}{\sqrt{2\pi}} \int_t^{\infty} e^{-u^2/2}\, du$$

Hence

$$(1.7) \qquad \alpha[\lambda|\phi(l)] = 2EG\left[\phi(l)\sqrt{\frac{s_1^2 + s_2^2}{1 + \lambda}}\right]$$

where the operator E stands for expected value.

$\alpha[\lambda|\phi(l)]$ can be expressed also in another way. For this purpose we shall write inequality (1.3) as follows:

$$(1.8) \qquad \left| \frac{\sqrt{2}\, y}{\sqrt{1 + \lambda}\,\sqrt{s_1^2 + (1/\lambda)s_2^2}} \right| \geqq \phi(l)\sqrt{\frac{2(s_1^2 + s_2^2)}{(1 + \lambda)[s_1^2 + (1/\lambda)s_2^2]}} = \frac{\sqrt{2}\,\phi(l)\,\sqrt{1 + l}}{\sqrt{(1 + \lambda)[1 + (l/\lambda)]}}$$

The expression

$$t = \frac{\sqrt{2}\, y}{\sqrt{1+\lambda}\,\sqrt{s_1^2 + (1/\lambda)s_2^2}} \tag{1.9}$$

can be seen to have the t-distribution with $2n$ degrees of freedom, where $n = N - 1$. Since $s_1^2 + (1/\lambda)s_2^2$ and l are independently distributed, t and l are also independently distributed. The distribution of l is the same as that of λF where F has the F-distribution with (n,n) degrees of freedom. Hence, it follows from (1.8) that

$$\alpha[\lambda|\phi(l)] = P\left\{|t| \geqq \frac{\sqrt{2}\,\phi(\lambda F)\,\sqrt{1+\lambda F}}{\sqrt{(1+\lambda)(1+F)}}\right\} \tag{1.10}$$

where t has the t-distribution with $2n$ degrees of freedom, F has the F-distribution with (n,n) degrees of freedom, and t and F are independent. For any constant c, let

$$H(c) = P\{|t| \geqq c\} \tag{1.11}$$

We then obtain from (1.10)

$$\alpha[\lambda|\phi(l)] = EH\left[\frac{\sqrt{2}\,\phi(\lambda F)\,\sqrt{1+\lambda F}}{\sqrt{(1+\lambda)(1+F)}}\right] \tag{1.12}$$

In this paper we shall be concerned mainly with the problem of a proper choice of the function $\phi(l)$. We shall impose the condition that the size of the critical region shall not exceed a prescribed value α_0, that is, $\phi(l)$ is to be chosen so that

$$\alpha[\lambda|\phi(l)] \leqq \alpha_0 \qquad \text{for any } \lambda \geqq 0 \tag{1.13}$$

From the class of functions $\phi(l)$ for which (1.13) is satisfied, we shall try to select one which makes the power of the test as large as possible in some sense. Of particular interest is the case when $\phi(l)$ equals a positive constant c. It will be shown in Sec. 2 that $\alpha(\lambda|c)$ takes its maximum value at $\lambda = 0$. It follows from (1.3) that $\alpha(0|c) = P\{|t_n| \geq c\}$, where t_n has the t-distribution with n degrees of freedom. Thus, if c is chosen so that $P\{|t_n| \geq c\} = \alpha_0$, condition (1.13) is fulfilled for $\phi(l) = c$. It can be shown that the critical region resulting from putting $\phi(l) = c$ is an asymptotically most powerful unbiased region. Thus, for large n, $\phi(l) = c$ will provide a satisfactory solution of our problem. For small n, however, the critical region corresponding to $\phi(l) = c$ is not very satisfactory, since other critical regions exist which are uniformly more powerful, as will be seen in Sec. 3.

In Sec. 3 the critical region corresponding to $\phi(l) = c - c'\dfrac{l}{(1+l)^2}$ is discussed, where c and c' are certain properly determined positive constants. This critical region has very nearly constant size even for small n. The difference $\alpha\left[\lambda|c - c'\dfrac{l}{(1+l)^2}\right] - \alpha\left[0|c - c'\dfrac{l}{(1+l)^2}\right]$ is shown to be of order

$1/n^2$, while the difference $\alpha(\lambda|c) - \alpha(0|c)$ is of the order $1/n$. A numerical example given in Sec. 3 suggests that even for $n = 5$ the function

$$\alpha\left[\lambda|c - c'\frac{l}{(1+l)^2}\right]$$

is nearly constant, while $\alpha(\lambda|c)$ varies considerably with λ for such small values of n. Clearly, since $c' > 0$, the critical region corresponding to

$$\phi(l) = c - c'\frac{l}{(1+l)^2}$$

is uniformly more powerful than the region corresponding to $\phi(l) = c$. While the region corresponding to $\phi(l) = c - c'\dfrac{l}{(1+l)^2}$ is certainly not the best possible one, it is a relatively simple one and in view of Theorem 3.2 it seems likely that no appreciably better critical regions exist even for small values of n.

Let F_0 be a given positive constant and let F^* be a random variable whose probability distribution is defined as follows:

$$P\{F^* < c\} = 1 \qquad \text{for } c > F_0 \tag{1.14}$$

$$P\{F^* < c\} = P\{F < c\} \qquad \text{for } c \leqq F_0 \tag{1.15}$$

where F has the F-distribution with (n,n) degrees of freedom. Let $\alpha^*[\lambda|\phi(l)]$ be the function we obtain by substituting F^* for F in (1.12), that is,

$$\alpha^*[\lambda|\phi(l)] = EH\left[\frac{\sqrt{2}\,\phi(\lambda F^*)\,\sqrt{1+\lambda F^*}}{\sqrt{(1+\lambda)(1+F^*)}}\right] \tag{1.16}$$

Clearly, since $0 \leqq H(u) \leqq 1$, $|\alpha^*[\lambda|\phi(l)] - \alpha[\lambda|\phi(l)]|$ can be made arbitrarily small by choosing F_0 sufficiently large.

It will be shown in Sec. 6 that for any F_0, no matter how large, and for any positive $\alpha < 1$, there exists exactly one analytic function $\phi_\alpha(l)$ which is not singular at $l = 0$ and for which $\alpha^*[\lambda|\phi_\alpha(l)]$ is exactly equal to α for all λ less than or equal to a certain positive value λ_0. Since $\phi_\alpha(l)$ is not singular at $l = 0$, it can be expanded in a power series in l. A method for computing step by step the successive coefficients of this power series is given in Sec. 4. If this power series can be extended by analytic continuation to cover the whole complex plane, $\alpha^*[\lambda|\phi_\alpha(\lambda)]$ can be shown to be equal to α over the entire range of λ. Even if for any finite F_0 there exists a function $\phi_\alpha(\lambda)$ such that $\alpha^*[\lambda|\phi_\alpha(\lambda)] = \alpha$ for all values of λ, the question still remains open whether such a function $\phi_\alpha(l)$ also exists in the limit case when $F_0 = \infty$.

2. Some properties of $\alpha(\lambda|c)$. Since the critical region corresponding to $\phi(l) = c$ is symmetric in the variables x_1 and x_2, it follows that

$$\alpha(\lambda|c) = \alpha(1/\lambda|c)$$

From (1.7) we obtain

$$\alpha(\lambda|c) = 2EG\left(c\sqrt{\frac{s_1^2 + s_2^2}{1+\lambda}}\right) \tag{2.1}$$

We put

$$ns_1^2 = \chi_1^2 \qquad \text{and} \qquad ns_2^2 = \lambda\chi_2^2 \tag{2.2}$$

Then χ_1^2 and χ_2^2 are independent random variables and each of them has the χ^2-distribution with n degrees of freedom. Hence it follows from (2.1) that

$$\begin{aligned} \alpha(\lambda|c) &= 2EG\left(\frac{c}{\sqrt{n}}\sqrt{\frac{\chi_1^2 + \lambda\chi_2^2}{1+\lambda}}\right) \\ &= E\left[G\left(\frac{c}{\sqrt{n}}\sqrt{\frac{\chi_1^2 + \lambda\chi_2^2}{1+\lambda}}\right) + G\left(\frac{c}{\sqrt{n}}\sqrt{\frac{\chi_2^2 + \lambda\chi_1^2}{1+\lambda}}\right)\right] \end{aligned} \tag{2.3}$$

We have

$$\begin{aligned} &\sqrt{2\pi}\,\frac{d}{d\lambda}\left[G\left(\frac{c}{\sqrt{n}}\sqrt{\frac{\chi_1^2 + \lambda\chi_2^2}{1+\lambda}}\right) + G\left(\frac{c}{\sqrt{n}}\sqrt{\frac{\chi_2^2 + \lambda\chi_1^2}{1+\lambda}}\right)\right] \\ &\quad = \frac{1}{2}\frac{c}{\sqrt{n}}\frac{\chi_1^2 - \chi_2^2}{(1+\lambda)^{\frac{3}{2}}}\left[\frac{1}{\sqrt{\chi_1^2 + \lambda\chi_2^2}}\exp\left(-\frac{1}{2}\frac{c^2}{n}\frac{\chi_1^2 + \lambda\chi_2^2}{1+\lambda}\right)\right. \\ &\qquad \left. - \frac{1}{\sqrt{\chi_2^2 + \lambda\chi_1^2}}\exp\left(-\frac{1}{2}\frac{c^2}{n}\frac{\chi_2^2 + \lambda\chi_1^2}{1+\lambda}\right)\right] \end{aligned} \tag{2.4}$$

We shall now show that the expression on the right side of (2.4) is negative when $\lambda < 1$ and $\chi_1^2 \neq \chi_2^2$. In fact, if $\chi_1^2 > \chi_2^2$ and $\lambda < 1$, we have $\chi_1^2 + \lambda\chi_2^2 > \chi_2^2 + \lambda\chi_1^2$. Hence

$$\left[\frac{1}{\sqrt{\chi_1^2 + \lambda\chi_2^2}}\exp\left(-\frac{1}{2}\frac{c^2}{n}\frac{\chi_1^2 + \lambda\chi_2^2}{1+\lambda}\right) - \frac{1}{\sqrt{\chi_2^2 + \lambda\chi_1^2}}\exp\left(-\frac{1}{2}\frac{c^2}{n}\frac{\chi_2^2 + \lambda\chi_1^2}{1+\lambda}\right)\right] < 0 \tag{2.5}$$

if $\lambda < 1$ and $\chi_1^2 > \chi_2^2$. If $\lambda < 1$ and $\chi_1^2 < \chi_2^2$, the difference in (2.5) is positive. Hence, the right member of (2.4) is negative when $\lambda < 1$ and $\chi_1^2 \neq \chi_2^2$. From this and (2.3) it follows that $\alpha(\lambda|c)$ is a strictly decreasing function of λ in the interval $0 \leqq \lambda \leqq 1$.

Our results may be summarized in the following theorem.

THEOREM 2.1: *The function $\alpha(\lambda|c)$ is strictly decreasing with increasing λ in the interval $0 \leqq \lambda \leqq 1$. Furthermore, we have $\alpha(\lambda|c) = \alpha(1/\lambda|c)$.*

It follows immediately from Theorem 2.1 that $\alpha(\lambda|c)$ takes its maximum value at $\lambda = 0$, and its minimum value at $\lambda = 1$. Let

$$v = \frac{1}{n}\frac{\chi_1^2 + \lambda\chi_2^2}{1+\lambda} \tag{2.6}$$

We expand $G(c\sqrt{v})$ in a finite Taylor series around $v = 1$. We have

$$(2.7)\quad G(c\sqrt{v}) = G(c) - \frac{(v-1)}{\sqrt{2\pi}}\frac{c}{2}e^{-\frac{1}{2}c^2} + \frac{1}{2\sqrt{2\pi}}(v-1)^2\left(\frac{c+c^3}{4}\right)e^{-\frac{1}{2}c^2}$$
$$- \frac{1}{6\sqrt{2\pi}}(v-1)^3\left(\frac{3c+2c^3+c^5}{8}\right)e^{-\frac{1}{2}c^2} + \frac{1}{24\sqrt{2\pi}}(v-1)^4\rho(\xi)$$

where ξ lies in the interval $[1,v]$ and

$$(2.8)\qquad \rho(\xi) = \left(\frac{15c}{16\xi^{\frac{7}{2}}} + \frac{9c^3}{16\xi^{\frac{5}{2}}} + \frac{3c^5}{16\xi^{\frac{3}{2}}} + \frac{c^7}{16\xi^{\frac{1}{2}}}\right)e^{-\frac{1}{2}c^2\xi} > 0$$

Clearly, for any positive δ the function $\rho(\xi)$ is bounded in the domain $\xi \geqq \delta$. We shall now show that $\rho(\xi)$ is bounded also when $\xi < \delta < 1$. Since ξ lies in the interval $[1,v]$, the inequality $\xi < \delta < 1$ implies that $v < \delta$. Hence, we merely have to show that $\rho(\xi)$ is bounded in the domain $v < \delta < 1$. From (2.7) we obtain

$$(2.9)\quad \rho(\xi) = \left\{\sqrt{2\pi}\,[G(c\sqrt{v}) - G(c)] + (v-1)\frac{c}{2}e^{-\frac{1}{2}c^2}\right.$$
$$\left. - \frac{1}{2}(v-1)^2\frac{(c+c^3)}{4}e^{-\frac{1}{2}c^2} + \frac{1}{6}(v-1)^3\frac{3c+2c^3+c^5}{8}e^{-\frac{1}{2}c^2}\right\}\frac{24}{(v-1)^4}$$

Clearly, the right-hand member of (2.9) is bounded in the domain $v < \delta < 1$. Hence, $\rho(\xi)$ is bounded in the whole domain of v. Since $E(v-1)^4 = 0(1/n^2)$ and since $\rho(\xi)$ is bounded, we have

$$(2.10)\qquad E[(v-1)^4\rho(\xi)] = 0\left(\frac{1}{n^2}\right)$$

Since $E(v-1)^3 = 0(1/n^2)$, taking expected values on both sides of (2.7), we obtain

$$(2.11)\quad \frac{1}{2}\alpha(\lambda|c) = G(c) + \frac{1}{n}\frac{1+\lambda^2}{(1+\lambda)^2}\frac{c+c^3}{4\sqrt{2\pi}}e^{-\frac{1}{2}c^2} + 0\left(\frac{1}{n^2}\right)$$
$$= G(c) + \frac{1}{n}\frac{c+c^3}{4\sqrt{2\pi}}e^{-\frac{1}{2}c^2}\left[1 - \frac{2\lambda}{(1+\lambda)^2}\right] + 0\left(\frac{1}{n^2}\right)$$

Hence we obtain

THEOREM 2.2: *The function $\alpha(\lambda|c)$ differs from the function*

$$(2.12)\qquad 2G(c) + \frac{1}{n}\frac{c+c^3}{2\sqrt{2\pi}}e^{-\frac{1}{2}c^2}\left[1 - \frac{2\lambda}{(1+\lambda)^2}\right]$$

by a quantity of the order $1/n^2$.

The equation

$$(2.13)\qquad \alpha(\lambda|c) - 2G(c) = 0\left(\frac{1}{n}\right)$$

is an immediate consequence of Theorem 2.2.

3. A critical region whose size differs from a constant by a quantity of the order $1/n^2$. It will be shown in this section that the size of the critical region $W(c,c')$ corresponding to $\phi(l) = c - c' \frac{l}{(1+l)^2}$, where c and c' are some properly chosen constants, differs from a constant by a quantity of the order $1/n^2$. We shall, furthermore, investigate the power function of the region $W(c,c')$. For this purpose we shall first prove the following lemma.

LEMMA 3.1: *Let $f(s_1^2,s_2^2)$ be a bounded function of s_1^2 and s_2^2 which admits continuous second-order partial derivatives in the neighborhood of the point*

$$(s_1^2,s_2^2) = (\sigma_1^2,\sigma_2^2)$$

Then

$$Ef(s_1^2,s_2^2) - f(\sigma_1^2,\sigma_2^2) = 0\left(\frac{1}{n}\right) \tag{3.1}$$

PROOF: Since the standard deviation of $s_i^2 (i = 1,2)$ is of the order $1/\sqrt{n}$, it follows from Chebyshev's inequality that, for any positive δ,

$$P\{\sigma_1^2 - \delta \leqq s_1^2 \leqq \sigma_1^2 + \delta,\ \sigma_2^2 - \delta \leqq s_2^2 \leqq \sigma_2^2 + \delta\} = 1 - 0\left(\frac{1}{n}\right) \tag{3.2}$$

Let s_1^{*2} and s_2^{*2} be two independently distributed random variables, and let the distribution of s_i^{*2} be equal to the conditional distribution of s_i^2 given that $\sigma_i^2 - \delta \leqq s_i^2 \leqq \sigma_i^2 + \delta$ $(i = 1, 2)$. Since $f(s_1^2,s_2^2)$ is bounded, it follows from (3.2) that

$$Ef(s_1^{*2},s_2^{*2}) - Ef(s_1^2,s_2^2) = 0\left(\frac{1}{n}\right) \tag{3.3}$$

Expanding $f(s_1^{*2},s_2^{*2})$ in a Taylor series around (σ_1^2,σ_2^2) and taking expected values, we obtain

$$E[f(s_1^{*2},s_2^{*2})] = f(\sigma_1^2,\sigma_2^2) + [E(s_1^{*2} - \sigma_1^2)] \frac{\partial f(\sigma_1^2,\sigma_2^2)}{\partial \sigma_1^2} + [E(s_2^{*2} - \sigma_2^2)] \tag{3.4}$$

$$\frac{\partial f(\sigma_1^2,\sigma_2^2)}{\partial \sigma_2^2} + \frac{1}{2} \sum_{j=1}^{2} \sum_{i=1}^{2} E[(s_i^{*2} - \sigma_i^2)(s_j^{*2} - \sigma_j^2)] \left. \frac{\partial^2 f(\sigma_1^2,\sigma_2^2)}{\partial \sigma_i^2\, \partial \sigma_j^2} \right|_{\sigma_i^2 = \bar{s}_i^2, \sigma_j^2 = \bar{s}_j^2}$$

where $\bar{s}_i^2$ lies in the interval $[\sigma_i^2, s_i^{*2}]$ $(i = 1, 2)$. It is known that

$$E|(s_i^2 - \sigma_i^2)(s_j^2 - \sigma_j^2)| = 0\left(\frac{1}{n}\right) \qquad (i, j = 1, 2) \tag{3.5}$$

From (3.2) and (3.5) it follows immediately that

$$E|(s_i^{*2} - \sigma_i^2)(s_j^{*2} - \sigma_j^2)| = 0\left(\frac{1}{n}\right) \qquad (i, j = 1,2) \tag{3.6}$$

To evaluate $E(s_i^{*2} - \sigma_i^2)$, let $g(s_i^2)$ denote the probability density function of s_i^2. Clearly,

$$(3.7)\quad 0 < \int_{|\sigma_i^2 - s_i^2| > \delta} (s_i^2 - \sigma_i^2) g(s_i^2)\, ds_i^2 < \int_{|\sigma_i^2 - s_i^2| > \delta} \frac{(s_i^2 - \sigma_i^2)^2}{\delta} g(s_i^2)\, ds_i^2 < \frac{1}{\delta} E(s_i^2 - \sigma_i^2)^2 = 0\left(\frac{1}{n}\right)$$

From this and (3.2) it follows easily that

$$(3.8)\qquad E(s_i^{*2} - \sigma_i^2) - E(s_i^2 - \sigma_i^2) = 0\left(\frac{1}{n}\right)$$

Hence, since $E(s_i^2 - \sigma_i^2) = 0$, we have

$$(3.9)\qquad E(s_i^{*2} - \sigma_i^2) = 0\left(\frac{1}{n}\right)$$

By assumption, the second derivatives of $f(s_1^2, s_2^2)$ are continuous in the neighborhood of (σ_1^2, σ_2^2). Hence, for sufficiently small δ,

$$\left.\frac{\partial^2 f(\sigma_1^2, \sigma_2^2)}{\partial \sigma_i^2 \partial \sigma_j^2}\right|_{\sigma_i^2 = \bar{s}_i^2, \sigma_j^2 = \bar{s}_j^2}$$

is bounded. Hence, we obtain from (3.4), (3.6), and (3.9)

$$(3.10)\qquad Ef(s_1^{*2}, s_2^{*2}) = f(\sigma_1^2, \sigma_2^2) + 0\left(\frac{1}{n}\right)$$

Lemma 3.1 follows from (3.3) and (3.10).

We are now able to prove the following theorem.

Theorem 3.1: *Let $\rho^*(l)$ be a bounded function of l which admits a continuous second derivative. Let, furthermore, $\beta[d, \lambda | \phi(l)]$ denote the power of the critical region* (1.3) *when*

$$(3.11)\qquad \left| \frac{\sqrt{N}\,(\mu_1 - \mu_2)}{\sqrt{\sigma_1^2 + \sigma_2^2}} \right| = d$$

that is, $\beta[d, \lambda | \phi(l)]$ is the probability that the observed sample will fall in the region (1.3) *when* (3.11) *holds.* (It is easy to verify that β is a function only of d and λ.) *Then*

$$(3.12)\quad \beta[d, \lambda | c + \frac{1}{n}\rho^*(l)] - \beta[d, \lambda | c] = -\frac{1}{\sqrt{2\pi}\, n}\rho^*(\lambda)[e^{-\frac{1}{2}(c-d)^2} + e^{-\frac{1}{2}(c+d)^2}] + 0\left(\frac{1}{n^2}\right)$$

Proof: Without loss of generality we may assume that $\sigma_1^2 = 1$ and $\sigma_2^2 = \lambda$. Let

$$w = \sqrt{\frac{s_1^2 + s_2^2}{1 + \lambda}}$$

Then

$$(3.13)\quad \beta\left[d,\lambda|c + \frac{1}{n}\rho^*(l)\right] = EG\left\{\left[c + \frac{1}{n}\rho^*(l)\right]w - d\right\} + EG\left\{\left[c + \frac{1}{n}\rho^*(l)\right]w + d\right\}$$

and

$$(3.14)\quad \beta(d,\lambda|c) = EG(cw - d) + EG(cw + d)$$

Now

$$(3.15)\quad G\left\{\left[c + \frac{1}{n}\rho^*(l)\right]w - d\right\} - G(cw - d) = -\frac{1}{\sqrt{2\pi}\,n}\,w\rho^*(l)e^{-\frac{1}{2}(cw-d)^2} + \frac{1}{2\sqrt{2\pi}}\frac{w^2}{n^2}[\rho^*(l)]^2ue^{-u^2/2}$$

where u lies in the interval

$$\left\{cw - d, \quad \left[c + \frac{1}{n}\rho^*(l)\right]w - d\right\}$$

Since $\rho^*(l)$ and $ue^{-u^2/2}$ are bounded functions we have

$$(3.16)\quad E\frac{w^2}{n^2}[\rho^*(l)]^2ue^{-u^2/2} = 0\left(\frac{1}{n^2}\right)$$

Applying Lemma 3.1, we see that

$$(3.17)\quad E\frac{w}{n}\rho^*(l)e^{-\frac{1}{2}(cw-d)^2} = \frac{\rho^*(\lambda)}{n}e^{-\frac{1}{2}(c-d)^2} + 0\left(\frac{1}{n^2}\right)$$

Hence

$$(3.18)\quad EG\left\{w\left[c + \frac{1}{n}\rho^*(l)\right] - d\right\} - EG(cw - d) = \frac{-1}{\sqrt{2\pi}\,n}\rho^*(\lambda)e^{-\frac{1}{2}(c-d)^2} + 0\left(\frac{1}{n^2}\right)$$

Similarly

$$(3.19)\quad EG\left\{\left[c + \frac{1}{n}\rho^*(l)\right]w + d\right\} - EG(cw + d) = -\frac{1}{\sqrt{2\pi}\,n}\rho^*(\lambda)e^{-\frac{1}{2}(c+d)^2} + 0\left(\frac{1}{n^2}\right)$$

Theorem 3.1 follows from (3.18) and (3.19).

Putting $d = 0$ and $\rho^*(l) = -c^*\dfrac{l}{(1 + l)^2}$, we obtain from Theorem 3.1, since $\beta[0,\lambda|\phi(l)] = \alpha[\lambda|\phi(l)]$,

$$(3.20)\quad \alpha\left[\lambda|c - \frac{c^*}{n}\frac{l}{(1 + l)^2}\right] - \alpha(\lambda|c) = \frac{\sqrt{2}}{\sqrt{\pi}\,n}c^*\frac{\lambda}{(1 + \lambda)^2}e^{-\frac{1}{2}c^2} + 0\left(\frac{1}{n^2}\right)$$

From Theorem 2.2 and (3.20) it follows that for

$$c^* = \frac{c + c^3}{2} \tag{3.21}$$

we have

$$\alpha\left[\lambda|c - \frac{c^*}{n}\frac{l}{(1+l)^2}\right] - \alpha\left[0|c - \frac{c^*}{n}\frac{l}{(1+l)^2}\right] = 0\left(\frac{1}{n^2}\right) \tag{3.22}$$

Since $P[l = 0] = 1$ when $\lambda = 0$,

$$\alpha\left[0|c - \frac{c^*}{n}\frac{l}{(1+l)^2}\right] = \alpha(0|c)$$

Hence

$$\alpha\left[\lambda|c - \frac{c^*}{n}\frac{l}{(1+l)^2}\right] - \alpha(0|c) = 0\left(\frac{1}{n^2}\right) \tag{3.23}$$

Let c' be the constant determined by

$$\alpha\left[\lambda|c - c'\frac{\lambda}{(1+\lambda)^2}\right] - \alpha(0|c) = 0 \qquad \text{for } \lambda = 1 \tag{3.24}$$

(It is easy to see that it is unique.) It follows from Theorem 3.1 that (3.23) remains valid if we replace $l/(1 + l)^2$ by $\lambda/(1 + \lambda)^2$, that is,

$$\alpha\left[\lambda|c - \frac{c^*}{n}\frac{\lambda}{(1+\lambda)^2}\right] - \alpha(0|c) = 0\left(\frac{1}{n^2}\right) \tag{3.25}$$

Equations (3.24) and (3.25) imply that

$$c' - \frac{c^*}{n} = 0\left(\frac{1}{n^2}\right) \tag{3.26}$$

It follows easily from Theorem 3.1 that (3.23) remains valid if we replace c^*/n by a function $\bar{c}(n)$ for which

$$\bar{c}(n) - \frac{c^*}{n} = 0\left(\frac{1}{n^2}\right)$$

In particular, we may substitute c' for c^*/n. Thus, we obtain

THEOREM 3.2: *We have*

$$\alpha\left[\lambda|c - c'\frac{l}{(1+l)^2}\right] - \alpha(0|c) = 0\left(\frac{1}{n^2}\right) \tag{3.27}$$

where the constant c' is determined by equation (3.24).

The value of c' can easily be obtained from a table of the t-distribution. In fact, for $\lambda = 1$,

$$\alpha\left[\lambda|c - c'\frac{\lambda}{(1+\lambda)^2}\right] = \alpha\left(1|c - \frac{c'}{4}\right) = P\left\{\left|\frac{\sqrt{N}\,(\bar{x}_1 - \bar{x}_2)}{\sqrt{s_1^2 + s_2^2}}\right| \geqq c - \frac{c'}{4}\right\}$$

$$= P\left\{|t_{2n}| \geqq c - \frac{c'}{4}\right\}$$

where t_{2n} has the t-distribution with $2n$ degrees of freedom. From a table of the t-distribution we can determine the value t^* for which

$$P\{|t_{2n}| \geqq t^*\} = \alpha(0|c) \tag{3.28}$$

Then

$$c' = 4(c - t^*) \tag{3.29}$$

According to Theorem 2.1, $\alpha(\lambda|c)$ is strictly decreasing with increasing λ in the interval $0 \leqq \lambda \leqq 1$. Hence, $\alpha[1|c - (c'/4)] = \alpha(0|c)$ implies that $c' > 0$. Thus, t^* must be less than c.

We shall now prove

THEOREM 3.3: *Let $\rho(l,n)$ be a bounded function of l and n such that the least upper bound of $\alpha[\lambda|c + (1/n)\rho(l,n)]$ with respect to λ does not exceed the least upper bound of $\alpha\left[\lambda|c - c'\dfrac{l}{(1+l)^2}\right]$ with respect to λ. Suppose that $\dfrac{\partial^2\rho(l,n)}{\partial l^2}$ exists and is a continuous and bounded function of l and a bounded function of n in any finite l-interval. Then $\beta[d,\lambda|c + (1/n)\rho(l,n)]$ cannot exceed*

$$\beta\left[d,\lambda|c - c'\frac{l}{(1+l)^2}\right]$$

except by a quantity of order $1/n^2$.

PROOF: It follows from Theorem 3.1 that

$$\beta\left[d,\lambda|c + \frac{1}{n}\rho(\lambda,n)\right] - \beta\left[d,\lambda|c + \frac{1}{n}\rho(l,n)\right] = 0\left(\frac{1}{n^2}\right) \tag{3.30}$$

and

$$\beta\left[d,\lambda|c - c'\frac{\lambda}{(1+\lambda)^2}\right] - \beta\left[d,\lambda|c - c'\frac{l}{(1+l)^2}\right] = 0\left(\frac{1}{n^2}\right) \tag{3.31}$$

Since

$$\alpha\left[\lambda|c - c'\frac{l}{(1+l)^2}\right] - \alpha\left[0|c - c'\frac{l}{(1+l)^2}\right] = 0\left(\frac{1}{n^2}\right)$$

and since the least upper bound of $\alpha[\lambda|c + (1/n)\rho(l,n)]$ with respect to λ does not exceed the least upper bound of $\alpha\left[\lambda|c - c'\dfrac{l}{(1+l)^2}\right]$, it follows that $\alpha[\lambda|c + (1/n)\rho(l,n)]$ cannot exceed $\alpha\left[\lambda|c - c'\dfrac{l}{(1+l)^2}\right]$ except by a quantity of order $1/n^2$. But then, because of equations (3.30) and (3.31), also $\alpha[\lambda|c + \dfrac{1}{n}\rho(\lambda,n)]$ cannot exceed $\alpha\left[\lambda|c - c'\dfrac{\lambda}{(1+\lambda^2)}\right]$ except by a quantity of order $1/n^2$. This in turn implies that $-\dfrac{1}{n}\rho(\lambda,n)$ cannot exceed $c'\dfrac{\lambda}{(1+\lambda)^2}$ except by a quantity of order $1/n^2$. Hence, also $\beta[d,\lambda|c + (1/n)\rho(\lambda,n)]$ cannot exceed $\beta\left[d,\lambda|c - c'\frac{\lambda}{(1+\lambda)^2}\right]$ except by a quantity of order $1/n^2$. This and equations (3.30) and (3.31) imply the validity of Theorem 3.3.

Theorem 3.3 shows that, if quantities of order $1/n^2$ are neglected, the critical region corresponding to $\phi(l) = c - c'\frac{l}{(1+l)^2}$ has the largest possible power as compared with any other region corresponding to functions $\phi(l)$ for which $\phi(l) - c = 0(1/n)$.

This still leaves the possibility open that a function $\phi(l)$ may exist for which $\phi(l) - c = 0(1/n)$ does not hold and which leads to a substantially greater power than $\phi(l) = c - c'\frac{l}{(1+l)^2}$. This is, however, rather unlikely in view of the fact that the power of the customary t-test when λ is known exceeds the power of the test corresponding to $\phi(l) = c$ only by a quantity of the order $1/n$. Thus, an optimum function $\phi(l)$ is likely to be in the neighborhood of the function $\phi(l) = c$.

Let $\gamma(d,\lambda)$ denote the power of the customary t-test when λ is known and the size of the critical region is equal to $\alpha(0|c)$. It can easily be verified that

$$\beta\left(d,1|c - \frac{c'}{4}\right) = \gamma(d,1)$$

Hence, because of (3.31), we have

$$\gamma(d,1) - \beta\left[d,1|c - c'\frac{l}{(1+l)^2}\right] = 0\left(\frac{1}{n^2}\right) \tag{3.32}$$

This shows that for $\lambda = 1$ the power of the test corresponding to

$$\phi(l) = c - c'\frac{l}{(1+l)^2}$$

cannot be improved except by a quantity of order $1/n^2$ even if we admit the use of functions $\phi(l)$ for which $\phi(l) - c = 0(1/n)$ does not hold.

As an example,[1] $\alpha\left[\lambda|c - c'\frac{l}{(1+l)^2}\right]$ has been computed when $n = 5$ and c has been chosen so that $\alpha(0|c) = .05$. For this purpose the formula (1.10) has been used, that is,

$$\alpha\left[\lambda|c - c'\frac{l}{(1+l)^2}\right] = P\left\{|t_{2n}| \geqq \frac{\sqrt{2}\sqrt{1+\lambda F}}{\sqrt{(1+\lambda)(1+F)}}\left[c - c'\frac{\lambda F}{(1+\lambda F)^2}\right]\right\}$$

when t_{2n} and F are independent, t_{2n} has the t-distribution with $2n$ degrees of freedom, and F has the F-distribution with (n,n) degrees of freedom. Lower

[1] The author is greatly indebted to Mr. George Carlton for carrying out these computations.

and upper limits for $\alpha\left[\lambda|c - c'\frac{l}{(1+l)^2}\right]$ were computed as follows: Let $F_0, F_1, F_2, \cdots, F_k$ be a set of values such that $F_0 = 0, F_k = \infty$, and

$$P\{F_{i-1} \leqq F \leqq F_i\} = \frac{1}{k} \qquad (i = 1, \cdots, k)$$

Let A_i be the maximum and B_i be the minimum of

$$\frac{\sqrt{2}\sqrt{1+\lambda F}}{\sqrt{(1+\lambda)(1+F)}}\left[c - c'\frac{\lambda F}{(1+\lambda F)^2}\right] \tag{3.33}$$

in the interval from F_{i-1} to F_i. Then

$$\underline{\alpha} = \frac{1}{k}\sum_{i=1}^{k} P\{|t_{2n}| \geqq A_i\} \tag{3.34}$$

is a lower limit, and

$$\bar{\alpha} = \frac{1}{k}\sum_{i=1}^{k} P\{|t_{2n}| \geqq B_i\} \tag{3.35}$$

is an upper limit of $\alpha\left[\lambda|c - c'\frac{l}{(1+l)^2}\right]$. These formulas yielded the limits for various values of $\sqrt{\lambda}$ as given in Table I.

TABLE I

$\sqrt{\lambda}$	0.0	0.1	0.2	0.3	0.4	0.5	0.6	0.7	0.8	0.9	1.0
$\underline{\alpha}$	.050	.047	.048	.049	.048	.047	.046	.046	.045	.044	.044
$\bar{\alpha}$	.050	.051	.053	.054	.053	.052	.050	.048	.047	.047	.046
$\alpha^* = \frac{1}{2}[\underline{\alpha} + \bar{\alpha}]$	.050	.048	.050	.051	.050	.049	.048	.047	.046	.045	.045

Since $\frac{l}{(1+l)^2} = \frac{1/l}{(1+1/l)^2}$, for reasons of symmetry we see that

$$\alpha\left[\lambda|c - c'\frac{l}{(1+l)^2}\right] = \alpha\left[1/\lambda|c - c'\frac{l}{(1+l)^2}\right]$$

The function (3.33) is a fairly smooth one. Thus, the mid-point between the two limits will be a good approximation to the exact value. As can be seen from Table I, $\alpha\left[\lambda|c - c'\frac{l}{(1+l)^2}\right]$ is nearly constant. On the other hand, the variation of $\alpha(\lambda|c)$ is considerable. It decreases from .05 to .0-- as λ increases from 0 to 1. The value of c' has been determined from equation

(3.24). Table I suggests that a slightly higher value of c' would give even better results.

4. Determination of $\phi(l)$ for which the first r derivatives of $\alpha^*[\lambda|\phi(l)]$ are equal to zero at $\lambda = 0$. In equation (1.16) $\alpha^*[\lambda|\phi(l)]$ was defined by the formula

$$\alpha^*[\lambda|\phi(l)] = EH\left[\frac{\sqrt{2}\,\phi(\lambda F^*)\,\sqrt{1+\lambda F^*}}{\sqrt{(1+\lambda)(1+F^*)}}\right] \tag{4.1}$$

where F^* is a random variable whose distribution is given as follows:

$$\begin{aligned} &P\{F^* < c\} = 1 \quad \text{for any } c \text{ greater than a given } F_0 \\ &P\{F^* < c\} = P\{F < c\} \quad \text{for any } c \leqq F_0 \end{aligned} \tag{4.2}$$

The random variable F has the F-distribution with (n,n) degrees of freedom. Thus, $\alpha^*[\lambda|\phi(l)]$ agrees with the expression for $\alpha[\lambda|\phi(l)]$ given in (1.12), except that F is replaced by F^*. Clearly, $\alpha^*[\lambda|\phi(l)] - \alpha[\lambda|\phi(l)]$ can be made arbitrarily small by choosing F_0 sufficiently large.

If the first r-derivatives of $\phi(\lambda)$ exist and are continuous functions of λ in the neighborhood of $\lambda = 0$, we may differentiate (4.1) at $\lambda = 0$ at least r times under the expected value sign, that is,

$$\left.\frac{d^i\alpha^*[\lambda|\phi(l)]}{d\lambda^i}\right|_{\lambda=0} = E\frac{d^i}{d\lambda^i}H\left[\frac{\sqrt{2}\,\phi(\lambda F^*)\,\sqrt{1+\lambda F^*}}{\sqrt{(1+\lambda)(1+F^*)}}\right]\Bigg|_{\lambda=0} \quad (i = 1, \cdots, r) \tag{4.3}$$

Let $c_0 = \phi(0)$ and let $i!\,c_i$ denote the ith derivative of $\phi(\lambda)$ at $\lambda = 0$. Clearly

$$\frac{d^i}{d\lambda^i}H\left[\frac{\sqrt{2}\,\phi(\lambda F^*)\,\sqrt{1+\lambda F^*}}{\sqrt{(1+\lambda)(1+F^*)}}\right]\Bigg|_{\lambda=0} \qquad (i = 1, \cdots, r) \tag{4.4}$$

does not depend on c_j for $j > i$ and is a linear function of c_i. Let $\rho_i(F^*)$ be the coefficient of c_i in the expression (4.4). One can easily verify that

$$\rho_i(F^*) = i!\,\frac{\sqrt{2}\,F^{*i}}{\sqrt{1+F^*}}\,\frac{dH(u)}{du}\Bigg|_{u=\frac{\sqrt{2}c_0}{\sqrt{1+F^*}}} \tag{4.5}$$

Since $dH(u)/du < 0$, we have

$$E\rho_i(F^*) < 0 \tag{4.6}$$

Thus, for any given values $c_0, c_1, \cdots, c_{i-1}$, the equation in c_i

$$E\frac{d^i}{d\lambda^i}H\left[\frac{\sqrt{2}\,\phi(\lambda F^*)\,\sqrt{1+\lambda F^*}}{\sqrt{(1+\lambda)(1+F^*)}}\right]\Bigg|_{\lambda=0} = 0 \tag{4.7}$$

has exactly one root. Hence, by solving successive linear equations we can determine step by step the constants $c_1, \cdots, c_r$ so that (4.7) is equal to zero for $i = 1, \cdots, r$. This shows the validity of the following theorem.

THEOREM 4.1: *For any given* $c_0 > 0$, *there exists exactly one sequence* $\{c_j\}$ ($j = 1, 2, \cdots$, ad inf.), *namely, the sequence* $\{c_j\}$ *determined uniquely by the condition that* (4.7) *holds for all integral values i, such that the first r-derivatives of* $\alpha^*[\lambda|Q_r(l)]$ *are equal to zero at* $\lambda = 0$, *where* $Q_r(l) = c_0 + c_1 l + c_2 l^2 + \cdots + c_r l^r$ ($r = 1, 2, \cdots$, ad inf.).

A power series $\sum_{i=0}^{\infty} a_i u^i$ will be said to be convergent if the radius of its circle of convergence is positive. If the radius of the circle of convergence is zero, the power series will be said to be divergent. We shall now prove the following theorem.

THEOREM 4.2: *Let* $\{c_j\}$ ($j = 1, \cdots$, ad inf.) *be the sequence determined by the equations* (4.7). *If* $Q(l) = \sum_{i=0}^{\infty} c_i l^i$ *is a convergent power series, there exists a positive value* λ_0 *such that* $\alpha^*[\lambda|Q(l)]$ *is constant over the interval* $0 \leqq \lambda \leqq \lambda_0$.

PROOF: Let

$$u = \frac{\sqrt{2}\,Q(\lambda F^*)\,\sqrt{1 + \lambda F^*}}{\sqrt{(1+\lambda)(1+F^*)}} \tag{4.8}$$

and

$$u_0 = \frac{\sqrt{2}\,Q(0)}{\sqrt{(1+F^*)}} \tag{4.9}$$

Since $Q(l)$ is a convergent power series, we see that $(u - u_0)$ can be expanded in a power series in λ which is uniformly convergent in λ and F^* over the domain $0 \leqq \lambda \leqq \lambda'$ and $0 \leqq F^* \leqq F_0$, where λ' is a positive number. Clearly, $\lambda' F_0$ is less than or equal to the radius of the circle of convergence of $Q(l)$. Since $H(u)$ can be expanded in a convergent power series in $(u - u_0)$, it follows that $H(u)$ can be expanded in a power series in λ,

$$H(u) = \sum_{i=0}^{\infty} \gamma_i(F^*)\lambda^i \tag{4.10}$$

which is uniformly convergent in λ and F^* over the domain $0 \leqq \lambda \leqq \lambda_0$ and $0 \leqq F^* \leqq F_0$, where λ_0 is a positive number. Clearly, $\lambda_0 F_0$ is less than or equal to the radius of the circle of convergence of $Q(l)$. Thus, for any $\lambda \leqq \lambda_0$, $Q(\lambda F^*)$ is defined over the whole range of F^*.

Because of the uniform convergence of the series in (4.10), we have, for any $\lambda \leqq \lambda_0$,

$$\alpha^*[\lambda|Q(l)] = EH(u) = \sum_{i=0}^{\infty} E[\gamma_i(F^*)]\lambda^i \tag{4.11}$$

But

$$i!E\gamma_i(F^*) = \left.\frac{d^i\alpha^*[\lambda|Q(l)]}{d\lambda^i}\right|_{\lambda_0} = 0 \qquad (i = 1, 2, \cdots, \text{ad inf.})$$

Hence, Theorem 4.2 is proved.

In Sec. 5 we shall prove several lemmas on divergent power series which will then be used in Sec. 6 to show that $Q(l)$ is a convergent power series.

Since $Q(l)$ is a convergent power series, it is possible to represent it in a power series in $l/(1+l)^2$, that is,

$$Q(l) = Q^*\left[\frac{l}{(1+l)^2}\right] = \sum_{i=0}^{\infty} c_i^* \frac{l^i}{(1+l)^{2i}} \tag{4.12}$$

The coefficients c_i^* can be determined as follows: We expand $Q^*[l/(1+l)^2]$ in a power series in l by expanding each term $c_i^* l^i/(1+l)^{2i}$ in a power series in l and summing these power series. Then the coefficient d_j of l^j in the expansion of $Q^*[l/(1+l)^2]$ does not depend on c_k^* for $k > j$ and is a linear function of c_j^*. After having determined the coefficients $c_0^*, c_1^*, \cdots, c_{j-1}^*$, the coefficient c_j^* can be computed from the linear equation $d_j = c_j$. Thus, putting $c_0^* = c_0$, the coefficients $c_1^*, c_2^* \cdots$, etc., can be determined step by step.

In the light of the results of Sec. 3, it seems preferable to use the expansion $Q^*[l/(1+l)^2]$ instead of $Q(l)$. Moreover, because of the symmetry of the variables x_1 and x_2, it is desirable that each term of the series should be symmetric in x_1 and x_2. This is fulfilled for $Q^*[l/(1+l)^2]$, since

$$\frac{l}{(1+l)^2} = \frac{1/l}{(1+1/l)^2}$$

5. Some lemmas on divergent power series. In this section we shall prove several lemmas on divergent power series. First we shall prove the following lemma.

LEMMA 5.1: *Let* $\{a_i\}$ $(i = 1, 2, \cdots)$ *be a sequence of positive numbers such that* $a_{i+1} \geqq a_i$ and $\lim_{i=\infty} a_i = \infty$.

Let
$$b_r = \frac{a_1}{a_{2r}} + \left(\frac{a_2}{a_{2r}}\right)^2 + \cdots + \left(\frac{a_r}{a_{2r}}\right)^r \qquad (r = 1, 2, \cdots)$$

Then zero is a limit point of the sequence $\{b_r\}$.

PROOF: We shall first show that there are infinitely many integral values r for each of which the following set of inequalities are fulfilled:

$$a_{2r} \geqq a_j[1 + (j)^{-\frac{1}{2}}] \qquad (j = 1, \cdots r) \tag{5.1}$$

Suppose that there are only a finite number of values r for which (5.1) holds. Then there exists a positive integer r_0 such that for any $r \geqq r_0$ the inequality (5.1) does not hold at least for one value $j \leqq r$. Hence, we have

$$a_{2r_0} < a_j[1 + (j)^{-\frac{1}{2}}]$$

for some value $j \leqq r_0$. Since

$$a_j[1 + (j)^{-\frac{1}{2}}] \leqq 2a_{r_0} \qquad \text{for } j \leq r_0 \tag{5.2}$$

we have

$$(5.3) \qquad a_{2r_0} < 2a_{r_0}$$

It follows from (5.2) and (5.3) that

$$(5.4) \qquad a_j[1 + (j)^{-\frac{1}{2}}] \leqq 2a_{r_0}[1 + (r_0 + 1)^{-\frac{1}{2}}] \qquad \text{for } j \leqq 2r_0$$

Hence, since for $r = 2r_0$ the inequality (5.1) does not hold for at least one value $j \leqq 2r_0$, we have

$$(5.5) \qquad a_{4r_0} < 2a_{r_0}[1 + (r_0 + 1)^{-\frac{1}{2}}]$$

In a similar way to that in which we obtained (5.4) and (5.5), we obtain

$$(5.6) \quad a_j[1 + (j)^{-\frac{1}{2}}] \leqq 2a_{r_0}[1 + (r_0 + 1)^{-\frac{1}{2}}][1 + (2r_0 + 1)^{-\frac{1}{2}}] \qquad \text{for } j \leqq 4r_0$$

$$(5.7) \qquad a_{8r_0} < 2a_{r_0}[1 + (r_0 + 1)^{-\frac{1}{2}}][1 + (2r_0 + 1)^{-\frac{1}{2}}]$$

In general, from the pair of equations

$$(5.8) \qquad a_j[1 + (j)^{-\frac{1}{2}}] \leq 2a_{r_0} \prod_{k=0}^{i-1} [1 + (2^k r_0 + 1)^{-\frac{1}{2}}] \qquad (j \leq 2^i r_0)$$

and

$$(5.9) \qquad a_{2^{i+1}r_0} < 2a_{r_0} \prod_{k=0}^{i-1} [1 + (2^k r_0 + 1)^{-\frac{1}{2}}] \qquad (i \geqq 1)$$

we obtain

$$(5.10) \quad a_j[1 + (j)^{-\frac{1}{2}}] \leq 2a_{r_0} \prod_{k=0}^{i} [1 + (2^k r_0 + 1)^{-\frac{1}{2}}] \qquad \text{for } j \leq 2^{i+1} r_0$$

$$(5.11) \qquad a_{2^{i+2}r_0} < 2a_{r_0} \prod_{k=0}^{i} [1 + (2^k r_0 + 1)^{-\frac{1}{2}}]$$

Thus, since (5.8) and (5.9) have beeen proved for $i \leqq 2$, we see by induction that they must hold for all integral values $i \geqq 1$. Since the product on the right side of (5.9) is a bounded function of i, we arrive at a contradiction to the assumption that $\lim_{j=\infty} a_j = \infty$. Hence we have proved that there are infinitely many values r for which (5.1) holds. Let $\{r_i\}$ ($i = 1, 2, \cdots$, ad inf.) be such a sequence of values r. Let j_0 be a fixed integer and let

$$(5.12) \quad b_{r,j_0} = \left(\frac{a_1}{a_{2r}}\right) + \left(\frac{a_2}{a_{2r}}\right)^2 + \cdots + \left(\frac{a_{j_0}}{a_{2r}}\right)^{j_0} + \sum_{j=j_0+1}^{\infty} \left\{\frac{a_j}{a_j[1 + (j)^{-\frac{1}{2}}]}\right\}^j$$

Clearly

$$(5.13) \qquad b_{r,j_0} \geq b_r \qquad \text{for } r = r_i \qquad (1 = 1, 2, \cdots, \text{ad inf.})$$

Since

$$\lim_{j=\infty} \left[\frac{1}{1 + (j)^{-\frac{1}{2}}}\right]^{(j)^{\frac{1}{2}}} = \frac{1}{e}$$

we have, for sufficiently large j_0,

$$\left[\frac{1}{1+(j)^{-\frac{1}{3}}}\right]^{(j)^{\frac{1}{3}}} < \frac{2}{e} \qquad \text{for } j \geq j_0$$

Thus,

$$\sum_{j=j_0+1}^{\infty} \left[\frac{1}{1+(j)^{-\frac{1}{3}}}\right]^{j} < \sum_{j=j_0+1}^{\infty} \left(\frac{2}{e}\right)^{j^{\frac{2}{3}}} \tag{5.14}$$

Since $(2/e)^{j^{\frac{2}{3}}} < 1/j^2$ for sufficiently large j, the series $\sum_j (2/e)^{j^{\frac{2}{3}}}$ is convergent. Hence, by choosing j_0 sufficiently large, we can make $\sum_{j=j_0+1}^{\infty} [1/1+(j)^{-\frac{1}{3}}]^j$ arbitrarily small. Hence, it follows from (5.12) that

$$\lim_{j_0=\infty} \limsup_{i=\infty} b_{r_i j_0} = 0 \tag{5.15}$$

Lemma 5.1 follows from (5.13) and (5.15).

Lemma 5.2: *Let* $\sum_{i=1}^{\infty} c_i\lambda^i$ *be a divergent power series and let* λ_r *be a positive value determined so that*

$$\max(|c_1|\lambda_r, \cdots, |c_{2r}|\lambda_r^{2r}) = 1$$

Let, furthermore,

$$S_r = \sum_{i=1}^{r} |c_i|\lambda_r^i$$

Then zero is a limit point of the sequence $\{S_r\}$ $(r = 1, 2, \cdots)$

Proof: Clearly, it is sufficient to prove Lemma 5.2 for the case when $c_i \geqq 0$ $(i = 1, 2, \cdots)$. Let $a_i = c_i^{1/i}$. Then

$$\lambda_r = \frac{1}{\max(a_1, \cdots, a_{2r})} \tag{5.16}$$

Because of the divergence of $\Sigma c_i\lambda^i$, we have

$$\limsup_{i=\infty} a_i = \infty \tag{5.17}$$

Let

$$A_i = \max(a_1, \cdots, a_i)$$

Then

$$\lim_{i=\infty} A_i = \infty$$

and

$$S_r = \frac{a_1}{A_{2r}} + \left(\frac{a_2}{A_{2r}}\right)^2 + \cdots + \left(\frac{a_r}{A_{2r}}\right)^r \leqq \sum_{j=1}^{r} \left(\frac{A_j}{A_{2r}}\right)^j \tag{5.18}$$

Lemma 5.2 follows from (5.18) and Lemma 5.1.

LEMMA 5.3: *Let* $\sum_{i=1}^{\infty} c_i\lambda^i$ *be a divergent power series and let* $\bar{\lambda}_r$ *be a positive value determined so that*

$$\max (|c_{r+1}|\bar{\lambda}_r^{r+1}, \cdots, |c_{2r}|\bar{\lambda}_r^{2r}) = 1$$

Let, furthermore,

$$\bar{S}_r = \sum_{i=1}^{r} |c_i|\bar{\lambda}_r^i$$

Then zero is a limit point of the sequence $\{\bar{S}_r\}$.

PROOF: Let

$$S_r = \sum_{i=1}^{r} |c_i|\lambda_r^i$$

where λ_r is defined by equation (5.16). It follows from Lemma 5.2 that there exists a sequence $\{r_i\}$ ($i = 1, 2, \cdots$, ad inf.) such that

$$\lim_{i=\infty} S_{r_i} = 0$$

For any i for which $S_{r_i} < 1$, we obviously have

$$\bar{\lambda}_{r_i} = \lambda_{r_i} \qquad S_{r_i} = \bar{S}_{r_i}$$

Hence Lemma 5.3 is proved.

LEMMA 5.4: *Let* $\Sigma c_i\lambda^i$ *be a divergent power series and let*

$$\lambda'_r = \left|\frac{1}{c_r}\right|^{1/r} \qquad (r = 1, 2, \cdots, \text{ad inf.})$$

Let, furthermore,

$$S'_r = \sum_{i=1}^{[r/2]} |c_i|\lambda_r'^i$$

where $[k]$ *denotes the largest integer* $\leqq k$. *Then there exists a sequence of integers* $\{r_j\}$ *such that*

$$\lim_{j=\infty} r_j = \infty, \quad \lim_{j=\infty} S'_{r_j} = 0 \tag{5.19}$$

and

$$|c_i|(\lambda'_{r_j})^i \leqq 1 \qquad \text{for } i \leqq r_j \text{ and for all } j \tag{5.20}$$

PROOF: For any positive integral k let $\bar{S}_k$ and $\bar{\lambda}_k$ be defined as in Lemma 5.3. For each positive integral k there exists an integral value m_k such that $k + 1. \leqq m_k \leqq 2k$ and $|c_{m_k}|(\bar{\lambda}_k)^{m_k} = 1$. Thus,

$$\lambda'_{m_k} = \bar{\lambda}_k \text{ and } S'_{m_k} \leqq \bar{S}_k \tag{5.21}$$

It follows from Lemma 5.3 that there exists a subsequence $\{k'\}$ of the sequence $\{k\}$ ($k = 1, 2, \cdots$, ad inf.) such that

$$\lim_{k=\infty} \bar{S}_{k'} = 0 \qquad \lim_{k=\infty} k' = \infty \tag{5.22}$$

and

$$\bar{S}_{k'} < 1 \qquad \text{for all values } k' \tag{5.23}$$

Let $r_k = m_{k'}$. Then it follows from (5.21) and (5.22) that

$$\lim_{k=\infty} r_k = \infty \qquad \text{and} \qquad \lim_{k=\infty} S'_{r_k} = 0 \tag{5.24}$$

From the definition of $\bar{\lambda}_k$ it follows that

$$|c_i|(\bar{\lambda}_k)^i \leqq 1 \qquad \text{for } k < i \leq 2k \tag{5.25}$$

Clearly, (5.23) implies that

$$|c_i|(\bar{\lambda}_{k'})^i < 1 \qquad \text{for } i \leqq k' \tag{5.26}$$

Hence, since $k' < m_{k'} \leqq 2k'$,

$$|c_i|(\bar{\lambda}_{k'})^i \leqq 1 \qquad \text{for } i \leqq m_{k'} \tag{5.27}$$

Since $\bar{\lambda}_{k'} = \lambda'_{m_{k'}}$ and $m_{k'} = r_k$, we obtain from (5.27)

$$|c_i|(\lambda'_{r_k})^i \leqq 1 \qquad \text{for } i \leqq r_k \tag{5.28}$$

Lemma 5.4 is a consequence of (5.24) and (5.28).

LEMMA 5.5: *Let* $a(u) = \sum_{i=0}^{\infty} a_i u^i$ *and* $p(\lambda) = \sum_{i=0}^{\infty} p_i \lambda^i$ *be two convergent power series with nonnegative coefficients. Let, furthermore,* $\{c_i\}$ ($i = 1, 2, \cdots$, ad inf.) *be a sequence of nonnegative numbers and* $\{r_i\}$ ($i = 1, 2, \cdots$, ad inf.) *a sequence of increasing positive integers for which the following conditions are fulfilled:*

$$c_i \lambda_r^i \leqq 1 \qquad \text{for } i \leqq r \tag{5.29}$$

$$\lim_{r=\infty} \sum_{i=1}^{[r/2]} c_i \lambda_r^i = 0 \tag{5.30}$$

where $\lambda_r = (1/c_r)^{1/r}$ *and* r *denotes a variable whose domain is restricted to the sequence* $\{r_j\}$. *For any* $i \leqq r$ *let* ρ_i *denote the coefficient of* λ^i *in the expansion of*

$$\sum_{j=2}^{\infty} a_j [T_r(\lambda) p(\lambda) - c_0 p_0]^j \tag{5.31}$$

in a power series in λ *where*

$$T_r(\lambda) = c_0 + c_1\lambda + \cdots + c_r\lambda^r \tag{5.32}$$

(It is clear that ρ_i does not depend on r in the domain $i \leq r$.) *Then*

(5.33) $$\lim_{r=\infty} \rho_i \lambda_r^i = 0$$

uniformly in i over the domain $i \leqq r$.

PROOF: Since the elements of the sequences $\{a_i\}$, $\{c_i\}$, and $\{p_i\}$ are nonnegative, the coefficients of the power series expansion of $T_r(\lambda)p(\lambda)$ are nonnegative, and consequently also $\rho_i \geqq 0$. Clearly

(5.34) $$\sum_{i=2}^{[r/2]} \rho_i \lambda_r^i \leqq \sum_{j=2}^{\infty} a_j[T_{[r/2]}(\lambda_r)p(\lambda_r) - c_0 p_0]^j$$

Because of (5.30), we have

(5.35) $$\lim_{r=\infty} [T_{[r/2]}(\lambda_r) - c_0] = 0$$

Since $\lim_{r=\infty} \lambda_r = 0$, we also have

(5.36) $$\lim_{r=\infty} [p(\lambda_r) - p_0] = 0$$

From (5.35) and (5.36) it follows that the right member of (5.34) converges to zero as $r \to \infty$. Hence

(5.37) $$\lim_{r=\infty} \sum_{i=2}^{[r/2]} \rho_i \lambda_r^i = 0$$

Since $\rho_1 = 0$, this proves (5.33) for $i \leqq [r/2]$.

Let

(5.38) $$U_r(\lambda) = T_{[r/2]}(\lambda)p(\lambda) - c_0 p_0$$

and

(5.39) $$V_r(\lambda) = [T_r(\lambda) - T_{[r/2]}(\lambda)]p(\lambda)$$

Consider the expression

(5.40) $$W_r(\lambda) = \sum_{j=2}^{\infty} \{ja_j[U_r(\lambda)]^{j-1}V_r(\lambda) + a_j[U_r(\lambda)]^j\}$$

It is clear that for $i \leqq r$ the coefficient of λ^i in the power series expansion of $W_r(\lambda)$ is equal to ρ_i. For any $i > [r/2]$ and $\leqq r$, let ρ_{ij} denote the coefficient of λ^i in the expansion of

(5.41) $$ja_j[U_r(\lambda)]^{j-1}V_r(\lambda) + a_j[U_r(\lambda)]^j \qquad (j = 2, 3 \cdots)$$

Since all coefficients of the power series expansions $a(u)$, $p(\lambda)$, and $T_r(\lambda)$ are nonnegative, we have, for any positive λ,

(5.42) $$\rho_{ij}\lambda^i \leqq ja_j[U_r(\lambda)]^{j-1}p(\lambda) \max (c_{[r/2]+1}\lambda^{[r/2]+1}, \cdots, c_r\lambda^r) + a_j[U_r(\lambda)]^j$$

From (5.29) and (5.42) we obtain

$$\rho_{ij}\lambda_r^i \leqq ja_j[U_r(\lambda)]^{j-1}p(\lambda) + a_j[U_r(\lambda)]^j \tag{5.43}$$

Hence, for any $i > [r/2]$ and $\leqq r$ we have

$$0 \leqq \rho_i\lambda_r^i = \sum_{j=2}^{\infty} \rho_{ij}\lambda_r^i \leq \sum_{j=2}^{\infty} \{ja_j[U_r(\lambda_r)]^{j-1}p(\lambda_r) + a_j[U_r(\lambda_r)^j]\} \tag{5.44}$$

Since $\lim_{r=\infty} U_r(\lambda_r) = 0, j > 1$ and $\Sigma a_i u^i$ is a convergent power series, the expression on the extreme right in (5.44) converges to zero as $r \to \infty$. Hence,

$$\lim_{r=\infty} \rho_i\lambda_r^i = 0 \qquad \text{for } [r/2] < i \leqq r \tag{5.45}$$

This convergence is uniform in i, since the expression on the extreme right in the inequality (5.44) is independent of i. Lemma 5.5 is a consequence of (5.37) and (5.44).

Lemma 5.6: *Let $f(x)$ be a function which can be expanded in a Taylor series around $x = x_0$, that is,*

$$f(x) = a_0 + \sum_{i=1}^{\infty} a_i(x - x_0)^i \tag{5.46}$$

where

$$a_0 = f(x_0) \qquad \text{and} \qquad a_i = \frac{1}{i!}\frac{d^if(x)}{dx^i}\bigg|_{x=x_0}$$

For any positive d, let $b_i(d)$ the maximum of $\frac{1}{i!}\left|\frac{d^if(x)}{dx^i}\right|$ in the domain

$$|x - x_0| \leqq d \qquad (i = 1, 2, \cdots, \text{ad inf.})$$

There exists a positive d such that the power series $\sum_{i=1}^{\infty} b_i(d)U^i$ is convergent.

Proof: We can assume without loss of generality that $x_0 = 0$. Clearly, $\bar{f}(x) = |a_0| + \sum_{i=1}^{\infty} |a_i|x^i$ is a convergent power series. Let r be its radius of convergence and let $d = r/2$. Let δ be a real number whose absolute value does not exceed d. We have

$$f(x + \delta) = a_0 + \sum_{i=1}^{\infty} a_i(x + \delta)^i = c_0(\delta) + \sum_{i=1}^{\infty} c_i(\delta)x^i \tag{5.47}$$

and

$$\bar{f}(x + d) = |a_0| + \sum |a_i|(x + d)^i = \bar{c}_0(d) + \sum_{i=1}^{\infty} \bar{c}_i(d)x^i \tag{5.48}$$

It is easy to see that for any δ for which $|\delta| \leqq d$ we have

$$\bar{c}_i(d) \geqq |c_i(\delta)| \tag{5.49}$$

But $c_i(\delta) = \frac{1}{i!} \frac{d^i f(x)}{dx^i}\bigg|_{x=\delta}$. Hence

$$\bar{c}_i(d) \geqq \frac{1}{i!} \left| \frac{d^i f(x)}{dx^i} \right|_{x=\delta} \quad \text{for } |\delta| \leqq d \tag{5.50}$$

Lemma 5.6 follows from (5.50) and the fact that $\sum_{i=1}^{\infty} \bar{c}_i(d)x^i$ is a convergent power series.

6. Proof that the power series $Q(l)$ defined in Sec. 4 is convergent. In Sec. 4 the coefficients c_i of $Q(l)$ were defined by the equations (4.7). We shall show here that $Q(l)$ is a convergent power series. For this purpose we shall assume that $Q(l)$ is divergent and we shall derive a contradiction from this assumption. Let

$$T_r(\lambda) = c_0 + |c_1|\lambda + \cdots + |c_r|\lambda^r \qquad (r = 1, 2, \cdots) \tag{6.1}$$

The functions $\sqrt{1 + \lambda F^*}$ and $1/\sqrt{1 + \lambda}$ can be expanded in power series thus:

$$\sqrt{1 + \lambda F^*} = 1 + \sum_{i=1}^{\infty} d_i \lambda^i F^{*i} \tag{6.2}$$

and

$$\frac{1}{\sqrt{1 + \lambda}} = 1 + \sum_{i=1}^{\infty} k_i \lambda^i \tag{6.3}$$

Let

$$D(\lambda) = 1 + \sum_{i=1}^{\infty} |d_i| F_0^i \lambda^i \tag{6.4}$$

and

$$K(\lambda) = 1 + \sum_{i=1}^{\infty} |k_i| F_0^i \lambda^i \tag{6.5}$$

Clearly, $D(\lambda)$ and $K(\lambda)$ are convergent power series. Consider the following two expressions

$$\sqrt{2}\, T_r(\lambda F_0) D(\lambda) K(\lambda) - \sqrt{2}\, c_0 \tag{6.6}$$

and

$$\frac{\sqrt{2}\, Q_r(\lambda F^*) \sqrt{1 + \lambda F^*}}{\sqrt{(1 + \lambda)(1 + F^*)}} - \frac{\sqrt{2}\, c_0}{\sqrt{1 + F^*}} \tag{6.7}$$

where $Q_r(l) = c_0 + c_1 l + \cdots + c_r l^r$. It follows easily from the definitions of $T_r(\lambda)$, $D(\lambda)$, and $K(\lambda)$ that

$$|\gamma_i(F^*)| \leqq \bar{\gamma}_i \qquad (i = 1, 2, \cdots \text{ ad inf.}) \tag{6.8}$$

when $\gamma_i(F^*)$ is the coefficient of λ^i in the power series expansion of (6.7) and $\bar{\gamma}_i$ is the coefficient of λ^i in the expression of (6.6). Let

$$H(u) = P\{|t_{2n}| < u\} \tag{6.9}$$

where t_{2n} has the t-distribution with $2n$ degrees of freedom. We expand $H(u)$ in a Taylor series around $u = u_0 = \sqrt{2}\, c_0/\sqrt{1 + F^*}$, that is,

$$H(u) = H(u_0) + \sum_{i=1}^{\infty} a_i(F^*)(u - u_0)^i \tag{6.10}$$

where

$$a_i(F^*) = \frac{1}{i!}\frac{d^iH(u)}{du^i}\bigg|_{u=u_0}$$

Let

$$\bar{H}(v) = |H(u_0)| + \sum_{i=1}^{\infty} |a_i(F^*)|v^i \tag{6.11}$$

Clearly, $\bar{H}(v)$ is a convergent power series. Since $\Sigma c_i\lambda^iF_0^i$ is a divergent power series, it follows from Lemma 5.4 that there exists a sequence

$$\{r_j\} \quad (j = 1, 2, \cdots)$$

of integers such that

$$|c_i|\lambda_r^iF_0^i \leqq 1 \qquad \text{for } i \leqq r \tag{6.12}$$

and

$$\lim_{r=\infty} \sum_{i=1}^{[r/2]} |c_i|\lambda_r^iF_0^i = 0 \tag{6.13}$$

where $\lambda_r = \dfrac{1}{F_0}\left|\dfrac{1}{c_r}\right|^{1/r}$ and r denotes a variable whose domain is restricted to the sequence $\{r_j\}$. Let $\bar{\rho}_i(F^*)$ be the coefficient of λ^i in the power series expansion of

$$\sum_{j=2}^{\infty} |a_j(F^*)|[\sqrt{2}\, T_r(\lambda F_0)D(\lambda)K(\lambda) - \sqrt{2}\, c_0]^j \qquad (i \leqq r) \tag{6.14}$$

and $\rho_i'(F^*)$ be the coefficient of λ^i in the expansion of

$$\sum_{j=2}^{\infty} a_j(F^*)\left[\frac{\sqrt{2}\, Q_r(\lambda F^*)\sqrt{1 + \lambda F^*}}{\sqrt{(1 + \lambda)(1 + F^*)}} - \frac{\sqrt{2}\, c_0}{\sqrt{1 + F^*}}\right]^j \tag{6.15}$$

Because of (6.8), we obviously have

$$|\rho_i'(F^*)| \leqq \bar{\rho}_i(F^*) \tag{6.16}$$

It follows from Lemma 5.5 that

$$\lim_{r=\infty} \bar{\rho}_r(F^*)\lambda_r^r = 0 \tag{6.17}$$

We shall now show that the convergence in (6.17) is uniform in F^*. For suppose that the convergence is not uniform. Then there exists a sequence $\{F_r^*\}$ such that

$$\liminf_{r=\infty} \bar{\rho}_r(F_r^*)\lambda_r^r > 0 \tag{6.18}$$

Since the domain of F^* is a finite and closed interval, there exists a subsequence $\{r'\}$ of the sequence $\{r\}$ such that $\lim_{r=\infty} F_{r'}^*$ exists. Denote this limit value by $\bar{F}$. Let $b_i(d)$ be the maximum of $|a_i(F^*)|$ in the interval $\bar{F} - d \leqq F^* \leqq \bar{F} + d$. It follows from Lemma 5.6 that there exists a positive value d such that the power series $\sum_{i=1}^{\infty} b_i(d)u^i$ is convergent. Let ρ_i^* be the coefficient of λ^i in the power series expansion of

$$\sum_{j=2}^{\infty} b_j(d)[\sqrt{2}\, T_r(\lambda F_0)D(\lambda)K(\lambda) - \sqrt{2}\, c_0]^j$$

Clearly

$$\rho_i^* \geqq \bar{\rho}_i(F^*) \tag{6.19}$$

for any F^* for which $\bar{F} - d \leqq F^* \leqq \bar{F} + d$. It follows from Lemma 5.5 that

$$\lim_{r=\infty} \rho_r^*\lambda_r^r = 0 \tag{6.20}$$

Hence, because of (6.19) we have

$$\lim_{r=\infty} \bar{\rho}_r(F^*)\lambda_r^r = 0 \tag{6.21}$$

uniformly in F^* over the interval $\bar{F} - d \leqq F^* \leqq \bar{F} + d$. But this is in contradiction to (6.18) and thus the convergence in (6.17) is proved to be uniform in F^*. From (6.16) we have

$$\lim_{r=\infty} \rho_r'(F^*)\lambda_r^r = 0 \tag{6.22}$$

uniformly in F^*. This implies that

$$\lim_{r=\infty} E\rho_r'(F^*)\lambda_r^r = 0 \tag{6.23}$$

The coefficients c_i ($i = 1, 2, \cdots$, ad inf.) of the power series $Q(l)$ have been determined so that

$$E\xi_j(F^*) = 0 \qquad (j = 1, \cdots, r) \tag{6.24}$$

where $\xi_j(F^*)$ is the coefficient of λ^j in the expansion of

$$H(u_0) + \sum_{j=1}^{\infty} a_j \left[\frac{\sqrt{2}\, Q_r(\lambda F^*)\sqrt{1 + \lambda F^*}}{\sqrt{(1 + \lambda)(1 + F^*)}} - \frac{\sqrt{2}\, c_0}{\sqrt{1 + F^*}}\right]^j \tag{6.25}$$

in a power series in λ. Let $\zeta_r(F^*)$ be the coefficient of λ^r in the expansion of

$$(6.26)\qquad \frac{\sqrt{2}\, Q_r(\lambda F^*)\, \sqrt{1 + \lambda F^*}}{\sqrt{(1+\lambda)(1+F^*)}}$$

in a power series in λ. From (6.23) and (6.24) it follows that

$$(6.27)\qquad \lim_{r=\infty} E\zeta_r(F^*)\lambda_r^r = 0$$

Let $\eta_r(F^*)$ be the coefficient of λ^r in the expansion of

$$(6.28)\qquad \frac{\sqrt{2}\, Q_{r-1}(\lambda F^*)\, \sqrt{1 + \lambda F^*}}{\sqrt{(1+\lambda)(1+F^*)}}$$

Let, furthermore, $\bar{\eta}_r$ be the coefficient of λ^r in the expansion of

$$(6.29)\qquad \sqrt{2}\, T_{r-1}(\lambda F_0) D(\lambda) K(\lambda)$$

Clearly

$$(6.30)\qquad |\eta_r(F^*)| \leqq \bar{\eta}_r$$

We have

$$(6.31)\qquad \bar{\eta}_r \lambda^r = \sum_{j=0}^{r-1} |c_j| F_0^j \lambda^j g_{r-j} \lambda^{r-j}$$

where g_m is the coefficient of λ^m in the expansion of $D(\lambda)K(\lambda)$. Since $g_m \geqq 0$, because of (6.12) we obtain from (6.31)

$$(6.32)\qquad \bar{\eta}_r \lambda_r^r \leqq \sum_{j=0}^{r-1} g_{r-j} \lambda_r^{r-j} \leqq D(\lambda_r)K(\lambda_r) - D(0)K(0)$$

Clearly

$$\lim_{r=\infty} [D(\lambda_r)K(\lambda_r) - D(0)K(0)] = 0$$

Hence

$$(6.33)\qquad \lim_{r=\infty} \bar{\eta}_r \lambda_r^r = 0$$

We have then, because of (6.30),

$$(6.34)\qquad \lim_{r=\infty} E\eta_r(F^*)\lambda_r^r = 0$$

From this and (6.27) we obtain

$$(6.35)\qquad \lim_{r=\infty} Ew_r(F^*)\lambda_r^r = 0$$

where $w_r(F^*)$ is the coefficient of λ^r in the expansion of

$$\frac{\sqrt{2}\, c_r F^{*r} \lambda^r \sqrt{1 + \lambda F^*}}{\sqrt{(1 + \lambda)(1 + F^*)}} \tag{6.36}$$

Clearly

$$w_r(F^*) = \frac{\sqrt{2}\, c_r F^{*r}}{\sqrt{1 + F^*}}$$

Since $P\{F^* = F_0\} > 0$, and since $|c_r| F_0^r \lambda_r^r = 1$, equation (6.35) cannot hold. Thus we arrive at a contradiction which proves that the power series

$$Q(l) = \sum_{i=0}^{\infty} c_i l^i$$

must be convergent.

NAME INDEX

(Boldface numbers indicate first page of article written in collaboration with Wald.)

SUBJECT INDEX